CARBONATITES

TITLES OF RELATED INTEREST

Boninites
A. J. Crawford (ed.)

Cathodoluminescence of geological materials
D. J. Marshall

Chemical fundamentals of geology
R. Gill

Crystal structures and cation sites of the rock-forming minerals
J. R. Smyth & D. L. Bish

The dark side of the Earth
R. Muir Wood

Deformation processes in minerals, ceramics and rocks
D. J. Barber & P. G. Meredith (eds)

Geology and mineral resources of West Africa
J. B. Wright

Igneous petrogenesis
B. M. Wilson

Image interpretation in geology
S. Drury

The inaccessible Earth
G. C. Brown & A. E. Mussett

The interpretation of igneous rocks
K. G. Cox et al.

Introduction to X-ray spectrometry
K. L. Williams

Komatiites
N. Arndt & E. Nisbet (eds)

Mathematics in geology
J. Ferguson

Metamorphism and metamorphic belts
A. K. Miyashiro

Perspectives on a dynamic Earth
T. R. Paton

Petrology of the igneous rocks
F. Hatch et al.

Petrology of the metamorphic rocks
R. Mason

Planetary landscapes
R. Greeley

A practical introduction to optical mineralogy
C. D. Gribble & A. J. Hall

Rheology of the Earth
G. Ranalli

Rutley's elements of mineralogy
C. D. Gribble

Simulating the Earth
J. Holloway & B. Wood

Statistical methods in geology
R. F. Cheeney

Volcanic successions
R. Cas & J. V. Wright

The young Earth
E. G. Nisbet

CARBONATITES
Genesis and Evolution

Edited by
Keith Bell
Ottawa–Carleton Geoscience Centre,
Department of Earth Sciences, Carleton University, Ottawa

London
UNWIN HYMAN
Boston   Sydney   Wellington

© Keith Bell, 1989
This book is copyright under the Berne Convention.
No reproduction without permission. All rights reserved.

Published by the Academic Division of
Unwin Hyman Ltd
15/17 Broadwick Street, London W1V 1FP, UK

Unwin Hyman Inc.,
8 Winchester Place, Winchester, Mass. 01890, USA

Allen & Unwin (Australia) Ltd,
8 Napier Street, North Sydney, NSW 2060, Australia

Allen & Unwin (New Zealand) Ltd in association with the Port Nicholson Press Ltd,
Compusales Building, 75 Ghuznee Street, Wellington 1, New Zealand

First published in 1989

British Library Cataloguing in Publication Data

Carbonatites.
1. Carbonatites
I. Bell, Keith
552'.2
ISBN 0-04-445068-0

Library of Congress Cataloging-in-Publication Data

Main entry under title:
Carbonatites: genesis and evolution.
Includes bibliographies and index.
1. Carbonatites—Congresses. I. Bell, Keith.
QE462.C36C36 1989 552'.3 88-33801
ISBN 0-04-445068-0 (alk. paper)

Typeset in 11 on 13 point Times by
Mathematical Composition Setters Ltd, Salisbury
and printed in Great Britain by
Butler & Tanner, Frome, Somerset

PREFACE

At the 1986 annual meeting of the Geological Association of Canada–Mineralogical Association of Canada–Canadian Geophysical Union in Ottawa a one-day special session was held on carbonatites. This volume is an outcome of that meeting and was originally intended as an up-to-date primer, a review of carefully selected topics for the general reader, covering research during the past decade or so. However, it became clear, both during and after the meeting, that several new research directions were being shaped and that these were going to have significant impact on future work. The idea of a book devoted solely to review papers was modified to accommodate these new findings, and out of this resulted the present volume.

It is now almost 20 years since the first comprehensive books on carbonatites were published, and 12 years have elapsed since the publication of the *Proceedings of the First International Symposium on Carbonatites*, the last major work on carbonatites. Despite this lengthy interval, most of the new and exciting data have been obtained only recently: new phase-equilibrium studies show the possibility of immiscibility for a wide compositional range of silicate melts; Nd, Pb, and Sr isotope studies clearly indicate the mantle as the source of carbonatite magmas; mantle metasomatism lurks in the background as a precursor to carbonatite genesis; and new information is being collected about the thermochemical and thermophysical properties of carbonate melts. Field work has now established several carbonatite and alkaline rock provinces associated with fold belts and rifted continental margins, well outside of stable cratonic areas normally associated with carbonatite activity. Extrusive carbonatites are more widespread than previously thought, illustrating, quite clearly, that low viscosity carbonate melts can form most of the typical volcanic products normally associated with silicate melts.

The papers contained in this volume are generally of two types. One includes several review articles, while the other presents more recent findings, marks relatively new directions, and adds still more controversy to the debate about the origin of carbonatite magmas. Early chapters document, for the most part, field relationships, chemistry, mineralogy, and mineral deposits associated with carbonatites. Isotope chemistry is covered in subsequent chapters, and the origin and diversification of carbonatite melts are left until last. These later chapters deal mainly with mantle processes and models, in particular metasomatism, the peridotite–C–H–O solidus, the role of asthenosphere and lithosphere in carbonatite genesis, and lithospheric thinning.

As with many other igneous rock types, nomenclature poses a problem in carbonatite petrology. There is still no widespread agreement concerning the classification of carbonatites. One classification scheme, recommended by the International Union of Geological Sciences Subcommission on the Systematics of Igneous Rocks, is based on modal mineralogy. An additional classification scheme presented in this volume is based on rock chemistry. Both schemes are evaluated, providing a useful starting point for deliberation and discussion. The naming of carbonatites in this volume, however, has been left to the discretion of the authors.

A deliberate attempt was also made to avoid the 'this is my favourite carbonatite' syndrome, although a volume about carbonatites would be incomplete without reviews of Phalaborwa and Oldoinyo Lengai. Both are important for quite different reasons. Phalaborwa is not only one of the oldest carbonatites known, but it also hosts one of the largest copper deposits on Earth. Of interest too, is the pyroxenitic material that makes up most of the complex at Phalaborwa, contrasting with the carbonatites that form less than 5% of the rocks observed at its surface. Oldoinyo Lengai, on the other hand, is a unique and spectacular example of an active carbonatite volcano that has erupted as recently as 1988. The significance of the natrocarbonatite lavas from Oldoinyo Lengai remains one of the most controversial issues in carbonatite research.

A common thread that runs through many of the papers is the role of alkali carbonate melts of the Lengai type in carbonatite genesis. The absence of abundant alkalies in most carbonatites has, in the past, been attributed to loss of alkalies by migration of fenitizing fluids from plutonic carbonatites, or by secondary processes such as weathering and recrystallization, particularly for carbonatites of extrusive origin. Recent chemical analyses of primary fragments in carbonatite pyroclastic deposits, and evaluation of textures in some extrusive carbonatites, convincingly show that few such rocks ever contained alkalies in the abundances seen in the Oldoinyo Lengai lavas. Carbonatites, it would seem, can be either alkali-rich or alkali-poor.

There is still no consensus about the origin of carbonatite magmas. Are they direct primary melts from the mantle? Are they derived by liquid immiscibility? Or are they the end-products of fractional crystallization of a silicate melt? In this volume each of these three models has been proposed by various authors, and, although this controversy will continue, most contributors favour the formation of carbonatites from differentiation of a carbonated silicate melt.

The chemical make-up of the mantle, particularly below the continents, and the nature of mantle fluids are areas that are presently demanding a great deal of attention. Carbonatites form one group of rocks that has much to say about the chemical and physical evolution of the subcontinental upper mantle, but there still remains doubt about whether the parental magma for carbonatites is of asthenospheric or lithospheric origin. Proposed models in this volume suggest that parental melts are derived from (a) depleted or metasomatized ancient lithosphere; (b) the asthenosphere with modification of melts as they migrate through the lithosphere; or (c) from both asthenosphere and lithosphere, interacting at the lithosphere–asthenosphere boundary. No matter which model is accepted, lithospheric thinning attributed to up-doming of asthenospheric material and associated rifting appears to have played a key role in producing the parental melts of carbonatites.

It is, however, the abundance of volatiles in carbonatites that makes them unique among igneous rocks, and herein lies their special position in the geological record. Mantle metasomatism and the near-surface analogue fenitization, are areas that are closely intertwined with carbonatite genesis and evolution. The stability of exotic minerals formed by metasomatism under mantle conditions, coupled with the possibility that carbonated silicate melts might well play a major role as one of the main, deep-seated metasomatic fluids, bring us closer towards an understanding of the role of volatiles in deep-seated processes.

There are recognizable gaps in this book, but these are the products of the lack of recent data. Fenitization has been touched on only briefly, and the detailed mineralogy and chemical compositions of the main carbonate minerals in carbonatites merit, but still await, thorough study. In addition, contributions from workers outside of Europe and North America were solicited but none was forthcoming.

In conclusion, this book is an account of where we are in carbonatite research, and how we arrived there. It is hoped that it will stimulate a resurgence of interest and research activity in carbonatites. Carbonatite magmas as a source of fluids, and carbonatites as a repository of Nb, Sr, and the rare earth elements (REEs), important in superconducting materials, make them especially topical and relevant. Not only do such rocks provide a very interesting way of sampling the mantle but they also hold many of the clues to understanding fluid migration and its evolution at both mantle and crustal levels. There is clearly much to be done. If this book serves to whet the research community's appetite for carbonatite research and augment the handful of workers involved in carbonatite endeavours, then I will consider the book a success and my time as editor well spent.

Director, Keith Bell
Ottawa–Carleton Geoscience Centre,
Ottawa, Canada

ACKNOWLEDGEMENTS

I would like to thank the Geological Association of Canada–Mineralogical Association of Canada–Canadian Geophysical Union for their help in sponsoring a one-day special session on carbonatites at their annual meeting in Ottawa. It was this meeting that provided the spark that started this volume. Particular thanks go to Roger Jones of Unwin Hyman whose handling of the volume was much appreciated.

All chapters in the volume were reviewed by the editor and at least two other referees. The following external referees are thanked for their thorough and, in many cases, rapid reviews: T. Andersen, R. L. Armstrong, W. R. Baragar, N. K. Basu, A. R. Berger, M. G. Best, T. Birkett, J. W. Card, K. D. Card, M. J. Carr, G. Y. Chao, W. Chesworth, L. Diamond, A. D. Edgar, R. E. Ernst, G. Faure, H. Gabrielse, D. H. Green, R. E. Harmer, G. Heiken, L. Hulbert, J. R. Ikingura, R. W. Kay, O. J. Kleppa, A. F. Koster van Groos, K. T. Kyser, M. B. Lambert, R. W. Le Maitre, B. E. Leake, G. Lofgren, J. Ludden, M. A. Menzies, R. H. Mitchell, J. M. Moore, B. O. Mysen, J. R. O'Neil, R. Parrish, A. R. Philpotts, R. G. Platt, P. J. Pollard, M. F. Roden, M. J. Sheridan, A. Streckeisen, R. P. Taylor, W. J. Verwoerd, R. H. Vollmer, E. B. Watson, R. F. Wendlandt, H. S. Yoder, and R. E. Zartman.

I would also like to thank G. Y. Chao and D. D. Hogarth, two of my colleagues at the Ottawa–Carleton Geoscience Centre. Their knowledge of the mineralogy of carbonatites and alkalic rocks in general contributed a great deal during the reviewing stages of the manuscripts. Finally, special thanks go to J. W. Card who acted as my 'general' reader. His thoroughness coupled with his invaluable insights added enormously to the final volume.

CONTENTS

Preface	vii
Acknowledgements	x
List of tables	xvi
List of contributors	xviii

1 Carbonatites: nomenclature, average chemical compositions, and element distribution *A. R. Woolley & D. R. C. Kempe* 1

	Abstract	1		Acknowledgements	13
1.1	Nomenclature	1		References	13
1.2	Average chemical composition	7			

2 The spatial and temporal distribution of carbonatites *A. R. Woolley* 15

	Abstract	15	2.5	Increase in carbonatite activity with time	33
2.1	Introduction	15	2.6	General features of the temporal and spatial distribution of carbonatites	34
2.2	Spatial and temporal distribution	16			
2.3	Tectonic considerations	31		Acknowledgements	34
2.4	Repetition of carbonatite activity	32		References	34

3 Field relations of carbonatites *D. S. Barker* 38

	Abstract	38	3.8	Metamorphosed carbonatites	56
3.1	Introduction	38	3.9	Hydrothermal dilatant veins and metasomatic replacement bodies associated with carbonatite-bearing plutons	57
3.2	Evidence for carbonatite magmas	39			
3.3	Carbonatite lavas	41			
3.4	Carbonatite tephra	44	3.10	Pseudocarbonatites	60
3.5	Carbonatites in small hypabyssal intrusions	47	3.11	Conclusions	61
				Acknowledgements	62
3.6	Carbonatites in plutons	50		References	63
3.7	Carbonatites associated with kimberlites	54			

4 Extrusive carbonatites and their significance *J. Keller* 70

	Abstract	70	4.5	Extrusive carbonatites from the Kaiserstuhl	79
4.1	Introduction	70			
4.2	Carbonatite volcanoes	71	4.6	Conclusions	85
4.3	Volcanological classification of carbonatites	76		Acknowledgements	86
				References	86
4.4	Natrocarbonatites versus calcite carbonatites	77			

5 Carbonatite magma: properties and processes *A. H. Treiman* — 89

	Abstract	89	5.5 Physical properties and processes	96
5.1	Introduction	89	5.6 Conclusions	101
5.2	Composition of carbonatite magma	90	Acknowledgements	102
5.3	Structure of carbonatite magmas	90	References	102
5.4	Thermochemical properties	91		

6 Pyrochlore, apatite and amphibole: distinctive minerals in carbonatite *D. D. Hogarth* — 105

	Abstract	105	6.4	Amphiboles	132
6.1	Introduction	105	6.5	Conclusions	140
6.2	Pyrochlore group	109		Acknowledgements	141
6.3	Apatite	121		References	141

7 Nature of economic mineralization in carbonatites and related rocks *A. N. Mariano* — 149

	Abstract	149	7.8	Fluorite	167
7.1	Introduction	150	7.9	Barite	168
7.2	Rare earth element mineralization	152	7.10	Vanadium	168
			7.11	Strontium	169
7.3	Niobium mineralization	156	7.12	Thorium and uranium	169
7.4	Apatite	160	7.13	Copper	170
7.5	Titanium mineralization	162	7.14	Conclusions	171
7.6	Stability of carbonatite minerals in laterites	164		Acknowledgements	172
7.7	Vermiculite	166		References	172

8 Nature and structural relationships of carbonatites from Southwest and West Tanzania *P. van Straaten* — 177

	Abstract	177	8.4	Spatial and temporal distribution	192
8.1	Introduction	177			
8.2	Carbonatites in Tanzania	178	8.5	Discussion and conclusions	194
8.3	The carbonatites of southwestern and western Tanzania	179		Acknowledgements	196
				References	197

9 Carbonatites in a continental margin environment – the Canadian Cordillera *J. Pell & T. Höy* — 200

	Abstract	200		core complexes, Omineca Crystalline Belt	206
9.1	Introduction	200			
9.2	Intrusive carbonatites in Palaeozoic rocks, Foreland Belt	202	9.5	Chemistry	211
			9.6	Tectonic implications	215
9.3	Intrusive carbonatites in Late Proterozoic rocks, Omineca Crystalline Belt	205	9.7	Conclusions	216
				Acknowledgements	217
9.4	Carbonatites associated with			References	217

10 Phalaborwa: a saga of magmatism, metasomatism, and miscibility
S. C. Eriksson — 221

	Abstract	221	10.6 Isotopic chemistry	242
10.1	Introduction	222	10.7 Origin of the complex	247
10.2	Geology	222	10.8 Conclusions	248
10.3	Age	230	Acknowledgements	249
10.4	Mineral chemistry	231	References	249
10.5	Whole-rock chemistry	238		

11 Sodium carbonatite extrusions from Oldoinyo Lengai, Tanzania: implications for carbonatite complex genesis
J. B. Dawson — 255

	Abstract	255	Acknowledgements	274
11.1	Introduction	255	References	275
11.2	Regional geological setting	256		
11.3	Oldoinyo Lengai, general geology	258		

12 Neodymium and Strontium isotope geochemistry of carbonatites
K. Bell & J. Blenkinsop — 278

	Abstract	278	12.5 The source region of carbonatites – lithosphere or asthenosphere?	293
12.1	Introduction	278		
12.2	Chemical background	279		
12.3	Strontium and Nd isotope chemistry of carbonatites	281	12.6 Conclusions	296
			Acknowledgements	297
12.4	Isotopic constraints on the evolution of carbonatite magmas	290	References	297

13 Stable isotope variations in carbonatites
P. Deines — 301

	Abstract	301	and oxygen isotope composition	337
13.1	Introduction	301		
13.2	Oxygen	306	13.5 Sulphur	342
13.3	Carbon	323	13.6 Summary	348
13.4	The relation between carbon		Acknowledgements	349
			References	350

14 Lead isotope relationships in carbonatites and alkalic complexes: an overview
S.-T. Kwon, G. R. Tilton & M. H. Grünenfelder — 360

	Abstract	360	14.5 Evolution models for Canadian complexes	379
14.1	Introduction	361		
14.2	North American alkalic complexes	362	14.6 Differentiation of U, Th, and Pb in mantle processes	381
14.3	Bulk Earth Pb isotopic evolution	374	14.7 The source of carbonatites	382
			14.8 Concluding statement	383
14.4	$^{206}Pb:^{204}Pb-^{87}Sr:^{86}Sr$ correlation	375	Acknowledgements	384
			References	384

15 The genesis of carbonatites by immiscibility B. A. Kjarsgaard & D. L. Hamilton 388

	Abstract	388	15.3 Discussion	398
15.1	Introduction	388	15.4 Summary	403
15.2	Evidence for silicate–carbonate immiscibility	389	Acknowledgements	403
			References	403

16 The behaviour of trace elements in the evolution of carbonatites D. L. Hamilton, P. Bedson & J. Esson 405

	Abstract	405	16.5 Effect of P on K_D	417
16.1	Introduction	405	16.6 Summary and applications	425
16.2	Methods	406	Acknowledgements	426
16.3	Effect of composition on K_D	411	References	427
16.4	Effect of T on K_D	413		

17 Diversification of carbonatite M. J. Le Bas 428

	Abstract	428	17.5 Primitive carbonatites	438
17.1	Introduction	428	17.6 Late-stage carbonatites	443
17.2	The choice of carbonatites	430	17.7 Summary	444
17.3	Basic assumptions	432	References	445
17.4	The nephelinite–carbonatite association	432		

18 Upper-mantle enrichment by kimberlitic or carbonatitic magmatism A. P. Jones 448

	Abstract	448	18.5 South African model	455
18.1	Introduction	448	18.6 Discussion	458
18.2	Mantle metasomatism	449	18.7 Conclusions	460
18.3	Mantle titanate minerals	451	Acknowledgements	461
18.4	Kimberlite signature	453	References	461

19 A model of mantle metasomatism by carbonated alkaline melts: trace-element and isotopic compositions of mantle source regions of carbonatite and other continental igneous rocks J. K. Meen, J. C. Ayers & E. J. Fregeau 464

	Abstract	464	19.5 Consequences of ijolite–peridotite interaction	478
19.1	Introduction	465		
19.2	Models of mantle metasomatism	467	19.6 Examples of materials derived from metasomatic mantle regions	489
19.3	Phase relations of peridotite–H_2O–CO_2	469	19.7 Summary	493
19.4	Experimental techniques	470	Acknowledgements	495
			References	495

20 Origin of carbonatites: evidence from phase equilibrium studies *P. J. Wyllie* — 500

	Abstract	500		low-temperature synthetic carbonatite magmas	514
20.1	Introduction	501	20.5	Immiscibility between silicate and carbonate liquids	520
20.2	Carbonate systems: precipitation of carbonates from melts	503	20.6	The formation of carbonate-rich melts in the mantle	523
20.3	Accessory minerals in synthetic carbonatite magmas	509	20.7	Kimberlites, nephelinites, and carbonatites	532
20.4	Relationships between high-temperature silicate melts and		20.8	The origin of carbonatites	537
				Acknowledgements	540
				References	540

21 Mantle metasomes and the kinship between carbonatites and kimberlites *S. E. Haggerty* — 546

	Abstract	546	21.5	Kimberlite and carbonatite genesis	553
21.1	Introduction	546	21.6	Conclusions	556
21.2	Carbonate in kimberlites and related rocks	547		Acknowledgements	557
21.3	Metasomatism	549		References	558
21.4	Metasome model	551			

22 Carbonatites, primary melts, and mantle dynamics *D. H. Eggler* — 561

	Abstract	561	22.4	Mantle dynamics	570
22.1	Introduction	562	22.5	Conclusions	575
22.2	Melting of carbonated peridotite	562		Acknowledgements	576
22.3	Primary melts and carbonatites	566		References	576
				Appendix	579

23 The origin and evolution of carbonatite magmas *J. Gittins* — 580

	Abstract	580	23.5	Alkalic carbonatite magma as a parental magma	594
23.1	Introduction	581			
23.2	Where do carbonatites come from?	582	23.6	Carbonatite lavas and pyroclastics: their bearing on carbonatite magma evolution	594
23.3	Assessment of magma derivation schemes	582	23.7	Carbonatite fluids and fenitization	597
23.4	Carbonatite magma evolution in the crust	589	23.8	Epilogue	599
				References	599

Index — 601

LIST OF TABLES

1.1	Carbonatites: averages and ranges of analyses	8
1.2	Rare earth element (REE) abundances in carbonatites (p.p.m.)	9
3.1	Criteria for magmatic heritage of carbonatite	39
3.2	Styles of occurrence of carbonatites	61
4.1	Chemical composition of selected extrusive carbonatites	79
5.1	Melting of carbonates: molar properties at 1 Kb pressure	91
5.2	Mixing in molten carbonates: regular solution parameters	94
5.3	Thermophysical properties of magmas near their liquidi	96
6.1	List of minerals reported from carbonatites	106
6.2	Carbonatite stages	108
6.3	Classification of the pyrochlore group	110
6.4	Some pyrochlore-group minerals from carbonatites	112
6.5	REE data of some pyrochlore-group minerals	119
6.6	Chemical composition of apatite from carbonatites and related rocks	122
6.7	REE distribution in apatite from some carbonatites	126
6.8	Chemical composition of typical amphiboles from carbonatite and related rocks	133
6.9	Maximum percentages of some constituents of amphiboles	139
7.1	Economic mineralization in carbonatites	150
7.2	Electron microprobe analyses of pyrochlores	158
7.3	Some examples of carbonatites with niobium-bearing laterites	165
8.1	Ages of carbonatites in SW and W. Tanzania	180
8.2	Size, shape, and orientation of carbonatites in SW and W. Tanzania	183
8.3	Chemical analyses of rocks from Ngualla carbonatite	186
8.4	Chemical analyses of carbonatites from Panda Hill and Sengeri Hill	187
8.5	Chemical analyses of rocks from the Mbalizi, Songwe Scarp, and Makonde carbonatites	190
8.6	Chemical analyses of rocks from the Sangu-Ikola and Nachendezwaya carbonatites	191
9.1	Carbonatites in British Columbia	203
9.2	Average chemical analyses of carbonatites from various British Columbia localities	212
10.1	Rb–Sr phlogopite isotopic data, Phalaborwa	230
10.2	Electron microprobe analyses of some representative minerals from Phalaborwa	232
10.3	Whole rock chemistry, Phalaborwa	239
11.1	Major- and trace-element contents and $^{87}Sr:^{86}Sr$ ratios of selected Oldoinyo Lengai silicate lavas, and a coeval olivine nephelinite from Oldoinyo Loolmurwak	264
11.2	Representative analyses of magmatic series plutonic blocks, Oldoinyo Lengai	266
11.3	Composition of Oldoinyo Lengai carbonatite lavas of the 1960 eruption, a ?pre-1917 altered natrocarbonatite lava, and the phases and matrix of the 1960 lavas	269
11.4	REE abundances of Oldoinyo Lengai lavas	270
12.1	Nd and Sr abundances in large-scale reservoirs	279

LIST OF TABLES

12.2	Summary of Sm, Nd, and Sr data from carbonatites	282
13.1	List of carbonatites and related rocks were used in the review	302
13.2	Mean carbon isotopic composition, standard deviation, and number of analyses of carbonatites and associated carbonate lavas	326
13.3	Sulphur isotopic composition of different carbonatite stages	347
14.1	U, Th, and Pb isotopic data for *c.* 1100-Ma-old alkalic complexes in the Canadian Shield	366
14.2	Summary of initial ratios for the alkalic complexes	375
14.3	Comparison of observed and calculated Pb ratios for Archaean galenas	377
14.4	BE Pb isotopic compositions	378
15.1	Run data from immiscibility studies	391
15.2	Electron microprobe data of quenched liquids from experimental runs	392
16.1	Composition of starting materials for trace-element studies	406
16.2	Trace-element content of charges for immiscibility studies	407
16.3	Major element analyses, wt%, of run products determined by electron microprobe	408
16.4	Measured K_D, silicate/carbonate, values for trace elements	410
18.1	Titanate mineral compared with average rock compositions	454
19.1	Compositions of materials employed in ijolite–harzburgite phase study	471
19.2	Results of experimental study on ijolite–harzburgite join	472
19.3	Compositions of melts and crystal assemblages obtained on the ijolite–harzburgite join at 20 Kb	475
19.4	Compositions of materials employed in saturation studies	476
19.5	Concentrations of TiO_2, ZrO_2, and P_2O_5 required to saturate melts in trace-element concentrating phases at 14 Kb	476
19.6	Calculated isotopic compositions of 2.75-Ga-old SCUM	481

LIST OF CONTRIBUTORS

J. C. Ayers, *Department of Geosciences, Pennsylvania State University, University Park, PA 16802, USA*

D. S. Barker, *Department of Geological Sciences, University of Texas, Austin, TX 78713-7909, USA*

P. Bedson, *Department of Geology, University of Manchester, Manchester, UK*

K. Bell, *Ottawa–Carleton Geoscience Centre, Department of Earth Sciences, Carleton University, Ottawa, Ontario K1S 5B6, Canada*

J. Blenkinsop, *Ottawa–Carleton Geoscience Centre, Department of Earth Sciences, Carleton University, Ottawa, Ontario K1S 5B6, Canada*

J. B. Dawson, *Department of Geology, University of Sheffield, Sheffield, UK*

P. Deines, *Department of Geosciences, Pennsylvania State University, University Park, PA 16802, USA*

D. H. Eggler, *Department of Geosciences, Pennsylvania State University, University Park, PA 16802, USA*

S. C. Eriksson, *Department of Geological Sciences, Virginia Polytechnic Institute and State University, Blacksburg, VA 24061, USA*

J. Esson, *Department of Geology, University of Manchester, Manchester, UK*

E. J. Fregeau, *Department of Geosciences, Pennsylvania State University, University Park, PA 16802, USA*

J. Gittins, *Department of Geology, University of Toronto, Toronto, Ontario M5S 1A1, Canada*

M. H. Grünenfelder, *Laboratory for Geochemistry and Mass Spectrometry, Federal Institute of Technology, CH-8092 Zurich, Switzerland*

S. E. Haggerty, *Department of Geology, University of Massachusetts, Amherst, MA 01003, USA*

D. L. Hamilton, *Department of Geology, University of Manchester, Manchester, UK*

D. D. Hogarth, *Ottawa–Carleton Geoscience Centre, Department of Geology, University of Ottawa, Ottawa, Ontario K1N 6N5, Canada*

T. Höy, *Geological Survey Branch, British Columbia Ministry of Energy, Mines and Petroleum Resources, Victoria, British Columbia, Canada*

A. P. Jones, *School of Geological Sciences, Kingston Polytechnic, Penrhyn Road, Kingston upon Thames, UK*

J. Keller, *Mineralogisch-Petrographisches Institut der Universität Freiburg, D-7800 Freiburg, FRG*

D. R. C. Kempe, *Department of Mineralogy, British Museum (Natural History), London, UK*

B. A. Kjarsgaard, *Department of Geology, University of Manchester, Manchester, UK*

S.-T. Kwon, *Department of Geological Sciences, University of California, Santa Barbara, CA 93106, USA*

M. J. Le Bas, *Department of Geology, University of Leicester, Leicester, UK*

A. N. Mariano, *Mineral Exploration Consultant, 48 Page Brook Road, Carlisle, MA 01741, USA*

J. K. Meen, *Department of Geosciences, Pennsylvania State University, PA 16802, USA*

J. Pell, *Department of Geological Sciences, University of British Columbia, Vancouver, British Columbia, Canada*

LIST OF CONTRIBUTORS

G. R. Tilton, *Department of Geological Sciences, University of California, Santa Barbara, CA 93106, USA*

A. H. Treiman, *Department of Geology, Boston University, Boston, MA 02215, USA*

P. van Straaten, *Department of Land Resource Sciences, University of Guelph, Guelph, Ontario N1G 2W1, Canada*

A. R. Woolley, *Department of Mineralogy, British Museum (Natural History), London, UK*

P. J. Wyllie, *Division of Geological and Planetary Sciences, California Institute of Technology, Pasadena, CA 91125, USA*

1
CARBONATITES: NOMENCLATURE, AVERAGE CHEMICAL COMPOSITIONS, AND ELEMENT DISTRIBUTION

A. R. WOOLLEY & D. R. C. KEMPE

ABSTRACT

A system of nomenclature is proposed in which carbonatites are named on the basis of carbonate mineralogy. To qualify as a carbonatite a rock must have > 50% carbonate minerals. Those dominantly composed of calcite are called calcite carbonatites (sövites and alvikites); and dolomite carbonatite, ankerite carbonatite, etc. are preferred to terms such as rauhaugite and beforsite. If the carbonate mineral, or minerals, have not been identified but a whole rock chemical analysis is available, carbonatites are divided according to the weight proportions of CaO, MgO, and $FeO + Fe_2O_3 + MnO$, such that 'calciocarbonatites' have > 80% CaO and 'magnesiocarbonatites' have $MgO > FeO + Fe_2O_3 + MnO$. Those with $FeO + Fe_2O_3 + MnO > MgO$ are called 'ferrocarbonatites'.

The average chemical compositions of carbonatites, which have been subdivided into calciocarbonatites, magnesiocarbonatites, and ferrocarbonatites, have been calculated from an initial data base of some 400 analyses. Analyses were excluded if they contained > 10% SiO_2, lacked values for CaO, MgO, FeO, Fe_2O_3 or MnO, reported 'loss on ignition' rather than CO_2, H_2O, etc., or provided only spectrographic determinations of trace elements. These criteria precluded the use of many REE (rare earth element) data. The greater number of analyses now available, and localities represented, counteracts the bias in earlier published averages by overrepresentation of data from some localities.

1.1 NOMENCLATURE

Despite Heinrich's (1966) review of the nomenclature and definitions of carbonatites, and the efforts of earlier researchers such as von Eckermann (1948) and Brögger (1921) in erecting comprehensive naming systems, rather less than half of the 400 or so samples that were analysed for this study had been adequately named. Commonly, these rocks are simply referred to as 'carbonatite', perhaps with mineral prefixes referring to the major silicate phases present, despite the fact that major subdivisions of the carbonatitic rocks were proposed and defined in the early days of carbonatite studies, notably by Brögger (1921). His proposed terms 'sövite' and 'rauhaugite' for calcitic and dolomitic carbonatites have been widely used, while von Eckermann's (1928, 1948) suggestions of 'alvikite' and 'beforsite' for their hypabyssal equivalents have also been widely adopted. The iron-rich carbonatites are sometimes given the prefix 'ferro-', but they also are comparatively ill-defined. Brögger (1921) also suggested a number of names for the carbonatitic rocks containing a high proportion of silicate minerals, but these have not been widely adopted, although the term 'silicocarbonatite' is commonly used.

The recommendations for the classification of carbonatites by the IUGS Subcommission on the Systematics of Igneous Rocks (Streckeisen 1980, p. 204–5) included the following:

'1 Carbonatites are igneous rocks, intrusive as well as extrusive, which contain more than 50% by volume of carbonate minerals.
2 The following classes of carbonatites are distinguished:
   (a) calcite–carbonatite (sövite, coarse-grained; alvikite, medium- to fine-grained);
   (b) dolomite–carbonatite (beforsite);
   (c) ferrocarbonatite (essentially composed of iron-rich carbonate minerals);
   (d) natrocarbonatite (essentially composed of sodium–potassium–calcium carbonates; at present, this rock is known only as the extrusive product of Oldoinyo Lengai volcano in N. Tanzania).
3 Carbonatites containing a mixture of various carbonate minerals, e.g. calcite and dolomite, are indicated by prefixes according to the established rules for quantitative composition of rocks at the 10–50–90% boundaries. The qualification dolomite-bearing, etc., may be used when the presence of the minor constituent (less than 10%) needs to be emphasized ...
4 Igneous rocks containing less than 10% carbonate minerals may be called calcite-bearing ijolite, dolomite-bearing peridotite, etc.
5 Igneous rocks containing between 10 and 50% carbonate minerals are called calcitic (or carbonatitic) ijolite, etc.
6 The terms melanocratic and leucocratic should not be applied when describing carbonatites.
7 Characteristic contents of other minerals are indicated by prefixes: apatite–pyrochlore–carbonatite, aegirine–calcite–carbonatite, magnetite–dolomite–carbonatite, etc.'

Although this system does not allow for a distinction to be made between coarser- and finer-grained dolomitic carbonatites, i.e. rauhaugites and beforsites, it is relatively comprehensive. However, in practice, apart from the sövites and alvikites, certain problems arise in applying the system (Woolley 1982). There are essentially four difficulties:

(a) Many carbonatites contain at least two carbonate species which are not always easy to distinguish, and may be intimately intergrown showing complex exsolution textures, as illustrated, for instance, in Heinrich (1966, Figs 7-4 to

7-6). Determining the modal proportions of such rocks, as well as those of fine grain size, may be extremely difficult.

(b) In addition to calcite and dolomite, ankerite, which is ill-defined, is also a common mineral in carbonatites, and there appears to be a complete solid solution series from dolomite through ferroan dolomite to ankerites which have a broad range of Fe:Mg ratios. This complexity is not allowed for in the IUGS classification.

(c) Siderite and magnesian siderite (breunnerite) are reported from some iron-rich carbonatites (Samoilov 1977, Kapustin 1980), and textures in many iron-rich carbonatites seem to indicate the former presence of iron-rich carbonates, which subsequently have altered to a less iron-rich carbonate phase plus iron oxides.

(d) Magnesite is also reported from carbonatites (Kapustin 1980, T. Deans, pers. comm.).

Because carbonatites are generally coarse rocks, their classification should be based on modal mineralogy, according to the recommendations of the IUGS Subcommission. However, it is generally necessary to use an electron microprobe to determine the carbonate species, and the common presence of complex intergrowths creates additional difficulties. Because the carbonate species are commonly not determined and may be complexly intergrown, many carbonatites are inadequately or inappropriately named. It is probable, for instance, that the dominant carbonate mineral in many rocks called rauhaugite or beforsite is, in fact, ankerite rather than dolomite. However, there has not, to date, been a detailed study of the full range of carbonate species in carbonatites, in spite of the fact that they are the major minerals in these rocks. Indeed, there are relatively few analyses of carbonatite carbonates in the literature at all.

Because of these difficulties, Woolley (1982) suggested a system of classification based on rock chemistry, using weight% CaO, MgO, and ($FeO + Fe_2O_3 + MnO$). Sövites and alvikites were defined as containing > 80% CaO, while the remainder were named depending on whether MgO or ($FeO + Fe_2O_3 + MnO$) were the greater: the former constituted the magnesiocarbonatites (rauhaugite and beforsite) and the latter the ferrocarbonatites.

For the purposes of the present paper over 400 chemical analyses from carbonatites have been collected (see Fig. 1.1) and, after selection for quality (as specified in the next section), used to test the system of nomenclature involving the proportions of CaO, MgO, and ($FeO + Fe_2O_3 + MnO$). The data from carbonatites originally described as sövite and alvikite, beforsite and rauhaugite, and ferrocarbonatite or siderite carbonatite, have been plotted on CMF diagrams (Figs 1.2 to 1.4). The sövites and alvikites (Fig. 1.2) form a distinct group; about 80% contain > 80% CaO. The rauhaugites and beforsites (Fig. 1.3) form a rather less compact, though nevertheless distinct, group, and the ferrocarbonatites (Fig. 1.4), although with somewhat fewer data, are also distinct. The boundaries suggested by Woolley (1982) have been drawn on Figures 1.2–1.4 and seem to reflect logical subdivisions of the data.

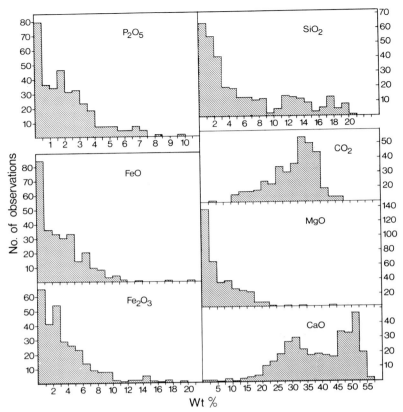

Figure 1.1 Frequency distribution diagrams for some major oxides for all acceptable carbonatite analyses collected. Note particularly the hiatus in the distribution for SiO_2 at about 10% and for CaO around 35–45%.

Analyses of carbonates from carbonatites, apart from calcites, are scarce, but some dolomites, ferroan dolomites, and ankerites are available (Barth & Ramberg 1966, Kapustin 1980). These analyses, together with unpublished data (D. D. Hogarth, pers. comm.), indicate that there is probably a continuous solid solution series in the shaded area of Figures 1.2–1.4. The area occupied by calcite analyses from carbonatites has not been indicated because it lies close to the CaO corner. The boundary of the sövite (alvikite) field at 80% CaO falls approximately midway between the concentrations of sövite (alvikite) and magnesium- and iron-rich carbonatites; this is also illustrated by the hiatus in the CaO frequency distribution (Fig. 1.1). Its position is also justified because it lies half way between the positions of calcite and the Mg–Fe carbonates in the CMF diagram.

There are a few analyses on Figures 1.3 and 1.4 which plot below the dolomite–ankerite compositional field. The one point near to the MgO corner (Fig. 1.3) represents data from a carbonatite from Lueshe (Meyer & Bethune 1958), that contains magnesite. Magnesite is also reported from Sallanlatva and Kovdor (Kapustin 1980). There is also a vague trend on Figures 1.3 and 1.4 towards the iron-bearing corner, which probably partly reflects the presence of Fe oxides, but

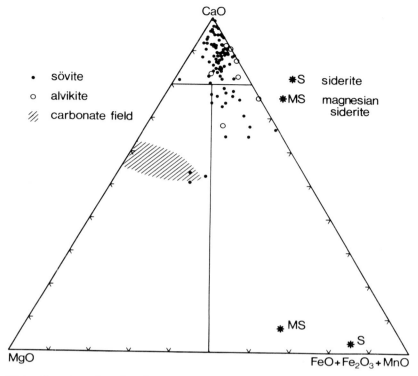

Figure 1.2 Plots of carbonatites described by authors as sövite and alvikite in terms of CaO–MgO–(FeO + Fe$_2$O$_3$ + MnO) (wt%), together with unspecified carbonatites that fall in the calciocarbonatite field. Dolomites, ferroan dolomites, and ankerites (Barth & Ramberg 1966, Kapustin 1980) plot in the shaded area. MS is a magnesian siderite (breunnerite) (Kapustin 1980) and S is siderite (Samoilov 1977) from carbonatites. Calcite analyses have not been plotted. Two of the rocks plotting close to the dolomite–ankerite zone were described as 'ankeritic sövite'. The field boundaries are discussed in the text.

which may also be caused by the presence of siderite or magnesian siderite. Analyses of magnesian siderite (breunnerite) (Kapustin 1980) and siderite (Samoilov 1977) from carbonatites are shown on Figures 1.2–1.4. It is possible that as more analyses become available limited solid solution between magnesite and siderite will be found in some magnesium- and iron-rich carbonatites.

It is suggested that carbonatites could be named according to the following system, which is similar to the IUGS system (Streckeisen 1980) and incorporates some changes suggested by Woolley (1982), together with further modifications:

(a) *General recommendation.* Carbonatites should be named according to their carbonate mineralogy; if this is not known then whole-rock chemistry should be used.

(b) *The term carbonatite.* The term 'carbonatite' should be restricted to those rocks with > 50% carbonate minerals.

(c) *Calcite, dolomite, ankerite carbonatites, etc.* If the carbonate species is identified then these mineral names should be used as a prefix. If more than one

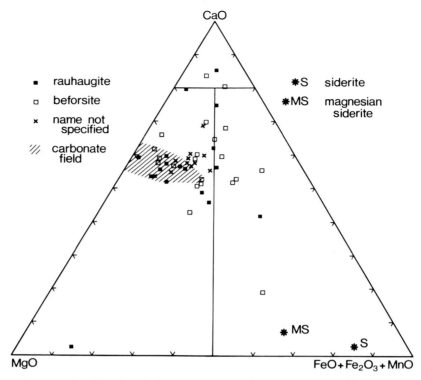

Figure 1.3 Data from carbonatites described as rauhaugite and beforsite, together with unspecified carbonatites which plot in the magnesiocarbonatite field. The rock plotting close to the MgO corner is a magnesite carbonatite from Lueshe. Shaded area, MS, and S as on Figure 1.2.

carbonate species is present then this should be reflected in the name, in the way recommended by Streckeisen (1980, p. 205 and Fig. 4). Use of such terms as sövite, alvikite, rauhaugite, and beforsite is not recommended.

(d) *Other characteristic minerals.* These may be indicated by prefixes, e.g. apatite–pyrochlore–calcite carbonatite, etc.

(e) *Rocks with 10–50% carbonate minerals.* Igneous rocks containing between 10 and 50% carbonate minerals may be called calcitic (or carbonatitic) ijolite, etc.

(f) *Rocks with < 10% carbonate minerals.* Igneous rocks with < 10% carbonate minerals may be called calcite-bearing ijolite, dolomite-bearing nepheline syenite, etc.

(g) *Calciocarbonatite, magnesiocarbonatite, and ferrocarbonatite.* If the carbonate minerals have not been determined but a chemical analysis is available, then carbonatites can be subdivided into three groups. If $CaO:CaO + MgO + FeO + Fe_2O_3 + MnO$ is greater than 0.8 then the rock is a calciocarbonatite; if it is less than 0.8 and $MgO > FeO + Fe_2O_3 + MnO$ it is a magnesiocarbonatite, but if $MgO < FeO + Fe_2O_3 + MnO$ it is a ferrocarbonatite.

(h) *Alkali-rich carbonatite (natrocarbonatite).* This term should be retained for rocks essentially composed of sodium–potassium–calcium carbonates.

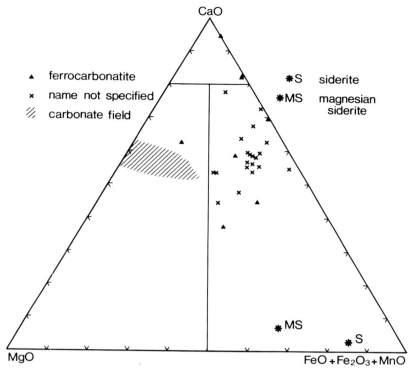

Figure 1.4 Data from carbonatites described as ferrocarbonatites, together with unspecified carbonatites which plot in the ferrocarbonatite field. Shaded area, MS, and S as on Figure 1.2.

1.2 AVERAGE CHEMICAL COMPOSITION

The average chemical compositions of igneous rocks are useful in helping with classification and for comparing one rock type with another. Gold (1963, 1966) and Heinrich (1966) have published averages using between 100 and 200 carbonatite analyses. Although useful, these averages have a number of deficiencies. First, all carbonatites were incorporated, including some with high values of silica, and secondly, nearly half the analyses in Gold's (1963) average came from the Alnö complex, and still made up about a quarter of the analyses used in his later compilation.

The effects of over-representation of one locality is particularly well shown by the SiO_2 values, because von Eckermann (1948) published many analyses of beforsite dykes from Alnö which commonly contain $> 20\%$, and in some cases $> 40\%$ SiO_2. Gold (1966) pointed out that the average SiO_2 value of the 65 available carbonatite analyses from Alnö is 18.88%, and as a result his initial average for SiO_2 (Gold 1963) was 12.10%, which was reduced to 9.58% in his 1966 compilation. The bias of over-represented occurrences was overcome by calculating an average based on averages for each individual occurrence, which removed the dominance of Alnö.

Table 1.1 Carbonatites: averages and ranges of analyses.

	Calciocarbonatite			Magnesiocarbonatite			Ferrocarbonatite		
	Av.	No.	Range	Av.	No.	Range	Av.	No.	Range
Wt%									
SiO_2	2.72	116	0.0–8.93	3.63	50	0.6–9.40	4.70	57	0.36–9.0
TiO_2	0.15	115	0.0–1.09	0.33	49	0.0–1.98	0.42	57	0.0–2.30
Al_2O_3	1.06	116	0.0–6.89	0.99	53	0.0–4.41	1.46	53	0.01–5.60
Fe_2O_3	2.25	97	0.0–9.28	2.41	48	0.0–9.57	7.44	50	0.46–17.84
FeO	1.01	91	0.0–4.70	3.93	47	0.0–10.40	5.28	50	0.0–20.28
MnO	0.52	119	0.0–2.57	0.96	54	0.02–5.47	1.65	57	0.23–5.53
MgO	1.80	122	0.0–8.11	15.06	54	9.25–24.82	6.05	58	0.10–14.50
CaO	49.12	118	39.24–55.40	30.12	53	20.8–47.0	32.77	58	9.20–46.43
Na_2O	0.29	102	0.0–1.73	0.29	44	0.0–2.23	0.39	46	0.0–1.52
K_2O	0.26	105	0.0–1.47	0.28	44	0.0–1.89	0.39	51	0.0–2.80
H_2O+	0.76	78	0.0–4.49	1.20	36	0.08–9.61	1.25	35	0.04–4.52
P_2O_5	2.10	119	0.0–10.41	1.90	51	0.0–11.30	1.97	54	0.0–11.56
CO_2	36.64	104	11.02–47.83	36.81	49	16.93–47.88	30.74	53	20.56–41.81
BaO	0.34	74	0.0–5.0	0.64	32	0.01–4.30	3.25	38	0.02–20.60
							0.80*	31	0.02–2.48
SrO	0.86	66	0.0–3.29	0.69	29	0.06–1.50	0.88	34	0.01–5.95
F	0.29	31	0.0–2.66	0.31	21	0.03–2.10	0.45	20	0.02–1.20
Cl	0.08	8	0.0–0.45	0.07	1	–	0.02	3	0.01–0.04
S	0.41	23	0.02–2.29	0.35	12	0.03–1.30	0.96	12	0.12–5.40
SO_3	0.88	15	0.02–3.87	1.08	13	0.06–2.86	4.14	21	0.06–17.10
							1.08†	14	0.06–3.00
Trace elements (p.p.m.)									
Li	0.1	1	–	–	–	–	10	1	–
Be	2.4	5	0.4–8	<5	1	–	12	1	–
Sc	7	7	0.6–18	14	3	10–17	10	2	9–14
V	80	31	0–300	89	9	7–280	191	16	56–340
Cr	13	10	2–479	55	9	2–175	62	8	15–135
Co	11	12	2–26	17	9	4–39	26	7	11–54
Ni	18	11	5–30	33	5	21–60	26	7	10–63
Cu	24	25	4–80	27	12	4–94	16	14	4–45
Zn	188	16	20–1120	251	10	15–851	606	9	35–1800
Ga	<5	1	–	5	1	–	12	1	–
Rb	14	6	4–35	31	4	2–80	–	–	–
Y	119	16	25–346	61	13	5–120	204	9	28–535
Zr	189	33	4–2320	165	14	0–550	127	13	0–900
Nb	1204	43	1–15000	569‡	18	10–3000	1292	17	10–5033
				1422	19	10–16780			
Mo	–	–	–	12	1	–	71	3	26–94
Ag	–	–	–	3.2	2	3.2–3.3	3.4	2	3.4–3.5
Cs	20	1	–	0.9	1	–	0.6	1	–
Hf	–	–	–	3.2	1	–	–	–	–
Ta	5	1	–	21	6	4–34	0.9	1	–
W	–	–	–	10	1	–	20	1	–
Au	–	–	–	–	–	–	12	2	10–15
Pb	56	6	30–108	89	7	30–244	217	4	46–400
Th	52	13	5–168	93	11	4–315	276	13	100–723
U	8.7	10	0.3–29	13	13	1–42	7.2	16	1–20

No. Number of analyses.
*   BaO average excluding seven rocks with exceptionally high values (9.6–20.6): see text.
† SO_3 average excluding seven rocks with exceptionally high values (5.49–17.10): see BaO above.
‡   Nb average excluding one rock which contains 16 780 p.p.m.

Table 1.2 Rare earth element (REE) abundances in carbonatites (p.p.m.).

	Calciocarbonatite			Magnesiocarbonatite			Ferrocarbonatite		
	Av.	No.	Range	Av.	No.	Range	Av.	No.	Range
La	608	31	90–1600	764	19	95–3655	2666	15	95–16883
Ce	1687	18	74–4152	2183	14	147–8905	5125	8	1091–19457
Pr	219	2	50–389	560	1	–	550	4	141–1324
Nd	883	8	190–1550	634	6	222–1755	1618	4	437–3430
Sm	130	2	95–164	45	6	33–75	128	4	30–233
Eu	39	2	29–48	12	6	3–20	34	4	11–78
Gd	105	2	91–119	–	–	–	130	4	31–226
Tb	9	2	9–10	4.5	6	0.9–8	16	4	4–36
Dy	34	2	22–46	–	–	–	52	4	11–105
Ho	6	2	3–9	–	–	–	6	4	1–9
Er	4	1	–	–	–	–	17	4	3–35
Tm	1	1	–	–	–	–	1.8	4	0.3–3
Yb	5	10	1.5–12	9.5	10	1–52	15.5	4	1–16
Lu	0.7	1	–	0.08	1	–	–	–	–

No. Number of analyses.

This approach also lowered the SiO_2 value to 5.82% (Gold 1966, Table 1, column III). The particular problem of carbonatites with high silica values has been reduced in the present paper by excluding carbonatites which contain > 10% SiO_2. The 10% figure is, to some extent, justified in terms of the frequency distribution diagram (Fig. 1.1) which indicates a break at this value. Some carbonatites with > 10% SiO_2 are undoubtedly hybrid rocks, but it is also clear that others are probably primary. It is unfortunate that the latter are excluded by the 10% SiO_2 criterion, but this was found to be the only practicable solution to this problem at present.

Averages have been calculated (Tables 1.1 and 1.2) from an initial database of over 400 analyses. However, many of these analyses were rejected according to a number of quality criteria. These include:

(a) analyses that did not include values for CaO, MgO, FeO, Fe_2O_3, and MnO, as these were necessary to enable classification of the rock;
(b) analyses giving total Fe as FeO;
(c) analyses with > 10% SiO_2;
(d) analyses quoting 'loss on ignition' rather than values for CO_2, H_2O, etc.;
(e) analyses in which two or more oxides were quoted in a combined form;
(f) phoscorites with exceptionally high values of P_2O_5 (commonly > 10%);
(g) analyses with exceptionally high values of BaO (up to 20.6%), Nb, and SO_3, but separate figures are given for these rocks (Table 1.1);
(h) analyses for which the trace elements were determined spectrographically (although some of these may have been accurate, none has been included because of the difficulty of deciding which are reliable);
(i) analyses of extrusive carbonatites. There is a large, and rapidly increasing, body of chemical data on extrusive carbonatites, both lavas and tuffs. Apart from the special case of the alkali-rich carbonatite of Oldoinyo Lengai, most of this extrusive material is calcitic. However, because these rocks have invariably

been subjected to weathering and alteration, none of them has been used in compiling the averages. Furthermore, some of these rocks may have been alkali-rich when extruded.

The problems of bias caused by the over-weighting of one or two complexes no longer remain in the calculated averages. This is principally because the Alnö data, formerly a problem, are now greatly reduced in their influence by the upper SiO_2 limit of 10% and the greater number of analyses now available from elsewhere. No other complex constitutes more than 5% of the total data. The major bias now results from the few analyses with exceptionally high values for certain elements such as Ba and Nb

Samoilov (1984) has collected an extensive amount of carbonatite chemical data, particularly for trace elements. Unfortunately, Samoilov's data could not be incorporated into the averages given here, because they do not include the information necessary for subdivision using the CMF diagram criteria.

1.2.1 Element distribution

Heinrich (1966) has discussed the abundance and some aspects of the distribution of major and minor elements in carbonatites, referring to his own data and to the average compositions of Pecora (1956) and Gold (1963, 1966). The notable features of the compilations presented in this paper (Tables 1.1 and 1.2) are briefly discussed below.

SiO_2. Silica increases through the series calciocarbonatite–magnesiocarbonatite–ferrocarbonatite. Late, drusy quartz is common in ferrocarbonatites and it has been suggested (e.g. Woolley 1969) that some of the silica present in carbonatites may be the result of desilicification of adjacent fenites. The reason for the much lower values than those previously reported has already been discussed.

TiO_2. Titania also increases through the series calciocarbonatite–magnesiocarbonatite–ferrocarbonatite. The averages obtained are somewhat lower than those reported by Gold (1966).

Al_2O_3. Averages for alumina are significantly lower than those previously reported, because of the exclusion of the high-silica carbonatites, which may include many aluminous silicates. There is little variation in Al_2O_3 contents among different carbonatite types.

Fe_2O_3. Iron – by definition – increases substantially through the series calciocarbonatite–magnesiocarbonatite–ferrocarbonatite. Only in the magnesiocarbonatites is $FeO > Fe_2O_3$, probably reflecting the presence of ankerite. In ferrocarbonatites much of the Fe is present as Fe oxides, including magnetite, but these rocks are almost invariably oxidized. Much of the Fe in calciocarbonatites is probably contained in magnetite. The $Fe^{2+}:Fe^{3+}$ ratios support such a conclusion.

MnO. Manganese characteristically increases through the series calciocarbonatite–magnesiocarbonatite–ferrocarbonatite, with very high values ($>5\%$) in some ferrocarbonatites. Some magnesiocarbonatites are manganese-rich. This reflects the increase in Mn through the sequence calcite, dolomite, ankerite, and siderite (Quon & Heinrich 1965).

Na$_2$O and K$_2$O. The major alkalis are very low, and have similar abundances in all intrusive carbonatites.

P$_2$O$_5$. Phosphorus is equally abundant and variable through all carbonatite types.

CO$_2$. The lower values found in ferrocarbonatites probably reflect both the greater proportion of non-carbonate minerals in these rocks, and the fact that CO_2 is lower in $FeCO_3$ than in $CaCO_3$.

BaO. High Ba values are one of the chemical characteristics of carbonatites, and there is a notable increase through the sequence calciocarbonatite–magnesiocarbonatite–ferrocarbonatite, with exceptionally high values in some ferrocarbonatites ($>20\%$). Seven of the collected analyses range from 9.6% to 20.6% BaO so that two averages are given in Table 1.1, one including and the other excluding this group. Rocks high in BaO are usually rich in SO_3 indicating that baryte is present.

SrO. Although a few SrO values $>5\%$ are found in the ferrocarbonatites, there is little variation in overall abundance among carbonatites. However, the BaO:SrO ratio varies from about 0.4 in the calciocarbonatites to 1 in the magnesiocarbonatites and ferrocarbonatites, but values >1 occur in barium-rich ferrocarbonatites. Generally, Sr is considered to be more abundant than Ba in carbonatites (e.g. Quon & Heinrich 1965, Gold 1966), but the present data indicate that this is only true for calciocarbonatites.

F and Cl. Fluorine is much more abundant than Cl in carbonatites and there is an increase in the F:Cl ratio through the sequence calciocarbonatite–magnesiocarbonatite–ferrocarbonatite although, as with many other elements, there is considerable variation between localities and individual samples. Fluorine is mainly held in fluorapatite, but fluorite is also an important phase in many carbonatites.

SO$_3$. The strong correlation between SO_3 and BaO has already been pointed out. Two SO_3 averages are given in Table 1.1. These include and exclude the same seven rocks which have exceptionally high SO_3, as well as BaO, values.

Co, Cr, Ni, and V. Cobalt, Cr, and V all increase, as would be expected, through the series calciocarbonatite–magnesiocarbonatite–ferrocarbonatite, but Ni shows a slight decrease in the ferrocarbonatites; this is probably a statistical aberration due to the few available data.

Cu. Values for Cu are not particularly high in carbonatites in general. The averages in Table 1.1 do not include data from Phalaborwa, South Africa, which would clearly enhance them significantly.

Nb. Niobium is one of the characteristic trace elements of carbonatites, and is principally concentrated in pyrochlore. Both the calciocarbonatite and ferrocarbonatite data indicate values of > 1200 p.p.m. while magnesiocarbonatites contain about 600 p.p.m. However, Nb ranges in magnesiocarbonatites up to 17 000 p.p.m. so that it is not clear if the lower average value for magnesiocarbonatites indicates a fundamental difference from other carbonatite types or is a statistical anomaly that reflects a limited database.

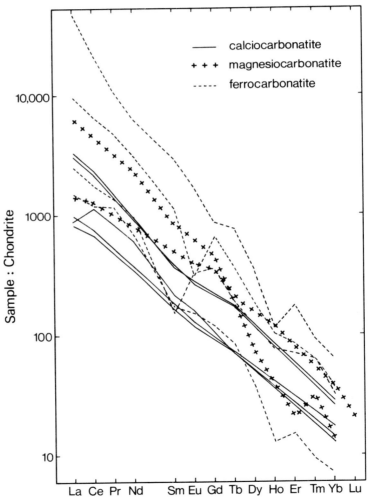

Figure 1.5 REE abundance patterns, normalized to chondrites (Wakita *et al.* 1971) for a range of carbonatites. Although there appears to be no systematic difference between the calciocarbonatites and magnesiocarbonatites, there is some indication of REE enrichment in average ferrocarbonatites. The steep patterns with very high, light REE abundances are typical of carbonatites.

Th and U. It has been shown at a number of localities (Heinrich 1966) that Th:U ratios in carbonatites are usually greater than unity and may be as high as 16 (Heier 1962). The Th:U ratios of the averages in Table 1.1 vary from 6 in the calciocarbonatites and 7 in the magnesiocarbonatites to nearly 40 in the ferrocarbonatites. In general, this reflects an increase in Th through the series.

The average values for Th and U quoted in Table 1.1 are very much lower than the average calculated by Gold (1966), although he included many more analyses. The reason for this difference is not clear, although much of the earlier data would have been determined spectrographically and hence excluded from the present study.

Rare earth elements (REEs). A useful review of the REEs in carbonatites is given by Cullers and Graf (1984), who point out that carbonatites contain the highest contents of REEs and highest LREE:HREE (La:Lu) ratios of any igneous rocks. It is particularly unfortunate that much of the REE data in the literature could not be used in this study because of the lack of accompanying information on major elements. It is apparent from Table 1.2 that La and Ce increase through the series calciocarbonatite–magnesiocarbonatite–ferrocarbonatite. For the remainder of the REEs, however, there is some variation, but generally the ferrocarbonatites contain the highest REE abundances. Some of the available data from the literature for carbonatites, classified according to the system recommended earlier in this paper, are shown on Figure 1.5. There is no consistent difference between the REE patterns for calciocarbonatites and magnesiocarbonatites. However, the ferrocarbonatites are, on the whole, richer in REEs, although the characteristic enrichment in the light REEs is common to all carbonatites.

Other elements. Most of the other trace elements given in Table 1.1 are based on too few analyses to be discussed in any detail, and it is clear that there are little or no high quality data for many elements.

ACKNOWLEDGEMENTS

We are extremely grateful to Dr M. J. Le Bas and Dr T. Deans for long discussions about this paper and for making numerous suggestions for its improvement. Dr A. C. Bishop also kindly read the paper and suggested a number of improvements. We are also indebted to Nick Goldman who gave valuable help and advice in the computer handling of the carbonatite analyses database. Valerie Jones drafted the figures.

REFERENCES

Barth, T. F. W. & I. B. Ramberg 1966. The Fen circular complex. In *Carbonatites*, O. F. Tuttle & J. Gittins (eds), 225–57. New York: Wiley.

Brögger, W. C. 1921. Die Eruptivgesteine des Kristianiagebietes. IV. Das Fengebiet in Telemark, Norwegen. *Videnskapsselskapets Skrifter. I. Mat-Naturv. Klasse* **1920**, No. 9.

Cullers, R. L. & J. L. Graf 1984. Rare earth elements in igneous rocks of the continental crust: predominantly basic and ultrabasic rocks. In *Developments in Geochemistry*, Vol. 2: *Rare earth element geochemistry*, P. Henderson (ed.), 237–74. Amsterdam: Elsevier.

Eckermann, H. von 1928. Dikes belonging to the Alnö-formation in the cuttings of the East coast railway. *Boletin, Geologiska Foreningens i Stockholm Forhandlingar*.

Eckermann, H. von 1948. The alkaline district of Alnö Island. *Sveriges Geologiska Undersokning*, No. 36.

Gold, D. P. 1963. Average chemical composition of carbonatites. *Economic Geology* **58**, 988–91.

Gold, D. P. 1966. The average and typical chemical composition of carbonatites. *International Mineralogical Association, Papers, Fourth General Meeting, Mineralogical Society of India*, 83–91.

Heier, K. S. 1962. A note on the U, Th and K contents in the nepheline syenite on Sternoy, North Norway. *Norsk Geologisk Tidsskrift* **42**, 287–92.

Heinrich, E. W. 1966. *The geology of carbonatites*. Chicago: Rand McNally.

Kapustin, Y. L. 1980. *Mineralogy of carbonatites*. New Delhi: Amerind Publishing Company. (translated from Russian)

Meyer, A. & P. de Bethune 1958. La carbonatite Lueshe (Kivu). *Bulletin du Service Géologique, Congo Belge et Ruanda-Urundi, Léopoldville* **8**(5), 1–19.

Pecora, W. T. 1956. Carbonatites: a review. *Bulletin of the Geological Society of America* **67**, 1537–56.

Quon, S. H. & E. W. Heinrich 1965. Abundance and significance of some minor elements in carbonatitic calcites and dolomites. *International Mineralogical Association, Papers, Fourth General Meeting, Mineralogical Society of India*, 29–36.

Samoilov, V. S. 1977. *Carbonatites (facies and conditions of formation)*. Moskva: Izd-vo Nauka. (in Russian)

Samoilov, V. S. 1984. *Geochemistry of carbonatites*. Moskva: Izd-vo Nauka. (in Russian)

Streckeisen, A. 1980. Classification and nomenclature of volcanic rocks, lamprophyres, carbonatites and melilitic rocks. IUGS Subcommission on the Systematics of Igneous Rocks. *Geologische Rundschau* **69**, 194–207.

Wakita, H., P. Rey & R. A. Schmitt 1971. Abundances of the 14 rare-earth elements and 12 other trace elements in Apollo 12 samples: five igneous and one breccia rocks and four soils. *Proceedings of the Second Lunar Science Conference*, 1319–29.

Woolley, A. R. 1969. Some aspects of fenitization with particular reference to Chilwa Island and Kangankunde, Malawi. *Bulletin of the British Museum (Natural History), Mineralogy* **2**(4), 189–219.

Woolley, A. R. 1982. A discussion of carbonatite evolution and nomenclature, and the generation of sodic and potassic fenites. *Mineralogical Magazine* **46**, 13–7.

2
THE SPATIAL AND TEMPORAL DISTRIBUTION OF CARBONATITES

A. R. WOOLLEY

ABSTRACT

About 330 carbonatites are now known world-wide. The majority are located in relatively stable, intra-plate areas but some are found near to plate margins and may be linked with orogenic activity or plate separation. They are commonly located on major lithospheric domes or are related to major lineaments, or both. Few carbonatites of Archaean age are known. Overall, there appears to have been an increase in carbonatitic activity with time, although this is episodic and appears to be temporally and spatially linked to major orogenic events. Carbonatites tend to form clusters or provinces, and in many areas there has been repetition of carbonatitic activity with time. In West and South Greenland, for instance, carbonatites were emplaced during four periods extending over 2500 Ma, while in the East African rifts late Proterozoic and early Palaeozoic carbonatites are juxtaposed with those of Cretaceous age. Localization of carbonatite activity over several time periods suggests lithospheric control.

2.1 INTRODUCTION

About 330 carbonatite occurrences are now known, compared with just under 200 listed some 20 years ago (Tuttle & Gittins 1966, Deans 1968). Many new discoveries are in areas previously unrepresented. Occurrences in Australia, for example, were not known 20 years ago; neither were the concentrations in Finland, nor those in Tamil Nadu and the Eastern Ghats of India. Other areas that formerly had only one or two examples are now known to be areas of major carbonatitic activity. It has been generally held that carbonatites are characteristic of the central parts of anorogenic continental areas, but the increase in the number of known occurrences now shows that many can be linked with orogenic activity. Furthermore, some carbonatites occur close to plate margins and their distribution is clearly related to plate movement (Garson 1984). Although only about half the known carbonatites have been dated, it is apparent that there has been a steady increase in carbonatitic activity with time. New geochronological information shows that in a number of areas carbonatitic magmatism was episodic and extended over hundreds of millions of years.

2.2 SPATIAL AND TEMPORAL DISTRIBUTION

2.2.1 Africa

Approximately half of the known carbonatites occur in Africa, with the majority concentrated in or close to the East African Rift, in a broad zone trending southwards from Kenya through Mozambique into South Africa (Fig. 2.1). There is a lesser, but significant concentration, along the southwest African coast. Generally these concentrations are associated with major faults, which may define rifts, but equally the faults may simply form sub-parallel sets or sub-radial patterns. In some areas of carbonatitic activity there is apparently a scarcity of large faults.

All the major carbonatite concentrations are associated with topographic highs, or swells, and it would appear that the faulting can often be interpreted as a response to crustal doming. Le Bas (1971) proposes about 18 major swells in Africa most of which are associated with carbonatites. Two major swells in Angola and Namibia

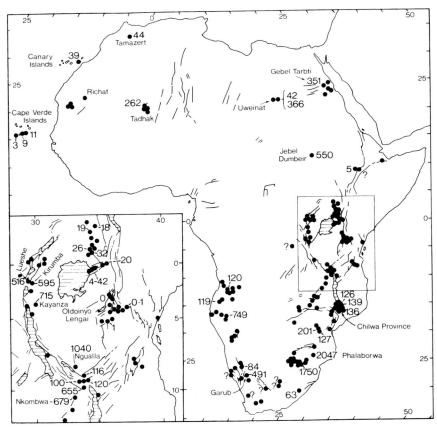

Figure 2.1 Distribution and dates in Ma of carbonatites in Africa. Major faults are indicated. Localities for which a carbonatitic origin is in some doubt are indicated by a question mark. Most of the dates represent averages of several determinations.

are truncated by the coast but their westerly extensions can be recognized in eastern South America. The largest swells are over 1000 km across, while smaller ones are measured in hundreds of kilometres; the Rungwe swell, for example, is 200×200 km. These large swells grade in to smaller ones which are generally associated with a single carbonatite centre. The Richat Dome in Mauritania, for instance, is some 38 km in diameter and contains a system of radial carbonatite dykes (Woolley *et al.* 1984), but the doming associated with individual carbonatite complexes such as Napak in Uganda and Tundulu in Malawi, is generally only one or two kilometres across, although superimposed on a regional swell.

Some of the African carbonatites define, or help to define, lineaments, the origins of which are unclear. Bailey (1961), for instance, defined a mid-Zambezi–Luangwa lineament along the Luangwa Rift in Zambia, into southwestern Tanzania and the Kenya coast. The carbonatites of Angola, together with the numerous associated alkaline complexes, were shown by Lapido-Loureiro (1973) to lie on a number of southwest-trending lines, and Marsh (1973) suggested that alkaline igneous complexes in Angola, Brazil, Namibia, and Uruguay define lineaments that can be correlated with transform faults in the South Atlantic. This conclusion is illustrated even more clearly by Prins (1981). A correlation of carbonatites and alkaline complexes in Egypt with transform faults in the Red Sea is suggested by Garson & Krs (1976).

Some carbonatites are concentrated where lineaments and/or fault systems cross. The clearest example is probably the Chilwa Province of southern Malawi and Mozambique, which lies at the junction of the east–west-trending Zambezi Rift and the north–south-trending southern part of the East African Rift (Fig. 2.1). A major concentration occurs in southwest Tanzania where the two branches of the East African Rift merge and are crossed by the northeast-trending Zambezi–Luangwa lineament.

A further type of structural setting is illustrated by the carbonatites and undersaturated alkaline complexes of the Tadhak area of Mali. These appear to be related to a well-defined rift which lies just west of a suture between the passive margin of the West African craton and the active margin of the continental mass to the east (Liegeois *et al.* 1983). The Tamazert carbonatite of Morocco, situated in the mountains of the High Atlas, appears to be unique among African carbonatites because of its association with a young orogenic belt.

The Canary, and Cape Verde Islands, situated just off the northwest African coast, are the only oceanic islands from which carbonatites have so far been described. The Canary Islands lie close to the edge of the continental shelf and appear to represent a westward continuation of Alpine structures in the Moroccan Atlas Mountains. The Cape Verde Islands, however, are probably truly oceanic (M. J. Le Bas, pers. comm.), so that these are the only carbonatites so far described that do not lie above continental lithosphere.

Only about one-quarter of the African carbonatites have been dated; these dates are shown in Figure 2.2. The oldest, Phalaborwa, in the Kaapvaal craton of South Africa, has an age of 2047 Ma (Eriksson 1984 and this volume) determined by the U–Pb method. There are ten or so further carbonatites known within this craton

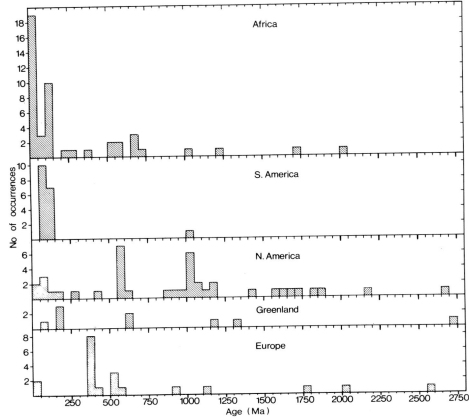

Figure 2.2 Frequency distribution diagram of carbonatite dates. When more than one age determination is available for a locality, an average has generally been plotted. No attempt has been made to recalculate all dates using the most recent decay constants.

and clearly much more is to be learned about this oldest group of African carbonatites.

Carbonatites between 139 and 116 Ma old are found in Angola, Zimbabwe, Malawi, and Tanzania and are generally considered synchronous with the breakup of Pangaea. It is probable that some of the Namibian, and probably also many of the undated occurrences in the western part of South Africa, belong to this group. Carbonatites of this age located in Malawi, Tanzania, and Zimbabwe are spatially associated with rifts, but undated carbonatites in Mozambique probably belong to this group.

It has generally been considered that in the East African Rift igneous activity becomes younger northwards. This is true in so far as present activity is confined to the northern half of the rift, but the apparent pattern has been much changed recently by the determination of Proterozoic ages for carbonatites and alkaline complexes in the central and western parts of the rift. In southwestern Tanzania there is a group of carbonatites with ages of 680 Ma to 730 Ma, while the Ngualla carbonatite is 1040 Ma (van Straaten 1986). The nearby Nkombwa carbonatite in Zambia gave a K–Ar date of 679 Ma (Bloomfield 1970).

Dates from Western Rift carbonatites include a K–Ar determination on biotite from Lueshe in Zaire of 516 Ma (Pouche 1979, quoted by von Maravic & Morteani 1980). Rb–Sr biotite dates from nepheline syenites and syenites belonging to the Kirumba carbonatite complex gave 555 Ma and 635 Ma (Cahen & Snelling 1966). Similar dates have been obtained from the Kayanza carbonatite along the Western Rift in Burundi (Kampunzu 1981, Tack *et al.* 1983). An average of 715 Ma for Kayanza is given in Figure 2.1. The Jebel Dumbeir carbonatite–nepheline syenite complex in the Sudan (Fig. 2.1), for which a Rb–Sr whole-rock isochron indicated an age of 550 Ma (Harris *et al.* 1983), may also belong to this group. In this context it is also worth noting that in Angola, of the three carbonatites dated, two are Cretaceous but one is Proterozoic in age (Lapido-Loureiro 1973). In southern Namibia a Rb–Sr mineral-whole-rock isochron on the Garub carbonatitic sill yielded a date of 491 Ma (Allsopp *et al.* 1979).

Two important points are indicated by this double grouping of dates. First, in southwestern Tanzania and Angola there is an intimate association of Proterozoic and Cretaceous carbonatites (see van Straaten, this volume). Secondly, the existence of carbonatites, and some alkaline complexes, spatially associated with the rift but Proterozoic in age, suggests the existence of some fundamental, long-lived lithospheric structure along the rift-valley system.

Gebel Tarbti is the only Egyptian carbonatite that has been dated, with nordmarkites from the complex yielding 351 Ma by the K–Ar method (Serencsits *et al.* 1979). However, the numerous alkaline complexes in the same area of Egypt give a spectrum of ages, and it is possible that the carbonatites will show the same range. It is tempting to interpret the concentration of Egyptian carbonatites close to the Red Sea coast as being structurally comparable with the provinces of Angola and Namibia and their relationship to the initial opening of the Atlantic Ocean. However, the date of 351 Ma from Gebel Tarbti seems much too old for it to be related to the initial fracturing and opening of the Red Sea. On the other hand, it could be that, as in the East African Rift, fracturing took place on an ancient and active lineament along which carbonatitic and alkaline igneous activity had long been focused.

In eastern Uganda and the Kavirondo Gulf area of western Kenya carbonatite activity started about 42 Ma and ceased about 4 Ma ago. Subsequent activity was concentrated in southern Kenya and northern Tanzania, and still continues, as spectacularly demonstrated by the carbonatite lava flow erupted recently from Oldoinyo Lengai (see Dawson, this volume).

A range of dates has also been obtained from the Uweinat carbonatite-bearing complex of Libya, with associated granites giving 42 Ma to 45 Ma but syenites 336 Ma (Vail 1976). It is not clear whether the carbonatite is consanguineous with either of these two events but if, as seems likely, it is genetically related to the syenites then it probably correlates with the Gebel Tarbti carbonatite of Egypt.

The carbonatites and undersaturated alkaline complexes of the Tadhak area of Mali are relatively isolated from other African occurrences (Fig. 2.1) and only the Tadhak carbonatite complex itself (262 Ma) has so far been dated (Liegeois *et al.* 1983). The Tamazert carbonatite of Morocco, for which dates of 44.8 Ma and

44.4 Ma have been obtained by the K–Ar and Rb–Sr methods respectively (Lancelot & Allègre 1974), correlates closely with a date of 38.6 Ma by K–Ar on aegirine from a syenite associated with a carbonatite on Fuerteventura, Canary Islands (Abdel-Monem et al. 1971). These dates support the contention that these islands are an extension of the Atlas Mountains and that the syenite is of similar age to the Tamazert carbonatite.

The widespread carbonatitic activity within the Cape Verde archipelago is possibly unique in its oceanic setting, and does not correlate in age with any other occurrences in West Africa. K–Ar dates of 11.1 Ma (Mitchell et al. 1983) and 10.3 Ma (Bernard-Griffiths et al. 1975) have been obtained for carbonatite from Maio, as well as 8.5 Ma and 9.8 Ma for Santiago (Bernard-Griffiths et al. 1975). A fission-track date on apatite from carbonatite on Brava indicates a much younger event at 3.4 Ma (Lancelot & Allègre 1974). Age data are not yet available for the carbonatites of Mauritania.

2.2.2 South America

The large concentration of carbonatites in southern Brazil has been known for many years, but recently several more have been discovered on the north side of the Amazon Basin. Together with a number of alkaline complexes, this Amazon group constitutes a major province (Fig. 2.3). Four carbonatites are known in the Amazon group, two in Brazil and one each in Guyana and Venezuela. A large ring structure that has been observed from the air in eastern Colombia may constitute a fifth. Of these carbonatites, only Mutum on the Brazil–Guyana border has been dated, and it yields a K–Ar date of 1026 Ma (Issler et al. 1975). However, alkaline complexes in the Amazon province dated by the Rb–Sr method range between 1800 Ma and 1500 Ma and it is probable that Mutum was emplaced within that time range. Other complexes of the same age may also lie in the Guyana craton hidden beneath the immensely thick laterite cover.

The carbonatites of southern Brazil comprise two groups. The more northerly (Fig. 2.3), in Goias and Minas Gerais, lies along an old, northwestward-trending lineament – the Alto-Paranaiba Uplift. The latter lies between the São Francisco Craton and the Paraná Basin, and is marked by gravimetric and magnetic highs. This group includes Araxá, Catalão, Salitre, and Tapira and most have similar ages, within the range 87 Ma to 70 Ma (Herz 1977). Eby & Mariano (1986) report an average date for Catalão I of 114 Ma using the fission-track method. The more southerly group, in the states of São Paulo, Santa Catarina, and Paraná, includes Anitapolis, Jacupiranga, and Lajes (Fig. 2.3), and contains both older and younger complexes in the range 131–65 Ma (Herz 1977, Eby & Mariano 1986). According to Eby & Mariano (1986), the carbonatites as a whole fall into three periods of emplacement that occurred 70–60 Ma, 90–80 Ma, and 135–110 Ma ago.

The more southerly group lies scattered across the Ponta Grossa Arch, and includes the Chiriguelo and adjacent Cerro Sarambí carbonatites of Paraguay (Fig. 2.3). Chiriguelo has been dated at 128 Ma by the K–Ar method on biotite (A. N. Mariano & M. D. Druecker, pers. comm.), which is similar to the ages obtained

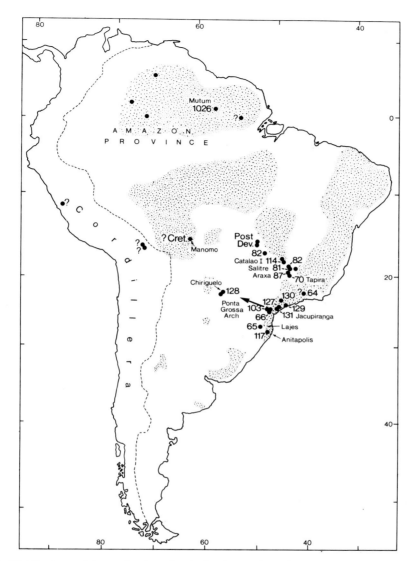

Figure 2.3 Distribution and dates of carbonatites in South America. Of the two major concentrations in Brazil, the more northerly lies along the northwest-trending Alto Paranaiba Uplift, and the southerly is distributed about the Ponta Grossa Arch; the axis of the latter is indicated by the arrow and extrapolates towards the Chiriguelo and Cerro Sarambí carbonatites in Paraguay. Question marks indicate localities for which a carbonatitic origin is in some doubt. Shaded areas indicate Precambrian cratons.

from the Brazilian carbonatites. Further to the northwest lies the large Manomo carbonatite of Bolivia, which has not been dated but which is thought also to be Cretaceous in age (Fletcher *et al.* 1981).

The extensive carbonatite and alkaline province of southern Brazil has been attributed by Herz (1977) and others to plume activity below a triple junction. The Minas Gerais–Goias group is thought to mark a failed arm. A pre-drift reconstruction of South America and Africa (Fig. 2.4), however, does not seem to lend support

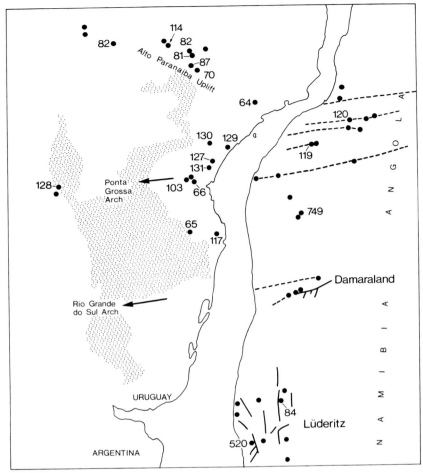

Figure 2.4 Pre-drift reconstruction of parts of South America and Africa to show the continuity of carbonatite provinces in Brazil and Angola. The dashed lines linking carbonatites in Angola are from Lapido-Loureiro (1973) and are collinear with the axis of the Ponta Grossa Arch, which is indicated by an arrow. The Rio Grande do Sul Arch is also shown. The shading indicates the distribution of the Paraná basalts in Brazil, Paraguay, and Uruguay.

to the presence of a triple junction, and the lack of a simple linear distribution for the complexes rules out either a migrating plume or a continental plate moving over a plume. On the other hand, the axis of the Ponta Grossa Arch does appear to parallel the lines of complexes distinguished by Lapido-Loureiro (1973) in Angola (Fig. 2.4). A westward continuation of the Lüderitz alkaline province in Namibia may continue into South America (see Fig. 2.4), and Marsh (1973) correlates some of the Namibian carbonatites with alkaline complexes in Uruguay. A possible continuation of the province into adjacent parts of Argentina is obscured by a cover of Quaternary deposits. It may be noteworthy that the Damaraland province of Namibia aligns with the Rio Grande do Sul Arch of southern Brazil (Fig. 2.4).

Dates for carbonatites on either side of the Atlantic are broadly coherent. The histograms of Figure 2.2 clearly indicate the correlation of the essentially Cretaceous carbonatite activity of South America with the major Cretaceous carbonatite event found over much of southern Africa. However, no equivalents have yet been found in South America to the complexes in Angola and South Africa dated at 749 Ma and 491 Ma.

2.2.3 North America

Carbonatites are widely distributed throughout North America (Fig. 2.5). Although most occur in the southern part of the Canadian Shield, some are scattered over the whole of the Precambrian craton and also along the central and southern Cordillera (Fig. 2.5). Mitchell (1973) obtained a common Pb age of 1523 Ma from Mountain Pass, and Lanphere (1964) obtained 1385 Ma and 1440 Ma by K–Ar and 1380 Ma and 1450 Ma by Rb–Sr from shonkinite of the same complex. More recent work by Dewitt *et al.* (1987) has yielded 1375 Ma for the carbonatite, by the Th–Pb method on monazite, with the associated alkaline igneous rocks emplaced between 1410 Ma and 1400 Ma. Dates from other carbonatites of western North America of 770 Ma, 570–511 Ma, 380–325 Ma, 99–90 Ma, 55–38 Ma, and Recent have also been found (Fig. 2.5). Most of these carbonatites constitute a very minor proportion of extensive provinces of alkaline rocks.

Special mention should be made of the carbonatites recently recognized in British Columbia, particularly in the vicinity of the Frenchman Cap gneiss dome, which are commonly sheet-like and laterally extensive (Pell 1986); the Mount Grace carbonatite, for example, extends over tens of kilometres and is extrusive. Most of these carbonatites have been metamorphosed during the Columbian orogeny (Pell 1986). Emplacement was in two episodes, the first at about 770 Ma and the second at 380–325 Ma (see Pell & Hoy, this volume).

Carbonatites are plentiful within the Canadian Shield, and although alkaline silicate rocks are also generally present, the carbonatites are volumetrically much more important than in the Cordillera. Kaminak Lake (Fig. 2.5) has been dated at 1820 Ma by K–Ar (Davidson 1970) but a Rb–Sr isochron gave a date of 2686 Ma (Wanless, in Currie 1976, p. 103). Big Spruce Lake in the Northwest Territories yielded dates of about 2180 Ma by Pb–Pb, Rb–Sr, and Sm–Nd (Cavell & Baadsgaard 1986, Cavell *et al.* 1986), while Castignon Lake in Quebec gave 1873 Ma by K–Ar (Dressler 1975). Nisikkatch Lake and Carb Lake are also possibly very old complexes but have not, so far, been dated.

A large and well-known concentration of carbonatites is aligned along the Kapuskasing High between the southern end of James Bay and the eastern end of Lake Superior (Fig. 2.5). The Kapuskasing High carbonatites fall into two groups of approximately 1900 Ma and 1100 Ma (Bell *et al.* 1987). Some of these complexes have been dated by at least two different methods and there is reasonable agreement. The younger group along the Kapuskasing structure, together with several carbonatites just to the west and those lying within the Grenville orogen, are clearly related spatially and temporally to the Grenville Orogeny. The distribution of these carbonatites up to several hundred kilometres from the margin of the belt is notable (Fig. 2.5). Similar distributions elsewhere will be described later.

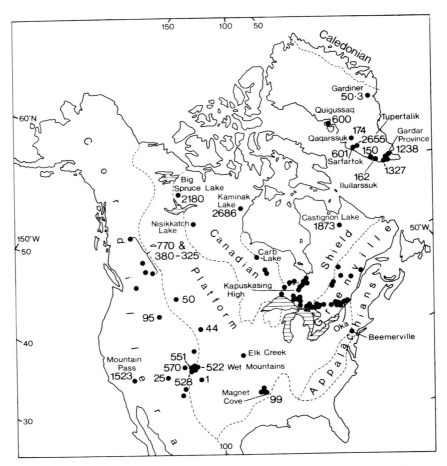

Figure 2.5 Distribution and dates of carbonatites in North America, including Greenland. The major thermotectonic provinces are indicated.

The Elk Creek carbonatite in Nebraska, USA, is overlain by Pennsylvanian limestones but has not been dated. It is, however, extrapolated along the southern extension of the Kapuskasing High and it may have been emplaced at the same time as either of the Kapuskasing groups, or it may be temporally associated with the Palaeozoic Wet Mountains carbonatites to the west (Fig. 2.5).

From the southern end of the Kapuskasing High a major concentration of carbonatites defines an approximately east–west line as far as Oka (Fig. 2.5), which is then continued into the Monteregian Hills. Along this line both of the Kapuskasing age groups are represented, together with a group around Lake Nipissing which includes the carbonatites of Callander Bay dated at 575–568 Ma (Currie 1976), and the Manitou Islands at 579–560 Ma (Lowdon et al. 1963, Gittins et al. 1967). Oka is much younger in age, yielding 109 Ma by Rb–Sr and 122 Ma (18 determinations) by fission-track dating (Gold et al. 1986). The Lake Nipissing complexes are matched in age by Arvida (Doig & Barton 1968), while the St. Honoré carbonatite should probably also be placed in this group because age determinations from it range from 656–629 Ma (Vallée & Dubuc 1970). These complexes (see Fig. 2.5) all lie to the west of the Appalachian orogen, in much the same way as the Kapuskasing High carbonatites lie to west of the Grenville Front.

The only carbonatite within the Appalachians is the minor occurrence at Beemerville in New Jersey (Fig. 2.5). Nepheline syenites associated with this carbonatite yield K–Ar and Rb–Sr dates that fall in the range 437–424 Ma (Zartman *et al.* 1967), similar to those associated with the Taconic orogeny.

2.2.4 *Greenland*

The dating of recently discovered carbonatites on the west coast of Greenland makes this one of the most remarkable carbonatite provinces of the world, illustrating as it does a very extensive range of ages within a relatively small area (Fig. 2.5). The Tupertalik intrusion yields K–Ar dates of about 2650 Ma (Larsen *et al.* 1983), making this one of the only three Archaean carbonatites known (the author has recently learned of a fourth newly discovered carbonatite in Canada that yields an Archaean date). In South Greenland, Rb–Sr whole rock isochrons on the Gardar Gronnedal-Ika and Igaliko intrusions yield dates of 1327 Ma and 1310–1167 Ma, respectively (Blaxland *et al.* 1978). It is perhaps noteworthy that the Gardar Province lies on a line extrapolated from the Kapuskasing High through the Castignon Lake carbonatite, a configuration that is little affected by moving Greenland into a pre-drift position (Fig. 2.5). The Grenville Front must originally have passed several hundred kilometres to the southeast of the Gardar Province, so that these essentially early Grenville intrusions have a similar spatial relationship to the Grenville Orogeny as do the carbonatites of Ontario and Quebec that lie west of the Grenville Front.

The Quigussaq (Umanak) carbonatite has been dated by K–Ar at 600 Ma (Larsen & Moller 1968) and Sarfartok at 603–599 Ma (Larsen *et al.* 1983), and thus both may be regarded as early Caledonian. The position of the western margin of the Caledonian Fold Belt in Greenland is indicated on Figure 2.5, and the location of Quigussaq and Sarfartok several hundred kilometres from the front is analogous to that of the St. Honoré and Arvida carbonatites of similar age in Canada with respect to the Appalachian front.

Another group of carbonatites in West Greenland, Qaqarssuk (176–169 Ma – Larsen *et al.* 1983), Frederikshaabs Isblink (150 Ma – Hansen 1981), and Iluilarssuk (162 Ma – Larsen & Moller 1968), have no obvious temporal equivalents elsewhere. The youngest complex that contains carbonatite in Greenland is the Gardiner intrusion in East Greenland (Fig. 2.5) which, at 50.3 Ma old (Gleadow & Brooks 1979), is one of the numerous Tertiary intrusions to be found in Greenland and elsewhere around the North Atlantic. It is not clear from its location whether Gardiner should be considered as part of the West Greenland carbonatite province, but even if excluded, the repetition of carbonatite activity along the west coast over a period of 2500 Ma makes this one of the most temporally extensive provinces of its kind in the world.

2.2.5 Europe

Northern Europe is another relatively small crustal segment in which carbonatites yield a broad spectrum of dates (Fig. 2.6). The oldest so far dated is Siilinjarvi in Finland, an elongated complex that has clearly been metamorphosed and tectonized. The Sm–Nd method yielded a figure of 2094 Ma (Basu & Puustinen 1982), but a U–Pb zircon date gave 2580 Ma (Patchett *et al.* 1981), making this complex closely comparable in age with Kaminak Lake and Tupertalik. The Laivajoki and Kortejarvi carbonatites of Finland have been dated by the K–Ar method at 2020 Ma

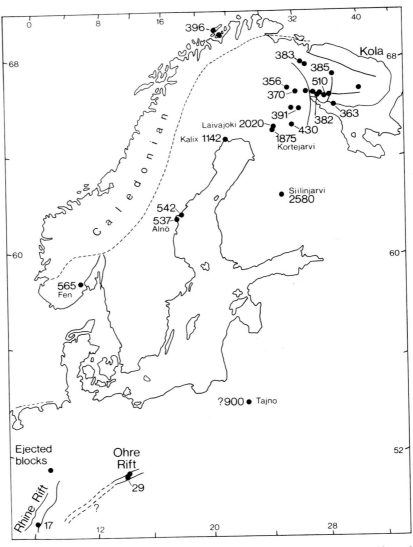

Figure 2.6 Distribution and dates of carbonatites in northern Europe. The Rhine and Ohre rifts and major faults in the Kola Peninsula are indicated.

(Vartiainen & Woolley 1974) and 1875 Ma (Kresten *et al.* 1977), but these complexes are probably metamorphosed, and are probably older.

Carbonatites are included among the dyke suite around Kalix in Sweden. Phlogopite from a breccia dyke gave a K–Ar date of 1142 Ma (Kresten *et al.* 1977). The significance of this essentially Grenville age is not clear, although rocks of Grenville age occur in southern Sweden. The Tajno carbonatite in Poland, detected beneath a sedimentary cover by drilling and thought to be about 900 Ma like other 'platform-type' intrusions in this part of the country (Dziedzic & Ryka 1983), is possibly related to the activity at Kalix.

The well-known Alnö and Fen carbonatite complexes have early Caledonian ages of 537 Ma (Kresten *et al.* 1977) and 565 Ma, respectively (Faul *et al.* 1959). The probable carbonatite intrusion of Arvika Bay, just north of Alnö, is also early Caledonian in age (Kresten *et al.* 1977). Dates in the range 540–480 Ma (Kononova & Shanin 1972) have been reported for Salmogorsk in the Kola Peninsula. The location of these four carbonatite complexes in the foreland to the Caledonian orogen (see Fig. 2.6) is essentially similar to that of the Caledonian carbonatites on the western side of the orogen in Greenland and Canada. This relationship was pointed out by Vartiainen & Woolley (1974) and contrasts with the view of Doig (1970) who linked the Caledonian complexes by a rift system extending across a pre-drift North American–European reconstruction.

Eight carbonatites in northern Finland and the Kola Peninsula yield dates between 430 Ma and 356 Ma and can be considered a late Caledonian group. These also lie in the Caledonian Foreland. The structural relationships of the carbonatites in northern Finland and the Kola Peninsula have been discussed by Vartiainen & Paarma (1979) who show that Kovdor, Turii, and the large Sokli carbonatite all lie on the Kandalaksha deep fracture, a lineament that can be traced from northern Finland through to the Polar Urals. This zone constitutes an ancient rift supposedly active since Archaean times. However, there are numerous other major fractures across the Kola Peninsula trending in several directions, reminiscent of the fault patterns across some of the alkaline provinces associated with doming in Africa.

The Kaiserstuhl carbonatite in the Rhine graben is 17 Ma old (Lippolt *et al.* 1963), while the Czechoslovakian carbonatites located in the Ohre rift (Fig. 2.6) are similar in age (L. Kopecky, pers. comm.). The relationship between the Rhine graben and the Alpine fold belt has been described by Illies & Greiner (1978), and once again the location and ages of the Kaiserstuhl and Ohre rift carbonatites point to a spatial and temporal relationship between carbonatite emplacement and orogenic activity.

2.2.6 USSR and Mongolia

Over 50 carbonatite occurrences are known within the USSR (Fig. 2.7). Those in the Kola Peninsula have already been referred to and comprise a major part of the northern European alkaline province (Vartiainen & Woolley 1974). Although alkaline complexes more than 2000 Ma old are known from Kola, all of the Kola carbonatites are late Caledonian or Hercynian in age. The oldest carbonatites

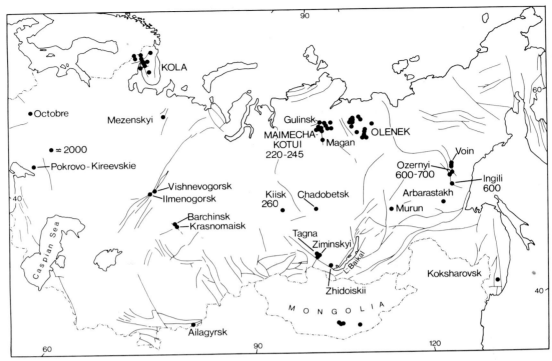

Figure 2.7 Distribution and dates of carbonatites in the USSR and Mongolia. Major faults are indicated.

known in the USSR are located in Ukraine and have ages in the range 2000–1900 Ma (Glevassky & Krivdik 1981, Bagdasarov *et al.* 1985).

The Maimecha-Kotui province (Fig. 2.7) lies on the northern part of the Siberian platform, and is related to rift structures in the basement (Borodin 1974). More than half of the alkaline complexes include carbonatite, of which the Gulinsk and Magan intrusions are particularly large and complex intrusions. All the dated complexes lie within the range 245–220 Ma (Borodin 1974).

The Aldan alkaline province, largely confined to the Aldan Shield is essentially divided into a northern and a southern part by the Stanov rift. Several other graben were also developed, particularly in Mesozoic times. There are more than 30 alkaline complexes in the province, but carbonatite occurs in only seven in the eastern part, where the complexes are emplaced in anticlinal structures. Carbonatite is the principal rock type in some complexes, covering, for instance, about 10 km^2 in the Arbarastakh complex and 8 km^2 in Ozernyi (Fig. 2.7). Borodin (1974) reports that most of the dated complexes fall in the range 300–250 Ma, but U–Pb dates indicate values from 700 Ma to 600 Ma for Ozernyi. Complexes associated with the Arbarastakh and Ingili massifs are 600 Ma old.

The group of carbonatites at Zhidoiskii, in the Sayan-Tuva province to the west of Lake Baikal (Fig. 2.7), have given U–Pb dates on pyrochlore of 450–410 Ma and 270–214 Ma (Borodin 1974); once again the area is characterized by large anticlinal structures and complex rift systems. To the north of Sayan, the anticlinorium of the

Enesei Ridge is divided by two systems of rifts, and alkaline complexes are concentrated close to the intersections of the two systems. The Kiisk carbonatite complex has been dated at 260 Ma (Borodin 1974).

To the south of Sayan in central Mongolia an east–west line of complexes includes four carbonatites (Samoilov & Kovalenko 1983) (Fig. 2.7).

It may be significant that, within the USSR, apart from the Kola, Maimecha-Kotui, and Olenek provinces, the principal concentrations of carbonatites and alkaline igneous complexes lie in faulted and rifted zones north of and adjacent to the central Asian and Pacific fold belts. There appears to be a broad temporal correlation of orogenic activity with the alkaline magmatic events in the stable, but heavily fractured, forelands to the north.

2.2.7 India–Pakistan–Afghanistan

About 20 carbonatites are known in the Eastern Ghats and Tamil Nadu provinces of southern India (Fig. 2.8). These are emplaced in Precambrian rocks but only Sevathur (Koratti), yielding 720 Ma (Deans & Powell 1968), and Jodipatti, yielding 700 Ma (Moralev *et al.* 1975), have been dated, both by the K–Ar method. From the descriptions of the alkaline silicate rocks from these provinces, some of the intrusions must have been metamorphosed, suggesting that older dates might be forthcoming. The Eppawala carbonatite of Sri Lanka has not been dated, but is likely to be part of the Tamil Nadu province.

The Amba Dongar carbonatite lies in the central part of the Deccan volcanic province; several small carbonatites also occur in the same area. Deans *et al.* (1973) obtained dates of 76 Ma and 61 Ma by the K–Ar method, showing that these carbonatites were coeval with the main period of Deccan volcanism (Kaneoka & Haramura 1973). North of Amba Dongar, and still within the area affected by the Deccan magmatism, is the Newania carbonatite, dated at 959 Ma by Deans & Powell (1968). This complex, along with Amba Dongar, constitutes yet another example of spatially related carbonatites emplaced at quite different times.

The same theme is continued by carbonatites in the Peshawar Plain alkaline province in northern Pakistan and Afghanistan (Fig. 2.8). The Loe Shilman and Malakand carbonatites have yielded dates of 31 Ma and 25 Ma, and 33 Ma and 24 Ma, respectively (Le Bas *et al.* in press). The Koga carbonatite has not been dated, but it is associated with syenites which are dated at 300 Ma and 50 Ma (D. R. C. Kempe & M. J. Le Bas, pers. comm.). This group of carbonatites, thought to be associated with rifting, is particularly interesting because of its location close to the northern margin of the Indian Plate, just south of the collision zone with the Eurasian Plate. The carbonatites lie on the concave side of a sharp bend in the edge of the Indian Plate, in a similar structural position to the Caledonian carbonatites of the Kola Peninsula and northern Finland. The Khanneshin carbonatite in southern Afghanistan does not appear to have been dated, but this and two other possible carbonatites in Afghanistan (Fig. 2.8), lie in a region of major faulting related to the collision of the Indian and Eurasian plates.

The relationship of the Afghanistan and Pakistan carbonatites to the northern

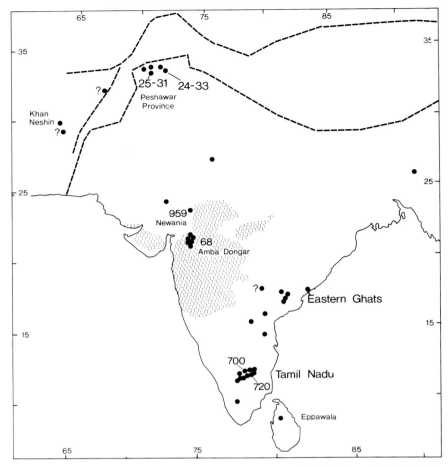

Figure 2.8 Distribution and dates of carbonatites in India, Pakistan, and Afghanistan. The dashed lines indicate the approximate limits of the Himalayan fold system. The shading shows the distribution of the Deccan basalts.

margin of the Indian Plate is remarkably similar to the structural situation of the four carbonatites of Mali, that were emplaced in the West African Craton but very close to its eastern margin (Liegeois *et al.* 1983). The structural similarity of these two areas, pointed out by Cahen *et al.* (1984, Fig. 21.8), is further emphasized by these carbonatite occurrences.

2.2.8 Other areas

Three carbonatites, and possibly a fourth, are known from Australia. K–Ar dating of the Mount Weld carbonatite in western Australia gave 2064 Ma (Goode 1981), while the Mud Tank carbonatite in the Strangways Range, which lies very close to the extensive Woolonga lineament, gave dates of 735 Ma, by the Rb–Sr method, and 732 Ma, by the U–Pb method (Black & Gulson 1978). Carbonatite dykes, associated with a lamprophyric dyke swarm in South Westland, South Island, New

Zealand, are located some 20 km from the Alpine fault and the swarm has been dated at 125 Ma (Wellman & Cooper 1971).

There appear to be at least eight intrusive carbonatite complexes in China, and, in addition, Ge & Zhongxin (1985) refer to some five 'marine facies volcano-sedimentary carbonatites'. In the absence of a full translation of their paper, however, the exact status of these carbonatites is not clear. Carbonatites are also known from Iran, Turkey, the Yemen, and possibly Italy.

2.3 TECTONIC CONSIDERATIONS

2.3.1 *Rift valleys, major faulting, and lithosphere doming*

Probably the single most widely held belief about the structural setting of carbonatites is that they are associated with rift valleys, and the carbonatites of East Africa, the Kaiserstuhl, and the Lake Nipissing area, among others, conform to this pattern. However, probably less than half the world's carbonatites occur in or near recognizable graben structures. The remainder are spatially related to major faults, of which a number of examples have been cited above. In some areas, e.g. the Kola Peninsula, the carbonatites are located at major fault intersections (Vartiainen & Paarma 1979).

In some areas, particularly in Africa, carbonatites and associated alkaline rocks are associated with structural domes. The domes are of various sizes, and it is apparent that much of the faulting, including rifting, is a response to the doming, and the carbonatites may be linked genetically in some way with these structures. Although the faulting associated with carbonatite provinces is relatively late, and probably a high-level response to doming, many of the lineaments associated with carbonatites have a geological history long pre-dating carbonatite emplacement. The alignment of transform faults in the South Atlantic with lines of alkaline complexes in Angola and Namibia (Marsh 1973) seems clearly to indicate a close genetic connection between magmatism and faulting.

2.3.2 *Major lineaments*

Reference has been made to several areas where carbonatite intrusions were emplaced along lines which vary in length from tens to several thousands of kilometres. Notable areas include the Kapuskasing group in Canada and the line extending from west of Lake Nipissing to Oka and continuing to the Monteregian Hills, the mid-Zambezi–Luangwa lineament in East Africa, and the lines of intrusions in Angola, Namibia, and South Africa. The carbonatites of East Africa, which are clearly associated with the Eastern and Western Rifts and the southerly continuation through Malawi and Mozambique, can also be considered as related to a major lineament, or lineaments, albeit rather complicated in detail.

Many lineaments undoubtedly persisted as active geological features over long periods of time. Along the Kapuskasing structure, for example, carbonatites range from about 1900 Ma to 1000 Ma (Bell *et al.* 1987) with a comparable age span along the Nipissing–Oka line. Similarly, the Proterozoic ages from carbonatites close to the East African Rift attest to the probable presence of some sort of older tectonic structure.

The correlation of lines of intrusion in southwest Africa with lines extrapolated from oceanic fracture zones across the continental margin along small circles, about a Cretaceous pole of rotation, has been described by Marsh (1973) and Prins (1981). Prins also pointed out that the spacing on the African continent of the linear igneous belts is nearly the same as that of many transform faults projected from the Mid-Atlantic Ridge. One possible inference from the southwest African occurrences is that the lines of intrusions lie along and above deep-seated fracture zones that probably extend into, if not to the base of, the lithosphere. That the linear distributions do not mark the passage of the continent over mantle plumes is an inference that is supported by the presence of intrusions of widely differing ages along some lineaments.

2.3.3 *Relationship to orogenic activity and plate margins*

Evidence cited in this paper shows that many carbonatites can be correlated in space and time with orogenic activity. This is significant not only because it must be taken into account in any general petrological model for carbonatites, but also because it is diametrically opposed to the generally accepted view that carbonatites are characteristic of stable continental, intra-plate areas. Although few carbonatites are located within the orogenic belts themselves, many occur in zones, up to several hundred kilometres wide, adjacent to the belts. This has been described for the North Atlantic area by Vartiainen & Woolley (1974), but may also hold for some of the older complexes in East Africa lying west of the Mozambique Belt, the young carbonatites of northern Pakistan, those of British Columbia, Canada, and possibly some of those of Mongolia and the southern USSR.

2.4 REPETITION OF CARBONATITE ACTIVITY

It is clear that in many areas there have been repeated periods of carbonatite emplacement. The best example of one such area so far recognized is South and West Greenland where there have been four distinct episodes within a period of 2500 Ma. Other examples include the carbonatites along the Kapuskasing structure and the Nipissing–Oka line in Canada, and the recent recognition of Pre-Cambrian and Palaeozoic carbonatites juxtaposed with much younger ones along certain sections of the East African Rift.

This periodic carbonatitic activity must indicate that the lithosphere plays an important role in the genesis of these rocks. The alternative hypothesis, involving mantle plumes, necessitates unlikely repeated impingement on the same area of lithosphere by asthenospheric material at intervals of some hundreds of millions of

2.5 INCREASE IN CARBONATITE ACTIVITY WITH TIME

Only about half the known carbonatites have been dated (see Fig. 2.9), mostly by the K-Ar method. In spite of the fact that the K-Ar dates are probably minimum limits for carbonatite activity, it is apparent from Figure 2.9 that the dates fall into groups which generally correspond to major orogenic and tectonic events. There is a middle Proterozoic group corresponding to the Hudsonian and Svecokarelian orogenies of North America and Europe respectively. A broad, major, later Proterozoic event corresponds to the Grenville orogeny. The peak between 750 Ma and 500 Ma includes both early Caledonian dates from both northern Europe and North America, as well as some African dates. A major period of carbonatite activity starting 200 Ma ago is perhaps associated with the breakup of Pangaea.

Although there are still many carbonatites to be dated, there is now sufficient information to suggest that Figure 2.9 reflects, with some certainty, the variation of carbonatite activity with time. The gradual increase in the number of carbonatite occurrences with time is undoubtedly real, and implies that the conditions necessary for the production of carbonatite were not only established by the late Archaean, but have become increasingly widespread with time.

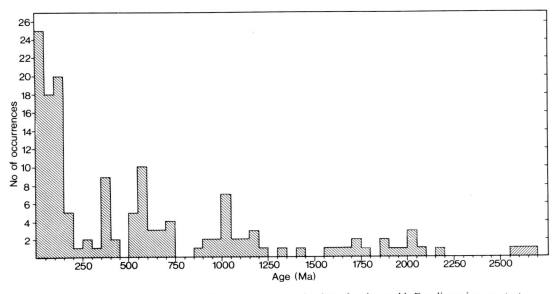

Figure 2.9 A frequency distribution diagram of carbonatite dates for the world. For discussion, see text.

2.6 GENERAL FEATURES OF THE TEMPORAL AND SPATIAL DISTRIBUTION OF CARBONATITES

The distribution in space and time of carbonatites has highlighted a number of general features important in any petrogenetic scheme for the origin of these rocks. These are:

(a) Many carbonatites appear to be related, in a general way, both spatially and temporally, to orogenic belts and to constructive and destructive plate margins.
(b) Carbonatites can be aligned with oceanic fracture zones.
(c) Carbonatites commonly occur on uplifted or domed areas which vary from tens to thousands of kilometres in diameter.
(d) Carbonatites are generally associated with major faulting and rifting related to doming.
(e) Carbonatite magmatism can be repeated over long periods of time.
(f) The generation of carbonatite magmas has increased with time.
(g) The necessary conditions for the production of carbonatite magmas were established by the end of the Archaean.

ACKNOWLEDGEMENTS

I am most grateful to Dr A. C. Bishop, Dr T. Deans, Dr D. R. C. Kempe, and the editor for critically reading the manuscript and making various suggestions for its improvement. Dr M. P. Orlova and Dr V. S. Samoilov kindly provided information on carbonatites of the USSR and Dr M. J. Le Bas made available age data on a carbonatite from Pakistan. Miss V. Jones drafted the figures.

REFERENCES

Abdel-Monem, A., N. D. Watkins & P. W. Gast 1971. Potassium–argon ages, volcanic stratigraphy, and geomagnetic polarity history of the Canary Islands: Lanzarote, Fuerteventura, Gran Canaria, and La Gomera. *American Journal of Science* **271**, 490–521.

Allsopp, H. L., E. O. Kostlin, H. J. Welke, A. J. Burger, A. Kroner & H. J. Blignault 1979. Rb–Sr and U–Pb geochronology of late Precambrian–early Palaeozoic igneous activity in the Richtersveld (South Africa) and southern South West Africa. *Transactions of the Geological Society of South Africa* **82**, 185–204.

Bagdasarov, Yu. A., S. N. Voronovskiy & L. V. Ovchinnikova 1985. Geologic position and radiometric age of a new carbonatite occurrence found in the area of the Kursk magnetic anomaly. *Doklady of the Academy of Sciences of the USSR, Earth Science Section* **282**(3), 84–7.

Bailey, D. K. 1961. The mid-Zambezi–Luangwa Rift and related carbonatite activity. *Geological Magazine* **98**, 277–84.

Bailey, D. K. 1977. Lithosphere control of continental rift magmatism. *Journal of the Geological Society of London* **133**, 103–6.

Basu, A. R. & K. Puustinen 1982. Nd-isotopic study of the Siilinjarvi carbonatite complex, eastern Finland, and evidence of early Proterozoic mantle enrichment. *Geological Society of America Abstracts with Programs* **14**, 140.

Bell, K., J. Blenkinsop, S. T. Kwon, G. R. Tilton & R. P. Sage 1987. Age and radiogenic isotopic systematics of the Borden carbonatite complex, Ontario, Canada. *Canadian Journal of Earth Sciences* **24**, 24–30.

Bernard-Griffiths, J., J-M. Cantagrel, C. A. Matos Alves, F. Mendes, A. Serralheiro & J. R. de Macedo 1975. Données radiométriques potassium–argon sur quelques formations magmatiques des Iles de l'archipel du Cap Vert. *Comptes Rendus Hebdomadaire des Séances de l'Académie des Sciences, Paris* **280**, 2429–32.

Black, L. P. & B. L. Gulson 1978. The age of the Mud Tank carbonatite, Strangways Range, Northern Territory. *BMR Journal of Australian Geology and Geophysics* **3**, 227–32.

Blaxland, A. B., O. van Breeman, C. H. Emeleus & J. G. Anderson 1978. Age and origin of the major syenite centres in the Gardar province of South Greenland: Rb–Sr studies. *Bulletin of the Geological Society of America* **89**, 231–44.

Bloomfield, K. 1970. Orogenic and post-orogenic plutonism in Malawi. In *African magmatism and tectonics*, T. N. Clifford & I. G. Gass (eds), 119–55. Edinburgh: Oliver & Boyd.

Borodin, L. S. 1974. *The principal provinces and formations of alkaline rocks.* Moscow: Izdatel'stvo Nauka. (in Russian)

Cahen, L. & N. J. Snelling 1966. *The geochronology of Equatorial Africa.* Amsterdam: North Holland.

Cahen, L., N. J. Snelling, J. Delhal & J. R. Vail 1984. *The geochronology and evolution of Africa.* Oxford: Clarendon Press.

Cavell, P. A. & H. Baadsgaard 1986. Geochronology of the Big Spruce Lake alkaline intrusion. *Canadian Journal of Earth Sciences* **23**, 1–10.

Cavell, P. A., H. Baadsgaard & R. St. J. Lambert 1986. Sm–Nd, Rb–Sr and U–Pb systematics of the Big Spruce Lake alkaline-carbonatite complex. *Geological Association of Canada–Mineralogical Association of Canada–Canadian Geophysical Union. Joint Annual Meeting, Ottawa. Program with Abstracts* **11**, 53.

Currie, K. L. 1976. *The alkaline rocks of Canada.* Geological Survey of Canada Bulletin, 239.

Davidson, A. 1970. *Precambrian geology, Kaminak Lake map-area, District of Keewatin.* Geological Survey of Canada, Paper 69-51.

Deans, T. 1968. World distribution of carbonatites in relation to volcanism. *Proceedings of the Geological Society of London* **1647**, 59–61.

Deans, T. & J. L. Powell 1968. Trace elements and strontium isotopes in carbonatites, fluorites and limestones from India and Pakistan. *Nature* **218**, 750–2.

Deans, T., R. N. Sukheswala, S. F. Sethna & S. G. Viladkar 1973. Metasomatic feldspar rocks (potash fenites) associated with the fluorite deposits and carbonatites of Amba Dongar, Gujarat, India. *Transactions of the Institution of Mining and Metallurgy, Section B, Applied Earth Science* **795**, 33–40.

Dewitt, E., L. M. Kwak & R. E. Zartman 1987. U–Th–Pb and $^{40}Ar/^{39}Ar$ dating of the Mountain Pass carbonatite and alkalic igneous rocks, S. E. Cal. *Geological Society of America, Abstracts with Programs* **19**, 642.

Doig, R. 1970. An alkaline rock province linking Europe and North America. *Canadian Journal of Earth Sciences* **7**, 22–8.

Doig, R. & J. M. Barton 1968. Ages of carbonatites and other alkaline rocks in Quebec. *Canadian Journal of Earth Sciences* **5**, 1401–7.

Dressler, B. 1975. Lamprophyres of the north-central Labrador Trough, Quebec, Canada. *Neues Jahrbuch für Mineralogie Stuttgart. Monatshefte* **1807**, 268–80.

Dziedzic, A. & W. Ryka 1983. Carbonatites in the Tajno intrusion (NE Poland). *Archiwum Mineralogiczne. Warszawa* **38**, 4–34.

Eby, G. N. & A. N. Mariano 1986. Geology and geochronology of carbonatites peripheral to the Paraná Basin, Brazil–Paraguay. *Geological Association of Canada–Mineralogical Association of Canada–Canadian Geophysical Union. Joint Annual Meeting, Ottawa. Program with Abstracts* **11**, 66.

Eriksson, S. C. 1984. Age of carbonatite and phoscorite magmatism of the Phalaborwa complex (South Africa). *Isotope Geoscience* **2**, 291–9.

Faul, H., P. L. D. Elmore & W. W. Brannock 1959. Age of the Fen carbonatite (Norway) and its relation to the intrusions of the Oslo region. *Geochimica et Cosmochimica Acta* **17**, 153–5.

Fletcher, C. J. N., J. D. Appleton, B. C. Webb & I. R. Basham 1981. Mineralization in the Cerro Manomo carbonatite complex, eastern Bolivia. *Transactions of the Institution of Mining and Metallurgy* **B90**, 37–50.

Garson, M. S. 1984. Relationship of carbonatites to plate tectonics. *Indian Mineralogist*, Sukheswala volume, 163–88.

Garson, M. S. & M. Krs 1976. Geophysical and geological evidence of the relationship of Red Sea transverse tectonics to ancient fractures. *Bulletin of the Geological Society of America* **87**, 169–81.

Ge, B. & Y. Zhongxin 1985. Carbonatites and related mineral resources. *Bulletin of the Institute of Mineral Deposits, Chinese Academy of Science* **13**, 1–195.

Gittins, J., R. M. Macintyre & D. York 1967. The ages of carbonatite complexes in eastern Canada. *Canadian Journal of Earth Sciences* **4**, 651–5.

Gleadow, A. J. W. & C. K. Brooks 1979. Fission track dating, thermal histories and tectonics of igneous intrusions in East Greenland. *Contributions to Mineralogy and Petrology* **71**, 45–60.

Glevassky, S. G. & S. G. Krivdik 1981. *Precambrian carbonatite complex of the Azov area*. Kiev: Naukova Dumka. (in Russian)

Gold, D., G. N. Eby, K. Bell & M. Vallée 1986. Carbonatites, diatremes, and ultra-alkaline rocks in the Oka area, Quebec. *Geological Association of Canada–Mineralogical Association of Canada–Canadian Geophysical Union, Joint Annual Meeting, Ottawa, Field Trip 21, Guidebook.*

Goode, A. D. T. 1981. Proterozoic geology of western Australia. In *Precambrian of the Southern Hemisphere*, D. R. Hunter (ed.), 105–203. Amsterdam: Elsevier.

Hansen, K. 1981. Systematic Sr-isotopic variation in alkaline rocks from West Greenland. *Lithos* **14**, 183–8.

Harris, N. B. W., A. E. R. O. Mohammed & M. Z. Shaddad 1983. Geochemistry and petrogenesis of a nepheline syenite–carbonatite complex from the Sudan. *Geological Magazine* **120**, 115–27.

Herz, N. 1977. Timing of spreading in the South Atlantic: information from Brazilian alkalic rocks. *Bulletin of the Geological Society of America* **88**, 101–12.

Illies, J. H. & G. Greiner 1978. Rhinegraben and the Alpine system. *Bulletin of the Geological Society of America* **89**, 770–82.

Issler, R. S., M. I. C. de Lima, M. B. G. de Montalvao & G. G. de Silva 1975. Intrusivas feldspatoididas no Craton Guianes. *Congresso Ibero-Americano de Geologia Economica* **11**, 363–82.

Kampunzu, A. B. 1981. *Le magmatisme du massif de Kahuzi (Kivu, Zaïre). Structure, pétrologie, signification et implication géodynamique*. Thèse des sciences, Université Lubumbashi, Zaïre.

Kaneoka, I. & H. Haramura 1973. K–Ar ages of successive lava flows from the Deccan Traps, India. *Earth and Planetary Science Letters* **8**, 229–36.

Kononova, V. A. & L. L. Shanin 1972. On possible application of nepheline for alkaline rock dating. *Bulletin Volcanologique* **35**, 251–64.

Kresten, P., I. Printzlau, D. Rex, H. Vartiainen & A. Woolley 1977. New ages of carbonatitic and alkaline ultramafic rocks from Sweden and Finland. *Geologiska Föreningens i Stockholm Förhandlingar. Stockholm* **99**, 62–5.

Lancelot, J. R. & C. J. Allègre 1974. Origin of carbonatitic magma in the light of the Pb–U–Th isotope system. *Earth and Planetary Science Letters* **22**, 233–8.

Lanphere, M. A. 1964. Geochronologic studies in the eastern Mojave Desert, California. *Journal of Geology* **72**, 381–99.

Lapido-Loureiro, F. E. 1973. Carbonatitos de Angola. *Memorias e Trabalhos. Instituto de Investigação Cientifica de Angola. Luanda* **11**, 1–242.

Larsen, L. M., D. C. Rex & K. Secher 1983. The age of carbonatites, kimberlites and lamprophyres from southern West Greenland: recurrent alkaline magmatism during 2500 million years. *Lithos* **16**, 215–21.

Larsen, O. & J. Moller 1968. K/Ar age determinations from western Greenland. I. Reconnaissance programme. *Rapport Grønlands Geologiske Undersogelse* **15**, 82–6.

Le Bas, M. J. 1971. Per-alkaline volcanism, crustal swelling, and rifting. *Nature* **230**, 85–7.

Le Bas, M. J., I. Mian & D. C. Rex, 1987. Age and nature of carbonatite emplacement in North Pakistan. *Geologische Rundschau* **76**, 317–23.

Liegeois, J. P., H. Bertrand, R. Black, R. Caby & J. Fabre 1983. Permian alkaline undersaturated and carbonatite province, and rifting along the West African craton. *Nature* **305**, 42–3.

Lippolt, H. J., W. Gentner & W. Wimmenauer 1963. Altersbestimmungen nach der Kalium–Argon-Methode an Tertiären Eruptivgesteinen Südwestdeutschlands. *Jahresheft. Geologisches Landesamt in Baden-Württemberg* **6**, 507–38.

Lowdon, J. A., C. H. Stockwell, H. W. Tipper & R. K. Wanless 1963. *Age determinations and geological studies*. Paper, Geological Survey of Canada. 62-17.

Maravic, H. von & G. Morteani 1980. Petrology and geochemistry of the carbonatite and syenite complex of Lueshe (N. E. Zaire). *Lithos* **13**, 159–70.

Marsh, J. S. 1973. Relationships between transform directions and alkaline igneous rock lineaments in Africa and South America. *Earth and Planetary Science Letters* **18**, 317–23.

Mitchell, J. G., M. J. Le Bas, J. Zielonka & H. Furnes 1983. On dating the magmatism of Maio, Cape Verde Islands. *Earth and Planetary Science Letters* **64**, 61–76.

Mitchell, R. H. 1973. Isotopic composition of lead in galena from the Mountain Pass carbonatite, California. *Nature* **241**, 17–18.

Moralev, V. M., S. N. Voronovskiy & L. S. Borodin 1975. New data on the age of carbonatite and syenite of southern India. *Doklady of the Academy of Sciences, U.S.S.R. Earth Science Section* **222**, 46–8.

Patchett, P. J., O. Kouvo, C. E. Hedge & M. Tatsumoto 1981. Evolution of continental crust and mantle, and mantle heterogeneity: evidence from Hf isotopes. *Contributions to Mineralogy and Petrology* **78**, 279–97.

Pell, J. 1986. Carbonatites in British Columbia: a review. *Geological Association of Canada–Mineralogical Association of Canada–Canadian Geophysical Union. Joint Annual Meeting, Ottawa. Programs with Abstracts.* **11**, 113.

Prins, P. 1981. The geochemical evolution of the alkaline and carbonatite complexes of the Damaraland igneous province, South West Africa. *Annale van die Universiteit van Stellenbosch. Serie A1(Geologiese)* **3**, 145–278.

Samoilov V. S. & V. I. Kovalenko 1983. *Alkaline complexes and carbonatites of Mongolia.* Moscow: Izdatel'stvo Nauka. (in Russian)

Serencsits, C. M., H. Faul, K. A. Foland, M. F. El Ramly & A. A. Hussein 1979. Alkaline ring complexes in Egypt: their ages and relationships to tectonic development of the Red Sea. *Annals of the Geological Survey of Egypt* **9**, 102–16.

Straaten P. van 1986. Some aspects of the geology of carbonatites in S. W. Tanzania. *Geological Association of Canada–Mineralogical Association of Canada–Geophysical Union of Canada. Joint Annual Meeting, Ottawa. Programs with Abstracts.* **11**, 140.

Tack, L., S. Dentish, J. P. Liegeois & P. de Paepe 1983. Age, Nd and Sr isotopic geochemistry of the alkaline plutonic complex of the Upper-Ruvuba, Burundi. *Résumé. 12th Colloquium on African Geology. Brussels.*

Tuttle, O. F. & J. Gittins 1966. *Carbonatites.* New York: John Wiley.

Vail, J. R. 1976. Location and geochronology of igneous ring-complexes and related rocks in north-east Africa. *Geologische Jahrbuch* **20B**, 97–114.

Vallée, M. & F. Dubuc 1970. The St-Honoré carbonatite complex, Quebec. *Transactions of the Canadian Institute of Mining and Metallurgy* **73**, 346–56.

Vartiainen, H. & H. Paarma 1979. Geological characteristics of the Sokli carbonatite complex, Finland. *Economic Geology* **74**, 1296–306.

Vartiainen, H. & A. R. Woolley 1974. The age of the Sokli carbonatite, Finland, and some relationships of the North Atlantic alkaline igneous province. *Bulletin of the Geological Society of Finland* **46**, 81–91.

Wellman, P. & A. Cooper 1971. Potassium–argon age of some New Zealand lamprophyre dykes near the Alpine fault. *New Zealand Journal of Geology and Geophysics* **14**, 341–50.

Woolley, A. R., A. H. Rankin, C. J. Elliott, A. C. Bishop & D. Niblett 1984. Carbonatite dykes from the Richat Dome, Mauritania, and the genesis of the dome. *Indian Mineralogist*, Sukheswala volume, 189–207.

Zartman, R. E., M. R. Brock, A. V. Heyl & H. H. Thomas 1967. K–Ar and Rb–Sr ages of some alkalic intrusive rocks from central and eastern United States. *American Journal of Science* **265**, 848–70.

3
FIELD RELATIONS OF CARBONATITES

D. S. BARKER

ABSTRACT

Carbonatites occur as intrusive, volcanic, hydrothermal, and replacement bodies. The existence of carbonate-rich magma is demonstrated by experiments and by volcanological observations. Evidence for magmatic carbonatites includes preservation of chilled contacts, lava flow surfaces, Pele's tears, and vesicles. Geothermometry gives permissive evidence, with temperature estimates in some carbonatites agreeing with experimentally determined solidus–liquidus ranges.

Carbonatite magma forms lava-flows and tephra, plugs, cone sheets, dykes, and rare sills, but apparently never large homogeneous plutons. Carbonatites have also been emplaced by plastic flow of solids, and perhaps as fluidized suspensions.

Late-stage magnesium- and iron-rich carbonatites have precipitated from hydrothermal fluids in dilatant fractures and have also been formed by metasomatic replacement of earlier carbonatites and silicate rocks. Late hydrothermal and metasomatic carbonatites should be discriminated from magmatic carbonatites, and from hydrothermal veins in which carbonate came from a non-magmatic source.

If carbonatite liquids are generated as primary magmas in the mantle, they should form earlier and ascend faster than silicate magmas. Most carbonatites are, however, emplaced after most of the much more voluminous silicate rocks with which they are associated. Such tardiness implies immiscible separation of carbonate liquid from fractionated silicate liquid within the crust.

Field relations confirm that both alkali-rich and alkali-poor carbonatite magmas exist, but have not contributed much toward settling the debate concerning which magma type is parental.

3.1 INTRODUCTION

Carbonatite is magmatic rock containing at least 50 modal% carbonate minerals, according to Streckeisen (1980). If the definition of carbonatite specifies that the carbon is juvenile, derived from an identifiable or suspected magma, not from wall rock or external fluids, then it is possible to include some subsolidus (hydrothermal and metasomatic) carbonate-rich rocks. This seems a useful extension of the definition, because abundant carbonate minerals can replace only slightly older magmatic carbonate-rich or silicate-rich rocks or fill dilatant voids within them; the carbon in these hydrothermal and metasomatic rocks was carried upward by magma, and has merely been recycled during subsolidus processes within localized

Table 3.1 Criteria for magmatic heritage of carbonatite.

Associated rocks
    feldspathoid-bearing (nephelinites, phonolites, nepheline syenites, and urtite-ijolite-melteigite series)
    melilite-bearing (melilitites, okaites, turjaites, uncompahgrites)
    kimberlites
    fenites (wall rocks metasomatically enriched in Fe and alkalis, and commonly depleted in Si)
Accessory minerals
    pyrochlore, melilite, nepheline, sodic clinopyroxene
Trace-element enrichment
    Sr, Ba, Zr, Nb, Th, REEs
Isotopic composition
    $\delta^{13}C = -1$ to $-9‰$ relative to PDB
    $\delta^{18}O = +6$ to $+12‰$ relative to SMOW
    initial $^{87}Sr:^{86}Sr < 0.706$

volumes of rock. The term carbonatite should extend to these late-stage rocks of igneous heritage, while excluding the many examples of altered igneous rocks in which secondary carbonate, even though constituting more than 50 modal%, was not carried in magma.

Any genetic definition of carbonatite must be combined with observational criteria (Table 3.1). Isotopic composition of carbon in carbonate minerals, as summarized by Deines in this volume, provides one criterion. Additional evidence for juvenile origin includes the association of most carbonatites with distinctive igneous rocks (nephelinites and phonolites or their coarser-grained equivalents, or melilite-rich rocks, or kimberlites), the presence of certain minerals, and enrichment in some trace elements. None of these pieces of evidence for juvenile carbon listed in Table 3.1 is absolutely incontrovertible when used alone.

3.2 EVIDENCE FOR CARBONATITE MAGMAS

Evidence for carbonate-rich liquids is drawn from experiments, volcanological observations, geothermometry, and structural and textural features. Magmatic textures are likely to be erased by plastic flow (Wenk *et al.* 1983) and by solution and reprecipitation (Fig. 3.1). 'One of the major problems encountered by all researchers on carbonatites is trying to distinguish primary magmatic textures from later replacement and secondary textures' (Mariano & Roeder 1983).

Few carbonate liquids quench to glass. Datta *et al.* (1964) reported experiments at 1 Kb that did produce glasses at 40–60 mole% $MgCO_3$ in the systems $K_2CO_3-MgCO_3$ and $CaCO_3-CaF_2-Ca(OH)_2-BaSO_4$. Jones & Wyllie (1983) formed glass by quenching at the liquidus temperature of approximately 650 °C at 1 Kb in the system $CaCO_3-CaF_2-Ca(OH)_2-BaSO_4-La(OH)_3$; without $BaSO_4$ and $La(OH)_3$ no glass formed. Liquids also did not quench to glass in the system $CaCO_3-Ca(OH)_2-La(OH)_3$ (Jones & Wyllie 1986). In systems of greater

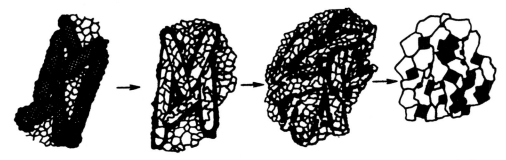

Figure 3.1 Carbonatite with primary magmatic texture (left) containing tabular calcite (cross-hatched) progressively recrystallizing to the right (after Zhabin 1978). Black represents non-carbonate phases.

pertinence to most carbonatites, $CaO-MgO-CO_2-H_2O$ (Boettcher *et al.* 1980) and $Na_2CO_3-K_2CO_3-CaCO_3$ (Cooper *et al.* 1975), no glass was found.

Keller (1981) recognized quenched droplets (Pele's tears) of carbonatite, 0.5–15 mm in diameter, at the Kaiserstuhl volcano, Germany, but these are holocrystalline. Chilled margins of carbonatites are not frequently reported (for examples, see Nash 1972, Fig. 8, and Le Bas 1977, Fig. 14.19 and Plate XVIIf), and are most convincing as quench features if crystals are elongated perpendicular to the contact. However, fine-grained margins could form by cataclasis.

Where primary igneous textures are preserved, calcite tends to form tabular crystals flattened perpendicular to the c-axis (Zhabin 1971, 1978), but dolomite forms rhombohedra (Gittins 1973). Treiman & Schedl (1983) pointed out the textural similarity of some carbonatites to spinifex-textured komatiites. Sutherland (1980) and Lapin & Vartiainen (1983) described orbicular and spherulitic textures involving olivine, phlogopite, magnetite, and calcite in carbonatites.

Most carbonate-rich systems yield non-polymerized ionic liquids (Treiman & Schedl 1983) of low viscosity and density. Katz-Lehnert & Keller (1987) emphasized the probable efficiency of crystal–liquid fractionation of dense oxide and silicate phases in such liquids; compositions of the fractionating phases differ strongly from the bulk composition of the magma.

Temperatures of crystallization can be estimated from mineral assemblages, homogenization temperatures of fluid inclusions, and oxygen isotope ratios. Although the magnetite–ilmenite geothermometer has rarely been applied to carbonatites, Secher & Larsen (1980) estimated 600–770 °C for crystallization of oxide pairs represented by the centres of discrete early-formed crystals in the Sarfartok carbonatite, W. Greenland. Treiman & Essene (1984) estimated a temperature of 640 °C at an assumed total pressure of 1 Kb for a late dyke at Oka, Quebec, containing an inferred eutectic assemblage of dolomite, calcite, periclase, apatite, and other phases. Unfortunately, most carbonatites do not contain mineral assemblages that permit well-constrained calculations of equilibration temperatures.

New calibration of the geothermometer based on the calcite–dolomite solvus (Anovitz & Essene 1987) has yet to be applied to carbonatites. Results from previous calibrations on the $CaCO_3-MgCO_3$ join only yield minimum temperatures reflecting the latest subsolidus loss of Mg from calcite to dolomite, except in those rare examples that contain coarsely 'perthitic' calcite hosts with dolomite lamellae wide enough for microprobe analyses. One sufficiently coarse intergrowth, from Goldray, Ontario, did give a minimum temperature of 885 °C (Gittins 1979).

Fluid inclusion studies also encounter obstacles, because calcite and dolomite tend to be open systems; 'most of the inclusions studied from carbonatites have been in apatite crystals' (Roedder 1984, p. 407). Romanchev (1972) and Romanchev & Sokolov (1980) reported consistently lower temperatures for homogenization of fluid inclusions in carbonatites (550–880 °C) than in the coexisting silicate rocks (760–1180 °C). Rankin (1977) reported a minimum temperature of 350 °C from apatite in the Tororo, Uganda, carbonatite, and Le Bas (1981) summarized results from apatites in carbonatites as 200–600 °C, but commonly 1000–1100 °C from apatites in the associated silicate rocks. Dawson (1966a) reported that the erupting natrocarbonatite lava of Oldoinyo Lengai was not incandescent, and therefore cooler than 600 °C, in agreement with fluid inclusion data of 550–560 °C (Romanchev & Sokolov 1980).

The fractionation of ^{18}O and ^{16}O has been applied as a geothermometer in carbonatites. Friedrichsen (1968) reported temperatures of 600–700 °C for carbonatites at Fen, Norway, assuming equilibration of calcite with water and an isotopic composition for the water. The ankerite–hematite pair gave a temperature of only 200 °C for an iron-rich late metasomatic carbonatite in the same complex. Conway & Taylor (1969) used the calcite–magnetite pair to estimate temperatures of 720–730 °C for Oka, Quebec, carbonatite and 595 °C for Magnet Cove, Arkansas, carbonatite. Suwa et al. (1975) estimated 380–800 °C for a range of hydrothermal to magmatic carbonatites at Mbeya, Tanzania, using calcite–dolomite pairs.

Extending to lower temperatures than the solidi of silicate magmas, estimates of crystallization temperatures for many carbonatites are consistent with solidus to liquidus temperature ranges in the experimentally studied systems $CaO-MgO-CO_2-H_2O$ (Fanelli et al. 1981), $CaO-MgO-SiO_2-CO_2-H_2O$ (Boettcher et al. 1980), $CaO-CO_2-H_2O$ (Wyllie & Tuttle 1960), and $Na_2CO_3-K_2CO_3-CaCO_3$ (Cooper et al. 1975).

3.3 CARBONATITE LAVAS

Carbonatite containing an alkali carbonate mineral is *natrocarbonatite*; carbonatite in which calcite is the only, or greatly dominant, carbonate is *sövite*. Names have also been given to carbonatites containing major amounts of dolomite and ankerite, but these will not be needed in this chapter.

The only carbonate-rich liquid seen to erupt was the natrocarbonatite of Oldoinyo Lengai, Tanzania (Dawson 1962a, 1962b, 1966a, Dawson et al. 1968, Donaldson et al. 1987). Oldoinyo Lengai is a 30 km³ cone made up of nephelinitic and phonolitic tephra and sparse interbedded lavas, topped by tephra and intracrater flows of natrocarbonatite. Carbonatites make up less than 1% of the volume of the cone; a few carbonatite tuffs are intercalated within the older silicate rocks. Ejected blocks of ijolite, sövite, and alkali-metasomatized basement rock are abundant. The 1960–66 natrocarbonatite lava within the summit crater formed flows of aa nearly 2 m thick and pahoehoe up to 30 cm thick (Dawson 1962b). More details are given by Dawson (this volume).

Gittins & McKie (1980) described the Oldoinyo Lengai lava as containing platey phenocrysts of nyerereite, $(Na_{0.82}K_{0.18})_2Ca(CO_3)_2$, and rounded phenocrysts of gregoryite, $(Na_{0.78}K_{0.05})_2Ca_{0.17}CO_3$, in a groundmass of the same two phases plus alabandite, MnS. Quiet effusion of low viscosity natrocarbonatite liquid ended in August 1966, with an explosive eruption of natrocarbonatite tephra accompanied by blocks of ijolite and melteigite (Dawson 1966a). Eruption of lava subsequently resumed (Dawson, this volume).

Oldoinyo Lengai silicate lavas do not form a simple fractionation sequence. 'The volcano is clearly not fed from a periodically tapped reservoir within which differentiation is proceeding. Rather we suggest that intermittent ascent of lava batches has occurred, with each batch evolving by differentiation and contamination along similar but not identical paths' (Donaldson *et al*. 1987).

The volcano Kerimasi (Hay 1983, Mariano & Roeder 1983), 12 km SE of Oldoinyo Lengai, is built of tephra and lava flows of nephelinite and carbonatite. The lava rimming the summit crater is porphyritic, with parallel euhedral calcite plates contained in a groundmass of euhedral calcite rhombs (average size 5 μm), as well as apatite. The calcite phenocrysts are compositionally zoned (Mariano & Roeder 1983) and are therefore considered primary, rather than replacing nyerereite or melilite.

The Quaternary Fort Portal volcanic field (Fig. 3.2) of southwest Uganda lies 750 km WNW of Oldoinyo Lengai (Nixon & Hornung 1973). Carbonatite lava escaped from a fissure on the NW flank of the tephra cone Kalyango to form a flow 1–5 m thick with a scoriaceous top. The flow covers an area of 0.3 km^2. Barker & Nixon (1983) concluded that the Fort Portal lava differs strongly from the natrocarbonatite of Oldoinyo Lengai; olivine, clinopyroxene, phlogopite, titanomagnetite, and tabular calcite occur in a groundmass of calcite, apatite, periclase, perovskite, and spurrite ($Ca_5(SiO_4)_2CO_3$). Monticellite forms rims on olivine and clinopyroxene. Phlogopite is the only phase with an alkali metal as an essential component, and the tabular calcite, although superficially resembling nyerereite in Oldoinyo Lengai natrocarbonatite, is zoned with respect to Sr and is interpreted as primary magmatic calcite. No coeval silicate magma was erupted in the Fort Portal volcanic field.

Khanneshin in southwest Afghanistan is a Pliocene to early Quaternary cone, deeply eroded, with a basal diameter of up to 8 km and a height of 700–750 m above its base (Abdullah *et al*. 1975, Vikhter *et al*. 1976, Alkhazov *et al*. 1978). A sövite core intrudes a cone of carbonatite tephra and lava, some containing ankerite and barite. Ring dykes and widespread radial dykes are also carbonatite. Minor leucite tephrites are reported as the only silicate rocks associated with the carbonatites.

At Qagssiarssuk in the Gardar province, SW Greenland (Stewart 1970), 1300-Ma-old carbonatite and nephelinite lavas are associated with small, compositionally

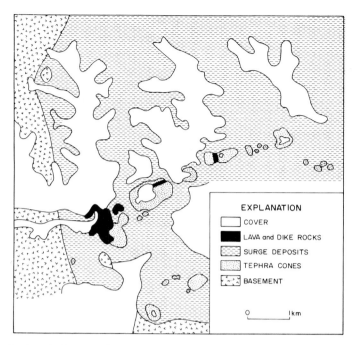

Figure 3.2 Generalized geologic map of SW part of the Fort Portal volcanic field, southwestern Uganda (after Nixon & Hornung 1973). All volcanic units are carbonatite.

similar, intrusive bodies and tephra. The carbonatite lavas contain apatite, calcite, chlorite, and hematite.

A vigorous debate concerns the significance of alkali-poor carbonatite lavas. Can both natrocarbonatite and sövite magmas be primary, and, if not, which is parental? Le Bas (1981, 1987) has assembled strong arguments that natrocarbonatite liquid separates immiscibly from melanephelinite magma and then, probably through a combination of crystal–liquid and vapour–liquid fractionation, evolves from alkali-rich through calcium-rich and ultimately to magnesium- and iron-rich, the opposite of the trend for silicate liquids. Twyman & Gittins (1987) and Gittins (this volume) proposed the reverse sequence; the parental magma is not nephelinitic but an olivine sövite liquid with around 8 wt% $Na_2O + K_2O$, generated by partial fusion of carbonate-bearing peridotite at a total pressure exceeding 27 Kb. Evolution by fractional crystallization produces more alkali-rich and hydrous compositions. If the residual liquid becomes water-saturated, an aqueous fluid separates and carries off much of the alkalis to metasomatize wall rocks. If the liquid does not become water-saturated, it is 'kept liquid by alkalis and halogens, and alkali loss is prevented by the absence of an aqueous phase. Alkalic carbonatite magmas are therefore late differentiates of a more normal mildly-alkalic sövite magma developed under low water fugacity', and 'are not to be considered parental to other carbonatite rock types' (Twyman & Gittins 1987).

Le Bas & Handley (1979) offered evidence that sövite and ijolite constitute an immiscible pair of liquids. Apatite, a liquidus phase in both magmas, has nearly the same composition in the carbonatite and ijolite, but apatite compositions change along diverging paths in more evolved silicate and carbonate liquids suggesting independent fractionation histories after unmixing.

At the Oka complex, Quebec, Treiman & Essene (1985) found identical liquidus phases (clinopyroxene, melilite, and perovskite) in sövite and silicate rocks again suggesting immiscibility. This relationship is also supported by identical initial $^{87}Sr:^{86}Sr$ ratios in carbonatite and silicate rocks, and by the presence of carbonate ocelli in silicate rock and, more rarely, of silicate rock in carbonatite. Furthermore, Treiman (1985) reported that a carbonatite at Oka containing calcite, monticellite, clinopyroxene, and olivine has negligible alkalis and caused no fenitization. At Magnet Cove, Arkansas, Nesbitt & Kelly (1977) found that fluid inclusions in monticellite from a sövite (also lacking a fenite halo) indicate a calcium-rich parental liquid.

These findings have not resolved the controversy over the primary nature of natrocarbonatite and sövite liquids. Some calcite-rich lavas are interpreted as leached natrocarbonatite; tabular calcite phenocrysts and microphenocrysts, previously thought to be replacements of melilite, are now considered pseudomorphs after nyerereite (Hay 1978, 1983, 1986, Hay & O'Neil 1983, Deans & Roberts 1984, Clarke & Roberts 1986) formed by incongruent dissolution of the alkali carbonate in meteoric water. Dawson et al. (1987) suggested that this mechanism may apply to all calcitic lavas and tephra. As evidence that calcite has replaced nyerereite at Oldoinyo Lengai, $\delta^{13}C$ and $\delta^{18}O$ in the calcite are both more positive than in magmatic calcite, and the calcite pseudomorphs are porous, rimmed by hydrated ferric oxides, and contain closely spaced lines of vacuoles following relict cleavage cracks parallel to the length of the original nyerereite. In the lavas and tuffs of Fort Portal, however, Barker & Nixon (1983) found no nyerereite or porous calcite after nyerereite, and concluded that the carbonatite liquid was very low in alkalis, precipitating calcite as the only carbonate phase and forming sövite lava.

Undoubtedly other carbonatite lavas await recognition, and all of the occurrences mentioned above deserve much more study. It appears that both natrocarbonatite and sövite liquids can erupt as lavas of low viscosity.

3.4 CARBONATITE TEPHRA

Deans & Roberts (1984) stated that volcanic carbonatites 'are now known to be worldwide, as at least 25 can be listed in 13 countries from Afghanistan to Zambia'. Nearly all are pyroclastic, rather than lava. Some of these are of great palaeontologic significance (Hay 1986) because of excellent fossil preservation in rapidly indurated airfall tuffs, especially of upland faunal and floral assemblages not represented in lacustrine and fluvial deposits.

At Oldoinyo Lengai, most carbonatite has been expelled as tephra. Dawson et al. (1968) reported that fine ash deposited on slopes was quickly reworked by wind; the particles must have been dry, although water vapour escaped from fumaroles at 50–80 °C (in some other occurrences of carbonatite tephra, cementation came soon after, or even during, deposition). The 1966 tephra were contaminated with lithic fragments and xenocrysts, and contained approximately 25 wt% SiO_2 and 9–11 wt% CO_2 (Dawson et al. 1968), in contrast with 0.05 wt% SiO_2 and at least 30 wt% CO_2 in the 1960 lava (Gittins & McKie 1980).

Dawson (1964a, b) pointed out that carbonate-cemented pyroclastic cones are numerous in northern Tanzania and that some, if not all, of the 'surface limestones' of that region are indurated carbonatite tephra.

Hay (1983) and Mariano and Roeder (1983) described calcite-cemented agglomerates and lapilli tuffs from Kerimasi. Many lapilli are carbonatite with calcite phenocrysts, some of which are interpreted as primary and others as pseudomorphs after nyereite. Other lapilli are melilitite; whether the melilitite and carbonatite lapilli are intermixed in the same layers is not specified. Hay (1983) listed possible pyroclastic flow deposits, in addition to airfall and ballistic tephra, at Kerimasi.

Hay (1978) and Hay & O'Neil (1983) described the Laetolil tuffs of northern Tanzania. Most are wind-reworked, but about 20% retain their primary airfall characteristics. The upper unit of the Laetolil Beds contains approximately 60 carbonatite and melilitite–carbonatite airfall layers, 1–5 cm thick, of microcrystalline calcite with biotite, clinopyroxene, melilite, and nepheline. The calcite has $\delta^{13}C$ averaging $-6.2‰$ relative to PDB, and in the range for magmatic carbonatites, but $\delta^{18}O$ averages $+26.0‰$ relative to SMOW, and is too high for magmatic carbonatites, suggesting low-temperature interaction with meteoric water. Hay & O'Neil concluded that the original natrocarbonatite tephra were first cemented by trona, and later by calcite and zeolites, as Na and K were leached away.

Pyatenko & Saprykina (1976) described Palaeozoic carbonatite (tephra?) from the Kola Peninsula, containing angular fragments of glassy melilitite. Clarke & Roberts (1986) also mention carbonatite tephra containing clasts of 'partly glassy trachyte ... in apparent pristine condition'. It seems unlikely that alkalis could be leached from the carbonatite matrix without concurrent devitrification of the melilitite and trachyte.

At Fort Portal, Uganda, nearly 50 Quaternary monogenetic tephra cones (Fig. 3.2) contain only carbonatite and lie along two east–northeast-trending arrays. Although airfall and surge deposits cover 140 km^2, their total volume is only about 0.25 km^3 (Nixon & Hornung 1973). Two kinds of tephra occur at Fort Portal. One is lapilli tuff containing round fragments, up to 15 mm in diameter, identical to the sövite lava described in the preceding section of this review, cemented by calcite (Barker & Nixon 1983). Lapilli are cored by phlogopite or, less commonly, clinopyroxene or fragments of basement rock. The second type of pyroclastic deposit is 'welded' flaggy tuff, with about 50% lithic fragments of the Toro-Ankolean basement, up to 2 m thick, draped on the topography as rapidly cemented, wet surge deposits. Crystallization of hydrous calcium silicates, analogous to the phases in Portland cement (Barker & Nixon 1983), occurred immediately after the tephra fell on wet ground. Thousands of phreatic 'blister mounds', up to 1.5 m high, formed after induration of the tephra had begun.

South of Fort Portal in the Katwe-Kikirongo and Bunyaruguru volcanic fields of western Uganda, Lloyd (1985) found that lapilli tuffs of olivine leucitite and related rocks are 'cemented by calcite, which has a trace element content indicative of a magmatic origin.'

In the Rufunza district, Zambia (Bailey 1966), tephra contain lapilli and lithic fragments of carbonatite, plus accidental fragments of Karroo and older rocks, cemented by ankerite. Like the carbonatite tephra at Fort Portal, those at Rufunza have no coeval silicate rocks.

At Tinderet, Kenya, Deans & Roberts (1984) described tuffs with carbonatite bombs that caused impact sag structures. Bombs and lapilli have tabular calcite that is commonly parallel or subparallel; some calcite is interpreted by Deans & Roberts as primary, and some as replacing nyererite. No glass is present. Ejected blocks of plutonic sövite are also reported. Skeletal titanomagnetite phenocrysts from the Tinderet tuffs show a distinctive habit, with growth pyramids, which Deans & Seager (1978) also recognized in carbonatite tephra from other localities, including Amba Dongar in India and Qagssiarssuk in SW Greenland.

Agglutinated carbonatite lapilli have been reported from the Kaiserstuhl volcano in the southern Rhine graben, in a layer 1–1.5 m thick (Keller 1981, this volume). In these lapilli, Hay & O'Neil (1983) found differences in Ba and Sr contents between calcite phenocrysts and microphenocrysts that indicate that some of the calcite was primary, and not secondary after nyererite. The lapilli have $\delta^{13}C$ values 2–3‰ lower and $\delta^{18}O$ values 7–8‰ higher than those found from intrusive carbonatites at Kaiserstuhl. These data suggest that the groundmass of the lapilli exchanged carbon and oxygen isotopes with groundwater at low temperature. The groundmass may have contained a high proportion of nyererite.

Other carbonate-rich tephra have been documented from Homa Mountain, Kenya (Clarke & Roberts 1986), northern Quebec, Canada (Dimroth 1970), the Kontozero district, Kola Peninsula, USSR (Pyatenko & Saprykina 1976), the Cape Verde Islands (Silva *et al.* 1981, Le Bas 1984), Khanneshin, Afghanistan (Abdullah *et al.* 1975, Vikhter *et al.* 1976, Alkhazov *et al.* 1978), Cerro Manomo, Velasco magmatic province, Bolivia (Fletcher & Litherland 1981), the Deckentuff of the Hegau region, Germany (Keller & Brey 1983, Brey & Keller 1984), Melkfontein in East Griqualand, southern Africa (Boctor *et al.* 1984), and a widespread unit, now metamorphosed, at Mt. Grace, British Columbia, Canada (Höy & Kwong 1986, Pell & Höy, this volume).

Ultramafic xenoliths have been reported in carbonatite tephra and associated silicate rocks at several localities, but never in carbonatite lavas or intrusions (Carswell 1980). At Lashaine volcano, northern Tanzania (Dawson *et al.* 1970), garnet and spinel lherzolites occur in carbonatite tephra overlying glassy ankaramite scoria; however, some xenoliths are coated with glassy lava, suggesting that they ascended with the silicate magma and later were recycled as accidental fragments in the carbonatite tephra.

The absence of ultramafic xenoliths from carbonatites can be attributed to the low viscosity and low density of carbonatite liquid, or to separation of an immiscible carbonatite liquid from mantle-derived silicate magma within a crustal reservoir. In both cases, ultramafic xenoliths would settle out. Diamonds have been reported from close to the intrusive Ngualla carbonatite, SW Tanzania (van Straaten 1986, this volume), but their relation to the carbonatite is unclear.

3.5 CARBONATITES IN SMALL HYPABYSSAL INTRUSIONS

Many carbonatites occur in swarms of parallel and radial dykes, ring dykes, cone sheets, diatremes, and subvolcanic pipes and plugs. Fewer references are cited here, in comparison with the preceding sections on carbonatite lavas and tephra, because the literature on volcanic carbonatites is less abundant and less familiar to most readers.

The dykes of the Maimecha-Kotui region, N. Siberia, as described by Zhabin & Cherepivskaya (1965), appear to be typical. These are vertical, up to 0.5 m wide, and consist of 60–80% euhedral tabular calcite phenocrysts set in a groundmass of apatite, calcite, and dolomite. Flow structure is generally well developed, and is expressed by parallelism of the calcite phenocrysts (Fig. 3.3).

A few of the many sövite dykes of the Kaiserstuhl contain skeletal and dendritic calcite and apatite, both elongated perpendicular to dyke walls (Katz & Keller 1981, Sommerauer & Katz-Lehnert 1985). The low-viscosity liquid is thought to have been supercooled, and to have lost heat rapidly.

In dykes of the McCloskey's Field area, NW of Ottawa, Canada (Hogarth 1966), dolomite forms euhedral rhombs in a groundmass of calcite. Wall rock is brecciated (Fig. 3.4) and fenitized; both processes are common, even adjacent to small carbonatite dykes.

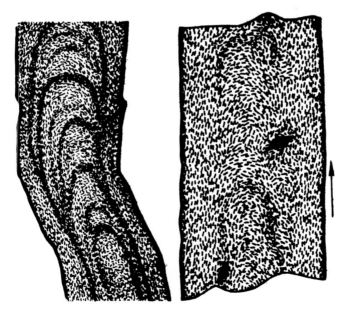

Figure 3.3 Carbonatite dykes with flow structure expressed by parallelism of tabular calcite, Guli complex, northern Siberia, USSR (after Zhabin 1971). Black represents wall rock inclusions. Dykes are 15 and 20 cm wide. Arrow indicates direction of flow.

Figure 3.4 Carbonatite (light) in brecciated and fenitized wallrock, McCloskey's Field, Quebec, Canada. Matchbook for scale is approximately 2.7 by 3 cm.

Phonolite dykes extend as far as 25 km from the Fen complex, Norway, although none apparently cuts the complex (Bergstøl 1979). Carbonatite dykes are much less abundant and widespread than the phonolite dykes, but are parallel to them, separated from the phonolite by screens of wall rock a few centimetres to several metres wide. The parallelism and proximity of the two rock types are important evidence for the contemporaneity of carbonatite and phonolite liquids, not otherwise demonstrated in the Fen complex.

At Alnö, Sweden (Kresten 1980), radial dykes of carbonatite were followed by carbonatite cone sheets in a complex array, dipping both inward and outward relative to the centre of the complex at angles varying from less than 30° to nearly vertical. A carbonatite ring dyke, 100–200 m wide, formed later than the radial dykes and cone sheets, is cut by ijolite and by still younger carbonatite cone sheets.

Carbonatite cone sheets at Homa Mountain, W. Kenya, filled a spiral fracture that is inferred to have grown in four stages over an ascending magma body 2–4 km below the surface (Bahat 1979).

Carbonatite commonly forms the matrix of breccia in diatremes (Dimroth 1970) and 'fluidized' dykes (Garson 1955, Sutherland 1980), and forms small plugs (McIver & Ferguson 1979). A plug of carbonatite containing calcite, magnetite, monticellite, and periclase is exposed at the centre of the summit crater of Kerimasi volcano (Mariano & Roeder 1983), and a plug of carbonatite 0.4 km in diameter lies at the centre of an ijolite intrusion about 3 km in diameter in the core of the deeply eroded, nephelinite volcano, Napak, in eastern Uganda (King & Sutherland 1966).

Sills of carbonatite are uncommon. Carbonatite-rich segregations occur in mafic silicate rocks in sills (Boctor & Boyd 1981), but tabular concordant bodies of carbonatite are very rare except as minor apophyses from dykes, and as metamorphosed carbonatite (§3.8 of this review), where their concordance may result from shear and flow.

Two large sills of carbonatite share perplexing features in their lateral extent and brecciated texture. The Kaluwe sill in Zambia (Bailey 1966), emplaced during folding, formed a body more than 12 km long and up to 250 m thick. The brecciated rock consists of apatite, calcite, oxidized magnetite, pyrochlore, and vermiculite. Wide variation in clast size defines layers up to several metres thick. Angular fragments of Karroo sedimentary rocks occur, in addition to the carbonatite clasts. No cogenetic silicate rocks are reported. The second sill lies a few kilometres north of the Amba Dongar carbonatite complex, India (Sukheswala & Borges 1975). This body, 11 km long and of variable but unspecified thickness, occurs along the contact between Cretaceous sandstone and the overlying Deccan basalt flows. The matrix (apatite, ankerite, calcite, garnet, magnetite, pyrochlore, sodic clinopyroxene, and titanite) surrounds clasts of the underlying sandstone, overlying basalt, and phonolite of unknown provenance. Horizons separating coarse and fine beds are deflected around large clasts. The contact between carbonatitic breccia and basalt is sharp, although the contact with the underlying sandstone is gradational. It appears that the sandstone was permeated by carbonate-rich fluid to a decreasing extent downward, although original sedimentary features, including cross-bedding, are still preserved.

Although a pyroclastic origin is inferred for a similar occurrence of concordant carbonatite breccia at Mt. Grace, British Columbia (§3.8), such an origin was rejected for the Zambian and Indian occurrences.

Some carbonatites form tabular bodies emplaced along faults. The Pollen carbonatite in Finnmark, Norway (Robins & Tysseland 1980), is 200 m wide by 1600 m long, and lies along a Caledonian thrust fault. The 170 m thick Loe Shilman sövite follows a thrust plane for at least 2.5 km astride the Afghanistan–Pakistan border (Le Bas *et al.* 1987). The Silai Patti sövite, in the same region, is 2–20 m thick and extends for at least 12 km along another thrust fault (Le Bas *et al.* 1987). In SW Tanzania, the Songwe scarp carbonatite (Brown 1964, van Straaten, this volume) is an array of lenses up to 30 m wide, reaching nearly 20 km along a normal fault; the fine-grained ankeritic carbonatite was emplaced before Cretaceous fault movement had stopped.

The evidence from hypabyssal intrusions indicates that carbonatite magma can be intruded forcefully, with vigorous release of gas, or can be extremely passive, only entering space prepared by stress fields independent of the ascending magma. Carbonatite is emplaced as liquids, nearly solid slurries, or gas–solid suspensions.

3.6 CARBONATITES IN PLUTONS

Most carbonatites occur in composite plutons with coeval silicate rocks. The latter are silica-undersaturated, usually mafic, the coarse-grained equivalents of the nephelinite and melilitite lavas associated with carbonatite lavas and tephra. Nephelinites associated with carbonatites are more evolved than those not associated with carbonatites (Le Bas 1987). They generally have lower Mg:Fe ratios and higher K:Na ratios; clinopyroxene is more abundant than olivine as a phenocryst phase. An antipathy between carbonatites and coeval silicate rocks containing calcic plagioclase was noted by Heinrich (1966) and Rock (1976).

Composite plutons that contain carbonatites are assemblages of plugs, dykes, ring dykes, and cone sheets, which may or may not share a common centre. Very few 'ring complexes' achieve radial symmetry. A typical complex is surrounded by a metasomatic (fenite) aureole that 'appears to have exploited the channelways created by shattering of the country rocks during vigorous magmatic emplacement' (Gittins 1978). Many complexes are elongated (Fig. 3.5) and contain, in addition to carbonatites, ultramafic rocks but lack nepheline syenite or fenite (Gittins 1978).

In the carbonatite-bearing plutons of the Homa Bay area, western Kenya (Le Bas 1977), the earliest carbonatites are calcitic and 'are typically coarse-grained and platey-textured, and have diffuse intrusive boundaries as if the carbonatite magma had soaked into the country rocks', but nevertheless are 'penetrative stock-like intrusions'. Cone sheets and dykes of finer sövite were subsequently emplaced and show 'sharp intrusive margins, commonly chilled and with internal flow structures; the geometry of the intrusive form shows them to be dilational'. These were followed by cone sheets and dykes of ankeritic carbonatite. The final stage was marked by veins of barite, calcite, fluorite, and quartz. At Homa Bay and in other complexes, fenitization usually can be attributed only to early carbonatites and to mafic silicate rocks (Gittins *et al.* 1975, Kresten & Morogan 1986).

The Jacupiranga carbonatite plug, Brazil (Gaspar & Wyllie 1987), consists of five intrusions. The first, third, and fourth are calcitic, the second contains both calcite and dolomite, and the fifth is dolomitic. Carbonatites emplaced during the second through the fifth intrusive events show trends of decreasing Mg:Fe in magnetite, while Mg and Mn increase and Ba and Nb decrease in the cores of phlogopites. The trends could record fractionation in parental magma 'from which successive batches of carbonatite were derived' (Gaspar & Wyllie 1987).

Flow structure in plutonic carbonatite (Garson 1955) is expressed by parallelism of xenoliths and by trains of apatite, oxides, and mafic silicates. These features, and contacts marked by abrupt changes in grain size and mineralogy, define nested arrays of conical bodies which dip steeply inward or outward.

At Fen, Norway, different textured bands of carbonatite and screens of fenite dip inward to suggest a focus 2–3 km below the present erosion surface. A gravity

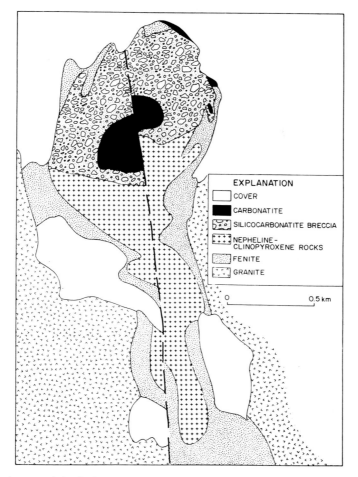

Figure 3.5 Geologic map of the Seabrook Lake complex, Ontario (generalized after R. P. Sage, pers. comm. 1987). Silicate-rich, carbonatite breccia is intruded by a core of dolomitic carbonatite.

survey of the Fen complex revealed a +23 mgal anomaly indicating a dense, probably ultramafic, root extending perhaps 15 km downward under the 2 km diameter complex (Ramberg 1973). Andersen (1986a) interpreted fluid inclusion data from apatites in the Fen carbonatites as indicating separation of carbonatite liquid from mafic silicate magma at a temperature exceeding 625 °C and a pressure above 4 Kb.

The Oka pluton, Quebec, Canada (Gold *et al.* 1986) is a double-ring complex with outward dipping carbonatite units in the southern ring and the periphery of the northern ring, but inward dipping bodies in the interior of the latter. Rhythmic layers (Fig. 3.6) are defined by varying amounts of biotite, clinopyroxene, magnetite, melilite, and monticellite, and are '1–5 cm thick, spaced 5–30 cm apart' (Gold 1969).

Carbonatites in other plutons have been well described by Vartiainen & Paarma

Figure 3.6 Rhythmic layers in carbonatite of the Bond zone, Oka, Quebec. View, approximately one metre wide, is of a block displayed on the campus of the Université de Montréal, Quebec, Canada.

(1979), Secher & Larsen (1980), Knudsen (1985), Basu & Mayila (1986), and van Straaten (this volume).

Although most carbonatites are emplaced in gneisses and granitic rocks, Bagdasarov & Buyakayte (1986) described three carbonatite plutons in the USSR which intruded marble and limestone. Intrusive contacts of carbonatite dykes against marble are sharp and linear, and, locally, carbonate wall rocks were stoped into carbonatite. The difference in behaviour of carbonate wall rock and carbonatite implies a great temperature difference. There was substantial exchange of Sr and C, as well as introduction of Ba and REEs into the wall rock. Other carbonatites that intruded sedimentary carbonate rocks have been described in the Cape Verde Islands (Stillman *et al*. 1982) and in the Ottawa region, Canada (Hogarth & Rushforth 1986).

Times of emplacement of carbonatites, relative to associated silicate rocks in the same complex, are fairly consistent. In general, the sequence from periphery to core, and from oldest to youngest, is nepheline syenite (if present), to nepheline–clinopyroxene rocks to carbonatites (Fig. 3.7). Kapustin (1986), however, lists the sequence as dunite and/or clinopyroxenite succeeded by melilite-bearing rocks, then nepheline–clinopyroxene rocks, nepheline syenite, and finally carbonatite.

Many carbonatites are cut by diatreme breccias (frequently described as lamprophyric or kimberlitic) and mafic to felsic dykes (Garson 1955, Egorov 1970, Kapustin 1971, Nash 1972, Brueckner & Rex 1980, Kresten 1980, Robins 1980, Robins & Tysseland 1980, Samoilov & Plyusnin 1982, Larsen *et al*. 1983). Carbonatite lavas and tephra are also commonly intercalated with, or overlain by,

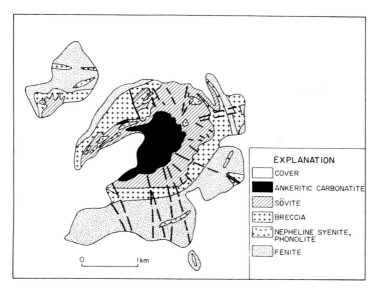

Figure 3.7 Geologic map of the Chilwa Island complex, Malawi (generalized after Garson 1966).

melilitites, nephelinites, phonolites, and tephrites (Stewart 1970, Pyatenko & Saprykina 1976, Keller 1981, Deans & Roberts 1984). These relations suggest that carbonate liquid coexisted with, or was survived by, silicate liquid.

There is only one clear case of carbonatite cut by a *pluton* of silicate rock. In the Mansouri ring complex in the Eastern Desert of Egypt (El Ramly & Hussein 1985), an arcuate carbonatite body and associated dykes are cut by a syenite pluton. Although the field relations show that the silicate rocks are younger than the carbonatite, it is likely that the carbonatite and syenite do not belong to the same magmatic episode, because K–Ar and Rb–Sr ages of alkalic rocks in this province range from Cambrian to Cretaceous.

Exceptions to the spatial, but not the temporal, sequence (with carbonatite normally at the centre) include the Prairie Lake complex, Ontario, Canada (Currie 1976), in which a core of silicate rocks is surrounded by a thin ring of younger carbonatite, and the Lueshe pluton, Zaire (Maravic & Morteani 1980), where a core of cancrinite syenite is surrounded by younger coarse sövite. A plug of still younger, fine-grained dolomitic carbonatite occupies the southeastern part of the pluton.

Carbonatite liquids have such low solidus temperatures, densities, and viscosities (Treiman, this volume) that they should form earlier, and ascend faster, than any cogenetic silicate magma. Nevertheless, carbonatites are emplaced later than most of the silicate rocks with which they are associated. One explanation for this is that carbonatite liquids have considerably deeper sources than the silicate liquids (Harris *et al.* 1983), but, for reasons given below, this seems unlikely.

The tardiness of carbonatite liquids in arriving at the level where their products are observed, strongly suggests that they have not come all the way from the upper mantle as independent and primary liquids (which should form early and rise rapidly), but have separated from parental carbonated silicate liquids closer to the surface. A current debate centres on the questions: Are ascending CO_2- and H_2O-rich liquids trapped at 65–80 km depths by an inflection in the peridotite solidus where they crystallize and give up their volatiles (Wyllie 1987)? Can such liquids rapidly ascend through fractures without significantly reacting with wall rock (Eggler 1987)? Or do they react with peridotite in the upper mantle (where carbonates are not stable in the peridotite assemblage) to lose H_2O and become enriched in CO_2 and alkalis (Meen 1987)?

It is improbable that natrocarbonatite liquid could retain its identity while ascending through silicate rocks of the upper mantle and crust. 'The standard way in the laboratory to dissolve granite or other silicate material is to add sodium carbonate and heat' (Koster van Groos & Wyllie 1963, Le Bas 1981).

Williams *et al.* (1986) used products of the radium and thorium disequilibrium series to demonstrate that the 1960–63 Oldoinyo Lengai natrocarbonatite liquid separated from a nephelinitic parent only 7–18 years before its eruption, and following shortly after the preceding major eruptive episode. These data suggest an origin by immiscibility at shallow depth, with between 2 and 20% of the parent extracted as carbonatite.

In summary, most carbonatites make up only a small fraction of the mass of exposed rock in a pluton, and appear to have coexisted as liquids with magmas (melilitites, nephelinites, phonolites or tephrites) that had already undergone protracted fractionation. Wide compositional gaps between carbonatites and the associated silicate rocks occur in most complexes. The few examples of silicate-rich carbonatites (e.g. Fen: Ramberg 1973) can be explained by mechanical mixing of nepheline–clinopyroxene cumulates from ijolitic magma with carbonatite, or by carbonate metasomatism of silicate rocks.

3.7 CARBONATITES ASSOCIATED WITH KIMBERLITES

Carbonate-rich rocks occur in small amounts within some kimberlites, and a lively debate has gone on for several years as to whether these should be called carbonatites (see Haggerty, this volume). The problem has not been clarified by the current tendency toward narrower definition of kimberlite (Skinner & Clement 1979) and the recognition of lamproite as a separate rock type (Scott Smith & Skinner 1984). Carbonate-rich rocks have yet to be reported with lamproites (Bergman 1987).

Dawson (1966*b*) stated that 'there is every gradation from kimberlites and alnoites to carbonatites', and Wyllie (1978) referred to the 'intimate association of kimberlite and carbonatite on a small scale'. On the regional scale, however, Lapin (1987) concluded that carbonatite complexes are concentrated in the margins of cratons, kimberlites in the interiors, and the two rarely intermingle. Gittins (1978) emphasized that carbonate-rich rocks in kimberlites differ from those associated with alkalic rocks in their smaller dimensions and simpler mineralogy. Le Bas (1984) concluded that 'carbonatite magma can also occur associated with kimberlites, but these carbonatites are alkali-poor and do not produce the characteristic fenitization'; furthermore, they lack the pyrochlore, sodic clinopyroxene, and rare-earth phosphates and carbonates that are characteristic of 'normal' carbonatites. However, McIver & Ferguson (1979) did report pyrochlore in some of the small plugs and dykes of calcitic rock cutting olivine melilitite and kimberlite at Saltpetre Kop, southwest Africa; ankeritic rocks also occur here, but apparently lack pyrochlore. Carbonate-rich rocks associated with kimberlites show much lower Ti, Zr, Nb, and REEs than those associated with alkalic rocks (Haggerty, this volume).

The most widely publicized occurrence of carbonate-rich rock in kimberlite is that of late dykes cutting kimberlite in the Premier Mine, South Africa (Fig. 3.8). These dykes contain about 50% calcite, with serpentine and magnetite, both pseudomorphous after olivine, and traces of apatite and phlogopite. Robinson (1975) interpreted them as products of residual liquid from crystallization of kimberlite magma. Suwa et al. (1975) reported that calcite from both the Premier carbonate-rich dykes and the kimberlite have slightly lower $\delta^{13}C$ and considerably higher $\delta^{18}O$ than primary calcite in carbonatites associated with alkalic rocks, and have exchanged oxygen with meteoric water. Scatena-Wachel et al. (1986) found that calcite in a Premier carbonate-rich dyke has an initial $^{87}Sr:^{86}Sr$ ratio of 0.7042, comparable to the value of 0.7045 obtained on clinopyroxene from the kimberlite.

Mitchell (1979) denied a genetic relationship between kimberlites and carbonatites. Although 'kimberlite magmas differentiate to carbonate-rich silica-poor residua', these carbonate-rich rocks should not be called carbonatites, in his view, in part because kimberlites do not occur in alkalic complexes and in part because kimberlite and carbonatite contain spinels and ilmenites of different compositions. If kimberlite and carbonatite represent a pair of immiscible liquids, the liquidus

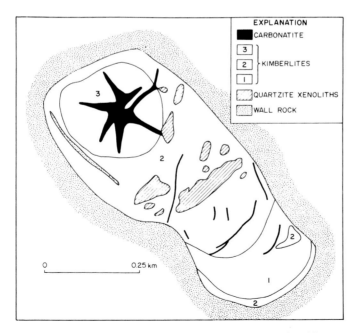

Figure 3.8 Geologic map of the 1085 foot (330 m) level, Premier Mine, South Africa (generalized after Anglo American Corporation 1969).

phases common to both, such as spinel and ilmenite, should have identical compositions in both rock types.

However, Gaspar & Wyllie (1984) demonstrated that ilmenites and spinels from the carbonate-rich rocks associated with kimberlites have compositions overlapping, or grading into, those of ilmenites and spinels from carbonatites associated with alkalic rocks. Gaspar & Wyllie state that 'consolidation of kimberlite magma produces a carbonate-rich residual liquid which precipitates ilmenite and spinel whose chemical characteristics may be indistinguishable from their counterparts in carbonatites, and distinct from those characteristic of kimberlites. The residual liquid precipitates carbonate which occurs in various forms, be it interstitial to kimberlite, or as ocelli, spherules, or globules, or as layers, or as discrete carbonate-rich dykes.' According to Gaspar & Wyllie, 'The parent magma of an alkalic rock association experiences a more complex history than does a kimberlite magma.'

Carbonate-rich rocks associated with kimberlites certainly merit the term carbonatite, if for no other reason than the repugnance of a new and distinctive rock name. The mineralogical data favour an origin for these carbonatites through crystal–liquid fractionation, not liquid immiscibility.

3.8 METAMORPHOSED CARBONATITES

Calcite easily recrystallizes and undergoes plastic flow. 'In contrast to most other rock forming minerals, calcite deforms plastically at room temperature provided that cleavage fracture is suppressed. Flow stresses in dolomite and silicates are an order of magnitude larger than in calcite' (Barber & Wenk 1979). Zhabin (1971, 1978) emphasized the almost inevitable erasure of primary calcite morphology in plutonic carbonatites. Remobilized carbonatites carrying boudinaged dykes of silicate rocks have been described by Garson (1955) and Gold & Vallée (1969).

Some carbonatites, emplaced before or during dynamic metamorphism of their enclosing rocks, are foliated or banded lenticular masses, still retaining their original distinctive magmatic mineral assemblages and trace element concentrations. Dismembered, but identifiable, fenite envelopes, and associated nepheline syenites and nepheline–clinopyroxene rocks, are commonly preserved. Examples include Mud Tank, 100 km NE of Alice Springs, Australia (Crohn & Moore 1984); Lonnie (Fig. 3.9) and Verity (Currie 1976), and others in British Columbia, Canada (see Pell & Höy, this volume); Tupertalik, Greenland (Larsen *et al.* 1983); Lillebukt on Stjernøy, Norway (Robins 1980, Skogen 1980); Breivikbotn on Sørøy, Norway (Sturt & Ramsay 1965), and Siilinjarvi, Finland (Puustinen 1971).

Although most metamorphosed carbonatites are sill-like, this geometry is rare among unmetamorphosed carbonatites. Probably the great ductility of calcite is the common explanation for this, but one occurrence, the Mt. Grace carbonatite of British Columbia (Hoy & Kwong 1986), is interpreted as a metamorphosed pyroclastic unit, forming a conformable layer averaging 3 m thick and extending for at least 25 km along strike, bedded, with clasts of albite–phlogopite rock and

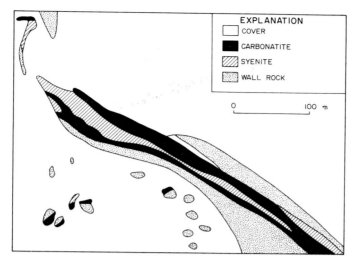

Figure 3.9 Geologic map of the Lonnie metamorphosed carbonatite, British Columbia (generalized after Currie 1976). Contacts and foliation are approximately vertical.

metapelite, in a fine-grained carbonate matrix (see Pell & Hoy, this volume). The layer is exposed in both limbs of a tight syncline, and enclosing metasediments are metamorphosed to sillimanite and kyanite grades. Relative to metasedimentary marble in the enveloping rocks, the Mt. Grace unit is strongly enriched in light rare earth elements (LREEs), Ba, Mn, Nb, and Sr.

3.9 HYDROTHERMAL DILATANT VEINS AND METASOMATIC REPLACEMENT BODIES ASSOCIATED WITH CARBONATITE-BEARING PLUTONS

There has been general agreement (for example, Kapustin 1971, 1982, 1987, Borodin 1978, Le Bas 1981, Sokolov 1985) that a sequence of emplacement from calcite- to dolomite- to ankerite-rich carbonatites occurs. Dolomite tends to be lacking at shallow levels, but is more abundant than calcite in deeply emplaced rocks (Le Bas 1977, 1984). Woolley's (1982) compilation of whole-rock analyses of carbonatites from different stages in several plutons showed calcium depletion accompanied by enrichment in either magnesium or iron.

Kapustin (1986) has recognized several different stages of carbonatite development. The earliest carbonatite is usually calcitic and forms plugs (the largest carbonatite bodies in most complexes) or nested cone sheets. Monticellite, melilite, or nepheline can be present (unlike in most later carbonatites), in addition to magnetite, apatite, olivine, phlogopite or biotite, amphibole, or clinopyroxene. A second stage consists of dykes, cone sheets, and shear zones in the earlier plugs, of calcite, dolomite, or both, commonly accompanied by pyrochlore, phlogopite or biotite, clinohumite, pyrrhotite, and richteritic amphibole. Both stages typically show sharp contacts and are rich in xenoliths of fenitized wall rock and magmatic silicate rocks of the complex. A third stage consists of dykes and dilatant veins of ankerite or dolomite, with or without calcite. A fourth stage (if present) is represented by thin hydrothermal veins, stockworks, and replacement bodies rich in calcite and dolomite or ankerite. In stages three and four, barite, chlorite, fluorite, hematite, pyrite, quartz, rutile, strontianite, and zeolites are among the typical accessory phases. With the exception of fluorite and pyrite, these minerals are rarely reported from stages one and two, and may be hallmarks of hydrothermal rather than magmatic origin. Whole-rock analyses show that Zr and Nb tend to decrease throughout the sequence, while Ba rises to a maximum in stage four. In calcite, Mg is highest in stage two, Fe and Mn in stage three. According to Kapustin (1982), Ba and Sr concentrations are relatively constant in the early carbonatites, and are lower than in the later units.

The small dilatant and replacement bodies of stages three and four are probably deuteric products. Magmatic calcite and dolomite of stages one and two are accompanied by silicates (forsteritic olivine, monticellite, melilite, nepheline, phlogopite/biotite, clinopyroxene, and amphibole) and oxides (magnetite, perovskite) that tend to react (Fig. 3.10) with water at lower temperatures to form the assemblage of dolomite or ankerite, quartz, zeolites, fluorite, chlorite, rutile, and hematite typical of stages three and four. Andersen (1984) states that 'reactions involving groundwater convecting in hydrothermal cells set up by the intrusives themselves may play a substantial part in the post-magmatic reequilibration of carbonatites.' Intense wall rock brecciation commonly extends farther than fenitization from carbonatite bodies, providing permeable zones that permit groundwater convection toward the carbonatites. Carbonatites of stages three and four structurally resemble the late fracture fillings and replacement bodies in granitic pegmatites (Černy 1982).

The effect of deuteric alteration in carbonatite can be seen at the Fen complex (Andersen 1984, 1986b, 1987), where 'rødberg', consisting mostly of calcite and dolomite, pervaded with dust-like hematite, formed along joints in ankeritic carbonatite.

Carbon and oxygen isotope ratios generally confirm exchange with meteoric water in late-stage carbonatites. Sheppard & Dawson (1973) and Suwa et al. (1975) showed that those carbonatites richer in dolomite, ankerite, fluorite, and barite commonly have higher $\delta^{18}O$ and constant or increased $\delta^{13}C$ than the earlier calcite-rich carbonatites.

A dyke and vein swarm at least 110 km long and 24–30 km wide on the South Island, New Zealand (Cooper 1986), is composed of approximately 1% carbonate-rich rock, 35% phonolites and trachytes, and the remainder lamprophyres. Strontium, Nd, and Pb isotope compositions for the carbonate-rich rocks are the same as for the associated silicate dyke rocks. Although high $\delta^{18}O$ values of carbonates indicate exchange with water or wall rock, $\delta^{13}C$ values are those of magmatic carbonatites (Blattner & Cooper 1974). Cooper (1986) concluded that the carbonate-rich veins are hydrothermal, but owe their distribution to a subjacent alkalic pluton.

Armbrustmacher (1979) distinguished two textural types of carbonatite in the

Figure 3.10 Biotite, surrounded and partly replaced by dolomite in sövite (calcite is stained), Cappelen quarry, Fen complex, Norway. Plane-polarized light. Longest dimension of the photomicrograph is 3 mm.

Wet Mountains, Colorado. Replacement carbonatites form tabular bodies, about 0.5 m wide and a few tens of metres long, have sharp contacts, and preserve textures typical of lamprophyre and syenite dykes in the area. The second group, less widespread, contains more calcite than dolomite, sparse ankerite and strontianite, and commonly barite, fluorite, hematite, and quartz. Some samples contain bastnaesite, pyrochlore, and other rare earth minerals. Armbrustmacher considered this second group to contain primary magmatic carbonatites, but the mineralogy strongly suggests a hydrothermal origin. In samples from both groups, $\delta^{13}C$ and $\delta^{18}O$ values are similar to those found in magmatic carbonatites, indicating that both types of carbonatite drew their carbon and oxygen from the same source, presumably an underlying magmatic body.

Hydrothermal dilatant veins and replacement bodies of carbonate-rich rock recycled from magmatic carbonatite differ from their precursors in containing a greater variety of carbonate phases, quartz, chlorite, and oxides (rutile, hematite) that indicate higher oxygen fugacity than the magnetite and ilmenite of earlier, magmatic, carbonatites. This same mineral assemblage, however, makes subsolidus carbonatites difficult to distinguish from hydrothermal and metasomatic rocks in which the carbonate came from a non-magmatic source.

3.10 PSEUDOCARBONATITES

Recrystallization of magmatic calcite, through subsolidus plastic flow, deuteric alteration, and solution-reprecipitation, leads to an evolutionary convergence of carbonatite textures toward those that are normally associated with hydrothermal calcite veins and metasedimentary carbonate rocks. These hydrothermal and metasedimentary products should be distinguished from those in which juvenile carbon is an essential component, but such discrimination can be difficult. Recognition of a magmatic heritage for carbonatites must be based on the associated igneous rocks, on the presence of fenites, and on the mineralogy, trace element composition, and isotopic ratios of the carbonate-rich rock (Table 3.1).

Magmatic carbonatites (Le Bas 1977, 1981) associated with melilite- and nepheline-bearing rocks and with fenites commonly contain one or more of apatite, phlogopite, magnetite, pyrochlore, and sodic clinopyroxene or amphibole, and usually are enriched in Ba, F, Nb, P, Sr, Th, Zr and REEs relative to remobilized marble and recrystallized limestone. Kapustin (1982) emphasized that carbonatites contain 'strontium and, to a smaller extent, barium in amounts very much larger than those found in other carbonate rocks of hydrothermal, metamorphic, or sedimentary origins.'

Carbonatites associated with kimberlites generally show less distinctive mineralogy and chemistry, and their magmatic origin must be inferred from field relations and from isotopic compositions of Sr, C, and O that are identical in carbonatite and the enclosing kimberlite.

As Gittins (1978) has pointed out, enrichment in Ba, F, Nb, P, Sr, Th, Zr, and REEs is not diagnostic, nor is the initial $^{87}Sr:^{86}Sr$ ratio, in discriminating between carbonatites on the one hand and hydrothermal veins or metasedimentary carbonates on the other. Phlogopite, forsteritic olivine, monticellite, periclase, and perovskite occur in marbles as well as in carbonatites. There seems to be no single decisive chemical or mineralogical criterion by which to establish a magmatic heritage for a carbonate-rich rock. Even carbonatite lava and tephra, although clearly magmatic, are widely considered to have undergone drastic compositional changes after eruption.

Garson *et al.* (1984) described carbonate-rich veins near the Great Glen Fault, Scotland. These contain calcite, sodic clinopyroxene and amphibole, barite, apatite, plagioclase, and pyrite, have high Sr, Ba, Nb, and REE contents, and are flanked by fenites and albitites. Garson *et al.* displayed reservations in terming these carbonatites. Initial $^{87}Sr:^{86}Sr$ ratios in two veins were 0.7095 and 0.7201.

Calcite-rich segregations have been reported in many mafic sills and dykes, and have been interpreted as vesicles filled with late carbonate precipitated from groundwater, as hydrothermal alteration products, and as products of immiscible carbonatite liquids. Kapustin (1985) described an analcime basalt sill containing veins and globules of calcite with barite, analcime, and traces of apatite, surrounded by unaltered basalt. The lack of alteration in the enclosing rock argues against a hydrothermal origin for the carbonate-rich bodies, but Sr, P, and Nb concentrations are lower than in most carbonatites.

Three layered kimberlitic sills at Benfontein, South Africa (Boctor & Boyd 1981), 1–1.5 m thick, contain carbonate-rich miniature diapirs in the lowermost sill, and 'blobs' and layers of carbonate near the tops of the middle and uppermost sills. The carbonate-rich bodies differ from those associated with alkalic rocks in having ilmenite that is Cr-rich, rather than Mn-rich, perovskite that is not strongly enriched in Nb, and spinel that is Cr-rich.

McGetchin *et al.* (1973) described a swarm of carbonate-rich dykes, consisting of kimberlitic breccia and lamprophyre, up to 30 cm wide and each only a few metres long, in Permian sandstone, siltstone, and shale within the Cane Valley diatreme, Utah. The dykes vary 'in strike, dip, and thickness along the strike in a more or less random way'. Dykes rich in carbonate are intercalated with those of kimberlitic breccia.

These examples illustrate the ambiguity that is encountered when some criteria listed in Table 3.1 are not satisfied.

3.11 CONCLUSIONS

Carbonatites display the full range of emplacement mechanisms shown by igneous silicate rocks (Table 3.2), from explosive volcanism to quiet effusion of low-viscosity liquid, from violently forceful injection into shattered wall rock to laminar flow into opportune space. The major differences from silicate magma emplacement are of scale and time. No known carbonatite body approaches the dimensions of large silicate intrusions. Carbonatites, where associated with magmatic silicate rocks, are

Table 3.2 Styles of occurrence of carbonatites.

Lavas
Tephra
Diatreme breccias
Small hypabyssal bodies (approximately in descending order of abundance)
  dykes
  cone sheets and ring dykes
  plugs
  sills
Composite plutons ('ring complexes')
  containing any combination of the above, usually with coeval magmatic silicate rocks (ultramafic or silica-undersaturated mafic to felsic)
Hydrothermal deposits in dilatant fractures
Metasomatic replacement bodies
Remobilized bodies
  any of the above, intensely recrystallized by subsolidus plastic flow

intruded or erupted late in the sequence and are strongly subordinate in mass to the associated silicate rocks.

Interpretation of carbonatite origins is hindered by the tendency of all calcite-rich rocks to recrystallize and to flow upward through more dense rock, and by the ease with which calcium carbonate is carried in hydrothermal fluids to fill fractures and to permeate and replace rocks. Deuteric alteration of magmatic carbonatites produces rocks containing carbonates richer in Mg, Fe, and Sr, with free silica, and with oxides that indicate higher oxygen fugacity than in carbonatite magmas. Such rocks, in which the carbon was derived from magma, should still be called carbonatites but distinguished from hydrothermal veins and alteration products, metasedimentary carbonate rocks, and clastic dykes.

Carbonatites in kimberlites show compositional gradation into their host, and usually lack the strong enrichment in trace elements typical of carbonatites that are associated with nepheline syenites, nepheline-clinopyroxene rocks, and melilite-bearing rocks. These relationships suggest that carbonatites in kimberlites are products of fractional crystallization, whereas those found with silica-undersaturated alkalic rocks either formed from an independent magma that ascended late, or by liquid immiscibility. Experimental evidence concerning immiscibility is discussed by Kjarsgaard & Hamilton (this volume). Some volcanic carbonatites, but not all, were alkali-rich, but intrusive carbonatites hold generally low amounts of Na and K, and then only in non-carbonate minerals. Alkali metasomatism accompanying many intrusive carbonatites indicates that a source of abundant alkalis was available either in the carbonatite magma or the associated silicate magma, or in both.

For carbonatite magmas, there are two competing hypotheses of generation and two alternative models of fractionation. Carbonatite magma may rise from the mantle as a primary liquid, or may separate immiscibly at shallow crustal levels from silica-undersaturated magma after prolonged differentiation. In either case, the carbonatite magma may initially be low in alkalis but fractionate to yield alkali-rich liquid as an extreme end-product, or may start alkali-rich, then lose its alkalis to the wall rocks to yield calcitic and dolomitic liquids.

There is little evidence for a primary carbonatite magma ascending directly from the mantle. It would be premature to conclude that alkali-rich carbonate liquid is parental to all calcium- and magnesium-rich carbonatites. Some carbonatite lavas were low in alkalis when erupted. The abundance of field and analytical evidence for immiscibility between calcitic liquids and silicate liquids appears stronger than experimental evidence against such immiscibility.

ACKNOWLEDGEMENTS

Comments on earlier drafts by M. J. Le Bas and A. H. Treiman were constructive, although neither can endorse all the conclusions, as is clear in their own chapters in this volume. J. G. Price, R. P. Taylor, and W. R. A. Baragar also made helpful suggestions.

REFERENCES

Abdullah, J., P. Bordet, J.-P. Carbonnel & J. Plas 1975. Sur l'existence d'un dôme recent de carbonatites dans le Régistan (Afghanistan du Sud). *Comptes Rendus Académie des Sciences Paris, Série D* **281**, 1801–4.

Alkhazov, V. Yu., Z. M. Atakishiyev & N. A. Azimi 1978. Geology and mineral resources of the early Quaternary Khanneshin carbonatite volcano (Southern Afghanistan). *International Geology Review* **20**, 281–5.

Andersen, T. 1984. Secondary processes in carbonatites: petrology of 'rødberg' (hematite–calcite–dolomite carbonatite) in the Fen central complex, Telemark (South Norway). *Lithos* **17**, 227–45.

Andersen, T. 1986*a*. Magmatic fluids in the Fen carbonatite complex, S.E. Norway. Evidence of mid-crustal fractionation from solid and fluid inclusions in apatite. *Contributions to Mineralogy and Petrology* **93**, 491–503.

Andersen, T. 1986*b*. Compositional variation of some rare earth minerals from the Fen Complex (Telemark, S.E. Norway): implications for the mobility of rare earths in a carbonatite system. *Mineralogical Magazine* **50**, 503–9.

Andersen, T. 1987. A model for the evolution of hematite carbonatites, based on whole-rock major and trace element data from the Fen complex, southeast Norway. *Applied Geochemistry* **2**, 163–80.

Anglo American Corporation of South Africa, Limited 1969. *The geology of Premier Mine*. Guidebook prepared for American Geological Institute International Field Institute.

Anovitz, L. M. & E. J. Essene 1987. Phase equilibria in the system $CaCO_3$–$MgCO_3$–$FeCO_3$. *Journal of Petrology* **28**, 389–414.

Armbrustmacher, T. J. 1979. Replacement and primary magmatic carbonatites from the Wet Mountains area, Fremont and Custer Counties, Colorado. *Economic Geology* **74**, 888–901.

Bagdasarov, Yu. A. & M. I. Buyakayte 1986. Isotopic data on carbonatite formation in carbonate sediments. *Geochemistry International* **22**, 30–8.

Bahat, D. 1979. Interpretation on the basis of Hertzian theory of a spiral carbonatite structure at Homa Mountain, Kenya. *Tectonophysics* **60**, 235–46.

Bailey, D. K. 1966. Carbonatite volcanoes and shallow intrusions in Zambia. In *Carbonatites*, O. F. Tuttle & J. Gittins (eds), 127–54. New York: Wiley.

Barber, D. J. & H. R. Wenk 1979. On geological aspects of calcite microstructure. *Tectonophysics* **54**, 45–60.

Barker, D. S. & P. H. Nixon 1983. Carbonatite lava and 'welded' air-fall tuff (natural Portland cement), Fort Portal field, western Uganda. *EOS. Transactions, American Geophysical Union* **64**, 896.

Basu, N. K. & A. Mayila 1986. Petrographic and chemical characteristics of the Panda Hill carbonatite complex, Tanzania. *Journal of African Earth Sciences* **5**, 589–98.

Bergman, S. C. 1987. Lamproites and other potassium-rich igneous rocks: a review of their occurrence, mineralogy and geochemistry. In *Alkaline igneous rocks*, J. G. Fitton & B. G. J. Upton (eds), 103–90. Geological Society of London, Special Publication 30.

Bergstøl, S. 1979. Tinguaite dikes adjacent to the Fen alkaline complex in Telemark, Norway. *Norsk Geologisk Tidsskrift* **59**, 115–24.

Blattner, P. & A. F. Cooper 1974. Carbon and oxygen isotopic composition of carbonatite dikes and metamorphic country rock of the Haast Schist terrain, New Zealand. *Contributions to Mineralogy and Petrology* **44**, 17–27.

Boctor, N. Z. & F. R. Boyd 1981. Oxide minerals in a layered kimberlite–carbonatite sill from Benfontein, South Africa. *Contributions to Mineralogy and Petrology* **76**, 253–9.

Boctor, N. Z., P. H. Nixon, F. Buckley & F. R. Boyd 1984. Petrology of carbonatite tuff from Melkfontein, East Griqualand, southern Africa. In *Kimberlites I: Kimberlites and related rocks*. Proceedings of the Third International Kimberlite Conference, J. Kornprobst (ed.), 75–82. New York: Elsevier.

Boettcher, A. L., J. K. Robertson & P. J. Wyllie 1980. Studies in synthetic carbonatite systems: solidus relationships for CaO–MgO–CO_2–H_2O to 40 kbar and CaO–MgO–SiO_2–CO_2–H_2O to 10 kbar. *Journal of Geophysical Research* **85**, 6937–43.

Borodin, L. S. 1978. Alkaline-ultramafic and carbonatite provinces of the USSR. In *Proceedings of the*

First International Symposium on Carbonatites, Poços de Caldas, Brazil, June 1976, 223–6. Brasilia: Brasil Departamento Nacional da Produção Mineral.

Brey, G. & J. Keller 1984. Die petrologische Charakterisierung des 'Deckentuff'–Magmas in Hegau. *Fortschritte der Mineralogie* **62**(1), 32–3.

Brown, P. E. 1964. The Songwe Scarp carbonatite and associated feldspathization in the Mbeya Range, Tanganyika. *Quarterly Journal of the Geological Society of London* **120**, 223–40.

Brueckner, H. K. & D. C. Rex 1980. K–Ar and Rb–Sr geochronology and Sr isotopic study of the Alnö alkaline complex, northeastern Sweden. *Lithos* **13**, 111–19.

Carswell, D. A. 1980. Mantle derived lherzolite nodules associated with kimberlite, carbonatite and basalt magmatism: a review. *Lithos* **13**, 121–38.

Černy, P. 1982. Anatomy and classification of granitic pegmatites. In *Short course in granitic pegmatites in science and industry*, P. Černy (ed.), 1–39. Mineralogical Association of Canada, Short Course Handbook 8.

Clarke, M. G. C. & B. Roberts 1986. Carbonated melilitites and calcitized alkalicarbonatites from Homa Mountain, western Kenya: a reinterpretation. *Geological Magazine* **123**, 683–92.

Conway, C. M. & H. P. Taylor, Jr. 1969. O^{18}/O^{16} and C^{13}/C^{12} ratios of coexisting minerals in the Oka and Magnet Cove carbonatite bodies. *Journal of Geology* **77**, 618–26.

Cooper, A. F. 1986. A carbonatitic lamprophyre dike swarm from the Southern Alps, Otago and Westland. In *Late Cenozoic volcanism in New Zealand*, I. E. M. Smith (ed.), 313–36. Royal Society of New Zealand Bulletin 23.

Cooper, A. F., J. Gittins & O. F. Tuttle 1975. The system Na_2CO_3–K_2CO_3–$CaCO_3$ at 1 kilobar and its significance in carbonatite petrogenesis. *American Journal of Science* **275**, 534–60.

Crohn, P. W. & D. H. Moore 1984. The Mud Tank carbonatite, Strangways Range, central Australia. *BMR Journal of Australian Geology and Geophysics* **9**, 13–18.

Currie, K. L. 1976. *The alkaline rocks of Canada*. Geological Survey of Canada, Bulletin 239.

Datta, R. K., D. M. Roy, S. P. Faile & O. F. Tuttle 1964. Glass formation in carbonate systems. *Journal of the American Ceramic Society* **47**, 153.

Dawson, J. B. 1962a. Sodium carbonate lavas from Oldoinyo Lengai, Tanganyika. *Nature* **195**, 1075–6.

Dawson, J. B. 1962b. The geology of Oldoinyo Lengai. *Bulletin Volcanologique* **24**, 349–87.

Dawson, J. B. 1964a. Carbonatitic volcanic ashes in northern Tanganyika. *Bulletin Volcanologique* **27**, 81–91.

Dawson, J. B. 1964b. Carbonate tuff cones in northern Tanganyika. *Geological Magazine* **101**, 129–37.

Dawson, J. B. 1966a. Oldoinyo Lengai – an active volcano with sodium carbonatite lava flows. In *Carbonatites*, O. F. Tuttle & J. Gittins (eds), 155–68. New York: Wiley.

Dawson, J. B. 1966b. The kimberlite–carbonatite relationship. *Mineralogical Society of India*, IMA volume, 1–4.

Dawson, J. B., P. Bowden & G. C. Clark 1968. Activity of the carbonatite volcano Oldoinyo Lengai, 1966. *Geologische Rundschau* **57**, 865–79.

Dawson, J. B., M. S. Garson & B. Roberts 1987. Altered former alkalic carbonatite lava from Oldoinyo Lengai, Tanzania: inferences for calcite carbonatite lavas. *Geology* **15**, 765–8.

Dawson, J. B., D. G. Powell & A. M. Reid 1970. Ultrabasic xenoliths and lava from the Lashaine volcano, northern Tanzania. *Journal of Petrology* **11**, 519–48.

Deans, T. & B. Roberts 1984. Carbonatite tuffs and lava clasts of the Tinderet foothills, western Kenya: a study of calcified natrocarbonatites. *Journal of the Geological Society of London* **141**, 563–80.

Deans, T. & A. F. Seager 1978. Stratiform magnetite crystals of abnormal morphology from volcanic carbonatites in Tanzania, Kenya, Greenland and India. *Mineralogical Magazine* **42**, 463–75.

Dimroth, E. 1970. Meimechites and carbonatites of the Castignon Lake complex, New Quebec. *Neues Jahrbuch für Mineralogie, Abhandlungen* **112**, 239–78.

Donaldson, C. H., J. B. Dawson, R. Kanaris-Sotiriou, R. A. Batchelor & J. N. Walsh 1987. The silicate lavas of Oldoinyo Lengai, Tanzania. *Neues Jahrbuch für Mineralogie, Abhandlungen* **156**, 247–79.

Eggler, D. H. 1987. Discussion of recent papers on carbonated peridotite, bearing on mantle metasomatism and magmatism: an alternative. *Earth and Planetary Science Letters* **82**, 398–400; and final comment, ibid., 403.

Egorov, L. S. 1970. Carbonatites and ultrabasic-alkaline rocks of the Maimecha-Kotui region, N. Siberia. *Lithos* **3**, 341–59.

El Ramly, M. F. & A. A. A. Hussein 1985. The ring complexes of the Eastern Desert of Egypt. *Journal of African Earth Sciences* **3**, 77–82.

Fanelli, M. F., N. Cava & P. J. Wyllie 1981. Co-precipitation of calcite and dolomite without portlandite at a new eutectic in $CaO-MgO-CO_2-H_2O$ at 2 kbar illuminates carbonatite petrogenesis. *EOS. Transactions, American Geophysical Union* **62**, 413.

Fletcher, C. J. N. & M. Litherland 1981. The geology and tectonic setting of the Velasco Alkaline Province, eastern Bolivia. *Journal of the Geological Society of London* **138**, 541–8.

Friedrichsen, H. 1968. Sauerstoffisotopen einiger minerale der Karbonatite des Fengebietes Sud Norwegen. *Lithos* **1**, 70–5.

Garson, M. S. 1955. Flow phenomena in carbonatites in southern Nyasaland. *Colonial Geology and Mineral Resources* **5**(3), 311–18.

Garson, M. S. 1966. Carbonatites in Malawi. In *Carbonatites*, O. F. Tuttle & J. Gittins (eds), 33–71. New York: Wiley.

Garson, M. S., J. S. Coats, N. M. S. Rock & T. Deans 1984. Fenites, breccia dykes, albitites, and carbonatitic veins near the Great Glen Fault, Inverness, Scotland. *Journal of the Geological Society of London* **141**, 711–32.

Gaspar, J. C. & P. J. Wyllie 1984. The alleged kimberlite–carbonatite relationship: evidence from ilmenite and spinel from Premier and Wesselton Mines and the Benfontein Sill, South Africa. *Contributions to Mineralogy and Petrology* **85**, 133–40.

Gaspar, J. C. & P. J. Wyllie 1987. The phlogopites from the Jacupiranga carbonatite intrusions. *Mineralogy and Petrology* **36**, 121–34.

Gittins, J. 1973. The significance of some porphyritic textures in carbonatites. *Canadian Mineralogist* **12**, 226–8.

Gittins, J. 1978. Some observations on the present status of carbonatite studies. In *Proceedings of the First International Symposium on Carbonatites, Poços de Caldas, Brazil*, June 1976, 107–15. Brasilia: Brasil Departamento Nacional da Produção Mineral.

Gittins, J. 1979. Problems inherent in the application of calcite–dolomite geothermometry to carbonatites. *Contributions to Mineralogy and Petrology* **69**, 1–4.

Gittins, J. & D. McKie 1980. Alkalic carbonatite magmas: Oldoinyo Lengai and its wider applicability. *Lithos* **13**, 213–15.

Gittins, J., C. R. Allen & A. F. Cooper 1975. Phlogopitization of pyroxenite: its bearing on the composition of carbonatite magmas. *Geological Magazine* **112**, 503–7.

Gold, D. P. 1969. The Oka carbonatite and alkaline complex. In *Guidebook for the geology of Monteregian Hills: Geological Association of Canada and Mineralogical Association of Canada, Annual Meeting, Montreal*, G. Pouliot (ed.), 43–62.

Gold, D. P. & M. Vallée 1969. *Field guide to the Oka area: description and itinerary*. Quebec Department of Natural Resources, Mines Branch, S-101.

Gold, D. P., G. N. Eby, K. Bell & M. Vallée 1986. Carbonatites, diatremes, and ultra-alkaline rocks in the Oka area, Quebec. *Geological Association of Canada–Mineralogical Association of Canada–Canadian Geophysical Union, Joint Annual Meeting, Ottawa, Field Trip 21, Guidebook*.

Harris, N. B. W., A. E. R. O. Mohammed & M. Z. Shaddad 1983. Geochemistry and petrogenesis of a nepheline syenite–carbonatite complex from the Sudan. *Geological Magazine* **120**, 115–27.

Hay, R. L. 1978. Melilitite–carbonatite tuffs in the Laetolil Beds of Tanzania. *Contributions to Mineralogy and Petrology* **67**, 357–67.

Hay, R. L. 1983. Natrocarbonatite tephra of Kerimasi volcano, Tanzania. *Geology* **11**, 599–602.

Hay, R. L. 1986. Role of tephra in the preservation of fossils in Cenozoic deposits of East Africa. In *Sedimentation in the African Rifts*, L. E. Frostick, R. W. Renant, I. Reid & J. J. Tiercelin (eds), 339–44. Geological Society of London, Special Publication 25.

Hay, R. L. & J. R. O'Neil 1983. Carbonatite tuffs in the Laetolil Beds of Tanzania and the Kaiserstuhl in Germany. *Contributions to Mineralogy and Petrology* **82**, 403–6.

Heinrich, E. W. 1966. *Geology of carbonatites*. Chicago: Rand McNally.

Hogarth, D. D. 1966. Intrusive carbonate rock near Ottawa, Canada. *Mineralogical Society of India*, IMA volume, 45–53.

Hogarth, D. D. & P. Rushforth 1986. Carbonatites and fenites near Ottawa, Ontario and Gatineau, Quebec. *Geological Association of Canada–Mineralogical Association of Canada–Canadian Geophysical Union, Joint Annual Meeting, Ottawa, Field Trip 9B, Guidebook*.

Höy, T. & Y. T. J. Kwong 1986. The Mount Grace carbonatite – An Nb and light rare earth element-enriched marble of probable pyroclastic origin in the Shuswap Complex, southeastern British Columbia. *Economic Geology* **81**, 1374–86.

Jones, A. P. & P. J. Wyllie 1983. Low-temperature glass quenched from a synthetic, rare earth carbonatite: implications for the origin of the Mountain Pass deposit, California. *Economic Geology* **78**, 1721–3.

Jones, A. P. & P. J. Wyllie 1986. Solubility of rare earth elements in carbonatite magmas, indicated by the liquidus surface in $CaCO_3-Ca(OH)_2-La(OH)_3$ at 1 kbar pressure. *Applied Geochemistry* **1**, 95–102.

Kapustin, Yu. L. 1971. *Mineralogy of carbonatites*. Moscow: Nauka. (English translation, 1980, New Delhi: Amerind Publishing.)

Kapustin, Yu. L. 1982. Geochemistry of strontium and barium in carbonatites. *Geochemistry International* **19**(2), 38–48.

Kapustin, Yu. L. 1985. A differentiated analcime basalt sill with segregated calcite. *International Geology Review* **27**, 964–76.

Kapustin, Yu. L. 1986. The origin of early calcitic carbonatites. *International Geology Review* **28**, 1031–44.

Kapustin, Yu. L. 1987. Magnesium metasomatism in early calcite-carbonatites. *International Geology Review* **29**, 193–206.

Katz, K. & J. Keller 1981. Comb-layering in carbonatite dykes. *Nature* **294**, 350–2.

Katz-Lehnert, K. & J. Keller 1987. Differentiation trends in carbonatite rocks of the Kaiserstuhl complex, SW Germany (abstract). *Terra Cognita* **7**, 397.

Keller, J. 1981. Carbonatitic volcanism in the Kaiserstuhl alkaline complex: evidence for highly fluid carbonatitic melts at the earth's surface. *Journal of Volcanology and Geothermal Research* **9**, 423–31.

Keller, J. & G. Brey 1983. The Hegau volcanic province, SW-Germany: olivine melilitites, phonolites, melilititic diatremes, carbonatites. *Terra Cognita* **3**, 143.

King, B. C. & D. S. Sutherland 1966. The carbonatite complexes of eastern Uganda. In *Carbonatites*, O. F. Tuttle & J. Gittins (eds), 73–126. New York: Wiley.

Knudsen, C. 1985. Investigation of the Qaqarssuk carbonatite complex, southern West Greenland. *Geological Survey of Greenland, Report* **125**, 34–40.

Koster van Groos, A. F. & P. J. Wyllie 1963. Experimental data bearing on the role of liquid immiscibility in the genesis of carbonatites. *Nature* **199**, 801–2.

Kresten, P. 1980. The Alnö complex: tectonics of dyke emplacement. *Lithos* **13**, 153–8.

Kresten, P. & V. Morogan 1986. Fenitization at the Fen complex, southern Norway. *Lithos* **19**, 27–42.

Lapin, A. V. 1987. The relationships between carbonatites and kimberlites and some problems of deep-seated magma formation. *International Geology Review* **28**, 955–64.

Lapin, A. V. & H. Vartiainen 1983. Orbicular and spherulitic carbonatites from Sokli and Vuorijarvi. *Lithos* **16**, 53–60.

Larsen, L. M., D. C. Rex & K. Secher 1983. The age of carbonatites, kimberlites and lamprophyres from southern West Greenland: recurrent alkaline magmatism during 2500 million years. *Lithos* **16**, 215–21.

Le Bas, M. J. 1977. *Carbonatite-nephelinite volcanism: an African case history*. New York: Wiley.

Le Bas, M. J. 1981. Carbonatite magmas. *Mineralogical Magazine* **44**, 133–40.

Le Bas, M. J. 1984. Oceanic carbonatites. In *Kimberlites I: Kimberlites and related rocks*. Proceedings of the Third International Kimberlite Conference, J. Kornprobst (ed.), 169–78. New York: Elsevier.

Le Bas, M. J. 1987. Nephelinites and carbonatites. In *Alkaline igneous rocks*, J. G. Fitton & B. G. J. Upton (eds), 53–83. Geological Society of London, Special Publication 30.

Le Bas, M. J. & C. D. Handley 1979. Variations in apatite composition in ijolitic and carbonatitic igneous rocks. *Nature* **279**, 54–6.

Le Bas, M. J., I. Mian & D. C. Rex 1987. Age and nature of carbonatite emplacement in North Pakistan. *Geologische Rundschau* **76**, 317–23.

Lloyd, F. E. 1985. Experimental melting and crystallization of glassy olivine melilitites. *Contributions to Mineralogy and Petrology* **90**, 236–43.

McGetchin, T. R., Y. S. Nikhanj & A. A. Chodos 1973. Carbonatite-kimberlite relations in the Cane Valley diatreme, San Juan County, Utah. *Journal of Geophysical Research* **78**, 1854–69.

McIver, J. R. & J. Ferguson 1979. Kimberlitic, melilititic, trachytic and carbonatite eruptives at Saltpetre Kop, Sutherland, South Africa. In *Kimberlites, diatremes, and diamonds: their geology, petrology, and geochemistry*. Proceedings of the Second International Kimberlite Conference, Vol. 1, F. R. Boyd & H. O. A. Meyer (eds), 111–28. Washington: American Geophysical Union.

Maravic, H. v. & G. Morteani 1980. Petrology and geochemistry of the carbonatite and syenite complex of Lueshe (N.E. Zaire). *Lithos* **13**, 159–70.

Mariano, A. N. & P. L. Roeder 1983. Kerimasi: a neglected carbonatite volcano. *Journal of Geology* **91**, 449–55.

Meen, J. K. 1987. Mantle metasomatism and carbonatites; an experimental study of a complex relationship. In *Mantle metasomatism and alkaline magmatism*, E. M. Morris & J. D. Pasteris (eds), 91–100. Geological Society of America Special Paper 215.

Mitchell, R. H. 1979. The alleged kimberlite–carbonatite relationship: additional contrary mineralogical evidence. *American Journal of Science* **279**, 570–89.

Nash, W. P. 1972. Mineralogy and petrology of the Iron Hill carbonatite complex, Colorado. *Geological Society of America Bulletin* **83**, 1361–82.

Nesbitt, B. E. & W. C. Kelly 1977. Magmatic and hydrothermal inclusions in carbonatite of the Magnet Cove complex, Arkansas. *Contributions to Mineralogy and Petrology* **63**, 271–94.

Nixon, P. H. & G. Hornung 1973. The carbonatite lavas and tuffs near Fort Portal, Western Uganda. *Overseas Geology and Mineral Resources* **41**, 168–79.

Puustinen, K. 1971. *Geology of the Siilinjarvi carbonatite complex, eastern Finland*. Bulletin Commission Géologique de Finlande 249.

Pyatenko, I. K. & L. G. Saprykina 1976. Carbonatite lavas and pyroclastics in the Paleozoic sedimentary volcanic sequence of the Kontozero District, Kola Peninsula. *Doklady Akademiya Nauk SSSR* **229**, 185–7. (English translation)

Ramberg, I. B. 1973. Gravity studies of the Fen complex, Norway, and their petrological significance. *Contributions to Mineralogy and Petrology* **38**, 115–34.

Rankin, A. H. 1977. Fluid-inclusion evidence for the formation conditions of apatite from the Tororo carbonatite complex of eastern Uganda. *Mineralogical Magazine* **41**, 155–64.

Robins, B. 1980. The evolution of the Lillebukt alkaline complex, Stjernøy, Norway. *Lithos* **13**, 219–20.

Robins, B. & M. Tysseland 1980. Fenitization of mafic cumulates by the Pollen carbonatite, Finnmark, Norway. *Lithos* **13**, 220.

Robinson, D. N. 1975. Magnetite–serpentine–calcite dykes at Premier Mine and aspects of their relationship to kimberlite and to carbonatite of alkalic carbonatite complexes. *Physics and Chemistry of the Earth* **9**, 61–70.

Rock, N. M. S. 1976. The role of CO_2 in alkali rock genesis. *Geological Magazine* **113**, 97–113.

Roedder, E. 1984. Fluid inclusions. *Mineralogical Society of America, Reviews in Mineralogy* **12**.

Romanchev, B. P. 1972. Inclusion thermometry and the formation conditions of some carbonatite complexes in East Africa. *Geochemistry International* **9**, 115–20.

Romanchev, B. P. & S. V. Sokolov 1980. Liquation in the production and geochemistry of the rocks in carbonatite complexes. *Geochemistry International* **16**, 125–35.

Samoilov, V. S. & G. S. Plyusnin 1982. The source of material for rare-earth carbonatites. *Geochemistry International* **19**(5), 13–25.

Scatena-Wachel, D. E., A. P. Jones, A. Mariano, R. W. Hinton & R. N. Clayton 1986. Strontium isotopes of kimberlites and carbonatites. *Terra Cognita* **6**, 202.

Scott Smith, B. H. & E. M. W. Skinner 1984. A new look at Prairie Creek, Arkansas. *Kimberlites I: Kimberlites and related rocks*. Proceedings of the Third International Kimberlite Conference, J. Kornprobst (ed.), 255–83. New York: Elsevier.

Secher, K. & L. M. Larsen 1980. Geology and mineralogy of the Sarfartoq carbonatite complex, southern West Greenland. *Lithos* **13**, 199–212.

Sheppard, S. M. F. & J. B. Dawson 1973. $^{13}C/^{12}C$, $^{18}O/^{16}O$ and D/H isotope variation in 'primary' igneous carbonatites. *Fortschritte der Mineralogie* **50**, 128–9.

Silva, L. C., M. J. Le Bas & A. H. F. Robertson 1981. An oceanic carbonatite volcano on Santiago, Cape Verde Islands. *Nature* **294**, 644–5.

Skinner, E. M. W. & C. R. Clement 1979. Mineralogical classification of southern African kimberlites. In *Kimberlites, diatremes, and diamonds: their geology, petrology, and geochemistry*. Proceedings

Skogen, J. H. 1980. The structural geology of the Lillebukt carbonatite, Stjernøy, Norway. *Lithos* **13**, 221.

Sokolov, S. V. 1985. Carbonates in ultramafite, alkali-rock, and carbonatite intrusions. *Geochemistry International* **22**(4), 150–66.

Sommerauer, J. & K. Katz-Lehnert 1985. Trapped phosphate melt inclusions in silicate–carbonate–hydroxyapatite from comb-layer alvikites from the Kaiserstuhl carbonatite complex (SW-Germany). *Contributions to Mineralogy and Petrology* **91**, 354–9.

Stewart, J. W. 1970. *Precambrian alkaline-ultramafic/carbonatite volcanism at Qagssiarssuk, South Greenland*. Geological Survey of Greenland Bulletin 84.

Stillman, C. J., H. Furnes, M. J. Le Bas, A. H. F. Robertson & J. Zielonka 1982. The geological history of Maio, Cape Verde Islands. *Journal of the Geological Society of London* **139**, 347–61.

Straaten, P. van 1986. Some aspects of the geology of carbonatites in S.W. Tanzania. *Geological Association of Canada – Mineralogical Association of Canada – Canadian Geophysical Union, Joint Annual Meeting, Ottawa. Program with Abstracts* **11**, 140.

Streckeisen, A. L. 1980. Classification and nomenclature of volcanic rocks, lamprophyres, carbonatites and melilitic rocks, IUGS Subcommission on the Systematics of Igneous Rocks, recommendations and suggestions. *Geologische Rundschau* **69**, 194–207.

Sturt, B. A. & D. M. Ramsay 1965. The alkaline province of the Breivikbotn area, Sørøy, northern Norway. *Norges Geologisk Undersøgelse* **231**, 6–142.

Sukheswala, R. N. & S. M. Borges 1975. The carbonatite-injected sandstones of Siriwasan, Chhota Udaipur, Gujarat. *Indian Journal of Earth Sciences* **2**, 1–10.

Sutherland, D. S. 1980. Two examples of fluidization from the Tororo carbonatite complex, southeast Uganda. *Proceedings of the Geologists' Association* **91**, 39–45.

Suwa, K., S. Oana, H. Wada & S. Osaki 1975. Isotope geochemistry and petrology of African carbonatites. *Physics and Chemistry of the Earth* **9**, 735–45.

Treiman, A. H. 1985. Low-alkali carbonatites in alkaline complexes: separate mantle sources for carbonate and alkalis? *Geological Society of America, Abstracts with Programs* **17**, 194.

Treiman, A. H. & E. J. Essene 1984. A periclase–dolomite–calcite carbonatite from the Oka complex, Quebec, and its calculated volatile composition. *Contributions to Mineralogy and Petrology* **85**, 149–57.

Treiman, A. H. & E. J. Essene 1985. The Oka carbonatite complex, Quebec: geology and evidence for silicate–carbonate liquid immiscibility. *American Mineralogist* **70**, 1101–13.

Treiman, A. H. & A. Schedl 1983. Properties of carbonatite magma and processes in carbonatite magma chambers. *Journal of Geology* **91**, 437–47.

Twyman, J. D. & J. Gittins 1987. Alkalic carbonatite magmas: parental or derivative? In *Alkaline igneous rocks*, J. G. Fitton & B. G. J. Upton (eds), 85–94. Geological Society of London Special Publication 30.

Vartiainen, H. & H. Paarma 1979. Geological characteristics of the Sokli carbonatite complex, Finland. *Economic Geology* **74**, 1296–306.

Vikhter, B. Ya., G. K. Yeremenko & V. M. Chmyrev 1976. A young volcanogenic carbonatite complex in Afghanistan. *International Geology Review* **18**, 1305–12.

Wenk, H.-R., D. J. Barber & R. J. Reeder 1983. Microstructures in carbonates. In *Carbonates: mineralogy and chemistry*, R. J. Reeder (ed.), 301–67. Mineralogical Society of America, Reviews in Mineralogy 11.

Williams, R. W., J. B. Gill & K. W. Bruland 1986. Ra–Th disequilibria systematics: Timescale of carbonatite magma formation at Oldoinyo Lengai Volcano, Tanzania. *Geochimica et Cosmochimica Acta* **50**, 1249–59.

Woolley, A. R. 1982. A discussion of carbonatite evolution and nomenclature, and the generation of sodic and potassic fenites. *Mineralogical Magazine* **46**, 13–17.

Wyllie, P. J. 1978. Silicate–carbonate systems with bearing on the origin and crystallization of carbonatites. In *Proceedings of the First International Symposium on Carbonatites, Poços de Caldas, Brazil*, June 1976, 61–78. Brasilia: Brasil Departamento Nacional da Produção Mineral.

Wyllie, P. J. 1987. Discussion of recent papers on carbonated peridotite, bearing on mantle metasomatism and magmatism. *Earth and Planetary Science Letters* **82**, 391–7; and response, ibid., 401–2.

Wyllie, P. J. & O. F. Tuttle 1960. The system $CaO-CO_2-H_2O$ and the origin of carbonatites. *Journal of Petrology* **1**, 1–46.

Zhabin, A. G. 1971. Primary textural-structural features of carbonatites and their metamorphic evolution. *International Geology Review* **13**, 1087–96.

Zhabin, A. G. 1978. Syngenesis and metamorphism of carbonatites. In *Proceedings of the First International Symposium on Carbonatites, Poços de Caldas, Brazil*, June 1976, 191–5. Brasilia: Brasil Departamento Nacional da Produção Mineral.

Zhabin, A. G. & G. Y. Cherepivskaya 1965. Carbonatite dikes as related to ultrabasic-alkalic extrusive igneous activity. *Doklady Akademiya Nauk SSSR, Earth Sciences Section* **160**, 135–8. (English translation)

4
EXTRUSIVE CARBONATITES AND THEIR SIGNIFICANCE

J. KELLER

ABSTRACT

Extrusive carbonatite lavas and volcanoclastic deposits with juvenile carbonatite clasts are now being recognized in many carbonatite complexes on a world-wide scale. Interpretations from these extrusive carbonatites add significantly to the observations made at Oldoinyo Lengai, Tanzania, the only active carbonatite volcano.

Carbonatite extrusive volcanism commonly indicates moderate to high explosivity of melts with extremely low viscosities. A wide range of volcanic features are recognized, including lava flows, widespread ashes, tuffs and tuff breccias, and phreatomagmatic deposits. Quite common and characteristic of carbonatite volcanic activity are tear-drop lapilli, juvenile carbonatite bombs, agglutinated lapilli tuffs, welded spatter, and agglomerates.

The quenched juvenile fragments in carbonatite pyroclastic deposits are important in constraining the physical properties and the original chemical composition of carbonatite melts. In spite of the unequivocal evidence for the existence of natrocarbonatite magma at Oldoinyo Lengai, volcanological, textural, and compositional evidence shows that for many Recent and older carbonatite extrusives primary crystallization is dominated by calcite. This conclusion argues against those models that suggest that calcite is generally, in most carbonatites, the secondary, post-solidification product of an alkali-rich carbonatite parent magma.

4.1 INTRODUCTION

The petrogenesis of carbonatite magma is still enigmatic. Whereas there is no doubt regarding the existence of carbonatite melts, there is less general agreement about the physical conditions, chemical composition, and phase relationships of the crystallizing carbonatite magma (Wyllie 1978, Le Bas 1981, Woolley 1982, Treiman & Schedl 1983, Le Bas 1987, Twyman & Gittins 1987).

One of the fundamental petrogenetic questions in carbonatite petrology is whether carbonatites represent direct partial melts from the upper mantle or whether they are generated during magmatic differentiation by liquid immiscibility or protracted fractionation (Koster van Groos & Wyllie 1968, Freestone & Hamilton 1980, Gittins & McKie 1980, Le Bas 1987, Bell & Blenkinsop 1987). Such problems are difficult to approach without knowing the composition of the carbonatitic melt phase.

The vast majority of carbonatites – whether from subvolcanic intrusions, high-level dykes, cone sheets, or volcanic extrusives – are calcitic, and less commonly dolomitic or ankeritic. This contrasts sharply with the recent products of the only active carbonatite volcano, Oldoinyo Lengai, Tanzania (Dawson 1962a, 1966, Nyamweru 1987), that has repeatedly produced sodium-dominated alkali carbonate lavas containing tabular Na–Ca-carbonate (nyerereite) and Na-carbonate (gregoryite) as liquidus phases (Cooper et al. 1975, McKie & Frankis 1977, Dawson et al. 1987).

An ongoing controversy involves the role of alkali-rich melts in the genesis of carbonatites (Gittins & McKie 1980, Keller 1981, Hay 1983, Deans & Roberts 1984, Twyman & Gittins 1987). If the chemical composition of the lavas from Oldoinyo Lengai is typical of most parent carbonatite magmas, then alkali-free carbonatites must reflect loss of alkalies from the melt during carbonatite evolution. If calcium carbonatites are calcified alkalicarbonatites of the lengaite type, as suggested by Deans & Roberts (1984) and Dawson et al. (1987), then large-scale chemical replacement after solidification is necessary. Analysis of unequivocal, rapidly chilled melt products should help clarify some of the problems associated with this controversy.

4.2 CARBONATITE VOLCANOES

Preservation of primary textures and compositions is more likely to occur in rapidly quenched, eruptive carbonatites than in their subvolcanic or plutonic counterparts. Extrusive carbonatites are now documented from a large number of occurrences, ranging in age from Precambrian and Palaeozoic to Recent. They appear wherever shallow-level carbonatite centres are exposed.

Carbonatite lava flows were first described by von Knorring & Du Bois (1961) from the Fort Portal field in western Uganda. These lavas are calcitic, vesicular, and fresh. Wyllie & Tuttle (1962) emphasized that the existence of these carbonatitic lavas is supported by the experimental results in the $CaO-CO_2-H_2O$ system. Nixon & Hornung (1973) gave a detailed account of the carbonatitic lavas and tuffs of the Fort Portal–Kasekere field pointing out that lava flows form only a small part of a wide variety of volcanic features and deposits. Nearly 50 tuff cones were described, that contain welded airfall tuffs, carbonatitic spatter, and, more significantly, lapilli tuffs with drop- and spindle-shaped, ovoid lapilli. A sample of tear-drop lapilli tuff from Saka Hill, Fort Portal (specimen no. C5484 of the Cambridge University collection), contains clear, tabular calcite microphenocrysts of primary origin, very similar in texture to the Pele's tear-drop lapilli from Henkenberg, Kaiserstuhl (Fig. 4.1) described by Keller (1981).

As early as 1966, Bailey emphasized the significance of extrusive carbonatites, and contrasted his observations at the Rufunsa volcanoes, Zambia, with the natrocarbonatitic composition of the recent lavas from Oldoinyo Lengai. Bailey concluded

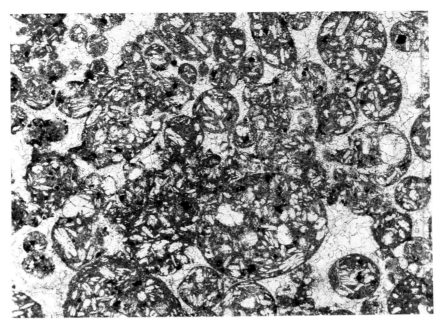

Figure 4.1 Tear-drop lapilli from Henkenberg (Kaiserstuhl). Microporphyritic lapilli supported by a matrix of secondary sparritic calcite, suggesting replacement of original fine-ash matrix. Plane-polarized light (PPL); sample width = 6 mm.

that the preservation of primary structures in the calciocarbonatitic pyroclastics of Chasweta and other Rufunsa carbonatite volcanoes ruled out original crystallization from a natrocarbonatite magma.

In the Homa Bay province of western Kenya (Le Bas 1977) extrusive carbonatites occur in several centres, including the Rangwa caldera (McCall 1963, Findlay & Rubie 1977), the Ruri carbonatite complexes (Dixon & Collins 1977), and the satellite vents of Homa Mountain (Clarke *et al.* 1977, Flegg *et al.* 1977, Clarke & Roberts 1986). All extrusive carbonatite volcanics of the Homa Bay area are spatially and temporally related to alvikite dykes corresponding to the C2, C3, and C4 carbonatitic stages of Le Bas (1977, 1981). These high-level dykes are probably feeders for the volcanic eruptions.

Tuffs of the Rangwa caldera contain abundant accretionary lapilli (Findlay & Rubie 1977, Le Bas 1977). They also show irregular pinch-and-swell structures and cross-bedding typical of base surge deposits of phreatomagmatic eruptions. Accretionary lapilli and bedding features, referred to as cross-bedding and ripple-marking, are reported from other carbonatite volcanoes (Bailey 1966, Deans & Roberts 1984). Nixon & Hornung (1973), Yeremenko *et al.* (1975), Clarke *et al.* (1977), and Hay (1983) refer briefly to possible phreatomagmatic eruptions associated with carbonatites. The carbonatitic maar-like craters of northern Tanzania (Dawson 1964a, b) were probably formed by phreatomagmatic activity. Although a systematic study of hydromagmatic eruptions and their deposits associated with carbonatite volcanoes has yet to be undertaken, it must be concluded from available descriptions that phreatomagmatic phenomena and pyroclastic surge deposits are widespread in carbonatite volcanism.

At Homa Mountain, the satellite vent of Got Chiewo consists of a well-preserved carbonatite cone composed of bedded, grey lapilli tuffs (Clarke & Roberts 1986). Porphyritic textures with platy, rectangular pseudomorphs made up of polycrystalline calcite, formerly believed to be melilite, are interpreted by Clarke & Roberts as pseudomorphs of nyereite, and the rock is considered to have originally crystallized as alkali carbonatite. Recent observations at Got Chiewo show that the grey carbonatite tuffs are composed of juvenile and accretionary lapilli set in an ash matrix of similar texture and composition to the lapilli themselves. The well-preserved crater rim is formed of densely agglutinated, flattened and welded spatter and lapilli tuff. Examples of these pyroclastic rocks are shown in Figures 4.2–4.4. Got Chiewo clearly demonstrates the highly fluid nature of erupting carbonatite lava. Welding and agglutination in the proximal crater-rim facies show that the juvenile pyroclasts were in a fluid or plastic state at the moment of deposition.

A similar, but very extensive deposit of densely agglutinated carbonatite agglomerate that probably resulted from vigorous lava fountaining occurs in the Ruri carbonatite complex, Kenya. On the southern flank of Minarot Valley in South Ruri (Dixon & Collins 1977), this agglomerate mantles the walls of the South Ruri caldera with a thick veneer of carbonatite spatter (Fig. 4.5).

Carbonatite tuffs from Kerimasi, Tanzania (Hay 1983) and from Tinderet, Kenya (Deans & Roberts 1984) have been proposed as the product of calcification of former natrocarbonatites. Deans & Roberts describe drop-shaped lapilli from

Figure 4.2 Agglutinated carbonatite spatter from the spatter rim of Got Chiewo carbonatite vent, Homa Mountain (Kenya). Sample HB12; PPL; sample width = 1.5 cm.

Figure 4.3 Agglutinated carbonatite spatter from same sample as in Figure 4.2, under cathodoluminescence, showing clear outlines of tabular pseudomorphs, now calcitic. PPL; sample width = 8 mm.

Figure 4.4 Shard texture of agglutinated carbonatite ash, pyroclastic-fall near-vent facies at Got Chiewo, Homa Mountain, Kenya, sample HB10. Composition as HB7 in Table 4.1. Large tabular pseudomorphs of polycrystalline calcite after nyerereite (?). PPL; sample width = 3.5 mm.

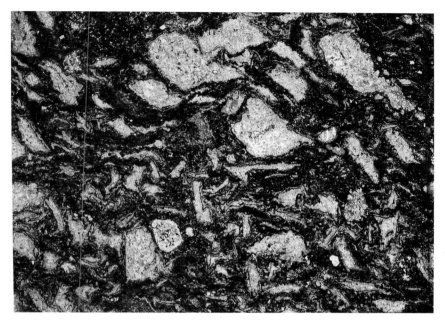

Figure 4.5 Pyroclastic texture of carbonatitic veneer deposit at South Ruri caldera, Kenya. Juvenile clasts are recrystallized calcite aggregates. Sample HB28.2. Composition as HB30 in Table 4.1. PPL; sample width = 1 cm.

Songhor (Tinderet) that are identical in texture and morphology to the Pele's tear-drop lapilli of calcite carbonatite from the Kaiserstuhl which represent melt fragments moulded by surface tension during sub-aerial eruptions (Keller 1981). Silva *et al.* (1981) and Silva and Ubaldo (1985) also compared globular carbonatitic tuffs from the Cape Verde Islands to the carbonatitic lava-spray products from the Kaiserstuhl area. Spherical tear-drop lapilli with primary calcite microphenocrysts, similar to the Kaiserstuhl examples, have also been documented from Kerimasi, Tanzania (Mariano & Roeder 1983). The widespread occurrence of juvenile melt-droplets in carbonatite volcanism is now well established.

Spherical fragments of different origin than the tear-drop lapilli have been described from pyroclastic carbonatite deposits. Accretionary lapilli, formed by ash accretion in phreatomagmatic eruptions, are recognized by the presence of concentric structures (e.g. Le Bas 1977). Sutherland (1980) has reported globular carbonatite formed during fluidization in vent-facies breccias from Tororo, Uganda, and spherical carbonatite lapilli have also been found in diatreme breccias (Bailey 1966). An example of cored pelletal lapilli from Hegau, Germany, with clasts coated by microporphyritic carbonatite, is shown in Figure 4.6. These fragments probably resulted from melt accretion around clasts in a fluidizing gas stream during diatreme formation.

The features exhibited by many carbonatite extrusives suggest that the erupted carbonatitic melts are extremely fluid (Keller 1981). Field observations of volcanic carbonatites thus agree with experimental results from simple carbonatite systems, which support the existence of low-viscosity carbonatite melts under very low

Figure 4.6 Cored pelletal lapilli from diatreme facies carbonatite, Hegau. Microporphyritic calcite carbonatite surrounding calcite megacrysts with apatite inclusions. Diameters of the two pelletal lapilli are 1.0 and 0.5 cm respectively; PPL.

pressures and temperatures down to about 550 °C. Krafft and Keller (in preparation) obtained temperatures of 495–500 °C on flowing natrocarbonatite lavas of Oldoinyo Lengai in June, 1988. Low temperature–low pressure melting has been demonstrated in both the alkali carbonatite systems Na_2CO_3–$CaCO_3$–K_2CO_3 (Cooper et al. 1975, Wyllie, this volume), and in the calcium carbonatite system CaO–CO_2–H_2O (Wyllie & Tuttle 1960, 1962). Addition of further components to the latter system (e.g. CaF_2, $Ca_3(PO_4)_2$, alkalies) would lower pressures and temperatures still further (Gittins & Tuttle 1964, Kuellmer et al. 1966, Wyllie 1966, 1978).

4.3 VOLCANOLOGICAL CLASSIFICATION OF CARBONATITES

A systematic study of volcanic phenomena and eruptive styles reflected by carbonatitic deposits is still much in its infancy. However, it is now well established that carbonatites exhibit many volcanic features, similar to those associated with gas-rich, low-viscosity silicate magmas. Extrusive products include lava flows, pyroclastic-fall deposits, surge deposits associated with phreatomagmatic eruptions, and possible pyroclastic-flow deposits.

Lava flows occur as pahoehoe- and aa-flows, reflecting variation in viscosities of the carbonatite magmas. Well-described examples are still few, and most are from Oldoinyo Lengai (Dawson 1962a, b, 1966) and Fort Portal (von Knorring & Du Bois 1961). Yeremenko et al. (1975) report amygdaloidal, alvikitic lavas from the Quaternary Khanneshin volcano in Afghanistan, and Pyatenko & Saprykina (1976) describe carbonatite lavas associated with tuffs of Palaeozoic age from the Kola

Peninsula, USSR. Stewart (1970) described highly vesicular, intensely carbonatized, amygdaloidal lava with scoriaceous flow tops from the Precambrian of southern Greenland. Of specific interest is the reference to structures resembling pillows (Stewart 1970).

Pyroclastic-fall deposits can be attributed to a variety of eruptive mechanisms. Violent, gas-rich explosions of carbonatitic magmas result in a high degree of fragmentation of country rocks and of the juvenile melt. Widespread ashes and tuff sheets have been documented, especially from Tanzania (e.g. Dawson 1964*b*, Hay & O'Neil 1983).

Drop-shaped lapilli and agglutinated splatter and agglomerates represent near-vent facies deposits of moderate explosive activity. This activity, compared to Hawaiian- and Strombolian-type eruptions associated with silicate magmas, is obviously a very common feature of carbonatite volcanism. Pyroclastic-fall deposits of shard-rich tuffs from Got Chiewo (Fig. 4.4) are produced by the violent disruption of a carbonatite melt by rapidly expanding gases. Agglomerates are common and are typically made up of agglutinated accumulations of carbonatite bombs. Increase in explosive intensity results in lava fountaining, that forms extensive, topography-mantling, spatter veneers.

Deposits of phreatomagmatic activity are characterized by the presence of accretionary lapilli and by cross-bedding features in carbonatite tuffs, interpreted as pyroclastic-surge deposits. Tuff-rings and maar-type explosion craters, such as those described by Dawson (1964*b*) from northern Tanzania, also point to widespread hydromagmatic activity.

Pyroclastic flow deposits from carbonatitic volcanoes have yet to be documented in a systematic way. Several authors use terms such as ignimbrite, nuée ardente, eutaxitic fiamme, and pumice to describe some of the extrusive products from carbonatite activity (e.g. Silva *et al.* 1981, Hay 1983). Further volcanological studies are needed to decide whether pyroclastic flows of the nuée ardente type are possible in carbonatite volcanism.

Diatreme breccias and vent agglomerates are important features in the deeper levels of carbonatite volcanoes (e.g. Bailey 1966, Dixon & Collins 1977, Keller & Brey 1983). Some similarities to kimberlite pipes are suggested by the presence of pelletal carbonatite lapilli, thought to represent juvenile melt material (Fig. 4.6). Lithic fragments, including carbonatites, are commonly rounded in the diatreme pipes, suggesting fluidization.

4.4 NATROCARBONATITES VERSUS CALCITE CARBONATITES

Nyerereite, $Na_2Ca(CO_3)_2$, and gregoryite, Na_2CO_3, both with K_2CO_3 in solid solution (Dawson 1962*a*, *b*, McKie & Frankis 1977, Gittins & McKie 1980), are the dominant products of the primary crystallization of alkali-rich carbonatite lavas of Oldoinyo Lengai. Both minerals are water-soluble and decompose under atmospheric conditions. It has been suggested by Dawson (1962*a*, *b*) and later by Hay (1983) and Deans & Roberts (1984) that nyerereite and gregoryite protocrysts will be pseudomorphosed, as they come into contact with percolating solutions, to polycrystalline calcite. This process is proposed as an explanation for the much commoner calcite carbonatites that contain very low alkali abundances. Evidence for this volume-by-volume replacement is based entirely on the interpretation of the tabular pseudomorphs seen in carbonatite extrusives. If such an exchange has taken place, it involved extensive mass transfer. One of the problems with this model is that many extrusive calcite carbonatites completely preserve their original textures, in spite of the proposed post-solidification replacement of large rock volumes by calcium-bearing aqueous solutions. Although fenitization and fluid inclusions associated with calcium carbonatite intrusions point to an elevated activity of alkalies in the fluid phase (Le Bas 1981, Woolley 1982), this alone cannot be taken as evidence for a melt of alkali carbonatitic nature.

The model of secondary calcification still lacks a quantitative geochemical treatment, and seems to imply minor changes in trace element abundances and isotopic compositions despite a volume exchange of about 40% if primary nyereite were replaced by calcite (Deans & Roberts 1984).

Table 4.1 shows chemical analyses from four samples of extrusive carbonatites from Kaiserstuhl, Germany and two samples from Kenya. The samples from Kenya contain tabular calcite crystals that were considered pseudomorphs of nyereite. Although there are significant differences between the two groups of samples (e.g. higher Ba, Ce, La, Mn, Nd, Y, and Zr in the Kenyan samples), both exhibit a typical carbonatite signature. There is little evidence to suggest that any of the major or trace elements have been removed. Chemical data for extrusive and related carbonatites given in Barber (1974) and Deans & Roberts (1984) support this conclusion. The high concentrations in the Kenyan samples of Ba, Sr, Y, and the REEs, along with high MnO and F are certainly not produced by secondary processes; they rather reflect the chemistry of late-stage carbonatites (Le Bas 1977).

Recrystallization, on the other hand, can be an important process in carbonatites, especially in the fine-grained, highly reactive, carbonatitic ashes. However, diagenetic recrystallization of fine-grained particles to coarser-crystalline aggregates in itself does not automatically imply large-scale replacement and chemical exchange. An example of complete recrystallization of carbonatitic ashes with chemical exchange of all characteristic trace elements, probably by meteoric solutions, is shown by the analysis from sample H37 given in Table 4.1.

Other attempts have been made to explain the common polycrystalline calcite pseudomorphs observed in many extrusive carbonatites. Replacement of melilite has been discussed as one possibility, but melilite pseudomorphs have rather diagnostic features and can be readily distinguished (e.g. Clarke & Roberts 1986). Polycrystalline recrystallization of a metastable calcium carbonate polymorph is yet another possibility. However, the latter is unlikely particularly when tabular pseudomorphs coexist with perfectly fresh, primary calcite crystals. Experimental results in the $CaO-CO_2-H_2O$ system (Wyllie & Tuttle 1960) show a field of primary portlandite ($Ca(OH_2)$), a mineral that forms clear, tabular plates. In the experiments, dissociation of portlandite occurred only at pressures < 100 bars. Thus, the replacement of primary portlandite by calcite, or the transformation of a metastable calcite

Table 4.1 Chemical composition of selected extrusive carbonatites.

	KB2	KB12	H31	H37	HB7	HB30
Wt%						
SiO_2	0.45	0.48	1.28	2.25	0.52	2.36
TiO_2	0.03	0.03	0.15	0.12	0.04	0.06
Al_2O_3	0.15	0.17	0.45	0.72	0.09	0.21
Fe_2O_3	0.98	0.94	2.47	0.73	2.91	2.47
MnO	0.41	0.35	0.37	0.02	1.70	3.01
MgO	0.36	0.33	0.75	0.48	0.46	0.62
CaO	52.60	52.90	51.10	53.00	49.69	47.91
Na_2O	0.04	0.06	0.09	0.00	0.20	0.24
K_2O	0.05	0.03	0.05	0.09	0.03	0.36
P_2O_5	1.56	1.21	1.57	0.13	0.88	0.69
H_2O	1.16	1.19	2.15	0.36	2.81	1.48
CO_2	39.84	40.51	38.95	41.74	36.64	33.18
F	0.54	0.54	0.19	0.04	1.22	3.57
Total	98.17	98.74	99.57	99.68	97.19	96.16
Trace elements (p.p.m.)						
V	80	80	60	30	106	81
Sr	5640	6230	3980	90	3246	3761
Ba	1360	1190	900	0	5986	6691
La	250	240	250	6	2622	3043
Ce	450	430	460	40	4589	5528
Nd	90	90	110	10	1077	1362
Y	30	30	40	4	232	430
Nb	540	510	470	40	453	430
Zr	15	60	50	20	76	45

KB2 and KB12 are from Kirchberg/Kaiserstuhl bulk lapilli tuff and isolated spherical lapilli respectively. H31 is from lapilli tuff from Henkenberg, Kaiserstuhl. All are calcium carbonatites. H37 is from a layer of completely recrystallized calcitic ash underlying the lapillistone at Henkenberg. HB7 is from Got Chiewo, Homa Mountain, and HB30 is spatter agglomerate from South Ruri.

polymorph, may be alternative explanations for the tabular pseudomorphs observed in many carbonatite extrusives.

4.5 EXTRUSIVE CARBONATITES FROM THE KAISERSTUHL

The Kaiserstuhl carbonatites belong to an alkaline complex that consists of olivine nephelinites, melilitites, tephrites, and phonolites. The complex has been described in detail by Wimmenauer (1966, 1974) and Keller (1984). Igneous activity of the Kaiserstuhl is of Miocene age and spans the range of 18–13 Ma. Keller (1984) has established a close genetic link between carbonatites and calcite-bearing melilite-hauyne nephelinites (bergalites) of the Kaiserstuhl area. Bergalites (Soellner 1913, Keller 1984) are similar in petrography and chemistry to okaites and turjaites. The Kaiserstuhl carbonatites occur as (a) subvolcanic sövite intrusions in the centre of the complex; (b) alvikitic dykes; and (c) extrusives.

A carbonatitic lapilli tuff from the Henkenberg locality has been described in detail by Keller (1981). This deposit consists almost entirely of juvenile, tear-drop shaped and sub-rounded, carbonatite lapilli, supported by secondary sparry calcite (Fig. 4.1). The lapilli bed, 1–1.5 m thick, overlies layers of fine-grained, white and very pure calcite ash and a polymict explosion breccia containing sövite clasts.

The lapilli are pyroclastic-fall, melt-droplets of a calcium carbonatitic magma which erupted explosively as lava fountains or some other lava-spray mechanism. The melt fragments, having had very low viscosities, were moulded during sub-aerial transport and some were still plastic on landing.

The extrusive activity at Kaiserstuhl is the surface expression of the emplacement of numerous high-level alvikitic dykes. The dykes, dominantly calcite carbonatites, commonly exhibit spectacular comb-quench borders with arborescent textures of skeletal calcite (Katz & Keller 1981). The alvikitic dykes and lapilli form a chemically homogeneous group, similar to the sövitic intrusives of the Kaiserstuhl.

It has been suggested that the tear-drop lapilli from the Kaiserstuhl largely preserve the chemical composition of a quenched carbonate liquid (Keller 1981). However, in a series of papers on the natrocarbonatite problem (Hay & O'Neil 1983, Deans & Roberts 1984, Dawson *et al.* 1987) it has been suggested that these lapilli represent natrocarbonatites in which nyerereite, now calcified, was an important or even the dominant primary phase.

Hay & O'Neil (1983) studied a sample from the Henkenberg lapillistone with the initial assumption that the tabular calcite phenocrysts and microphenocrysts are possible pseudomorphs after nyerereite. The lack of any recrystallization in the monocrystalline phenocrysts, however, coupled with chemical arguments led Hay & O'Neil (1983) to accept the primary nature of the calcite phenocrysts and microphenocrysts. It was then assumed that the supposed sodium carbonates were originally contained in the microcrystalline matrix of the lapilli. There is, however, no evidence that the groundmass originally contained any substantial amounts of alkali carbonates. The distinct carbon and oxygen isotopic compositions of the extrusive volcanics in the Kaiserstuhl complex can be explained by magmatic fractionation and only very minor recrystallization of the fine-grained calcitic groundmass in contact with meteoric water (Hubberten *et al.* 1988). However, the textural, petrographic, and chemical evidence argues strongly against large-scale chemical transport and replacement processes in the Kaiserstuhl rocks.

A second example of carbonatitic lapillistone from the Kaiserstuhl occurs at the Kirchberg locality, about 1.5 km SE of Henkenberg. Although lapilli deposits from Kirchberg and Henkenberg have striking similarities, they are separate volcanic units, and were erupted from different vents. The Kirchberg lapilli provide additional and complementary evidence for the primary igneous composition that was suggested already for the lapilli deposit from Henkenberg (Keller 1981).

A major difference between the Henkenberg and Kirchberg extrusive units is the lack of secondary, sparritic calcite cement in the Kirchberg lapilli tuff. This difference can be seen by comparing Figures 4.1 and 4.8.

A large part of the Kirchberg lapilli tuff consists of perfectly spherical, juvenile melt droplets. Lapilli diameters generally reach a maximum size of 2 cm (Fig. 4.7), and grade into ash-sized fragments which have identical porphyritic to microporphyritic textures (Fig. 4.8). The matrix-supported lapilli tuff shows a high degree of

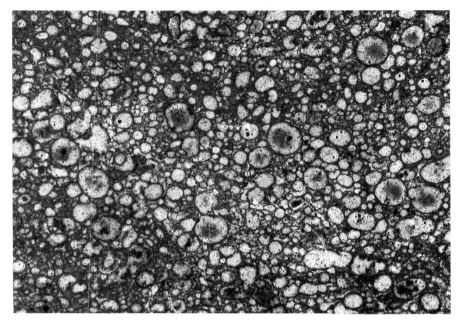

Figure 4.7 Carbonatitic lapilli tuff from Kirchberg (Kaiserstuhl), sample KB2. PPL; sample width = 4.7 cm.

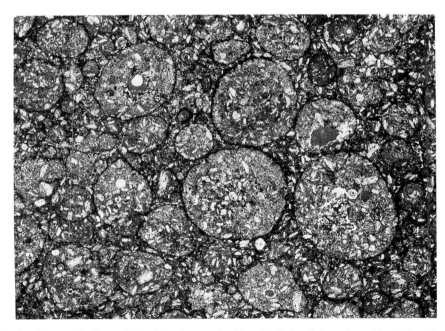

Figure 4.8 Lapilli tuff Kirchberg KB2 with microporphyritic spherical and drop-shaped lapilli in ash matrix of identical texture and composition. PPL; sample width = 1.5 cm.

primary compaction, and in some layers the ejectamenta were still plastic and deformed during accumulation. Such agglutinated and welded lapilli beds (see Fig. 4.9) grade into welded carbonatite spatter which is very common in African carbonatite volcanoes (Nixon & Hornung 1973, Clarke & Roberts 1986). Figures 4.2 and 4.4 show textures typical of African agglutinated extrusive carbonatites.

The similar chemical compositions of the isolated spherical lapilli and the bulk lapilli tuff (Table 4.1) support the proposed juvenile nature of the whole deposit, and show the lack of any selective replacement in the more reactive ash matrix.

Vesicularity of the spherical lapilli is another distinctive feature of the Kirchberg extrusive carbonatites. As commonly observed in basaltic Pele's tears, exsolution of gases results in vesicular or even spongy textures, especially within the centres of the tear-drop lapilli. This vesicular texture can be observed in some of the lapilli shown in Figure 4.7.

The preservation of these delicate vesicular textures, and the absence of secondary calcite inside the vesicles, are important in assessing what role, if any, was played by secondary processes. On the basis of these observations it seems that fluid migration must have been very minor, and it is difficult to envisage a volume-by-volume replacement of large amounts of Na by Ca in these rocks.

Parts of the Kirchberg lapilli tuff are so well sorted that closely packed lapilli occur without any ash matrix. The lapilli form an open, clast-supported framework with hollow interstices. Again, and in contrast to the Henkenberg material, there is no secondary calcite matrix.

Figure 4.9 Agglutinated lapilli with plastic deformation in Kirchberg lapilli bed. Sample KB10.2. PPL; sample width = 9 mm.

The obvious lack of cementation in the open-framework spaces and gas vesicles, in addition to the preservation of delicate walls of quenched carbonatite between vesicles, are compelling evidence against aqueous fluids precipitating calcite into these rocks. Even minor solution and redeposition within the lapillistone is difficult to accept on the basis of the textural evidence. The preservation of accidental clasts of fresh, silicate glass, despite the tendency for alkaline silicate glass to alter in the presence of CO_2-rich pore water, is further evidence against migrating solutions.

Additional evidence for a primary origin for the calcite at Kaiserstuhl can be deduced from carbonatitic lava-blocks of bomb size, embedded within the lapilli tuff at Kirchberg. Both lapilli and lava bombs have similar porphyritic textures, but differ in crystallinity and phenocryst size. The chemical compositions of the lava bombs are identical to those from lapilli in the same deposit, and the carbon and oxygen isotopic compositions fall in the restricted field of extrusive carbonatites of the Kaiserstuhl (Hubberten *et al.* 1988). Therefore, highly porphyritic lava bombs and microporphyritic lapilli represent juvenile components of the same magma. The importance of porphyritic textures as a primary magmatic feature in carbonatites has been emphasized by Gittins (1973) and Zhabin (1978). Clear calcite phenocrysts, tabular on {0001}, make up about 50% of the bombs (see Figs. 4.10 & 4.11). In thin section, the tabular crystals are sub-angular laths (up to 0.6 mm in size) with rounded edges indicating the development of rhombohedral and pyramidal crystal faces. Basal sections show the tablets to have pseudohexagonal forms. The matrix

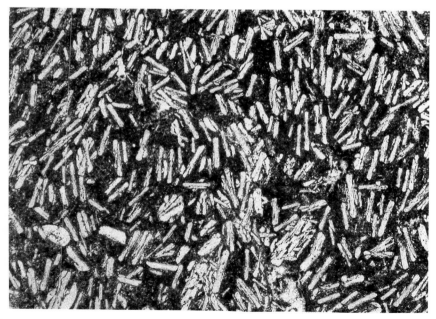

Figure 4.10 Trachytoidal texture of microphenocrysts of calcite tabular on {0001} in lava bomb from Kirchberg. Sample KB325b. PPL; sample width = 8 mm.

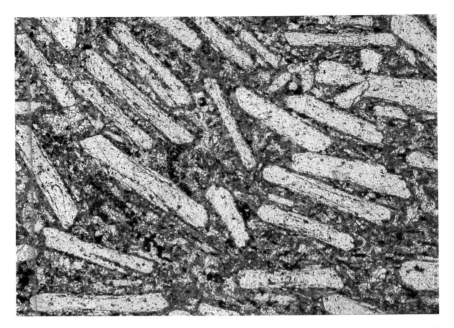

Figure 4.11 Tabular calcite phenocrysts from lava bomb, Kirchberg. Sample KB326.6. Monocrystalline calcite with minute inclusions of mainly dusty magnetite. PPL; sample width = 1.5 mm.

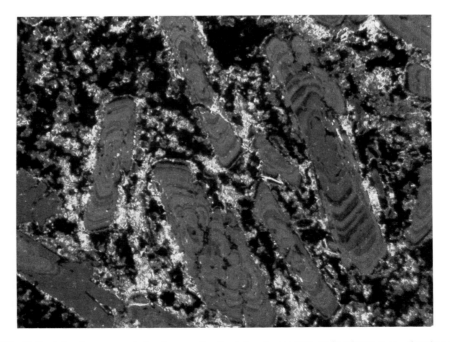

Figure 4.12 Cathodoluminescence photomicrograph of calcite phenocrysts as in Figure 4.11, showing primary magmatic zonation of calcite. PPL; sample width = 1.3 mm.

consists of fine-grained calcite and magnetite dust, the result of rapid chilling of the carbonatite melt. Any low-temperature recrystallization in such a fine-grained, granular groundmass can only have happened on a very limited scale.

Microprobe traverses from core to rim of the calcite phenocrysts illustrated in Figure 4.11 failed to reveal any detectable Na in the calcite phenocrysts nor any sodium-bearing phases associated with the rims. Thus, precipitation of calcite persisted throughout solidification of these carbonatite bombs, without any significant alkali enrichment in the residual melt.

The primary igneous nature of calcite phenocrysts in the bombs is further shown by conspicuous oscillatory zoning, observed using cathodoluminescence (Fig. 4.12). Zoning takes the form of alternating orange to orange–brown bands, a feature attributed to variation in Mn^{2+} contents of the calcite phenocrysts (Mariano 1978). Zoning of this type can only be the result of magmatic crystallization, and closely resembles the zoning of primary magmatic calcite phenocrysts from a carbonatite lava from Kerimasi, described by Mariano & Roeder (1983). Zoned phenocrysts of this type cannot be produced by secondary processes, nor can they survive recrystallization or chemical replacement.

The igneous nature of the calcite phenocrysts, their abundance in the bulk rock and the lack of Na enrichment during crystallization are clear indications that crystallization of these carbonatites was dominated by calcite. All the evidence obtained so far points to a calciocarbonatitic composition for the erupting magma.

4.6 CONCLUSIONS

Extrusive carbonatites show a wide variety of volcanological features that reflect the physical properties and the chemical composition of erupting carbonate magmas.

Petrographic examination of lapilli from two occurrences in the Kaiserstuhl complex rules out significant alteration of these rocks after volcanic quenching. The suggestion that extrusive calcium carbonatites originally crystallized as natrocarbonatites similar to the modern lavas of Oldoinyo Lengai, and the proposal that carbonatites, in general, are produced by replacement of original nyerereite or gregoryite is not supported by examples from the Kaiserstuhl complex. Tear-drop lapilli, lapilli tuffs made up of spherical melt fragments, and larger carbonatite lava bombs, exhibit textural features which rule out any chemical replacement and re-organization during a diagenetic, post-solidification process. The loss of significant amounts of Na accompanied by the introduction of Ca by migrating pore fluids is unlikely, particularly when primary pyroclastic textures, such as open inter-clast frameworks and open vesicles in lapilli, are preserved.

Porphyritic lapilli and lava bombs with tabular calcite phenocrysts from the Kaiserstuhl complex provide strong evidence that magmatic crystallization in these rocks was governed by the precipitation of calcite from a melt of calcium carbonatitic composition. It is concluded that the former presence of alkali carbonates in calcite carbonatites can not be assumed on the basis of interpretation of pseudomorphs. A careful analysis of textures as well as a quantitative treatment of the distribution and possible redistribution of major and trace elements are essential for the evaluation of the original magmatic composition.

ACKNOWLEDGEMENTS

This paper has benefitted from discussions with many friends and colleagues during excursions to the Kaiserstuhl. Special thanks are due to John Gittins and Mike Le Bas for an invitation to the stimulating Homa Bay Carbonatite Workshop held in 1987. Keith Bell is thanked for his encouragement, patience, and editorial guidance. Martin Clarke was an inspiring guide through Homa Mountain. Research is supported by Deutsche Forschungsgemeinschaft. S. O. Agrell provided two specimens from Fort Portal. Thanks for technical assistance are given to Erika Lutz, Hildegard Schlegel, Hildegard Dewald, and Klaus Fesenmeier. Ulrich Koberski helped with the cathodoluminescence and Kerstin Katz-Lehnert with the XRF analysis. Ulrike Mues became expert at word processing during the preparation of this manuscript. Grant Heiken, M. F. Sheridan, and J. W. Card gave valuable help as reviewers of this paper.

REFERENCES

Bailey, D. K. 1966. Carbonatite volcanoes and shallow intrusions in Zambia. In *Carbonatites*, O. F. Tuttle & J. Gittins (eds), 127–54. New York: Interscience.

Barber, C. 1974. The geochemistry of carbonatites and related rocks from two carbonatite complexes, south Nyanza, Kenya. *Lithos* **7**, 53–63.

Bell, K. & J. Blenkinsop 1987. Nd and Sr isotopic compositions of East African carbonatites: Implications for mantle heterogeneity. *Geology* **15**, 99–102.

Clarke, M. G. C. & B. Roberts 1986. Carbonated melilitites and calcitized alkali carbonatites from Homa Mountain, western Kenya: A reinterpretation. *Geological Magazine* **123**, 683–92.

Clarke, M. C. G., A. M. Flegg, D. S. Sutherland & M. J. Le Bas 1977. Homa Mountain IV: Clastic deposits and late volcanism. In *Carbonatite–nephelinite volcanism: an African case history*, M. J. Le Bas (ed.), 246–54. London: Wiley.

Cooper, A. F., J. Gittins & O. F. Tuttle 1975. The system Na_2CO_3–K_2CO_3–$CaCO_3$ at 1 kilobar and its significance in carbonatite petrogenesis. *American Journal of Science* **275**, 534–60.

Dawson, J. B. 1962a. Sodium carbonate lavas from Oldoinyo Lengai, Tanganyika. *Nature* **195**, 1075–6.

Dawson, J. B. 1962b. The geology of Oldoinyo Lengai. *Bulletin Volcanologique* **24**, 349–87.

Dawson, J. B. 1964a. Carbonate tuff cones in northern Tanganyika. *Geological Magazine* **101**, 129–37.

Dawson, J. B. 1964b. Carbonatitic ashes in northern Tanganyika. *Bulletin Volcanologique* **27**, 81–92.

Dawson, J. B. 1966. Oldoinyo Lengai – an active volcano with sodium carbonatite lava flows. In *Carbonatites*, O. F. Tuttle & J. Gittins (eds), 155–68. New York: Interscience.

Dawson, J. B., M. S. Garson & B. Roberts 1987. Altered former alkalic carbonatite lava from Oldoinyo Lengai, Tanzania: Inferences for calcite carbonatite lavas. *Geology* **15**, 765–8.

Deans, T. & B. Roberts 1984. Carbonatite tuffs and lava clasts of the Tinderet foothills, western Kenya: A study of calcified natrocarbonatites. *Journal of the Geological Society London* **141**, 563–80.

Dixon, J. A. & B. A. Collins 1977. The carbonatitic complex of North and South Ruri. In *Carbonatite–nephelinite volcanism: an African case history*, M. J. Le Bas (ed.), 169–98. London: Wiley.

Findlay, A. L. & D. C. Rubie 1977. Kisingiri V: the Rangwa carbonatitic caldera. In *Carbonatite–*

nephelinite volcanism: an African case history, M. J. Le Bas (ed.), 101–8. London: Wiley.

Flegg, A. M., M. C. G. Clarke, M. J. Le Bas & D. S. Sutherland 1977. Homa Mountain III: Peripheral carbonatite centres. In *Carbonatite–nephelinite volcanism: an African case history*, M. J. Le Bas (ed.), 233–45. London: Wiley.

Freestone, I. C. & D. L. Hamilton 1980. The role of liquid immiscibility in the genesis of carbonatites – an experimental study. *Contributions to Mineralogy and Petrology* **73**, 105–17.

Gittins, J. 1973. The significance of some porphyritic textures in carbonatites. *Canadian Mineralogist* **12**, 226–8.

Gittins, J. & D. McKie 1980. Alkalic carbonatite magmas: Oldoinyo Lengai and its wider applicability. *Lithos* **13**, 213–15.

Gittins, J. & O. F. Tuttle 1964. The system CaF_2–$Ca(OH)_2$–$CaCO_3$. *American Journal of Science* **262**, 66–75.

Hay, R. L. 1983. Natrocarbonatite tephra of Kerimasi volcano, Tanzania. *Geology* **11**, 599–602.

Hay, R. L. & J. R. O'Neil 1983. Carbonatite tuffs in the Laetolil beds of Tanzania and the Kaiserstuhl in Germany. *Contributions to Mineralogy and Petrology* **82**, 403–6.

Hubberten, H. W., K. Katz-Lehnert & J. Keller 1988. Carbon and oxygen isotope investigations in carbonatites and related rocks from the Kaiserstuhl, Germany. *Chemical Geology* **70**, 257–74.

Katz, K. & J. Keller 1981. Comb-layering in carbonatitic dykes. *Nature* **294**, 350–2.

Keller, J. 1981. Carbonatitic volcanism in the Kaiserstuhl alkaline complex: Evidence for highly fluid carbonatitic melts at the earth's surface. *Journal of Volcanology and Geothermal Research* **9**, 423–31.

Keller, J. 1984. Der jungtertiäre Vulkanismus Südwestdeutschlands: Exkursionen im Kaiserstuhl und Hegau. *Fortschritte der Mineralogie* **62**(2), 2–35.

Keller, J. & G. Brey 1983. The Hegau volcanic province, SW Germany: Olivine melilitites, phonolites, melilitic diatremes, carbonatites. *Terra Cognita* **3**, 143.

Knorring, O. von & C. G. B. Du Bois 1961. Carbonatitic lava from Fort Portal area in western Uganda. *Nature* **192**, 1064–5.

Koster van Groos, A. F. & P. J. Wyllie 1968. Liquid immiscibility in the join $NaAlSi_3O_8$–Na_2CO_3–H_2O and its bearing on the genesis of carbonatites. *American Journal of Science* **266**, 932–67.

Kuellmer, F. J., A. P. Visocky & O. F. Tuttle 1966. Preliminary survey of the system barite–calcite–fluorite at 500 bars. In *Carbonatites*, O. F. Tuttle & J. Gittins (eds), 353–64. New York: Wiley.

Le Bas, M. J. 1977. *Carbonatite–nephelinite volcanism: an African case history*. London: Wiley.

Le Bas, M. J. 1981. Carbonatite magmas. *Mineralogical Magazine* **44**, 133–40.

Le Bas, M. J. 1987. Nephelinites and carbonatites. In *Alkaline igneous rocks*, J. G. Fitton & B. G. J. Upton (eds), 53–83. Geological Society Special Publication 30.

McCall, G. J. H. 1963. A reconsideration of certain aspects of the Rangwa and Ruri carbonatite complexes in western Kenya. *Geological Magazine* **100**, 181–5.

McKie, D. & E. J. Frankis 1977. Nyerereite: A new volcanic carbonate mineral from Oldoinyo Lengai. *Zeitschrift für Kristallographie* **145**, 73–95.

Mariano, A. N. 1978. The application of cathodoluminescence for carbonatite exploration and characterization. In *Proceedings of the First International Symposium on Carbonatites, Poços de Caldas, Brazil*, June 1976, 39–57. Brasilia: Brasil Departamento Nacional da Produção Mineral.

Mariano, A. N. & P. L. Roeder 1983. Kerimasi: A neglected carbonatite volcano. *Journal of Geology* **91**, 449–55.

Nixon, P. H. & G. Hornung 1973. The carbonatitic lavas and tuffs near Fort Portal, Western Uganda. *Overseas Geology and Mineral Resources* **41**, 168–79.

Nyamweru, C. 1987. Oldoinyo Lengai (Tanzania): Eruptive history 1983–June 1987. *SEAN Bulletin* **12**, (8), 2–5, and (10), 2.

Pyatenko, I. K. & L. G. Saprykina 1976. Carbonatite lavas and pyroclastics in the Palaeozoic sedimentary sequence of the Kontozero district, Kola peninsula. *Akademia Nauk SSSR Doklady, Earth Science Section* **229**, 185–7.

Silva, L. C. & L. M. Ubaldo 1985. Considerações geológicas e petrogenéticas sobre os tufos carbonatiticos globulares da estrutura alcalino-carbonatitica do Norte de Santiago, arquipélago de Cabo Verde. *Garcia de Orta, série de Geologia* **8**, 1–6.

Silva, L. C., M. J. Le Bas & A. H. F. Robertson 1981. An oceanic carbonatite volcano on Santiago, Cape Verde Islands. *Nature* **294**, 644–5.

Soellner, J. 1913. Über Bergalith, ein neues melilithreiches Gestein aus dem Kaiserstuhl. *Mitteilungen der badischen geologischen Landesanstalt* **7**, 415–66.

Stewart, J. W. 1970. *Precambrian alkaline-ultramafic/carbonatite volcanism at Quassiarssuk, South Greenland*. Grönlands Geologiske Undersögelse Bulletin 84.

Sutherland, D. S. 1980. Two examples of fluidisation from the Tororo carbonatite complex, southeast Uganda. *Proceedings of the Geological Association* **91**, 39–45.

Treiman, A. H. & A. Schedl 1983. Properties of carbonatite magma and processes in carbonatite magma chambers. *Journal of Geology* **91**, 437–47.

Twyman, J. D. & J. Gittins 1987. Alkalic carbonatite magmas: parental or derivative? In *Alkaline igneous rocks*, J. G. Fitton & B. G. J. Upton (eds), 85–94. Geological Society Special Publication 30.

Wimmenauer, W. 1966. The eruptive rocks and carbonatites of the Kaiserstuhl, Germany. In *Carbonatites*, O. F. Tuttle & J. Gittins (eds), 183–204. New York: Wiley.

Wimmenauer, W. 1974. The alkaline province of Central Europe and France. In *The alkaline rocks*, H. Sörensen (ed.), 238–71. London: Wiley.

Woolley, A. R. 1982. A discussion of carbonatite evolution and nomenclature, and the generation of sodic and potassic fenites. *Mineralogical Magazine* **46**, 13–17.

Wyllie, P. J. 1966. Experimental studies of carbonatite problems: The origin and differentiation of carbonatite magmas. In *Carbonatites*, O. F. Tuttle & J. Gittins (eds), 311–52. New York: Wiley.

Wyllie, P. J. 1978. Silicate–carbonate systems with bearing on the origin and crystallization of carbonatites. In *Proceedings of the First International Symposium on Carbonatites, Poços de Caldas, Brazil*, June 1976, 61–78. Brasilia: Brasil Departamento Nacional da Produção Mineral.

Wyllie, P. J. & O. F. Tuttle 1960. The system $CaO-CO_2-H_2O$ and the origin of carbonatites. *Journal of Petrology* **1**, 1–46.

Wyllie, P. J. & O. F. Tuttle 1962. Carbonatitic lavas. *Nature* **194**, 1269.

Yeremenko, G. K., B. Ya. Vikhter, V. M. Chmyrev & K. Khamidi 1975. Quaternary volcanic carbonatite complex in Afghanistan. *Akademia Nauk SSSR Doklady, Earth Science Section* **223**, 53–5.

Zhabin, A. G. 1978. Syngenesis and metamorphism of carbonatites. In *Proceedings of the First International Symposium on Carbonatites, Poços de Caldas, Brazil*, June 1976, 191–5. Brasilia: Brasil Departamento Nacional da Produção Mineral.

5
CARBONATITE MAGMA: PROPERTIES AND PROCESSES

A. H. TREIMAN

ABSTRACT

Carbonatite magmas are ionic liquids, composed of individual ions bound without covalency. The anions and the cations may each be considered as independent solutions (Temkin model), which behave according to the regular solution model. The heat of melting for $CaCO_3$, derived from 1 Kb liquidus phase equilibria, is 32.5 kjoule/g.f.w., similar in magnitude to heats of melting for alkali carbonates. Regular solution properties, based on liquidus phase equilibria, show that the solutions among carbonate and most unpolymerized anions are sensibly ideal; solution of carbonate and polymerized aluminosilicate is strongly non-ideal, with concomitant liquid immiscibility. Cation solutions are not ideal, as dissimilar species of cations tend to cluster; consequently, formation of mixed-cation compounds is favoured. The speciations of water and sulphur are complex. Because ionic species in carbonatite magmas are not bound covalently, the magmas have very low viscosity, similar to that of water. The low viscosity implies that flow of carbonatites will be rapid and turbulent, and that large crystals will not grow while suspended in the magma. The large crystals in many carbonatites may represent dendritic crystals grown *in situ* on the walls of a magma chamber.

5.1 INTRODUCTION

Carbonatites are igneous rocks, replete with all of the structures and textures characteristic of basaltic or granitic rocks (Barker, this volume). As with more common igneous rocks, understanding the origins and petrology of carbonatites depends on understanding the nature of carbonatite magmas and the processes which can occur in carbonatite magma bodies. Two classes of basic data are needed to treat carbonatites as igneous rocks: thermochemical (related to melting, solution, and crystallization) and thermophysical (related to flow styles and speeds, and to motion of crystals). The basic thermochemical parameters include heats, entropies, and volumes of fusion for the various components of carbonatite magma, solution parameters for the components in magma, heat capacities, and ionic diffusivities. The basic thermophysical parameters include volume, viscosity, thermal diffusivity, thermal expansion, and compressibility. The purpose of this paper is to derive and review some of these basic fundamental data on carbonatite magmas.

Knowledge of the thermochemical parameters has come slowly, in part because of the late date at which carbonatites were recognized to be magmatic, and in part because of uncertainty about the compositions of carbonatite magmas (e.g. Le Bas 1981, Treiman 1985, Gittins this volume). Some data are available in the engineering literature, and other data may be derived from the many phase-equilibrium studies of carbonatite-like compositions in the literature. But much of the thermochemistry of carbonatite magma is unknown.

The thermophysical properties of carbonatite magma have not been measured directly, in great part because many carbonatite-like compositions are not stable at 1 bar pressure. These properties have been inferred from those of related synthetic systems (e.g. Treiman & Schedl 1983), which extrapolation is reasonable because all of the synthetic compositions have very similar thermophysical properties. Given estimates of physical and thermal properties of carbonatite magma, one can apply standard fluid dynamic relations to determine speeds of crystal settling in carbonatites, and rates and styles of magmatic flow. Crystallization processes, such as quenching and growth of dendritic crystals, may be important in carbonatites, but are not yet amenable to quantification.

5.2 COMPOSITION OF CARBONATITE MAGMA

Knowing the possible compositions of carbonatite magmas is essential in inferring the chemical and physical properties of the magmas. It is generally agreed that alkali and alkaline earth carbonates and hydroxides are essential components of the magmas, with lesser quantities of transition metal cations, aluminium, silicon, phosphorus, fluorine, etc. There is little agreement on the proportions of these components (e.g. Gittins & McKie 1980, Le Bas 1981, Dawson *et al.* 1987). To me, it seems likely that carbonatite melts span a wide range of compositions (Treiman 1985), and the results here are intended to be as general as possible.

5.3 STRUCTURE OF CARBONATITE MAGMAS

Carbonate–hydroxide melts, and presumably carbonatite magmas, have relatively simple structures and thermochemistry. Such melts are *ionic liquids* (or fused salts), composed of discrete anions and cations bound together by ionic forces (e.g. Zarzycki 1962, Sundheim 1964, Lumsden 1966, Kleppa 1977, 1981). Covalent bonding is unimportant in most ionic melts, and there is little or no polymerization of anions; this is especially true for carbonate melts, in which there is no tendency for carbonate ions to polymerize. Much of the thermochemical and structural complexity of silicate melts stems from the prevalence of silicate polymerization (e.g. Mysen *et al.* 1982); these complexities are absent in ionic melts.

An accurate simplification of ionic liquids is to treat them as independent interpenetrating solutions of cations and of anions. This is the quasi-lattice or Temkin (1945) model, which is justifiable because enormous energy is needed to exchange a cation surrounded by anions for an anion surrounded by anions (Blander 1964). The model's predictions are in reasonable accord with experimental studies of ionic liquids. Other models of ionic liquids (e.g. conformal solution) may be more accurate, but most of the data available on carbonate systems are not sufficiently precise to justify such treatment.

5.4 THERMOCHEMICAL PROPERTIES

The essential thermochemical data for carbonatite melts and their synthetic analogues are the heats, entropies, and volume changes on melting of the various chemical components of the melts, and the effects of solution on these properties. For salts of alkali metals (Na, K) most of the data are available from direct measurements at 1 bar (summarized in Janz et al. 1979, Selman & Maru 1981). These 1-bar values may be extrapolated to geologically reasonable pressures (1 Kb) using P-T melting curves (Klement & Cohen 1975). The extrapolated values are given in Table 5.1.

Data for salts of the alkaline earth metals (Ca, Mg) are more difficult to acquire, because most such salts devolatilize at 1 bar before melting. Two approaches have been used to investigate these melts: extrapolation from solutions that *are* stable at 1 bar (e.g. Førland 1955), and interpretation of liquidus phase equilibria (e.g. Flood et al. 1949, Bradley 1962). The latter approach is particularly useful for carbonatite-like melts because high-pressure liquidi have been determined in many relevant systems. Much of the data presented here is derived from phase equilibria.

Table 5.1 Melting of carbonates: molar properties at 1 Kb pressure.

Compound	T_f K	$\Delta \bar{H}_f^0$ kJ/g.f.w.	$\Delta \bar{S}_f^0$ J/g.f.w.-K	$\Delta \bar{C}_p^0$ J/g.f.w.-K	$\Delta \bar{V}_f^0$ cm^3/g.f.w.
Na$_2$CO$_3$	1145[1]	30.1	26.3	-8.0	4.2
K$_2$CO$_3$	1200[1]	28.2	23.5	-1.1	4.6
Li$_2$CO$_3$	1003[2]	45.0	45.0	-7.6	2.4
CaCO$_3$	1583[3]	32.5	20.5	—	2.6
MgCO$_3$	1750[4]	~ 35	~ 20	—	0.7

Heats and entropies of alkali carbonates extrapolated from 1-bar values: Janz et al. (1979), Selman and Maru (1981).

T_f as measured; volumes from high-pressure phase equilibria. Data for alkaline earth carbonates described in text.

[1] Koster van Groos & Wyllie (1966).
[2] Klement & Cohen (1975).
[3] after Wyllie & Tuttle (1960).
[4] extrapolated from Irving & Wyllie (1975).

5.4.1 *Derivation from phase equilibria*

The heat of melting (or enthalpy of fusion) for a compound may be derived from binary liquidus equilibria, if the thermochemical characteristics of the melt and crystals are known. In this case, use of liquidus equilibria to obtain heats of melting is equivalent to use of freezing point depression (Lewis & Randall 1961), a standard thermochemical technique. Heats of melting derived here and in Førland (1955) and Bradley (1962) are from systems in which the solid phase has no solid solutions (i.e. its activity is unity). Similarly, these workers have assumed that the melts may be represented either as ideal or regular solutions.

The regular solution model is commonly adequate in describing multicomponent melt solutions where all species have similar bonding characteristics (e.g. non-polar solutes in a non-polar solvent) and where there is no complexation. In the regular solution model of an ionic liquid, cations (or anions) do not interact with other species exactly as with their own. Each cross-species interaction involves an excess energy, which is independent of the bulk composition of the solution. This last assumption is not strictly true for alkali carbonate melts (Andersen & Kleppa 1976) and probably not strictly true for carbonatite-like melts, but is close enough for our purposes. In ionic liquids, the Temkin model implies that the cation and anion solutions may each be considered as a separate regular solution.

In the regular solution model, two terms contribute to the free energy of mixing ($\Delta \bar{G}_M$): an entropy term from random mixing of ions, and an enthalpy term from heat released (or absorbed) on association of unlike ions. For a binary melt solution between components A and B, the regular solution model predicts the free energy of mixing to be:

$$\Delta \bar{G}_M = + RT(X_A \ln X_A + X_B \ln X_B) + X_A X_B \cdot W \tag{5.1}$$

(Lewis & Randall 1961), where Xs are gram-formula-weight fractions of the components in solution and W is an empirical interaction parameter (here assumed to be independent of temperature and pressure). The first term of this equation is the entropy contribution from random mixing, and the second is the enthalpy contribution from ion association. The thermochemical activities, α, of components in regular solutions are simply:

$$RT \ln \alpha = RT \ln X + (1 - X)^2 \cdot W. \tag{5.2}$$

If W is zero, there is no heat associated with mixing, and the solution is ideal. If W is greater than zero, heat is expended in mixing the solution and it is energetically favourable for similar ions to be adjacent; liquid immiscibility becomes possible. If W is less than zero, heat is released on mixing and it is energetically favourable for dissimilar ions to be adjacent; in this case, formation of intermediate compounds is favoured.

Following Lewis & Randall (1961), the liquidus surface in a binary phase diagram may be modelled in terms of the heat of melting and the regular solution parameter, W. For example, in the binary system A–B, the location of the A-saturated liquidus ($T, X_{A,melt}$) is given by:

$$RT \ln X_{A,melt} + X_{B,melt}^2 \cdot W = \frac{T - T_{f,A}}{T_{f,A}} \Delta \bar{H}_{f,A}^0, \tag{5.3}$$

where $T_{f,A}$ is the melting temperature of pure A and $\Delta \bar{H}^0_{f,A}$ is the heat of melting for pure A. This equation assumes that the solid phase is pure A, and that the change in heat capacity on melting is negligible compared to the heat of fusion. The former assumption is valid for some systems (e.g. $CaCO_3$–$Ca(OH)_2$); the latter assumption appears to be valid (to $\pm 1\%$) for carbonate systems (unpublished data).

5.4.2 Heats of melting

Heats of melting for carbonate components of carbonatite magmas are given in Table 5.1, along with estimates of their changes in volume, entropy, and heat capacity on melting. Heats of melting for alkali carbonates are extrapolated from 1-bar values. The heat of melting of $CaCO_3$ is inferred from 1-Kb binary liquidi: $CaCO_3$–Na_2CO_3 (Cooper *et al.* 1975); $CaCO_3$–K_2CO_3 (Cooper *et al.* 1975); $CaCO_3$–$Ca(OH)_2$ (Wyllie & Tuttle 1960); and $CaCO_3$–CaF_2 (Gittins & Tuttle 1964, Kuellmer *et al.* 1966 [500 bars]); details of the calculation are to be published. The heat of fusion inferred here is a few kilojoules/gram-formula-weight smaller than that inferred by Flood *et al.* (1952) from vapour pressure measurements. The source of the disagreement is not yet clear. The heat of melting for $MgCO_3$ is derived from the 1-bar liquidus in $MgCO_3$–K_2CO_3 (Ragone *et al.* 1966), and the melting temperatures at 1 bar and 1 Kb are extrapolated from the high-pressure experiments of Irving & Wyllie (1975). Neither the 1-bar liquidus nor the melting temperature are known precisely, so the derived heat of fusion is highly uncertain. Thermochemical data for melting of dolomite, $CaMg(CO_3)_2$, cannot yet be estimated; there are no available binary equilibria in which dolomite does not take solid solutions.

The entropy changes on melting (Table 5.1) are calculated directly from the heat of melting:

$$\Delta \bar{S}^0_f = \Delta \bar{H}^0_f / T_f \qquad (5.4)$$

because melting of the pure phase is isothermal. The volume changes on melting ($\Delta \bar{V}^0_f$, Table 5.1) may be derived from density measurements, or from experimentally determined polybaric liquidi using the Clausius–Clapeyron relation:

$$(dP/dT)_{liquidus} = \Delta \bar{S}^0_f / \Delta \bar{V}^0_f. \qquad (5.5)$$

Where data are available (e.g. Na_2CO_3), both methods yield comparable volume changes of melting.

5.4.3 Heats of mixing

Because carbonatite magmas are mixed carbonate melts (Le Bas 1981), their solution properties are important for understanding their chemistry and physical properties. Details of the mixing properties of carbonate (or other ionic) melts may be complex (see Kleppa 1981), but most data relevant to carbonatites are so imprecise that we need consider only the simplest mixing models. In general, mixing in ionic melt systems is close to ideal, except where there are large differences in cation size or in bond character. Both the cation and anion solutions of carbonate–hydroxide melts are nearly regular, and the regular solution model is inadequate only in explaining high-quality calorimetric data (e.g. Kleppa & Julsrud 1980). As above, non-ideality in a regular solution is represented as an enthalpy (heat) of mixing term, W; there are assumed to be no entropy or volume changes beyond those of the ideal solution. In an ideal melt solution W is zero; deviations from zero imply greater and greater non-ideality, leading either to formation of intermediate compounds ($W < 0$) or liquid immiscibility ($W > 0$).

Regular solution parameters, W, for alkali and alkaline earth cations in carbonate melts are summarized in Table 5.2. In general, values of W are negative and moderately large. Negative Ws, suggesting association of different cation species, are consistent with the existence of many mixed-cation carbonate compounds. Such compounds include $CaNa_2(CO_3)_2$ (nyerereite), $CaK_2(CO_3)_2$ (fairchildite), and $MgK_2(CO_3)_2$ (Ragone et al. 1966).

There are few or no data available on the behaviour of other cations in carbonatite magmas. Limited phase-equilibrium evidence on Ba and La in carbonatites (Kuellmer et al. 1966, Jones & Wyllie 1986) do not suggest gross non-ideality. The regular solution W for Ca^{2+} and La^{3+} may be negative, as minerals like bastnaesite, parisite, and synchysite are not uncommon in carbonatites. Essentially nothing is known of other important cations in carbonatites, including Fe, Ti, Nb, Ta, Zr, and Cu.

Table 5.2 Mixing in molten carbonates: regular solution parameters.

Ions	W kJ/(g.f.w.)^2-K	Counter ion	Ref.
Ca^{2+}-Na^+	-6	CO_3^{2-}	*
Ca^{2+}-K^+	-12	CO_3^{2-}	*
Ca^{2+}-Li^+	-2.5	CO_3^{2-}	1
Mg^{2+}-K^+	-25	CO_3^{2-}	*
Na^+-K^+	-5.6	CO_3^{2-}	2
Na^+-Li^+	-11.2	CO_3^{2-}	2
CO_3^{2-}-OH^-	-1	Ca^{2+}	*
CO_3^{2-}-F^-	-3.5	Ca^{2+}	*
CO_3^{2-}-F^-	~ -2	K^+	3
CO_3^{2-}-Cl^-	-1.7	Na^+	3
CO_3^{2-}-SO_4^{2-}	0	Na^+	4
CO_3^{2-}-O^{2-}	~ 0	Na^+	5
CO_3^{2-}-O_2^{2-}	~ 0	Na^+	5

*Unpublished data.
1. Lumsden 1966.
2. Average from Andersen & Kleppa (1976).
3. From phase diagrams in Levin et al. (1969).
4. Flood et al. (1952).
5. Selman & Maru (1981).

Regular solution parameters, W, for anions (Table 5.2), are near zero, implying that anion solutions in carbonate melts are nearly ideal: 'It is well-known that mixed anion–common cation fused salts often are very nearly ideal solutions' (Kleppa & Julsrud 1980). Data in the table come from interpretation of liquidus phase equilibria and from direct measurement at 1 bar. Of particular importance is the electrochemical study by Flood *et al.* (1952), which showed that solution of carbonate and sulphate is ideal. Near-ideality of the anion solution is consistent with both the lack of liquid immiscibility in carbonate–simple anion solutions, and the paucity of mixed-anion carbonate minerals from high-temperature settings.

Limited data are available for anions not shown in Table 5.2. High-pressure liquidi involving sulphide and carbonate show no gross non-ideality (Helz & Wyllie 1979), but have not been rigorously analysed. Mixing of carbonate and orthophosphate (PO_4^{3-}) is probably also near ideality, although there are no data available. Mixing of silicate and carbonate is complicated, as discussed next.

Aluminosilicate anions. Oxyanions of silicon and aluminium are important in carbonatite petrogenesis because almost all carbonatites contain silicate or aluminosilicate minerals, and some carbonatites may form by liquid immiscibility with aluminosilicate magmas (Freestone & Hamilton 1980, Treiman & Essene 1985, Kjarsgaard & Hamilton, this volume). Solution of carbonate and small silicate anions is probably close to ideality, although direct data are lacking. Near-ideality may be inferred by analogy with the ideal mixing of carbonate and sulphate (Flood *et al.* 1952) and the ideal mixing of oxide and silicate (e.g. Waseda & Toguri 1978). The interaction parameter, W, between carbonate and orthosilicate (SiO_4^{4-}) may be negative because the mixed-anion mineral spurrite ($Ca_5(SiO_4)_2CO_3$) is a liquidus phase in $CaCO_3$–Ca_2SiO_4 (Wyllie & Haas 1965). Similarly, the existence of tilleyite ($Ca_5Si_2O_7(CO_3)_2$) may reflect a negative W for mixing of carbonate and sorosilicate ($Si_2O_7^{6-}$), even though tilleyite is not a liquidus phase. Mixing of aluminate and carbonate anions may also be ideal, but is of little significance as aluminate minerals (e.g. spinel) are exceedingly rare in carbonatites.

Melt solution of carbonate and large-polymer aluminosilicate anions is very far from ideality, as shown by the existence of immiscibility between carbonate and aluminosilicate magmas. This liquid immiscibility covers a wide range of natural and synthetic compositions (Koster van Groos & Wyllie 1966, Freestone & Hamilton 1980, Treiman & Essene 1985, Kjarsgaard & Hamilton, this volume, and many others). Even in compositions without immiscibility, carbonate anions tend to form clusters which exclude polymerized silicate (Mysen & Virgo 1980). The degree of non-ideality is a function of the composition of the silicate melt, both the degree of polymerization and the cation species present (Mysen & Virgo 1980, Fine & Stolper 1985). Although there has been study of the thermochemistry of carbonate solution in aluminosilicate liquids (above references), there are no data on the solution of aluminosilicates in carbonate liquids.

Water. Despite the acknowledged importance of water in the genesis of carbonatites (Wyllie & Tuttle 1960, Le Bas 1981, Treiman & Essene 1984), little is known of the behaviour of water in carbonatite magmas. It is likely that water will be mostly ionized in a strongly polarizing environment like a carbonate melt, and that hydroxyl may be the most important ionic species (hydroxyl is discussed above). Other species will form according to reactions such as:

$$H_2O + CO_3^{2-} \rightleftharpoons OH^- + HCO_3^-; \tag{5.6}$$

$$H_2O \rightleftharpoons OH^- + H^+; \tag{5.7}$$

$$OH^- \rightleftharpoons O^{2-} + H^+; \tag{5.8}$$

$$H_2O + H^+ \rightleftharpoons H_3O^+. \tag{5.9}$$

In general, ionized species will be strongly favoured over neutral water molecules, but the abundances of species at equilibrium will be dependent on the oxidation state (Eh) and hydrogen content (pH) of the melt. It is likely that mixing among these anions and carbonate is close to ideality, but the abundances of individual species cannot yet be calculated.

5.5 PHYSICAL PROPERTIES AND PROCESSES

To model physical processes in carbonatite intrusions, the physical and thermal properties of carbonatite magmas must be known. The most important properties include viscosity, density, ion diffusion coefficients, thermal expansion, heat of melting, heat capacity, and thermal diffusivity. Heats of melting and mixing properties are discussed above; density has been calculated by Nesbitt & Kelly (1977) and estimated by Treiman & Schedl (1983).

Table 5.3 shows the inferred and calculated properties of carbonatite magma compared to those of basalt and rhyolite magmas. Most of the values for carbonatite magma have been estimated from those of alkali carbonate melts

Table 5.3 Thermophysical properties of magmas near their liquidi.

		Carbonatite[1]	Basalt[2]	Rhyolite
Viscosity (η)	poise	5×10^{-2}	5×10^{2}	$\sim 10^{8}$
Density (ρ)	gm cm^{-3}	2.2	2.7	2.5
Thermal expansion (α)	10^{-4} K^{-1}	2.3	0.25	0.25
Heat of fusion (L)	J gm^{-1}	250	400	225
Heat capacity (C_p)	J(gm K)$^{-1}$	2.0	1.2	1.2
Thermal diffusivity ($\varkappa$)	cm^2 sec^{-1}	4×10^{-3}	8×10^{-3}	7×10^{-3}
Thermal conductivity (k)	J(cm K sec)$^{-1}$	0.015	0.025	0.020
Cation diffusivity (D)	cm^2 sec^{-1}	5×10^{-5}	5×10^{-8}	10^{-10}

1. Treiman & Schedl 1983.
2. Bartlett 1969, Huppert & Sparks 1980, Dunn 1986, Scarfe 1986.

(Treiman & Schedl 1983), because no direct measurements have been made. The analogy between natural carbonatite magma and alkali carbonate melt is likely to be best for properties dependent on the structure of the melt (e.g. viscosity and ion diffusion coefficients). The viscosity of carbonatite magma is likely to be somewhat greater than that of alkali carbonate melt because of the greater electrostatic attractions between divalent (and greater) cations in natural magmas compared to monovalent cations in the synthetic melts. The latent heat of fusion is also likely to be greater to account for dissolved silicate. It must be remembered that carbonatite magmas span a range of compositions, and the values tabulated in Table 5.3 must be taken as guides, not gospel.

The viscosity of carbonatite magma may be affected strongly by content of aluminosilicate anions and of water. There are no data on the viscosities of aluminosilicate-bearing carbonate melts, so one can make an analogy with silica-bearing oxide melts. At low concentrations, silica is almost all in the form of orthosilicate anions, SiO_4^{4-}; melts with these anions behave as ionic liquids (Masson *et al.* 1970). At greater concentrations of silica, polymerized silicate anions will be dominant; the presence of these anions, particularly chain or sheet or framework anions will increase viscosity significantly. However, extensively polymerized aluminosilicate anions are partitioned into silicate melt coexisting immiscibly with the carbonate melt (e.g. Koster van Groos & Wyllie 1966, 1968, 1973, Wendlandt & Harrison 1979, Freestone & Hamilton 1980, Treiman & Essene 1985, Kjarsgaard & Hamilton, this volume). The carbonate melt will thus contain little silica or alumina, and will remain fluid.

Water dissolved in carbonatite magma will probably tend to decrease viscosity slightly. In an ionic melt (like a carbonatite magma) it is likely that water will be ionized completely (or almost so); the resultant ions will behave just as ions in the melt solution. If the carbonatite magma contains alumina and silica, the effect of added water is not obvious. Water might tend to polymerize or depolymerize the silicate anions, depending on the concentrations of silicate and hydroxyl (and other) species in the melt:

$$Si_2O_7^{6-} + H_2O \rightleftharpoons 2SiO_3(OH)^{3-}; \qquad (5.10)$$

or

$$2SiO_4^{4-} + H_2O \rightleftharpoons Si_2O_7^{6-} + 2OH^-. \qquad (5.11)$$

More work is needed to understand the relative importance of these and other such equilibria in carbonatite melts.

5.5.1 *Crystal settling*

Crystal settling has often been invoked in the formation and differentiation of carbonatites (Wyllie & Tuttle 1960, Wyllie 1966, Nesbitt & Kelly 1977). The theory of settling rates has been studied extensively, and applied to carbonatites by Treiman

& Schedl (1983). Settling velocities for common carbonatite minerals are quite rapid: up to 60 cm/sec for centimetre-sized oxide (or sulphide) grains. Upward flow rates in the cores of kilometre-sized intrusions are only of the order of 1 cm/sec. Crystals larger than 0.5 mm could not be suspended by the flow, and would settle to the bottom of the chamber. Settling velocities would be reduced if the magma were rich in phenocrysts.

Crystal settling could be effective in differentiation of carbonatites, but it is not clear how the resultant cumulates could be recognized. Layered (?) rocks with original grain sizes approximately 1 mm (the largest suspensible size) may be indicative of crystal accumulation. Possible crystal cumulates cited by Wyllie (1966) are from dykes and small bodies (30 m), not larger intrusions. Evidence cited by Nesbitt & Kelly (1977) for crystal settling may be explained by dendritic crystallization (see below).

Related to crystal settling is the question of phenocrysts in carbonatite magma (Gittins 1973). If crystals settle so quickly, how are porphyritic textures formed? Under slow cooling, phenocrysts would settle out and it is possible that some porphyritic carbonatites are crystal cumulates. Dykes of porphyritic carbonatite may have been emplaced as crystal mushes, and not as fluids to crystallize *in situ*. It is also possible that 'porphyritic textures' in carbonatites may not reflect magmatic processes at all, but low-temperature or phreatic alteration. Phenocrysts of dolomite should be particularly suspect, because euhedral dolomite crystals form readily during diagenesis of carbonate rock. Williams *et al.* (1954, Fig. 119A) present a superb drawing of euhedral diagenetic dolomite in a fine-grained limestone.

5.5.2 *Flow*

The flow of carbonatite magmas is qualitatively different from that of basalt or granite magmas, because carbonatite magma is much less viscous (Table 5.3; Treiman & Schedl 1983). Carbonatite magma is as viscous as water or light machine oil, while basalt is as viscous as tar or cold molasses. In carbonatite intrusions, magma convection is turbulent, even violently turbulent; the Rayleigh number for flow in typical intrusions exceeds that of laminar flow by more than nine orders of magnitude (Treiman & Schedl 1983). Only for thin, horizontal sills is laminar convection possible. Carbonatite magma flows rapidly in plug-shaped intrusions; downward flow near the walls may be 1 m/sec. Velocity decreases as the intrusion cools, but may still be 20 cm/sec after 1000 years.

5.5.3 *Cooling and crystallization*

When cooling is not so rapid as to cause quenching, feathery or elongate *dendritic* crystals (e.g. Fig. 5.1) will form around any available nuclei. Dendritic crystallization is important in metallurgy and in the origins of many basaltic rocks (e.g. Taubeneck & Poldervaart 1960, Wager & Brown 1968, Pyke *et al.* 1973, Berg 1980, Naslund 1984). Dendritic carbonate crystals (and related elongate needles) are common in small carbonatite dykes (Girault 1966, Katz & Keller 1981, Treiman &

Figure 5.1 Comb-layering of calcite crystals with interstitial silicate and oxide minerals; rock is 8 cm long. From monticellite–diopside–olivine–pyrochlore carbonatite, open pit A-2, Oka carbonatite complex, Canada.

Figure 5.2 Blebs of silicate and oxide minerals, 'fingerprint' texture, in a 10 cm cleavage fragment of calcite. Sample is part of a metre-sized calcite crystal, interpreted as a filled calcite dendrite in which the blebs crystallized from melt trapped among the dendrite's branches. From monticellite–diopside–olivine–pyrochlore ore carbonatite, open pit A-2, Oka carbonatite complex, Canada.

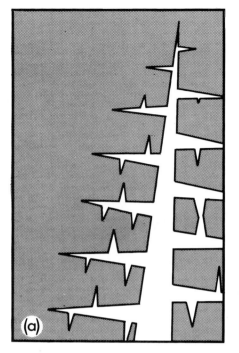 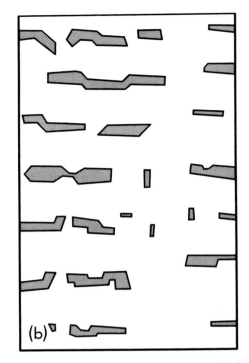

Figure 5.3 Development of 'fingerprint' textures during growth of a dendritic crystal. (a) Dendritic crystal of calcite (white) in carbonatite magma (shaded); (b) coalescence of the dendrite arms producing a large single crystal of calcite (white) with aligned pockets of magma (shaded). The magma pockets will crystallize to form aligned blebs of non-carbonate minerals.

Schedl 1983). Elongate needles of calcite may form *comb-layering* textures (Katz & Keller 1981, and Fig. 5.1) akin to those of spinifex olivine or pyroxene in komatiites (Pyke *et al.* 1973).

In larger carbonatite intrusions, feathery dendrites apparently do not occur; instead 'fingerprint' textures are found (Fig. 5.2), marked by aligned and bifurcating blebs of silicate and oxide minerals within large crystals of a carbonate mineral. I interpret these textures to have formed by infilling of large dendritic crystals. During growth of a dendrite (Fig. 5.3a), its branches will ramify to form smaller branches and twigs. Twigs from adjacent branches will grow and eventually touch and coalesce; because the twigs are part of the same crystal, coalescence poses no structural difficulties. As the twigs coalesce, cells of magma are blocked off from the body of the intrusion. The cells eventually crystallize to form blebs of oxide and silicate (and other) minerals aligned within a single crystal of carbonate (Fig. 5.3b). Figure 5.2 shows an example of aligned blebs in a single calcite crystal from the Oka carbonatite complex, Quebec, Canada. Similar textures occur in rocks from Magnet Cove (Arkansas, USA) and Phalaborwa (South Africa); it is possible that filled dendritic crystals are common in larger carbonatite intrusions.

Recognition of dendritic or filled dendritic textures is important in understanding the petrology and geochemistry of carbonatites. For example, the sample shown in

Figure 5.2 is from a zone mined for Nb (in pyrochlore); dendritic crystallization may have played a role in localization of the ore. Another example is a study of melt inclusions in silicate minerals of the Magnet Cove carbonatite (Nesbitt & Kelly 1977). The rock was found to be much richer in calcite than the inclusions, and it was hypothesized that significant settling of silicate minerals had occurred after entrapment of the inclusions but before solidification of the rock. An alternative explanation is that the inclusions' host grains formed in melt cells in a filled dendrite (Fig. 5.3b).

5.6 CONCLUSIONS

Mixing of cations and anions in carbonate melts can be modelled as regular solutions, at least to the accuracy of available phase equilibria. The anion solutions are nearly ideal, except for those with polymerized aluminosilicates. Non-ideality of carbonate–polymer mixing is shown by immiscibility between carbonate and silicate magmas; many carbonatites may in fact originate by liquid immiscibility (Treiman & Essene 1985, Kjarsgaard & Hamilton, this volume). Cation solutions in carbonate melts deviate from ideality, in that unlike cations tend to associate.

Considerable work remains before quantitative models of carbonatite magma can be formulated. This study has considered only binary systems of a few of the common cations and anions in carbonatite magmas, and only in a limited pressure range. A few published phase equilibria may yet be interpreted to good effect, but many more calorimetric, phase equilibrium, and electrochemical studies are needed. The most pressing need is for data on magnesium-bearing carbonate melts. After Ca, Mg is the most abundant alkaline earth in carbonatites, but data on its melting and solution behaviour are almost non-existent. It may be possible to interpret high-pressure phase equilibria in systems with Ca- and Mg-carbonate (e.g. Byrnes & Wyllie 1981), but to do so requires models of solid carbonate activity coefficients applicable at super-solvus conditions. Although activity models are available, it is not proven that they are applicable at liquidus temperatures. More work is needed to clarify the behaviour of water in carbonatite magmas, and particularly the interaction of water, carbonate, and silicate. Similarly, the behaviour of S in carbonatite magma is likely to be complex, because it may occur in oxidation states ranging from sulphide to disulphide (S_2^{2-}) to sulphate. The redox behaviour of S cannot be considered independent of other equilibria involving Fe, carbonate, and H_2O. The most daunting task is understanding the behaviour of silicate and aluminosilicate compounds in carbonate melts. It seems likely that solution between carbonate and orthosilicate or sorosilicate anions is nearly ideal, but the existence of liquid immiscibility between carbonate and aluminosilicate melts (e.g. Kjarsgaard & Hamilton, this volume) proves that mixing with large-polymer species is far from ideal. Important considerations are the abundances of various aluminosilicate polymer species, the effects of different cations on polymerization, the effect of CO_2 pressure on silicate polymerization, and the effects of various hydrogen-bearing ion species.

Physically, carbonatite magmas are significantly different from silicate magmas; although the same processes may occur in both, the relative importances of such processes may be different. Treiman & Schedl (1983) presented a preliminary investigation of physical processes in carbonatite magmas, but much more remains to be done. It is important to verify the inferred low viscosity (near water) of carbonatite magma, and determine the effects of temperature and composition on its viscosity. The low viscosity requires that crystals will settle rapidly out of carbonatite magmas, and thus that the grain size of cumulates will be small. The low viscosity also suggests that flow (convective or otherwise) in carbonatite intrusions is turbulent, and may be very rapid. Similarly, carbonatite magma may be able to travel rapidly through very small cracks (McKenzie 1985) and thus be collected from regions of very small degrees of partial melting. Finally, the low viscosity of carbonatite magmas is related to rapid transport of ions within the magmas; such rapid transport may allow the growth of large, single crystals, dendrites, on to the walls of intrusions.

ACKNOWLEDGEMENTS

Reviews of this chapter by K. Bell, O. Kleppa, G. Lofgren, and B. Mysen are appreciated. The early collaboration with A. Schedl was indispensable. Without the help of P. Strother, the plates here would not have been completed. I am particularly grateful to K. Bell for organizing the symposium on carbonatites at the 1986 GAC–MAC–CGU meeting and organizing this book.

REFERENCES

Andersen, B. K. & O. J. Kleppa 1976. Enthalpies of mixing in binary liquid alkali carbonate mixtures. *Acta Chemica Scandinavia* **A30**, 751–8.

Bartlett, R. W. 1969. Magma convection, temperature distribution, and differentiation. *American Journal of Science* **287**, 1067–82.

Berg, J. H. 1980. Snowflake troctolite in the Hettasch intrusion, Labrador: Evidence for magma mixing and supercooling in a plutonic environment. *Contributions to Mineralogy and Petrology* **72**, 339–51.

Blander, M. 1964. Thermodynamic properties of molten salt solutions. In *Molten salt chemistry*, M. Blander (ed.), 127–237. New York: Interscience.

Bradley, R. S. 1962. Thermodynamic calculations on phase equilibria involving fused salts. Part I. General theory and application to equilibria involving calcium carbonate at high pressure. *American Journal of Science* **260**, 374–82.

Byrnes, A. P. & P. J. Wyllie 1981. Subsolidus and melting relations for the join $CaCO_3$–$MgCO_3$ at 10 kilobars. *Geochimica et Cosmochimica Acta* **45**, 321–8.

Cooper, A. F., J. Gittins & O. F. Tuttle 1975. The system Na_2CO_3–K_2CO_3–$CaCO_3$ at 1 kilobar and its significance in carbonatite petrogenesis. *American Journal of Science* **275**, 534–60.

Dawson, J. B., M. S. Garson & B. Roberts 1987. Altered former alkalic carbonatite lava from Oldoinyo Lengai, Tanzania. Inferences for calcite carbonatite lavas. *Geology* **15**, 765–8.

Dunn, J. T. 1986. Diffusion in silicate melts: An introduction and literature review. In *Short course on silicate melts*, C. M. Scarfe (ed.), 57–92. Toronto: Mineralogical Association of Canada.

Fine, G. & E. Stolper 1985. The speciation of carbon dioxide in sodium aluminosilicate glasses. *Contributions to Mineralogy and Petrology* **91**, 105–21.

Flood, H., T. Førland & B. Roald 1949. The equilibrium $CaCO_{3(melt)} = CaO_{(s)} + CO_2$. The activity coefficients of calcium carbonate. *Journal of the American Chemical Society* **71**, 572–5.

Flood, H., T. Førland & K. Motzfeld 1952. On the oxygen electrode in molten salts. *Acta Chemica Scandinavia* **6**, 257–69.

Førland, T. 1955. An investigation of the activity of calcium carbonate in mixtures of fused salts. *Journal of Physical Chemistry* **59**, 152–6.

Freestone, I. C. & D. L. Hamilton 1980. The role of liquid immiscibility in the genesis of carbonatites. *Contributions to Mineralogy and Petrology* **73**, 105–17.

Girault, J. 1966. Genèse et géochimie de l'apatite et la calcite dans les roches liées au complexe carbonatitique et hyperalcalin d'Oka (Canada). *Bulletin de la Société française de Minéralogie et Cristallographie* **89**, 469–513.

Gittins, J. 1973. The significance of some porphyritic textures in carbonatites. *Canadian Mineralogist* **12**, 226–8.

Gittins, J. & D. McKie 1980. Alkalic carbonatite magmas: Oldoinyo Lengai and its wider applicability. *Lithos* **13**, 213–15.

Gittins, J. & O. F. Tuttle 1964. The system CaF_2–$Ca(OH)_2$–$CaCO_3$. *American Journal of Science* **262**, 66–75.

Helz, G. R. & P. J. Wyllie 1979. Liquidus relationships in the system $CaCO_3$–$Ca(OH)_2$–CaS and the solubility of sulfur in carbonatite magmas. *Geochimica et Cosmochimica Acta* **43**, 259–65.

Huppert, H. E. & R. S. J. Sparks 1980. The fluid dynamics of a basaltic magma chamber replenished by influx of hot, dense ultrabasic magmas. *Contributions to Mineralogy and Petrology* **75**, 279–89.

Irving, A. J. & P. J. Wyllie 1975. Solidus and melting relationships for calcite, magnesite and the join $CaCO_3$–$MgCO_3$ to 36 kilobars. *Geochimica et Cosmochimica Acta* **39**, 35–53.

Janz, G. J., C. B. Allen, N. P. Bansal, R. M. Murphy & R. P. T. Tomkins 1979. *Physical properties data compilations relevant to energy storage. II Molten salts: data on single and multi-component systems*. Nat. Stand. Ref. Data Ser. NSRDS-NBS 61, Part II. Washington, DC: US Government Printing Office.

Jones, A. P. & P. J. Wyllie 1986. Solubility of rare earth elements in carbonatite magmas, indicated by the liquidus surface in $CaCO_3$–$Ca(OH)_2$–$La(OH)_3$. *Applied Geochemistry* **1**, 95–102.

Katz K. & J. Keller 1981. Comb-layering in carbonatite dykes. *Nature* **294**, 350–2.

Klement, W., Jr. & L. H. Cohen 1975. Solid–solid and solid–liquid transitions in K_2CO_3, Na_2CO_3, and Li_2CO_3: investigations to > 5 kbar by differential thermal analysis; thermodynamics and structural correlations. *Berichte Bunsengesellschaft* **79**, 327–34.

Kleppa, O. J. 1977. Thermodynamic properties of molten salt solutions. In *Thermodynamics in Geology*, D. G. Fraser (ed.), 279–99. Boston: D. Reidel.

Kleppa, O. J. 1981. Thermodynamics of simple molten salt mixtures. In *Thermodynamics of minerals and melts*, R. C. Newton, A. Navrotsky & B. J. Wood (eds), 179–88. New York: Springer.

Kleppa, O. J. & S. Julsrud 1980. Thermodynamics of charge-unsymmetrical anion mixtures I. The liquid systems AF–A_2SO_4. *Acta Chemica Scandinavia* **A34**, 655–65.

Koster van Groos, A. F. & P. J. Wyllie 1966. Liquid immiscibility in the system Na_2O–Al_2O_3–SiO_2–CO_2 at pressures up to 1 kilobar. *American Journal of Science* **264**, 234–55.

Koster van Groos, A. F. & P. J. Wyllie 1968. Liquid immiscibility in the join $NaAlSi_3O_8$–Na_2CO_3–H_2O and its bearing on the genesis of carbonatites. *American Journal of Science* **266**, 932–67.

Koster van Groos, A. F. & P. J. Wyllie 1973. Liquid immiscibility in the join $CaAl_2Si_2O_8$–$NaAlSi_3O_8$–Na_2CO_3–H_2O. *American Journal of Science* **273**, 465–87.

Kuellmer, F. J., A. P. Visocky & O. F. Tuttle 1966. Preliminary survey of the system barite–calcite–fluorite at 500 bars. In *Carbonatites*, O. F. Tuttle & J. Gittins (eds), 353–64. New York: Wiley and Sons.

Le Bas, M. J. 1981. Carbonatite magmas. *Mineralogical Magazine* **44**, 133–40.

Levin, E. M., C. R. Robbins & H. F. McMurdie 1969. *Phase diagrams for ceramicists*. Columbus: The American Ceramic Society.

Lewis, G. N. & M. Randall 1961. *Thermodynamics*, 2nd edn (revised by K. J. Pitzer & L. Brewer). New York: McGraw-Hill.

Lumsden, J. 1966. *Thermodynamics of molten salt mixtures*. London: Academic Press.

McKenzie, D. 1985. The extraction of magma from the crust and mantle. *Earth and Planetary Science Letters* **74**, 81–91.

Masson, C. R., I. B. Smith & S. G. Whiteway 1970. Activities and ionic distributions in liquid silicates: applications of polymer theory. *Journal of the Iron and Steel Institute* **210**, 89–96.

Mysen, B. O. & D. Virgo 1980. Solubility mechanisms of carbon dioxide in silicate melts: a Raman spectroscopic study. *American Mineralogist* **65**, 885–99.

Mysen, B. O., D. Virgo & F. A. Siefert 1982. The structure of silicate melts: Implications for chemical and physical properties of natural magma. *Reviews of Geophysics and Space Physics* **20**, 353–83.

Naslund, H. R. 1984. Petrology of the upper border series of the Skaergaard intrusion. *Journal of Petrology* **25**, 185–212.

Nesbitt, B. E. & W. C. Kelly 1977. Magmatic and hydrothermal inclusions in carbonatites of the Magnet Cove complex, Arkansas. *Contributions to Mineralogy and Petrology* **63**, 271–94.

Pyke, D. R., A. J. Naldrett & O. R. Eckstrand 1973. Archaean ultramafic flows in Munro Township, Ontario. *Bulletin of the Geological Society of America* **84**, 955–78.

Ragone, S. E., R. K. Datta, D. M. Roy & O. F. Tuttle 1966. The system potassium carbonate–magnesium carbonate. *Journal of Chemical Physics* **70**, 3360–1.

Scarfe, C. M. 1986. Viscosity and density of silicate melts. In *Short course on silicate melts*, C. M. Scarfe (ed.), 36–56. Toronto: Mineralogical Association of Canada.

Selman, J. R. & H. C. Maru 1981. Physical chemistry and electrochemistry of alkali carbonate melts, with special reference to molten-carbonate fuel cells. In *Advances in molten salt chemistry*, Vol. 4, G. Mamantov & J. Braunstein (eds), 159–389. New York: Plenum Press.

Stolper, E. & G. Fine 1985. The speciation of carbon dioxide in sodium aluminosilicate glasses. *Contributions to Mineralogy and Petrology* **91**, 105–21.

Sundheim, B. R. (ed.) 1964. *Fused salts*. New York: McGraw-Hill.

Taubeneck, W. H. & A. Poldervaart 1960. Geology of the Elkhorn Mountains, northeastern Oregon, part 2: Willow Lake intrusion. *Bulletin of the Geological Society of America* **71**, 1295–332.

Temkin, M. 1945. Mixtures of fused salts as ionic solutions. *Acta Physicochimica USSR* **20**, 411–20.

Treiman, A. H. 1985. Low alkali carbonatites in alkaline complexes: separate sources for carbonate and alkalis? *Geological Society of America Abstracts* **17**, 194.

Treiman, A. H. & A. Schedl 1983. Properties of carbonatite magma and processes in carbonatite magma chambers. *Journal of Geology* **91**, 437–47.

Treiman, A. H. & E. J. Essene 1984. A periclase–dolomite–calcite carbonatite from the Oka complex, Quebec, and its calculated volatile composition. *Contributions to Mineralogy and Petrology* **85**, 149–57.

Treiman, A. H. & E. J. Essene 1985. The Oka carbonatite complex, Quebec: Geology and evidence for liquid immiscibility. *American Mineralogist* **70**, 1101–13.

Wager, L. R. & G. M. Brown 1968. *Layered igneous rocks*. San Francisco: Freeman.

Waseda, Y. & J. M. Toguri 1978. The structure of the molten $FeO-SiO_2$ system. *Transactions, Metallurgical Society of AIME* **9b**, 595–601.

Wendlandt R. F. & W. J. Harrison 1979. Rare earth partitioning between immiscible carbonate and silicate liquids and CO_2 vapor: results and implications for the formation of light rare earth enriched rocks. *Contributions to Mineralogy and Petrology* **69**, 400–19.

Williams, H., F. J. Turner & C. M. Gilbert 1954. *Petrography*. San Francisco: Freeman.

Wyllie, P. J. 1966. Experimental studies of carbonatite problems: The origin and differentiation of carbonatite magmas. In *Carbonatites*, O. F. Tuttle & J. Gittins (eds), 311–52. San Francisco: Freeman.

Wyllie, P. J. & J. L. Haas, Jr. 1965. The system $CaO-SiO_2-CO_2-H_2O$: 1. Melting relationships with excess vapor at 1 kilobar pressure. *Geochimica et Cosmochimica Acta* **29**, 871–92.

Wyllie, P. J. & O. F. Tuttle 1960. The system $CaO-CO_2-H_2O$ and the origin of carbonatites. *Journal of Petrology* **1**, 1–46.

Zarzycki, J. 1962. High-temperature X-ray diffraction studies of fused salts: structure of molten alkali carbonates and sulphates. *Discussions of the Faraday Society* **32**, 38–48.

6
PYROCHLORE, APATITE AND AMPHIBOLE: DISTINCTIVE MINERALS IN CARBONATITE

D. D. HOGARTH

ABSTRACT

Two hundred and eighty minerals are known from carbonatite. Prominent are the pyrochlore, apatite, and amphibole groups that occur in both intrusive and extrusive rock.

Pyrochlore-group minerals include Ba, Ca–Na, Ce, Pb, Sr, Th, and U members of the pyrochlore subgroup. Uranpyrochlore commonly contains 5–15 wt% Ta_2O_5 and, in some deposits, is in separate zones from low-uranium, low-tantalum pyrochlore. Primary pyrochlore normally has La:Yb > 100:1, and some pyrochlores, especially uranpyrochlore, have pronounced positive Ce anomalies. On weathering, A ions and F are progressively removed, Nb:Ta decreases, and the mineral becomes hydrated.

Apatite includes fluorapatite, hydroxylapatite, carbonate-fluorapatite, and carbonate hydroxylapatite. The Sr:Mn ratio is commonly > 10:1, but varies widely. During carbonatite development, F and Sr increase in apatite. Rare earth elements (REEs) are normally high in apatite and can attain 8.3 wt%. The La:Yb ratio is > 100:1 and commonly > 200:1. Some varieties show pronounced negative Ce anomalies, others negative Eu anomalies. On weathering, Sr and light rare earth elements (LREEs) are removed, F increases (to overfilling of the normal F site) and the mineral becomes carbonated (with 2–4 wt% CO_2).

In carbonatites, early amphiboles are calcic (magnesio-hastingsite and edenite), late are alkalic (magnesio-arfvedsonite–richterite–winchite–magnesio-riebeckite). Magnesio-kataphorite is comparatively rare, and actinolite and tremolite have yet to be confirmed from carbonatites. Chemical data are consistent with tetrahedral Fe^{3+} and exsolution in the alkali amphiboles. Some amphibole crystals have a 'normal' zonal scheme (magnesium- and calcium-rich cores, iron- and alkali-rich rims), while others have a complex 'reverse' zonal scheme.

6.1 INTRODUCTION

Essentials of the mineralogy of carbonatites have been brought together under one cover by Kapustin (1980), where 135 mineral species have been noted and the majority described in detail. Up to 1987, some 280 species have been reported, the exact number dependent on definitions of the words 'mineral' and 'carbonatite'. Included in the 280 are 44 weathering products, 10 species unique (for carbonatites) to effusive carbonatites of Afghanistan, Africa, Germany, and USSR and 33 species unique to the dawsonite-rich, silicocarbonatites of St. Michel, Montreal Island, Quebec, Canada (Sabina 1979, Simpson 1980, A. P. Sabina, pers. comm).

Table 6.1 List of minerals reported from carbonatites.

Name	Type	Name	Type	Name	Type
ACMITE	P, F	CALZIRTITE	P	edenite	P
AEGIRINE-AUGITE	P	cancrinite	P	elpidite	F
aeschynite-(Ce)	P	carbocernaite	P	endellite	F
AKERMANITE	P	carbonate-apatite	P, W	epidote	P
alabandite	P, V	catapleiite	P	eudialyte	P
ALBITE	P, V, F	celadonite	F	euxenite-(Ce)	P
ALLANITE-(Ce)	P	CELESTITE	P, F	fenghuanite	P
almandine	F	cerianite	P, W	fergusonite-β-(Ce)	P
alstonite	P	ceriopyrochlore	P	ferrikataphorite	P
ANALCIME	F	cerite	P	ferrimolybdite	W
anatase	P, F, W	cerussite	W, F	ferrohornblende	P
ancylite-(Ce)	P	chabazite	P	fersmite	P
ANDRADITE	P	chalcocite	P	florencite-(Ce)	P
anglesite	W	CHALCOPYRITE	P	fluocerite-(La)	W
anhydrite	P	chalcosiderite	W	FLUORAPATITE	W, P, V, F
ANKERITE	P	CHAMOSITE	W	FLUORITE	P, F, V
antimony	P	charoite	P	FORSTERITE	P
aragonite	P, V, W	chondrodite	P	franconite	F
arsenopyite	P	churchite	W	galena	P, F
azurite	W	clinochlore	P	gearksutite	P
baddeleyite	P, F, W	clinohumite	P	geikielite	P
BARIOPYROCHLORE	P, W	collinsite	P	GIBBSITE	W
BARITE	P, F, W	COLUMBITE	P	glauberite	P
barytocalcite	P	cordylite	P	GOETHITE	W, F
BASTNAESITE	P	crandallite	P	gold	P
beidelite	W	cristobalite	F	gorceixite	W
benstonite	P	crocoite	F	goyazite	P, W
betafite	P	cryolite	P, F	graphite	P, F
BIOTITE	P, V, F	cubanite	P	GREGORYITE	V
bobierrite	P	cuprite	W	GYPSUM	P, W, F
bonshtedtite	P	cuspidine	V	halite	F
bornite	P	dachiardite	F	HALLOYSITE	W, F
boulangerite	P	daqingshanite-(Ce)	P	harmotome	F
bournonite	P	DAWSONITE	F	hastingsite	P
bravoite	P	DELVAUXITE	W	HEMATITE	P, V, W, F
britholite-(Ce)	P	diamond	P	hexahydrite	W
brookite	P, F	dickite	F	hisingerite	P
brucite	P, V, W	digenite	P	hochelagaite	F
burbankite	P	DIOPSIDE	V, P, F	hollandite	W
calciobetafite	P	DOLOMITE	P, V, F	huanghoite-(Ce)	P
CALCITE	P, V, W, F	doyleite	F	humboltine	F
calkinsite-(Ce)	P	dresserite	F	hydrocerussite	F
hydrodresserite	F	monohydrocalcite	W	röntgenite-(Ce)	P
hydrotalcite	P	MONTICELLITE	P	rozenite	W, F

Table 6.1 (contd).

Name	Type	Name	Type	Name	Type
ilmenite	P	MONTMORILLONITE	W, F	rutile	P, W, F
ilmenorutile	F	montroyalite	F	sabinaite	F
isokite	W	mordenite	F	sahamalite-(Ce)	P
jamesonite	P	muscovite	P	saponite	W
JAROSITE	W	nahcolite	F	schorlomite	P
kaersutite	V	natrofairchildite	P	schröckingerite	W
kalicinite	V	natrojarosite	F	scolecite	P
kalipyrochlore	P	natrolite	P	sellaite	P
KAOLINITE	W, F	nepheline	P	serpentine	P, W
khanneshite	V	niocalite	P	shortite	P
kutnahorite	P	nontronite	W	siderite	P, F
labuntsovite	P	nordstrandite	F	siderotile	W
lamprophyllite	P	norsethite	P	silver	P
lanthanite-(Ce)	P	NYEREREITE	V	smithsonite	W
larnite	P	opal	W	smythite	F
lavenite	P	orthoclase	P, F	sodalite	P
lazurite	P	pachnolite	P	sphalerite	P, F
lepidocrocite	W	palygorskite	P	spinel	P
liebigite	W	pargasite	P	spurrite	V
linnaeite	P	parisite-(Ce)	P	stilpnomelane	P
litharge	W	pectolite	P	strengite	W
lorenzenite	P	pentlandite	P	strontianite	P, F
lueshite	P	periclase	P, V	strontiodresserite	F
maghemite	W	perovskite	P, V	sulphur	W, F
magnesio-arfvedsonite	P	phillipsite	P	svanbergite	P
magnesio-ferrite	V	PHLOGOPITE	P, V	synchysite-(Ce)	P, F
magnesio-hastingsite	P	platinum	W	talc	P, F
magnesio-kataphorite	P	plumbopyrochlore	P	tazheranite	P
magnesio-riebeckite	P	prehnite	P	tengerite-(Y)	W
magnesite	P	prosopite	P	tenorite	W
MAGNETITE	P, V, F	pseudorutile	F	tetrahedrite	P
malachite	W	PYRITE	P, V, F	thaumasite	V
manasseite	P	PYROCHLORE	P, V, F	thenardite	P, F
marcasite	P, F	PYROLUSITE	W	thermonatrite	V
mesolite	P	pyrope	P	thomsonite	P
meta-autunite	W	pyrophanite	P	thorbastnaesite	F
microcline	P	pyrrhotite	P, F	thorianite	P
millerite	P	quartz	P, W, F	thorite	P
mitridatite	W	ralstonite	P	titanite	P, V
moissanite	P	rhabdophane-(Ce)	W	torbernite	W
molybdenite	P, F	rhodochrosite	P	trona	W
molybdite	W	RICHTERITE	P	uranpyrochlore	P, V
MONAZITE-(Ce)	P, W	riebeckite	P	valleriite	P
VERMICULITE	P, W	romanechite	W	variscite	W
vesuvianite	P	wavellite	W	wurtzite	F
viitaniemiite	F	weberite	P	xenotime-(Y)	W
vinogradovite	P	WELOGANITE	F	zircon	P
vivianite	P	wöhlerite	P	zirkelite	P
wadeite	P	wollastonite	P	zoisite	P
		wulfenite	P		

Names in capitals represent common minerals; in lower case, uncommon or rare minerals. Carbonatite types are: F, silicocarbonatite from Montreal Island, Canada; P, plutonic carbonatite; V, volcanic carbonatite; W, weathering products.

Kapustin (1973) described 98 minerals that characterize weathered crusts overlying carbonatites, including several Mn oxides, phosphates of the crandallite group, Fe sulphates, and a number of hydrous Mg silicates.

Lists of minerals found in carbonatites are given by Heinrich (1966) and descriptions, complete to about 1970, are presented by Kapustin (1980). Table 6.1 includes minerals from these two works and, in addition, attempts to update the listing. Mineral nomenclature and spellings follow Fleischer (1987). Abundance is a subjective rating, with common minerals listed in capitals, less common minerals listed in lower case. Some 'common' minerals characterize an individual but distinct carbonatite but are very rare or unknown elsewhere, e.g. gregoryite and nyerereite from Oldoinyo Lengai, Tanzania, and dawsonite and weloganite from Montreal Island, Canada.

Although carbonates are the most abundant minerals in carbonatite, the amount of data is disproportionately small. Chemical trends, especially in calcite, dolomite, and ankerite, have been discussed by Sokolov (1985). Early carbonatites are rich in calcite, which commonly contains considerable Sr (e.g. up to 2.0% at Oka: Pouliot 1970). Dolomite normally precipitates after calcite, and ankerite after dolomite. This ankerite may contain considerable Mn (e.g. 1.7–2.0% in various carbonatites of USSR; Samoylov 1977). Calcite may persist into and beyond the ankerite stage

Table 6.2 Carbonatite stages, modified after Sokolov (1985).

Stage	Typical silicates	Typical carbonates	Geochemical specialization of carbonates	Apatite	Pyrochlore	Amphibole
Early I	Aeg.-augite, phlogopite, K-feldspar, andradite, wollastonite, monticellite	Sr – calcite	Ca, Sr			
II	Aeg.-augite, forsterite, clinohumite, albite, tetraferriphlogopite	Sr-calcite, dolomite	Ca, Mg, Sr			
III	Tetraferriphlogopite, serpentine, aeg.-augite, clinohumite	calcite, ankerite, alstonite, strontianite	Ca, Mg, Fe, Sr, Na, Ba			
IV Late	Aegirine, chlorite, sericite, quartz, phlogopite	manganoan calcite, siderite, ankerite, parisite, bastnaesite, magnesite	Mn, Fe, Ca, Mg, Ce, Sr, Ba, F			
Post-carbonatite	Aegirine, zeolites, muscovite, prehnite	calcite, manganoan calcite, barytocalcite	Ca, Mn, Ba, Fe			

and commonly contains less Sr than earlier calcite (e.g. 0.1–0.3% at Gatineau, Quebec, Canada; Hogarth *et al.* 1985). At this stage Sr is strongly partitioned into apatite and barite. If REEs and F have not been removed in early apatite, they may precipitate as REE fluocarbonates at a late stage (as at Mountain Pass, California; Olson *et al.* 1954). Syn- or post-ankerite siderite occurs in several carbonatites, but the mineral is high in Mg (e.g. 5.2% Mg at Iron Hill, Colorado; Nash 1972*a*).

Most minerals listed in Table 6.1 are either rare or atypical of average carbonatite and, because of this, only the well-documented pyrochlore, apatite, and amphibole mineral groups are discussed in this chapter. Pyrochlore is typical of many carbonatites and its presence at once distinguishes carbonatite from limestone and marble. Apatite is characteristic of most early carbonatites and amphibole is common, though non-essential, in the intermediate stages of carbonatite formation (Table 6.2).

Table 6.2, modified after Sokolov (1985), depicts a continuous transition through stages I–IV, each with its own chemistry, Ca → Mg → Fe + Mg → Mn + Fe. The final stage ('post-carbonatite') is characterized by late-stage calcite veins that cut carbonatite and associated silicate rocks. It is equivalent to the 'zeolitic carbonatite facies' of Samoylov (1977) and 'late stage IV carbonatite' of Kapustin (1980).

Apatite is most abundant in Stage 1 and persists to late Stage II. After Stage II apatite is normally present as secondary apatite only, derived from primary apatite by solution and reprecipitation. Pyrochlore appears near the end of Stage I and persists into Stage III. Amphiboles are ubiquitous in all carbonatites and are even present in post-carbonatite veins. All three mineral groups have also been noted from effusive carbonatite. All occur together in carbonatite lava near volcanic vents at Catanda, Angola (Silva 1973), in carbonatitic tuff near Fort Portal, Uganda (Kapustin & Polyakov 1982), and in extrusive carbonatite in the Kanneshin Mountains, Afghanistan (Vikhter *et al.* 1975).

6.2 PYROCHLORE GROUP

6.2.1 *Structure, formula, and classification*

Pyrochlore, ideally $8[NaCaNb_2O_6F]$, is cubic with space group Fd3m and has the structure type $E8_1$ in the Structurbericht (Gaertner 1930). The group cell-formula can be written $A_{16-x}B_{16}O_{48}(O,OH,F)_{8-y} \cdot zH_2O$, where x and y are vacant sites in the unit cell and x, y, and z are non-rational. In carbonatites, x attains a maximum of at least 13.25, y a maximum of 3.8, and z a maximum of 25.4. A atoms are As, Ba, Bi, Ca, Cs, K, Mg, Mn, Na, Pb, REEs, Sb, Sn, Sr, Th, U, Y; B atoms are Nb, Ta, Ti, V. The assignment of Al, Fe, Si, and Zr is uncertain. Members with much more than 5–10% U_3O_8 or ThO_2 (e.g. uranpyrochlore, uranmicrolite, and betafite) are normally metamict, although very young uranium-rich members may be crystalline.

Bonshtedt-Kupletskaya's (1966) classification of the pyrochlore group is based on

Table 6.3 Classification of the pyrochlore group; species found in carbonatites are in capitals.

A ions characteristic of species		Pyrochlore subgroup Nb + Ta > 2Ti Nb > Ta	Microlite subgroup Nb + Ta > 2Ti Ta ⩾ Nb	Betafite subgroup 2Ti ⩾ Nb + Ta
Na + Ca but no other A-atom > 20% total atoms		PYROCHLORE	microlite	CALCIOBETAFITE
One or more A-atoms, other than Na or Ca, >20% total A-atoms	K	KALIPYROCHLORE		
	Cs		cestibtantite	
	Sn		stannomicrolite	
	Ba	BARIOPYROCHLORE	bariomicrolite	
	Sr	(UN-NAMED)		
Species named by most abundant A-atom, other than Na or Ca	REE ΣCe > ΣY	CERIOPYROCHLORE		
	REE ΣY > ΣCe	yttropyrochlore		yttrobetafite
	Bi		bismutomicrolite	
	Sb			stibiobetafite
	Pb	PLUMBOPYROCHLORE	plumbomicrolite	plumbobetafite
	U	URANPYROCHLORE	uranmicrolite	BETAFITE
	Th	(UN-NAMED)		

REE = lanthanides + Y; ΣCe = (La → Eu), ΣY = (Gd → Lu) + Y. For purposes of species definition REE counts as one atom.

dominance or equivalence of Nb, Ta, Ti, and A-ion characterization, and results in 17 species. However, the limits of variation of A ions are not precisely defined.

The Pyrochlore Subcommittee published its conclusion and recommendations in 1977 (Hogarth 1977). The classification was modelled after the classification of Bonshtedt-Kupletskaya. Three subgroups were suggested: pyrochlore Nb + Ta > 2Ti, Nb > Ta; microlite Nb + Ta > 2Ti, Ta ⩾ Nb; betafite 2Ti ⩾ Nb + Ta. Within the subgroups, species were defined as:

(a) Na–Ca members – Na and Ca, but no other A-atom to exceed 20% of the A atoms;
(b) other members – one or more A atoms, other than Na or Ca, to exceed 20 atomic % of the A atoms present.

The updated classification scheme is presented in Table 6.3.

Recently, Kuz'menko (1984) proposed a more elaborate classification, which would have genetic significance in the description of mineral deposits. However, in the author's view, the advantages of the Kuz'menko scheme are outweighed by its complexity. The term hatchettolite of Kuz'menko, equivalent to tantalian uranpyrochlore in the IMA nomenclature, seems to have wide usage amongst Soviet mineralogists.

6.2.2 Compositional limits

Most analyses of pyrochlore from carbonatite correspond to pyrochlore (sp.), with the formula $(Ca, Na...)_{15.7-16}(Nb,Ta,Ti)_{16}O_{48}(OH,F)_8 \cdot 0-3H_2O$, i.e. a formula approaching stoichiometry. Pyrochlore species, identified from carbonatites, are shown in Table 6.3 and selected compositions are given in Table 6.4. Minerals such as ceriopyrochlore and betafite are uncommon in carbonatite, and kalipyrochlore and the Sr and Th members are very rare. Kalipyrochlore has been identified from one locality only (Lueshe, Zaire; Van Wambeke 1978). The greenish white type (analysis 1, Table 6.4) is characterized by high K (52% of the A ions) and a large deficit of A ions (83% vacancy in this site). Potassium-rich members of the group are not restricted to carbonatite. Bariopyrochlore, first described from carbonatite of Panda Hill, Tanzania ('pandaite'; Jäger et al. 1959), is also known from carbonatite at Mrima Hill, Kenya (Deans 1966) and Araxá, Brazil (analysis 2, Table 6.4; Filho et al. 1984). Late pyrochlore commonly contains a few per cent BaO. Ceriopyrochlore is known from sövite from a niobium orebody at Oka, Quebec, Canada (analysis 4, Table 6.4; Kalogeropoulos 1977). Cerium-rich pyrochlore from Kaiserstuhl, Germany ('ceriopyrochlore'; Van Wambeke 1980) gives ΣCe as 18% of A ions if the formula is calculated on 16 B ions, and is best regarded as cerian pyrochlore. Plumbopyrochlore, found in sövite from Oka by Kalogeropoulos (1977), has 29% of the A ions as Pb and 12% of the A sites are vacant (analysis 5, Table 6.4).

Although betafite and calciobetafite are not typical of carbonatite because of insufficient Ti, examples of betafite from eastern Siberia have been reported, e.g. analysis 7, Table 6.4; occurrence 1 of Pozharitskaya & Samoylov (1972). In Canada, betafite has been identified from sövite boulders near Prairie Lake, Ontario (Sage 1983d). Betafite has been tentatively identified from the Loe Shilman carbonatite, Pakistan, on the basis of partial chemical analyses (Rahman 1980), but the X-ray diffraction intensities suggest uranpyrochlore rather than betafite. Betafite is a common mineral in alkali amphibole veins and the phlogopite–apatite matrix of breccias associated with carbonatite near Meech Lake, Quebec, Canada (Hogarth 1959, 1961). Calciobetafite has been described from sövite of the Kovdor Massif, USSR (Borodin et al. 1973).

An analysis of the Sr member of the pyrochlore group from an unidentified carbonatite in Siberia (Gaidukova 1966) is given in Table 6.4, analysis 3. If P_2O_5, along with combined CaO, is subtracted as apatite, Sr makes up 47% of the A ions and 80% of the A sites are left unfilled. This is 'strontiopyrochlore', an unofficial name used by Kuz'menko (1984). This mineral has also been reported from rauhaugite near St. André, Quebec, Canada (Gold et al. 1986).

Other pyrochlores with high Sr, from carbonatite or closely associated rocks include: strontian bariopyrochlore (Panda Hill, Tanzania, 6.4% SrO; Jäger et al. 1959); strontian pyrochlore (Iron Hill, USA, 6.94% SrO; Nash 1972; Nkombwa Hill, Zambia, 3.8% SrO; Deans 1966; Napak, Uganda, 2.02% SrO; Deans 1966; St. André, 4.20% SrO; Gold et al. 1986); strontian uranpyrochlore (Sebl'yavr Massif, USSR, 4.11% SrO; Subbotin et al. 1985); and strontian betafite (in breccia, Meech Lake, 2.01% SrO; Hogarth 1961).

According to Efimov et al. (1985) 20 analyses of pyrochlore from carbonatite (occurrence not specified) averaged 0.90% SrO; 15 early pyrochlores averaged 0.38% SrO; and three late pyrochlores averaged 2.46%. A number of pyrochlores

Table 6.4 Some pyrochlore-group minerals from carbonatites.

Wt%	1	2	3	4	5	6	7	8	9	10	11	12
Nb_2O_5	80.05	63.42	67.06	47.2	43.26	51.45	40.2	40.13	31.55	68.18	61.78	46.1
Ta_2O_5	0.11	0.15	1.40	2.46	1.73	3.46	4.2	13.91	25.60	1.08	7.22	2.2
WO_3	0.033	–	–	–	–	–	–	–	–	–	–	–
V_2O_5	4.12	2.30	4.46	6.69	4.80	5.00	13.45	3.21	6.26	2.13	2.32	4.38
TiO_2	0.37	–	0.48	0.66	0.41	0.16	3.42	0.20	1.46	0.02	0.02	–
ZrO_2	0.06	0.10	–	–	–	–	–	–	–	–	–	–
SnO_2	–	–	–	–	–	–	–	–	–	–	–	–
SiO_2	–	–	0.50	3.84	2.84	0.00	–	0.48	–	–	–	–
ThO_2	0.17	2.34	0.83	1.43	3.15	–	3.72	15.28	16.02	<0.05	<0.05	<0.05
U_3O_{8T}	0.08	–	0.61	–	–	24.79	27.8	9.48	–	<0.05	3.05	18.9
Al_2O_3	–	–	0.67	–	–	–	–	0.53	–	–	–	–
Fe_2O_{3T}	0.21	2.63	1.22	3.01	2.69	3.06	–	2.68	1.49	<0.01	0.02	6.12
MnO	0.06	0.13	–	1.56	1.18	0.15	–	–	–	<0.02	<0.02	–
MgO	0.11	–	–	–	–	–	–	–	–	–	–	–
PbO	0.02	0.42	–	–	22.04	–	–	–	–	–	–	0.71
La_2O_3	0.13	–	–	2.22	–	0.77	–	–	–	0.25	0.12	0.135
Ce_2O_3	0.25	–	–	18.90	10.68	3.01	–	–	–	0.43	0.235	0.31
Pr_2O_3	0.045	–	–	–	–	–	–	–	–	–	–	–
Nd_2O_3	0.045	–	–	1.62	–	0.58	–	–	–	–	0.08	0.12
Gd_2O_3	0.01	–	–	–	–	–	–	–	–	–	–	–
Y_2O_3	0.022	3.29	–	–	–	–	–	–	–	–	–	–
ΣREE_2O_3	–	–	2.11	–	–	–	–	–	1.00	–	–	–
CaO	0.13	0.44	4.08	9.80	6.56	6.33	–	6.37	1.77	16.04	14.54	15.5
SrO	1.73	–	5.60	–	–	0.21	–	0.44	0.95	0.65	0.56	0.5
BaO	0.38	16.51	0.72	–	–	–	–	–	–	–	–	–
Na_2O	0.58	–	0.25	–	–	1.11	3.73	–	0.82	7.81	7.54	0.55
K_2O	2.76	–	0.07	–	–	–	–	–	–	–	–	0.07
H_2O_T	8.37	8.50	5.81	2.21	1.84	–	–	7.47	11.35	–	–	5.12
F	0.11	–	0.71	0.85	1.13	–	–	0.90	2.38	4.62	3.76	0.60
Total	99.96	100.23	96.58	102.31	102.31	100.08	96.52	101.17	100.65	101.34	101.32	101.37

Names, occurrences, and references of above analyses (names in quotes are not approved by IMA but appear in Soviet literature). Oxygen equivalents of F have been subtracted. 1, *kalipyrochlore*, Lueshe carbonatite, Zaire (Van Wambeke 1978); 2, *bariopyrochlore*, Araxá, Brazil (Filho *et al.* 1984); 3, '*strontiopyrochlore*' [3.06% P_2O_5 not included in total], Sayan, USSR (Gaidukova 1966); 4, *ceriopyrochlore*, St. Lawrence Mine, Oka, Quebec (Kalogeropoulos 1977); 5, *plumbopyrochlore*, St. Lawrence Mine, Oka, Quebec (Kalogeropoulos 1977); 6, *uranpyrochlore*, St. Lawrence Mine, Oka, Quebec (Petruk & Owens 1975); 7, *betafite*, eastern Siberia (Pozharitskaya & Samoylov 1972); 8, '*thoriopyrochlore*', Siberia (Gaidukova *et al.* 1963); 9, *tantalian uranpyrochlore*, Upper Sayansk Massif (Kapustin 1973); 10, *pyrochlore*, centre of zoned crystal, Blue River, British Columbia (Hogarth, unpublished); 11, *pyrochlore*, rim of zoned crystal, Blue River, British Columbia (Hogarth, unpublished); 12, *uranpyrochlore*, Meech Lake, Quebec (Hogarth, unpublished). Dashes indicate: not analysed.

from carbonatite are on record as containing 1–2% SrO. Strontian pyrochlore is not restricted to carbonatite but has also been reported from fenite and albitite.

Tantalum, an important element in some carbonatite pyrochlores, reaches 26.9% Ta_2O_5 in uranoan pyrochlore from early sövite from an unidentified locality in eastern Siberia (Pozharitskaya & Samoylov 1972). Twenty analyses from this occurrence averaged 18.7% Ta_2O_5 (Nb:Ta = 1.7:1) and 13.5% U_3O_8. Considerable Ta is present in pyrochlore from the Bailundo carbonatite, Angola (Lapido-Loureiro 1973), originally described as microlite by Aires-Barros (1968).

High tantalum (5–15% Ta_2O_5) pyrochlore-group minerals have also been reported from carbonatite at:

Canada: Crevier (Laplante 1980);
Finland: Sokli (Lindqvist & Rehtijärvi 1979);
India: Koratti, Tamil Nadu (Borodin *et al.* 1971);
USSR: Dalbykha, Maimecha-Kotui (Bagdasarov & Danilin 1982),
 Kola Peninsula (Kukharenko 1965a, Kirnarskii *et al.* 1968,
  Lebedeva *et al.* 1973, Semenov 1977, Kapustin 1980),
 Sebl'yavr (Subbotin *et al.* 1985),
 'Siberia' (Gaidukova *et al.* 1963, Bagdasarov 1974, Kapustin 1980),
 Tien-Shan (Efimov *et al.* 1980), and
 Turii Peninsula (Lapin 1975).

According to Kapustin (1974), in USSR 'a number of zones of Ta-rich uranpyrochlore, containing 10–20% Ta_2O_5, have been investigated as potential sources of tantalum. Individual deposits have been shown to contain 10 000 tonnes of Ta_2O_5' (translated by D. D. Hogarth).

A Th member of the pyrochlore group, containing 15.28% ThO_2, has been described by Gaidukova & Zdorik (1962) and Gaidukova *et al.* (1963) from a carbonatite in 'Siberia' ('Sayan' in Semenov (1977); analysis 8, Table 6.4). Calculations show that 24% of the *A* ions present are Th, but 39% of the *A* site is vacant. This mineral was unofficially named 'thoriopyrochlore' by Kuz'menko (1984).

Sixteen pyrochlore-group minerals from a carbonatite of eastern Siberia average 7.5% and contain up to 12.0% ThO_2 (Pozharitskaya & Samoylov 1972). Some of the highest Th values were from low-uranium (< 2% U_3O_8) pyrochlore but some were from uranpyrochlore. Another thorian pyrochlore, with 11.19% ThO_2, was recorded from a carbonatite from Kola (Kirnarskii *et al.* 1968). A thorian calciobetafite (6.15% ThO_2) from sövite from the Kovdor Massif, was described by Borodin *et al.* (1973). A pyrochlore concentrate from Oka with 8.5% ThO_2 by radiometric analysis (Nickel 1962) may represent the metamict variety, with 7.23% ThO_2 and 1.83% U_3O_8 by X-ray fluorescence (Hogarth 1961). According to Efimov *et al.* (1985) 20 carbonatite pyrochlores, mainly early (from unspecified localities), averaged 1.80% ThO_2. Pyrochlore from the Aley carbonatite, British Columbia, Canada contains 3.44% ThO_2 in the core and 0.26 ThO_2 on the rim of crystals (Mäder 1987).

Thorium increases during weathering. Thus pyrochlore from alluvium at Vuoyarvi contains 5.55% ThO_2 (Kapustin 1973).

Uranpyrochlore and uranoan pyrochlore are typical of many carbonatites. Uranoan pyrochlores involve two types of ionic substitution:

(a) $Ca^{2+} + 2Nb^{5+} \rightarrow U^{4+} + 2Ti^{4+}$ ('betafite scheme'); and
(b) $2Ca^{2+} \rightarrow U^{4+} + \square$ ('uranpyrochlore scheme').

The latter scheme is important in all uranoan pyrochlores and uranpyrochlores and results in a deficit of A cations (Borodin & Nazarenko 1957, Hogarth 1961). The highest U_3O_8 content is normally about 28% (a partial analysis of uranpyrochlore from Crevier gave 37.6% U_3O_8; Laplante 1980) and the attendant deficit of A ions suggests undersaturation of U. This explains why free uraninite is never associated with pyrochlore-group minerals. Non-metamict, tantalum-rich uranpyrochlore has recently been identified from the Fort Portal (Uganda) tuff (D. D. Hogarth, unpublished data).

Tin, normally restricted to pegmatitic pyrochlore, is also known in pyrochlore from sövite at Oka, which contains up to 3.43% SnO_2 (Kalogeropoulos 1977).

The roles of Al, Fe, Si, and Zr in pyrochlores are uncertain. High values (about 11%) of SiO_2 were reported in pyrochlore (var. chalcolamprite and endeiolite) from syenite in Greenland (Flink 1898, 1901) but at least some silica is thought to be present as silicate impurities (Hogarth 1977). Lesser amounts of SiO_2 are reported from carbonatitic pyrochlore minerals. Electron microprobe imaging of pyrochlore from the USSR showed Si concentrated as flecks in several concentric zones. The Si, considered a silicate, was not included in the analyses (Polezhaeva & Strel'nikova 1980). The ion Si^{4+} substituting for the much larger Nb^{5+} in sixfold co-ordination, should be looked at critically. Similarly, high-aluminium pyrochlore needs further investigation.

Iron is present up to several per cent in some pyrochlore-group minerals from carbonatite, and is generally thought to be present in the crystal structure (e.g. Palache et al. 1944, Bonshtedt-Kupletskaya 1967). For reasons of charge and ionic radius, Fe^{2+} is normally placed in A and Fe^{3+} in B, but there is no compelling evidence for this assignment. Perrault (1968) obtained good agreement of observed and calculated X-ray intensities of five pyrochlores from Oka, by assuming all Fe as Fe^{3+} in B. However, the maximum Fe in his specimens was 1.35% total Fe. Comparison of observed and calculated intensities with all Fe in A and all Fe in B in five heated and unheated pyrochlores and betafites (trace to 2.67% total Fe) by Hogarth (1961), did not solve the problem.

Zirconium has been reported up to several per cent in pyrochlore from carbonatite, but many of these analyses are suspect. Lebedeva et al. (1973) compared chemical and microprobe analyses from the same specimens and found that chemically analysed pyrochlores contained much higher ZrO_2, which was attributed to minute zircon and baddeleyite inclusions. Pyrochlore from the Arbarastakh (USSR) carbonatite gave 30.46% ZrO_2 by chemical analysis, but 3.0% ZrO_2 by electron microprobe.

6.2.3 *Some chemical correlations and generalizations*

The positive correlation of Ta and U in pyrochlore-group minerals (see Kukharenko 1965a) generally holds true in carbonatite, and this has led most Soviet geologists to consider 'hatchettolite' as a separate mineral species, defined as a pyrochlore mineral containing considerable Ta and U. Uranpyrochlore, however, from the Sol'bel'der sövite, Tuva, contains 18.84% U_3O_8, and only 0.39% Ta_2O_5 (Semenov 1977). Semenov notes that Ta normally also correlates positively with Fe, H, Ti, and Zr, and negatively with Ca, F, Na, and Nb. Most carbonatite pyrochlores are characterized by high Ba, Sr, Th, Zr, and low Pb, REEs, Sn compared with pyrochlores from nepheline syenites and alkali granites (Efimov *et al.* 1985). Early carbonatites are enriched in Ta, Th, Ti, U, Zr, and late carbonatites in Ba, Pb, Sr (Efimov *et al.* 1985). Most late, tantalum-poor carbonatite pyrochlores are enriched in OH and H_2O at the expense of F (Kapustin 1980).

Indeed, pyrochlore compositions are so variable that it is difficult to generalize. Uranium–tantalum-rich, titanium-poor pyrochlores do seem to characterize certain carbonatites, whereas barium-, potassium-, and strontium-rich pyrochlores seem restricted to late-generation carbonatite. Titanium-rich pyrochlores are rare in carbonatite.

6.2.4 *Regional variation in pyrochlore composition*

Regional variations of pyrochlore composition have shown that, in eastern Siberian carbonatites, tantalian uranpyrochlore ('hatchettolite') occurs in the outer zones and pyrochlore in the inner zones, although, in some places, uranpyrochlore is associated with pyrochlore (Bagdasarov 1972, Pozharitskaya & Samoylov 1972). Bagdasarov (1974) showed that in a carbonatite complex of Siberia some Nb and Ta orebodies were separated, one from the other, but some were zoned with tantalum-rich ore (Nb:Ta < 12:1) concentrated in the centre of the orebodies and niobium-rich ore (Nb:Ta > 20:1) on the periphery. In some orebodies, niobium-rich zones were found near the contact and tantalum-rich zones at a distance from pyroxene–nepheline rocks. Efimov *et al.* (1985) demonstrated that pyrochlore from carbonatites in the Urals, within a strip 120 km long and up to 6 km wide, have large chemical variations. Uranium and U:Th ratios are much higher in the axial zones than in the contact zones but Nb and Nb:Ta ratios are higher in the contact zones. The authors also concluded that pyrochlore from linear carbonatites is richer in Ti and U but poorer in Ta and Zr than in stock-like carbonatites. However, this last conclusion was based on limited data.

Laplante (1980) showed that, at Crevier, pyrochlore from syenite was low in U and Ta, compared with tantalian uranpyrochlore from carbonatite, and the data plot in two distinct areas in a Nb–Ta–Ti diagram, with pyrochlore and perhaps uranpyrochlore showing compositional trends (Fig. 6.1). Six analyses of pyrochlore from Turii Peninsula, USSR (Lapin 1975) can be fitted to the two trends delineated by Laplante (1980). Compositions of Oka pyrochlore do not follow these trends.

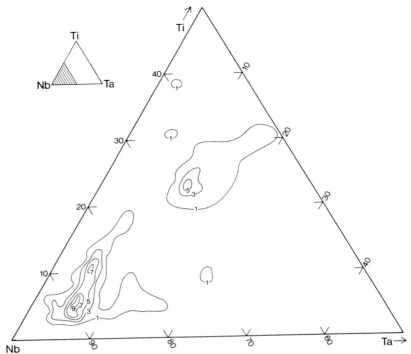

Figure 6.1 Contoured atomic distribution of 281 microprobe analyses of pyrochlore-group minerals from the Crevier Pluton, Quebec, Canada, on a Nb–Ta–Ti diagram. The niobium-rich mode corresponds to pyrochlore from syenite; the titanium–tantalum-rich mode corresponds to uranpyrochlore from carbonatite. Diagram from Laplante (1980); published with permission of R. Laplante (Ressources Aiguebelle, Rouyn-Noranda, Quebec) and Ecole Polytechnique, Montréal, Quebec.

6.2.5 Zoned crystals

Some pyrochlore crystals (e.g. Blue River, Oka, and Meech Lake) have radioactive zone(s) near the rims (Hogarth 1959). Some zones are sharp and probably represent growth discontinuities; other zones are diffuse and may represent replacement. Microprobe analyses of pyrochlore crystals from Mbeya, Tanzania, with calcium-depleted and thorium-enriched rims, may also reflect replacement (Veen 1963).

Some pyrochlore crystals show repetitive radioactive zoning (Mineev & Razenkova 1962, Vronskii & Basina 1963). Pyrochlore from an unidentified carbonatite shows repetitive zoning, with the most radioactive zone close to the edge of the crystal (Chistov & Denisov 1971). The authors conclude that 'the composition of ore-forming solutions experiences multiple and commonly rhythmic oscillation during carbonatite formation and that radioactive elements increase during the final stages of ore formation. Columbitization post-dates the breakdown of crystal structure in zones enriched in these elements' (translation by D. D. Hogarth).

These findings contrast markedly with data given by Lebedeva et al. (1973) who found uranium-rich rims but tantalum-rich cores in pyrochlore crystals from the Kovdor carbonatite. Polezhaeva & Strel'nikova (1980) concluded that Ta, Th, Ti,

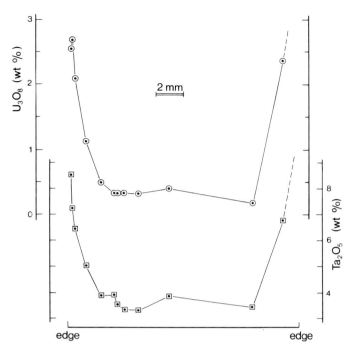

Figure 6.2 Analyses across a pyrochlore grain from the Blue River (Verity Property) carbonatite, British Columbia. Crystal supplied from specimen 1224, National Mineral Collection, Geological Survey of Canada. Analyst, P. Jones, Carleton University.

and U were progressively depleted during carbonatite development. Pyrochlore from the Sarfartôk carbonatite, Greenland has higher U (5%) in the dark central parts than the lighter margins of crystals (< 1%; Secher & Larsen 1980). Pyrochlore from the Urals is characterized by a decrease in U but an increase in Ta, from core to periphery (Efimov et al. 1985).

Petruk & Owens (1975) found some pyrochlore from Oka was enriched in U in narrow zones near the outer parts of the crystals. Kalogeropoulos (1977) found that the rims were consistently higher in Ca, F, and Na than the cores, but the cores were higher in Ta and U than the rims. Cerium and Fe were not consistently distributed between core and rim. Data from strongly colour-zoned pyrochlore from the Blue River carbonatite, British Columbia are shown in Figure 6.2 and two analyses are given in Table 6.4 (analyses 10 & 11). The more deeply coloured zone(s) are enriched in Ta and U; these elements are highest in zones at or near the rim.

While the bulk of evidence seems to support the evolution from low-uranium pyrochlore to tantalian uranpyrochlore during carbonatite development (Chistov & Denisov 1971, Laplante 1980), reversals are possible.

6.2.6 Rare earth element (REE) distribution

Pyrochlore-group minerals are potential REE carriers and some varieties, like those from Oka, contain large amounts of these elements. A notable feature is the

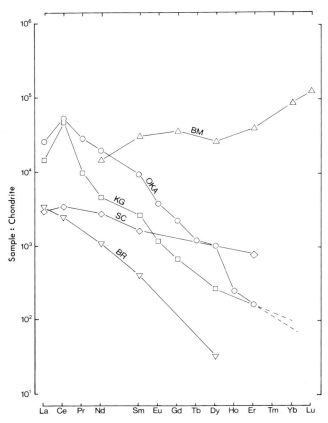

Figure 6.3 REE abundances of pyrochlore-group minerals normalized against the chondritic values of Boynton (1984). BM, betafite from granite pegmatite, Betafo, Madagascar (Junge *et al.* 1983). BR, radioactive pyrochlore from carbonatite, Blue River (Verity Property), British Columbia, Canada; analyst J. Loop, University of Ottawa. KG, radioactive pyrochlore from carbonatite, Badlock, Kaiserstuhl, Germany (Van Wambeke 1964). OKA, cerian pyrochlore from carbonatite, sample OL-7, Oka, Quebec, Canada (Eby 1975). SC, betafite crystal from skarn, Bancroft (Silver Crater Property), Ontario, Canada; analyst, J. Loop, University of Ottawa.

pronounced LREE enrichment (see Fig. 6.3) which reflects the REE distribution in carbonatites. The La:Yb ratio, commonly > 100:1 in carbonatite pyrochlore, is about 7:1 in betafite from the Silver Crater skarn (Hogarth unpublished data) and < 0.1:1 in betafite from the Betafo granite pegmatite (Junge *et al.* 1983).

Cerium anomalies can be expressed as δCe, where δCe = Ce analysed : Ce interpolated between La and Nd. A positive anomaly ($\delta Ce > 1:1$) characterizes pyrochlore-group minerals from carbonatite (e.g. Semenov 1963, Balashov & Pozharitskaya 1968; see also Table 6.5) but also appears in pyrochlore and betafite from skarn deposits such as Silver Crater (Fig. 6.3), fenite (Borodin & Barinskii 1961), nepheline syenite (Borodin & Barinskii 1961), and aegirine syenite (Lin *et al.* 1973) and is apparently controlled structurally. Eby (1975) attributes this anomaly to the result of oxidation of Ce^{3+} to Ce^{4+}, the tetravalent ion being more easily accommodated in the pyrochlore structure than the trivalent ion. However, Kapustin (1966) has noted that this anomaly can also be explained by La depletion.

Table 6.5 REE data of some pyrochlore-group minerals.

Location	Specimen	δCe	δEu	Nd:La	La:Yb	Method	Reference
USSR	1st generation pyr.	1.6(3)	0.4(2)	2.2(3)	110(1)	unknown	Kapustin (1980)
USSR	2nd generation pyr.	2.05(3)	0.3(2)	1.8(3)	–	unknown	Kapustin (1980)
USSR	3rd generation pyr.	1.9(3)	0.3(2)	1.2(3)	75(1)	unknown	Kapustin (1980)
USSR	4th generation pyr.	2.2(1)	0.8(1)	1.2(1)	–	unknown	Kapustin (1980)
USSR	uranpyrochlore	1.9(1)	0.4(1)	2.9(?)	40(?)	XRF	Borodin & Barinskii (1961)
USSR	pyrochlore	1.9(10)	0.7(10)	1.8(10)	50	XRF	Borodin & Barinskii (1961)
Eastern Siberia	uranpyrochlore	2.9(1)	–	1.4(1)	100(1)	unknown	Zdorik (1966)
Eastern Siberia	pyrochlore	1.3(1)	–	1.7(1)	100(1)	unknown	Zdorik (1966)
Mbalizi, Tanzania	pyrochlore	2.6(5)	0.1(1)	0.9(5)	400(3)	XRF	Pentel'kov & Matrosova (1978)
Kaiserstuhl, FRG	pyrochlore	2.9(5)	0.5(1)	0.8(5)	100(5)	XRF	Van Wambeke (1964)
Oka, Canada	pyrochlore No. OL-7	2.2(1)	0.8(1)	1.5(1)	100(1)	XRF	Eby (1975)
Oka, Canada	pyrochlore No. G13-8	2.5(1)	1.0(1)	1.1(1)	100(1)	XRF	Eby (1975)
Oka, Canada	pyrochlore rims	2.5(3)	–	1.5(3)	–	EM	Kalogeropoulos (1977)
Oka, Canada	pyrochlore cores	2.8(3)	–	0.7(3)	–	EM	Kalogeropoulos (1977)
Blackburn, Canada	pyrochlore	2.1(5)	–	0.9(5)	–	EM	unpublished data
Blue River, Canada	pyrochlore	1.1(1)	–	0.6(1)	–	DCP	this chapter
Meech Lake, Canada	uranpyrochlore	1.2(1)	–	0.9(1)	–	DCP	this chapter

δCe = Ce analysed: Ce interpolated (La → Nd); δEu = Eu analysed: Eu interpolated (Sm → Gd). Numbers in parentheses indicate number of analyses considered. Analytical methods: DCP, direct-current plasma; EM, electron microprobe; XRF, X-ray fluorescence. Dashes indicate insufficient data for calculation.

Pyrochlores from carbonatite commonly have δCe > 2:1 but for some carbonatite pyrochlores, such as those of Blue River, Meech Lake, and Zdorik's 'eastern Siberian carbonatite', the Ce anomaly is weak or undetectable (Table 6.5).

Europium anomalies can be expressed as δEu, where δEu = Eu analysed : Eu interpolated between Sm and Gd. A negative anomaly (δEu < 1:1; Table 6.5) varies from strong at Mbalizi (about 0.1:1) to barely apparent at Oka (about 1:1). Possibly Eu^{2+} was selectively partitioned into diopside which precipitated from the carbonatite magma, or possibly Eu^{2+} was selectively removed by diopside or plagioclase in the wall rock.

Neodymium : La ratios in pyrochlore-group minerals are large in carbonatite (1.5–3.0:1) but comparatively small in nepheline syenite and syenite (0.4–0.5:1; Borodin & Barinskii 1961). The new data extend Nd : La from carbonatite to lower values (Table 6.3) which may overlap those of pyrochlore from nepheline syenite and syenite. Early pyrochlore may be characterized by high Nd : La ratios (c. 2:1), late pyrochlore by low Nd : La ratios (c. 1:1). Zoned pyrochlore from Oka shows a similar trend, with cores showing distinctly higher Nd : La ratios than rims. A plot of

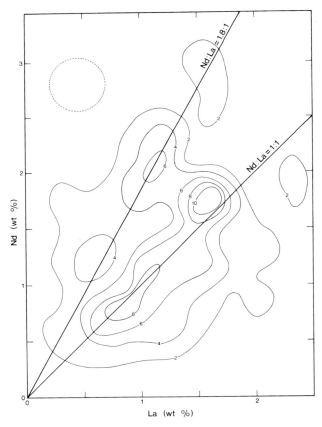

Figure 6.4 Contoured La–Nd distribution from 101 microprobe analyses of pyrochlore from sövite at the St. Lawrence Mine, Oka, Quebec. Note modes at Nd : La = 1:1 and 9:5. Analyses are from Kalogeropoulos (1977). Contours have been positioned according to the number of data points within a circle (shown on diagram).

Nd–La analyses from Oka (Kalogeropoulos 1977) shows a bimodal distribution (Fig. 6.4) which may reflect two pulses of mineralization.

6.2.7 Effects of weathering

Pyrochlore is commonly concentrated in eluvium above carbonatite. The chemical effects of hypergene alteration have been discussed by Kapustin (1973). During the weathering process La, Nb, Ta, Th, Zr, U, and H_2O increase, whereas Ca, F, HREEs, and Na decrease. In the most advanced stage of weathering U is also removed. In general, as weathering progresses A ions are depleted, the Nb : Ta ratio decreases and the mineral becomes hydrated. In the final stage, the mineral breaks down to a mixture of oxides, including $(Nb,Ta)_2O_5 nH_2O$ and La_2O_3. Analysis 9, Table 6.4, represents weathered pyrochlore.

Kapustin (1980) notes that pyrochlore, being brittle, is easily fragmented and dispersed in alluvial processes but 'columbitized' pyrochlore (a late-stage replacement) does not degrade.

Potassium and Ba enrichment, accompanied by hydration and a general loss of A ions and F, may be produced by certain types of weathering (cf. analyses 1 & 2, Table 6.4).

6.3 APATITE

6.3.1 Introduction

The apatite group is characteristic of carbonatites, commonly making up 2–5% by volume. Concentrations are particularly high in contact zones (Kapustin 1980). Rocks containing essential amounts of magnetite and apatite were termed 'phoscorite' by Russell *et al.* (1954) and 'kamforite' by Borodin *et al.* (1973). Examples are known from Phalaborwa, South Africa (Hanekom *et al.* 1965, Eriksson, this volume); Nemegos, Canada (Hodder 1961); and Kovdor, USSR (Rimskaya-Korsakova 1965). Mineral compositions, trace-element compositions, and fluid and solid inclusions suggest a genetic relationship between the magnetite–apatite bodies and carbonatites (Sokolov 1983). Primary and secondary apatite are concentrated in iron-rich residual soil derived from weathering of carbonatite. Examples occur at Glenover, South Africa; Sukulu, Uganda; Rufunsa Gorge, Zambia; and a number of other occurrences in Africa, Brazil, Canada, and USSR (Deans 1966, Kapustin 1973, Sage 1983*d*).

The apatite group contains the bulk of REEs in some carbonatites, most of the F in early- and middle-stage carbonatites, and much of the Sr in late-stage carbonatites. Fleischer (1987) lists 17 minerals in the apatite group, defined as those species with apatite structure and containing essential $(PO_4)^{3-}$, $(AsO_4)^{3-}$, $(VO_4)^{3-}$, or $(CO_3)^{2-}$ anions. In carbonatites, the most common species, fluorapatite, is characterized by appreciable substitution of all elements in the ideal formula – $Ca_{10}(PO_4)_6F_2$. Four species only, fluorapatite, hydroxylapatite, carbonate–fluorap-

Table 6.6 Chemical composition of apatite from carbonatites and related rocks.

Wt%	1	2	3	4	5	6	7	8	9	10	11	12
La_2O_3	0.771	5.6	n.a.	0.13	0.00	0.14	n.a.	0.12	0.24	0.04	0.05	n.a.
Ce_2O_3	1.500	14.4	n.a.	0.23	0.00	0.37	n.a.	0.34	0.59	0.26	0.18	n.a.
Pr_2O_3	0.246	1.9	1.47	0.02	0.00	n.a.	0.48	n.a.	n.a.	n.a.	n.a.	n.a.
Nd_2O_3	0.503	8.0	n.a.	0.08	0.00	n.a.	n.a.	0.16	0.25	0.22	n.a.	n.a.
Sm_2O_3	0.058	1.5	n.a.	0.01	0.00	n.a.	n.a.	0.09	n.a.	n.a.	n.a.	n.a.
ThO_2	0.03	5.62	0.13	n.a.	n.a.	0.0	n.a.	n.a.	n.a.	n.a.	n.a.	n.a.
Y_2O_3	0.075	0.4	n.a.	0.02	n.a.	n.a.	n.a.	0.02	0.02	0.11	0.01	n.a.
FeO_T	0.091	0.13	0.23	0.09	0.21	0.0	0.40	0.09	0.08	0.15	0.03	0.10
MnO	0.048	n.a.	0.01	0.03	0.00	n.a.	n.a.	0.01	0.01	0.07	0.01	n.a.
MgO	n.a.	0.20	0.29	0.45	0.29	0.1	0.10	0.09	0.09	0.00	0.21	0.14
BaO	n.a.	n.a.	n.a.	n.a.	0.04	0.00	n.a.	0.08	0.01	0.05	0.00	n.a.
SrO	0.906	n.a.	0.21	0.50	0.16	0.32	4.72	2.92	1.82	0.07	0.52	n.a.
CaO	51.76	28.84	53.63	54.24	54.27	54.8	51.04	52.33	53.05	54.58	55.01	55.35
Na_2O	0.07	0.21	0.25	0.14	0.04	0.00	n.a.	0.19	0.25	0.10	n.a.	n.a.
K_2O	n.a.	n.a.	0.01	0.00	0.01	n.a.	n.a.	0.02	0.02	0.00	0.10	n.a.
SiO_2	1.79	12.28	1.02	0.00	0.00	1.13	n.a.	0.02	0.04	0.04	0.25	n.a.
P_2O_5	37.19	16.96	40.24	42.80	38.11	38.7	40.20	40.09	41.04	39.67	40.76	39.55
SO_3	1.26c	n.a.	0.02	n.a.	0.00	0.456	n.a.	0.19	0.17	0.08	n.a.	n.a.
CO_2	0.19c	n.a.	0.82	n.a.	3.50	0.87	0.00	1.6	n.a.	1.33c	0.34	1.80
F	2.65	2.10	1.87	2.02	2.30	0.93	3.40	3.57	3.34	3.50	3.10	3.79
Cl	0.01	n.a.	0.10	0.00	0.02	0.20	0.05	0.00	0.01	0.02	n.a.	0.1
H_2O^+	0.46c	0.54	0.38	0.50	2.32	1.28c	0.59	0.12	0.18c	0.11c	0.12	0.53
H_2O^-	n.a.	n.a.	n.a.	0.00	n.a.	n.a.	0.00	n.a.	n.a.	n.a.	0.52	0.30
Rem.	0.183	1.56	0.99	0.05	0.44	n.a.	0.10	n.a.	n.a.	0.03	0.09	0.05
	99.79	100.24	101.67	101.31	101.71	99.30	101.08	101.96	101.21	100.43	100.95	101.71
O = F	1.12	0.88	0.81	0.85	0.97	0.44	1.44	1.50	1.41	1.47	1.31	1.62
Total	98.67	99.36	100.86	100.46	100.74	98.86	99.64	100.46	99.80	98.96	99.64	100.09

1, *Silicon-rich fluorapatite*, Oka, Quebec (Roeder et al. 1987). Rem. is Eu_2O_3, 0.019; Gd_2O_3, 0.089; Tb_2O_3, 0.069; Dy_2O_3, 0.003; As_2O_3, 0.003. 2, *Britholite*, Oka, Quebec (Hughson & Sen Gupta 1964). Rem. is Gd_2O_3, 0.9; $HREE_2O_3$, 0.1; Al_2O_3, 0.47; TiO_2, 0.09. 3, *Carbonate-fluorapatite*, Kovdor, USSR (Bulakh 1984). Rem. is insol., 0.81; TiO_2, 0.07; Al_2O_3, 0.11. 4, *Fluorapatite*, Sallanlatva, USSR (Rimskaya-Korsakova 1965). Rem. is Al_2O_3, 0.04; Gd_2O_3, 0.01. 5, *Carbonate-hydroxylfluorapatite* ('*staffelite*'), Kovdor, USSR (Rimskaya-Korsakova 1965). Rem. is CuO, 0.44. 6, *Carbonate-hydroxylapatite*, Kaiserstuhl, Germany (Sommerauer & Katz-Lehnert 1985). 7, *Strontian fluorapatite*, Nkombwa Hill, Zambia (Deans 1966). Rem. is insol., 0.10. 8, *Strontian carbonate-fluorapatite*, Gatineau, Quebec (Hogarth 1988). 9, *Strontian fluorapatite*, Gatineau, Quebec (Hogarth et al. 1987). 10, *Carbonate(?)-fluorapatite*, Gatineau, Quebec (Hogarth et al. 1985). Rem. is UO_2, 0.03. 11, *Fluorapatite*, Rangwa, Kenya (Prins 1973). Rem. is ZrO_2, 0.09. 12, *Carbonate-fluorapatite*, Busumbu, Uganda (Davies 1947). Rem. is insol., 0.05. n.a. = not analysed; c = calculated.

atite, and carbonate–hydroxylapatite, are known from carbonatite. Another isostructural species, britholite, while found in carbonatite, is a silicate and therefore not included in Fleischer's tabulation.

Apatite forms early in carbonatite and can persist into late-stage mineralization. Thus at Gatineau, Quebec, Canada, dolomite–calcite carbonatite contains 0.59–1.64% P_2O_5 or 1.5–4.0% apatite (Hogarth *et al.* 1985). The late carbonatite of Mush Khuduk, Outer Mongolia, is equally rich in apatite (Samoylov & Kovalenko 1983). This late-generation apatite is fluorapatite or carbonate–fluorapatite, rich in REEs and Sr ('podolite' of Soviet geologists). However, apatite may dissolve and reprecipitate as stalactitic-to-banded, carbonate-hydroxyl-fluorapatite ('francolite' of Rimskaya-Korsakova, 'staffelite' of Kapustin), low in REEs and Sr. In the USSR and Finland apatite cements clasts and fills breccias (Rimskaya-Korsakova 1965, Kapustin 1973, 1977, 1980, Rimskaya-Korsakova *et al.* 1979). There seems to be only one analysis of apatite from volcanic carbonatite presently available, viz. hydroxyl(?)-fluorapatite from Kerimasi, Tanzania (Roeder *et al.* 1987). The only unusual features of this apatite are relatively high contents of FeO (0.158%) and Cl (0.04%), and weak LREE enrichment.

Table 6.6 presents apatite and britholite analyses from carbonatite. Analysis 2 is britholite from a britholite–vermiculite vein that cuts carbonatite, whereas analysis 5 is a carbonate-hydroxyl-fluorapatite that forms cement in a breccia. Analysis 12 represents carbonate–fluorapatite from residual soil overlying carbonatite. All other specimens are from carbonatite *sensu stricto*.

6.3.2 *Structure of apatite*

The unit cell of fluorapatite contains $Ca_{10}P_6O_{24}F_2$ and the mineral has the space group symmetry $P6_3/m$ (Bragg *et al.* 1965). The structure is described as type VIII, C9 by Wyckoff (1957). Calcium occupies two distinct positions in the structure: Ca_I with a multiplicity of four, co-ordinated with nine oxygen atoms, and Ca_{II} with a multiplicity of six, co-ordinated with six O and one F. Each O is bonded to one P, one Ca_I and two Ca_{II} atoms (Beevers & McIntyre 1946). Hydroxylapatite and carbonate–apatite have modified fluorapatite structures.

In chlorapatite, the symmetry degenerates to monoclinic (pseudohexagonal) $P2_1/a$ (Young 1966, Prener 1967, Hounslow & Chao 1970). Miscibility of monoclinic with hexagonal apatite is strictly limited (Hounslow & Chao 1970, Taborszky 1972*a*, Hogarth 1988). In addition, the monoclinic phase is metastable and readily changes to the hexagonal form during alteration (Taborszky 1972*b*). With the possible exception of a single example from Phalaborwa, the domain of chlorapatite lies outside of the range of compositions of apatite from carbonatite.

In this review, Ca' will signify all ions in the Ca sites, P' all ions in the P site, and F' all ions in the F site. In Table 6.6, analyses 1, 6, 9, and 10, H_2O has been calculated by assuming two F' ions in the standard formula unit. In analysis 10, CO_2 was calculated on the basis of ten Ca' ions and then the P' ions, including C, were brought to six. The resulting charge on P' + Ca' ions was 49.7 (v. 50

theoretical). In analysis 1, P' was adjusted to six, based on ten Ca' ions. The ions C^{4+} and S^{6+} were then resolved by assuming the charge on P' + Ca' = 50.

6.3.3 Ca' ions

High Sr and low Mn are well-known characteristics of apatite from carbonatite. Preliminary data indicate that the Sr:Mn ratio is at least 50:1 in apatite from carbonatite but < 0.2:1 in apatite from gneiss and granite pegmatite (Brasseur *et al.* 1962). Although Sr:Mn in carbonatite is undoubtedly high (normally > 10:1), this ratio may vary greatly due to the abundance of Sr. For example, the ratio is approximately 1:1 for secondary carbonate–fluorapatite from Gatineau (*c.* 0.1% Sr, analysis 10, Table 6.6), 20:1 for fluorapatite from Oka (0.5% Sr; analysis 1, Table 6.6) and > 100:1 for primary fluorapatite from Gatineau (> 1.5% Sr, analysis 9, Table 6.6). On the other hand, fluorapatite from granite pegmatite north of Timmins, Ontario has Sr:Mn 4.4:1 (0.15% Sr; Roeder *et al.* 1987). Figure 6.5 shows generalized Sr–Mn relationships of apatite from carbonatite, granite pegmatite, phosphorite, and regional skarn.

At the Kovdor carbonatite, Sr in apatite increases from early to late carbonatite (0.23 → 0.70% SrO; Rimskaya-Korsakova *et al.* 1979), a generalization that appears to hold world-wide. Strontium enters calcite in early carbonatite but in the late-stage (ankerite–calcite) Sr is concentrated in barite, fluorapatite, and strontianite. Carbonate-hydroxyl-fluorapatite ('francolite'), filling breccias, contains only 0.17% SrO. Kapustin (1982) has shown that Sr increases from an average of 0.33% in

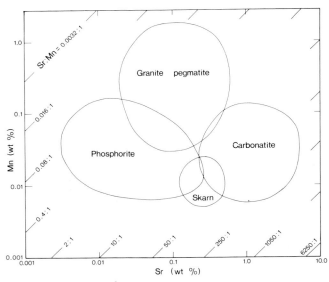

Figure 6.5 Mn–Sr fields of apatite from skarn, phosphorite, granite pegmatite, and carbonatite from world-wide localities, based on 30 analyses of apatite from skarn, 17 analyses from phosphorite, 28 analyses from granite pegmatite, and 62 analyses from carbonatite. Manganoan apatite from granite pegmatite, with > 1% Mn, has been omitted for simplicity.

fluorapatite from early carbonatite, to 0.90% in carbonate–fluorapatite from late carbonatite. Strontium isotope studies of the carbonate-hydroxyl-fluorapatite breccias at Kovdor, indicate hydrothermal deposition involving groundwater (Landa *et al.* 1983).

Barium in apatite increases from 0.04% in apatite from early carbonatite to 0.19% in apatite from late carbonatite (Kapustin 1977, 1982).

Total REE abundances in apatite from carbonatite show great variation and range from < 1% ΣREE in some carbonatites to 8.3% in fluorapatite from the Chernigov carbonatite, Ukraine (Vil'kovich & Pozharitskaya 1982). Apatite from Mush Khuduk, Outer Mongolia (Samoylov & Kovalenko 1983), with ΣREE 13.5%, may represent impure apatite.

Fluorapatite from Oka is particularly rich in REEs with total REE values that range from 1.3% to 7.7% (Girault 1966, Eby 1975, Mariano 1985; analysis 1, Table 6.6). The No. 5 zone of the Chernigov carbonatite contains apatite averaging 6.6% ΣREE (Zhukov *et al.* 1973, Lapitskii *et al.* 1974, Marchenko *et al.* 1980, 1984, Vil'kovich & Pozharitskaya 1982). On the other hand, much primary apatite from carbonatite is low in REEs, e.g. Iron Hill, Alnö and many Soviet examples (Nash 1972b, Kapustin 1977). Secondary carbonate-hydroxyl-fluorapatite contains very little REEs (analysis 5, Table 6.6; Rimskaya-Korsakova *et al.* 1979). Kapustin (1977) maintains that, with the exception of secondary, carbonate-hydroxyl-fluorapatite ('staffelite'), ΣREE remained relatively constant and low during carbonatite development. Rimskaya-Korsakova *et al.* (1979) noted that, at Kovdor, total REEs increase from early to late carbonatite (0.17 → 0.70 ΣREE$_2$O$_3$ + Y$_2$O$_3$).

Much work has been done on REE distribution, and a tabular listing of 939 REE analyses of apatite was given by Fleischer & Altschuler (1982) and summarized by Fleischer & Altschuler (1986). Rare earth characteristics are summarized in Table 6.7. The high ratio of LREE:HREE is characteristic of carbonatites. Most apatite specimens from carbonatites have La:Yb = 150–200:1 but some from Oka, and Gatineau, have ratios of 300:1 and more. Typical chondrite-normalized plots are given in Figure 6.6.

Apatite from carbonatite in the Kola Peninsula and Siberia shows an increase in HREEs, from early to middle stages of development (Balashov & Pozharitskaya 1968), the reverse of the trend outlined by Rimskaya-Korsakova *et al.* (1979), a feature ascribed to changes in alkalinity of the carbonatite magma. Secondary apatite from the youngest carbonatites near Gatineau, relatively rich in Nd, with respect to La and Ce (Hogarth 1983), may reflect a bias towards the HREEs. Such a trend may be typical of secondary apatite. 'Botryoidal apatite in laterite overlying carbonatite in Western Australia', has ΣREEs of 0.18% and La:Yb = 0.03:1 (Roeder *et al.* 1987). The only analysed apatite from volcanic carbonatite, fluorapatite from Kerimasi, Tanzania, has ΣREEs of 0.52% and also has weak LREE enrichment (La:Yb = 19:1; Roeder *et al.* 1987).

Mariano & Ring (1975) and Mariano (1978) have shown that apatite from carbonatite, unlike apatite from common igneous and metamorphic rocks, luminesces blue with cathode-ray excitation, due to the emission by Eu^{2+} at 410–430 μm. Moroshkin & Gorobets (1985) determined Eu^{2+}:total Eu from luminescence

Table 6.7 REE distribution in apatite from some carbonatites.

No.	Reference	Area(s)	Method	ΣREE(%)	La:Yb	δCe	δEu
1	Eby (1975; BOK-7)	Oka, Canada	XRF	2.4	310	0.8	0.6
2	Eby (1975; OI-1)	Oka, Canada	XRF	3.0	270	1.0	0.5
3	Mariano (1985)	Bond zone, Oka, Canada	NAA	7.7*	565	1.0	1.0
4	Hogarth (unpub.)	Gatineau Park, Canada	DCP	0.5*	170	0.9	0.6
5	Hogarth (1988, AP 12A)	Perkins, Canada	DCP	0.8*	450	1.0	1.0
6	Vil'kovich & Pozharitskaya (1982; 16)	Chernigov, Zone 5, Ukraine	NAA	8.3*	550	1.0	1.2
7	Kapustin (1977)	various localities, USSR	OES	0.5	100	1.0	1.0
8	Landa et al. (1983)	various localities, USSR	OES	0.2	200	0.9	1.0
9	Puchelt & Emmermann (1976; 10)	Kaiserstuhl, Germany	NAA	0.8*	140	0.35	1.0
10	Puchelt & Emmermann (1976; 11)	Kaiserstuhl, Germany	NAA	0.7*	210	0.35	1.1

*Involves extrapolation and/or interpolation. Abbreviations: DCP, direct-current plasma; NAA, neutron activation; OES, optical emission spectroscopy; XRF, X-ray fluorescence. δCe = measured Ce; interpolated Ce, δEu = measured Eu: interpolated Eu.

characteristics, which they considered a suitable indicator of redox conditions. The Eu^{2+} : total Eu values for apatite from carbonatite of the Kola Peninsula are grouped around 0.51:1. However, apatite from carbonatite of eastern Siberia averaged 0.35:1, a value in the range of most igneous and metamorphic rocks.

Philpotts (1970) assumed equal interphase partitioning of Eu^{2+} and Sr^{2+} for two minerals in equilibrium, and calculated $Eu^{2+}:Eu^{3+}$ by measuring Sr and Eu in each phase. Eby (1975) applied this method to calcite and apatite from Oka carbonatite and found virtually all Eu to be Eu^{3+}. No attempt has been made to compare the two methods on apatite from the same rock.

Negative Ce and Eu anomalies occur in apatite from carbonatite (Eby 1975, Puchelt & Emmermann 1976). These are reflected in δCe (measured Ce : interpolated Ce) and δEu values (measured Eu : interpolated Eu) listed in Table 6.7. Values between 0.75 and 0.5 can be regarded as weak negative anomalies and those < 0.5 are strong negative anomalies. In Figure 6.6, a Ce anomaly is evident in curve 9, and Eu anomalies in curves 2 and 4. Eby (1975), Puchelt & Emmermann (1976), and Kovalenko *et al.* (1982) explain these negative Eu anomalies by exclusion of Eu^{2+}

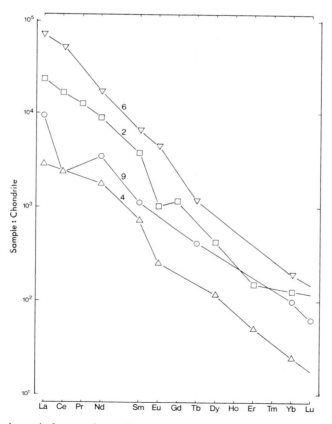

Figure 6.6 REE abundances in four apatite specimens (numbered as in Table 6.7). Analyses of Lu (analysis 6), Tb and Ho (analysis 2) have been omitted for simplicity. Normalization is based on chondritic values of Boynton (1984).

from the apatite structure. However, rejection of Eu^{2+} is not in agreement with crystallochemical principles (bond strength and ionic radius) and some specimens showing strong Eu^{2+} cathodoluminescence have no perceptible Eu anomaly. Puchelt & Emmermann (1976) contend that negative Ce anomalies may be due to extraction of Ce by another phase (such as allanite, monazite, or titanite). The large negative Ce anomalies found by these authors at Kaiserstuhl have not been reported elsewhere.

In apatite from phoscorite (early), calcitic carbonatites (intermediate), and dolomitic carbonatites (late), Rimskaya-Korsakova *et al.* (1979) found that La was strongly depleted in the earliest phoscorites, and Sm, Eu and Gd depleted in stage II and III carbonatites. No attempt was made to explain this behaviour.

Le Bas & Handley (1979), on the basis of chemical analyses of apatite from carbonatite and pyroxenite–ijolite–urtite from western Kenya, suggested an origin of carbonatite by fractional crystallization of an ijolite magma and evolution as a pair of immiscible, conjugate liquids. The latest crystallate from the carbonatite fluid would be notably enriched in REEs. Two separate lineages were also suggested by Landa *et al.* (1983) who compared average apatite analyses from glimmerite rocks of the ijolite series, calcitic carbonatite, and phoscorite. The authors concluded that the trends suggest apatite crystallized from two different but genetically related systems, namely phoscorite–carbonatite and ijolite–glimmerite. Geothermometry and paragenetic studies, however, suggest that the two series progressed *towards* and *converged* at ijolite.

Iron (normally 0.0X% Fe) and Mg (0.0X–0.X% MgO) also appear to substitute for Ca in the apatite structure. Sodium is commonly present at 0.X% in primary apatite and 0.0X% in secondary apatite (analysis 5, Table 6.6) from carbonatite. Apatite from 'alkalic ultrabasic rock and carbonatite' (Petrov & Zhuravel 1971) averages 0.36% Na, distinctly higher than the value compiled for apatite from common igneous and metamorphic rocks. It is also higher than in apatite from marble and regionally metamorphosed skarn, and approximately the same as for primary apatite from carbonatite and fenite near Gatineau, Quebec (Hogarth 1988). Marchenko *et al.* (1975), without defining compositional limits, note that the high Na content of (primary) apatite is distinct from that of other rock types.

There is some evidence that Ce, Na, and Sr are *ordered* within specific Ca sites. Klevtsova & Borisov (1964) assigned Na^+ and Ce^{3+} to Ca_I, and Sr^{2+} to Ca_{II} in the apatite structure of belovite, ideally $_I(Na_2Ce_2)_{\Sigma=4II}Sr_6(PO_4)_6(OH)_2$. However, re-assignment of Na^+ to Ca_{II} and Ce^{3+} to Ca_I would lower oversaturation of the positive bond on O (by applying Pauling's second rule). Borisov & Klevtsova (1964) suggest that, at low concentrations, all three elements, Ce, Na, and Sr, belong with Ca_{II}. Khudolozhkin *et al.* (1972) assigned Sr and Ce to Ca_{II}, and Na to Ca_I. The last two proposals would result in excess oversaturation of positive bonds. Perfect bond-charge balance in britholite is attained with the following assignment: $_ICe_{4II}(Ce_2Ca_4)_{\Sigma=6}(SiO_4)_6(OH)_2$.

The abundance of U and Th is low in apatite from carbonatite. Puchelt & Emmermann (1976) report 23 and 15 p.p.m U, and 18 and 22 p.p.m. Th for two specimens from Kaiserstuhl. Bell *et al.* (1987) report up to 1.79 p.p.m. U and 14.1

p.p.m. Th from apatite in the Borden carbonatite, Canada. Bulakh (1984) reports 0.13% ThO_2 in apatite from phoscorite of Kovdor. This is the highest Th content in apatite from carbonatite and closely related rocks known to the author. Apatite from eight specimens from the Oka carbonatite gave 1.6–105 p.p.m. U by neutron activation (Gold *et al.* 1986).

An unusual constituent in apatite from carbonatite breccia at Kovdor (analysis 5, Table 6.6) is Cu, presumably present essentially as Cu^{2+} and responsible for the blue–green colour of the specimen. In addition, Rimskaya-Korsakova (1965) reports spectrographic analyses of 30–8000 p.p.m. Cu, 300–500 p.p.m. Ba, and 1–5 p.p.m. Mn. Pozharitskaya (1962) records up to 100 p.p.m. Pb in fluorapatite from stage 1 carbonatite, and up to 500 p.p.m. Be and 100 p.p.m. Pb (spectrographic analyses) from carbonate–fluorapatite of post-carbonatite veins at an unidentified carbonatite in Siberia.

6.3.4 P' ions

Probably all carbonatite apatites contain appreciable C in their structure, because they have been precipitated from a carbonate-rich fluid. Late-generation apatite is particularly rich in C (up to several per cent CO_2; e.g. analyses 5, 8, 12, Table 6.6). McConnell (1973) entertains replacement of $(PO_4)^{3-}$ groups by $(O_4H_4)^{4-}$. Borneman-Starynkevich & Belov (1953) have suggested that C is co-ordinated with three O and one (F, OH), thereby explaining F in excess of available F' positions (e.g. analysis 12, Table 6.6). The two groups, PO_4 and CO_3F, have identical charge.

In carbonatite apatite Si is variable but commonly in high concentration (0.X–X% SiO_2). Typical are the values quoted by Petrov & Zhuravel (1971) and Bulakh (1984), Sommerauer & Katz-Lehnert (1985) and those of this chapter (analyses 1, 3 & 6, Table 6.6). These values overlap those of apatite from common igneous and metamorphic rocks (Petrov & Zhuravel 1971; Hogarth 1988).

Sulphur is present in small amounts (0.0X–0.X%, rarely X% SO_3). Its abundance is low in apatite from the Kola Peninsula (Rimskaya-Korsakova 1965) but reaches a maximum of 2.98% SO_3 in an analysis of Petrov & Zhuravel (1971) used in their compilation of apatite from alkalic ultrabasic rock and carbonatite. In analysis 1, Table 6.6, 0.5% S^{6+} is required for charge balance.

Niobium is present (0.13–0.16% Nb_2O_5) in three apatite specimens from pyrochlore-rich carbonatite at Iron Hill, Colorado (Nash 1972). However, these abundances may be due to inclusions of Nb-rich minerals. Khodyreva (1972) determined 0.105% Nb_2O_5 and 0.001% Ta_2O_5 in apatite from a residual carbonatite cap in eastern Siberia but concluded that these were probably due to impurities. An identical conclusion was reached by Anastasenko *et al.* (1978), who analysed apatite (0.00016–0.0005% Ta) from carbonatite of the Turii Peninsula. Niobium in fluorapatite from a uranpyrochlore-bearing zone in Gatineau Park, Quebec is below analytical detection (0.002% Nb; D. D. Hogarth, unpublished data).

Zirconium (480–1580 p.p.m.) was found in apatite from four African carbonatites (Prins 1973) and it was suggested that Zr might replace P.

Arsenic appears to be very low. Fluorapatite from carbonatite of Western

Australia (occurrence unspecified), Kerimasi, Blue River, and Oka ranged from 0.001 to 0.0002% by proton probe (Roeder et al. 1987). Late apatite from Gatineau contains 0.03% As (Hogarth et al. 1985).

6.3.5 F' ions

Normally, F predominates over all F' ions. In many carbonate–apatites it exceeds two atoms per formula unit and the 'extra' F has been assumed to replace O co-ordinated with C in the P site. On the other hand, some apatites from Kaiserstuhl are particularly low in F, with values as low as 0.55% (Sommerauer & Katz-Lehnert 1985). Low-fluorine apatite is also recorded from the Ozernaya Varaka carbonatite (0.93% F), and from phoscorite at Kovdor (0.88% F) and Essei, USSR (0.62% F) (Rimskaya-Korsakova 1965, Borodin et al. 1973).

McConnell (1973) has suggested that C is present in the structure as (a) subvertical CO_3 groups, with C replacing P; and (b) horizontal CO_3 groups, with C positioned on a triad axis. Groups (a) and (b) are in the ratio 3 : 1 and the charge is balanced by H_3O on the triad axis replacing Ca_I. This model has the advantage of explaining the positive correlation of birefringence with C-content, but it is inconsistent with the relative ionic radii of H_3O^+ and Ca^{2+}, and it fails to explain the 'extra' F in some analyses of carbonate-apatite.

Rimskaya-Korsakova et al. (1979) traced the enrichment of F in apatite during development of carbonatite at Kovdor. Apatite from the earliest phoscorite averaged 0.93% F, from carbonatite of stages I and II 0.47%, from stages III and IV 1.80%, and from post-ore breccias 2.60%.

In contrast to F, Cl is low and commonly below the detection limit of present analytical methods. Kaiserstuhl apatite has a relatively large percentage of Cl (up to 0.20% in carbonatite; Sommerauer & Katz-Lehnert 1985). Du Toit (1931) reported chlorapatite (6.45% Cl) from a specimen collected near April Kop, Phalaborwa and, 'although fluorine was tested for by a delicate method, only a trace was found'. Subsequent search has failed to verify either chlorapatite or high-chlorine fluorapatite at Phalaborwa.

A method has been devised (Korzhinskiy 1981) to determine f_{HF} and f_{HCl} of the fluid phase during carbonatite emplacement, if F and Cl in apatite, total P, T, and f_{H_2O} are known. The principle has been applied to carbonatite near Perkins, Quebec by Hogarth et al. (1987). Geothermometers based on partition of F between coexisting apatite and biotite have been proposed (Stormer & Carmichael 1971, Chernysheva et al. 1976, Ludington 1978).

6.3.6 Coupled substitution

Apatite from carbonatite involves at least five coupled substitution mechanisms:

1 $Ca^{2+} + Ca^{2+} \rightarrow Na^+ + REE^{3+}$
2 $Ca^{2+} + P^{5+} \rightarrow REE^{3+} + Si^{4+}$
3 $P^{5+} + P^{5+} \rightarrow (Si, C)^{4+} + Si^{6+}$

4 $P^{5+} + O^{2-} \rightarrow (Si, C)^{4+} + (OH, F)^-$
5 $P^{5+} + F^- \rightarrow C^{4+} + \square$

Scheme 1, the 'belovite scheme', is common in apatite from carbonatite and is dominant in the compositions of apatites 4 and 8, Table 6.6. It characterizes the replacement in apatite from the Gatineau area, Quebec (Hogarth *et al.* 1985). Scheme 2, the 'britholite scheme', is far less common in apatite from carbonatite but is the dominant scheme for specimen 1, Table 6.6, from the 'Britholite Zone', Oka. The remaining $(Si, C)^{4+}$ in this specimen can be paired against Si^{6+} in scheme 3, to restore charge balance. Apatite 6 is mainly an example of scheme 4 and, subordinately, schemes 1 and 2. Scheme 4 requires additional water for OH (which would improve the total by 0.26%). Analysis 5 leads to a deficit in F' ions which can be explained by scheme 5. The other analyses in Table 6.6 can be treated in a similar manner. Coupled substitution of Kaiserstuhl apatite is considered in detail by Sommerauer & Katz-Lehnert (1985).

6.3.7 Zoned crystals

Apatite crystals from both carbonatite and fenite in the Gatineau area, Quebec, are strongly zoned, commonly with REE and Na enrichment near the rim, Sr and F enrichment in the core. A microprobe traverse across a crystal was described by Hogarth *et al.* (1985). Data for another crystal from the same carbonatite are illustrated in Figure 6.7. Zonation is shown by bright-blue cathodoluminescence of the border zone, presumably mainly due to the effect of Eu^{2+} (Mariano & Ring 1975, Mariano 1978), and a 'dead' core. The zonal sequence differs from that of the

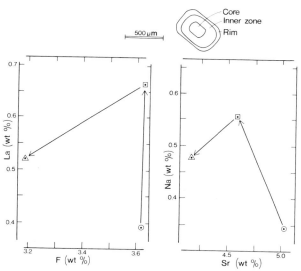

Figure 6.7 Fluorapatite grain showing three distinct zones from dolomite–calcite–barite carbonatite of the Gatineau area, Quebec. Arrows indicate direction of crystal growth. Diagram is based on 40 step-count analyses along the length of the grain. Analyst, T. N. Solberg, Virginia Polytechnic Institute.

apatite from Blackburn, Ontario; this apatite is normally enriched in F on the rim and REEs and Sr are enriched in the core (Hogarth *et al.* 1988).

6.4 AMPHIBOLES

6.4.1 Introduction

Amphiboles occur throughout the entire carbonatite range. The group appears sporadically in early calcite–clinopyroxene carbonatite (Stage I), reaches a maximum in 'tetraferriphlogopite' carbonatite (Stage III) and persists into the post-carbonatite, calcite-vein stage. During this development, amphiboles vary from early calcic aluminous to late alkalic aluminium free. The amphibole group is also found in some phoscorite and as phenocrysts in certain volcanic carbonatites.

Reviews of amphiboles from carbonatite have been given by Bulakh (1965), Kovalenko (1968), Samoylov *et al.* (1974), Samoylov & Gormasheva (1975), Samoylov (1977), Fabriès (1978), and Kapustin (1980).

For comparative purposes and classification, amphibole formulae have been calculated, or recalculated, into 13 $(C + Z)$ cations (wet-chemical analyses) or 13 $(C + Z)$ cations and 46 positive charges (microprobe analyses) in the standard formula unit $A_{0-1}B_2C_5Z_8O_{22}W_2$, that permits calculation of oxides into wt% FeO and Fe_2O_3 (Neumann 1976, Droop 1987). The IMA nomenclature (Leake 1978) with its recommended order of cation assignment is used throughout. Twelve analyses given in Table 6.8, are typical of amphiboles from carbonatite and closely related rocks. Included in the Table are atomic parameters m (100 Mg : ΣC ions), the 'magnesium index' of Samoylov (1977), mf (Mg : [Mg + Fe^{2+}]), used in the IMA classification, n (100 [Na + K] : [Na + K + Ca]), the 'alkalinity index' of Samoylov (1977), and o (100 Fe^{3+} : total Fe), a semiquantitative oxidation parameter.

6.4.2 General composition and classification

The following amphiboles have been identified from carbonatite: calcic amphiboles – magnesio-hastingsite, magnesian hastingsite, edenite, edenitic hornblende; sodic-calcic amphiboles – magnesio-katophorite, pargasite, ferroan pargasite, winchite, richterite; sodic amphiboles – magnesio-riebeckite, riebeckite, magnesio-arfvedsonite, arfvedsonite. Two unusual (one-occurrence) amphiboles are winchite (Jans & de Bethune 1968) and arfvedsonite (Sage 1983*e*).

Other amphiboles reported without chemical analyses are: actinolite (optically determined), tremolite (optically determined), kaersutite (optically determined), ferro-hornblende ('barkevikite', optically determined), glaucophane and ferri-katophorite. The names katophorite, eckermannite, 'hornblende', and hastingsite, along with chemical analyses, have appeared in the literature, but identification was based on pre-1978 definitions.

The most thorough investigations of amphiboles from carbonatite were made by Samoylov *et al.* (1974), Samoylov & Gormasheva (1975), and Samoylov (1977). In

Table 6.8 Chemical composition of typical amphiboles from carbonatite and related rocks.

Stage	I					II				III, IV		Post-carb.	
	1	2	3	4	5	6	7	8	9	10	11	12
Wt%												
SiO_2	37.81	45.2	45.87	48.38	52.14	54.37	55.86	54.52	53.69	54.37	56.70	54.08
TiO_2	1.72	1.42	1.78	1.48	0.74	0.24	0.07	0.10	0.08	0.09	0.46	0.26
Al_2O_3	17.33	8.06	7.52	6.39	4.03	1.80	0.10	0.43	0.76	0.81	0.57	0.77
Cr_2O_3	n.a.	n.a.	0.08	n.a.	n.a.	0.00	n.a.	n.a.	0.03	0.00	n.a.	n.a.
Fe_2O_3	4.52	0.00c	8.48c	5.56	6.71	4.06	7.12c	5.13c	9.05c	12.63c	8.87c	18.24
FeO	4.04	9.78c	7.90c	9.82	6.40	3.01	0.00c	0.55c	5.26c	1.42c	0.00c	6.00
MnO	0.11	0.24	0.18	0.16	0.33	0.12	0.07	0.07	0.06	0.13	0.12	0.13
MgO	16.36	15.0	12.96	13.07	15.52	19.19	20.42	21.16	15.72	16.64	19.49	10.01
BaO	n.a.	n.a.	n.a.	n.a.	0.00	0.00	0.00	0.00	n.a.	0.05	n.a.	n.a.
SrO	n.a.	n.a.	n.a.	n.a.	0.03	0.00	0.06	0.01	n.a.	0.00	n.a.	n.a.
CaO	12.90	11.5	7.02	6.55	4.49	5.92	3.32	5.07	1.64	0.52	4.32	0.35
Na_2O	3.47	4.41	5.49	6.74	7.31	6.53	5.88	6.17	8.60	8.96	6.02	6.44
K_2O	0.73	0.50	0.68	0.82	0.76	1.05	4.27	2.84	1.62	1.67	0.98	0.66
F	n.a.	n.a.	n.a.	0.41	0.41	1.35	2.05	1.70	n.a.	3.22	n.a.	n.a.
Cl	n.a.	n.a.	n.a.	n.a.	n.a.	n.a.	0.00	0.01	n.a.	0.00	n.a.	n.a.
H_2O^+	n.a.	n.a.	n.a.	1.38	1.60	1.3	1.15c	1.29c	n.a.	0.57c	n.a.	n.a.
H_2O^-	n.a.	n.a.	n.a.	0.11	n.a.	n.a.	n.a.	n.a.	n.a.	n.a.	n.a.	n.a.
L.O.I.	1.24	n.a.	n.a.	n.a.	n.a.	n.a.	n.a.	n.a.	n.a.	n.a.	n.a.	3.26
Total	100.23	96.1	97.96	100.87	100.20	98.94	100.37	99.05	96.51	101.08	97.53	100.20
O = F, Cl				0.17	0.17	0.57	0.86	0.72		1.36		
				100.70	100.03	98.37	99.51	98.33		99.72		
o	50	0	49	34	49	55	100	89	61	89	100	73
mf	0.88	0.73	0.75	0.70	0.81	0.92	1.00	0.99	0.84	0.95	1.00	0.75
m	70	67	57	57	66	82	86	90	68	71	81	44
n	36	43	60	67	76	69	82	74	91	97	74	97
Method	WC	EM	EM	WC	WC	XRF EM, WC	EM	EM	EM	EM	EM	WC

1, *Magnesio-hastingsite*, Turii peninsula, USSR; associated with pyroxene, biotite, and calcite (Samoylov 1977, No. 24, p. 61). 2, *Edenite*, Sokli, Finland; in phoscorite (Vartiainen 1980). 3, *Magnesio-katophorite*, Goldray carbonatite, Ontario; ave. of 2, with magnetite and calcite (Sage 1983c). 4, *Magnesio-katophorite*, Turii peninsula, USSR; with pyroxene, biotite, and calcite (Samoylov 1977, No. 2, p. 108). 5, *Richterite*, Iron Hill, Colorado; with pyroxene and calcite (Larsen 1942). 6, *Richterite*, Homa Bay, Kenya; with pyroxene and calcite (Sutherland 1969). 7 *Potassium richterite*, Gatineau, Quebec; with zoned crystal with ferriphlogopite, fluorapatite, and calcite (Hogarth *et al.* 1987). 8, *Potassium richterite*; rim of crystal No. 7. 9, *Magnesio-arfvedsonite*, Argor carbonatite, Ontario; ave. of 2, with calcite (Sage 1983a). 10, *Fluormagnesio-arfvedsonite*, Gatineau, Quebec; with aegirine, barite, calcite, ankerite, fluorapatite, and monazite (Hogarth, unpublished). 11, *Magnesio-riebeckite*, Cargill carbonatite, Ontario; with magnetite, fluorapatite, calcite, and ankerite (Allen 1972). 12, *Magnesio-riebeckite*, Turii peninsula, USSR; from 'zeolitic facies carbonatite'. Total includes 0.06% P_2O_5 (Samoylov 1977, No. 3, p. 206).

Abbreviations: $m = 100$ Mg : ΣC ions, $mf = $ Mg : (Mg + Fe^{2+}), $n = 100$ (Na + K) : (Na + Ca + K), $o = 100$ Fe^{3+} : total Fe (all atomic). EM, electron microprobe; WC, wet chemical; XRF, X-ray fluorescence; c denotes calculated value; n.a., not analysed.

the last publication, Samoylov considered 96 amphibole analyses but, of these, 75 only can be regarded as derived from carbonatite proper and the closely related rocks phoscorite and glimmerite. His most important conclusions are summarized below:

(a) Early amphiboles have considerable Al (per cents) mainly as $_{IV}$Al (as much as 2.6 atoms per formula unit; commonly > 2 atoms per formula unit). $_{VI}$Al is comparatively minor. 'Hastingsite' (= magnesio-hastingsite) is the type mineral; 'katophorite' (= magnesio-kataphorite) is comparatively rare and forms at lower temperatures, tending to replace magnesio-hastingsite. Analyses 1 and 4, Table 6.8 are characteristic.
(b) Most high-aluminium amphiboles are also rich in Ti.
(c) The aluminium-rich amphiboles are best developed in the apical parts of the complexes.
(d) Amphiboles developed in middle- to late-stage carbonatites are low-aluminium and can be divided into richterite–magnesio-arfvedsonite ($A + B > 2.70$ cations) and 'riebeckite' (= magnesio-riebeckite–magnesio-arfvedsonite; $A + B < 2.70$ cations) rows, a complete series extending from one row, through magnesio-arfvedsonite, to the other (Fig. 6.8). The richterite–magnesio-arfvedsonite row is the most common and characterizes the middle to late stage of carbonatite development.
(e) Soda amphiboles from carbonatite contain more Mg ($m = 42–100$) than those from agpaitic rocks (normally $m = 0–10$).
(f) The general trend for richterite–arfvedsonite involves an increase in alkalis and Fe with carbonatite development. Some Fe^{3+} may enter the tetrahedral sites.
(g) Compositions of most low-aluminium amphiboles fall within a narrow triangle, bounded by riebeckite–magnesio-arfvedsonite–richterite, on a $_{VI}R^{3+}$ v. $R^+ + R^{2+}$ plot (Fig. 6.8).

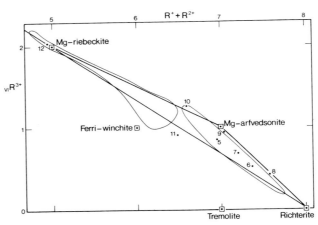

Figure 6.8 $_{VI}R^{3+}$ v. $R^+ + R^{2+}$ plot of low-aluminium amphiboles. Approximate fields of Samoylov & Gormasheva (1975) are outlined. Low-aluminium amphiboles of Table 6.8 (analyses 5–12) are shown.

(h) Tremolite and actinolite are not found in carbonatites. The lowest alkalinity index, n, for low-aluminium amphiboles is 62 (in richterite from calcite–dolomite at Kovdor). Low-aluminium amphiboles with $n < 80$ are very rare in carbonatites.

(i) Ti is not important in low-aluminium amphiboles and rarely attains 4 wt%.

(j) The order of formation of low-aluminium amphiboles is richterite → magnesio-arfvedsonite → magnesio-riebeckite. The very latest amphibole is transitional from riebeckite to magnesio-riebeckite (e.g. analysis 12, Table 6.8) from post-carbonatite veins (the 'zeolitic carbonatite facies' of Samoylov).

6.4.3 Comments on the compositional trends outlined by Samoylov

The principal amhibole trends for carbonatite are outlined on the $Ca + {}_{IV}Al$ v. $Si + Na + K$ plot in Figure 6.9. With the exception of rare edenite, magnesio-katophorite, and magnesio-riebeckite, analyses fall into two domains:

(a) high-aluminium amphiboles in the field extending from the hypothetical compound $NaCa_2(Mg_3Fe_2^{3+})(Si_5Al_3)O_{22}(OH)_2$ (symbol 'Hal') to near edenite, implying the substitution $_cFe^{3+} + {}_zAl^{3+} \rightarrow {}_cMg^{2+} + {}_zSi^{4+}$; and

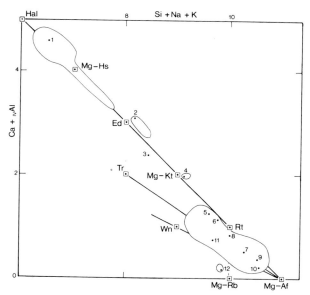

Figure 6.9 $Ca + {}_{IV}Al$ v. $Si + Na + K$ plot, after Fabriès (1978), of amphiboles from carbonatites. Approximate fields are outlined. Data taken from the following sources: *winchite–richterite–magnesio-arfvedsonite* field, 61 analyses (Fabriès 1978); *magnesio-riebeckite* field, two analyses from Samoylov (1977); *magnesio-katophorite* field, three analyses (Samoylov 1977); *edenite* field, two analyses (Vartiainen 1980); *magnesio-hastingsite* field, 23 analyses (Samoylov 1977), eight 'xenocrystic amphibole' analyses (McCormick & Heathcote 1987), one 'hornblende' (Gold 1966). The 12 points derived from data in Table 6.8 are superimposed. Mineral abbreviations: Ed, edenite; Mg-Af, magnesio-arfvedsonite; Mg-Hs, magnesio-hastingsite; Mg-Kt, magnesio-katophorite; Mg-Rb, magnesio-riebeckite; Tr, tremolite; Wn, winchite; Hal, $NaCa_2(Mg_3Fe_2^{3+})(Si_5Al_3)O_{22}(OH)_2$ (hypothetical compound).

(b) low-aluminium amphiboles extending from near magnesio-arfvedsonite towards tremolite, implying major substitution of the type $_A\text{Na}^+ + 2_B\text{Na}^+ + _c\text{Fe}^{3+} \rightarrow {_A}\square + 2_B\text{Ca}^{2+} + _c\text{Mg}^{2+}$.

Compositions of amphiboles in Table 6.8 fit well on the proposed carbonatitic trends and, with the exception of analyses 2, 3, 4, and 12, all lie within the two main compositional fields. In terms of carbonatite development, magnesio-hastingsite (e.g. amphibole 1) would be expected to form early, edenite–magnesio-katophorite (amphiboles 2–4) somewhat later, richterite (amphiboles 5–8) later still, and magnesio-arfvedsonite and magnesio-riebeckite (amphiboles 10, 11) last of all. No analysed amphiboles from carbonatite have tremolite–actinolite compositions, minerals that have been tentatively identified in many studies.

Seventy-five analyses of Samoylov have been contoured for distribution of Ca and Fe in Figure 6.10. Superimposed on this diagram are data from the 12 amphiboles of Table 6.8, eight xenocrystic amphiboles of McCormick & Heathcote (1987), and a 'hornblende' of Gold (1966).

Amphibole compositions fall into two main fields: low-calcium corresponding to low-aluminium; and high-calcium, corresponding to high-aluminium (Samoylov &

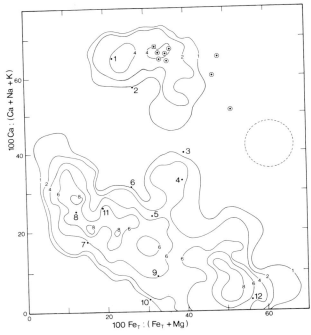

Figure 6.10 Contoured 100 Ca : (Ca + Na + K) v. 100 Fe : (Fe + Mg) plot for low-aluminium, low-calcium amphiboles (64 analyses) from Samoylov (1977) and high-aluminium, high-calcium amphiboles (11 analyses) from Samoylov (1977). Plots of the 12 analyses of Table 6.8 (points), eight analyses of 'xenocrystic amphiboles' from McCormick & Heathcote (1987; double circles), and one analysis of 'hornblende' (Gold 1966; square) are superimposed for comparison. Contours have been positioned according to the number of data points within a circle (shown on diagram).

Gormasheva 1975). An ill-defined area with intermediate Ca and Fe values, corresponding to magnesio-kataphorite, tends to bridge the two major domains.

The low-calcium domain has a general negative Fe–Ca slope and extends from magnesio-riebeckite and most magnesio-arfvedsonites (lower right), to richterite (upper left). Amphibole 11 is very different from amphibole 12 (also apparent in Figs 6.8 & 6.9). Both are magnesio-riebeckite but amphibole 11 has definite links with magnesio-arfvedsonite and especially richterite.

Teal (1979) notes that, in the Firesand River Complex, Canada, richterite occurs in glimmerite and the peripheral parts of carbonatite, whereas magnesio-arfvedsonite characterizes the central part of the carbonatite. A similar relationship occurs in carbonatite and glimmerite near Meech Lake, Quebec (D. D. Hogarth, unpublished data). Potassium richterite is also known from glimmerite in the Siilinjarvi Complex, Finland (Puustinen 1972) and from Gatineau, Quebec (analyses 7, 8; Table 6.8).

In contrast to low-calcium amphiboles, the high-calcium domain is tightly constrained, though based on a limited number (11) of analyses. Analysis 2 (Table 6.8), 'hornblende' of Gold (1966), and five of the eight xenolithic amphiboles analysed by McCormick & Heathcote (1987) also fall within this domain.

6.4.4 Tetrahedral iron

A number of authors have assigned Fe^{3+} to tetrahedral sites in alkali and sodic-calcic amphiboles, notably the sodic-calcic amphiboles from carbonatite of the Kursk Magnetic Anomaly, USSR, where up to 0.5 atoms per formula unit were assigned in this manner (Kurenkina 1976). Hawthorne (1983) points out that the sole evidence for this assignment is a low total of other tetrahedral ions in the calculated formula, and the analyses may, themselves, be at fault. However, the normal environment of carbonatites of middle-stage development (high oxidation, low Al activity) is favourable for the introduction of tetrahedral Fe in silicates. Many reported amphiboles bearing tetrahedral Fe are associated with 'tetraferriphlogopite', a variety of phlogopite with well-established $_{IV}Fe^{3+}$ (Hogarth et al. 1970). In this regard, amphiboles 7 and 8, Table 6.8 (with calculated $_{IV}Fe^{3+}$ 0.08 and 0.14 atoms per formula unit, respectively), associated with 'tetraferriphlogopite' and normal phlogopite, have been shown by Mössbauer spectroscopy to contain tetrahedral Fe (Hogarth et al. 1987).

6.4.5 Potassium and F, and a possible compositional gap

Although most amphiboles from carbonatite or closely related rock contain little K, some contain > 0.5 atoms K per formula unit, and therefore warrant the modifier potassium before their name (Leake 1978), viz. potassium ferroan pargasite and potassium magnesian hastingsite (McCormick & Heathcote 1987); potassium magnesio-arfvedsonite (Teal 1979, Secher & Larsen 1980, Mian & Le Bas 1986); potassium richterite (Larsen 1942, Teal 1979, Hogarth et al. 1987, and analyses 7, 8, Table 6.8). These varieties must reflect significant K in the parent magma.

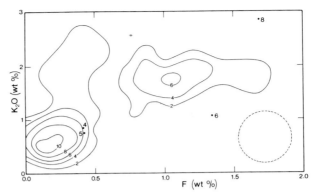

Figure 6.11 Contoured K₂O v. F (wt%) diagram for 45 analyses of low-aluminium amphiboles. The compilation comprises all amphiboles for which fluorine analyses were recorded by Samoylov (1977). Superimposed are plots from analyses 4, 5, 6, and 8 of Table 6.8 (points). Contours have been positioned according to the number of data points within a circle (shown on the diagram).

In fenite near Cantley, Canada, potassium-, fluorine-rich magnesio-arfvedsonite was found near magnesio-riebeckite, comparatively rich in Fe and poor in K and F. The distinct compositions were ascribed to different palaeosomes (Hogarth & Lapointe 1984). Similar magnesio-riebeckite near Gatineau, Canada, forms sharply defined overgrowths on magnesio-arfvedsonite, suggesting a compositional gap between the two minerals or precipitation of an overgrowth at lower temperatures (Hogarth *et al.* 1987). It is also known that some arfvedsonite from agpaite in southwest Greenland is composed of a submicroscopic intergrowth of two amphiboles (Sahama 1956).

The possible occurrence of two coexisting amphiboles in carbonatite can be investigated by contouring a K₂O – F diagram (Fig 6.11). Bimodal distributions are apparent in Figures 6.10 and 6.11. The modes match magnesio-arfvedsonite and magnesio-riebeckite trends for the amphiboles from the Cantley fenite (Hogarth & Lapointe 1984). The data are consistent with magnesio-arfvedsonite and magnesio-riebeckite representing immiscible solids. Carbonatites may have formed at temperatures near the solvus dome whereas the Cantley fenites may have formed well below the solvus.

6.4.6 *Minor elements*

Of the various constituents listed in Table 6.9, TiO_2 alone attains one per cent. Titanium has its highest concentration in early, aluminium-rich amphiboles. Amphiboles from west-central Arkansas (e.g. analysis 5, Table 6.9) are unusually titanium-rich, even for magnesio-hastingsite. The amphibole from Valentine Township, Ontario (analysis 6) is even more unusual because it is an aluminium-poor member (0.03% Al_2O_3) containing a large amount of Ti (> 2% TiO_2). Kaersutite, $NaCa_2(Mg_4Ti)(Si_6Al_2)(O_{23}OH)$, has been reported as phenocrysts in carbonatite lava at Fort Portal, Uganda (Kapustin & Polyakov 1982), and kaersutite and riebeckite have been described from carbonatite tuff, in addition to unidentified

Table 6.9 Maximum percentages of some constituents of amphiboles.

No.		Wt%	Method	Locality	Mineral	Rock	Reference
1	Nb_2O_5	0.034	OS	Siberia	Mg-Af	Cbt	Pozharitskaya (1962)
2	Nb_2O_5	0.03	ED	Eastern Siberia		Cbt	Khodyreva (1972)
3	Ta_2O_5	0.0016	ED	Eastern Siberia		Cbt	Khodyreva (1972)
4	ZrO_2	0.06	OS	Afrikanda, USSR		Cbt	Kukharenko (1965b)
5	TiO_2	3.71	EM	Oppelo, Arkansas	Mg-Hs	Cbt	McCormick & Heathcote (1987)
6	TiO_2	2.11	EM	Valentine Township, Ontario	Af	Cbt	Sage (1983e)
7	Cr_2O_3	0.26	EM	Firesand River, Ontario	Mg-Af	Cbt	Teal (1979)
8	BaO	0.12	WC	Iron Hill, Colorado	Mg-Af	Cbt	Larsen (1942)
9	SrO	0.06	EM	Gatineau, Quebec	Rct	Glim	This paper
10	MnO	0.58	WC	Lower Sayansk Massif, USSR	Mg-Rb	Cbt	Samoylov (1977)
11	V	0.1	OS	Siberia	Rb(?)	P(?)	Pozharitskaya (1962)
12	Cu	0.03	OS	Alkalic massif, Kola Pen.	Hbl	Pxt	Bulakh (1965)
13	Be	0.03	OS	Alkalic massif, Kola Pen.	Hbl	Pxt	Bulakh (1965)
14	Sc	0.003	OS	Alkalic massif, Kola Pen.	Hbl	Pxt	Bulakh (1965)
15	Sn	0.005	OS	Siberia	Mg-Af	Cbt	Pozharitskaya (1962)
16	Co	0.005	OS	Siberia	Mg-Af	Cbt	Pozharitskaya (1962)
17	Ga	0.005	OS	Siberia	Mg-Af	Cbt	Pozharitskaya (1962)
18	Be	0.005	OS	Siberia	Mg-Af	Cbt	Pozharitskaya (1962)

Abbreviations Method: ED, electrodialysis + acid solution; EM, electron microprobe; OS, optical spectroscopy; WC, wet chemical. *Minerals:* Af, arfvedsonite; Hbl, 'hornblende' (as given); Mg-Af, magnesio-arfvedsonite; Mg-Hs, magnesio-hastingsite; Mg-Rb, magnesio-riebeckite; Rb, riebeckite (given as 'crocidolite', identification based on optical properties); Rct, richterite. *Rocks:* Cbt, carbonatite; Glim, glimmerite; P, post-carbonatite vein; Pxt, pyroxenite.

amphiboles, in the lava of Catanda, Angola (Silva 1973). No chemical analyses were given. 'Amphibole' has been recorded in volcanic carbonatite at Khanneshin Mountains, Afghanistan (Vikhter *et al.* 1975), Kaiserstuhl, Germany (Keller 1981), and Tinderet, Kenya (Deans & Roberts 1984).

Manganese oxide, normally < 0.25%, reaches 0.6% in amphibole from the Lower Sayansk carbonatite, USSR (analysis 10, Table 6.9).

Chromium is virtually absent in amphiboles from carbonatite, but at Firesand River, Ontario it is consistently above detection limit (0.03% Cr_2O_3; Teal 1979). This reflects the whole-rock composition of 580 p.p.m. in a sövite analysis and 900 p.p.m. in a silicocarbonatite analysis (Sage 1983b). Average Cr values in amphiboles from carbonatite are higher than those from glimmerite (Teal 1979).

Barium and Sr have been analysed in amphiboles from carbonatite in small quantities. The maximum Ba known to the writer is from Iron Hill (Table 6.9, analysis 8). Both Ba and Sr are consistently above background in amphiboles from five carbonatites near Gatineau (which include the carbonatite of analysis 9). These large ions probably belong to the *A* site.

Khodyreva (1972) analysed 35 aliquots of an amphibole from a carbonatite from eastern Siberia. Niobium and, to a lesser extent, Ta gave similar values from all aliquots, which suggests that Nb and Ta substitute isomorphously. Kuz'menko (1961) suggested that these elements enter amphibole through the coupled substitution

$Al^{3+} + 2Fe^{3+} \rightarrow (Nb, Ta)^{5+} + 2Mg^{2+}$. The co-ordination of Nb was not specified but there are difficulties assigning Nb to either octahedral or tetrahedral sites (Parker & Fleischer 1968).

6.4.7 Zoned amphiboles

Zoned amphiboles from carbonatite are of two types: normal and reversed. Normal zoning progresses from a magnesium- and calcium-rich core to an iron- and alkali-rich rim. An extreme case is found in amphiboles of the Sarfartôk carbonatite, Greenland (Secher & Larsen 1980) where richterite is mantled by magnesio-arfvedsonite. The variation from core to rim is mainly $Ca^{2+} + (Mg^{2+}, Fe^{2+}) \rightarrow Na^+ + Fe^{3+}$, indicating that Na and Fe not only increase in the later amphibole but crystallization proceeded from reduction to oxidation, a zonation similar to that observed in amphiboles from carbonatite in the Gatineau Park, Quebec, Canada (D. D. Hogarth, unpublished data).

The amphiboles from carbonatite of west-central Arkansas (McCormick & Heathcote 1987) must be regarded as having complex, reversed zoning, because they have iron-rich cores and magnesium-rich rims. Offsetting the Fe $\rightarrow$ Mg sequence is the zonation $_cAl^{3+}$ (core) $\rightarrow$ $_cFe^{3+}$ (rim), and oxidation is greater in the later amphiboles. Early amphiboles (recalculated to 13 cations and 46 charges) are hastingsitic-to-pargasitic; late amphiboles are hastingsitic. McCormick & Heathcote (1987) conclude that the early, xenocrystic amphiboles with variable composition formed during a period of rapid fluid evolution and that later amphiboles equilibrated with the magma fluid, accounting for their similar compositions.

A more complicated reversed zonal scheme is shown by amphibole analyses 7 and 8 in Table 6.8. The core of this amphibole is relatively rich in K and total Fe but a broad zone near the rim is relatively rich in Ca and Mg. Sodium varies little in these zones and the major change during crystallization can be explained as $_AK^+ + _BNa^+ + _cFe^{3+}$ (core) $\rightarrow$ $_ANa^+ + _BCa^{2+} + _c(Mg, Fe)^{2+}$ (inner rim). Crystallization has progressed with reduction. At the extreme rim (not represented in Table 6.8), there is an abrupt increase in Na, Fe^{3+}, and F at the expense of Ca, Mg, and OH that can be expressed as $_BCa^{2+} + _c(Mg, Fe)^{2+} + _wOH^-$ (inner rim) $\rightarrow$ $_BNa^+ + _cFe^{3+} + _wF^-$ (outer rim). This suggests that crystallization returned to the normal scheme, with increase in oxidation.

6.5 CONCLUSIONS

Each of the three mineral groups is characterized by a wide range of ionic substitution and, therefore, their compositions offer promise in understanding the chemical processes involved in their formation. In addition, REE substitution in apatite and Ta–U substitution in pyrochlore, have economic implications. The chemical trends of these elements during carbonatite development, and distribution of apatite and pyrochlore, both regionally and locally, should be studied in detail.

The chemistry of the amphibole group in carbonatites is well established and immiscibility within the alkali amphiboles is suspected. One possible control here is oxidation, with the more highly oxidized magnesio-riebeckite exsolved on the rims of magnesio-arfvedsonite, though disequilibrium precipitation is equally plausible. Each alternative awaits further testing.

An obvious avenue for future research would be to extend the coverage to include other carbonatite minerals, such as carbonates, micas, pyroxenes, and oxides. Oxidation trends in pyroxenes and micas could be correlated with $Eu^{3+}:Eu^{2+}$ ratios in coexisting apatite, $Fe^{3+}:Fe^{2+}$ ratios in coexisting amphiboles, and the type of Fe–Ti oxide assemblage.

The basic understanding of the structural chemistry of amphibole, apatite, and pyrochlore groups alone is far from satisfactory. For example, tetrahedral Fe, known to be present in amphiboles from fenites, is only suspected in amphiboles from carbonatites, and the role of Fe in pyrochlore minerals is unknown. In both mineral groups, Fe is a candidate for Mössbauer spectroscopy.

Clearly, we have come a considerable distance in the understanding of carbonatite minerals but much remains to be done.

ACKNOWLEDGEMENTS

The author is grateful for the following help in this study: Peter Jones (Carleton University) and Todd Solberg (Virginia Polytechnic Institute) for electron microprobe analyses; John Loop (University of Ottawa) for REE analyses by DCP; Edward Hearn (University of Ottawa) for drafting, and Julie Hayes (University of Ottawa) for typing this chapter. The manuscript was improved after reviews by Keith Bell, Tyson Birkett, J. W. Card, G. Y. Chao, and B. E. Leake.

REFERENCES

Aires-Barros, L. 1968. Sobre a presença de microlite no minério ferro-titanado ocorente na estrutura anelar carbonítica do Bailundo (Angola). *Garcia de Orta* **16**, 375–9.

Allen, C. 1972. *The petrology of the ultramafic suite associated with the Cargill alkaline ultramafic carbonatite complex, Kapuskasing, Ontario.* M.Sc. thesis, University of Toronto, Toronto.

Anastasenko, G. F., A. G. Bulakh, P. A. Vaganov & A. I. Bulnaev 1978. Mineralogy and geochemistry of tantalum in kataforites and carbonatites of the Turii Peninsula (Murmansk region). In *Mineralogy i Paragenezisy Mineralov,* P. M. Tatarinov (ed.), 38–46. Leningrad: Nauka. (in Russian)

Bagdasarov, Yu. A. 1972. Application of nuclear-physical methods for the study of rock, ore and mineral samples. Spatial distribution of uranium and thorium in pyrochlore crystals from carbonatites. *Trudy, Vsesoyuznyi Nauchno-Issledovatel'skii Institut Yadernoi Geofiziki i Geokhimii* **13**, 60–5. (in Russian)

Bagdasarov, Yu. A. 1974. Types of tantalum–niobium ore and some characteristics of their distribution in carbonatites. *Geologiya Rudnykh Mestorozhdenii* **16**, (5), 15–24. (in Russian)

Bagdasarov, Yu. A. & E. L. Danilin 1982. The Dalbykha carbonatite massif. *Doklady Akademii Nauk SSR* **267**, (6), 1440–4. (in Russian)

Balashov, Yu. A. & L. K. Pozharitskaya 1968. Factors governing the behavior of rare-earth elements in the carbonatite process. *Geochemistry International* **5**, 271–88.

Beevers, C. A. & D. B. McIntyre 1946. The atomic structure of fluorapatite and its relation to that of tooth and bone material. *Mineralogical Magazine* **27**, 254–7.

Bell, K., J. Blenkinsop, S. T. Kwon, G. R. Tilton & R. P. Sage 1987. Age and radiogenic isotopic systematics of the Borden carbonatite complex, Ontario, Canada. *Canadian Journal of Earth Sciences* **24**, 24–30.

Bonshtedt-Kupletskaya, E. M. 1966. On the question of systematics of the pyrochlore–microlite group. *Zapiski Vsesoyuznogo Mineralogicheskogo Obshchestva* **95**, 134–44. (in Russian)

Bonshtedt-Kupletskaya, E. M. 1967. The pyrochlore group. In *Mineralii*, Vol. 2 (3), F. Chukhrov (ed.), 145–81. Moscow: Nauka. (in Russian)

Borisov, S. V. & R. F. Klevtsova 1964. Crystallochemistry of some minerals of the apatite group. *Rentgenografiya Mineral'nogo Syr'ya* **4**, 62–6. (in Russian)

Borneman-Starynkevich, I. D. & N. V. Belov 1953. On carbonate–apatite. *Doklady Akademii Nauk SSSR* **90**, 89–92. (in Russian).

Borodin, L. S. & R. L. Barinskii 1961. Rare earth assemblages in pyrochlores from alkalic ultrabasic massifs and carbonatites. *Geochemistry* **10**, 517–24.

Borodin, L. S. & I. I. Nazarenko 1957. Chemical composition of pyrochlore and isomorphic replacements in the $A_2B_2X_7$ molecule. *Geokhimiya* **4**, 278–95. (in Russian with English summary)

Borodin, L. S., A. V. Lapin & A. G. Kharchenkov 1973. *Redkometal'nye kamafority. Formatsiya apatit-forsterit-magnetitovykh porod b shchelochno-ultraosnovnykh i karbonatitovykh massivakh*. Moscow: Nauka. (in Russian)

Borodin, L. S., V. Gopal, V. M. Moralev, V. Subramanian & V. Ponikarov 1971. Precambrian carbonatites of Tamil Nadu, South India. *Journal of the Geological Society of India* **12**, 101–12.

Boynton, W. V. 1984, Cosmochemistry of the rare earth elements: meteorite studies. In *Rare earth element geochemistry. Developments in geochemistry*, Vol. 2, P. Henderson (ed.), 63–114. Amsterdam: Elsevier.

Bragg, Sir L., G. F. Claringbull & W. H. Taylor 1965. *Crystal structures of minerals*. London: G. Bell and Sons.

Brasseur, H., P. Herman & A. Hubaux 1962. Apatites de l'est du Congo et du Ruanda. *Annales de la Société géologique de Belgique* **85**, (2), 61–85.

Bulakh, A. G. 1965. Amphibole group. In *Kaledonskii kompleks ul'traosnovnykh, shchelochnykh porod i karbonatitov Kol'skogo Poluostrova i severnoi Karelia*, A. A. Kukharenko (ed.), 453–63. Moscow: Nedra. (in Russian)

Bulakh, A. G. 1984. On reproducibility of results of chemical analyses of minerals and validity of their formulae (with apatites as example) *Mineralogicheskii Zhurnal* **6**, (4), 87–92. (in Russian)

Chernysheva, Ye. A., L. L. Petrov & L. V. Chernyshev 1976. Distribution of fluorine between coexisting phlogopite and apatite in carbonatites and the behavior of fluorine in the carbonatite-forming process. *Geochemistry International* **13**, (3), 14–22.

Chistov, L. B. & A. F. Denisov 1971. Compositional zoning in pyrochlore crystals from carbonatites. *Nauchnye Trudy, Gosudarstvennyi Nauchno-Issledovatel'skii Proektnyi Institut Redkometallicheskoi Promyshlennosti* **25**, 102–7. (in Russian)

Davies, K. A. 1947. The phosphate deposits of the Eastern Province, Uganda. *Economic Geology* **42**, 137–46.

Deans, T. 1966. Economic mineralogy of African carbonatites. In *Carbonatites*, O. F. Tuttle & J. Gittins (eds), 385–413. New York: Wiley.

Deans, T. & B. Roberts 1984. Carbonatite tuffs and lava clasts of the Tinderet foothills, western Kenya: a study of calcified natrocarbonatites. *Journal of the Geological Society of London* **141**, 563–80.

Droop, G. T. R. 1987. A general equation for estimating Fe^{3+} concentration in ferromagnesian silicates and oxides from microprobe analyses using stoichiometric criteria. *Mineralogical Magazine* **51**, 431–5.

Du Toit, A. L. 1931. The genesis of the pyroxenite–apatite rocks of Palabora, eastern Transvaal. *Transactions of the Geological Society of South Africa* **34**, 107–27.

Eby, G. N. 1975. Abundance and distribution of the rare-earth elements and yttrium in the rocks and minerals of the Oka carbonatite complex, Quebec. *Geochimica et Cosmochimica Acta* **39**, 597–620.

Efimov, A. F., E. M. Es'kova, S. I. Lebedeva, and V. Ya. Levin 1985. Typochemistry of accessory pyrochlore in rocks of an alkaline complex in the Urals. *Geokhimiya* **1985**, (2), 201–8. (in Russian)

Efimov, A. F., E. I. Semenov, A. N. Akramov, V. D. Dusmatov & G. V. Lyubomilova 1980. The pyrochlore mineral group in alkalic rocks of eastern Karategin (southern Tien-Shan). *Doklady Akademii Nauk Tadzhikskoi SSR* **23**, (3), 158–61. (in Russian)

Fabriès, J. 1978. Les types paragénétiques des amphiboles sodiques dans les roches magmatiques. *Bulletin de Minéralogie* **101**, 155–65.

Filho, A. I., P. R. A. dos S. Lima & O. M. de Souza 1984. Aspects of the geology of the Barreiro carbonatitic complex, Araxá, MG, Brazil. In *Carbonatitic complexes of Brazil: Geology*, A. B. Da Silva (ed.), 19–44. São Paulo: Companhia Brazileira de Metallurgia e Mineração.

Fleischer, M. & Z. S. Altschuler 1982. *The lanthanides and yttrium in minerals of the apatite group: a review.* U. S. Geological Survey Open File Report 82-783.

Fleischer, M. & Z. S. Altschuler 1986. The lanthanides and yttrium in minerals of the apatite group – an analysis of the available data. *Neues Jahrbuch für Mineralogie, Monatshefte*, 467–80.

Fleischer, M. 1987. *Glossary of mineral species.* Tucson: Mineralogical Record.

Flink, G. 1898. Berattelse om en mineralogisk resa i Syd-Grønland sommaren 1897. *Meddelelser om Grønland* **14**, 221–62.

Flink, G. 1901. On the minerals from Narsarsuk on the Firth of Tunugdliarfik in southern Greenland. *Meddelelser om Grønland* **24**, 7–180.

Gaertner, H. R. von 1930. Die Kristallstrukturen von Loparit und Pyrochlor. *Neues Jahrbuch für Mineralogie, Beilage-Band* **61**, Abt. A, 1–30.

Gaidukova, V. S. 1966. Pyrochlore and calcium aeschynite from carbonatites. *Geologiya Mestorozhdenii Redkikh Elementov* **30**, 72–6. (in Russian)

Gaidukova, V. S. & T. B. Zdorik 1962. Mineralogy of rare elements in carbonatites. *Geologiya Mestorozhdenii Redkikh Elementov* **17**, 86–117. (in Russian)

Gaidukova, V. S., L. I. Polupanova & T. I. Stolyarova 1963. Hatchettolite from the Siberian carbonatites. *Mineral'noe Syr'e (Vsesoyuznyi Nauchno-Issledovatel'skii Institut Mineral'nogo. Syr'ya)* **7**, 86–95. (in Russian)

Girault, J. 1966. Genèse et géochimie de l'apatite et de la calcite dans les roches liées au complex carbonatitique et hyperalcaline d'Oka (Canada). *Bulletin de la Société français de Minéralogie et Cristallographie* **89**, 496–513.

Gold, D. P. 1966. The minerals of the Oka carbonatite and alkaline complex, Oka, Quebec. *Mineralogical Society of India*, International Mineralogical Association vol., 109–25.

Gold, D. P., G. N. Eby, K. Bell & M. Vallée 1986. Carbonatites, diatremes, and ultra-alkaline rock of the Oka area, Quebec. *Geological Association of Canada–Mineralogical Association of Canada–Canadian Geophysical Union, Joint Annual Meeting*, Ottawa, Field Trip 21, Guidebook.

Hanekom, H. J., C. M. v. H. van Staden, P. J. Smit & D. R. Pike 1965. *The geology of the Palabora igneous complex.* Memoir of the Geological Survey of South Africa 54.

Hawthorne, F. C. 1983. The crystal chemistry of the amphiboles. *Canadian Mineralogist* **21**, 173–480.

Heinrich, E. W. 1966. *The geology of carbonatites.* Chicago: Rand McNally.

Hodder, R. W. 1961. *Alkaline rocks and niobium deposits near Nemegos, Ontario.* Geological Survey of Canada, Bulletin 70.

Hogarth, D. D. 1959. *A mineralogical study of pyrochlore and betafite.* Ph.D. Thesis, McGill University, Montreal.

Hogarth, D. D. 1961. A study of pyrochlore and betafite. *Canadian Mineralogist* **6**, 610–33.

Hogarth, D. D. 1977. Classification and nomenclature of the pyrochlore group. *American Mineralogist* **62**, 403–10.

Hogarth, D. D. 1983. *Rare earth element minerals from four carbonatites of Templeton Township, Quebec.* '83 Mineralogical Society of America Symposium on Alkaline Complexes, 21–2. (abstract)

Hogarth, D. D. 1988. Chemical composition of fluorapatite and associated minerals from skarn near Gatineau, Quebec. *Mineralogical Magazine* **52**, 347–58.

Hogarth, D. D. & P. Lapointe 1984. Amphibole and pyroxene development in fenite from Cantley, Quebec. *Canadian Mineralogist* **22**, 281–95.

Hogarth, D. D., F. F. Brown & A. M. Pritchard 1970. Biabsorption, Mössbauer spectra, and chemical investigation of five phlogopite samples from Quebec. *Canadian Mineralogist* **10**, 710–22.

Hogarth, D. D., R. Hartree, J. Loop & T. N. Solberg 1985. Rare-earth element minerals in four carbonatites near Gatineau, Quebec. *American Mineralogist* **70**, 1135–42.

Hogarth, D. D., G. Y. Chao & M. G. Townsend 1987. Potassium and fluorine-rich amphiboles from the Gatineau area, Quebec. *Canadian Mineralogist* **25**, 739–53.

Hogarth, D. D., P. Rushforth & R. H. McCorkell 1988. The Blackburn carbonatites near Ottawa, Ontario: dykes with fluidized emplacement. *Canadian Mineralogist* **26**, 377–90.

Hounslow, A. W. & G. Y. Chao 1970. Monoclinic chlorapatite from Ontario. *Canadian Mineralogist* **10**, 252–9.

Hughson, M. R. & J. G. Sen Gupta 1964. A thorian intermediate member of the britholite–apatite series. *American Mineralogist* **49**, 937–51.

Jäger, E., E. Niggli & A. H. van der Veen 1959. A hydrated barium–strontium pyrochlore in a biotite rock from Panda Hill, Tanganyika. *Mineralogical Magazine* **32**, 10–25.

Jans, H. & P. de Bethune 1968. The alkalic amphibole of the Lueshe carbonatite. *Papers and Proceedings of the 5th General Meeting. International Mineralogical Association* (London), 312–14.

Junge, W., J. Knoth & R. Rath 1983. Chemische und optische Untersuchungen von komplexen Titan-Niob-Tantalaten (Betafiten). *Neues Jahrbuch für Mineralogie, Abhandlungen* **147**, 169–83.

Kalogeropoulos, S. I. 1977. *Geochemistry and mineralogy of the St. Lawrence pyrochlore deposit, Oka, P.Q.* M.Sc. thesis, Queen's University, Kingston.

Kapustin, Yu. L. 1966. Geochemistry of rare-earth elements in carbonatites. *Geochemistry International* **3**, 1054–64.

Kapustin, Yu. L. 1973. *Mineralogiya Kory Vyvetrivaniya Karbonatitov*. Moscow: Nedra. (in Russian)

Kapustin, Yu. L. 1974. Carbonatites as a major rare metal source. In *Geologiya Syr'evye Resursy Redkikh Elementov v SSSR, Tezisy Dokladovma Vsesoyuznom Soveshchanii*, 2nd edn, G. I. Gorbunov (ed.), 44–6. Apatity (USSR): Akademiya Nauk. (in Russian)

Kapustin, Yu. L. 1977. Distribution of Sr, Ba and the rare earths in apatite from carbonatite complexes. *Geochemistry International* **14**, (4), 71–80.

Kapustin, Yu. L. 1980. *Mineralogy of carbonatites*, D. K. Biswas (trans.). New Delhi: Amerind.

Kapustin, Yu. L. 1982. Geochemistry of strontium and barium in carbonatites. *Geochemistry International* **19**, (2), 38–48.

Kapustin, Yu. L. & A. I. Polyakov 1982. Volcanic carbonatites of east Africa. *Zapiski Vsesoyuznogo Mineralogicheskogo Obshchestva* **111**, 639–55. (in Russian)

Keller, J. 1981. Carbonatitic volcanism in the Kaiserstuhl alkaline complex: evidence for highly fluid carbonatite melts at the earth's surface. *Journal of Volcanology and Geothermal Research* **9**, 423–31.

Khodyreva, A. I. 1972. Distribution of tantalum and niobium in weathering crust minerals of carbonatites of eastern Siberia. In *Voprosy Mineralogii Gornykh Porod i Rud Vostochnoi Sibiri*, B. M. Smakin (ed.), 90–6. Irkutsk: Siberian Division, Institute of Geochemistry. (in Russian)

Khudolozhkin, V. O., V. S. Urusov & K. I. Tobelko 1972. Ordering of Ca and Sr in cation positions in the hydroxylapatite–belovite isomorphous series. *Geochemistry International* **9**, 827–33.

Kirnarskii, Yu. M., B. V. Afanas'ev & Yu. P. Men'shikov 1968. Accessory thorian betafite from carbonatite. *Materialy po Mineralogii Kol'skogo Poluostrova* **6**, 258–64. (in Russian)

Klevtsova, R. F. & S. V. Borisov 1964. The crystal structure of belovite. *Journal of Structural Chemistry* **5**, 137–8. Summarized in *Structure Reports* **29**, 374.

Korzhinskiy, M. A. 1981. Apatite solid solutions as indicators of the fugacity of HCl° and HF° in hydrothermal fluids. *Geochemistry International* **18**, (3), 44–60.

Kovalenko, V. I. 1968. On the chemical composition, properties, and mineral paragenesis of riebeckite and arfvedsonite. *Papers and Proceedings of the 5th General Meeting. International Mineralogical Association* (London), 261–84.

Kovalenko, V. I., V. S. Antipin, N. V. Vladykin, Ye. V. Smirnova & Yu. A. Balashov 1982. Rare-earth distribution coefficients in apatite and behavior in magmatic processes. *Geochemistry International* **19**, (1), 171–83.

Kukharenko, A. A. 1965a. Pyrochlore group. In *Kaledonskii kompleks ul'traosnovnykh, shchelochnykh porod i karbonatitov Kol'skogo Poluostrova i severnoi Karelii*, A. A. Kukharenko (ed.), 345–62. Moscow: Nedra. (in Russian)

Kukharenko, A. A. 1965b. Geochemical characteristics of alkaline-ultrabasic rocks of the Caledonian complexes. In *Kaledonskii kompleks ul'traosnovnykh, shchelochnykh porod i karbonatitov Kol'skogo Poluostrova i severnoi Karelii*, A. A. Kukharenko (ed.), 547–644. Moscow: Nedra. (in Russian)

Kurenkina, I. E. 1976. Paragenetic associations in vein carbonatites of the Kursk Magnetic Anomaly. In *Problemy Geneticheskoi Informatsii v Mineralogii*, M. V. Fishman (ed.), 107–8. Syktyvkar, USSR: Akad. Nauk SSSR. (in Russian)

Kuz'menko M. V. 1961. The geochemistry of tantalum and niobium. *International Geology Review* **3**, 9–25.

Kuz'menko, M. V. 1984. Aspects of the systematics and typical chemistry of tantaloniobates of the pyrochlore group. In *Tipokhimizm Mineralov Granitnykh Pegmatitov*, M. V. Kuz'menko (ed.), 5–32. Moscow: IMGRE. (in Russian)

Landa, E. A., N. I. Krasnova, A. N. Tarnovskaya & Yu. P. Shergina 1983. The distribution of rare earths and yttrium in apatite from alkali-ultrabasic and carbonatite intrusions and the origin of apatite mineralization. *Geochemistry International* **20**, (1), 77–87.

Lapido-Loureiro, F. E. de V. 1973. Carbonatitos de Angola. *Memórias e Trabalhos do Instituto de Investigação Cientifica de Angola*, 11.

Lapin, A. V. 1975. Geological characteristics of mineral paragenesis of apatite–magnetite rock and carbonatite of the central massif of the Turii Peninsula. In *Metasomatizm i Rudoobrazovanie*, Yu. V. Kazitzyn (ed.), 167–76. Moscow: Nedra. (in Russian)

Lapitskii, Ye. M., E. N. Kachanov, M. V. Mitkeev & A. I. Nikonov 1974. Carbonatites of the northwest Sea of Azov region. *Geologicheskii Zhurnal* **34**, (2), 144–5. (in Russian)

Laplante, R. 1980. *Etude de la minéralisation en Nb–Ta–U du complexe igné alcalin de Crevier, Compté Roberval, Lac St. Jean, P. Q.* M.Sc. thesis, Ecole Polytechnique, Montréal.

Larsen, E. S. 1942. *Alkalic rocks of Iron Hill, Gunnison County, Colorado*. U. S. Geological Survey, Professional Paper 197A.

Leake, B. E., 1978. Nomenclature of amphiboles. *Canadian Mineralogist* **16**, 501–20.

Le Bas, M. J. & C. D. Handley 1979. Variation in apatite composition in ijolitic and carbonatitic igneous rocks. *Nature* **279**, 54–6.

Lebedeva, S. I., V. P. Bytov, L. S. Dubakina & K. V. Yurkina 1973. Microprobe determination of the mode of deposition of trace elements in zircon-bearing pyrochlores and hatchettolites. In *Issledovaniya v oblasti rudnoi mineralogii*, M. S. Bezsmertnaya (ed.), 133–45. Moscow: Nauka (in Russian)

Lin, T.-S., W.-H. Hung & C.-M. Shih 1973. Ceruranpyrochlore – a new variety of pyrochlore. *Geochimica* **1**, (1), 39–45. (in Chinese).

Lindqvist, K. & P. Rehtijärvi 1979. Pyrochlore from the Sokli carbonatite complex, northern Finland. *Bulletin of the Geological Society of Finland* **51**, 81–93.

Ludington, S. 1978. The biotite–apatite geothermometer revisited. *American Mineralogist* **63**, 551–3.

McConnell, D. 1973. *Apatite: its crystal chemistry, mineralogy, utilization, and geologic and biologic occurrences*. New York: Springer Verlag.

McCormick, G. R. & R. C. Heathcote 1987. Mineral chemistry and petrogenesis of carbonatite intrusions, Perry and Conway counties, Arkansas. *American Mineralogist* **72**, 59–66.

Mäder, U. K. 1987. *The Aley carbonatite complex, northern Rocky Mountains, British Columbia (94B/5)*. British Columbia Ministry of Energy, Mines and Petroleum Resources, Paper 1987-1, 283–8.

Marchenko, Ye. Ya., A. I. Chashka, P. N. Storchak, O. D. Timoshenko & E. M. Lapitskiy 1975. Typomorphism of apatite from carbonatite. *Doklady, Academy of Sciences, Earth Science Section* **223**, 138–40.

Marchenko, E. Ya., G. K. Eremenko, N. F. Rusakov, O. V. Dovgal & O. O. Zonova 1984. Apatite from carbonatites in the southwestern Chernigov area of the Sea of Azov region. *Dopovidi Akademii Nauk Ukrains'koi RSR. Seriya B: Geologichni, Khimichni ta Biologichni Nauki*, (2), 22–5 (in Ukrainian)

Marchenko, E. Ya., S. I. Kirikilitsa, P. N. Storchak, G. G. Kon'kov & V. I. Vasenko 1980. Apatite from carbonatites of the Ukrainian Shield. *Mineralogicheskii Zhurnal* **2**, (4), 59–66. (in Russian)

Mariano, A. N. 1978. The application of cathodoluminescence for carbonatite exploration and characterization. In *Proceedings of the First International Symposium on Carbonatites, Poços de Caldas, Brazil*, June 1976, 39–57. Brasilia: Brasil Departamento Nacional da Produção Mineral.

Mariano A. N. & P. J. Ring 1975. Europium-activated cathodoluminescence in minerals. *Geochimica et Cosmochimica Acta* **39**, 649–60.

Mariano, C. 1985. *The paragenesis of the Bond Zone of the Oka carbonatite complex, Oka, Quebec*. B.Sc. thesis, Bucknell University, Lewisburg.

Mian, I. & M. J. Le Bas 1986. Sodic amphiboles in fenites from the Loe Shilman carbonatite complex, NW Pakistan. *Mineralogical Magazine* **50**, 187–97.

Mineev, D. A. & N. I. Razenkova 1962. On zoned crystals of pyrochlore from the Vishnevey Gor. *Zapiski Vsesoyuznogo Mineralogicheskogo Obshchestva* **91**, 89–93. (in Russian)

Moroshkin, V. V. & B. S. Gorobets 1985. Experimental estimation of comparative Eh conditions of formation of minerals based on the ratio of valence forms of europium in apatite. *Mineralogicheskii Sbornik (Lvov)* **39**, (2), 21–6. (in Russian)

Nash, W. P. 1972a. Mineralogy and petrology of the Iron Hill carbonatite complex, Colorado. *Geological Society of America Bulletin* **83**, 1361–82.

Nash, W. P. 1972b. Apatite–calcite equilibria in carbonatites: chemistry of apatite from Iron Hill, Colorado. *Geochimica et Cosmochimica Acta* **36**, 1313–19.

Neumann, E.-R. 1976. Two refinements for calculating the structural formulae for pyroxenes and amphiboles. *Norsk Geologisk Tidsskrift* **56**, 1–6.

Nickel, E. H. 1962. *Compositional variations in pyrochlore and niobian perovskite from a niobium deposit in the Oka district of Quebec*. Canada Mines Branch, Technical Bulletin 31.

Olson, J. C., D. R. Shawe, L. C. Pray & W. N. Sharp 1954. *Rare-earth mineral deposits of the Mountain Pass district, San Bernardino County, California*. U.S. Geological Survey, Professional Paper 261.

Palache, C., H. Berman & C. Frondel 1944. *A system of mineralogy of James Dwight Dana and Edward Salisbury Dana*, 7th edn., Vol. 1. New York: Wiley.

Parker, R. L. & M. Fleischer 1968. *Geochemistry of niobium and tantalum*. U.S. Geological Survey, Professional Paper 612.

Pentel'kov, V. G. & T. I. Matrosova 1978. Pyrochlore from the Mbalizi carbonatites, Tanzania. *Doklady, Academy of Sciences, Earth Science Section* **241**, 160–4.

Perrault, G. 1968. La composition et la structure cristalline du pyrochlore d'Oka, P.Q. *Canadian Mineralogist* **9**, 383–402.

Petrov, P. A. & R. F. Zhuravel 1971. Comparative characteristics of the chemical composition of apatite from different rock types. *Trudy Sibirskogo Nauchno-Issledovatel'skogo Instituta Geologii, Geofiziki i Mineral'nogo Syr'ya* **108**, 119–27. (in Russian)

Petruk, W. & D. R. Owens 1975. Electron microprobe analyses for pyrochlores from Oka, Quebec. *Canadian Mineralogist* **13**, 282–5.

Philpotts, J. A. 1970. Redox estimation from a calculation of Eu^{2+} and Eu^{3+} concentrations in natural phases. *Earth and Planetary Science Letters* **9**, 257–68.

Polezhaeva, L. I. & L. A. Strel'nikova 1980. Microprobe studies of pyrochlore as a source of new information on their typomorphism. In *Metody Izucheniya Sostava i Svoistv Mineralov i Gornykh Porod Kol'skogo Poluostrova*, I. V. Bel'kov (ed.), 92–8. Apatity (USSR): Akademii Nauk. (in Russian).

Pouliot, G. 1970. Study of carbonatitic calcites from Oka, Quebec. *Canadian Mineralogist* **10**, 511–40.

Pozharitskaya, L. K. 1962. Mineralogical–petrological peculiarities of carbonatite massifs. *Geologiya Mestorozhdenii Redkikh Elementov* **17**, 70–86. (in Russian).

Pozharitskaya, L. K. & V. S. Samoylov 1972. *Petrologiya, Mineralogiya i Geokhimiya Karbonatitov Vostochnoi Sibiri*, L. M. Salikov (ed.). Moscow: Nauka. (in Russian).

Prener, J. S. 1967. The growth and crystallographic properties of calcium fluor- and chlorapatite crystals. *Journal of the Electrochemical Society* **114**, 77–83.

Prins, P. 1973. Apatite from African carbonatites. *Lithos* **6**, 133–44.

Puchelt, H. & R. Emmermann 1976. Bearing of rare earth patterns of apatites from igneous and metamorphic rocks. *Earth and Planetary Science Letters* **31**, 279–86.

Puustinen, K. 1972. Richterite and actinolite from the Siilinjarvi carbonatite complex, Finland. *Bulletin of the Geological Society of Finland* **44**, (1), 83–6.

Rahman, M. A. 1980. Betafite in carbonatite from Khyber Agency, N.W.F.P. *Pakistan Journal of Scientific Research* **32**, 5–8.

Rimskaya-Korsakova, O. M. 1965. Apatite group. In *Kaledonskii kompleks ul'traosnovnykh, shchelochnykh porod i karbonatitov Kol'skogo Poluostrova i severnoi Karelii*, A. A. Kukharenko (ed.), 507–19. Moscow: Nedra. (in Russian)

Rimskaya-Korsakova, O. M., N. I. Krasnova & L. N. Kopylova 1979. Typochemism of apatites in the Kovdor complex deposit. *Mineralogiya i Geokhimiya* **6**, 58–70. (in Russian)

Roeder, P. L., D. MacArthur, Xin-Pei Ma, G. R. Palmer & A. N. Mariano 1987. Cathodoluminescence and microprobe study of rare earth elements in apatite. *American Mineralogist* **72**, 801–11.

Russell, H. D., S. A. Hiemstra & D. Groeneveld 1954. The mineralogy and petrology of the carbonatite at Loolekop, eastern Transvaal. *Transactions of the Geological Society of South Africa* **57**, 197–208.

Sabina, A. P. 1979. *Minerals of the Francon Quarry (Montreal Island): a progress report*. Geological Survey of Canada, Paper 79-1A, 115–20.

Sage, R. P. 1983a. *Geology of the Argor carbonatite complex*. Ontario Geological Survey, Open File Report 5396.

Sage, R. P. 1983*b* *Geology of the Firesand River carbonatite complex.* Ontario Geological Survey Open File Report 5403.

Sage, R. P. 1983*c*. *Geology of the Goldray carbonatite complex.* Ontario Geological Survey, Open File Report 5404.

Sage, R. P. 1983*d*. *Geology of the Prairie Lake carbonatite complex.* Ontario Geological Survey, Open File Report 5412.

Sage, R. P. 1983*e*. *Geology of the Valentine Township carbonatite complex.* Ontario Geological Survey, Open File Report 5418.

Sahama, T. G. 1956. Optical anomalies in arfvedsonite from Greenland. *American Mineralogist* **41**, 509–12.

Samoylov, V. S. 1977. *Karbonatity (fatsii i usloviya obrazovaniya).* Moscow: Nauka. (in Russian)

Samoylov, V. S. & G. S. Gormasheva 1975. Alkali amphiboles of carbonatites and genetically related rocks. *Zapiski Vsesoyuznogo Mineralogicheskogo Obshchestva* **104**, 145–59. (in Russian)

Samoylov, V. S. & V. I. Kovalenko 1983. *Alkaline rocks and carbonatite complexes from Mongolia.* Joint Soviet-Mongolian Scientific Exploration Geological Expedition. Trudy 35. (in Russian)

Samoylov, V. S., G. S. Gormasheva & E. A. Chernysheva 1974. Low-aluminum alkalic amphiboles of carbonatites. *Ezhegodnik* – 1973; Yearbook Institute of Geochemistry, Irkutsk, 189–93. (in Russian, English abstract)

Secher, K. & L.-M. Larsen 1980. Geology and mineralogy of the Sarfartôq carbonatite complex, west Greenland. *Lithos* **13**, 199–212.

Semenov, E. I. 1963. *Mineralogiya Redkikh Zemel'*. Moscow: Akademii Nauk. (in Russian)

Semenov, E. I. 1977. *Tipokhimizm Mineralov Shchelochnykh Massivov.* Moscow: Nedra. (in Russian)

Silva, M. V. S. 1973. *Estrutura Vulcânico-Carbonatitica da Catanda (Angola)*. Boletim Serviços de Geologia e Minas de Angola, No. 24, 5–14.

Simpson, R. H. F. 1980. *A petrologic and geochemical study of the Francon silicocarbonatite sills, Montreal, Quebec.* B.Sc. thesis, University of Ottawa, Ottawa.

Sokolov, S. V. 1983. The genetic unity of the apatite–magnetite ores and carbonatites in alkali-ultrabasic intrusions. *Geochemistry International* **20**, 86–98.

Sokolov, S. V. 1985. Carbonates in ultramafic alkali-rock, and carbonatite intrusions. *Geochemistry International* **22**, 150–66.

Sommerauer, J. & K. Katz-Lehnert 1985. A new partial substitution mechanism of CO_3^{2-}/CO_3OH^{3-} and SiO_4^{4-} for the PO_4^{3-} group in hydroxyapatite from the Kaiserstuhl alkaline complex (SW-Germany). *Contributions to Mineralogy and Petrology* **91**, 360–8.

Stormer, J. C. & I. S. E. Carmichael 1971. Fluorine–hydroxyl exchange in apatite and biotite: a potential igneous geothermometer. *Contributions to Mineralogy and Petrology* **31**, 121–31.

Subbotin, V. V., Yu. M. Kirnarskii, G. S. Kurbatova, L. A. Strel'nikova & G. F. Subbotina 1985. Composition of apatite-bearing rocks of the central zone of the Sebl'yavr Massif. In *Petrologiya i Minerageniya Shchelochnykh, Shchelochno-Ul'traosnovnykh i Karbonatitovykh Kompleksov Karelo-Kol'skovo Regiona*, T. N. Ivanova (ed.), 61–9. Apatity (USSR): Akademii Nauk. (in Russian)

Sutherland, D. 1969. Sodic amphiboles and pyroxenes from fenites in east Africa. *Contributions to Mineralogy and Petrology* **24**, 114–35.

Taborszky, F. K. 1972*a*. Chemismus und Optik der Apatite. *Neues Jahrbuch für Mineralogie Monatshefte*, 79–91.

Taborszky, F. K. 1972*b*. Das problem der Cl-Apatite. *Lithos* **5**, 315–24.

Teal, S. E. 1979. *The geology and petrology of the Firesand River carbonatite complex, northwestern Ontario.* M.Sc. thesis, McMaster University, Hamilton.

Vartiainen, H. 1980. *The petrography, mineralogy and petrochemistry of the Sokli carbonatite massif, northern Finland.* Bulletin of the Geological Survey of Finland 313.

Veen, A. H. van der 1963. *A study of pyrochlore.* Verhandelingen van het Koninklijk Nederlands geologisch mijnbouwkundig Genootschap, Geologische Serie 22.

Vikhter, B. Ya., G. K. Eremenko & V. M. Umyrev 1975. A young volcanic carbonatite complex in Afghanistan. *Sovetskaya Geologiya* **1975**, (10), 107–16. (in Russian)

Vil'kovich, R. V. & L. K. Pozharitskaya 1982. Rare earth elements in calcite and apatite of the Chernigov zone. *Geokhimiya*, 511–18. (in Russian)

Vronskii, A. V. & V. A. Basina 1963. Use of X-ray photography for the study of zoned pyrochlore. *Nauchnye Trudy Irkutskii Gosudarstvennyi Nauchno – Issledovatel'skii Institut Redkikh Metallov* **11**, 44–7. (in Russian)

Wambeke, L. Van 1964. Géochimie minéral des carbonatites du Kaiserstuhl. In *Les roches alkalines et les carbonatites du Kaiserstuhl*, L. Van Wambeke (ed.), 65–91. Euratom 1827 d, f, e.

Wambeke, L. Van 1978. Kalipyrochlore, a new mineral of the pyrochlore group. *American Mineralogist* **63**, 528–30.

Wambeke, L. Van 1980. Latrappite and ceriopyrochlore, new minerals for the Federal Republic of Germany. *Neues Jahrbuch für Mineralogie, Monatshefte* no. 4, 171–4.

Wyckoff, R. W. G. 1957. *Crystal structures*. New York: Interscience.

Wyllie, P. J., K. G. Cox & G. M. Bigger 1962. The habit of apatite in synthetic systems and igneous rocks. *Journal of Petrology* **3**, 238–43.

Young, R. A. 1966. Dependence of apatite properties on crystal structural details. *Transactions of the New York Academy of Science, Ser. 2* **29**, 949–59.

Zdorik, T. B. 1966. Evolution of rare-earth mineralization in carbonatite of an eastern Siberian deposit. *Geologiya Mestorozhdenii Redkikh Elementov* **30**, 121–31. (in Russian)

Zhukov, G. V., V. A. Varkhotov, I. I. Sakhatskii & Yu. P. Goryaev 1973. Discovery of phosphate mineralization in the western Sea of Azov region. *Geologicheskii Zhurnal* **33**, (5), 150–2. (in Russian)

7
NATURE OF ECONOMIC MINERALIZATION IN CARBONATITES AND RELATED ROCKS
A. N. MARIANO

ABSTRACT

Carbonatites continue to be a major source of Nb, phosphate, and REEs (rare earth elements).

Carbonatite REE mineralization in established or potential ore quantities, includes ancylite, bastnaesite-type minerals, britholite, crandallite-group minerals, and monazite. A genetic classification of REE deposits associated with carbonatites is as follows:

(a) primary, from carbonatite melts (bastnaesite and parisite – Mountain Pass, USA);
(b) hydrothermal (bastnaesite and monazite – Bayan Obo, China; Wigu Hill, Tanzania; Karonge, Burundi); and
(c) supergene, developed in carbonatite-derived laterites (Araxá and Catalão I, Brazil; Cerro Impacto, Venezuela; Mrima, Kenya; Mt. Weld, Australia).

Pyrochlore is the most important Nb mineral in carbonatites. Other Nb minerals can occur that crystallize with, or are independent of pyrochlore, and may be potential ores for Nb. Some carbonatite pyrochlores contain Ta abundances at economic levels. Examples include the carbonatite bodies of the Shuswap Terrain (Canada), Crevier (Canada), Fen (Norway), and Ngualla (Tanzania). In regions of extreme lateritic weathering of carbonatites, pyrochlore is totally destroyed and Nb partitions into supergene REE minerals of the crandallite group.

Apatite is the most important mineral presently being mined in carbonatite complexes, and its phosphate content is higher than that of marine phosphorites.

Anatase deposits, the decalcification weathering products of perovskite in pyroxenites from the carbonatite complexes of Goiás and Minas Gerais, Brazil, constitute some of the largest TiO_2 ore bodies in the world. Although minor deuteric alteration in fresh pyroxenite can convert perovskite to anatase, field and chemical evidence establishes a direct correlation between weathering and anatase development.

Other commodities associated with carbonatites include barite, Cu, fluorite, Sr, V, Th, and U.

ECONOMIC MINERALIZATION

7.1 INTRODUCTION

Carbonatites originate in the upper mantle and, along with genetically related alkaline rocks, characteristically display geochemical enrichments in Ba, Nb, REEs, Sr, Ta, Th, U, and Zr. Within the Earth's crust carbonatites form relatively small bodies with such distinctive mineral and geochemical features that they can still be distinguished even after tectonic disruption or regional metamorphic activity.

Some of the best known, or potential, ore deposits for Cu, Nb, REEs, fluorite, phosphate, and vermiculite are associated with carbonatites. They include the virtually 'inexhaustable' Nb reserves of Araxá, the bastnaesite deposit of Mountain Pass, and the Cu and multi-commodity deposit of Phalaborwa, South Africa. Table 7.1 gives examples of deposits associated with carbonatites.

This chapter deals with some aspects of Nb, REE, and Ti mineralization in carbonatites and genetically related rocks that have not been treated in previous publications. It draws from field and laboratory examination of rocks from more than 150 carbonatite complexes in Africa, Asia, Australia, Europe, North America, and South America. Much of the author's unpublished information is derived from confidential company reports and reports to the United Nations Revolving Fund for Natural Resources Exploration. Although little has been written about carbonatites and mineralization, tribute is paid to T. Deans of the British Overseas Geological Survey for his pioneering work on the economic mineralogy of carbonatites, which laid the foundations for all work in this field.

Table 7.1 Economic mineralization in carbonatites.

Commodity	Locality	Mineralization	Reference
REE	Mountain Pass, USA	Bastnaesite in carbonatite, open-pit mining, reserves at end of 1986 — 40 mt (million tons) of 7.67 wt% REO (rare earth oxides)	J. O. Landreth (pers. comm., Molycorp Inc. 1987)
REE	Bayan Obo, Inner Mongolia, China	Hydrothermal bastnaesite and monazite associated with iron ore, proven reserves 37 mt of ore	Mining Annual Review (1986)
REE	Gakara-Karonge, Burundi	Hydrothermal veins of high-grade bastnaesite and minor monazite, 1978 output of 404 t (tons) bastnaesite	J. Brinckmann (pers. comm. 1983)
REE	Araxá, Minas Gerais, Brazil	Supergene REE minerals in laterite, 800 000 metric tons of 13.5 wt% REO	Souza & Castro (1968)
Niobium	Araxá, Minas Gerais, Brazil	Pyrochlore in laterite, 460 mt avg. 3.0 wt% Nb_2O_5	Engineering and Mining Journal (1986)
Niobium	Catalão I, Goiás, Brazil	Pyrochlore in laterite, 4.2 mt of ore, avg. 1.20 wt% Nb_2O_5; a second body contains 22 mt of ore, avg. 1.07 wt% Nb_2O_5	Pinkuss & Guimarães (1981)
Niobium	Catalão II, Goiás, Brazil	Pyrochlore in carbonatite, reserves 1.4 mt of 2.19 wt% Nb_2O_5	Geologia do Brasil (1984)

Table 7.1 (cont).

Commodity	Locality	Mineralization	Reference
Niobium	Tapira, Minas Gerais, Brazil	Pyrochlore in carbonatite, 126 mt of ore with 1.2 wt% Nb_2O_5	Geologia do Brasil (1984)
Niobium	St. Honoré, Quebec, Canada	Pyrochlore in carbonatite, 12.2 mt of ore, 0.66 wt% Nb_2O_5	Canadian Mines Handbook (1987–1988)
Niobium	Oka, Quebec, Canada	Pyrochlore in carbonatite, 25.4 mt, avg. 0.44 wt% Nb_2O_5	Canadian Mines Handbook (1975–1976)
Niobium	Sarfartôk, Greenland	Pyrochlore veins in carbonatite, 0.1 mt of 15 wt% Nb_2O_5	Secher (1986)
Phosphate	Tapira, Minas Gerais, Brazil	Primary apatite in carbonatite and pyroxenite, 266 mt of 8 wt% P_2O_5 proved	Engineering and Mining Journal (1986)
Phosphate	Araxá, Minas Gerais, Brazil	Secondary apatite with barite and ferric iron oxides, 94 mt of 15 wt% P_2O_5 proved	Engineering and Mining Journal (1986)
Phosphate	Patrocinio, Minas Gerais, Brazil	Primary magmatic apatite in carbonatite and pyroxenite, 200 mt reserves	Geologia do Brasil (1984)
Phosphate	Catalão I, Goiás, Brazil	Primary and secondary apatite in carbonatite and pyroxenite, 112 mt of 5–17 wt% P_2O_5	Geologia do Brasil (1984)
Phosphate	Jacupiranga, São Paulo, Brazil	Primary apatite in carbonatite, 500 000 t/y of 35 wt% P_2O_5	Engineering and Mining Journal (1986)
Phosphate	Anitapolis, Santa Catarina, Brazil	Primary apatite in pyroxenite and glimmerite, and residual apatite, 0.5 mt of 35 wt% P_2O_5	Geologia do Brasil (1984)
Phosphate	Ipanema, São Paulo, Brazil	Primary apatite in glimmerite, reserves 28 mt at 7.5 wt% P_2O_5	Geologia do Brasil (1984)
Phosphate	Phalaborwa, South Africa	Primary apatite in phoscorite, pyroxenite and carbonatite, reserves in pyroxenite 600 mt of 7 wt% P_2O_5	Mew (1980)
Phosphate	Cargill, Ontario, Canada	Eluvial apatite, 60 mt of ore grading 20.2 wt% P_2O_5	Canadian Minerals Yearbook (1985)
Phosphate	Martinson Lake, Ontario, Canada	Eluvial apatite, 57 mt of ore grading 23 wt% P_2O_5	Canadian Minerals Yearbook (1985)
Phosphate	Sukulu, Uganda	Residual apatite in weathered carbonatite, 202 mt of 12.8 wt% P_2O_5	Mew (1980)
Titanium	Tapira, Minas Gerais, Brazil	Anatase in laterite, 136 mt of 19.86 wt% TiO_2	Anuário Mineral Brasileiro (1986)
Titanium	Catalão I, Goiás, Brazil	Anatase in laterite measured, indicated 125 mt of 10 wt% TiO_2	Carvalho (1974)

(continued)

Table 7.1 (cont).

Commodity	Locality	Mineralization	Reference
Titanium	Patrocinio Area, Minas Gerais, Brazil: includes Serra Negra, Salitre I and Salitre II	Anatase in laterite, 260 mt of 26.49 wt% TiO_2	Anuário Mineral Brasileiro (1986)
Titanium	Serra de Maicuru, Pará, Brazil	Anatase in laterite, 500 mt of TiO_2	Revista Minerios (1987)
Titanium	Powderhorn Complex, Gunnison Co., Colorado USA	Perovskite in pyroxenite, 419 mt of 12 wt% TiO_2	Denver Post, 25 Feb., 1976
Vermiculite	Phalaborwa, South Africa	Vermiculite in pyroxenite-pegmatoid, production is 180 000 t/y of concentrate with 90% vermiculite	Verwoerd (1986)
Vermiculite	Rainy Creek Complex, Libby, Montana, USA	Vermiculite in pyroxenite, 6500 short tons/day processing capacity	Engineering and Mining Journal 1987/1988
Vermiculite	Ipanema, São Paulo, Brazil	Vermiculite in glimmerite, reserves 4.7 mt	Geologia do Brasil (1984)
Fluorite	Amba Dongar, India	Hydrothermal overprint on fenites at contact between carbonatite and country rock, 11.6 mt of 30 wt% CaF_2	Deans *et al.* (1973)
Fluorite	Okorusu, Namibia	Hydrothermal overprint on fenites at contact between carbonatite and country rock, 7–10 mt of 35 wt% CaF_2	Van Zijl (1962)
Fluorite	Mato Preto, Paraná, Brazil	Hydrothermal in carbonatite, 4.3 mt of 58 wt% CaF_2	Anuário Mineral Brasileiro (1986)
Fluorite	Tchivira, Quilengues, Angola	Hydrothermal infillings in brecciated fenite at contact between ijolite and carbonatite	Mariano (unpublished data 1980)

7.2 REE MINERALIZATION

A distinctive aspect of carbonatites is the presence of anomalously high REE contents, and for some exploration geologists Sr and REEs are the most reliable geochemical indicators of igneous carbonates.

The REEs can be used to discriminate between carbonates of sedimentary and igneous origin. Turekian (1971) shows ΣREE of 15 p.p.m. for crustal carbonates, and Haskin *et al.* (1966) show limestone and marble ΣREE values of between 7 and 175 p.p.m. In general, the ΣREE content in sedimentary carbonates rarely exceeds 200 p.p.m. In contrast, the ΣREE in carbonatites range from ~500 to

> 10 000 p.p.m. (Loubet *et al.* 1972, Eby 1975, Möller *et al.* 1980). Unpublished whole rock, mass spectrographic analyses for carbonatites from more than 150 complexes indicate an average ΣREE of ~2000 p.p.m. The major carbonate minerals in all carbonatites contain anomalous quantities of REEs. Carbonatite calcite and dolomite are therefore potential sources for REEs during hydrothermal and weathering processes. Discrete REE minerals also occur in carbonatites and apatite can also be a major source of REEs. The list of independent REE minerals in carbonatites and their genetically related alkaline igneous rocks is extensive, and is probably best summarized by Kapustin (1980). The best known are bastnaesite and monazite.

7.2.1 Magmatic REE minerals

In carbonatite complexes there may be several phases of carbonatite activity. Primary REE minerals have not been found in the early members, but appear to have been introduced by late-stage magmatic or hydrothermal activity. The bastnaesite–parisite mineralization at Mountain Pass is a rare case where REE minerals are primary phases in a late-stage carbonatite. Although the deposit is described as hydrothermal by Vlasov (1966) and Semonov (1974), the texture and mineral paragenesis show that bastnaesite and parisite from the main Sulfide Queen ore body are primary magmatic minerals. Experiments by Jones & Wyllie (1983), designed to synthesize a fluid similar to the Mountain Pass primary carbonatite magma, suggested that bastnaesite and calcite could precipitate together. The Mountain Pass deposit is unique in representing the largest known accumulation of primary REE minerals in carbonatite. On the basis of replacement textures and associated mineralogy, only a small portion of the REE mineralization at Mountain Pass is considered hydrothermal.

In Kangankunde Hill, Malawi (Deans 1966) late-stage ankeritic carbonatite is associated with primary monazite.

7.2.2 Hydrothermal REE minerals

Rare earth element minerals in most carbonatites are precipitated from hydrothermal solutions. The temperatures may vary from a few hundred degrees Celsius to almost ambient conditions. Thermal springs, commonly associated with carbonatite complexes, generate fluids that permeate carbonatite masses and easily dissolve carbonates, apatite, and sulphides, and become increasingly enriched in Ba, F, SO_4^{2-}, Sr, REEs, and Th. In most carbonatites, hydrothermal REE minerals occur in veinlets or as interstitial fillings, and appear as fine-grained polycrystalline clusters, commonly associated with barite, fluorite, hematite, quartz, strontianite, and sulphides. If phosphate is available, monazite is formed. In the absence of phosphate, minerals commonly found include ancylite [$SrCe(CO_3)_2(OH)\cdot H_2O$], bastnaesite [$(REE)(CO_3)F$], britholite [$(REE,Ca)_5(SiO_4, PO_4)_3(OH,F)$], parisite [$(REE)_2Ca(CO_3)_3F_2$], and synchysite [$(REE)Ca(CO_3)_2F$].

Although the system CO_2–CaO–P_2O_5–REE–H_2O has not been thoroughly studied, it has been shown that REE phosphates preferentially precipitate in aqueous systems. Recht & Ghassemi (1970) studied La salts as a means of removing phosphate from wastewater. They used La^{3+} because of availability and cost, but concluded that any Ce-free REE mixture would work equally well. In their study of a highly alkaline aqueous bicarbonate and phosphate system at ambient and higher temperatures, they found a linear relationship between phosphate removal and amount of added La. The precipitate was exclusively $La^{3+}(PO_4)^{3-}$ without the formation of any REE carbonate. The findings agree favourably with the paragenesis of REE minerals in some geologic environments, because it explains why apatite is never found as primary inclusions in REE carbonate minerals, and why REE carbonate minerals never occur as primary inclusions in monazite. Apatite associated with monazite is always interstitial, and has probably crystallized from a solution or melt with excess P_2O_5, after all of the available REE had been used up. The absence of primary apatite in REE carbonate-mineralized occurrences such as Mountain Pass, Wigu Hill, and Itapirapuá, São Paulo, Brazil, is consistent with this model.

An example of selective REE preference for phosphate is shown in fresh carbonatite from the central cores of the Araxá and Catalão I complexes. Low-temperature, hydrothermal solutions, rich in REEs, have permeated fresh carbonatite along fractures, and reacted with apatite to produce pseudomorphs of polycrystalline monazite and minor quartz. The absence of any Ce anomalies supports reducing conditions during monazite formation.

Harrison & Watson (1984), in their study on the solubility and dissolution kinetics of apatite in felsic melts at 850–1500 °C and at 8 Kb, allowed apatite partly to dissolve in melts containing 0–10 wt% H_2O. Among other findings, they showed that Ca has a transport rate that is generally two orders of magnitude faster than P. Such data may help to explain the selective removal of Ca from apatite in the presence of REE-rich hydrothermal solutions and why P remains relatively immobile.

Although hydrothermal REE deposits derived from, or contained in, carbonatites are numerous, most consist of small, high-grade occurrences that do not contain sufficient tonnage to constitute an economic deposit.

One deposit of high-grade bastnaesite and monazite from a possible carbonatite has been mined on a small scale at Karonge. It consists of almost pure bastnaesite–monazite veins that cut quartzites and schists. Deans (1966) and van Wambeke (1977) point out that the chemistry of Karonge reflects a hydrothermal origin from a carbonatite source. One argument against such an origin has been the apparent absence of alkaline rocks or minerals (Aderca & Van Tassel 1971), but bastnaesite veins containing feldspar with brilliant red, Fe^{3+} activated cathodoluminescence (CL) have been observed (unpublished data, A. N. Mariano) similar to that encountered in fenite associated with alkaline rocks and carbonatites (Mariano 1988). Although no carbonatite is known in Karonge, a carbonatite-alkaline igneous complex does exist at Matongo-Bandaga, approximately 50 km north of Karonge (Tack & De Paepe 1980). The Karonge bastnaesite veins probably represent an example where REEs have been mobilized over some distance by hydrothermal solutions and redeposited in high concentrations.

The REE mineralization at Bayan Obo includes relatively coarse, granular bastnaesite and monazite with quartz and fluorite. Geochemistry and mineralogy support a carbonatite link (Zhou et al. 1980), and the presence of large quantities of quartz and fluorite in the ore rocks suggests hydrothermal activity. This may be one of the largest hydrothermal deposits genetically linked with mantle-derived igneous rocks.

7.2.3 Supergene REE mineralization

When carbonatites are subjected to chemical weathering, the carbonate minerals are easily dissolved and, as a result, Ca and Mg can be selectively removed from the system, leaving behind the less mobile elements. This occurs even in the initial stages of chemical weathering. For example, outcrop carbonatite above the Manny Zone at Oka, Quebec, Canada, contains yellow patches of earthy monazite and strontianite produced from the selective leaching and removal of Ca and Mg from apatite, calcite, and dolomite. The less mobile REEs and Sr combine with PO_4^{3-} and CO_3^{2-} respectively to form secondary monazite and strontianite (unpublished data, A. N. Mariano). At Magnet Cove, Arkansas, USA, Rose et al. (1958) also report earthy monazite as a weathering product of carbonatite. Their description and characterization of the earthy monazite is almost identical to supergene monazite encountered in carbonatite laterites of tropical climates.

Sövite and beforsite, barren of REE minerals, from Araxá, yield values of between 0.28 and 0.5 wt% REEs (unpublished data, A. N. Mariano). In contrast, the laterite derived from Araxá carbonatite contains large quantities of supergene REE minerals. Laterite from the Área Zero in the northeast quadrant of Araxá (800 000 metric tons of proven reserves) contains an average grade of 13.5 wt% REE oxides (Souza & Castro 1968), while laterite from the central core of Araxá, where niobium is mined, has an average of 7 wt% monazite and 5 wt% gorceixite/goyazite (Jaffe & Selchow 1960). The development of extensive supergene REE mineralization in laterites above carbonatites does not require independent REE minerals to be present in the carbonatite source. Rather, the source of the REEs is carbonates and apatite. Large quantities of supergene REE minerals can be produced in tropical climates with moderate to high rainfall, and in those complexes where interior drainage and a basin-type topography traps residuum from decalcified carbonatite. Such conditions exist at Mrima, Araxá, Catalão I, Morro dos Seis Lagos (Brazil), and Cerro Impacto (Venezuela); Twareitau Mountain (Guyana); and in certain parts of Bonga Mountain (Angola); and Chiriguelo (Paraguay).

The most common supergene REE mineral in carbonatites is monazite. It is abundant because most carbonatites contain accessory apatite, and REEs and PO_4^{3-} are liberated from carbonates and apatite during the weathering process to form monazite. In circular carbonatite complexes, supergene monazite is concentrated in central areas, whereas crandallite-group, REE minerals are concentrated in the outer parts where alkaline silicate rocks and country rocks can contribute Al.

Supergene monazite occurs as submicroscopic crystallites, with individual crystal dimensions of about 20 nm; they form porous, greenish-yellow, spongy aggregates that may be amenable to beneficiation from laterite. Monazite can also occur as euhedral crystals that originally crystallized as hexagonal (REE)PO$_4$.nH$_2$O. With subsequent dehydration this forms anhydrous, monoclinic monazite similar to the type commonly encountered in the laterites of Araxá, Catalão I, and Mt. Weld.

Supergene monazite shows a higher Y and mid-atomic number REE content than primary monazite, a condition that holds for all of the supergene REE minerals. In some carbonatite complexes, both supergene and hydrothermal REE mineralization include xenotime or churchite as very fine-grained botryoidal or fibrous masses, easily distinguished from the other REE minerals even in the earthy form, because of their specific cathodoluminescence spectra. Hydrothermal and supergene processes can produce HREE enrichment. In view of the current interest in Y and HREEs for superconductors and permanent magnets, the natural enrichment of Y in carbonatites should be of interest to exploration geologists.

Other REE minerals commonly found in carbonatite laterites include bastnaesite, parisite, synchysite, and crandallite-group minerals, florencite [(REE)Al$_3$(PO$_4$)$_2$-(OH)$_6$], gorceixite [BaAl$_3$(PO$_4$)(PO$_3$OH)(OH)$_6$], and goyazite [SrAl$_3$(PO$_4$)$_2$-(OH)$_5$.H$_2$O]. Rhabdophane [(REE)PO$_4$.H$_2$O] is less commonly encountered than monazite or the crandallite-group minerals.

Almost invariably during the development of supergene REE minerals, Ce^{3+} is oxidized to Ce^{4+} and separated from the other REEs, commonly forming cerianite (CeO$_2$) which is usually inextricably associated with another Ce-depleted mineral.

Supergene REE mineralized laterites overlying carbonatite bodies may constitute the largest source of REEs in the world. There is also evidence that natural separation of the REEs during hydrothermal and weathering activities can produce enrichment of mid-atomic number REEs and Y in carbonatites. The successful beneficiation of REE ore and development of viable, separated REE products from laterites will require technical innovations.

7.3 NIOBIUM MINERALIZATION

Carbonatites and alkaline rocks constitute the world's largest source of Nb, and the literature about Nb in these environments is extensive (Vlasov 1966). Niobium may proxy for other elements in several rock-forming minerals (e.g. perovskite, rutile, titanite), or may occur as a major element in some accessory minerals (e.g. ferrocolumbite, pyrochlore, wöhlerite). The presence of Nb minerals or enrichment in Nb is a good criterion for carbonatite identification. Because some carbonatites contain low Nb concentrations (Bowden 1968), the absence of anomalous Nb in carbonatite exploration programmes does not rule out the presence of carbonatite.

Niobium is usually concentrated in intermediate pulses of carbonatite activity (e.g. Araxá and Catalão I). In contrast, early carbonatites and very late pulses enriched in REEs are relatively impoverished in Nb. Examples of REE-enriched carbonatites with anomalously low Nb contents include Mountain Pass, Wigu Hill, Itapirapuá, and Adiounedj, Mali (unpublished data, A. N. Mariano).

7.3.1 *Major niobium minerals*

Pyrochlore is the only mineral that is mined for Nb in carbonatites. Loparite [$(Ce, Na, Ca)_2(Ti, Nb)_2O_6$] is mined in the USSR in agpaitic nepheline syenite complexes, but not in carbonatite bodies (Vlasov 1968). Latrappite and niobian perovskite can crystallize together with pyrochlore in carbonatites, e.g. Oka, but the perovskite minerals were not recovered during pyrochlore extraction at Oka (Gold *et al.* 1967).

At least four other Nb minerals can occur in sufficient quantities to affect or dominate the Nb content of a carbonatite. They include ferrocolumbite [$FeNb_2O_6$], fersmite [$(Ca, REE, Na)(Nb, Ta, Ti)_2(O, OH, F)_6$], niocalite [$Ca_4NbSi_2O_{10}(O, F)$], and wöhlerite [$NaCa_2(Zr, Nb)Si_2O_8(O, OH, F)$].

Pyrochlore The crystal chemistry of pyrochlore in carbonatites has been reviewed by Hogarth (this volume). Recent interest in the pyrochlore structure as a host for the storage of high-level, nuclear waste has prompted more studies into the crystal chemistry and chemical exchange mechanisms in pyrochlore (Chakoumakos 1984, Lumpkin & Ewing 1985).

Pyrochlore most commonly occurs as disseminated octahedra in carbonatite. One such example is the euhedral pyrochlores dispersed in sövite at the Hydro Quarry in Fen. Although this type of occurrence usually constitutes low-grade, sub-economic mineralization, one exception is the Bonga carbonatite where coarse-grained disseminated pyrochlore locally makes up more than 2 vol% of the sövite.

Primary pyrochlore also occurs in veins that cut earlier carbonatite. It may be present in ore quantities commonly accompanied by apatite, biotite, dolomite, magnetite, and phlogopite. Pyrochlore veins are characterized by brecciation. The veins may be as much as 10 Ma later than the host carbonatite (Eby & Mariano 1986). The vein pyrochlores differ chemically from pyrochlore found disseminated in the earlier carbonatite. The best examples of transgressive pyrochlore mineralization can be seen at Araxá and Catalão I. Although both carbonatites have been lateritized and have undergone some eluvial enrichment, fresh carbonatite below the laterite at Araxá has revealed high-grade, pyrochlore mineralization (Silva 1985).

Most alteration in pyrochlore is deuteric, involving predominantly *A*-site cation exchange. At Wigu Hill disseminated pyrochlore in rauhaugite shows replacement of Na and Ca by Ba and REEs. No secondary REE minerals occur in this rock. In fresh carbonatite drill core from the centre of the Araxá complex, the pyrochlores are the normal Na–Ca types, but microfractures within the pyrochlore grains show incipient alteration to bariopyrochlore.

There are many examples of carbonatites where pyrochlore has crystallized directly from hydrothermal fluids. Examples include Araxá and Catalão I, Chiriguelo, and Panda Hill, Tanzania. These pyrochlores crystallized with quartz, fluorite, and bastnaesite-type minerals, and occur as fine-grained euhedral, bright-yellow octahedra, rich in Pb. Unlike the larger, primary pyrochlores in these complexes, they are restricted in size (50–200 μm), and differ radically in composition (Table 7.2).

Although the Araxá laterite is fairly thick, the bariopyrochlore has not been significantly affected by chemical weathering. A chondrite-normalized REE distribution plot for Araxá bariopyrochlore has a positive Ce anomaly, suggesting that the pyrochlore has not been altered during lateritic weathering.

In many carbonatites pyrochlores are low in Ta (Bakes *et al.* 1964), a point that was also emphasized by Heinrich (1966). However, Hogarth (this volume) shows that several carbonatite pyrochlores have high Ta values. The Ta contents of pyrochlores may vary considerably within a single complex, although late-stage pyrochlores are usually enriched in Ta. Yefimov *et al.* (1985) state that pyrochlores from elongate carbonatite-bearing complexes differ chemically from those of carbonatites in ring complexes. The former have elevated Ti and U but lower Ta and

Table 7.2 Electron microprobe analyses of pyrochlores.

	1	2	3	4	5
Wt%					
Nb_2O_5	40.41	63.51	56.35	44.41	36.52
Ta_2O_5	12.82	0.69	0.00	17.94	17.83
TiO_2	2.65	3.74	3.42	3.31	7.42
ZrO_2	0.40	0.75	1.04	0.15	0.49
SiO_2	2.69	0.16	0.47	0.46	3.39
ThO_2	1.47	2.38	0.86	0.11	0.10
U_3O_8	16.23	0.17	0.03	14.83	18.38
Al_2O_3	0.40	0.43	0.06	0.01	0.05
Fe_2O_3	1.40	1.18	0.30	0.29	3.88
MnO	0.33	0.00	0.00	0.07	0.02
MgO	0.09	0.03	0.00	0.02	0.00
PbO	9.62	0.89	21.70	0.14	3.07
Ce_2O_3	2.06	6.50*	6.09*	0.85	1.20*
CaO	3.02	0.41	0.07	9.59	3.38
SrO	1.19	0.00	0.08	0.28	0.00
BaO	3.36	15.50	6.13	0.14	2.27
Na_2O	0.05	0.04	0.10	6.26	0.00
K_2O	0.10	0.02	0.13	0.04	0.00
F	0.55	0.23	NA	3.02	NA
	98.84	96.63		101.92	
$-O = F$	-0.23	-0.10		-1.26	
Total	98.61	96.53	96.83	100.66	98.00

1. Pyrochlore – Hydro Quarry sövite: Fen, Norway
2. Bariopyrochlore – laterite; orebody: Araxá, Brazil
3. Plumbopyrochlore – laterite; orebody: Araxá, Brazil
4. Pyrochlore – rauhaugite; Bone Creek, Blue River, BC, Canada
5. Pyrochlore – rauhaugite; Ngualla, Tanzania
NA – not analysed; *represents total REEs.

Zr than the latter. This observation, however, does not seem to hold for North American carbonatites. Consistently high Ta pyrochlores are found in carbonatite sills of the Blue River area of British Columbia, Canada, as well as the linear Crevier carbonatite of Quebec (Bergeron 1980).

Ferrocolumbite In most carbonatites, columbite has been described as a replacement mineral after pyrochlore (Heinrich 1966, Kapustin 1980), but in many cases texture, morphology, and chemistry suggest a primary magmatic origin. On the basis of texture, ferrocolumbite and pyrochlore, found in a fine-grained, apatite rauhaugite from the Fir Claim in Blue River (unpublished data, A. N. Mariano), clearly precipitated together. The Blue River pyrochlores are unusually high in Ta (Table 7.2). In contrast, Blue River ferrocolumbites show no detectable Ta. Replacement of pyrochlore to ferrocolumbite would require removal of Ta. Chemically this is unlikely and, furthermore, there are no secondary Ta minerals in the rocks. In unweathered carbonatite from Blue River, Fen, and Ngualla most ferrocolumbites are isolated or intergrown with pyrochlore and are not replacement products.

Ferrocolumbite commonly occurs as a primary Nb mineral in some carbonatites. Examples include the Vipeto area of the Fen complex; the Mill Claim, Bone Creek Claim, and Fir Claim of Blue River; and Gem Park, Colorado, USA. In these carbonatites ferrocolumbite occurs as subhedral prisms containing minor amounts of Ti and Mn. Carbonatite ferrocolumbites, unlike those of alkali granites and pegmatites, are opaque and rarely show red internal reflection and translucency related to Ta and Mn (Ramdohr 1969). In general, ferrocolumbites from carbonatites are very low in Ta and Mn.

Fersmite Fersmite is reported as a replacement mineral after pyrochlore (Heinrich 1966, Kapustin 1980), but evidence from some carbonatites suggests that this may not always be the case.

In the Blue River carbonatite, fersmite occurs as isolated crystals adjacent to pyrochlore or ferrocolumbite, or as intergrowths with pyrochlore. No evidence can be seen for a secondary origin for fersmite.

Niocalite Niocalite, the major Nb mineral in sövite in some areas of the Bond Zone at Oka (Nickel et al. 1958), contains 16–19 wt% Nb_2O_5 (Gold et al. 1967). Where encountered with sövite in the Bond Zone, niocalite is accompanied by trace amounts of pyrochlore and latrappite (Mariano 1985). Bulk sample analysis of Bond Zone sövite from the high niobium-bearing areas show that niocalite contributed 50 wt% of the Nb content. As a silicate phase with a relatively low Nb content, niocalite is an economically deleterious mineral in the Bond Zone.

Wöhlerite Although wöhlerite is well documented in alkalic igneous rocks (Sørensen 1974, Kapustin 1980), it has until recently escaped recognition in carbonatites. In the Prairie Lake carbonatite of Ontario, Canada, wöhlerite occurs as a primary mineral in sövite, silicocarbonatites, pyroxenites, ijolites, urtites, melteigites, and malignites. Wöhlerite can reach 7 vol% in some of the silicocarbonatites and malignites, and in such cases accounts for virtually all the Nb found in these rocks (A. N. Mariano & P. L. Roeder, unpublished data). Using CL, wöhlerite has been found in carbonatites at Prairie Lake and Oka (Canada), Tchivira (Angola), Muri Mountains (Guyana), and In Imanal (Mali). The Nb_2O_5 content in wöhlerites from six localities was found to range from 12 to 16 wt% (P. L. Roeder, pers. comm.).

Wöhlerite is invariably associated with other Nb minerals. At Oka, wöhlerite with 15% Nb_2O_5 is associated with pyrochlore and latrappite.

7.3.2 Niobium in rock-forming minerals

The distribution of Nb in carbonatites can be extensive if perovskite and titanite are present as rock-forming minerals. Although considerable amounts of Nb can occur in rutile and brookite, these minerals do not occur in large quantities in carbonatites. Niobium is contained in perovskite in carbonatite complexes peripheral to the Paraná Basin of Brazil. The complexes Araxá, Catalão I, Salitre, Serra Negra, and Tapira are grouped together by Ulbrich & Gomes (1981). With the exception of Araxá, pyroxenites associated with these complexes contain vast quantities of perovskite with tenths of a wt% Nb. In contrast, perovskite is virtually absent in Araxá where Nb is concentrated in pyrochlore in the core carbonatite in quantities that dwarf all other known Nb deposits in the world.

In some carbonatites the amount of Nb in perovskite may be substantial, whereas the same mineral in associated silicate rocks usually contains significantly lower Nb abundances. This is well demonstrated at Oka where the carbonatites contain latrappite [$(Ca, Na)(Nb, Ti, Fe)O_3$], a variety of perovskite. Niobium minerals or Nb-rich perovskite are absent from the silicate rocks (Eby 1971). At Magnet Cove, perovskite in the silicate rocks contains little Nb whereas sövite from the Kimsey quarry contains niobian perovskite.

An example of Nb in titanite is shown in the Howard Creek carbonatite near Blue River, British Columbia. Titanite, occurring in large quantities in a magnesio-hastingsite amphibolite, has an average Nb value of ~0.7 wt%. The associated carbonatites are essentially devoid of pyrochlore. In contrast, the carbonatites in the nearby Verity, Paradise, and Bone Creek areas contain pyrochlore but associated amphibolites contain only trace amounts of titanite (unpublished data, A. N. Mariano).

7.4 APATITE

On a world level, the monetary value of phosphate is greater than any other commodity derived from carbonatite mining. In 1987 the income from carbonatite apatite from Brazil and South Africa alone exceeded that derived from the copper production at Phalaborwa.

Most of the world's phosphate production comes from marine phosphorites which yield an average BLP (bone phosphate of lime – $Ca_3(PO_4)_2$) of 70 wt% of the beneficiated product. In contrast, carbonatite apatites produce up-graded phosphatic material containing 79–88% BLP. The higher average BLP content makes carbonatite-derived phosphates more desirable in some markets, particularly those of western Europe.

Primary apatite is probably the most widespread accessory mineral in early- and intermediate-stage carbonatites and associated alkaline rocks. In some carbonatites, glimmerites, ijolites, pyroxenites, and silicocarbonatites the apatite content can exceed > 50 vol%; in such cases the rock is termed an igneous phosphorite. Monomineralic apatite rock occurs in many carbonatites. Examples include Mt. Weld, Catalão II and Salitre I (Brazil), the Holla Farm area of the Fen complex, and the central core area of the Bailundo carbonatite in Angola. At the In Imanal carbonatite complex, phosphorite was recognized by Mariano (1983) and mapped by Sauvage & Savard (1985).

The Khibina massif of the Kola Peninsula, containing the largest known accumulation of monomineralic apatite rock, consists predominantly of nepheline syenites with minor carbonatite (J. Gittins, pers. comm.). The Kovdor alkali-ultramafic complex of the Kola Peninsula also has minor carbonatite, and contains ores of apatite, baddeleyite, magnetite, and vermiculite similar to Phalaborwa but without copper (Smirnov 1977, Notholt 1979).

Although apatite is primary in most carbonatites, eluvial accumulations can occur. Karst-developed apatite deposits associated with carbonatite include Cargill in Canada (Sandvik & Erdosh 1977) and Matongo-Bandaga (United Nations Development Program 1979). Initial exploration at Matongo-Bandaga established 5.1 million tons of 13% P_2O_5.

Apatite can be destroyed during the weathering cycle. At Araxá and Catalão I, bariopyrochlore is mined from laterite in the centre of the complexes, but primary apatite is totally decomposed. This also holds true for the anatase deposits in Brazil, where weathering has decalcified perovskite and has also destroyed most of the apatite from the original pyroxenites. In Araxá apatite is mined approximately 1 km from a niobium mine where apatite is absent. The apatite ore, usually described as residual with secondary collophanic phosphate (Grossi Sad & Torres 1978), contains large quantities of optical quality, light-green hydrothermal barite. Much of the ore consists of ovoid apatite grains with overgrowths of secondary apatite. The persistence of apatite ore in the outer annulus at Araxá and its absence in the central core is a result of the degree of lateritic weathering. Apatite fission-track dates from the centre and the northwest edge of the complex are similar, suggesting that hydrothermal activity was closely related to carbonatite emplacement (Eby & Mariano 1986).

Apatite from carbonatites and alkaline igneous rocks contains high REE abundances. Some apatite in sövite from the Bond Zone at Oka averages > 7 wt% REEs (Mariano 1985), and apatite from alkali syenites of Pajarito Mountain, New Mexico contains > 19 wt% REE oxides (Roeder *et al.* 1987). In general, however, carbonatite apatites rarely exceed 2 wt% REEs, and at present REEs are not a by-product of phosphate mining in carbonatites.

7.5 TITANIUM MINERALIZATION

Titanium is an important element in most carbonatites and alkaline complexes where it may form its own minerals or may substitute for other elements in rock-forming minerals. The mineralogy of Ti in carbonatites is diverse but only few minerals contain Ti in potentially economic grade and tonnage.

Although titanite is not considered to be a Ti ore, it occurs in vast quantities in some carbonatites and some genetically related igneous rocks. For example, at Howard Creek a carbonatite-related, magnesio-hastingsite amphibolite contains large quantities of coarse-grained euhedral titanite. The amphibolite contains >100 million tons of TiO_2 averaging 15% in grade (unpublished work, A. N. Mariano & J. A. Gower).

In hydrothermal deposits associated with carbonatites rutile and brookite may be locally abundant. They have usually crystallized together with carbonates, fluorite, quartz, and low-temperature feldspars, as void fillings and as disseminations in coexisting igneous rocks, fenites, and country rock. At Magnet Cove several rutile and brookite hydrothermal deposits occur (Peterson 1966), and one body, the Magnet Cove Rutile Co. deposit, was mined for a short time.

All members of the perovskite group occur, mostly as accessory minerals in carbonatites, alkaline rocks, and fenites. With the exception of loparite, $(Ce, Na, Ca)_2(Ti, Nb)_2O_6$, from the agpaitic nepheline syenite complex of the Lovozero Massif in the Kola Peninsula (Smirnov 1977), perovskite has never been exploited as an ore mineral.

Perovskite has been recognized in several carbonatites from Minas Gerais and Goiás in Brazil (Tröger 1928, Palache *et al.* 1944), and recently, as a consequence of phosphate and niobium exploration, several carbonatites were found to contain large quantities of Ti.

At least four carbonatites in the eastern periphery of the Paraná Basin include units with ore accumulations of TiO_2 in the form of anatase, a product of weathering of perovskite. These Cretaceous carbonatites include Salitre I and II, Serra Negra, and Tapira, all in Minas Gerais, and Catalão I in Goiás. Anatase is derived in these cases from pyroxenite, one of the earliest rocks to crystallize.

In the circular structures at Catalão I, Serra Negra and Tapira perovskite-bearing pyroxenites form annuli between central carbonatite cores and surrounding, domed Precambrian sediments which are locally fenitized and silicified. Salitre I and II, satellites of Serra Negra, occur on the southeast edge of the complex. Salitre I is oval-shaped, and consists of pyroxenite and a small, apatite-rich, carbonatite body. Salitre II, a circular plug covering an area of approximately 2.5 km^2, is almost exclusively pyroxenite transected by thin carbonatite veins and minor, late lamprophyres. Doming of peripheral country rock, massive fenite caps, absence of extrusive units, and preservation of overlying sediments, indicate that these carbonatites did not breach their contemporary land surface.

In some areas pyroxenite grades into glimmerite and the presence of inclusions of apatite, calcite, magnetite, and salite in phlogopite indicates the primary nature of the glimmerite. Magnetite and perovskite occur as cumulus aggregates, disseminations and bands that alternate with apatite, calcite, and pyroxene layers. Apatite and magmatic calcite are always present. The apatite content varies considerably, and in some areas the perovskite-rich, Ti orebodies converge into apatite-rich, phosphate orebodies. Perovskite is restricted to the pyroxenites and some peridotite bands that occur between the pyroxenites and country rock. Perovskite is absent, or rare, in the carbonatite.

7.5.1 In-situ *decalcification of perovskite to anatase*

Intense lateritic weathering of the pyroxenites from Catalão I, Salitre I and II, Serra Negra, and Tapira is extensive and varies in depth between 40 and 120 metres. The breakdown of apatite, calcite, and salite is accompanied by decalcification of perovskite. Perovskite from all of the pyroxenites contains low, but significant, amounts of Nb, REEs, and Th. Abundances of these elements in perovskite are similar to those found in perovskite from pyroxenites of the Powderhorn carbonatite of Gunnison County, Colorado (Nash 1972). Perovskites from Tapira contain ~0.5 wt% Nb, ~1 wt% REEs, and ~0.05 wt% Th.

When these perovskites are decalcified, the Nb, REEs, and Th are incorporated in the newly formed anatase. For example, anatase concentrates from Tapira contain 0.1–0.2% Nb_2O_5, ~1% REEs and 0.08% Th. Perovskite and anatase data from Salitre II show very similar patterns on a chondrite-normalized, REE distribution plot, suggesting that anatase was derived from perovskite. In the Fazenda Boa Vista area of the Tapira complex, perovskite appears as idiomorphic octahedra, and in the weathered zone these are pseudomorphosed by anatase (Cassedanne & Cassedanne 1973).

In the pyroxenite bodies, anatase is concentrated in the zone of weathering, and at deeper levels residual cores of perovskite are found within the anatase grains. Below the zone of weathering only fresh perovskite is encountered. Electron microscopy shows that anatase commonly displays a platey texture. Rare earth element minerals occur as submicroscopic inclusions in the anatase or as overgrowths produced by weathering. They include brockite, cerianite, monazite, and several of the crandallite-group minerals.

Anatase mineralization, because of its origin, is amenable to open-pit mining. Grade and tonnage data for anatase in Brazilian carbonatite complexes, shown in Table 7.1, demonstrate that these are major TiO_2 resources. Pilot plant operations have been established for all of the complexes by various Brazilian firms. At Tapira a 92–94 wt% TiO_2 concentrate is currently being stockpiled. This type of concentrate is ideally suited to pigment manufacture because of the relatively low content of Cr, Fe, and V. The Brazilian anatase resources are large enough to influence the world supply and demand balance for Ti. Pyroxenites with perovskite mineralization similar to the Minas Gerais and Goiás pyroxenites are also known from other areas, including the Powderhorn carbonatite complex (Nash 1972). In addition, drill core associated with a deep-seated magnetic and gravity anomaly, near Rison in south central Arkansas, consists of major perovskite in pyroxenite; and in the USSR, perovskite–titanomagnetite–apatite mineralization occurs in pyroxenites of the Nizhnesayan Massif (Nichipurenko *et al.* 1980). In the Amazon area 500 million tons of anatase are reported in laterite from the alkaline complex of Maicuru in Pará, Brazil (Revista Minerios 1987, Issler *et al.* 1975). Residual perovskite–magnetite boulders have been found in laterite above the Maraconaí complex also in Pará (D. L. Mathias, pers. comm.). In a more recent study, the Maraconaí complex was found to contain economic quantities of anatase (Fonseca & Rigon 1984).

7.6 STABILITY OF CARBONATITE MINERALS IN LATERITES

Carbonatites in areas of combined high humidity and high temperatures can, over a long period, show extreme lateritic weathering so that only the most resistant primary minerals survive. Magnetite, pyrochlore, and ilmenite are only slightly affected during laterite formation in the Cretaceous carbonatites of Minas Gerais and southern Goiás, with the result that these minerals abound in the laterites. Although this type of weathering extends to depths of 300 m (Silva *et al.* 1979), it is still only an intermediate stage of lateritic weathering. In both the Lueshe and Bingo carbonatites of Zaire, magnetite and pyrochlore are essentially preserved in spite of extensive, deep weathering. At Mrima, Kenya, the weathered carbonatite residuum also contains pyrochlore and magnetite. In contrast, the breakdown of minerals from carbonatites in Amazonas-type environments is far greater than that found in the laterites of Africa, southern Brazil, and Paraguay.

The decomposition of minerals in some carbonatites is not necessarily seen in associated alkalic igneous rocks or fenites. At Cerro Impacto, carbonatite has been completely weathered but well-preserved apatite, feldspars, and other minerals can be found in weathered fenites. In the Muri Mountains, apatite, calcite, feldspar, and pyrochlore are totally preserved in nepheline syenites and fenites, whereas the nearby Twareitau Mountain, a suspected carbonatite, is totally lateritized.

The highly altered nature of carbonatite minerals in such environments is probably linked to high concentrations of H_2CO_3 in meteoric waters and, to a lesser extent, the development of H_2SO_4 from accessory sulphides. In addition, many carbonatite bodies tend to develop areas of internal drainage.

These conditions which hold, more or less, for all carbonatites do not explain the stability of pyrochlore and magnetite in Araxá-type, and their instability in Cerro Impacto-type, laterites. A possible explanation may be age differences. In Table 7.3 a list of some carbonatites with niobium-bearing laterites is given along with their ages. Most are characterized by large carbonatite-derived laterite covers that contain high Nb_2O_5 contents.

The mineralogy and geochemistry of carbonatite laterites are a complex function of age, depth of weathering, structure, geomorphology, tectonics, and the physicochemical nature of the pre-lateritized rock units. In Lower Palaeozoic laterites incipient breakdown of pyrochlore is indicated by leaching of elements contained in the A and Y sites. According to Es'kova & Nazarenko (1960), hydration of pyrochlore causes selective leaching and removal of Ca, Na, REEs (A sites), and F (Y sites). Leached pyrochlores are usually very friable and, as a result, laterite with

Table 7.3 Some examples of carbonatites with niobium-bearing laterites.

Carbonatite	Description	Degree of weathering	Age of complex
Araxá, Minas Gerais, Brazil	Deep weathering, circular structure-central depression	Moderate: magnetite and pyrochlore preserved	81 Ma (Hasui & Cordani 1968)
Bonga, Angola	Mountain with laterite in central depression; shallow, peripheral laterites at base; large internal karst system	Moderate: magnetite and pyrochlore preserved	112 Ma (Matos Alves 1968)
Catalão I, Brazil	Deep weathering, circular structure with central depression	Moderate: magnetite and pyrochlore preserved	113 Ma (Eby & Mariano 1986)
Lueshe, Zaire	Hilly and dissected local areas of deep weathering	Intermediate: some pyrochlore stripped of A-site cations; most pyrochlore friable and pulverable	516 Ma (Maravic & Morteani 1980)
Matongo-Bandaga, Burundi	Hilly and dissected, local areas of shallow laterite and local karst areas with preserved primary apatite	Moderate: magnetite and pyrochlore preserved	739 Ma (Tack et al. 1983)
Ngualla, Tanzania	Hilly and dissected, local areas of shallow laterite	Moderate: magnetite preserved, pyrochlore strongly altered	1040 Ma (van Straaten 1986)
Twareitau Mountain, Guyana	Hilly and dissected, very deep lateritic weathering	Extreme: magnetite and pyrochlore destroyed; pyrochlore in fresh nepheline syenites and fenites	Adjacent mountain, Mutum, dated 1026 Ma (Teixeira et al. 1976)
Mt. Weld, Australia	Lateritic cover with moderate depth	Extreme: magnetite and pyrochlore destroyed at shallow level; intermediate levels show pyrochlore with stripped A-site cations; fresh carbonatite below weathering horizon shows normal pyrochlore	2064 Ma (Goode 1981)

good Nb assays may lose most of the Nb during beneficiation, or simply by washing and cleaning. These observations are based on the study of laterites from Nb-mineralized carbonatites that range in age from 80 Ma to > 2000 Ma. Magnetite is more stable in old laterites than is pyrochlore, but in the most intense weathering profiles of old laterites (> 500 Ma) even magnetite becomes unstable (e.g. Cerro Impacto, Morro dos Seis Lagos, Twareitau, and the upper zones of the laterite at Mt. Weld).

In carbonatites that have undergone extreme lateritic weathering, the mineralogy changes with depth. In the upper profile, the laterite consists mostly of insoluble ferric iron oxides, aluminum oxides, clays, supergene monazite, crandallite-group minerals rich in Ba, Nb, REEs and Sr, and cerianite. In this zone all primary minerals have disappeared with the exception of trace amounts of zircon and rutile. Vadose water or low temperature, hydrothermal solutions can introduce quartz, and caliche-type calcite and supergene apatite into the laterite. The calcite and apatite can be easily distinguished from primary minerals by their CL emission spectra (Mariano 1988). Both minerals contain Sm^{3+}, Dy^{3+}, and Eu^{3+}, but Mn^{2+} is absent. In the development of REE-bearing minerals usually the LREEs are dominant, but in some cases the HREEs and Y are separated from the LREEs and selectively concentrated due to their higher migratory capacity (Vlasov 1966). In such cases, xenotime (YPO_4) or churchite ($YPO_4.2H_2O$) may occur in appreciable quantities (unpublished data, A. N. Mariano).

As lateritic weathering diminishes with depth, magnetite is encountered, followed by partly decomposed pyrochlore octahedra consisting of a pyrochlore core surrounded by an outer rim of a crandallite-type mineral, rich in Nb. At deeper levels fresh pyrochlore appears, and eventually primary apatite becomes prominent. With continued depth, partly decomposed calcite or dolomite is encountered followed by fresh unweathered carbonatite.

Positive Nb anomalies in the Morro dos Seis Lago laterite were attributed to rutile and brookite (Justo & Sousa 1984). In the case of the Cerro Impacto laterites the Nb source was not identified (Aarden *et al.* 1973). The amount of rutile and brookite present is not enough to account for the high Nb contents of the laterites, which range from tenths of a per cent to > 2.5 wt%. Although some Nb may be contained in iron oxide phases, the presence of Nb-bearing crandallite-group minerals in laterites from Morro dos Seis Lagos, Cerro Impacto, Twareitau, and Mt. Weld, where pyrochlore has been destroyed, at least partly accounts for the high Nb values of the laterites.

7.7 VERMICULITE

Carbonatite and genetically related silicate rocks commonly contain high concentrations of biotite or phlogopite, and many carbonatites and pyroxenites grade into glimmerites. Biotite or phlogopite concentrations in carbonatite complexes are sometimes attributed to metasomatic fluids, but more commonly they are products of primary magmatic crystallization, or reaction products unrelated to fenitization.

In the Jacupiranga carbonatite of São Paulo, Brazil, sövite contains large xenoliths of pyroxenite and ijolite strongly altered at the margins to phlogopite, magnetite, and other mafic minerals. Magnetite concentrations in bands that alternate with phlogopite rule out fenitization as a process in this case, because magnetite does not crystallize under conditions of strong alkali metasomatism (Mariano 1988).

Under certain conditions, such as surface weathering or circulating groundwaters, biotite and phlogopite may be transformed to vermiculite, and as a result vermiculite prospects are numerous in carbonatite environments.

The world's largest vermiculite deposit at Rainy Creek near Libby, Montana is associated with an alkaline-ultramafic igneous complex where glimmerites and biotite-rich pyroxenites have been altered to vermiculite and hydrobiotite. Although carbonatites have not been found at Rainy Creek, Boettcher (1967) suggests that fenites in the complex are derived from fenitizing fluids that may have migrated from a deep-seated carbonatite. The formation of the world's second largest vermiculite deposit at Phalaborwa in pegmatoidal pyroxenites, is attributed entirely to surface weathering of phlogopite (Palabora Mining Company 1976).

Vermiculite occurrences are also documented for Cargill (Vos 1981), Catalão I (Carvalho 1974), Mud Tank, Australia (Moore 1984), and Tapira (Grossi Sad & Torres 1971).

7.8 FLUORITE

Accessory fluorite is common in carbonatites and alkaline rocks. Fluorite concentrations of economic significance in carbonatites, however, are confined to late hydrothermal activity. Fluorite occurs as euhedral crystals lining solution cavities or voids in breccias. Other indications of hydrothermal activity include the presence of quartz and the low crystallization temperatures (100–150°C) revealed by fluid inclusion studies on fluorites from Amba Dongar, India, and Okorusu, Namibia (Roedder 1973).

Carbonatite-derived solutions that travel long distances and eventually deposit fluorite in overlying sediments may give little or no geochemical evidence of derivation from a carbonatite source (Deans 1978). The spatial association of the fluorite deposits of the Illinois–Kentucky region with Hick's Dome and several carbonate-bearing explosion breccias in the same general area, is a strong indication that they may be derived from a carbonatite source.

Fluorite is currently being mined at Amba Dongar (Deans et al. 1973) and was mined for a short period at Okorusu (Verwoerd 1986). The Mato Preto carbonatite of Paraná, Brazil, contains about 4.35 million tons of 58% CaF_2 (Anuário Mineral Brasileiro 1986). The Bayan Obo iron-REE mine contains vast quantities of fluorite intimately associated with bastnaesite. According to Hou et al. (1987) the REEs were probably introduced by exhalative hot-spring emanations into a marine basin. In contrast, Zhou et al. (1980) maintain that the mineralization is associated with carbonatite. The strong LREE enrichment of the apatite, bastnaesite, and monazite, and the Fe^{3+} activation shown in the CL emission spectra for some of the feldspars at Bayan Obo support a carbonatite origin (unpublished data, A. N. Mariano).

7.9 BARITE

Although barite can be found in all stages of carbonatite evolution, it is usually concentrated in late-stage intrusions where it can crystallize directly from a carbonate melt (e.g. Mountain Pass), or from low-temperature hydrothermal fluids (e.g. Araxá, Chiriguelo, Mrima).

At Mountain Pass, barite is both primary and, to a lesser extent, hydrothermal, forming veins that cut the carbonatite. Primary barite at Mountain Pass is found in sövite or rauhaugite and is associated with calcite, dolomite, bastnaesite, and parisite. Vein barite at Mountain Pass can be found associated with calcite, quartz, and strontianite.

Sövite from the centre of the Chiriguelo complex contains minor biotite and about 1% disseminated anhedral barite. In contrast, local laterite accumulations near the sövite contain veins of euhedral barite and quartz that cut residual layering in the laterite (A. N. Mariano, unpublished data). Veins rich in euhedral barite of optical quality are associated with laterites that overlie carbonatite at Araxá where total reserves are 463 million tons of 20% barite (Silva 1985). Barite from the laterites at Chiriguelo and Araxá probably precipitated from low-temperature groundwater solutions.

7.10 VANADIUM

Probably the two most important vanadium-bearing minerals in carbonatites are aegirine and magnetite. The V content of aegirine from various rocks at Magnet Cove averages 0.1 wt% (Erickson & Blade 1963), while an aegirine-augite from the Libby alkaline-ultramafic complex (Larsen & Hunt 1914) contains 2.71 wt% V. The Matongo-Bandaga carbonatite contains aegirine from an aegirine sövite (unpublished data, A. N. Mariano) with an average value of 1.08 wt% V (S. E. Haggerty, pers. comm.).

According to Heathcote & Owens (1981) the V ore deposit at Potash Sulphur Springs, Arkansas is the product of weathering of an altered aegirine pyroxenite. The ore occurs in an aegirine and sanidine fenite zone near the contact between alkaline igneous rocks and country rock. A saprolitic cover was developed that contains a complicated mixture of V minerals including navajoite [$V_2O_5 \cdot H_2O$], montroseite [$(V, Fe)O(OH)$], fervanite [$Fe_4(VO_4)_4 \cdot 5H_2O$], and hewettite [$CaV_6O_{16} \cdot 9H_2O$]. Vanadium is also suspected to be trapped by clays (D. R. Owens, pers. comm.). The deposit was reported to contain about 5 million tons of 0.6 wt% V and was mined only for V. Vanadium abundances in magnetite from some carbonatites range between 0.06 and 0.54 wt% (Prins 1972). In areas of extensive chemical weathering residual laterites may show some enrichment in V. The Cerro Impacto laterite yields V abundances between 0.08 and 0.20 wt% (unpublished data, A. N. Mariano). However, many carbonatite-derived laterites are not associated with V enrichments. Such examples include Morro dos Seis

Lagos, Araxá, Catalão I, Chiriguelo, Twareitau, Bonga, Mrima, Ngualla, and Mt. Weld (unpublished data, A. N. Mariano).

The absence of V enrichment in carbonatite laterites probably reflects V^{5+} mobility in the secondary environment. Vanadium is similar to U in the weathering cycle and neither of these elements forms residual enrichments in areas of strong oxidation.

7.11 STRONTIUM

Carbonatites are notably enriched in Sr, and in exploration programmes Sr is one of the best geochemical indicators for carbonatite. In addition to the high Sr content in the major rock-forming carbonate minerals and apatite, Sr may also crystallize directly from carbonatite melts as strontianite or celestite. At Mountain Pass, barite contains appreciable Sr (Olson *et al.* 1954). Barite and celestite also form exsolution intergrowths (unpublished data, A. N. Mariano).

Strontianite accompanies REE minerals in hydrothermal deposits (Bear Lodge, Wyoming) and supergene crusts (Magnet Cove and Oka) and is also found in fenites from Mountain Pass, Chiriguelo, Panda Hill, and Gem Park (unpublished data, A. N. Mariano).

Strontianite occurs in potential economic grade and tonnage at Kangankunde (Garson & Morgan 1978) and Ondurakorume, Namibia (Verwoerd 1986).

7.12 THORIUM AND URANIUM

Thorium and U are enriched in carbonatites and alkaline igneous rocks relative to crustal abundances. In carbonatites Th is usually more abundant than U. Woolley (this volume) reports average values for U as 8.7 p.p.m. (calciocarbonatite), 13 p.p.m. (magnesiocarbonatite), and 7.2 p.p.m. (ferrocarbonatite). Average values for Th are 52 p.p.m. (calciocarbonatite), 93 p.p.m. (magnesiocarbonatite), and 276 p.p.m. (ferrocarbonatite). Because of the high concentration of Th and U, radiometric signatures can distinguish carbonatites from non-igneous carbonate rocks.

The most important Th- and U-bearing minerals in carbonatites are perovskite, pyrochlore, the REE minerals, and thorite. Many other minerals occurring in trace quantities also contain Th and U. For example, uranothorianite is a by-product at Phalaborwa.

Hogarth (this volume) shows that there is considerable variation in U : Th ratios in pyrochlores. However, Th-rich pyrochlores are rare in carbonatites, and for those carbonatites mined for pyrochlore, e.g. Araxá, Catalão I, Fen, Oka and St. Honoré (Canada), the actinide elements are not very abundant. At Araxá, Fen, and Oka, U and Th enter slag produced from the aluminothermic reduction used for producing ferroniobium. Very high U values have been found in pyrochlores from Blue River, Fen, Ngualla (Table 7.2), Crevier (Bergeron 1980), and Prairie Lake (unpublished data, A. N. Mariano). All of these, with the exception of Prairie Lake, also contain high Ta.

In carbonatites which contain perovskite in major quantities, radiometric anomalies are common. This is mainly due to Th in the perovskite that usually averages a few hundred p.p.m. Examples include Catalão I, Salitre, Serra Negra, Tapira, Powderhorn, and Kerimasi (Tanzania). Decalcification of perovskite usually produces a slight enrichment of Th in the resulting anatase.

Thorium and U are also found in REE minerals, although actinide minerals such as thorite are actually rare at Mountain Pass, Wigu Hill, and Kangankunde. Thorium values for monazite from Kangankunde and Mrima, and bastnaesite from Mountain Pass of 0.07 wt%, 0.11 wt% and < 0.09 wt%, respectively (Deans 1966), are representative of Th abundances in REE minerals from carbonatites. Monazite from carbonatites is always much lower in Th than monazite from pegmatites and mineralized veins associated with granite. In carbonatite complexes thorite is usually associated with REE minerals of hydrothermal origin, and in rödbergite dykes. Anomalous thorite mineralization is also found associated with REE minerals in the fenite rim of Chiriguelo (unpublished data, A. N. Mariano).

Most carbonatites of the Amazon area in South America were located by aerial surveys. They occur as prominent topographic expressions associated with radiometric anomalies. In spite of formidable logistic problems, some of these complexes were explored in considerable detail, primarily for U. However, because of the intense oxidation in these areas and the highly mobile nature of U, it is unlikely that any concentrations would be found in this type of environment. The high radioactive signature of the laterites covering these carbonatites is from residual Th.

Despite anomalous Th and U contents in carbonatites there is no indication, currently, that either of these elements is concentrated in sufficient quantities to form economic deposits.

7.13 COPPER

The only significant copper mineralization in carbonatites is at Phalaborwa (see Eriksson, this volume) where a vertical pipe of phoscorite, with an elliptical cross-section, occurs within a micaceous pyroxenite body (Palabora Mining Company 1976). The phoscorite pipe is associated with a banded carbonatite unit. Both were subjected to fracturing and subsequent infilling by a late transgressive carbonatite. Copper mineralization consists predominantly of early disseminated bornite introduced between the two carbonatite emplacement periods. Later copper mineralization formed after emplacement of the transgressive carbonatite unit and was concentrated along fractures in the latter, predominantly as chalcopyrite. Copper minerals include chalcocite, cubanite, cuprite, malachite, and valleriite $[4(Fe, Cu)S \cdot 3(Mg, Al)(OH)_2]$.

The Cu mineralization at Phalaborwa is unique. As a late emplacement event it may have resulted from residual hydrothermal solutions (Aldous 1986). Although late-stage sulphide mineralization is encountered in many carbonatites, none has shown the same degree of mineralization as that observed at Phalaborwa.

Other commodities including baddeleyite and precious metals (gold and platinoids) are not treated in this chapter. As by-products at Phalaborwa, their economic extraction is dependent on Cu mining.

In some carbonatites (Jacupiranga, Sukulu, Chiriguelo) relatively pure sövites are the source of calcite for cement and agricultural lime which are of particular value to developing countries. Another bulk commodity is hematite ore that has been mined from rödbergites of Fen (Saether 1957) and Kalkfeld, Namibia (Van Zijl 1962).

7.14 CONCLUSIONS

Although carbonatites probably constitute the best source rocks for REEs and Nb, the largest production and economic value results from apatite mining in Europe, Brazil, and South Africa, followed by Cu mining from the Phalaborwa deposit.

Rare earth mineralization in carbonatites can be classified into three categories. The first is primary magmatic crystallization. The Mountain Pass deposit is the only known example, and bastnaesite and parisite are the only ore minerals. The second is hydrothermal mineralization where ancylite, monazite, or bastnaesite-type minerals form replacement veins in carbonatite or related rocks, and are usually accompanied by barite, fluorite, quartz, and strontianite. Wigu Hill and Karonge are examples. The third category is supergene mineralization produced by weathering of carbonatite bodies. The source is not from primary REE minerals but from the chemical breakdown of calcite, dolomite, and apatite.

The only economic Nb mineral in carbonatites is pyrochlore, although other Nb minerals are at least locally enriched and may constitute the dominant source of Nb. They include ferrocolumbite, fersmite, niobian-perovskite, niocalite, and wöhlerite.

Although pyrochlore is a common accessory mineral of magmatic crystallization in carbonatites, transgressive veins that intrude earlier barren carbonatites are the best source of economic grade mineralization. Examples of this type include Araxá and Catalão I.

Infrequently, pyrochlore of hydrothermal origin is associated with fluorite, quartz, and REE minerals. Hydrothermal pyrochlore can be distinguished from magmatic pyrochlore by composition, mineral association, and textural evidence.

In the past, carbonatite pyrochlores were believed to be impoverished in Ta. Many occurrences are now known of pyrochlores with appreciable Ta contents that may constitute attractive economic sources.

In carbonatite laterites older than 500 Ma incipient breakdown of pyrochlore is demonstrated by A-site cation leaching. Prolonged lateritic weathering of carbonatites in the Amazon area destroys pyrochlore and magnetite, and some of the Nb is partitioned into REE-bearing crandallite-group minerals.

In several carbonatite complexes in Brazil, lateritic weathering of pyroxenites rich in perovskite has produced large tonnages of high-grade TiO_2 soils consisting of anatase that is a decalcification product of perovskite. These deposits will have a profound impact on the world Ti market.

Apatite deposits constitute the largest bulk mining in carbonatites. They have a higher phosphate content than marine phosphorites and the future outlook for increased world mining of carbonatite apatite is good.

From the descriptions of the various deposits covered in this chapter it can be seen that the mineral deposits associated with carbonatites can be produced by a variety of ways that include magmatic, hydrothermal, and exogenic processes. At Araxá and Phalaborwa all three processes have produced ore accumulations of various minerals. It may be concluded that carbonatite complexes provide the locus for several metal deposits that are presently of strategic importance, containing elements which are otherwise largely dispersed in the Earth's crust.

ACKNOWLEDGEMENTS

The author gratefully acknowledges the suggestions, advice, and criticisms of this paper made by T. L. Armbrustmacher, K. A. Grace, P. J. Pollard, P. L. Roeder, and W. J. Verwoerd. Particular appreciation is extended to K. Bell for his patience and guidance in the preparation of the manuscript. Finally, special thanks are given to M. D. Druecker, J. Gittins, M. Marchetto, and T. C. James for many stimulating geologic discussions concerning economic minerals in carbonatites.

REFERENCES

Aarden, H. M., J. M. I. de Arozena, P. Moticska, J. G. Navarro, J. Z. Pasquali & R. S. G. Sifontes 1973. *El complejo geologico del area del Impacto Districo Cedeño, Estado Bolivar, Venezuela.* Internal Report Ministerio de Minas e Hidrocarburos, Direccion de Geologia.

Aderca, B. M. & R. Van Tassel 1971. Les gisements de terres rares de la Karonge. *Académie Royale Sciences d'Outre Mer, Classe Société Naturelles et Médicales*, N.S. **XVIII**, (5), 4–117.

Aldous, R. T. H. 1986. Copper-rich fluid inclusions in pyroxenes from the Guide copper mine, a satellite intrusion of the Palabora Igneous Complex, South Africa. *Economic Geology* **81**, 143–55.

Anuário Mineral Brasileiro 1986. Divisão de Economia Mineral-DNPM Setor de Autarquias Norte Quadro ol-Bloco B 70.040 Brasilia (DF)-Brasil.

Bakes, J. M., P. G. Jeffery & J. Sandor 1964. Pyrochlore minerals as a potential source of reactor-grade niobium. *Nature* **204**, 867–8.

Bergeron, A. 1980. *Petrography and geochemistry of the Crevier igneous alkaline complex and the metasomatized country rocks.* M.Sc. thesis, University of Quebec, Chicoutimi.

Boettcher, A. L. 1967. The Rainy Creek alkaline-ultramafic igneous complex near Libby, Montana I: Ultramafic rocks and fenite. *Journal of Geology* **75**, 526–53.

Bowden, P. 1968. Trace elements in carbonatites and limestones. *Nature* **219**, 716.

Canadian Minerals Yearbook 1985. Canadian Government Publishing Center, Ottawa.

Canadian Mines Handbook 1975–1976 edn. Northern Miner Press, Toronto.

Canadian Mines Handbook 1987–1988 edn. Northern Miner Press, Toronto.

Carvalho, W. T. De 1974. *Recursor minerais do complexo ultrámafico-alcalino de Catalão I, Goiás.* Metais De Goias S/A – METAGO Goiânia, Goiás, Brasil.

Cassedanne, J. P. & O. Cassedanne 1973. Note sur l'anatase de Tapira (Minas Gerais, Brésil). *Bulletin Société Française de Minéralogie et Cristallographie* **96**, 316–18.

Chakoumakos, B. C. 1984. Systematics of the pyrochlore structure type, ideal $A_2B_2X_6Y$. *Journal of Solid State Chemistry* **53**, 120–9.

Deans, T. 1966. Economic mineralogy of African carbonatites. In *Carbonatites*, O. F. Tuttle & J. Gittins (eds), 385–413. New York: Wiley.

Deans, T. 1978. Mineral production from carbonatite complexes. In *Proceedings of the First International Symposium on Carbonatites*, Poços de Caldas, Brazil, June 1976, 123–33. Brasilia: Brasil Departamento Nacional da Produção Mineral.

Deans, T., R. N. Sukheswala, S. F. Sethna & S. G. Viladkar 1973. Metasomatic feldspar rocks (potash fenites) associated with the fluorite deposits and carbonatites of Amba Dongar, Gujarat, India. *Institution of Mining and Metallurgy Transactions Section B* **81**, B1–B9.

Denver Post, 25 February 1976. Buttes Gas and Oil Company announcement.

Eby, G. N. 1971. *Rare-earth, yttrium and scandium geochemistry of the Oka carbonatite complex, Oka, Quebec*. Ph.D. thesis, Boston University, Boston.

Eby, G. N. 1975. Abundance and distribution of the rare earth elements and yttrium in the rocks and minerals of the Oka carbonatite complex, Quebec. *Geochimica et Cosmochimica Acta* **39**, 597–620.

Eby, G. N. & A. N. Mariano 1986. Geology and geochronology of carbonatites peripheral to the Paraná Basin, Brazil–Paraguay. *Geological Association of Canada–Mineralogical Association of Canada–Canadian Geophysical Union, Joint Annual Meeting, Ottawa. Programs with Abstracts* **11**, 66.

Engineering and Mining Journal 1986. International Directory of Mining. New York: McGraw Hill.

Engineering and Mining Journal 1987/1988. International Directory of Mining. New York: McGraw Hill.

Erickson, R. L. & L. V. Blade 1963. *Geochemistry and petrology of the alkalic igneous complex at Magnet Cove, Arkansas*. United States Geological Survey Professional Paper 425, United States Government Printing Office, Washington, DC.

Es'kova, E. M. & I. I. Nazarenko 1960. Pyrochlore from the Vishnevye Mountains, its paragenetic associations and chemical composition. *Trudy Institute of Mineralogy Geochemistry and Crystallochemistry of Rare Elements* **4**, 33–50.

Fonseca, L. R. & J. C. Rigon 1984. *Ocorrências de titânio no complexo ultramáfico-alcalino de Maraconaí no estado do Pará*. Anais do XXXIII Congresso Brasileiro de Geologia, Rio de Janeiro.

Garson, M. S. & D. J. Morgan 1978. Secondary strontianite at Kangankunde carbonatite complex, Malawi. *Institution of Mining and Metallurgy Transactions Section B* **87**, B70–B73.

Geologia Do Brasil 1984. República federativa do Brasil, Ministério das minas e energia, Departamento Nacional da Produção Mineral, Brasilia.

Gold, D. P., M. Vallée & J. P. Charette 1967. Economic geology and geophysics of the Oka alkaline complex, Quebec. *Canadian Mining and Metallurgical Bulletin* **60**, 1131–44.

Goode, A. D. T. 1981. Proterozoic geology of Western Australia. In *Precambrian of the Southern Hemisphere*, D. R. Hunter (ed.), 105–203. Amsterdam: Elsevier.

Grossi Sad, J. H. & N. Torres 1971. *Geologia e recursos minerais de complexo de Tapira, M. G. Brazil*. Belo Horizonte DNPM 3.° Distrito, Geologia e Sondagens Ltda.

Grossi Sad, J. H. & N. Torres 1978. Geology and mineral resources of the Barreiro complex, Araxá Brazil. In *Proceedings of the First International Symposium on Carbonatites*, Poços de Caldas, Minas Gerais, Brazil, June 1976, 307–12. Brasilia: Brasil Departamento Nacional da Produção Mineral.

Harrison, M. T. & E. B. Watson 1984. The behavior of apatite during crustal anatexis: equilibrium and kinetic consideration. *Geochimica et Cosmochimica Acta* **48**, 1467–77.

Haskin, L. A., T. R. Wildeman, F. A. Frey, K. A. Collins, C. R. Keedy & M. A. Haskin 1966. Rare earths in sediments. *Journal of Geophysical Research* **71**, 6091–105.

Hasui, Y. & U. G. Cordani 1968. *Idades potássio-argônio de rochas eruptivas mesozoica do oeste mineiro e sul de Goiás*. Anais Do XXII Congresso Brasileiro De Geologia, Belo Horizonte, 139–43.

Heathcote, R. C. & D. R. Owens 1981. Formation of vanadium ore at Potash Sulphur Springs, Arkansas. Geological Society of America, *Abstracts with Programs* **13**, 470.

Heinrich, E. W. 1966. *The geology of carbonatites*. Chicago: Rand McNally.

Hou, Z., Y. Ren, Q. Meng, L. J. Drew, R. L. Erickson & E. C. T. Chao 1987. Characteristics of the Bayan Obo rare earth bearing iron deposit of Inner Mongolia, China. *Geological Society of America, Abstracts with Programs* **19**, (7), 708.

Issler, R. S., M. I. C. Lima, R. M. G. Montalavão & G. G. Silva 1975. Magmatismo alcalino no cráton Guianês. *Anais X, Conferência Geologia Interguianas* **1**, 103–22.

Jaffe, H. W. & D. H. Selchow 1960. *Mineralogy of the Araxá Columbium Deposit*. Union Carbide Ore Company Research Report, 4. Research Center, Tuxedo, N.Y.

Jones, A. P. & P. J. Wyllie 1983. Low-temperature glass quenched from a synthetic, rare earth carbonatite: implications for the origin of the Mountain Pass Deposit, California. *Economic Geology* **78**, 1721–3.

Justo, L. J. E. C. & M. M. de Sousa 1984. *Jazida de nióbio do Morro dos Seis Lagos Anais Z.°* Simpósio Amazônico, Manaus 8 – Brasil.

Kapustin, Yu. L. 1980. *Mineralogy of carbonatites*, D. K. Biswa (trans.). New Delhi: Amerind.

Larsen, E. S. & F. W. Hunt 1914. Zwei vanadinhaltige aegirine von Libby, Montana. *Zeitschrift für Kristallographie* **53**, 209.

Loubet, M., M. Bernat, M. Javoy & C. J. Allègre 1972. Rare earth contents in carbonatites. *Earth and Planetary Science Letters* **14**, 226–32.

Lumpkin, G. R. & R. C. Ewing 1985. Natural pyrochlores: analogues for actinide host phases in radioactive waste forms. In *Scientific basis for nuclear waste management*, vol. VIII, C. M. Janzen, J. A. Stone & R. C. Ewing (eds), 647–54. Materials Research Society Symposia Proceedings 44, Materials Research Society, Pittsburgh.

Maravič, H. V. & G. Morteani 1980. Petrology and geochemistry of the carbonatite and syenite complex of Lueshe (N.E. Zaire). *Lithos* **13**, 159–70.

Mariano, A. N. 1983. *A petrographic description of carbonatite and related rocks from the Adrar des Iforas, Mali*. Report to the United Nations Revolving Fund for Natural Resources Exploration.

Mariano, A. N. 1988. Some further geologic applications of cathodoluminescence. In *Cathodoluminescence of geological materials*, D. J. Marshall (ed.). London: Unwin Hyman.

Mariano, C. 1985. *The paragenesis of the Bond Zone of the Oka carbonatite complex, Oka, Quebec*. B.Sc. thesis, Bucknell University, Lewisburg.

Matos Alves, C. A. 1968. *Estudo geológico e petrológico do maciço alcaline-carbonatitico do Quicuco*. Lisboa: Junta de Investigacoes do Ultramar.

Mew, M. C. 1980. *World survey of phosphate deposits*, 4th edn, M. C. Mew (ed.). The British Sulphur Corporation Ltd, Parnell House, London.

Mining Annual Review 1986. *Mining Journal of London*.

Möller, P., G. Morteani & F. Schley 1980. Discussion of REE distribution patterns of carbonatites and alkaline rocks. *Lithos* **13**, 171–9.

Moore, D. H. 1984. The Mud Tank vermiculite prospect, Alice Springs 1 : 250,000 Sheet area SF 53/14 Northern Territory Government, Department of Mines and Energy, Northern Territory Geological Survey Records (unpubl.).

Nash, W. P. 1972. Mineralogy and petrology of the Iron Hill Carbonatite Complex, Colorado. *Geological Society of America Bulletin* **83**, 1361–82.

Nichipurenko, V., O. Nichipurenko & L. Zhirova 1980. Perovskite–titanomagnetite–amphibole rocks of the Nizhnesayan Massif. *Geologiya i Geofizika* **5**, 60–8.

Nickel, E. H., J. F. Rowland & J. A. Maxwell 1958. The composition and crystallography of niocalite. *Canadian Mineralogist* **6**, 264–72.

Notholt, A. J. G. 1979. The economic geology and development of igneous phosphate deposits in Europe and the USSR. *Economic Geology* **74**, 339–50.

Olson, J. C., D. R. Shawe, L. C. Pray & W. N. Sharp 1954. *Rare-earth mineral deposits of the Mountain Pass District, San Bernardino County, California*. United States Geological Survey Professional Paper 261, United States Government Printing Office, Washington, DC.

Palabora Mining Company Limited & Mine Geological and Mineralogical Staff 1976. The geology and the economic deposits of copper, iron, and vermiculite in the Palabora Igneous Complex: A brief review. *Economic Geology* **71**, 177–92.

Palache, C., H. Berman & C. Frondel 1944. *Dana's System of Mineralogy*, 7th edn, Vol. 1. New York: Wiley.

Peterson, E. C. 1966. *Titanium resources of the United States*. Bureau of Mines Information Circular 8290.

Pinkuss, M. L. & H. Guimarãés 1981. *Mining ore preparation and ferro-niobium production at Mineracão Catalão*. International Symposium-Niobium 81, San Francisco.

Prins, P. 1972. Composition of magnetite from carbonatites. *Lithos* **5**, 227–40.

Ramdohr, P. 1969. *The ore minerals and their intergrowths*. New York: Pergamon Press.

Recht, H. L. & M. Ghassemi 1970. *Phosphate removal from wastewaters using lanthanum precip-*

itation. Report for the Federal Water Quality Adm. Dept. of the Interior Program #17010 EFX. *Revista Minerios* 1987. **12**, (125), 22.

Roedder, E. 1973. Fluid inclusions from the fluorite deposits associated with carbonatite of Amba Dongar, India and Okorusu, South West Africa. *Institution of Mining and Metallurgy Transactions Section B* **82**, B35–B39.

Roeder, P. L., D. MacArthur, Xin-Pei Ma, G. L. Palmer & A. N. Mariano 1987. Cathodoluminescence and microprobe study of rare-earth elements in apatite. *American Mineralogist* **72**, 801–11.

Rose, H. J., Jr, L. V. Blade & M. Ross 1958. Earthy monazite at Magnet Cove, Arkansas. *American Mineralogist* **43**, 995–7.

Saether E. 1957. The alkaline rock province of the Fen area in southern Norway. *Det Kongelige Norske Videnskabers Selskabs Skrifter* NRI, 1–150.

Sandvik, P. O. & G. Erdosh 1977. Geology of the Cargill phosphate deposit in northern Ontario. *Canadian Mining and Metallurgical Bulletin* **69**, 90–6.

Sauvage, J. F. & R. Savard 1985. Les complexes alcalins sous-saturés à carbonatites de la région d'In Imanal (Sahara Malien): une presentation. *Journal of African Earth Sciences* **3**, 143–9.

Secher, K. 1986. Exploration of the Sarfartôq carbonatite complex, southern West Greenland. *Rapport-Grønlands Geologiske Undersogelse* **128**, 89–101.

Semenov, E. I. 1974. Economic mineralogy of alkaline rocks. In *The alkaline rocks*, H. Sørensen (ed.). New York: Wiley.

Silva, A. B. da, 1985. *Araxá, uma reserva inesgotavel de nióbio. Contribuições à Geologia e à Petrologia*, 175–9. Dedicado à memória de Djalma Guimarães, Publicado sob o patrocinio da Companhia Brasileira de Metalurgia e Mineracão, Núcleo de Minas Gerais Sociedade Brasileira de Geologia, Belo Horizonte, Brasil.

Silva, A. B. da, M. Marchetto & O. M. DeSouza 1979. Geology of the Araxá (Barreiro) Carbonatite, Brazil. *Geological Association of Canada–Mineralogical Association of Canada, Joint Annual Meeting, Quebec, Program with Abstracts* **4**, 79.

Smirnov, V. I. 1977. *Ore deposits of the USSR*, Vol. III. San Francisco: Pitman.

Sørensen, H. 1974. *The alkaline rocks*. New York: Wiley.

Souza, J. M. & L. D. Castro 1968. *Geologia do Deposito de Terras Raras, Nióbio e Urânio da Área Zero, Araxá*. Anais Do XXII Congresso Brasileiro de Geologia, Belo Horizonte.

Straaten, P. van 1986. Some aspects of the geology of carbonatites in S.W. Tanzania. *Geological Association of Canada–Mineralogical Association of Canada–Canadian Geophysical Union, Joint Annual Meeting, Ottawa. Program with Abstracts*, **11**, 140.

Tack, L. & P. De Paepe 1981. *Le massif intrusif de syenites foidales de la Haute-Ruvubu, Burundi. Musée Royal de l'Afrique Central Tervuren (Belgique) Rapport Annuel*, 1980, 173–8.

Tack, L., S. Deutsch, J. P. Liegeois & P. De Paepe 1983. *Age, Nd and Sr isotopic geochemistry of the alkaline plutonic complex of the Upper-Ruvubu, Burundi, Résumé.* 12th Colloquium of African Geology, Brussels.

Teixeira, W., M. A. S. Basei & C. C. G. Tassinari 1976. *Geochronologia das Folhas Tumucumaque e Santarém*. Congresso Brasileiro de Geologia, 29° Resumo das Communicacões. Belo Horizonte, Sociedade Brasileiro Geologia, Bulletin 1, 193.

Tröger, E. 1928. Alkaligesteine aus der Serro do Salitre im westlichen Minas Gerais, Brasilien. *Centralblatt für Mineralogie* **1928A**, 202–4.

Turekian, K. K. 1971. *Encyclopedia of science and technology*. New York: McGraw-Hill.

Ulbrich, H. H. G. J. & C. B. Gomes 1981. Alkaline rocks from continental Brazil: A review. *Earth Science Reviews* **17**, 135–54.

United Nations Development Program 1979. *Synthèse des travaux de recherches sur les mineralisations associées à des carbonatites de Matongo-Bandaga*. Juin 1979. BDI/77/003.

Verwoerd, W. J. 1986. Mineral deposits associated with carbonatites and alkaline rocks. In *Mineral deposits of southern Africa*, vols I & II, C. P. Anhaeusser & S. Maske (eds), 2173–91. Johannesburg: Geological Society of South Africa.

Vlasov, K. A. 1966. *Geochemistry and mineralogy of rare elements and genetic types of their deposits.* Vol. I: *Geochemistry of rare element deposits.* Jerusalem, Israel program for scientific translations.

Vlasov, K. A. 1968. *Geochemistry and mineralogy of rare elements and genetic types of their deposits.* Vol. III: *Genetic types of rare-element deposits.* Jerusalem, Israel program for scientific translations.

Vos, M. A. 1981. *Industrial Minerals of the Cargill Complex.* Summary of Field Work, 1981 Ontario Geological Survey, Miscellaneous Paper 100, 224–9.

Wambeke, L. Van 1977. The Karonge rare earth deposits, Republic of Burundi; New mineralogical–geochemical data and origin of the mineralization. *Mineralium Deposita* **12**, 373–80.

Yefimov, A. F., Ye. M. Yes'kova, S. I. Lebedeva & V. Ya. Levin 1985. Type compositions of accessory pyrochlore in a Ural alkali complex. *Geochemistry International* **22**, (6), 68–75.

Zhou, Z., Li Gongyuan, Sung Tongyun & Liu Yuguan 1980. On the geological characteristics and the genesis of the dolomitic carbonatites at Bayan Obo, Inner Mongolia. *Geological Review* **26**, (1), 35–42.

Zÿl, P. J. Van 1962. The geology, structure and petrology of the alkaline instrusions of Kalkfeld and Okorusu and the invaded Damara rocks. *Annale Universiteit van Stellenbosch Serie A* **37**, (4), 237–339.

8
NATURE AND STRUCTURAL RELATIONSHIPS OF CARBONATITES FROM SOUTHWEST AND WEST TANZANIA

P. VAN STRAATEN

ABSTRACT

Three major periods of carbonatite–alkaline intrusions are recognized in Tanzania: Late Proterozoic (750–680 Ma), Cretaceous (120–100 Ma), and Cenozoic (40–0 Ma). Virtually all carbonatites were emplaced outside the Archaean Tanzania Craton and their emplacement is spatially associated with repeatedly rejuvenated fault and shear zones.

With the exception of Ngualla, all Proterozoic complexes were intruded either during or shortly before the initial phase of the Pan-African thermotectonic rejuvenation. The Cretaceous carbonatites intruded normal faults formed during a tensional stress field well after the main post-Karroo, pre-Cretaceous rifting phases and the breakup of Gondwanaland.

Three rock types are common, and can occur within the same complex. These are, in descending order of volume and age: sövite, magnesiocarbonatite, and ferrocarbonatite. Less common are varieties rich in apatite, fluorite, barite, and rare earth element (REE) minerals.

The intermittent association of carbonatites within a restricted area of the Earth's crust in East and Central Africa, over a long time-span, and the spatial relationships to structural features, suggest that carbonatite magmatism is related to recurrent reactivation of older structures, and argues for control by crustal rather than mantle processes.

8.1 INTRODUCTION

Two prominent features of East African geology are the great number of carbonatites and alkaline complexes, and the presence of continental rifting. This has led geologists to infer a genetic relationship between the two, a view that is largely based on the observation that most carbonatites and alkaline complexes in East Africa occur in or close to the Cenozoic Rift System. However, any relationship to recent rifting is not as straightforward as it may appear. Indeed, many authors provide alternative models, some of which involve a link between alkaline magmatism and orogeny rather than taphrogenesis (Gittins *et al.* 1967,

Woolley, this volume). Others link carbonatite magmatism with large-scale up-doming (Le Bas 1980, 1987) or with tectonically quiet periods (Sørensen 1974). Such a lack of consensus provides sufficient justification for re-examining the relationship of carbonatite magmatism to various geological processes such as orogenesis, rifting, and epeirogenesis. The carbonatites of southwestern and western Tanzania and their structural setting provide some interesting observations that may help towards resolving these problems.

8.2 CARBONATITES IN TANZANIA

More than 15 carbonatites are known in Tanzania (Fig. 8.1). Three distinct clusters occur, separated by several hundred kilometres. These are:

(a) the north Tanzania cluster, W. and SW of Arusha;
(b) the east-central Tanzania cluster, S. of Morogoro; and
(c) the southwest Tanzania cluster, W. of Mbeya.

A few carbonatites occur outside these clusters, and among them are Sangu-Ikola, on the shores of Lake Tanganyika, and Makonde, east of Lake Nyasa.

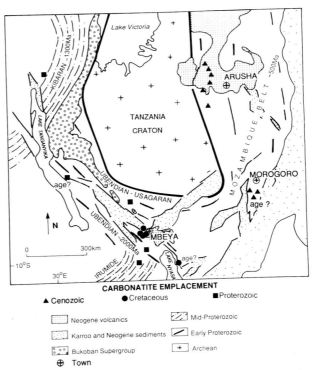

Figure 8.1 Distribution of carbonatites in Tanzania. Carbonatite magmatism is restricted to zones outside the Archaean Tanzania craton. Proterozoic carbonatites occur at the W. and SW side of the Tanzania craton, Cretaceous carbonatites are restricted to SW Tanzania, and Cenozoic carbonatites occur only along the eastern margin of the Tanzania craton, at the southern end of the Eastern Rift.

The carbonatites in northern Tanzania occur at the southern end of the Eastern (Gregory) Rift, and most are associated with alkaline volcanism. The carbonatites occur either as plugs or as lava flows and carbonatitic pyroclastics (Gallapo, Hanang, and Oldoinyo Lengai). All are of Tertiary to Recent age.

Little is known about occurrences in east-central Tanzania. The Wigu Hill and Maji ya Weta carbonatites, approximately 100 km south of Morogoro, form complex dykes injected into brecciated country rock. Although not directly associated with alkaline volcanic rocks, the carbonatites are considered part of the volcanic province between the Uluguru Mountains and the Rufiji River (Stockley 1943, Spence 1957, Kreuser 1984, Hankel 1987). No radiometric dates are available from the Wigu and Maji ya Weta carbonatites, but a carbonatite dyke from the nearby Sumbadzi River yielded an Eocene age (Hankel 1987).

8.3 THE CARBONATITES OF SOUTHWESTERN AND WESTERN TANZANIA

Nine carbonatite complexes are known to occur in SW and W. Tanzania (Fig. 8.2), and the geology of these complexes illustrates the variety of features associated with carbonatite magmatism.

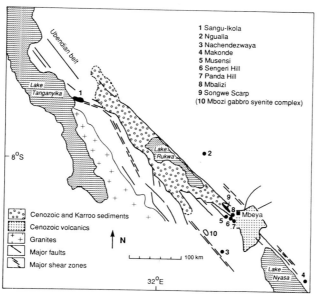

Figure 8.2 Location of carbonatites and alkaline complexes in SW and W. Tanzania in relation to major shear and fault zones.

8.3.1 Ages of carbonatite magmatism in SW and W. Tanzania

Two major episodes of carbonatite–alkaline intrusions are recognized in SW and W. Tanzania: Proterozoic and Cretaceous (Table 8.1). Of the carbonatites that have been dated, one is Proterozoic and four are Cretaceous.

Table 8.1 Ages of carbonatites in SW and W. Tanzania.

Name of complex	Age	Method of age dating	Reference
Ngualla	1040 ± 40 Ma	K–Ar, biotite	1
Nachendezwaya	probably Proterozoic, associated with the 685 ± 62 Ma (Rb–Sr) Songwe-Ilomba Hill complex	geological field relationships	2, 3, 4
Sangu-Ikola	probably Proterozoic, post-Bukoban – pre-Karroo	geological field relationships	5
Mbalizi	122 ± 8; 118 ± 9 Ma	K–Ar, phlogopite	6
Panda Hill	113 ± 6 Ma	K–Ar, mica concentrate	7
Sengeri Hill	probably same age as Panda Hill carbonatite	geological field relationships	8, 9
Musensi	101 ± 12; 96 ± 9 Ma	K–Ar, biotite	10
Songwe Scarp	100 ± 10 Ma	K–Ar, K–feldspar	10
Makonde	unknown		11

References: 1, Cahen & Snelling (1966); 2, Horne (1959); 3, MacFarlane (1966); 4, Ray (1974); 5, Coetzee (1962); 6, Pentelkov & Voronovskiy (1979); 7, Snelling (1965); 8, Fick & van der Heyde (1959); 9, RUDIS (1980); 10, Miller & Brown (1963); 11, Haldemann & Harpum (1952).

The Ngualla carbonatite of SW Tanzania yields a biotite K–Ar age of 1040 ± 40 Ma (Cahen & Snelling 1966). Recent isotope investigations by groups at Carleton University, Ottawa, and the University of California, Santa Barbara, using Rb–Sr, Sm–Nd, and U–Pb methods confirm the Proterozoic age, but also indicate a fairly complex evolution for the Ngualla carbonatite complex. The radiometric ages of the Nachendezwaya and Sangu-Ikola carbonatites have not yet been established but field relationships indicate a Proterozoic age.

Mesozoic carbonatites occur in East Africa south of 8°S latitude, in Malawi, Mozambique, Tanzania, Zambia, and Zimbabwe. In Tanzania all known Mesozoic carbonatites occur close to Mbeya, in the southwest of the country (Fig. 8.3). The four Mesozoic carbonatites of SW Tanzania are of Cretaceous age. They are: Panda Hill, Musensi, Mbalizi, and Songwe Scarp. The Sengeri Hill carbonatite, close to Panda Hill, is thought to be also of Cretaceous age. The age of the Makonde carbonatite east of Lake Nyasa is unknown.

The Proterozoic carbonatite–alkaline complexes can be viewed in the wider

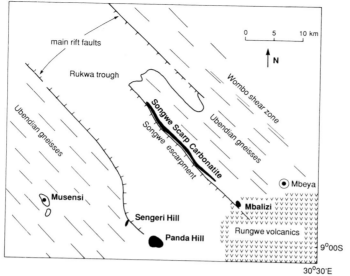

Figure 8.3 Geological sketch map with location of Cretaceous carbonatites near Mbeya, SW Tanzania.

context of Proterozoic carbonatites and alkaline complexes in East and Central Africa. Two major clusters are recognized. A northern cluster is found in Zaire and Burundi, and a southern cluster in the border area of Tanzania, Zambia, and Malawi. The carbonatite–alkaline complexes in Zaire and Burundi include Lueshe with a biotite K–Ar age of 516 ± 26 Ma (Maravič & Morteani 1980), Kirumba with Rb–Sr ages of 665 ± 33 Ma and 555 ± 17 Ma respectively (Cahen & Snelling 1966), and the Upper Ruvubu complex with foid-bearing syenites and the associated Matongo carbonatite. A U–Pb zircon date of 739 ± 7 Ma was obtained from this syenite and a Rb–Sr rehomogenization event at 699 ± 13 Ma has been documented (Tack *et al.* 1984).

Tack & De Paepe (1981) linked the two clusters of carbonatite–alkaline complexes together to form a curvilinear zone, over 1000 km long, which stretches from Malawi and Zambia through Tanzania into Zaire and Burundi. The carbonatite that links these two clusters is the elongated Sangu-Ikola complex, on the eastern shore of Lake Tanganyika (Coetzee 1962, van Straaten 1983). It was probably intruded during the late Proterozoic, after deposition of the Sanyika sediments (correlated with Bukoban sediments) and prior to Karroo sedimentation (Coetzee 1962).

Proterozoic carbonatites and alkaline complexes at the southern end of this zone occur in the border area of NE Zambia, N. Malawi, and SW Tanzania. In SW Tanzania they include the Mbozi gabbro-syenite complex dated at 745 ± 25 Ma (Cahen & Snelling 1966) and the Nachendezwaya carbonatite complex, which lies 3 km from the Songwe-Ilomba Hill syenite complex of Malawi. The Nkombwa carbonatite of NE Zambia, intruded at 680 ± 25 Ma (Snelling *et al.* 1964) and the Songwe-Ilomba Hill complex of northern Malawi dated at 685 ± 62 Ma (Ray 1974) are found in the same general area.

8.3.2 Field relationships

The carbonatites of southwest and west Tanzania intruded the *c.* 2000-Ma-old Ubendian mobile belt. Most intruded NW-striking gneisses and amphibolites while Ngualla intruded Proterozoic felsic volcanics, and Musensi intruded volcanic rocks and pyroxenites.

The main feature of carbonatite emplacement in this part of Tanzania is, however, the proximity of the complexes to repeatedly reactivated shear and fault zones (Fig. 8.2). The Nachendezwaya carbonatite occurs close to a major shear zone that can be traced at least some 600 km from N. Malawi to Lake Tanganyika (Ray 1974, Daly 1985). The steeply dipping Sangu-Ikola carbonatite is also associated with a major shear zone (Coetzee 1962, McConnell 1972). The Cretaceous Panda Hill and Sengeri Hill carbonatites intruded a fault zone on the SW side of the Rukwa trough, while the Mbalizi and Songwe Scarp carbonatites penetrated a fault zone at the NE side of the same trough (Fig. 8.3). The Makonde carbonatite occurs in a shear zone east of Lake Nyasa.

8.3.3 Size and shape

The Proterozoic carbonatites vary in size and shape (Table 8.2). Circular or oval shaped carbonatites include Ngualla (Fig. 8.4), Panda Hill (Fig. 8.5), and possibly the largely hidden Mbalizi and Musensi carbonatites. Diameters range from 3 km (Ngualla) to only a few metres (Musensi).

The Nachendezwaya and Sangu-Ikola carbonatites are elongate. With a surface area of 14 km², the Sangu-Ikola carbonatite on the shores of Lake Tanganyika is the most extensive carbonatite in Tanzania (Coetzee 1962, van Straaten 1983). It consists of three intrusive bodies the longest of which is 18 km long and up to 1.5 km wide.

The linear carbonatites include Makonde, Sengeri Hill, and Songwe Scarp carbonatites. The longest, dyke-like carbonatite is the NW-striking, 50 m wide Songwe Scarp carbonatite exposed over a length of 18 km (Brown 1964).

8.3.4 Fenite zones

Metasomatic alteration differs in nature and extent around the various carbonatite intrusions. The carbonatites of Ngualla, Panda Hill, Sengeri Hill, Mbalizi, and Songwe Scarp are all associated with potassium-rich fenites that contain K-feldspar and phlogopite. All of the fenites are fractured and veined.

At the edge of the Mbalizi carbonatite phlogopite-rich fenites also occur, and these are associated with abundant apatite. Abnormally high concentrations of fluorapatite and pyrochlore are found in the fenitized roof portion of the Panda Hill carbonatite, indicating that phosphorus-, fluorine-, and niobium-rich fenitizing fluids, perhaps trapped in the cupola, permeated the overlying country rocks.

Only the fenites associated with the Nachendezwaya and Sangu-Ikola carbonatites are sodium-rich. Widespread development of albitites and aegirine–albite rocks can

Table 8.2 Size, shape, and orientation of carbonatites in SW and W. Tanzania

Name of complex	Size	Shape	Orientation	Reference
Ngualla	3 km diameter	circular		1, 2
Nachendezwaya	400 m long, 25–90 m wide	lenticular	NW-SE	3
Sangu-Ikola (three separate bodies)	(a) 18 km long, 100–1500 m wide (b) 3000 × 700 m (c) 1000 × 200 m	all lenticular	NW-SE	4
Mbalizi	1100 × 400 m	oval		5, 6
Panda Hill	1.5 km diameter	circular		7, 8, 9, 10, 11
Sengeri Hill	2000 × 100 m	linear	NE-SW	9, 10
Musensi	outcrop only few m²	probably circular		12
Songwe Scarp	18 km × 50 m	linear	NS-SE	13
Makonde	140 × 7 m	linear	NW-SE	14

References: 1, James (1956a, b); 2, Williamson Diamonds (1970); 3, Horne (1959); 4, Coetzee (1962); 5, Mtuy et al. (1987); 6, van Straaten (1987); 7, James (1954); 8, Fawley & James (1955); 9, Fick & van der Heyde (1959); 10, RUDIS (1980); 11, Basu & Mayila (1986); 12, Miller & Brown (1963); 13, Brown (1964); 14, Haldemann & Harpum (1952).

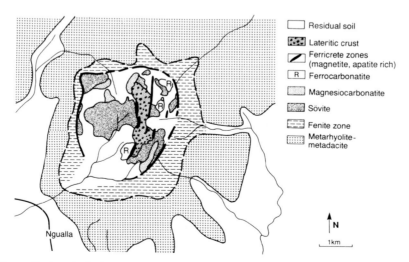

Figure 8.4 Generalized geological map of the Ngualla carbonatite, modified after Williamson Diamonds (1970).

be seen at the contact of the Sangu-Ikola carbonatite with amphibole gneisses. It is noteworthy that these two (probably Proterozoic) carbonatites are elongate and intruded into or near extensive shear zones. The carbonatites associated with potassium-rich fenite zones, the plug-like Ngualla carbonatite, and all of the Mesozoic carbonatites intruded brittle fault zones.

8.3.5 Structure and composition

Detailed geological field studies of, and new chemical and mineralogical analyses from, carbonatites of SW and W. Tanzania reveal both variation and similarity in chemistry and structure. The Tanzanian complexes include few silicate rocks and can include all or some of the main carbonatite rock types: sövite, magnesiocarbonatite and ferrocarbonatite (Woolley 1982). In some carbonatites all three rock types occur in the same complex. The two clear examples are Ngualla and Panda Hill. Both carbonatites are zoned with a sövitic core, surrounded by successively more magnesium- and iron-rich carbonatite rock types.

The Ngualla and Panda Hill complexes contain all major types of carbonatites which are in descending order of volume and age: sövite, magnesiocarbonatite, and ferrocarbonatite.

In the Ngualla complex, sövite is the most common carbonatite rock type. The sövite, concentrated in the central part of the complex, is mainly white and massive, at places very coarse-grained containing dendritic apatite up to 4 cm long. Towards the margins of the complex, the sövite is less massive, and is cut by later carbonatite intrusions, in particular veins of magnesiocarbonatite.

The brownish weathering, magnesiocarbonatite is pervasive throughout the sövitic carbonatite, except in the central part, occurring mainly in veins and schlieren. Networks of intersecting magnesiocarbonatite veins make up > 50% of some outcrops.

Ferrocarbonatite forms the youngest carbonatite rock at Ngualla. Dull- to purplish-red in colour, it occurs mainly in the northeastern and southeastern portions of the complex as scattered large massive pods. Ankerite is the predominant mineral, and large xenoliths of magnesiocarbonatite and earlier carbonatite rocks are common.

Dykes and veins of various compositions intruded the carbonatite at different stages throughout its development. Mafic dykes, up to 20 cm wide, consisting of calcite, biotite, pyroxene and/or amphibole, crosscut the sövite but are transected by later magnesiocarbonatites.

The Panda Hill carbonatite (Fig. 8.5), one of the best described carbonatite complexes in East Africa, displays a zonal arrangement of rock types similar to Ngualla. According to James (1954), the inner ring of Panda Hill is largely sövite surrounded by magnesiocarbonatite. Dykes of varying composition, including ferrocarbonatites, occur near the outer margins.

Fawley & James (1955) interpreted the Panda Hill carbonatite as the central plug of a much larger volcanic complex. Based on geological field relationships and drilling evidence, Fick & van der Heyde (1959) suggested that the exposed

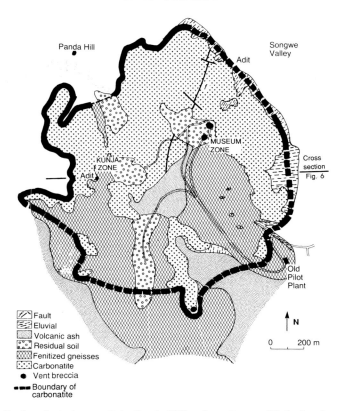

Figure 8.5 Generalized geological map of the Panda Hill carbonatite, modified after James (1954) and RUDIS (1980).

carbonatite is part of a high-level intrusion. Borehole data and current field investigations indicate that the central part of the carbonatite is largely covered by fenitized country rock, the fenites forming a shallow-dipping roof cap above the sövite (Fig. 8.6). Sövite was emplaced early in the sequence, ferrocarbonatite late.

Other common features that characterize many Tanzanian carbonatites include ferrocarbonatite breccias, possibly vent breccias, with rounded and angular fenitized basement fragments set in a ferrocarbonatite matrix. Breccias of this type have been observed at Panda Hill, Sengeri Hill, Musensi, and carbonatites of central-east Tanzania. It is possible that these breccias might be expressions of a late volatile-rich stage.

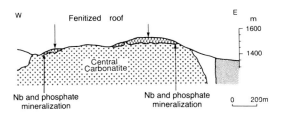

Figure 8.6 E–W cross-section of the Panda Hill carbonatite.

The chemical evolution of the sövite, magnesiocarbonatite, and ferrocarbonatite is shown by the behaviour of certain major and trace elements. Chemical analysis from rock types from Ngualla Panda Hill, and Sengeri Hill are shown in Tables 8.3 and 8.4 and some data are given in Figures 8.7 and 8.8. Trace element analyses from Ngualla show that Sr is most abundant in the sövite. Barium, along with the rare earth elements (REEs), especially the light REEs, are concentrated in the ferrocarbonatites. The chondrite-normalized REE signatures of early- or late-stage carbonatites from Ngualla show a strong fractionation (Fig. 8.9) from sövite (La:Yb, 50–90) to late ferrocarbonatite (La:Yb, >3000). The late stages of the Ngualla carbonatite intrusion yield the highest concentration of light REEs. The P_2O_5 data show bimodal distribution with greatest concentrations occurring in the sövites and the late-stage, magnetite–apatite veins.

Fine-grained, blue-grey, and commonly magnesium-rich carbonatite dykes cut sövites of the Panda Hill and Mbalizi carbonatites. The Songwe Scarp carbonatite is made up almost entirely of this rock type.

Detailed mapping of the Mbalizi carbonatite shows that the main part of the exposed carbonatite is a coarse-grained, phosphatic sövite that cuts a central

Table 8.3 Chemical analyses of rocks from Ngualla.

	Fenite	Central sövite	'Alnöite'	Magnesio-carb.	Ferro-carb.	Apatite-magnetite veins
*						
Wt%						
SiO_2	61.63	3.45	6.36	19.86	0.92	5.57	0.71
Al_2O_3	15.43	0.17	0.36	1.21	0.13	0.20	2.87
Fe_2O_3	6.23	2.89	5.34	18.80	3.99	12.13	36.69
MgO	2.01	3.34	4.76	13.08	21.00	1.94	1.06
CaO	0.37	46.56	49.24	21.94	28.01	45.17	31.49
Na_2O	1.60	0.52	0.75	0.52	0.34	0.32	0.40
K_2O	10.62	0.25	0.20	0.09	0.07	0.11	0.07
TiO_2	0.57	0.10	0.29	5.23	0.01	0.00	4.97
MnO	0.16	0.17	0.20	0.30	0.19	0.32	0.39
P_2O_5	0.01	6.79	12.76	0.19	3.81	0.14	19.99
LOI	1.37	35.76	19.74	18.78	41.53	34.10	1.35
Trace elements (p.p.m.)						
Rb	212	9	6	40	9	3	5
Sr	99	4120	2510	1160	4010	1120	900
Y	22	19	27	18	12	7	27
Zr	113	192	380	184	171	38	500
Nb	77	28	93	108	132	7	190
Ba	534	360	620	510	550	7200	46100
Ce	387	240	245	135	46	3400	450
Nd	73	175	250	74	42	1350	410
La	65	240	163	85	54	2370	280

*XRF analyses, McMaster University, Hamilton, Canada.
LOI, Loss on ignition.

Table 8.4 Chemical analyses of carbonatites from Panda Hill and Sengeri Hill.

	Panda Hill			Sengeri Hill	
	Sövite museum zone	Fenite museum zone	Apatite zone Kunja area		
*Wt%					
SiO_2	6.98	26.85	11.69	10.85	18.92
Al_2O_3	0.09	0.60	0.57	1.20	5.1
Fe_2O_3	1.93	8.56	20.50	9.31	7.64
MgO	4.72	2.14	2.06	12.03	11.39
CaO	46.14	35.44	38.78	26.22	22.00
Na_2O	0.22	0.23	0.21	0.10	0.12
K_2O	0.07	0.47	0.08	1.24	4.36
TiO_2	0.03	0.32	0.87	0.48	0.76
MnO	0.43	0.37	0.09	0.49	0.42
P_2O_5	6.85	22.06	22.98	4.14	0.25
LOI	32.54	2.96	2.16	33.94	29.04
Trace elements (p.p.m.)					
Rb	11	n.a.	17	17	87
Sr	8100	4200	2980	6250	5800
Y	35	n.a.	29	26	33
Zr	1480	n.a.	1470	1350	1080
Nb	6174	10150	4640	227	648
Ba	343	4700	1690	2690	2050
Ce	791	n.a.	1960	893	895
Nd	302	n.a.	990	446	311
La	402	n.a.	850	443	603

* XRF analyses, McMaster University, Hamilton, Canada.
LOI, Loss on ignition; n.a., not analysed.

pyroxenite. Much of the sövite is rich in dendritic apatite. Data from apatite–phlogopite rock in the fenite zone, the pyroxenite and sövite of Mbalizi are shown in Table 8.5. It is noteworthy that the chondrite-normalized REE distribution pattern of the pyroxenite differs little from that of the sövite (Fig. 8.10) indicating either a cogenetic relationship or a strong chemical overprinting of the pyroxenite by the later sövite.

The Mbalizi carbonatite is intruded by the fine-grained Songwe Scarp carbonatite (SSC) which lies along strike with the Mbalizi carbonatite (Fig. 8.3). The Songwe Scarp carbonatite is a 50 m-thick dyke-like body, intruded along the post-Karroo and pre-Cretaceous fault zone over a distance of at least 18 km. Three distinct events are associated with the dyke: widespread feldspathization of the country rock by potassium-rich fluids, injection of the carbonatite along the old fault zone, and the late-stage brecciation, manifested by a red 'siliceous breccia', which veined and silicified the carbonatite (Brown 1964).

Field relationships suggest a common origin for both the Mbalizi and Songwe Scarp carbonatites. However, both have quite different textures and chemical

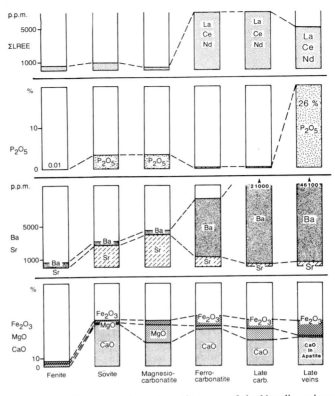

Figure 8.7 Selected element distribution from the main stages of the Ngualla carbonatite intrusion.

signatures. The fine-grained, Songwe Scarp carbonatite is enriched in heavy REEs and has Y values up to five times higher than those of the Mbalizi carbonatite.

The chondrite-normalized REE plots of data from both Mbalizi and Songwe Scarp carbonatites show two quite different trends. The REE distribution pattern of the Songwe Scarp dyke displays a much shallower slope than the Mbalizi REE pattern. Whether these REE signatures reflect two different magmas, or are caused

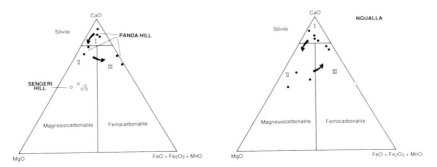

Figure 8.8 Plots of carbonatite analyses from the three main carbonatite rock types of Ngualla, Panda Hill, and Sengeri Hill carbonatites. The sequential order of intrusion is from old to young: I, sövite; II, magnesiocarbonatite; III, ferrocarbonatite. Classification after Woolley (1982).

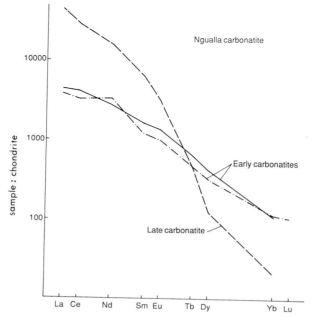

Figure 8.9 Chondrite-normalized REE plot for selected rock types of the Ngualla carbonatite. (Analyses by instrumental neutron activation, McMaster University, Hamilton, Canada.)

by other processes, such as wall–rock interaction (Moeller *et al.* 1980), or accumulation of volatiles, remains unclear.

New chemical data from Makonde, Nachendezwaya, and Sangu-Ikola are given in Tables 8.5 and 8.6. It is worth noting that the REE values from Sangu-Ikola (Fig. 8.11) are relatively low in comparison to other carbonatites. The chemical signature of Makonde previously described as 'apatite limestone' (Haldemann & Harpum 1952) is characteristic of carbonatites. This is also seen in the chondrite-normalized REE distribution pattern (Fig. 8.11).

8.3.6 *Mineralization*

The carbonatites of SW and W. Tanzania have been investigated by mining companies for their Nb, REE, base metal, fluorite, Ba, diamond, and phosphate potential. Most carbonatites from this area contain < 0.1% Nb_2O_5; highest Nb contents are reported from Panda Hill (0.3–1% Nb_2O_5). A pilot plant for the recovery of pyrochlore from the upper portions of the Panda Hill carbonatite was set up in the early 1960s, but full-scale mining did not develop (RUDIS 1980).

Rare earth element and Y mineralization are reported from the Songwe Scarp carbonatite and Ngualla (van Straaten 1987). Values of up to 0.12% Y have been determined from weathered materials of the Songwe Scarp carbonatite and REE values up to 4% from residual soils of Ngualla. Highest REE concentrations seem to characterize the late-stage carbonatites, and overlying residual soils.

Table 8.5 Chemical analyses of rocks from the Mbalizi, Songwe Scarp, and Makonde carbonatites.

	Mbalizi			Songwe Scarp		Makonde
	Fenite	Pyroxenite	Sövite	Ferrocarb.	Magnesiocarb.	Magnesiocarb.
*						
Wt%						
SiO_2	29.60	26.42	1.54	13.50	13.39	5.06
Al_2O_3	8.66	5.54	0.09	1.79	2.58	0.51
Fe_2O_3	20.41	11.74	3.22	10.34	4.84	5.38
MgO	4.51	8.73	1.38	8.77	10.04	6.12
CaO	12.99	23.18	50.06	27.58	29.31	40.65
Na_2O	0.22	0.70	n.d.	0.28	0.10	0.43
K_2O	2.49	4.88	0.06	1.61	2.71	0.23
TiO_2	4.87	1.64	n.d.	1.61	1.10	0.11
MnO	0.19	0.23	0.19	0.24	0.22	0.23
P_2O_5	7.17	5.25	8.60	7.95	9.89	6.41
LOI	8.89	11.69	34.75	26.34	25.82	34.88
Trace elements (p.p.m.)						
Rb	n.a.	99	21	20	23	40
Sr	n.a.	2560	6180	2470	3010	10700
Y	n.a.	43	132	350	420	66
Zr	n.a.	872	880	138	195	470
Nb	n.a.	301	980	162	155	17
Ba	n.a.	260	400	3970	430	1090
Ce	n.a.	701	677	630	440	788
Nd	n.a.	346	251	480	120	370
La	n.a.	244	282	310	270	333

*XRF analyses, McMaster University, Hamilton, Canada.
LOI, Loss on ignition; n.a., not analysed; n.d., not detected.

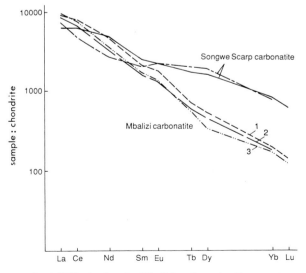

Figure 8.10 Chondrite-normalized REE plot for the Mbalizi carbonatite (1, central pyroxenite; 2, 3, sövite) and the Songwe Scarp carbonatite. (Analyses by instrumental neutron activation, McMaster University, Hamilton, Canada.)

Table 8.6 Chemical analyses of rocks from the Sangu-Ikola and Nachendezwaya carbonatites.

*	Nachendezwaya			Sangu-Ikola		
	Sövite	Sövite	Sövite	Sövite	Magnesiocarb.	Ferro-carb.
Wt%						
SiO_2	0.45	0.23	2.93	2.59	0.49	3.26	3.45
Al_2O_3	0.03	0.02	0.12	0.22	0.28	0.07	0.08
Fe_2O_3	1.66	2.06	4.36	3.84	4.76	2.83	7.75
MgO	4.83	7.10	3.31	3.31	16.68	17.23	6.32
CaO	49.71	46.57	46.72	48.14	30.90	30.80	43.03
Na_2O	0.04	0.05	0.07	0.04	0.02	0.11	0.05
K_2O	0.01	0.01	0.01	0.06	0.18	0.22	0.01
TiO_2	0.02	0.01	0.10	0.08	0.01	0.02	0.14
MnO	0.25	0.34	0.27	0.23	0.37	0.38	0.19
P_2O_5	2.85	0.07	12.10	7.16	3.57	4.53	0.17
LOI	40.15	43.54	30.01	34.32	42.73	40.55	38.81
Trace elements (p.p.m.)							
Rb	<1	<1	<1	6	7	8	<1
Sr	4230	4205	1050	3170	2650	2520	960
Y	10	10	6	22	20	23	7
Zr	163	161	308	920	513	549	216
Nb	7	7	19	16	19	501	8
Ba	377	348	299	84	13	14	250
Ce	287	240	200	176	185	165	182
Nd	85	81	77	68	87	76	68
La	290	195	78	78	82	80	90

*XRF analyses, McMaster University, Hamilton, Canada.
LOI, Loss on ignition.

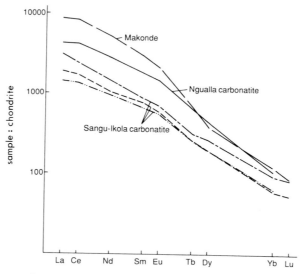

Figure 8.11 Chondrite-normalized REE plot for selected carbonatites: Makonde, Ngualla, and three samples of Sangu-Ikola. (Analyses by instrumental neutron activation, McMaster University, Hamilton, Canada.)

Diamonds in uneconomic quantities were discovered at Ngualla by Williamson Diamonds (1970), but the diamond-bearing source rock has never been identified. It is interesting to note that no kimberlite indicator minerals have so far been found.

Late-stage veins in carbonatites commonly contain base metals, Ba, F, and P. Despite intensive surveys by various mining companies, no economically interesting base metal mineralization has been found in the carbonatites of SW and W. Tanzania. Uneconomic quantities of Pb–Zn–Cu minerals are reported from Sangu-Ikola and Ngualla carbonatites.

Fluorite occurs in small veinlets and as void fillings at Ngualla, Panda Hill, and Sengeri Hill. Barite veins, several metres wide, occur at Ngualla (Williamson Diamonds 1970, van Straaten 1987).

Apatite concentrations are found in three environments: (a) in sövite/fenite contact zones, (b) in late-stage magnetite–apatite veins; and (c) in residual soils.

Accumulations of fluorapatite are reported by the Tanzania–Canada Agrogeology project from the sövite/fenite contact zone of Panda Hill and Mbalizi carbonatites. Detailed investigations at Panda Hill show that apatite mineralization is spatially related to the shallow dipping carbonatite/fenite roof zone. In the Panda Hill, Ngualla, and Mbalizi carbonatites, concentrations of long prismatic apatites are found in coarse-grained sövites similar to the dendritic textures described by Treiman (this volume). Late-stage magnetite–apatite veins are found at Ngualla where they are up to 20 m wide, several hundred metres long, and contain 12–35% P_2O_5. At Panda Hill apatite–magnetite veins are cut by carbonatite. Extensive accumulations of phosphates (up to 20% P_2O_5) have been discovered in residual soils overlying the Ngualla and Panda Hill carbonatites.

8.4 SPATIAL AND TEMPORAL DISTRIBUTION

Several observations from Tanzania indicate that the spatial distribution of carbonatite intrusions is determined by crustal structures. Of note is the virtual absence of carbonatite–alkaline complexes within the Archaean Tanzania Craton. In Tanzania, carbonatite magmatism is restricted to zones close to the margins of the craton (Figs 8.1, 8.12), generally within zones underlain by Proterozoic mobile belts.

The second observation is that the carbonatites occur either along extensive linear and curvilinear zones or in clusters. The Proterozoic carbonatites fall along a zone that stretches from northern Malawi to Burundi and eastern Zaire. The Mesozoic carbonatites occur in clusters close to post-Karroo faults and fault intersections.

The third observation is that the carbonatites were intruded into mechanically weak structures in Proterozoic belts. These structural anisotropies are mainly shear zones, repeatedly rejuvenated and probably extending to deep levels within the crust. Major shear zones in southwestern and western Tanzania were rejuvenated during the Irumide event, the Pan-African episode, and during post-Karroo times (Brown 1962, McConnell 1972, Ueda *et al.* 1975, Daly 1985). Two of the three Proterozoic carbonatites were intruded along structurally well-defined, northwest-

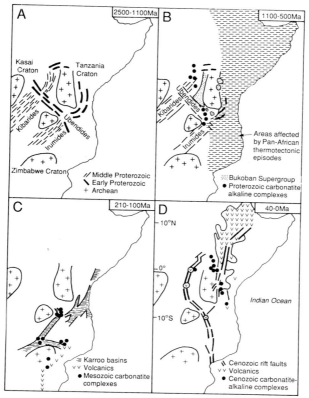

Figure 8.12 Carbonatites of East Africa in space and time. (A) The Tanzania craton is surrounded by Early and Middle Proterozoic mobile belts. No carbonatite magmatism is known. (B) Late Proterozoic carbonatites were intruded along a curvilinear zone into mobile belts outside the Archaean cratons. (C) Mesozoic carbonatites were intruded along post-Karroo fault zones and fault intersections. No Mesozoic carbonatites are known north of 8°S. (D) Cenozoic carbonatites were intruded mainly along the eastern side of the Tanzania craton. No Cenozoic carbonatites are known south of 8°S.

striking shear zones, and all Cretaceous carbonatites intruded zones of recurrent rejuvenation of shear and fault zones.

The temporal distribution of carbonatites in East Africa is marked by three major magmatic episodes (Fig. 8.13). Most carbonatites from East Africa were emplaced during the Late Proterozoic, mid-Mesozoic, and the Cenozoic. In southwestern and western Tanzania, carbonatites and associated alkaline complexes were intruded only during the Proterozoic and Mesozoic. No Cenozoic carbonatites are reported from the Western Rift south of the equator. With the exception of Ngualla, all Late Proterozoic carbonatites and alkaline complexes were emplaced between 750 and 650 Ma, a time that coincides with the initial phase of the Pan-African thermotectonic episode.

All known Mesozoic carbonatites of south-central Africa, other than Shawa in Zimbabwe, are of Cretaceous age. The carbonatites of southern Malawi were intruded between 139 and 130 Ma, Chishanya in Zimbabwe at 127 Ma, and the Tanzanian carbonatites between 120 and 100 Ma.

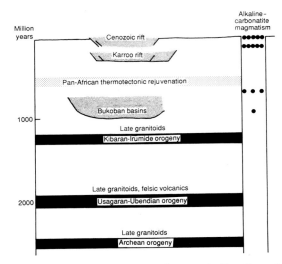

Figure 8.13 Major magmatic and tectonic events in East-Central Africa.

8.5 DISCUSSION AND CONCLUSIONS

The common association of carbonatite magmatism and rifting is largely based on studies of the East African Rift System where the two are closely associated (Baker *et al.* 1972, Logatchev 1974, Mitchell & Garson 1976).

Most current compilations stress the spatial relationship between carbonatites and major structures. Commonly, however, the time of magma emplacement is overlooked. The generalizations of Mitchell & Garson (1976), who matched the various rift systems with Cenozoic or older carbonatitic centres, have to be reassessed in order to account for the age differences between intrusion and rift formation.

Some workers correlate carbonatite events with orogenies. Gittins *et al.* (1967) point to a possible temporal relationship between carbonatite magmatism and major periods of orogenic deformation in eastern Canada, and Woolley (this volume) emphasizes the importance of orogenic activity and carbonatite magmatism.

A contrary view to the rift-carbonatite and orogenic belt-carbonatite association has been developed by Le Bas (1980, 1987), who relates alkaline magmatism in East Africa to large-scale regional up-doming. He describes the spatial distribution of carbonatites in East Africa in terms of an 'East African Alkaline Province'. Kimberlites occur in the central part of a concentrically zoned 'province', followed by carbonatites, olivine-poor nephelinites and ijolites in the adjacent zone, and alkali basaltic volcanism at the periphery. However, Le Bas has grouped together magmatic episodes of different ages, and has combined Cenozoic carbonatite events with Mesozoic and even Precambrian kimberlites.

Others claim that there is no direct genetic relationship between carbonatite–alkaline magmatism and rifting or orogenic events. Sørensen (1974), for instance, stresses that alkaline magmatism developed during periods of tectonic quiescence, while Bailey (1961, 1974, 1977) emphasized lithospheric control of carbonatite emplacement in East and Central Africa and stressed that carbonatites were intruded along recurrently opening fractures.

Obviously there are many different geodynamic models that attempt to explain carbonatite magmatism in space and time. In western and southwestern Tanzania, the spatial relationships of the Late Proterozoic carbonatites are characterized by alignment of carbonatites along persistently rejuvenated deep structures that mark extensive curvilinear zones. The time of carbonatite intrusion is bracketed by the final stages of a period of relative tectonic quiescence (deposition of the Bukoban succession) and the start of the Pan-African thermotectonic event (Fig. 8.13). Apart from the shallow basins of Bukoban age there is little evidence of palaeorifts. Bukoban basins at the western side of the Tanzanian craton show some signs of late Proterozoic rifting, including stratiform copper mineralization and widespread eruption of basaltic flows and dyke intrusions, but lack the typical normal faults (Halligan 1962) and higher subsidence rates associated with continental rifting.

The nature of the Pan-African thermotectonic event is much debated. Several authors follow the interpretation of Watson (1976), Martin & Porada (1977), Priem *et al.* (1979), and Gabert (1984) who interpret the Pan-African event as essentially non-orogenic with broad-scale, mantle diapirism causing stretching and thinning of the crust on a regional scale. Other authors, like Kröner (1984), interpret the Pan-African event in terms of plate tectonics and related orogenic activity. Although the cause of the Pan-African episode remains uncertain, it appears that carbonatite magmatism was closely associated with the initiation of this extensive thermotectonic event.

The time–space relationships of the Mesozoic carbonatites are clearer. East African, Mesozoic carbonatite magmatism is related to tensional stress. Most Cretaceous carbonatites of SW Tanzania were intruded into fault zones associated with post-Karroo rifting. They were emplaced at the very late stages of post-Karroo rifting or during a tectonically quiet period between two major rift events, the post-Karroo–pre-Cretaceous and the Cenozoic. The carbonatites were emplaced long after the breakup of Gondwanaland (Fig. 8.14).

The East African carbonatite magmatism of Mesozoic age is certainly not associated with crustal shortening or with orogenesis. The nearest Mesozoic fold belt, the Cape Fold Belt, lies more than 2500 km to the south and was formed during the Early Triassic.

Along the Eastern branch of the East African Rift system, mainly in Uganda, Kenya, and northern Tanzania, recurrence of carbonatite volcanism took place during the Cenozoic. This magmatism is again associated with recent tensional events, rather than orogenesis.

The conclusion is that the Late Proterozoic and Cretaceous carbonatites and alkaline complexes in southwestern and western Tanzania intrude persistent shear and fault zones. The intrusion of the Precambrian carbonatites seems to be associated with early phases of the Pan-African thermotectonic event, the cause of which is still not understood. All East African carbonatites of Mesozoic and younger ages were related to extensional episodes and not to orogenic activities.

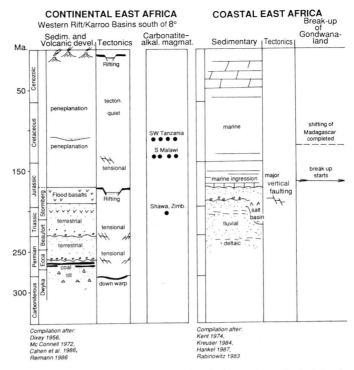

Figure 8.14 The age of carbonatite–alkaline magmatism in relation to the geological development of continental and coastal East Africa.

Deep-reaching, persistent fault and shear structures outside the cratonic blocks seem to have been preferred channelways for carbonatite volcanism. Such anisotropies were suitable for recurrent structural reactivation, including rifting.

ACKNOWLEDGEMENTS

This study is based on geological field work by members of the Tanzania–Canada Agrogeology project, funded by the Tanzanian government and the International Development Research Centre, Canada. My Tanzanian colleagues E. P. Mchihiyo, P. Masao, F. T. Mtuy, O. W. Kiranga, W. Z. Mbasha, and G. J. Mbawala are thanked for the support during the successful geological field work 1985–87, *asante sana*.

Special thanks go to Dr W. Chesworth for his invaluable advice and suggestions during field work and the preparation of this paper. I am also indebted to P. S. Smith for the REE analyses and D. Crabtree for assisting with thin-section work and data plotting.

Thanks are also due to the three reviewers for their constructive suggestions. I also appreciate the help of M. Metcalf who typed the manuscript and D. Wilson and D. Irvine for drafting.

REFERENCES

Bailey, D. K. 1961. The mid-Zambezi-Luangwa rift and related carbonatite activity. *Geological Magazine* **98**, 277–84.

Bailey, D. K. 1974. Continental rifting and alkaline magmatism. In *The alkaline rocks*, H. Sørensen, (ed.), 148–59. New York: Wiley.

Bailey, D. K. 1977. Lithosphere control of continental rift magmatism. *Journal of the Geological Society of London* **133**, 103–6.

Baker, B. H., P. Mohr & L. A. J. Williams 1972. *Geology of the eastern rift system of Africa*. Geological Society of America Special Paper 136.

Basu, N. K. & A. Mayila 1986. Petrographic and chemical characteristics of the Panda Hill carbonatite complex, Tanzania. *Journal of African Earth Sciences* **5**, 589–98.

Brown, P. E. 1962. The tectonic and metamorphic history of the Pre-Cambrian rocks of the Mbeya region, South-West Tanzania. *Quarterly Journal of the Geological Society of London* **118**, 295–317.

Brown, P. E. 1964. The Songwe Scarp carbonatite and associated feldspathization. *Quarterly Journal of the Geological Society of London* **120**, 223–40.

Cahen, L. & N. J. Snelling 1966. *The geochronology of equatorial Africa*. Amsterdam: North-Holland.

Cahen, L., N. J. Snelling, J. Delhal, J. R. Vail, M. Bonhomme & D. Ledent 1986. *The geochronology and evolution of Africa*. Oxford: Clarendon Press.

Coetzee, G. L. 1962. *The origin of the Sangu carbonatite complex and associated rocks, Karema depression, Tanganyika Territory, E. Africa*. Ph.D. thesis, University of Wisconsin, Madison.

Daly, M. C. 1985. The intracratonic Irumide Belt of Zambia and its bearing on collision orogeny during the Proterozoic of Africa. In *Collision Tectonics*, M. P. Coward & A. Ries (eds), 30–42. London: Special Publication of the Geological Society of London.

Dixey, F. 1956. *The East African Rift system*. Colonial Geology and Mineral Resources. Bulletin Supplement No. 1.

Fawley, A. P. & T. C. James 1955. A pyrochlore carbonatite, southern Tanganyika. *Economic Geology* **50**, 571–85.

Fick, L. J. & C. van der Heyde 1959. Additional data on the geology of the Mbeya carbonatite. *Economic Geology* **54**, 843–72.

Gabert, G. 1984. Structural lithological units of Proterozoic rocks in East Africa, their base, cover and mineralisation. In *African Geology*, J. Klerkx & J. Michot (eds), 11–22. Tervuren: Musée Royal de l'Afrique Centrale.

Gittins, J., R. M. MacIntyre & D. York 1967. The ages of carbonatite complexes in Eastern Canada. *Canadian Journal of Earth Sciences* **4**, 651–5.

Haldemann, E. G. & J. R. Harpum 1952. A preliminary account of the carbonate rocks of Central Ukisi, Njombe District (Tanganyika), Abstract. *Colonial Geology and Mineral Resources* **3**, 243–5.

Halligan, R. 1962. *The Proterozoic rocks of western Tanganyika*. Bulletin of the Geological Survey of Tanganyika, 34.

Hankel, O. 1987. Lithostratigraphic subdivision of the Karroo rocks of the Luwegu Basin (Tanzania) and their biostratigraphic classification based on microfloras, macrofloras, fossil woods and vertebrates. *Geologische Rundschau* **76**, 539–65.

Horne, R. G. 1959. Nachendezwaya Carbonatite. *Records of the Geological Survey of Tanganyika* **9**, 37–9.

James, T. C. 1954. *Preliminary report on geological investigations at Panda Hill, Mbeya district*. Geological Survey of Tanganyika.

James, T. C. 1956a. Carbonatite investigations. *Records of the Tanganyika Geological Survey* **IV**, 1954, 20.

James, T. C. 1956b. *Carbonatites and rift valleys in East Africa*. Geological Survey of Tanganyika unpublished report TCJ 134.

Kent, P. 1974. Continental margin of East Africa – A region of vertical movement. In *The geology of continental margins*, C. A. Burk & C. L. Drake (eds), 313–20. New York: Springer.

Kreuser, T. 1984. Karroo basins in Tanzania. In *African geology*, J. Klerkx & J. Michot (eds), 231–45. Tervuren: Musée Royal de l'Afrique Centrale.

Kröner, A. 1984. Late Precambrian plate tectonics and orogeny: a need to redefine the term Pan-

African. In *African Geology*, J. Klerkx & J. Michot (eds), 23–8. Tervuren: Musée Royal de l'Afrique Centrale.

Le Bas, M. J. 1980. The East African Cenozoic magmatic province. *Accademia Nazionale dei Lincei, Atti dei convegni Lincei* **47**, 111–22.

Le Bas, M. J. 1987. Nephelinites and carbonatites. In *Alkaline igneous rocks*, J. G. Fitton & B. G. J. Upton (eds), 53–83. London: Geological Society Special Publication 30.

Logatchev, N. A. 1974. Magmatism-tectonics relationship during the Kenya rift development and the prime cause of strongly and mildly alkaline volcanic suite formation. In *Afar between continental and oceanic rifting*, Vol. II, A. Pilger & A. Roesler (eds), 96–106. Stuttgart: Schweitzerbart.

McConnell, R. B. 1972. Geological development of the rift system of eastern Africa. *Geological Society of America Bulletin* **83**, 2549–72.

MacFarlane, A. 1966. *The geology of Tunduma and Itumba districts*. Quarter Degree Sheet 258 and 258S. Mineral Resources Division of Tanzania.

Maravič, H. V. & G. Morteani 1980. Petrology and geochemistry of the carbonatite and syenite complex of Lueshe (NE Zaire). *Lithos* **13**, 159–70.

Martin, H. & H. Porada 1977. The intercratonic branch of the Damara Orogen in Southwest Africa. II. Discussion of relationships with the Pan-African mobile belt system. *Precambrian Research* **5**, 339–57.

Miller, J. A. & P. E. Brown 1963. The age of some carbonatite activity in south-west Tanganyika. *Geological Magazine* **100**, 276–9.

Mitchell, A. H. G. & M. S. Garson 1976. Mineralization of plate boundaries. *Minerals Science and Engineering* **8**, 129–69.

Moeller, P., G. Morteani & F. Schley 1980. Discussion of REE distribution patterns of carbonatites and alkalic rocks. *Lithos* **13**, 171–9.

Mtuy, F. T., E. P. Mchihiyo, P. Masao & P. van Straaten 1987. *Phosphate mineralization associated with the Panda, Mbalizi and Songwe Scarp carbonatites, Mbeya region, Tanzania*. Abstract, 14th Colloquium on African Geology, Berlin, 18–22 Aug. 1987, CIFEG occ. publ. 1987/12.

Pentelkov, V. G. & S. N. Voronovskiy 1979. Radiometric age of the Mbalizi carbonatite, Tanzania and correlation with other carbonatites of the Rukwa-Malawi rift zone. *Akademiia Nauk SSSR Doklady Earth Science Sections* **235**, 92–4.

Priem, H. N. A., N. A. I. M. Boelrijk, E. H. Hebeda, E. A. Th. Verdurman & R. H. Verschure 1979. Isotopic age determinations on granitic and gneissic rocks from the Ubendian-Usagaran system in Southern Tanzania. *Precambrian Research* **9**, 227–39.

Rabinowitz, P. D. 1983. The separation of Madagascar and Africa. *Science* **220**, 67–9.

Ray, G. E. 1974. The structural and metamorphic geology of northern Malawi. *Journal of the Geological Society of London* **130**, 427–40.

Reimann, K. U. 1986. Prospects for oil and gas in Zimbabwe, Zambia and Botswana. *Episodes* **9**, 95–100.

RUDIS 1980. *Prefeasibility study. Fertilizer Raw Materials*. Ljubljana, Yugoslavia: RUDIS.

Snelling, N. J. 1965. Age determination on three African carbonatites. *Nature* **205**, 491.

Snelling, N. J., E. Hamilton, A. Drysdall & C. Stillman 1964. A review of age determinations from Northern Rhodesia. *Economic Geology* **59**, 961–81.

Sørensen, H. (ed.), 1974. *The alkaline rocks*. New York: Wiley.

Spence, J. 1957. The geology of part of the Eastern Province of Tanganyika. *Bulletin of the Geological Survey of Tanganyika*, 28.

Stockley, G. M. 1943. The geology of the Rufiji District, including a small portion of the Northern Kilwa District (Matumbi Hills). *Tanganyika Notes and Records* **16**, 7–28.

Straaten, P. van 1983. *Interim report on results from the Sangu-Ikola carbonatite*. Report ESAMRDC/83/Tech/26, Eastern and Southern African Mineral Resources Development Centre, Dodoma, Tanzania.

Straaten, P. van 1987. *Mineral resources potential in SW Tanzania*. Abstract, 14th Colloquium on African Geology, Berlin, 18–22 Aug. 1987, CIFEG occ. publ. 1987/12, 175–6.

Tack, L. & P. De Paepe 1981. Le massif intrusif de syenites foidales de la Haute-Ruvubu, Burundi. *Musée Royal de l'Afrique Centrale Tervuren (Belgique) Rapport Annuel*. 1980, 173–8.

Tack, L., P. De Paepe, S. Deutsch & J. P. Liegeois 1984. The alkaline plutonic complex of the upper Ruvubu (Burundi): geology, age, isotopic geochemistry and implications for the regional geology of the Western Rift. In *African Geology*, J. Klerkx & J. Michot (eds), 91–114. Tervuren: Musée Royal de l'Afrique Centrale.

Ueda, Y., I. Matsuzaka & K. Suwa 1975. *Potassium–argon age determinations on the Tanzanian igneous and metamorphic rocks.* First Preliminary Report of African Studies, Nagoya University, Japan, 76–80.

Watson, J. V. 1976. Vertical movements in Proterozoic structural provinces. *Philosophical Transactions of the Royal Society of London, Series A* **280**, 629–40.

Williamson Diamonds (1970). *Ngualla.* Internal report, Geological Survey of Tanzania, Dodoma.

Woolley, A. R. 1982. A discussion of carbonatite evolution and nomenclature, and the generation of sodic and potassic fenites. *Mineralogical Magazine* **46**, 13–17.

9
CARBONATITES IN A CONTINENTAL MARGIN ENVIRONMENT – THE CANADIAN CORDILLERA

J. PELL & T. HÖY

ABSTRACT

In western Canada, carbonatites, syenite gneisses, and related alkaline rocks are present in a broad zone which parallels the Rocky Mountain Trench. These rocks were emplaced in the continental margin sedimentary sequence prior to the orogenesis which resulted in the formation of the Canadian Cordillera. Their emplacement was related to periodic extension or rifting episodes along the western margin of ancestral North America.

Within this broad zone, three distinct spatial, geologic, and geochronologic belts can be identified. Alkalic rocks and carbonatites in the Foreland Belt are high-level intrusions of Devono-Mississippian age hosted in Lower to Middle Palaeozoic strata. They tend to be large and elliptical in shape, have significant alteration halos and are enriched in F, Nb, and REEs (rare earth elements). Carbonatites in the Omineca Belt, immediately west of the Rocky Mountain Trench, are deeper-level Devono-Mississippian intrusions hosted in late Precambrian strata. They are thin, discontinuous, concordant intrusions with only narrow fenite zones, and are commonly less enriched in F, Nb, and REEs than carbonatites in the Foreland Belt. Extrusive and intrusive carbonatites and nepheline syenite gneisses occur in cover sequences, presumed to be late Proterozoic to early Palaeozoic, above core complexes in the Omineca Belt. Syenite gneisses intruding the base of the succession are late Proterozoic whereas extrusive carbonatite, higher in the sequence, is interpreted to be Eocambrian to early Cambrian. Intrusive carbonatites in this sequence are similar to the extrusive carbonatite and are believed to be approximately the same age. They are high-level intrusive bodies, parallel to bedding, associated with extensive fenitic aureoles and moderate REE enrichment.

9.1 INTRODUCTION

An interesting alkaline igneous province is present in the eastern part of the Canadian Cordillera. It comprises carbonatites, nepheline and sodalite syenites, some ijolite series rocks, one kimberlite locality, and numerous ultramafic and lamprophyric diatreme breccias, all of which intruded the Cordilleran miogeoclinal succession prior to the deformation and metamorphism associated with the Jura-

Cretaceous orogeny. These rocks have important implications because they were emplaced in a continental margin environment as opposed to the more common cratonic or shield association (Heinrich 1966, Dawson 1980) and they also detail the subsequent effects of orogenesis. This chapter presents a broad overview of these somewhat enigmatic rocks, most of which have had little detailed petrologic work.

In British Columbia, carbonatites and related rocks are found in a broad zone parallel to and encompassing the Rocky Mountain Trench (Pell 1987a). They occur in three discrete areas (Fig. 9.1): in the Foreland Belt, east of the Rocky Mountain Trench; along the eastern edge of the Omineca Belt; and in the vicinity of the Frenchman Cap dome, a core gneiss complex, also within the Omineca Belt. Carbonatites in the Foreland Belt include the Aley carbonatite complex (Mader 1986, 1987; Pell 1986a), the Bearpaw Ridge sodalite syenite (Pell 1985, 1986b), the Ice River syenite and carbonatite complex (Currie 1975, 1976a) and the Rock Canyon Creek fluorite and REE prospect, which is a carbonatite-related deposit (Hora & Kwong 1986, Pell & Hora 1987). Three of the intrusions within this belt are subcircular to elliptical in plan, have extensive metasomatic and/or contact metamorphic alteration halos and are hosted in Middle Cambrian to Middle Devonian

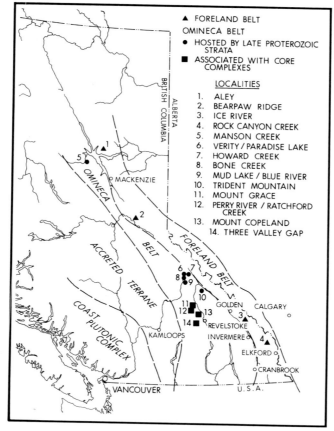

Figure 9.1 Index map showing carbonatite localities in the Canadian Cordillera (from Pell 1987a).

miogeoclinal rocks. The Rock Canyon Creek 'showing' comprises an elongate zone of metasomatic alteration with no known associated intrusive rocks (Pell & Hora 1987). During orogenesis, the intrusions within this belt were subjected to sub-greenschist facies metamorphism. The obvious effects of the deformation are minor; the intrusions appear to have behaved as rigid bodies and were simply rotated, tilted and/or transported eastward in thrust slices.

Carbonatites along the eastern margin of the Omineca Belt are found westward from the Rocky Mountain Trench for 50 km or more. All the intrusions within this belt are hosted in late Precambrian (Upper Proterozoic) to early Cambrian metasedimentary rocks. They form foliated, sill-like bodies that have been multiply deformed and metamorphosed to amphibolite facies during the Middle Mesozoic orogeny. These carbonatites have thin sodic pyroxene and amphibole-rich fenitic margins. The belt comprises carbonatites associated with monzonites and some syenites in the Manson Creek area (Rowe 1958, Currie 1976a, Pell 1985), carbonatites associated with sodalite syenites and some urtites in the Blue River area, including the Bone Creek, Howard Creek, Mud Lake, Paradise Lake, Verity localities (Rowe 1958, Currie 1976a, White 1982, Pell 1985) and nepheline and sodalite syenites at Trident Mountain (Currie 1976a, Perkins 1983, Pell 1986b).

The most westerly area contains extrusive and intrusive carbonatites and syenite gneiss bodies in a mixed paragneiss succession of uncertain age along the margins of the Frenchman Cap gneiss dome (McMillan 1970, McMillan & Moore 1974, Höy & Kwong 1986, Höy & Pell, 1986) in the core of the Omineca Belt. The intrusive and extrusive alkaline rocks in this area are conformable bodies that were deformed and metamorphosed to amphibolite facies during orogenesis. The Mount Copeland syenite gneiss (Fyles 1970, Currie 1976b) occurs along the southern margin of the gneiss dome, whereas the Mount Grace carbonatite tuff, intrusive carbonatites, and intrusive syenite gneisses occur along the western flank (McMillan 1970, McMillan & Moore 1974, Höy & Kwong 1986, Höy & Pell 1986, Höy 1987).

9.2 INTRUSIVE CARBONATITES IN PALAEOZOIC ROCKS, FORELAND BELT

9.2.1 Introduction

Carbonatites, associated syenites and alkaline ultramafic silicate rocks are found in a number of localities in the Foreland Belt of the Canadian Cordillera (Fig. 9.1). They were emplaced in Devonian to Mississippian time into continental margin platformal carbonate and clastic rocks of Middle Cambrian to Lower Devonian age which were deposited along the western margin of the North American continent (Pell 1987a; Pell et al., in preparation; and Table 9.1). Potassium–argon dates of mica separates from the Aley carbonatite complex (K. Pride, pers. comm. 1986) in the northern Rocky Mountains (Fig. 9.1) are 339 ± 12 Ma and 349 ± 12 Ma. Recent U–Pb data obtained from zircons, titanite, and ^{40}Ar-^{39}Ar dates on hornblende from the Ice River complex in the southern Rocky Mountains (Fig. 9.1) yield a

Table 9.1 Carbonatites in British Columbia.

Igneous suite	Location	Host succession, age	Metamorphism	Dating method	Age of intrusion
Foreland Belt					
Aley complex carbonatite	Northern Rocky Mountains	Kechika Formation Cambro-Ordovician	sub-greenschist	K-Ar, micas	339 ± 12 Ma[1,2] 349 ± 12 Ma[1,2]
Ice River complex syenites, etc.	Southern Rocky Mountains	Chancellor & Ottertail Fms, McKay Grp Upper Cambrian to Cambro-Ordovician	sub-greenschist to lower greenschist	U-Pb, zircon	368 ± 4 Ma[3]
Rock Canyon Creek prospect	Southern Rocky Mountains	Basal Devonian Unit Middle Devonian	sub-greenschist		?
Eastern Omineca Belt					
Lonnie carbonatite	Manson Creek area, Omineca Belt	Wolverine Complex late Proterozoic	lower amphibolite facies (garnet zone)	U-Pb, zircon	c. 370 Ma[4]
Vergil carbonatite	Manson Creek area, Omineca Belt	Wolverine Complex late Proterozoic	lower amphibolite facies (garnet zone)	U-Pb, zircon	c. 370 Ma[4]
Paradise Lake nepheline syenite	Blue River area, Omineca Belt	Horsethief Creek Group Hadrynian (late Proterozoic)	upper amphibolite facies (kyanite zone)	U-Pb, zircon	c. 370 Ma[4]
Verity carbonatite	Blue River area, Omineca Belt	Horsethief Creek Group Hadrynian (late Proterozoic)	upper amphibolite facies (kyanite zone)	U-Pb, zircon	c. 323 Ma[5]
Trident Mountain nepheline syenite	Mica Creek area, Omineca Belt	Horsethief Creek Group Hadrynian (late Proterozoic)	upper amphibolite facies (kyanite-sillimanite zone)	U-Pb, zircon	378 ± 7 Ma[5]
Frenchman Cap area					
Mount Grace extrusive carbonatites	Frenchman Cap, Omineca Belt	'mantling gneisses' ? late Proterozoic to early Palaeozoic	upper amphibolite facies (sillimanite zone)	relative, interpreted	Eocambrian[6,7]
Perry River intrusive carbonatites	Frenchman Cap, Omineca Belt	'mantling gneisses' ? late Proterozoic to early Palaeozoic	upper amphibolite facies (sillimanite zone)	relative, interpreted	Eocambrian[6,7]
Mount Copeland syenite gneiss	Frenchman Cap, Omineca Belt	'mantling gneisses' ? late Proterozoic to early Palaeozoic	upper amphibolite facies (sillimanite zone)	U-Pb, zircon	773 Ma[8]

References: 1, K. Pride, pers. comm.; 2, Mader (1986); 3, Parrish et al. (1987); 4, Pell et al. (in preparation); 5, R. R. Parrish, unpublished information; 6, this study; 7, Höy & Godwin (1988); 8, Okulitch et al. (1981).

c. 370 Ma emplacement age (Parrish *et al.* 1987). During the Columbian Orogeny these carbonatites were regionally metamorphosed to lower greenschist grade, tilted, weakly deformed, and transported eastward in thrust slices.

9.2.2 Description of carbonatites

The carbonatites and alkaline rock complexes in the Foreland Belt are relatively large intrusions; the Aley and Ice River complexes are characteristic of this belt. The Aley complex, a zoned sub-circular intrusion, is approximately 3.5 km in diameter and consists of a rauhaugite core surrounded by 'amphibolite' (in part fenitized alkaline silicate rocks). Calcite carbonatite dykes and carbonatites with high REE contents occur within the core. An extensive contact aureole of recrystallized carbonate rocks up to 500 m wide surrounds the amphibolite. Carbonatite dykes enriched in REE-carbonate minerals intrude the aureole (Mader 1986, 1987, Pell 1986*a*, 1987*a*). The mineralogy of the Aley complex is extremely varied (Mader 1987); the most abundant minerals of the rauhaugites include apatite, dolomite, magnetite, phlogopite, pyrite, zircon, and the niobium minerals columbite, fersmite, and pyrochlore. Sövite dykes generally contain sodic amphiboles in addition to the accessory minerals of the rauhaugites. The carbonatite dykes are ankeritic and contain barite, bastnaesite, fluorite, pyrite, and a variety of other REE-carbonate minerals. The Aley complex is economically interesting as extensive zones contain between 0.6 and 0.8% Nb_2O_5 (K. Pride, pers. comm.).

The Ice River complex (Fig. 9.1) is an arcuate shaped, zoned alkaline ultramafic intrusion exposed over an area of approximately 30 km^2. Two distinct intrusive series are present within the complex: an early, rhythmically layered feldspar-free intrusion of ijolite, jacupirangite, and urtite, which is cored by a carbonatite plug and cut by carbonatite dykes rich in mafic silicates and oxides; and a cross-cutting, zoned syenitic series which is associated with zeolitic and feldspathic carbonatites (Currie 1975). The enclosing sedimentary rocks are hornfelsed and skarned, and locally have undergone Na-metasomatism (Currie 1975).

The carbonatites are volumetrically minor but important constituents of the Ice River complex. They occur in a number of localities and display considerable lithologic variation. Black-, buff-, red-, and white-weathering varieties are present in the layered ultramafic series. The black carbonatites commonly occur as dykes containing elemental carbon and tetranatrolite near their margins, with aegirine, berthierine, biotite, calcite, edingtonite, ilmenite, perovskite, siderite, minor sphalerite, and traces of pyrite throughout. The red-weathering carbonatite is compositionally similar to the black, but contains fewer non-carbonate minerals and both siderite and zeolites are absent. The coarse-grained buff- and white-weathering carbonatites are composed of calcite with accessory aegirine, apatite, and pyrite ± ilmenite, magnetite, phlogopite, and pyrochlore (Currie 1975). The carbonatites associated with the syenites differ from those in the ultramafic series; the only silicate minerals present are feldspars, rare phlogopite, and zeolites (analcime, edingtonite, and natrolite). Minor mineral phases include apatite, barite, ilmenite, pyrite, and rutile (Currie 1975).

9.3 INTRUSIVE CARBONATITES IN LATE PROTEROZOIC ROCKS, OMINECA CRYSTALLINE BELT

9.3.1 Introduction

Intrusive carbonatites and related rocks are located in three areas within the Omineca Crystalline Belt; near Manson Creek, in the Blue River area and on Trident Mountain (Fig. 9.1). They are hosted by Upper Proterozoic strata, the Wolverine Complex in the Manson Creek area (Lang et al. 1946), and the Horsethief Creek Group of the Windermere Supergroup at Blue River and Trident Mountain (Wheeler 1965, Campbell 1967). These host rocks consist predominantly of psammites, semipelitic schists, and marbles. Carbonatites intruded the lower portion of the miogeoclinal succession in Devono-Mississippian time (Table 9.1), at the same time as those now exposed in the Foreland Belt (Fig. 9.1). U–Pb ages on zircons from the Manson Creek area are 350 ± 10 Ma and 370 ± 20 Ma and from the Blue River area, 325 ± 25 Ma, 328 ± 30 Ma, and 350 ± 5 Ma (Pell 1987a, Pell et al. in preparation). Both the carbonatites and host rocks were subsequently deformed and regionally metamorphosed to lower amphibolite grade in the Manson Creek area and to middle amphibolite grade (kyanite zone) in the Blue River area.

9.3.2 Description of carbonatites

Carbonatites in the Omineca Belt are concordant, sill-like intrusions; they vary in thickness from a few centimetres to approximately 7 m and are up to 1 km long. The carbonatites are commonly associated or interlayered with alkalic rocks such as nepheline and sodalite syenites, monzonites, and urtites or melteigites. They are deformed, foliated, and locally outline outcrop-scale folds (Fig. 9.2). Compositional banding, defined by apatite- and amphibole-rich layers parallel to the margins of the carbonatites and the regional schistosity, is also commonly present; this is considered to be an original igneous texture.

In the Blue River area, olivine-pyroxene-phlogopite sövites, amphibole, and phlogopite-bearing rauhaugites and biotite sövites are recognized. Pyroxenes present are mainly diopsides; amphiboles may be actinolite, richterite, soda-tremolite, or tremolite. Phlogopites have either normal or reverse pleochroism. In the Manson Creek area, aegirine sövites and biotite sövites are present. Accessory minerals in both localities include apatite, magnetite, and pyrochlore and, less commonly, microcline and plagioclase, with trace amounts of columbite, ilmenite, and zircon (with crystals up to 3 cm in size). Allanite, baddeleyite, monazite, and rutile have also been reported (Pell 1987a).

Fenitization is relatively minor adjacent to these carbonatites. In the Blue River area, mafic pyroxene-amphibole fenites, 1–30 cm thick, separate the carbonatites from the host metasedimentary rocks. In the Manson Creek area, the alteration halos are only slightly wider. Pyroxene–amphibole fenites, interlayered with the carbonatites and the metasedimentary rocks within 10 m of the intrusions, contain

Figure 9.2 Bedding-parallel carbonatite layer in Late Proterozoic Horsethief Creek Group metasedimentary rocks outlining an F_2 fold, Howard Creek area, Omineca Belt.

aegirine and arfvedsonite which presumably replaced biotite, the original mafic silicate mineral (Pell 1987a).

9.4 CARBONATITES ASSOCIATED WITH CORE COMPLEXES, OMINECA CRYSTALLINE BELT

9.4.1 Introduction

Carbonatites, both extrusive and intrusive, are associated with metamorphosed syenitic rocks on the margins of Frenchman Cap dome, one of a series of domal structures in the Monashee Complex of the Omineca Belt (Wheeler 1965, McMillan 1973, Höy 1987). The Frenchman Cap dome is cored by paragneiss and orthogneiss of probable Lower Proterozoic age (R. L. Armstrong, pers. comm.), and mantled by a succession of highly deformed and metamorphosed paragneiss, calc-silicate gneiss, quartzite, and marble (Fig. 9.3), originally sediments deposited in a shallow marine, platformal environment. This succession, referred to as the mantling gneisses, hosts the carbonatites and intrusive syenites. It is structurally overlain by metamorphic rocks of the Selkirk allochthon. The entire sequence has undergone intense polyphase deformation and high-grade regional metamorphism that spanned a period from late Jurassic(?) to perhaps Palaeocene (see summary by Okulitch 1984).

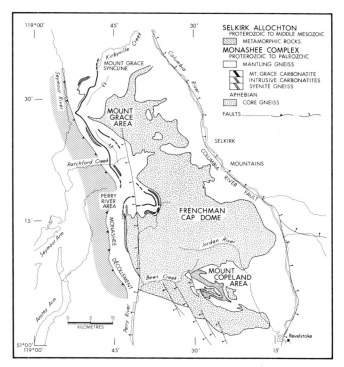

Figure 9.3 Tectonic map of the Frenchman Cap dome, Omineca Belt, southeastern British Columbia (modified from Read & Brown 1981) showing localities of carbonatites and syenite gneisses.

The age of the mantling paragneiss succession is not known with any certainty. It has been correlated with Eocambrian and Lower Palaeozoic platformal rocks in the Kootenay Arc to the east (Wheeler 1965, Fyles 1970, Höy & McMillan 1979), with Middle Proterozoic Belt–Purcell rocks (Brown & Psutka 1979) and with older rocks (Brown & Read 1983). An approximate U–Pb date of 773 Ma has been obtained from zircons from the Mount Copeland syenite gneiss which intrudes the basal part of the succession in the Jordan River area (Okulitch *et al.* 1981), and an early Cambrian age is indicated by Pb–Pb galena data from the stratiform Cottonbelt lead–zinc deposit higher in the succession (Höy & Godwin, 1988). These data are consistent with a late Proterozoic to early Palaeozoic age for the cover metasedimentary succession, although the lowermost strata intruded by nepheline syenites could be much older.

The ages of carbonatites and other alkalic rocks in the Frenchman Cap area may span a considerable range. The Mount Copeland syenite gneiss (Fyles 1970, Currie 1976*b*) is apparently Upper Proterozoic in age (Okulitch *et al.* 1981; Table 9.1). The extrusive Mount Grace carbonatite (Höy & Kwong 1986, Höy & Pell 1986) occurs much higher in the succession, only 100 m below the Cottonbelt deposit and, like the Cottonbelt deposit, is probably early Cambrian or Eocambrian in age. Intrusive carbonatites occur at two stratigraphic levels, near the base of the cover succession (Perry River carbonatites) and above the Mount Grace carbonatite (Ratchford

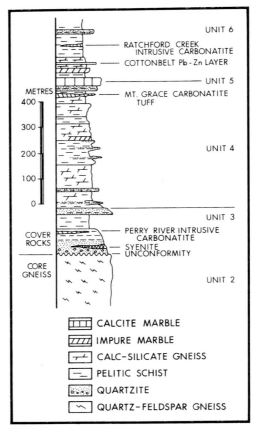

Figure 9.4 The mantling gneiss succession, Frenchman Cap dome, showing the location of the Mount Grace carbonatite tuff, intrusive carbonatites, and syenite gneisses (after Höy 1987).

Creek or Ren carbonatite, see Fig. 9.4). The age of these intrusive carbonatites is bracketed only by a late Proterozoic to early Palaeozoic age for the host succession and a later Mesozoic age for regional metamorphism and deformation. The spatial association of the intrusive carbonatites with the Mount Grace carbonatite, as well as petrographic and geochemical similarities, suggest that they may be of about the same age as the extrusives.

9.4.2 Intrusive carbonatites

The intrusive carbonatites, although occurring at different stratigraphic levels (Fig. 9.4), have many similarities. They form zones of intimately intermixed carbonatite, fenite, and paragneiss up to several tens of metres thick and several kilometres in length. Although these zones appear to be concordant on a local scale, they cut across stratigraphy on a regional scale. They have been internally deformed and metamorphosed to amphibolite facies (sillimanite zone) together with the country rocks, producing a zone of granoblastic marbles and gneisses.

Carbonatites occur as thick, buff-weathering, foliated, and laminated layers; thin, discontinuous swirled lenses; and small, coarse-grained pods with calcite cores and biotite–amphibole margins within the volumetrically more abundant fenites. They generally comprise 70–90% calcite or dolomite, with variable amounts of apatite, phlogopite, and sodic amphibole; phlogopites may display reverse pleochroism. Aegirine, chalcopyrite, ilmenite, magnetite, molybdenite, plagioclase, pyrite, pyrochlore, pyrrhotite, titanite, and zircon may be present as accessory minerals (Höy & Pell 1986).

Fenitization is intense and widespread around the carbonatite lenses. Three types are distinguished, both chemically and mineralogically (Höy 1987). The most prominent is mafic pyroxene–amphibole fenite; it is commonly interlayered with thin bands of albite-rich fenite. Layering in these fenites mimics bedding and foliation in adjacent and interlayered metasedimentary rocks. A third prominent fenite type comprises predominantly potassic feldspar and albite. It resembles a syenite but commonly retains the well-bedded nature of the protolith sedimentary layers and may have gradational contacts with other fenites. Locally it has remobilized, producing rheomorphic fenites with sharp intrusive contacts. The interlayering of the types of fenites and of fenites with paragneisses suggests that fenitization selectively replaced the layered rocks and affected strata of different compositions in different ways.

9.4.3 *The Mount Grace extrusive carbonatite*

The Mount Grace carbonatite is a marble layer that averages 2–5 m in thickness and has been traced and extrapolated for at least 100 km along strike length (Höy & Kwong 1986, Höy & Pell 1986, Höy 1987). Its contacts with overlying and underlying calcareous gneisses are generally sharp, but may grade through a distance of approximately 1 m into grey-weathering, massive to thin-bedded sedimentary marble. In contrast to intrusive carbonatites, fenitized margins are absent. The Mount Grace carbonatite is interpreted to be a pyroclastic carbonatite that was deposited on shallow marine tidal flats (Höy & Kwong 1986). Evidence includes its great lateral extent, its distinctly layered nature, its occurrence at a specific stratigraphic horizon, and its lack of contact alteration.

In the field, the Mount Grace carbonatite can be easily recognized by its unusual medium-brown weathering colour, by the number and variety of included clasts, and by its mineralogy. Grains of dark-brown phlogopite, colourless apatite, and needles of green sodic amphibole weather in relief. Pyrrhotite, pyrochlore, and zircon are locally-developed accessory minerals, and barite, monazite, and strontianite are present in trace amounts (Höy 1987). The carbonatite has been deformed and metamorphosed, producing a foliated to massive, commonly coarse-grained marble.

The Mount Grace carbonatite is commonly internally bedded with a layer or several layers of 'blocky' tephra interbedded with finer grained, massive or laminated carbonatite (Fig. 9.5). The blocky tephra layer contains three distinctive types of matrix-supported clasts. These are small granular albitite clasts up to 3 cm

Figure 9.5 Carbonatite tuff and agglomerate, Mount Grace extrusive, Kirbyville Creek area. Darker coloured, lithic fragment-rich agglomerate layers are interbedded with lighter coloured sedimentary marble layers.

in diameter, consisting of pure albite or albite with variable amounts of phlogopite; 'syenite' clasts, 1–10 cm in diameter, consisting of potassic feldspar with variable amounts of apatite, calcite, plagioclase, and rare feldspathoids; and large rounded to sub-rounded heterolithic clasts up to 20 cm in diameter. The albitite and 'syenite' clasts are interpreted to be pieces of fenite similar to the albite and potassic feldspar–albite fenites that are associated with intrusive carbonatite in the Perry River area to the south, and to albite fenites described by Le Bas (1981). The lithic clasts generally consist of gneiss, quartzite, and schist derived primarily from the underlying core gneisses. Clasts are generally randomly distributed throughout the blocky tephra, but in some layers they are concentrated in the central portion or occasionally graded with clast size increasing up-section.

The thickness of the Mount Grace carbonatite and the size and variety of included clasts vary considerably along strike, with both generally increasing to the north. Southern exposures of the carbonatite layer are approximately 1 m thick and clasts average 2–3 cm in diameter. Further north, the carbonatite is greater than 8 m thick and includes a thick section of mixed, coarse, blocky tephra and fine-grained tuff at the base, overlain by interlayered metasedimentary marble and fine-grained tuff. It continues to thicken northward, increasing to greater than 20 m. Thin-bedded, fine-grained tuff layers interbedded with calc-silicate gneiss and marble layers are overlain by thick, coarse, blocky tephra layers. Large gneissic blocks and blocks of 'syenitic' material, several metres across, occur throughout the coarse sections. A few thin, fine-grained tuff layers occur in the immediately overlying metasedimentary rocks.

The correlation between thickness and size of included clasts and the systematic change in these parameters in restricted areas suggest proximity to volcanic vents. Volcanism began with minor, intermittent deposition of fine ash, followed with increasing intensity by eruption of thick accumulations of coarse, blocky tephra. Explosive activity appears to have ceased abruptly, but was followed by minor, intermittent, and local deposition of ash.

9.5 CHEMISTRY

Chemical analyses (Table 9.2) of a limited number of intrusive carbonatites in the Canadian Cordillera suggest the presence of two types; calcite carbonatites ranging in composition from sövite to ferrocarbonatite, and magnesiocarbonatites (Fig. 9.6). Sövites containing abundant amphibole, ilmenite, magnetite, phlogopite, pyroxene, and pyrite fall into the ferrocarbonatite range. The Mount Grace carbonatite tuff is predominantly sövitic in composition (Höy 1987). On the basis of major elements, the carbonatites from the different tectonic belts are indistinguishable.

Analysed carbonatites from British Columbia are enriched in Mn and Sr, and most are enriched in Ba (Table 9.2); average concentrations of these elements are several thousand p.p.m., typical of carbonatites world-wide. Carbonatites and related metasomatic alteration zones in the Rocky Mountains Foreland Belt are

Table 9.2 Average chemical compositions of carbonatites from various British Columbia localities.

Belt	Foreland Belt						Eastern Omineca Belt			Frenchman Cap area		
	1	2	3	4	5	6	7	8	9	10	11	12
Wt%												
SiO_2	2.35	1.62	3.98	1.20	4.65	1.78	9.37	5.23	5.55	3.69	8.72	3.19
TiO_2	0.06	0.03	0.03	0.02	0.07	0.02	0.47	0.53	0.17	0.45	0.10	0.05
Al_2O_3	0.31	0.15	0.44	0.03	1.50	1.70	4.91	0.67	0.20	0.66	2.38	0.45
Fe_2O_3†	3.96	2.56	11.25	9.00	1.85	2.80	4.44	8.43	6.84	2.21	4.10	2.96
MnO	0.40	0.32	3.71	1.75	0.36	1.05	0.40	0.23	0.34	0.33	0.41	0.42
MgO	16.34	3.42	9.60	12.70	0.35	12.35	1.45	5.02	15.96	1.26	4.77	16.46
CaO	30.82	45.73	26.37	30.6	48.71	32.05	41.87	42.03	29.88	47.37	40.75	32.58
Na_2O	0.21	0.23	0.96	0.13	0.25	0.07	1.71	0.13	0.11	0.33	1.01	0.15
K_2O	0.12	0.10	0.05	0.11	0.98	0.24	1.54	0.30	0.17	0.24	0.71	0.10
LOI	39.26	37.92	37.93	44.10	41.08	32.73	31.46	29.06	36.00	34.82	32.87	40.03
P_2O_5	3.73	3.56	0.11	0.01	0.03	1.04	1.66	5.17	2.25	0.30	0.40	3.12
Trace elements (p.p.m.)												
Sr	1380	4268	3100	3600	2550	6858	6693	3143	3024	1950	5233	4576
Ba	32	343	42150	1100	625	18300	1194	307	360	1529	3283	657
Zr	84	277	338	nd	170	203	25	588	92	208	103	99
La	310	315	2480	650	480	3933	373	248	156	2007	570	471
Ce	750	710	5985	1000	640	5567	671	539	323	2795	922	877
Nd	240	n.a.	2300	820	800	2033	210	229	129	1186	323	250
No.	4	3	2	1	2	2	3	3	5	5	6	3

No., Number of analyses; n.a., not analysed; LOI, loss on ignition.
Samples from Foreland Belt: 1, Average Aley complex rauhaugite (analyses from Mader 1987, Pell 1987); 2, Average Aley complex sövite (analyses from Mader 1987, Pell 1987); 3, Average REE-enriched dyke, Aley complex (analyses from Mader 1987); 4, Rauhaugite, Ice River complex (Currie 1975); 5, Average sövite, Ice River complex (analyses from Currie 1975); 6, Average intensely metasomatized carbonate, Rock Canyon Creek (Pell 1987).
Samples from Eastern Omineca Belt: 7, Average sövite, Manson Creek (analyses from Pell 1987); 8, Average sövite, Blue River area (analyses from Pell 1987); 9, Average rauhaugite, Blue River area (analyses from Pell 1987).
Samples from Frenchman Cap Area: 10, Average intrusive sövite, Perry River area (analyses from Höy 1987); 11, Average extrusive Mount Grace carbonatite tuff (analyses from Höy & Kwong 1986); 12, Average intrusive rauhaugite, Ratchford Creek area (analyses from Höy 1987).
†, total Fe.

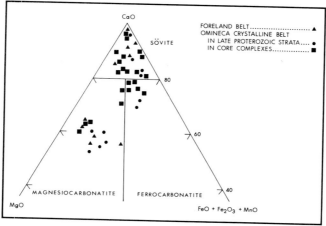

Figure 9.6 Carbonatite classification diagram (after Woolley 1982). Analyses of carbonatites of the Foreland Belt include the Aley complex (Mader 1987, Pell 1987a) and the Ice River complex (Currie 1975). Carbonatites of the Omineca Belt, hosted in late Proterozoic strata, include those in the Blue River and Manson Creek areas (Pell 1987a). Carbonatites associated with core complexes (Frenchman Cap dome) in the Omineca Belt include intrusive and extrusive carbonatites from the Mount Grace, Perry River, and Ratchford Creek areas (Höy & Kwong 1986, Höy 1987).

commonly associated with significant fluorine anomalies (Pell 1987a). The Aley carbonatite is significantly enriched in Nb, whereas others have average Nb_2O_5 contents of less than 0.2% (Pell 1987a). Nb:Ta ratios for most of the carbonatites in British Columbia are typical of carbonatites elsewhere, generally greater than 100:1; however, the Blue River carbonatites have ratios of approximately 5:1.

Concentrations of the REEs are highly variable, but the carbonatites tend to show enrichment in the LREEs. The highest total REE concentrations and the greatest degree of LREE enrichment are shown by carbonatites in the Foreland Belt (Fig. 9.7). Samples from the Rock Canyon Creek showing may contain in excess of 2.5% total REE oxides (Pell 1987a). Carbonatites from the Frenchman Cap area are moderately enriched in REEs; the Mount Grace extrusive carbonatite has total REE concentrations that range from approximately 600 p.p.m. to greater than 8000 p.p.m., with average values of 1000–2000 p.p.m. Rare earth element concentrations in the Ren carbonatite are similar to those in the Mount Grace carbonatite. The other intrusive carbonatites in the Frenchman Cap area have highly variable REE contents, ranging up to 13 000 p.p.m. (based on only four analyses). Carbonatites hosted by Upper Proterozoic metasedimentary strata in the Omineca Belt, north of Frenchman Cap dome, contain the lowest overall concentrations of REEs and are the least enriched in the LREEs (Fig. 9.7).

Compositional variations of fenites can be displayed readily on an $Na_2O + K_2O$–$MgO + Fe_2O_3$ total–CaO plot (Fig. 9.8). The diagram distinguishes between three major types of fenite. Albite fenites, characterized by high sodic content, are most common as clasts within the extrusive Mount Grace carbonatite. Potassic feldspar–albite fenites have high Na and K, and appreciable Fe + Mg contents. They are similar in composition to syenites and occur adjacent to intrusive carbonatites in

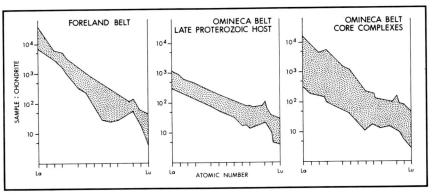

Figure 9.7 Chondrite-normalized REE plots of British Columbia carbonatites. All analyses by neutron activation. Samples from the Foreland Belt are from REE-enriched dykes, Aley complex, and Rock Canyon Creek metasomatic zone (Pell 1987a). Samples from carbonatites hosted in late Proterozoic strata, Omineca Belt, are from the Blue River and Manson Creek areas (Pell 1987a). Samples from extrusive and intrusive carbonatites associated with core complexes, Omineca Belt, are from the Mount Grace, Perry River, and Ratchford Creek areas (Höy & Pell 1986, Höy 1987). Chondrite normalizing factors are from H. Wakita & D. Zellmar (unpublished) quoted in Henderson (1984).

the Frenchman Cap dome. Pyroxene–amphibole fenites have high Fe and Mg content; they also occur adjacent to intrusive carbonatites in the Frenchman Cap area and are the most common type of fenite associated with carbonatites elsewhere in the Omineca Belt.

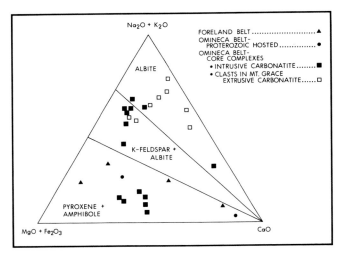

Figure 9.8 $MgO + Fe_2O_3(T) - Na_2O + K_2O - CaO$ plot of fenite compositions from various British Columbian carbonatite suites. Those in the Foreland Belt are from the outer amphibolite ring of the Aley carbonatite (Mader 1987, Pell 1987a); mafic fenites are associated with carbonatites hosted by late Proterozoic rocks in the Omineca Belt, Blue River, and Manson Creek areas (Pell 1987a); in the Frenchman Cap area fenite is associated with intrusive carbonatites at Perry River and occurs as clasts (albitite) in the Mount Grace extrusive carbonatite (Höy 1987).

9.6 TECTONIC IMPLICATIONS

The emplacement of carbonatites and other alkaline rocks in the Canadian Cordillera appears to be related, in part, to extension and/or rifting along the western continental margin which both initiated and modified the Cordilleran miogeocline. Sedimentological and stratigraphic evidence indicates that the western continental margin was tectonically active throughout much of the Proterozoic and Palaeozoic. It does not appear to have behaved entirely in a passive manner and therefore may not be strictly analogous to the present-day Atlantic margin as early workers proposed (e.g. Stewart 1972, Stewart & Poole 1974); rather, it appears that several superimposed 'passive margin-type' sequences are present as a result of periodic extensional activity (Thompson et al. 1987, Pell & Simony 1987).

The earliest dated event recorded by alkaline activity in western Canada is represented by the Mount Copeland syenite of late Proterozoic age (c. 750 Ma); it may record extension or rifting of the North American craton and initiation of the late Proterozoic Windermere basin. Diabase dykes and sills of similar age (700 Ma) in northern Canada (Armstrong et al. 1982) also record extension preceding Windermere deposition.

Extrusion of the Mount Grace carbonatite and intrusion of shallow-level carbonatites, accompanied by the formation of extensive zones of fenitization, probably occurred in Eocambrian to early Cambrian time (Höy & Godwin 1988). These rocks occur in a cover succession above core gneisses of the Frenchman Cap area. Emplacement of the alkalic rocks may have coincided with foundering of an extensive, Lower Cambrian platform to the east. This period is interpreted by many workers as the time of continental rifting along the western continental margin (Bond & Kominz 1984, Devlin & Bond 1984). In the southwestern United States, carbonatites of Eocambrian to early Cambrian age are reported from a number of localities; for example, the McLure Mountain carbonatite–alkalic complex, the Gem Park and the Iron Hill carbonatite complexes in Colorado, and the Lobo Hills, New Mexico syenite and carbonatite (see summary in McLemore 1987). Although these intrusions are structurally inboard of the Mount Grace carbonatite, their emplacement may be related to the same large-scale extensional tectonic event.

A number of periods of Palaeozoic extension are inferred along the western continental margin (Pell 1987a, Pell et al. in preparation). The earliest event is Ordovician–Silurian and is recorded by the emplacement of diatremes and alkaline basalts in the southern Rocky Mountains of British Columbia (Pell 1987a, b). Carbonatites of approximately the same age are found in the Lemitar Mountains of New Mexico (McLemore 1987), but none are known from the Canadian Cordillera. A second period of alkaline activity on the western margin of North America during the Palaeozoic occurred in early Devonian times. Ultrabasic lamprophyres and minettes in south-central British Columbia and some diatremes in southern British Columbia were emplaced at this time (Pell, in preparation). Diatreme breccias in the Yukon Territories (for example, the Mountain Diatreme, Godwin & Price 1987) are of the same age. In a more continental setting, early Devonian kimberlites are reported from the Colorado–Wyoming state line district (McCallum *et al.* 1975, Hausel *et al.* 1979).

A third Palaeozoic extensional event at the end of the Devonian (*c.* 350 Ma) resulted in intrusion of carbonatites into the miogeoclinal succession in the Foreland and Omineca belts. Aillikite diatremes and dykes were also emplaced at this time (Pell 1987*a, b*). The tectonic instability resulting from the major Devono-Mississippian extensional event is also evident in the stratigraphic record; volcanic rocks (some peralkaline in composition), syn-sedimentary block faults, and chert-pebble conglomerates are reported from the mid-Devonian to early Mississippian sequence in the northern and central Canadian Cordillera (Gordey 1981, Mortensen 1982, Gordey *et al.* 1987) as well as in the southern Canadian Cordillera (Wheeler 1965). The Devono-Mississippian extension was synchronous with, or slightly post-dated, compression to the south that was associated with the Antler Orogeny. This suggests that extensional tectonics may not have resulted simply from continental rifting, but may be a result of a more complex scenario. An incipient continental back-arc rift may have developed as localized obduction (and possibly subduction?) occurred to the south and outboard. Alternatively, extensional basins may have resulted from strike-slip faulting outboard of the preserved margin of the miogeocline, as proposed by Eisbacher (1983) and Gordey *et al.* (1987).

The last Palaeozoic extensional event is inferred from the presence of Permo-Triassic kimberlite in the southern Canadian Rocky Mountains (Grieve 1982, Hall *et al.* 1986, Pell 1987*a, b*). It was followed by a compressional regime in late Jurassic to early Tertiary time as the Pacific margin was telescoped by collisional tectonics. Alkaline igneous rocks within the miogeoclinal succession on the western margin were deformed, metamorphosed, and transported eastward within thrust sheets. Their present distribution near the Rocky Mountain Trench is due to original location along a rifted continental margin and not to later tectonics.

9.7 CONCLUSIONS

Carbonatites and syenite gneisses present in three belts parallel to the Rocky Mountain Trench were emplaced in the miogeoclinal prism prior to the late Jurassic–early Tertiary Orogeny. Alkalic rocks and carbonatites in the Foreland Belt are high-level Devono-Mississippian intrusions hosted in Lower to Middle Palaeozoic strata. They tend to be large and elliptical in shape and have significant alteration halos. Carbonatites in the Omineca Belt, immediately west of the Rocky Mountain Trench, are deeper-level Devono-Mississippian intrusions hosted in late Precambrian strata. They are thin, discontinuous, concordant intrusions with only narrow fenite zones. The most western belt includes extrusive and intrusive carbonatites and nepheline syenite gneisses in cover sequences above core complexes in the Omineca Belt. Host strata are presumed to be late Proterozoic to early Palaeozoic in age; syenite gneisses intruding the base of the succession are late Proterozoic (Okulitch *et al.* 1981, Scammell 1986) whereas the extrusive carbonatite, which is higher in the sequence, is interpreted to be Eocambrian to early

Cambrian (Höy & Godwin 1988). Intrusive carbonatites in this sequence are similar to the extrusive carbonatite and are believed to be approximately the same age. They are bedding-parallel, high-level intrusive bodies with extensive fenitic aureoles.

Carbonatites in the Foreland Belt are interpreted to have been emplaced at shallower structural levels than the Devono-Mississippian carbonatites in the Omineca Belt. They are associated with extensive zones of fenite and enrichment of F, Nb, and REEs, perhaps due to a higher volatile content of the magmas. Those in the Omineca Belt have narrower fenite zones and are commonly less highly enriched in F, Nb, and REEs.

Emplacement of carbonatites and alkalic rocks record extensional regimes in the Canadian Cordillera in late Proterozoic (approximately 750 Ma ago), Eocambrian to Lower Cambrian, Ordovician–Silurian, early Devonian, late Devonian, and perhaps Permo-Triassic time. These periods of extension are related to rifting or incipient rifting events which led to the initiation and modification of the Cordilleran miogeocline.

ACKNOWLEDGEMENTS

We would like to acknowledge logistical and financial support of the Canada/British Columbia Mineral Development Agreement; the Natural Sciences and Engineering Research Council; and the Geological Survey Branch, British Columbia Ministry of Energy, Mines, and Petroleum Resources. We would also like to extend thanks for support and information supplied by personnel of various companies that are active in the exploration and development of carbonatite properties in British Columbia; in particular the geologists of Cominco Ltd and Duval Ltd, and C. Graf of Active Minerals Ltd. K. Pride, C. Godwin, R. R. Parrish, and Z. D. Hora provided helpful discussions; W. J. McMillan and J. M. Newell (Geological Survey Branch, British Columbia Ministry of Energy, Mines, and Petroleum Resources), R. R. Parrish and H. Gabrielse (Geological Survey of Canada), and K. Bell and J. M. Moore (Carleton University) reviewed and improved the manuscript.

REFERENCES

Armstrong, R. L., G. H. Eisbacher & P. D. Evans 1982. Age and stratigraphic–tectonic significance of Proterozoic diabase sheets, Mackenzie Mountains, northwestern Canada. *Canadian Journal of Earth Sciences* **19**, 316–23.

Bond, G. C. & M. A. Kominz 1984. Construction of tectonic subsidence curves for the Early Paleozoic miogeocline, Southern Canadian Rocky Mountains: Implications for subsidence mechanisms, age of breakup and crustal thinning. *Geological Society of America Bulletin* **95**, 155–73.

Brown, R. L. & J. F. Psutka 1979. Stratigraphy of the east flank of Frenchman Cap dome, Shuswap Complex, British Columbia. Current Research Part A, Geological Survey of Canada Paper 79-1A, 35–6.

Brown, R. L. & P. B. Read 1983. Shuswap terrane of British Columbia: a Mesozoic 'Core Complex'. *Geology* **11**, 164–8.

Campbell, R. B. 1967. Canoe River, British Columbia. *Geological Survey of Canada* Map 15–1967.

Currie, K. L. 1975. *The geology and petrology of the Ice River alkaline complex*. Geological Survey of Canada Bulletin 245.

Currie, K. L. 1976a. *The alkaline rocks of Canada*. Geological Survey of Canada Bulletin 239.

Currie, K. L. 1976b. *Notes on the petrology of nepheline gneisses near Mount Copeland, British Columbia*. Geological Survey of Canada Bulletin 265.

Dawson, J. B. 1980. *Kimberlites and their xenoliths*. New York: Springer.

Devlin, W. J. & G. C. Bond 1984. Syn-depositional tectonism related to continental breakup in the Hamill Group, northern Selkirk Mountains, British Columbia. *Geological Association of Canada —Mineralogical Association of Canada, Joint Annual Meeting, London, Program with Abstracts* **9**, 57.

Eisbacher, G. H. 1983. Devonian–Mississippian sinistral transcurrent faulting along the cratonic margin of western North America: A hypothesis. *Geology* **11**, 7–10.

Fyles, J. T. 1970. *The Jordan River area, near Revelstoke, British Columbia*. British Columbia Department of Mines and Petroleum Resources Bulletin 57.

Godwin, C. I. & N. J. Price 1988. Geology of the Mountain Diatreme kimberlite, north-central Mackenzie Mountains, District of Mackenzie, N.W.T. In *Mineral deposits of the northern Canadian Cordillera*. Canadian Institute of Mining and Metallurgy Special Volume, (in press).

Gordey, S. P. 1981. *Stratigraphy, structure and tectonic evolution of the Southern Pelly Mountains in the Indigo Lake area, Yukon Territory*. Geological Survey of Canada Bulletin 318.

Gordey, S. P., J. G. Abbott, D. Templeman-Kluit & H. Gabrielse 1987. 'Antler' clastics in the Canadian Cordillera. *Geology* **15**, 103–7.

Grieve, D. A. 1982. 1980 – Petrology and chemistry of the Cross Kimberlite (82J/2). *British Columbia Ministry of Energy, Mines and Petroleum Resources, Geology in British Columbia, 1977–1981*, 34–41.

Hall, D. C., H. Helmstaedt & D. J. Schulze 1986. *The Cross Diatreme: A kimberlite in a young orogenic belt*. 4th International Kimberlite Conference Extended Abstracts, Geological Society of Australia Abstract Series, No. 16, 30–2.

Hausel, W. D., M. E. McCallum & T. L. Woodzick 1979. *Exploration for diamond-bearing kimberlite in Colorado and Wyoming: an evaluation of exploration techniques*. Geological Survey of Wyoming, Report of Investigations 19.

Heinrich, E. W. 1966. *The geology of carbonatites*. Chicago: Rand McNally.

Henderson, P. (ed.) 1984. *Developments in geochemistry*. Vol. 2: *Rare earth element geochemistry*. New York: Elsevier.

Hora, D. Z. & Y. T. J. Kwong 1986. *Anomalous rare earth elements (REE) in the Deep Purple and Candy Claims, British Columbia*. British Columbia Ministry of Energy Mines and Petroleum Resources Geological Fieldwork 1985, Paper 1986-1, 241–2.

Höy, T. 1987. *Geology of the Cottonbelt lead–zinc–magnetite layer, carbonatites and alkalic rocks in the Mount Grace area, Frenchman Cap dome, southeastern British Columbia*. British Columbia Ministry of Energy, Mines and Petroleum Resources Bulletin 80.

Höy, T. & C. I. Godwin 1988. Significance of a galena lead isotope Cambrian date for the stratiform Cottonbelt deposit in the Monashee complex, southeastern British Columbia. *Canadian Journal of Earth Sciences* **25**, No. 9.

Höy, T. & Y. T. J. Kwong 1986. The Mount Grace carbonatite: A niobium and light rare earth element-enriched marble of probable pyroclastic origin in the Shuswap Complex, Southeastern British Columbia. *Economic Geology* **81**, 1374–86.

Höy, T. & W. J. McMillan 1979. *Geology in the vicinity of Frenchman Cap gneiss dome*. British Columbia Ministry of Energy, Mines and Petroleum Resources, Geological Fieldwork 1978, 25–30.

Höy, T. & J. Pell 1986. *Carbonatites and associated alkalic rocks, Perry River and Mount Grace areas, Shuswap Complex, southeastern British Columbia*. British Columbia Ministry of Energy, Mines and Petroleum Resources Geological Fieldwork 1985, Paper 1986-1, 69–87.

Lang, A. H., J. E. Armstrong & J. B. Thurber 1946. *Manson Creek*. Geological Survey of Canada, Map 867A.

Le Bas, M. J. 1981. Carbonatite magmas. *Mineralogical Magazine* **44**, 133–40.

McCallum, M. E., D. H. Eggler & L. K. Burns 1975. Kimberlite diatremes in northern Colorado and southern Wyoming. *Physics and Chemistry of the Earth* **9**, 149–62.

McLemore, V. T. 1987. Geology and regional implications of carbonatites in the Lemitar Mountains, central New Mexico. *Journal of Geology* **95**, 255–70.

McMillan, W. J. 1970. West Flank, Frenchman Cap gneiss dome, Shuswap Terrane, British Columbia. In *Structure of the Canadian Cordillera*, J. O. Wheeler (ed.), 99–106. Geological Association of Canada, Special Paper No. 6.

McMillan, W. J. 1973. *Petrology and structure of the west flank, Frenchman Cap dome, near Revelstoke, British Columbia*. Geological Survey of Canada Paper 71-29.

McMillan, W. J. & J. M. Moore 1974. Gneissic alkalic rocks and carbonatites in Frenchman Cap gneiss dome, Shuswap Complex, British Columbia. *Canadian Journal of Earth Sciences* **11**, 304–18.

Mader, U. K. 1986. *The Aley carbonatite complex*. M.Sc. thesis, University of British Columbia, Vancouver.

Mader, U. K. 1987. *The Aley carbonatite complex, northern Rocky Mountains, British Columbia*. British Columbia Ministry of Energy, Mines and Petroleum Resources Geological Fieldwork 1986, Paper 1987-1, 283–8.

Mortensen, J. K. 1982. Geological setting and tectonic significance of Mississippian felsic metavolcanic rocks in the Pelly Mountains, southeastern Yukon Territory. *Canadian Journal of Earth Sciences* **19**, 8–22.

Okulitch, A. V. 1984. The role of the Shuswap Metamorphic Complex in Cordilleran tectonism: a review. *Canadian Journal of Earth Sciences* **21**, 1171–93.

Okulitch, A. V., W. D. Loveridge & R. W. Sullivan 1981. *Preliminary radiometric analyses of zircons from the Mount Copeland syenite gneiss, Shuswap metamorphic complex, British Columbia*. Report of Activities Part A, Geological Survey of Canada Paper 81-1A, 33–6.

Parrish, R. R., S. Heinrich & D. Archibald 1987. *Age of the Ice River complex, southeastern British Columbia*. Geological Survey of Canada Paper 87-2, 33–7.

Pell, J. 1985. *Carbonatites and related rocks in British Columbia*. British Columbia Ministry of Energy, Mines and Petroleum Resources, Geological Fieldwork 1984. Paper 1985-1, 85–94.

Pell, J. 1986a. *Carbonatites in British Columbia: The Aley property*. British Columbia Ministry of Energy, Mines and Petroleum Resources, Geological Fieldwork 1985. Paper 1986-1, 275–7.

Pell, J. 1986b. *Nepheline syenite gneiss complexes in British Columbia*. British Columbia Ministry of Energy, Mines and Petroleum Resources, Geological Fieldwork 1985, Paper 1986-1, 255–60.

Pell, J. 1987a. *Alkaline ultrabasic rocks in British Columbia: Carbonatites, nepheline syenites, kimberlites, ultramafic lamprophyres and related rocks*. British Columbia Ministry of Energy, Mines and Petroleum Resources, Open File Report 1987-17.

Pell, J. 1987b. *Alkalic ultrabasic diatremes in British Columbia: Petrology, geochronology and tectonic significance*. British Columbia Ministry of Energy, Mines and Petroleum Resources, Geological Fieldwork 1986, Paper 1987-1, 259–72.

Pell, J. & D. Z. Hora 1987. *Geology of the Rock Canyon Creek fluorite/rare earth element showing, southern Rocky Mountains*. British Columbia Ministry of Energy, Mines, and Petroleum Resources, Geological Fieldwork 1986, Paper 1987-1, 255–8.

Pell, J. & P. S. Simony 1987. New correlations of Hadrynian strata, south central British Columbia. *Canadian Journal of Earth Sciences* **24**, 302–13.

Perkins, M. J. 1983. *Structural geology and stratigraphy of the northern Big Bend of the Columbia River, Selkirk Mountains, southeastern British Columbia*. Ph.D. thesis, Carleton University, Ottawa.

Read, P. B. & R. L. Brown 1981. Columbia River fault zone: Southeastern margin of the Shuswap and Monashee complexes, southern British Columbia. *Canadian Journal of Earth Sciences* **18**, 1127–45.

Rowe, R. B. 1958. *Niobium (columbium) deposits of Canada*. Geological Survey of Canada, Economic Geology Series 18.

Scammell, R. J. 1986. *Stratigraphy, structure and metamorphism of the north flank of the Monashee Complex, southeastern British Columbia: A record of Proterozoic extension and Phanerozoic crustal thickening*. M.Sc. thesis, Carleton University, Ottawa.

Stewart, J. H. 1972. Initial deposits in the Cordilleran geosyncline: Evidence of late Precambrian (< 850 my) continental separation. *Geological Society of America Bulletin* **83**, 1345–60.

Stewart, J. H. & F. G. Poole 1974. Lower Paleozoic and uppermost Precambrian miogeocline, Great Basin, western United States. In *Tectonics and sedimentation*, W. R. Dickinson (ed.), 28–57. Society of Economic Paleontologists and Mineralogists Special Publication 22, Tulsa, Oklahoma.

Thompson, R. I., E. Mercier and C. Roots 1987. Extension and its influence on Canadian Cordilleran passive-margin evolution. In *Continental extensional tectonics*, M. P. Howard, J. F. Dewey, & P. L. Hancock, (eds), 409–17. Geological Society Special Publication 28.

Wheeler, J. O. 1965. *Big Bend map area, British Columbia*. Geological Survey of Canada Paper 64-32.

White, G. P. E. 1982. *Notes on carbonatites in central British Columbia*. British Columbia Ministry of Energy, Mines and Petroleum Resources, Geological Fieldwork 1981, Paper 1982-1, 68–9.

Woolley, A. R. 1982. A discussion of carbonatite evolution and nomenclature, and the generation of sodic and potassic fenites. *Mineralogical Magazine* **46**, 13–17.

10
PHALABORWA: A SAGA OF MAGMATISM, METASOMATISM AND MISCIBILITY

S. C. ERIKSSON

ABSTRACT

The Phalaborwa Complex is composed of copper-bearing carbonatite and phoscorite set in a large body of clinopyroxenites and syenites. Cumulus textures, intrusive relations, and chemistry point to magmatic crystallization of apatite, clinopyroxene, and phlogopite as the main mechanism in forming most of the complex. Clinopyroxenes from cumulus rocks have Fe:(Fe + Mg) = 0.07–0.29 and low Al_2O_3 ($\leqslant 0.5$ wt%). The low Al is attributed to high alkalinity in the magma. Metasomatism has affected some rocks, to form clinopyroxenes with a wide range in Fe:(Fe + Mg) ratios of 0.12–0.66 and greater enrichment in acmite. Mica ranges from magnesium-rich phlogopite to magnesium-rich biotite [(Fe:(Fe + Mg) = 0.04–0.52)]. Olivine (Fo_{79}–Fo_{91}) in phoscorite and carbonatite is low in NiO ($\leqslant 0.06$ wt%). Phlogopites in phoscorite and carbonatite show a separate crystallization trend from those found in clinopyroxenite, suggesting two separate magmas.

Rb–Sr analyses of phlogopite from clinopyroxenite, phoscorite, and carbonatite yield an age of 2012 ± 19 Ma, which is within experimental error of the previously determined age of 2047 Ma for the carbonate rocks. The age of the complex is considered to be 2047 +11/−8 Ma. Initial ^{87}Sr:^{86}Sr ratios of the carbonatites (0.70393–0.71022) are different from those of clinopyroxenites (0.71140–0.71282), suggesting that differentiation alone was not responsible for carbonatite formation. Initial ^{143}Nd:^{144}Nd ratios (0.50961–0.50977) also characterize the source for the carbonatites as enriched in large ion lithophile (LILE) elements relative to bulk Earth. $\delta^{18}O$ (+7.7‰ to +8.6‰) and $\delta^{13}C$ (−5.1‰ to −3.6‰) for carbonatites and $\delta^{18}O$ (+7.26‰ to +7.53‰) for clinopyroxenes indicate wide variation in isotopic composition, and this is attributed to the mixing of at least two magmas. The Pb is highly radiogenic (^{206}Pb:^{204}Pb = 19.026 ± 5 to 34.655 ± 9; ^{207}Pb:^{204}Pb = 15.935 ± 5 to 17.993 ± 4), and a high μ source is indicated.

Satellite pipes concentrated on or near the periphery of the main complex range from homogeneous bodies to highly brecciated bodies with multiple intrusions. The breccias are interpreted as forming at the upper part of the intrusion with passive infilling of magma lower in the pipe. Whole-rock chemistry of two satellite pipes, Kgopoeloe and Spitskop, supports this model and shows magma modification by interaction with country rock. The high total REE (rare earth element) and LREE contents as well as K:Na ratios of the satellite syenites also characterize the main complex.

10.1 INTRODUCTION

The Phalaborwa Complex is unique. Well known for its Cu mineralization, it set world records in mining circles in 1980 when over 521 000 tons of material were loaded and hauled in one 24-hour period. It is also the largest open-cast mining operation in Africa and second largest in the world. Aside from the variety of minerals extracted for Cu, P, U, and Zr, the large quantity of material and simple metallurgical techniques used at Phalaborwa allow Au, Ag, and Pt to be recovered from anode slimes and to be its second biggest source of income.

Although many geologists in the late nineteenth century recognized the 'limestone' and copper-stainings of Phalaborwa, Mellor (1906) first postulated an origin involving modification of granitic magma by pre-existing limestone. Very early studies focused on the nature of the contact between the feldspar-bearing pyroxenites of the complex and the surrounding syenites and granites.

Hanekom *et al.* (1965) documented work on the complex prior to 1965; this earlier data, mainly descriptive, dealt with the carbonatite and phoscorite which constitute less than 5% of the surface area of the complex. Various clinopyroxenites make up most of the complex, and syenites both rim the main complex and form prominent pipes concentrated outside the complex. Later workers recognized the importance of the numerous silicate rocks in unravelling the petrogenesis of the entire rock suite. This chapter presents unpublished work of the author since 1975 and emphasizes other published and unpublished work since that time. Previously unpublished geologic data, along with mineral chemistry, trace element, isotopic ratio, palaeomagnetic and fluid inclusion data, show that no simple process alone is at work in this complex. It appears that several magmas interacted with each other and their derived fluids.

10.2 GEOLOGY

10.2.1 Introduction

The Phalaborwa Complex is intruded into Archaean granites, gneisses, quartzites, granulites, amphibolites, and talc- and serpentine-schists (Hall 1912*a*, Shand 1931, Brandt 1948, Hanekom *et al.* 1965). The main complex (Fig. 10.1) is an elongate, irregularly shaped body covering about 16 km^2. Gravity data (Hanekom *et al.* 1965) indicate a pipe-like body, pitching 76°–80° to the east, and the pyroxenite extends to a depth of at least 5 km. Diamond drilling by the Palabora Mining Company indicates that the carbonatite is at least 1500 m deep (Palabora Mining Company, pers. comm.).

There has been a great deal of controversy over contact relations, starting with Hall (1912*a, b*) who postulated the presence of a 'Palabora' granite as a distinct rock type surrounding the complex. This pink granite, less siliceous toward the contact with the complex, grades into a feldspar-rich rock containing pyroxene, which he attributed to differentiation of a single magma. Shand (1931) considered the feldspar-bearing pyroxenite primary. Gevers (1948) disregarded Shand's inter-

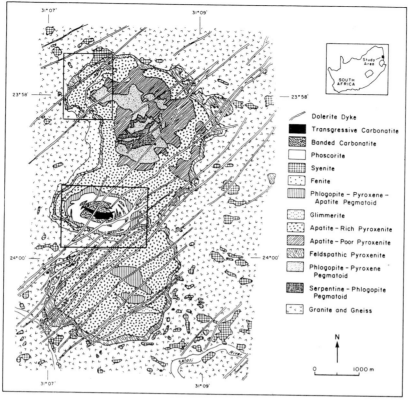

Figure 10.1 Geology of the Phalaborwa Complex (after Hanekom *et al.* 1965, Fourie 1981, Eriksson *et al.* 1985). Plugs surrounding the main complex are referred to as syenite. Recent mapping of plugs shows more complexity than previously published (Hanekom *et al.* 1965, Frick 1975). Boxes show localities of Figures 10.2 and 10.3.

pretation on the basis of 'schlieren' of clinopyroxene in syenite, quartz syenite, and 'Palabora' granite, and supported Hall's view that the feldspathic pyroxenite represents contamination of pure pyroxene rock with granite. Hanekom *et al.* (1965) re-investigated the outer boundary of the pyroxenite and concluded that feldspathic pyroxenite is best developed where pyroxenite is associated with younger syenite intrusions near or on the pyroxenite contact. The intrusive nature of the pyroxenite is demonstrated by transgressive pyroxenite dykes within granite-gneiss. Hanekom and his co-workers interpreted the feldspathic pyroxenite as a mixture of pyroxenite magma and Archaean granite.

The extent of metasomatism is difficult to establish. In the north, equigranular rocks, up to 5 km from the contact, contain dark-green pyroxenes. Whether the pyroxenes were present prior to the intrusion of the main complex, are related to emplacement of numerous syenite plugs, or are caused by metasomatism from the main complex is difficult to determine. However, at the contact with the main complex, country rock contains relict plagioclase feldspar, suggesting that metasomatism is not intense around the Phalaborwa Complex. Certainly, brecciation is absent around the main complex.

Studies of the feldspathic pyroxenite and its relation to the country rock (Fourie 1981, Eriksson 1982, Eriksson *et al.* 1985) indicate that this rock is magmatic rather than metasomatic. Factors supporting the magmatic origin are:

(a) no one fenite type is present around all of the complex;
(b) in the Foskor area, feldspathic pyroxenite does not grade into fenite;
(c) in the vicinity of the carbonatite, the relatively thin sequence of feldspathic pyroxenite is in contact with a micaceous, gneissic white granite;
(d) feldspathic pyroxenite extends into the pipe more extensively than previously noted;
(e) petrography and cathodoluminescence of the feldspar (Eriksson 1982) indicate none of the characteristics of metasomatic feldspar (Strauss & Truter 1950, Heinrich & Moore 1970, Mariano 1978); and
(f) the initial $^{87}Sr:^{86}Sr$ ratios of the feldspathic pyroxenite are similar to those from the majority of clinopyroxenites in the complex (Eriksson 1982).

10.2.2 Clinopyroxenities

The most recent account of the clinopyroxenites (Eriksson *et al.* 1985) includes a detailed map of the northwestern portion of the main complex (Fig. 10.2). Clinopyroxenites make up about 70% of the main complex and their genesis is one of the most important problems at Phalaborwa. These rocks grade from massive pyroxenite (very little mica with or without apatite) to glimmerite (phlogopite-rich rock with apatite, < 25% clinopyroxene). Pegmatites are common and a variety of textures make their origin debatable. Simple cumulus textures, 'inch-scale' layering, flow textures, cross-cutting relationships, and inclusions of one rock type in another point to simultaneous crystallization of clinopyroxene, phlogopite, and apatite. Monomineralic assemblages of pyroxene and apatite are attributed to crystal settling. Flow differentiation contributed to separation and accumulation of phlogopite into glimmerite, and multiple surges of magma created a wide variety of cross-cutting and gradational contacts (Eriksson *et al.* 1985). It seems that phlogopite precipitated from an ultrabasic magma rather than resulting from metasomatic alteration of clinopyroxene or olivine (Eriksson 1982, Eriksson *et al.* 1985). Apatite also appears to be a primary precipitate.

10.2.3 Phoscorite

Phoscorite, a locally named rock, consists of olivine, apatite, and magnetite confined to the central region containing carbonate rocks (Fig. 10.3). Hanekom *et al.* (1965) quoted an average modal composition of 25% apatite, 18% carbonate, 35% magnetite, and 22% combined serpentine, olivine, and mica. Mineral proportions vary considerably from 100% magnetite to 100% olivine, with minerals forming rough, vertical banding that parallels the shape of the body. Lombaard *et al.* (1964) described pure olivine rock up to 15 m wide in the eastern portion of the phoscorite. Carbonatite occurs in irregular patches and lenses in phoscorite, and

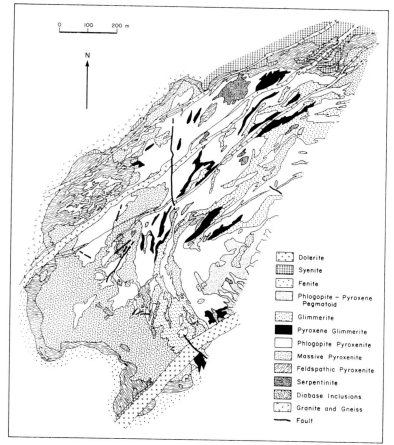

Figure 10.2 Detailed geology of clinopyroxenites and associated rocks in the northwestern corner of the main Phalaborwa Complex (after Fourie 1981, Eriksson *et al.* 1985). Foskor mine takes up most of the area shown on this map.

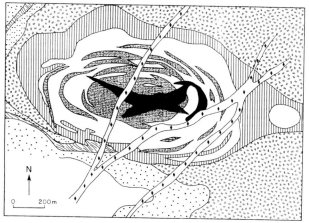

Figure 10.3 Detailed geology of the central carbonatite-bearing region of the main complex. See Figure 10.1 for legend. Diamond pattern shows position of dolerite dyke. The Palabora Mining Company open-cast, copper mine encompasses this area.

these become larger and more numerous inwards. From the description of Lombaard *et al*. (1964), both the inner and outer contacts of phoscorite appear gradational, but Palabora Mining Company (PMC) personnel (pers. comm.) describe the outer contact as abrupt. Current mining in the contact area has revealed cross-cutting relationships with abundant fragments of micaceous pyroxenite surrounded by carbonate and serpentine veins. A major feature of phoscorite is phlogopite with reverse pleochroism, that occurs as masses surrounding olivine and as euhedral books.

Olivine grains, up to several centimetres in length, are serpentinized to different degrees. There is no decrease in serpentinization with depth (PMC Mining Staff, pers. comm.), and highly serpentinized olivine may coexist with almost fresh olivines. Some olivines are rimmed with phlogopite or chondrodite. Hennig-Michaeli (1968) noted that clinohumite is commonly present. Bornite, chalcopyrite, and pentlandite, present within olivines as minute blebs and in fluid inclusions, are considered the products of early sulphide liquids (Van Rensburg 1965, Aldous 1980). Euhedral olivine is also poikilitically enclosed in chalcopyrite with bornite exsolution.

At least two generations of magnetite exist. Primary magnetite forms anhedral to euhedral grains that range from < 1 mm to tens of centimetres across. Hennig-Michaeli (1968) noted ilmenite-rich rims on some magnetite grains. Van Rensburg (1965) and Hanekom *et al*. (1965) observed ulvospinel and ilmenite exsolution in the magnetite. Late magnetite replaces baddeleyite in places (Van Rensburg 1965) and rims sulphides (Hennig-Michaeli 1968).

Apatite occurs as short, stubby crystals a few millimetres in length, or as larger, anhedral masses. The {0001} cleavage is pronounced, and crystals may occur bent and fractured. No primary fluid inclusions have been found in the apatite (Aldous 1980), in contrast to apatites found in many other carbonatites (Rankin 1975, 1977). Apatites do, however, contain minute elongate sulphide rods parallel to the *c* axis of the apatite (Aldous 1980). This crystallographic control of bornite–chalcopyrite intergrowths (10% chalcopyrite, 90% bornite solid-solution) suggests exsolution within the apatites and a copper-, and sulphur-rich environment (Aldous 1980).

Carbonate minerals are invariably present in the phoscorite. Calcite occurs as small, disseminated grains, in veinlets and monomineralic lenses, and can make up several per cent of the rock. Baddeleyite occurs as black, prismatic crystals up to a few centimetres in length (Hiemstra 1955) and, in contrast to baddeleyite from numerous other carbonatites, is extremely iron-rich (A. N. Mariano, pers. comm.). Clinopyroxene is rarely present in phoscorite, suggesting that pyroxene in phoscorite is probably xenocrystic. Bornite and chalcocite are the dominant sulphides and appear late in the crystallization sequence (Russell *et al*. 1954, Lombaard *et al*. 1964, Van Rensburg 1965, Hennig-Michaeli 1968, Palabora Mining Company 1976, Aldous 1980).

10.2.4 Carbonatites

The Palabora Mining Company (1976) recognizes two main periods of carbonatite emplacement (Fig. 10.3). Magnetite-rich carbonatite maintains the elliptical, concentric zoning of the phoscorite and is referred to as banded carbonatite. Early workers (Lombaard *et al*. 1964) noted several generations of banded carbonatite, each similar in mineralogy and texture, a feature now attributed to remobilization. Current usage classifies all the carbonatites with fine magnetite bands as banded carbonatite. The division into banded carbonatite and transgressive carbonatite is based primarily on gross textural features and sulphide mineralogy.

Calcite in the banded carbonatite contains up to 7.5% $MgCO_3$ (Lombaard *et al*. 1964) and commonly contains exsolution lamellae of dolomite (Van Rensburg 1965). Some of the banded carbonatite, whose banding is not concordant to the phoscorite–carbonatite contact, contains interstitial lenses of dolomite (Van Rensburg 1965, Aldous 1980). Olivine, rare in the banded carbonatite, is commonly fractured and partially replaced by phlogopite, monazite, or clinohumite. Serpentinization is minimal. Magnetite, found as discrete, idiomorphic grains 1 mm to 1 cm in length, is aligned parallel to the outer contact of the banded carbonatite, giving it a finely banded appearance. A later generation of magnetite forms rims on the sulphides and also lies in phlogopite cleavage planes. Bornite, the main sulphide mineral, occurs as disseminated grains, droplets enclosed within olivine, magnetite, and calcite (Van Rensburg 1965), as massive patches, and as lenses parallel to primary magnetite banding.

The younger, transgressive carbonatite member occupies the central portion of the centre plug and radiates primarily along fractures oriented at N70W and N70E (Fig. 10.3). Magnetite and silicate phases occur as disoriented clots and lenses. Veins of transgressive carbonatite within the banded carbonatite, the phoscorite, and the massive, micaceous, and feldspathic pyroxenites are common. Two curvilinear intrusions of transgressive carbonatite, in the eastern portion of the banded carbonatite, persist with depth and contain a larger than normal proportion of mica (Palabora Mining Company 1976).

Carbonate of the transgressive carbonatite is more magnesium-rich than that of banded carbonatite, and contains up to 14% $MgCO_3$ (Hennig-Michaeli 1968). Subordinate dolomite occurs as discrete grains. Phlogopite, rare in banded carbonatite, is abundant in the transgressive carbonatite and always exhibits reverse pleochroism similar to that found in phoscorite. Some phlogopites are strained, and phenocrysts are commonly zoned.

Apatite, more abundant in the transgressive carbonatite than in the banded carbonatite (Hennig-Michaeli 1968), occurs as disseminated anhedral grains, idiomorphic crystals, and more rarely as elongate crystals with parallel alignment. The latter, several centimetres long and 2–3 mm in diameter, do not occur in the banded carbonatite.

The magnetite content is the same in both types of carbonatite. In the transgressive carbonatite, magnetite forms blebs and idiomorphic crystals, from 1–2 mm to tens of centimetres in diameter. Magnetite makes up from < 1% to > 90% of a rock over distances of about a metre.

Exsolution lamellae of ilmenite in magnetite have been reported (Lombaard *et al*. 1964, Hanekom *et al*. 1965, Van Rensburg 1965), and Hennig-Michaeli (1968) documented isolated ilmenite grains associated with magnetite in both types of carbonatite. Chalcopyrite, the major sulphide in the transgressive carbonatite, contains exsolution lamellae of bornite and cubanite, and occurs as disseminated grains and massive blebs along fracture planes. Olivine is very rare in the transgressive carbonatite but clinohumite and chondrodite appear as crystal aggregates from < 1 mm to a few centimetres in size.

10.2.5 Economic geology

Phoscorite and carbonatite host the orebodies in the Phalaborwa Complex in which copper sulphides, magnetite, baddeleyite, apatite, and uranoan thorianite are the major economic minerals. Vermiculite is recovered economically from the central phlogopite–serpentine body of the northern lobe.

The transgressive carbonatite is the rock type richest in sulphides, and averages 1 wt% Cu contained in chalcopyrite and cubanite. Mineralized lenses occur along numerous fractures concordant with the shape of the transgressive carbonatite. In contrast, banded carbonatite contains only scattered blebs of sulphides aligned with major minerals, and sulphide distribution is unrelated to fractures. Banded carbonatite contains the same sulphides as the transgressive carbonatite but in much lower proportions. Sulphides, primarily bornite, in the phoscorite also occur in non-structurally controlled blebs and appear to replace the constituent minerals, mainly calcite. All sulphides crystallized later than apatite and magnetite. Valleriite occurs as a late-stage phase along shear planes and fractures.

Magnetite is zoned in both quantity and 'quality'. Magnetite, with up to 4 wt% TiO_2, is most abundant (25–50%) in the phoscorite. Titanium decreases towards the centre of the complex, and magnetite in the transgressive carbonatite contains >1% TiO_2. Apatite is present in economic proportions only in the phoscorite. Zirconium, contained in baddeleyite, is richest in phoscorite and is also recovered from the carbonatites. Minor Ni, Au, Pt group metals, Ag, Se, Te, Th, and U are also recovered from the complex.

The evidence is equivocal for a hydrothermal or magmatic origin for the sulphides (Hanekom et al. 1965, Heinrich 1970), but Aldous (1980) presented a model of automhetasomatic mineralization in which the sulphides were closely associated with phoscorite and carbonatite emplacement. Fracturing, brecciation, and recrystallization associated with late stages of intrusion, redistributed some of the sulphides.

10.2.6 Satellite pipes

Satellite bodies of syenite, granite, and feldspathic pyroxenite are concentrated on or near the periphery of the main complex (Fig. 10.1) but some occur well away from the complex, several kilometres to the northeast and northwest. These intrusions form distinct, pipe-like bodies, and range from a simple, homogeneous rock type, suggesting passive emplacement, to complex structures involving multiple intrusions with several generations of brecciation. The rock types include feldspathic pyroxenite, alkali syenite, alkali quartz syenite, alkali granite and trachyte.

The Guide Copper Mine, adjacent to Kitchener's Kop syenite, 4 km NW of the main complex, has a U-shaped outcrop covering approximately 15 000 m^2 (Hanekom *et al.* 1965). The feldspathic pyroxenite closely resembles that of the main complex but contains abundant bornite interstitial to clinopyroxenes (Eriksson 1985). The microcline is extremely fresh and cathodoluminescence indicates a high Fe^{3+} content. The clinopyroxenes exhibit oscillatory zoning (Eriksson 1985) and contain abundant, primary fluid inclusions of an aqueous silicate liquid which crystallizes pyroxene and K-feldspar on cooling (Aldous 1986).

The Kgopoeloe pipe, 3 km north of the main complex, is 150 m in diameter and consists of pink equigranular granite, with a dominant, central portion composed of horizontally reposing, pink granitic breccia fragments set in a matrix of darker, alkali quartz syenite. Gneisses, schists, and granites of the country rock are present as large xenoliths within the granite. Breccia fragments are small and closely packed at the breccia–granite contact, and increase in size and interfragment-distance both inward and downward. The tabular fragments can reach up to 0.5 m in thickness and 2–3 m in length. The overall, horizontal orientation and tabular nature of these fragments are distinctive.

At the Spitskop pipe, brecciation is limited to a 1–2 m thick zone between an outer syenite forming a low rise above the general landscape and the vertical syenite pipe. A distinct ring of alkali quartz syenite encloses the pipe. Syenite, made up of acicular aegirine poikilitically enclosed in microcline, grades inward to an alkali quartz syenite. A distinctive flow texture is marked in the alkali quartz syenite by tabular crystals of orthoclase.

The syenite pipes of Kgopoeloe and Spitskop (see map in Hanekom *et al.* 1965) represent two styles of intrusion. The highly brecciated character of Kgopoeloe reflects explosive activity, whereas Spitskop contains little breccia and shows evidence of multiple magmatic intrusion.

Several modes of origin have been proposed for the syenitic magmatism. Brandt (1948) suggested that the alkali granite magma was passively intruded and preceded explosive brecciation. Subsequent magmatism involved at least five distinct types of syenites. Frick (1975) concluded, on the basis of petrography and major element chemistry, that although there are gross textural differences among the syenites, they represent differentiates from an alkali basalt magma that was parental to the main complex.

10.2.7 *Dykes*

Several dykes of various size cut the entire complex. These were thought to be Karroo dolerite (Hanekom *et al.* 1965) until Briden (1976) suggested an age of 1900 Ma based on palaeomagnetic data. More detailed work (J. C. Briden, pers. comm.) confirms that most cross-cutting dykes in the Phalaborwa complex are Precambrian.

Taljaard (1936) documented a medium-grained, melilite–augite rock collected in the Phalaborwa area, but there is no information about its precise location. Shand (1931) 'collected a number of dark basaltic-looking rocks but all turned out to be ordinary plagioclase basalts.' Subsequently, Eriksson (1982) investigated dolerites and diabases in the Foskor pit, along with new occurrences of dark, fine-grained rocks at the western periphery of the PMC mining area. Petrographic analysis showed them to be dolerites (Eriksson 1982). Inclusions of large, fenitized dolerite xenoliths within the main complex as well as truncated dolerite dykes, indicate dolerite magmatism prior to the formation of the complex.

10.3 AGE

The main Phalaborwa Complex is generally regarded as at least 2000 Ma old. Analysis of several uranoan thorianites and baddeleyites from phoscorite and carbonatite (Eriksson 1984) yielded an isochron of $2047 + 11/- 8$ Ma, similar to the

Table 10.1 Rb-Sr phlogopite isotopic data.

Sample No.	Rock type	Rb (p.p.m.)	Sr (p.p.m.)	$^{87}Rb:^{86}Sr$ (atomic)	$^{87}Sr:^{86}Sr$ (atomic)
P1C	Transgressive carbonatite	722.3	27.88	90.77	3.304
P134-1006K	Phoscorite	746.0	30.78	85.41	3.137
		758.2	36.27	69.84	2.709
F16P	Micaceous pyroxenite	782.2	13.02	340.7	11.177
P28P	Micaceous pyroxenite	614.1	24.67	86.65	3.1987
		606.9	23.26	90.76	3.342
F100P*	Micaceous pyroxenite	781.5	58.81	41.46	1.908
P1R64-2395P	Glimmerite	806.7	11.25	516.7	16.761
F102X*	Glimmerite fragment in pyroxenite	827.2	19.93	175.2	5.912
P29P	Glimmerite	590.4	294.6	5.681	0.871
		624.5	42.90	45.22	1.8601
		629.9	51.62	38.03	1.8961
		660.4	188.0	10.03	0.9442
		662.7	17.94	134.3	4.630
		651.5	30.07	74.96	3.1429
		616.2	7.197	1211	41.98
		545.5	8.917	390.9	14.06
		518.1	488.1	2.988	0.7983
		624.1	20.76	111.5	4.0486

$\lambda = 1.42 \times 10^{-11}$ (Steiger & Jäger 1977); error in ages for Figure 10.4 calculated using in-run precision of $^{87}Sr:^{86}Sr$ ratios which are less than 0.1%. *Measured at University of Texas at Dallas by M. Halpern. Sample P29P measured several times with different washing procedures (for details see Eriksson 1982).

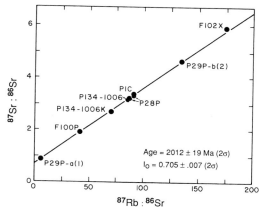

Figure 10.4 Rb–Sr isochron of phlogopite from the main complex. The 2012 Ma date is within error of the U–Pb date of 2047 Ma based on U minerals in carbonatite and phoscorite. Numbers refer to samples listed in Table 10.1.

previously quoted value of 2060 Ma (Holmes & Cahen 1956). Because baddeleyite and uranoan thorianite occur exclusively in carbonate-bearing rocks, the age of the major silicate magmatism is defined less clearly.

Unpublished Rb–Sr analyses (Table 10.1) of phlogopite from clinopyroxenites, carbonatite, and phoscorite plotted on Figure 10.4, yield an isochron of 2012 ± 19 Ma (2σ). U–Pb analysis by ion microprobe (Eriksson, unpublished data) of a single zircon grain occurring in glimmerite near carbonatite confirms that the carbonatite formed c. 2050 Ma ago, and that the glimmerite zone of the major silicate rock types was coeval with carbonatite magmatism.

Palaeomagnetic data on clinopyroxenites from the main complex show an identical pole position to gabbros from the Bushveld Igneous Complex (Briden 1976), known to be of similar age to the Phalaborwa complex. Thus, there is unequivocal evidence that the clinopyroxenites, phoscorite, and carbonatites were intruded over a relatively short time-span at about 2047 Ma.

The age of the syenites, both in the main complex and as satellites, is less certain. Palaeomagnetic data indicate a duration of magmatism of up to 200 Ma (Briden 1976, Morgan & Briden 1981). One syenite pipe has a pole position identical to the 2047 Ma clinopyroxenite of the main complex, whereas a syenite adjacent to the northern periphery of the main complex has a pole position similar to that of some large dolerite dykes which cut the complex, considered to be Mashonaland dolerites dated at 1950 Ma.

10.4 MINERAL CHEMISTRY

10.4.1 *Clinopyroxenes*

Representative analyses of clinopyroxenes are listed in Table 10.2. Pyroxenes from Phalaborwa form two groups: those in rocks with cumulus textures and those in

Table 10.2 Electron microprobe analyses of some representative minerals from Phalaborwa.

	Clinopyroxenes					Micas				Olivine		Magnetite
	F15M	F100P	P4P	P120P	P120P	P374N	F100P	P29P	P19K	P2506C	P374N	P3C	P42K	LK675C

Wt%														
SiO$_2$	54.10	54.20	53.70	53.50	54.50	52.00	41.90	41.80	40.30	41.50	36.20	40.80	40.40	0.07
TiO$_2$	0.00	0.15	0.08	0.05	0.00	0.03	0.88	1.50	0.25	0.02	4.32	n.a.	n.a.	0.00
Al$_2$O$_3$	0.26	0.00	0.14	0.07	0.00	2.02	11.00	10.80	11.30	6.74	13.10	n.a.	n.a.	0.00
FeO	4.02	2.72	8.27	8.78	4.04	12.05	6.39	6.39	3.22	11.88	20.80	19.80	12.00	93.90
MnO	0.00	0.04	0.25	0.17	0.13	0.49	0.08	0.08	0.03	0.09	0.18	—	—	0.16
NiO	n.a.	n.a.	n.a.	n.a.	n.a.	n.a.	—	—	—	—	—	0.02	0.03	—
MgO	16.20	17.20	14.10	13.20	16.30	11.10	26.10	23.10	26.30	26.50	10.80	49.60	47.70	n.a.
CaO	24.80	24.80	23.40	23.20	24.50	21.00	0.00	0.00	0.06	0.00	0.0	0.04	0.03	n.a.
Na$_2$O	0.28	0.00	0.58	0.91	0.43	1.35	0.06	0.00	0.06	0.00	0.00	n.a.	n.a.	n.a.
K$_2$O	n.a.	n.a.	n.a.	n.a.	n.a.	n.a.	10.30	10.10	10.30	10.40	9.29	n.a.	n.a.	n.a.
F	n.a.	n.a.	n.a.	n.a.	n.a.	n.a.	n.a.	n.a.	0.60	—	n.a.	n.a.	n.a.	n.a.
Cl	n.a.	n.a.	n.a.	n.a.	n.a.	n.a.	n.a.	n.a.	0.00	n.a.	—	n.a.	n.a.	n.a.
Total	99.66	98.81	100.52	99.88	99.90	100.04	96.70	93.77	91.76	97.05	94.69	100.26	100.16	94.13

Cations per formula unit based on: 6 oxygens | 22 oxygens | 4 oxygens

	F15M	F100P	P4P	P120P	P120P	P374N	F100P	P29P	P19K	P2506C	P374N	P3C	P42K	
Si	1.988	1.995	1.987	1.994	1.995	1.957	5.925	6.025	5.888	6.041	5.616	0.995	0.999	—
Al	0.011	0.000	0.006	0.003	0.000	0.089	1.839	1.839	1.944	1.157	2.399	—	—	—
Ti	0.000	0.004	0.002	0.001	0.000	0.000	0.094	0.162	0.027	0.002	0.505	—	—	—
Fe^{3+}	0.020	0.000	0.042	0.066	0.031	0.099	—	—	—	—	—	—	—	—
Fe^{2+}	0.104	0.084	0.215	0.208	0.093	0.270	0.758	0.971	0.393	1.447	2.711	0.200	0.248	—
Mn	0.000	0.001	0.008	0.005	0.004	0.016	0.009	0.004	0.004	0.010	0.024	0.000	0.000	—
Mg	0.888	0.941	0.777	0.733	0.889	0.623	5.505	4.916	5.72	5.745	2.491	1.810	1.767	—
Ca	0.976	0.976	0.928	0.927	0.962	0.847	0.000	0.000	0.000	0.000	0.000	0.001	0.001	—
Na	0.020	0.000	0.042	0.066	0.031	0.098	0.015	0.001	0.017	0.016	0.021	—	—	—
K	—	—	—	—	—	—	1.852	1.857	1.920	1.939	1.843	—	—	—
Fe:(Fe+Mg)	0.12	0.12	0.25	0.27	0.12	0.37	0.12	0.16	0.06	0.20	0.52	0.10	0.12	

Sample numbers: F15M, massive pyroxenite; F100P, P4P, P120P, P29P, micaceous pyroxenite; P19K, P42K, phoscorite; P3C, P2506C, LK675C, carbonatite; P374N, fenite. Details of microprobe data in Eriksson (1982).

rocks with a wide variety of textures including fenitized country rock. The pyroxenes can also be divided into two groups on the basis of chemistry.

Group I pyroxenes are chemically homogeneous, whereas Group II pyroxenes contain considerable intra-, and inter-grain variation within a single specimen. Group I pyroxenes are found in various clinopyroxenites in the northwestern and central portions of the complex, while Group II pyroxenes are limited to the micaceous clinopyroxenites, phoscorite, and fenitized country rocks.

Pyroxene analyses are shown on the Wo-En-Fs system in Figure 10.5. Group I pyroxenes are diopsides and salites with $Fe:(Fe + Mg)$ ratios between 0.07–0.29, where Fe represents total Fe. Within any one grain or thin section there is little change in $Fe:(Fe + Mg)$ or the Wo component. Group II pyroxenes vary considerably in Fe enrichment and have $Fe:(Fe + Mg)$ ratios between 0.12–0.66. Generally, the Wo component is fairly constant. Group II pyroxenes show both normal and reverse zonation. One extreme example has reverse chemical zonation of $Fe:(Fe + Mg) = 0.66-0.21$ along with marked reverse colour zonation. The wide variation in Group II chemistry can be attributed to changing conditions during metasomatism or variation in palaeosome chemistry.

The $Fe:(Fe + Mg)$ ratios of Group I pyroxenes are comparable to groundmass diopsides from kimberlites (Emeleus & Andrews 1975, Dawson *et al.* 1977) and clinopyroxenes from kimberlite xenoliths (Dawson *et al.* 1977). The limited $Fe:(Fe + Mg)$ ratios of the Group I pyroxenes suggest restricted Fe enrichment. At the Square Top intrusion, New South Wales, $Fe:(Fe + Mg)$ ratios of pyroxenes are slightly variable (0.24 at the lower contact to 0.32 at the exposed top), and this was attributed to limited FeO enrichment in pyroxenes due to the increasing oxidation state of successive liquid fractions limiting the FeO available for the clinopyroxenes (Wilkinson 1966). Presnall (1966) presented a mechanism of a buffering system for H_2O-rich basaltic magmas; hydrogen diffusion to wall-rock buffers the oxygen fugacity of the magma thereby limiting the range in Fe:Mg ratios.

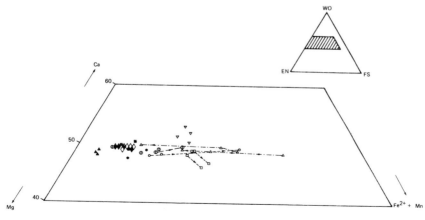

Figure 10.5 Composition of clinopyroxenes from the main complex plotted in the Wo–En–Fs system. Completely filled symbols represent data from magmatic clinopyroxenes (Group I); remainder are from clinopyroxenes affected by metasomatism. Arrows within box indicate compositional zones within single grains.

Eriksson (1982) plotted numerous analyses of clinopyroxenes relative to their geographic position in the complex and showed that there was no systematic increase in Fe:(Fe + Mg) ratios. This she attributed to multiple surges of magma.

Pyroxenes from alkaline rocks are known for their high acmite components, but Figure 10.6 shows the acmite component to be extremely low and restricted in the Group I pyroxenes. The two trends of Na enrichment, suggested by the Group II pyroxenes, may be due to variation in palaeosome or fluid chemistry.

Aluminium, Ti, and Cr contents are extremely low in all of the Phalaborwa pyroxenes. Chromium is less than the detection limit, TiO_2 is < 0.29 wt% and Al_2O_3 is < 0.5 wt% in the magmatic pyroxenes. Pyroxenes from the fenitized country rock contain up to 2.0 wt% Al_2O_3.

Group I pyroxenes are characterized by high Ca, low Fe:(Fe + Mg), and low Al_2O_3 contents. It has been suggested that the Ca content of a clinopyroxene depends on the SiO_2 activity in a magma rather than its alkalinity or Ca content (Smith & Lindsley 1971, Gibb 1973, Carmichael *et al.* 1974). Most basic and ultrabasic rocks from systems of low SiO_2 content have Al_2O_3 contents much higher than the pyroxenes from Phalaborwa (Shonkin Sag = 1.3–1.9%, Nash & Wilkinson

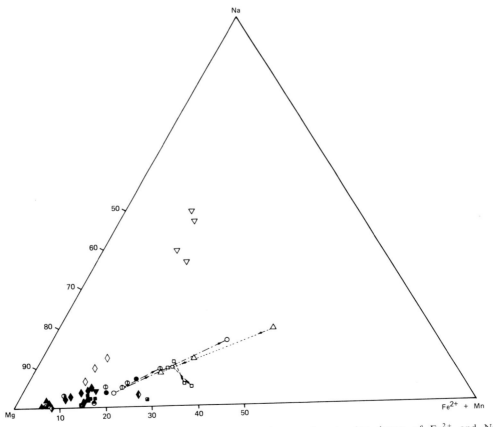

Figure 10.6 Compositions of clinopyroxenes from the main complex showing degree of Fe^{2+} and Na^+ enrichment. Symbols as in Figure 10.5.

1970; Ilimaussaq = 2.5%, Larsen 1976; eastern Uganda = 2.9–5.9%, Tyler & King 1967). Alkalic pyroxenite xenoliths (Dawson & Smith 1973) with similar mineralogy and texture to the micaceous and massive pyroxenites from Phalaborwa have much higher Al_2O_3 contents, while groundmass diopsides from kimberlite (Dawson et al. 1970) contain 3–6% Al_2O_3. Pyroxenes with comparable high Ca and low Al do occur in extremely silica-undersaturated melteigites and ijolites in Uganda (Tyler & King 1967, Edgar 1979), in madupite from Leucite Hills (Carmichael 1967), and in the Square Top intrusion (Wilkinson 1966), although the Al_2O_3 contents of these rocks can also be very high (7.32–14%). The extreme Al compositions may reflect the peralkalinity of the parental magma where high concentrations of K^+ ions affect the melt structure by inhibiting Ca–Al and Ca–Ti bonds (Edgar et al. 1976, Barton & Hamilton 1978).

Chromium is extremely low in the Phalaborwa pyroxenes considering their high Mg number (Mg : (Mg + Fe)). In a basaltic magma with 200–300 p.p.m. Cr and a $K_D = 4$ (Schreiber 1976), clinopyroxenes should have a concentration of 800–1200 p.p.m. Cr. Thus, there is a suggestion that some chromium-rich phase may have been removed from the magma at Phalaborwa at an earlier stage. Magnetite with $K_D = 10$ (Cox et al. 1979) is a likely candidate.

10.4.2 Mica

Micas at Phalaborwa show more textural, optical and chemical variation than any of the other minerals in the complex. Most micas in the main complex are phlogopite, although biotite occurs in the fenites. Isolated patches of hydrobiotite are present (Hanekom et al. 1965) and probably reflect late hydrothermal alteration. Vermiculite, a major economic mineral at Phalaborwa, is confined largely to surface outcrops and decreases in abundance dramatically with depth.

Fe : (Fe + Mg) ratios (0.04–0.52) indicate a wide range of composition from Mg-rich phlogopite to Mg-rich biotite (Table 10.2). The phlogopite exhibits both normal and reverse pleochroism; the latter is widespread in phlogopite from carbonatites (Rimsaite 1969, Suwa & Aoki 1975) and other ultrabasic rocks (Boettcher 1966, 1967, Aoki 1974, Emeleus & Andrews 1975, Dawson & Smith 1977, Smith et al. 1978). Mica compositions (Table 10.2) show that tetrahedral sites in micas contain 7.176–8.000 cations of Si, Al, and Ti. Significant Fe^{3+} is present in tetrahedral co-ordination in some micas and significant amounts cause the reverse pleochroism.

Phlogopite from Phalaborwa with distinct reverse pleochroism is found in phoscorite and carbonatite and only rarely in pyroxenites. Micas with tetrahedral sites completely filled with (Al + Si) occur only in pyroxenites. Previous studies have attributed phlogopite with reverse pleochroism to bulk magma chemistry (Rimsaite 1969, Nash 1972, Gittins et al. 1975, Suwa & Aoki 1975). Likewise, at Phalaborwa, crystallization of phlogopite from an Al-depleted carbonatitic magma would produce further Al-deficiency, creating micas with reverse pleochroism. Aldous (1980) analysed phlogopite with reverse pleochroism from fluid inclusions in olivine from the Phalaborwa Complex and showed that the mica contains almost no Al_2O_3.

This could imply that the phlogopite crystallized from a magma very low in Al, that the phlogopite is indeed magmatic, and that the olivine did not crystallize at the level of the complex presently exposed.

Micas show two trends in chemistry (Fig. 10.7), one with increasing Al with $Fe:(Fe + Mg)$ and one with decreasing Al with increasing $Fe:(Fe + Mg)$. The trend of increasing Al with $Fe:(Fe + Mg)$ in phlogopite from clinopyroxenites may reflect the increase in Al in silicate magma as Al-deficient phlogopite and Al-poor clinopyroxene crystallize. High $Fe:(Fe + Mg) = 0.51$ and high $TiO_2 = 4.32\%$ occur in biotite from metasomatized country rock.

Micas that show an inverse correlation between Al and $Fe:(Fe + Mg)$ occur in phoscorite and carbonatite. As Fe^{3+} substitutes for the Al, tetrahedral Fe shows a concomitant increase with increasing $Fe:(Fe + Mg)$. Correlation of tetrahedral Fe with $Fe:(Fe + Mg)$ was also noted by Smith *et al.* (1978) in micas from Type II kimberlites. They noted that the correlation of Fe^{3+} with $Fe:(Fe + Mg)$ may reflect increasing f_{O_2} in a magma due to loss of volatiles to surrounding rocks. High H_2O pressure decreases the stability field of magnetite (Eggler & Burnham 1973) and could explain the lack of magnetite in the pyroxenites, and the presence of magnetite only in rocks containing phlogopite with reverse pleochroism. Smith *et al.* (1978) also tentatively suggested, based on the data of Rimsaite (1969) and Bass *et al.* (1974), that carbonatite magmas give rise to micas with tetrahedral sites completely filled with Si and Al. This is not the case at Phalaborwa.

Phlogopite also forms reaction rims around olivine. These micas are Al-depleted and show reverse pleochroism. A more Al-rich composition with normal pleochroism is found for micas associated with clinopyroxenite (Fig. 10.7). Boettcher (1967) noted a similar feature in micas from the Rainy Creek Igneous Complex. The reaction:

$$\text{olivine} + \text{liquid 1} \rightarrow \text{phlogopite} + \text{liquid 2}$$

produces phlogopite during cooling of an ultrabasic liquid (Edgar *et al.* 1976), and micas found rimming olivine in phoscorite may represent a reaction of olivine with magma. Clinopyroxene is also present as a reaction product at Phalaborwa. In the possible reaction (Modreski & Boettcher 1973):

$$\text{forsterite} + \text{liquid} \rightarrow \text{phlogopite} + \text{diopside} + \text{vapour}$$

the type of magma involved may influence whether phlogopite alone or phlogopite plus clinopyroxene are formed.

10.4.3 Olivine

Phalaborwa olivines have compositions that range from Fo_{79} to Fo_{91} (Table 10.2). Most grains are unzoned. Banded carbonatite has olivine with the highest Fo content, phoscorite has the lowest, and no olivine has been found in the transgressive carbonatite.

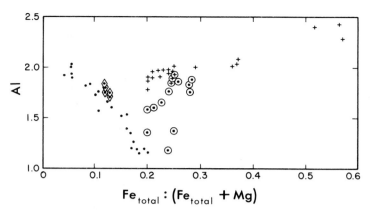

Figure 10.7 Compositions of micas from the main complex plotted as number of Al cations per unit formula versus $Fe_{total}:(Fe_{total} + Mg)$. Crosses (+), micas with normal pleochroism; dots (●), micas with reverse pleochroism; dots with diamond (◊), micas from 'inch-scale' layering; dots with circle (⊙), micas from olivine reaction rim.

The olivines are extremely low in NiO (< 0.06 wt%). Such low concentrations and lack of correspondence between Ni and Fo content is a significant feature in Phalaborwa samples, particularly when olivines of composition Fo_{91} in some ultramafic rocks can contain up to 0.28 wt% NiO (Duke & Naldrett 1978). One xenocrystic olivine (Fo_{87}) from a Phalaborwa clinopyroxenite contains 0.28 ± 0.03 wt% NiO, which suggests that it probably crystallized from a magma different from that which formed the phoscorite and carbonatite.

The very low NiO content of the Fo-rich olivines suggests that Ni has been removed from the magma. The model of Duke & Naldrett (1978) involved separation of olivine and molten sulphide from mafic and ultramafic liquids; under appropriate conditions olivines of low Ni content at Fo_{89-91} can be produced. This model is difficult to accept for the Phalaborwa olivines, because of the high Cu concentrations in the phoscorite and carbonatites. Not only are copper sulphides abundant in these rocks, but reports of copper sulphides as inclusions in olivines (Van Rensburg 1965, Aldous 1980) point to Cu being present during olivine crystallization. The distribution coefficient of Cu between olivine and sulphide melt, $K_D = 178$, is thought to be similar in mafic and ultramafic melts (Rajamani & Naldrett 1978). This strong partitioning of Cu into sulphide melt precludes significant amounts of sulphide liquid acting as a sink for Ni in Phalaborwa. However, Hart & Davis (1978) have shown that partitioning of Ni between olivine and liquid is highly dependent on melt composition; in some cases, the distribution coefficient of Ni between olivine and liquid can be very high.

Magnetite fractionation might produce liquids low in Ni without significantly lowering the MgO content of a magma. Depending on temperature and f_{O_2} (Lindstrom 1976) the K_D can be as high as 77, and removal of small amounts of magnetite from a liquid would reduce the Ni content dramatically.

10.4.4 Magnetite

Magnetites from Phalaborwa have not been well studied but available analyses show that they are relatively pure, show very little compositional variation, and have extremely low Cr_2O_3 and low TiO_2 contents (Hanekom *et al.* 1965, Van Rensburg 1965, Hennig-Michaeli 1968, Aldous 1980, Eriksson 1982). A typical analysis is given in Table 10.2.

Magnetite, from analysed concentrates from mining operations (Palabora Mining Company 1976), shows a decrease from 4.0% TiO_2 in the outer phoscorite to less than 0.10% TiO_2 in banded carbonatite. The 4.0% TiO_2 suggests the presence of ilmenite in the mineral concentrate. Aldous (1986) shows that magnetite from phoscorite contains more MgO (0.65–1.45%) than magnetites from banded carbonatite (0.35–1.01% MgO).

The compositions of primary oxide minerals depend on bulk chemistry of the magma, depth of emplacement, and oxygen fugacity of the magma. Haggerty (1976) notes that in highly undersaturated rocks where the activity of SiO_2 and f_{O_2} are dominating parameters, TiO_2 is commonly depleted in the oxides. The TiO_2 contents in magnetite from the Phalaborwa carbonatite are lower than those from the phoscorite and may indicate a decrease in f_{O_2} from phoscorite to carbonatite or may simply reflect early-formed phases, that removed titanium. Haggerty (1976) also notes that magma composition rather than pressure determines the magnetite Mg content. Decrease in Mg from phoscorite to carbonatite may, therefore, reflect a variation in magma composition.

10.4.5 Amphibole

Amphibole was found in only one sample associated with the main complex, a fenitized country rock. Its composition $\square_{0.5}(K_{0.3}Na_{0.2})(Ca_{1.2}Na_{0.8})(Mg_{3.3}Fe^{2+}_{1.2}Fe^{3+}_{0.5})Si_8O_{22}(OH)_2$ indicates that the components are richterite, $Na(NaCa)(Mg_5)Si_8O_{22}(OH)_2$, and winchite, $\square(NaCa)(Mg_4Fe^{3+})Si_8O_{22}(OH)_2$, with some substitution of Fe^{2+} and K. Amphibole is one of the few minerals containing significant Na in the main complex at Phalaborwa, and coexisting pyroxene is the most Na-rich of those analysed. Low Ti (0.07% TiO_2) and Al (0.33% Al_2O_3) characterize amphibole and pyroxenes of the main complex.

10.5 WHOLE-ROCK CHEMISTRY

10.5.1 Main complex

Chemical analyses of clinopyroxenites are listed in Table 10.3. These rocks are cumulates and the major element chemistry does not reflect liquid compositions. However, major elements show wide variation in oxide abundances over very short distances in the Phalaborwa Complex, supporting the cumulative processes inferred from textural information. Variation in P_2O_5 content (0.48–11.44%) is attributed to apatite accumulation.

Table 10.3 Whole rock chemistry.

	Main Complex							Satellite Bodies							
	F12F	F67F	F5P	F28P	F25M	F38M	G1F	K19m	K21m	K12g	K17g	S2d	S21r	S3r	S13c	S17c
Wt%																
SiO_2	54.41	39.89	39.24	45.86	50.50	48.11	53.18	59.44	58.84	76.19	68.74	61.19	62.47	61.01	64.55	63.00
TiO_2	0.14	0.21	0.40	0.36	0.18	0.15	0.19	0.31	0.84	0.16	0.02	1.04	1.10	0.97	1.27	0.85
Al_2O_3	1.00	1.55	3.78	6.02	0.70	0.45	3.61	11.78	10.65	13.15	18.12	14.96	8.42	8.16	10.61	12.72
Fe_2O_3	0.38	0.81	1.89	2.73	1.91	1.02	3.33	3.35	4.70	0.94	0.09	3.87	9.96	7.82	9.43	7.33
FeO	2.82	2.31	1.78	4.02	4.05	3.35	4.88	1.17	1.50	0.00	n.a.	2.45	2.28	2.65	0.45	0.03
MnO	0.05	0.29	0.14	0.20	0.22	0.35	0.15	0.15	0.32	0.00	0.45	0.06	0.56	0.41	0.09	0.10
MgO	13.72	12.76	17.08	20.57	14.77	14.04	9.56	2.50	3.06	0.41	0.23	0.88	3.74	4.95	0.57	0.77
CaO	22.04	28.41	21.69	10.83	23.88	25.54	18.70	6.66	7.93	1.06	0.08	0.55	0.92	1.93	0.63	0.44
Na_2O	0.61	0.58	0.00	0.00	0.17	0.69	1.06	1.85	1.79	3.60	1.84	1.66	2.65	3.36	2.34	1.92
K_2O	0.65	1.14	3.53	5.82	0.18	0.12	3.24	8.52	8.19	4.50	10.96	12.99	7.38	7.80	8.99	11.07
P_2O_5	4.18	11.44	8.97	0.48	2.72	5.44	1.29	0.78	0.69	0.11	0.15	0.26	0.23	0.58	0.14	0.34
H_2O^-	0.06	0.04	0.09	0.09	0.17	0.04	0.00	0.03	0.47	0.07	0.06	0.00	0.16	0.05	0.21	0.16
LOI	0.65	0.72	1.81	2.37	0.56	0.81	0.86	3.73	1.97	0.26	0.17	0.78	1.12	0.71	0.47	0.39
Total	100.71	100.15	100.40	99.35	100.01	100.01	100.05	100.27	100.95	100.45	100.91	100.69	100.99	100.40	99.75	99.12
Mg #	89	88	90	85	82	85	68	52	49	46	84	21	37	48	10	17
Trace Elements (p.p.m.)																
Sr	1039	2084	870	289	874	1362	781	1314	1644	352	81	46	96	878	50	58
Rb	22	71	302	485	10	6	120	370	284	133	463	420	262	262	329	406
Ba	57	126	161	500	2	BDL	641	2026	2254	1086	1468	3822	460	720	401	440
Ce	373	1213	646	21	356	760	189	366	365	46	39	340	438	355	215	591
La	132	627	208	20	127	378	62	200	208	17	29	138	293	128	111	308
Cu	n.d.	n.d.	n.a.	114	94	81	1804	n.a.	n.a.	n.a.	n.a.	22	n.a.	13	9	7
Ni	44	36	87	48	110	50	43	221	24	n.d.	5	18	12	3	7	9
Zn	30	16	32	97	43	16	66	16	45	n.d.	5	n.d.	90	184	2	n.a.
Nb	4	3	n.a.	8	3	5	3	125	201	20	n.d.	86	405	454	59	74
Zr	55	6	21	12	19	18	162	6	19	6	3	28	26	20	658	626
Y	44	93	62	12	35	54	25	262	815	130	13	270	1167	603	11	63
V	23	n.a.	10	68	47	n.a.	8	39	59	8	n.a.	n.a.	n.a.	n.a.	n.a.	n.a.

Sample numbers: F12F, F67F, G1F, feldspathic pyroxenite; F5P, F28P, micaceous pyroxenite; F25M, massive pyroxenite; K19m, K21m, breccia matrix; K12g, K17g, granite; S2d, dyke; S21r, S3r, ring syenite; S13c, S17c, centre syenite. Details of XRF analyses in Eriksson (1982). n.a., Not analysed; n.d., not detected.

Within the feldspathic, and massive clinopyroxenites, the only phase containing Fe and Mg is clinopyroxene. Magnesium numbers are consistently high (76–90) and show no systematic change with geographic location. Feldspathic pyroxenites are generally more Mg-rich (82–89) than massive pyroxenites (76–89). Among the micaceous pyroxenites, pegmatoidal varieties (see Fig. 10.3) have the highest value (90). Multiple surges of fresh magma would account for the spatial variation in Mg-number. A notable feature of the major element chemistry of the Phalaborwa pyroxenites is low TiO_2 values. The average value, 0.23%, is well below the concentration for most alkaline rocks.

Strontium in clinopyroxenites averages 1032 p.p.m. Feldspathic pyroxenites have the highest Sr content (average = 1175 p.p.m.) and micaceous pyroxenites the lowest (average = 150 p.p.m.). The strong correlation of Sr with P_2O_5 reflects compositional control by cumulus apatite. Strontium contents of clinopyroxenes, measured by isotope dilution of leached samples, range from 318–553 p.p.m. (Ericksson 1982). These concentrations are extremely high, even for such Ca-rich pyroxenes, although similar concentrations in clinopyroxenes from granular nodules in kimberlites have been reported (Shimizu 1975). Using a distribution coefficient of $K_D = 0.22$ (Sun *et al.* 1974), the liquid coexisting with the Phalaborwa pyroxenes would have 909–2045 p.p.m. Sr. The highest Sr concentration found in the literature for a rock believed to represent a liquid composition is 2195 p.p.m. and this was measured from the chilled lower margin of the Shonkin Sag laccolith (Nash & Wilkinson 1970). Experimental work (Koster van Groos 1975) suggests that Sr fractionates into a silicate melt in equilibrium with carbonatite magma. Given the close association of the clinopyroxenites with apatite-rich carbonatite (950–6300 p.p.m. Sr) at Phalaborwa, the magma forming the clinopyroxenites was probably very rich in Sr.

Rare earth elements in the Phalaborwa clinopyroxenites, carbonatites, and phoscorite are enriched in LREEs with over 2000 times chondrite abundances (Aldous 1980, Eriksson 1982). Aldous (1980) shows a fractionation trend among apatites with total REE content highest in the carbonatites and the LREEs progressively higher in phoscorite, pyroxenite, and carbonatite. Apatite controls the total REEs in these rocks.

One enigmatic aspect of Phalaborwa is the extremely low Nb content of both clinopyroxenites and carbonatites (Eriksson 1982). Deans (1966) noted that Nb is minor or absent in complexes where carbonatite is subordinate to syenites, pyroxenites, and dunites and is minimal in late-stage REE-, Sr-, and Ba-carbonatites. At Phalaborwa, it is possible that Nb partitioned into a fluid phase and reached upper levels of the intrusion, subsequently eroded. However, Nb substitutes readily for Ca and has been shown to be an intrinsic part of carbonatite magmatism (Watkinson 1970).

Zirconium concentrations of the magma may be inferred from the chemistry of the cumulates (Table 10.3). Clinopyroxenites have values from 6 to 55 p.p.m. Zr. Assuming that the Zr is in the clinopyroxene and using a $K_D = 0.10–0.12$ (Irving 1978, Pearce & Norry 1979), clinopyroxenes crystallized from magmas of

16–460 p.p.m. Zr. This is not high compared to other ultrabasic liquids (melteigite from Fen contains 0.1% ZrO, potassic lavas from Leucite Hills contain 0.28% ZrO, Carmichael 1967). At Phalaborwa, zircon is extremely rare and is found in only one area of the phoscorite–clinopyroxenite contact. Zircon does not form in clinopyroxenites and the relatively low concentration points to a silicate liquid which was not highly differentiated. In contrast, Zr concentrations in phoscorite are sufficiently high to form baddeleyite.

10.5.2 Satellites

Whole-rock analyses of the country rock granite (samples K12g and K17g), breccia fragments, and the syenite (K19m and K21m) from Kgopoeloe are given in Table 10.3. Samples K12g and K17g illustrate leaching of granitic breccia fragments by syenite. Sample K17g is much lower in SiO_2, Na_2O, and CaO and higher in K_2O and Al_2O_3 than the unfenitized granite (K12g), reflecting replacement of plagioclase by potassium feldspar, the removal of quartz and the breakdown of biotite. Trace element analyses indicate removal of Sr and Zr in these rocks with increase in Rb and La. Bulk assimilation is not important in modifying the syenitic magma.

Chemical compositions of the outer ring syenite (samples S3r and S21r) and the inner cumulus syenite (samples S13c and S17c) can be modelled by crystallizing aegirine, feldspar, and apatite in the proportions of 65:35:1 with a change to crystallization of 95% feldspar and 5% aegirine when 60% of the magma remains. These proportions correspond extremely well to the modal proportions of minerals found in these rocks.

The magma that formed the satellite pipes was only slightly silica-undersaturated. To derive such a liquid from a magma similar in composition to that of the main complex, the liquid must have contained at least 54–55% SiO_2. It is doubtful whether clinopyroxene of the main complex could have crystallized from such a silica-rich composition. The pyroxene chemistry in the satellites (Eriksson 1982) shows the same low Al_2O_3 abundances found in pyroxenes from the main complex. The enriched total REEs and LREEs and high K:Na ratio also characterize both magmas. Overall, the magma that produced the satellite bodies is much higher in SiO_2, Zr, Fe:(Fe+Mg), and Na:K and much lower in Ba and Sr than the magma that produced the clinopyroxenites of the main complex. There is no indication that the clinopyroxenites of the main complex and the syenites of the satellites crystallized from the same magmas in spite of obvious similarities between the two.

The clinopyroxenite of the Guide Copper Mine sample, G1F, is more similar in composition to the feldspathic pyroxenites of the main complex. The low Al content of the calcium-rich clinopyroxene, along with identical initial $^{87}Sr:^{86}Sr$ ratios (Eriksson 1982) and the abundant inclusions of copper sulphides in clinopyroxenes (Aldous 1986) suggest a close tie between the clinopyroxenite of Guide Copper Mine and the main complex.

10.6 ISOTOPIC CHEMISTRY

The Phalaborwa Complex provides a unique opportunity for examining the isotopic characteristics of multiple melting events in old, subcontinental mantle. Unlike kimberlites which may mechanically incorporate fragments of the crust, or basalts which may be contaminated chemically, carbonatites can rise quickly through the crust, thus minimizing any contamination. The isotopic systems Rb–Sr, Sm–Nd, U–Pb, O, C, and S have been studied to determine the relationship among the rock types, origin of mineralization, and the nature of the magma source.

10.6.1 Rb–Sr

The Sr isotopic compositions of the clinopyroxenites and carbonatites were used to determine the genetic relationship between the carbonatite and the silicate-bearing rocks of the complex (Eriksson 1982). The problematic origin of the feldspathic pyroxenite was also addressed.

Analyses of whole rock and mineral separates of clinopyroxene and apatite from carbonatites, and mineral separates from the clinopyroxenites are plotted in Figure 10.8. Most of the Phalaborwa carbonatites have initial $^{87}Sr:^{86}Sr$ ratios that lie between 0.70393 and 0.70680; one sample has an initial $^{87}Sr:^{86}Sr$ ratio of 0.71022. Clinopyroxenes from clinopyroxenites have initial $^{87}Sr:^{86}Sr$ ratios of 0.71140–0.71242 and one xenocrystic clinopyroxene in carbonatite has an initial $^{87}Sr:^{86}Sr = 0.70529$. Apatites have initial $^{87}Sr:^{86}Sr$ ratios of 0.71008–0.71234.

Coexisting phases from some rocks have different Sr isotopic compositions (see Fig. 10.8). In one carbonatite sample the initial $^{87}Sr:^{86}Sr$ ratio from a clinopyroxene xenocryst is 0.0008 higher than the carbonates, while coexisting apatite and clinopyroxene from a massive clinopyroxenite differ by 0.0004. These differences are well outside the limits of experimental uncertainty.

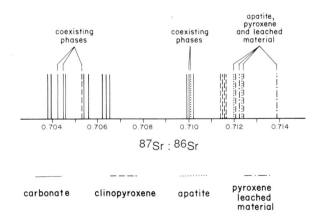

Figure 10.8 Summary of initial $^{87}Sr:^{86}Sr$ ratios for rocks and minerals from the Phalaborwa Complex. Analyses of minerals from the same rock and the material leached from a clinopyroxene sample are shown as coexisting phases. Each line represents one analysis.

Several methods were used in Eriksson's study (1982) to ascertain the role of pre- or post-crystallization contamination. Weathering and groundwater contamination were addressed by:

(a) leaching experiments on clinopyroxene;
(b) analysis of carbonatite that contained different amounts of phlogopite;
(c) correlation of isotopic compositions with rock mineralogy; and
(d) analysis of O and C isotopic compositions.

Leaching of a clinopyroxene was used to establish whether a contaminant, rich in radiogenic Sr derived from hydrothermal fluids or groundwater, could have contributed to the highly radiogenic nature of the clinopyroxenes. Two leaches (6N HCl for 1 hour, and 6N HCl for 6 hours in an ultrasonic bath) gave different ratios (Fig. 10.8). The more drastic treatment yielded a leach that had an initial $^{87}Sr:^{86}Sr$ ratio identical to that of the clinopyroxene. Both samples of the leachate contained 56–59 p.p.m. Sr compared to the 550 p.p.m. Sr of the clinopyroxene, and it was concluded that material adhering to clinopyroxene surfaces was not a contributor to the high initial $^{87}Sr:^{86}Sr$ ratios of the clinopyroxenes. These studies (Eriksson 1982, Allsopp & Eriksson 1986) show that although small variations in $^{87}Sr:^{86}Sr$ ratios ($\leqslant 0.0003$) could be caused by secondary processes, the differences of $\geqslant 0.002$ could not.

Contamination of a carbonatite or ultramafic magma to produce the high initial $^{87}Sr:^{86}Sr$ ratios in the clinopyroxenites is more difficult to evaluate. Although there is correspondence between Sr content and initial $^{87}Sr:^{86}Sr$ ratios (Fig. 10.9), the high concentrations of Sr in the carbonatites (4200–5000 p.p.m.), apatites (2600–14 000 p.p.m.), and clinopyroxenes (320–550 p.p.m.) would require prohibitively large amounts of crustal contamination. More complex models, such as preferential removal of Sr from country rocks, are difficult to quantify but are considered unlikely due to the rapid ascent of these magmas. In addition, carbonatites from other localities (Bell *et al.* 1982, Grünenfelder *et al.* 1986, Bell &

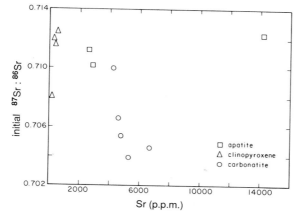

Figure 10.9 Initial $^{87}Sr:^{86}Sr$ ratios versus Sr concentrations for carbonatites and minerals.

Blenkinsop 1987, Tilton *et al.* 1987, Nelson *et al.* 1988) have not exhibited the extreme variations in Sr isotopic composition observed in the Phalaborwa samples. Hence, the differences in initial $^{87}Sr:^{86}Sr$ ratios in the Phalaborwa Complex are considered as primary, resulting from variations in source regions. The isotopic data shown in Figure 10.9 require the mixing of at least two carbonatite magmas, one with an initial $^{87}Sr:^{86}Sr$ ratio $\leqslant 0.7039$ and another with a high initial $^{87}Sr:^{86}Sr$ ratio $\geqslant 0.710$.

10.6.2 Stable isotopes

Four whole-rock carbonatites have $\delta^{18}O$ of $+7.7$ to $+8.6‰$ and $\delta^{13}C$ of -5.1 to $-3.6‰$, and three clinopyroxenes from clinopyroxenites have $\delta^{18}O$ of $+7.26$ to $+7.53‰$. Clinopyroxenes, known to be extremely resistant to deuteric alteration and change in oxygen isotopic composition, can be used to calculate the $\delta^{18}O$ composition of the magma that formed the cumulus clinopyroxenites. Because this magma is thought similar to many potassium-rich lavas relatively low in SiO_2 (Eriksson 1982), a $\delta^{18}O$ fractionation of $0.25‰$ (Taylor *et al.* 1979, Taylor *et al.* 1984) was used and this yielded whole-rock $\delta^{18}O$ values of $+7.5$ to $+7.7$, comparable to the $\delta^{18}O$ values of the Phalaborwa carbonatites.

Plots of the type shown in Figures 10.9 and 10.10 have been attributed to simple binary mixing (Bell & Powell 1969, Taylor *et al.* 1979, Taylor 1980, Vollmer & Norry 1983). Ferrara *et al.* (1985), using data from a potassium-rich volcanic suite, characterized a mantle-derived magma with a $\delta^{18}O$ of about $+6$ or $+7‰$ and a $^{87}Sr:^{86}Sr$ of about 0.710; this supports the contention of Hawkesworth & Vollmer (1979) that high $\delta^{18}O$ values of $+7$ to $+8‰$ exist in the subcontinental upper mantle. Therefore, the 'high' end member for the Phalaborwa magmas

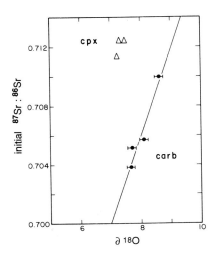

Figure 10.10 Initial $^{87}Sr:^{86}Sr$ ratios versus $\delta^{18}O$ values for clinopyroxenes and carbonatites. The solid line can be projected to a $^{87}Sr:^{86}Sr$ ratio of 0.7015 to calculate the $\delta^{18}O$ value for one of the end members, assuming simple binary mixing of two carbonatite magmas.

($\delta^{18}O = 8.6‰$) is only slightly higher than that previously reported for subcontinental mantle (Hawkesworth & Vollmer 1979).

Two features of the Phalaborwa data point to the mixing of two carbonatite magmas rather than mixing of a carbonatite magma with continental crust. First, a correlation between $\delta^{18}O$ and $\delta^{13}C$ observed for the Phalaborwa rocks would probably not occur with crustal contamination because the highly metamorphosed granitic basement does not contain significant carbonate-bearing rocks that could generate high $\delta^{13}C$ values. Secondly, in many alkaline magmas with high Sr contents contaminated by crustal rocks, significant changes in $\delta^{18}O$ would be expected with no concomitant changes in $^{87}Sr:^{86}Sr$ ratios (Hawkesworth & Vollmer 1979, Ferrara et al. 1985). At Phalaborwa both the O and Sr isotopic compositions show marked variations.

If a least-squares analysis is applied to data from the four carbonatites, shown in Figure 10.10, and the line is extrapolated to a $^{87}Sr:^{86}Sr$ ratio similar to that of oceanic island basalts at 2050 Ma (estimated $^{87}Sr:^{86}Sr = 0.7015$), a $\delta^{18}O$ value of $+7.2‰$ is obtained for the source. This end member is more enriched in $\delta^{18}O$ than many ocean island basalts (Hawaii $\delta^{18}O = +5.4 \pm 0.4‰$), alkali basalts ($6.2 \pm 0.5‰$), mid-ocean ridge basalts ($5.7 \pm 0.5‰$), continental tholeiites (5.5 and 6.5‰) (Kyser et al. 1982) and other carbonatites (Taylor et al. 1967, Deines this volume). The other end member has a $\delta^{18}O$ of $> +8.5$, which is extremely enriched in ^{18}O compared to most volcanic rocks but only slightly more enriched than the sources proposed for some potassium-rich lavas.

10.6.3 Nd–Sr

Epsilon Sr and Nd values plotted on an anti-correlation plot chemically characterize the mantle source with respect to depletion or enrichment of LIL elements relative to bulk Earth. The Nd and Sr isotopic data from Phalaborwa (Fig. 10.11) indicate a source region enriched in the LIL elements relative to bulk Earth. A significant variation in initial $^{143}Nd:^{144}Nd$ ratios occurs (0.50961 ± 5 to 0.50977 ± 4) for three carbonatite samples. The array of isotopic data shown in Figure 10.11 has been attributed to contamination by crustal rocks, mixing of heterogeneous mantle sources, or partial melting of a layered mantle. The carbonatite Nd and Sr isotopic compositions from Phalaborwa are significantly different from those of carbonatites from the USA (Tilton et al. 1987), Canada (Bell et al. 1982), and East Africa (Bell & Blenkinsop 1987), and lie in the enriched quadrant in Figure 10.11. Potassium-rich lamproites from western Australia and Spain (Nelson et al. 1986) exhibit similar isotopic compositions, and are thought to reflect sources highly enriched in LIL elements, a feature attributed to metasomatized mantle, rather than crustal contamination (Nelson et al. 1986).

Smith (1983) proposed two distinct sources for South African kimberlites: an undepleted to slightly depleted source for Group I kimberlites, which he tentatively correlated with the asthenosphere, and a second, enriched source for Group II kimberlites. This source for Group II kimberlites is less enriched than the most extreme Phalaborwa sample. Mixing of carbonatite magmas from two isotopically

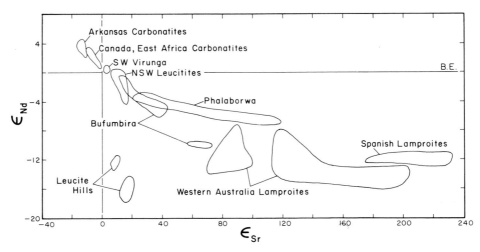

Figure 10.11 $\varepsilon_{Nd}(T)$ versus $\varepsilon_{Sr}(T)$ for a variety of alkaline rocks. Data from Hawkesworth & Vollmer 1979, Vollmer & Norry 1983, Grünenfelder et al. 1986, Nelson et al. 1986, Bell & Blenkinsop 1987. B.E., bulk Earth. Values used are $^{143}Nd:^{144}Nd = 0.51264$, $^{147}Sm:^{144}Nd = 0.1936$, $^{87}Sr:^{86}Sr = 0.7045$, $^{87}Rb:^{86}Sr = 0.0827$.

and chemically distinct sources similar to those that produced the kimberlite could generate the array of Phalaborwa Nd and Sr data.

Other evidence (Erlank et al. 1980) shows that the mantle beneath southern Africa has been heterogeneous through time and that mantle enrichment is common. Strontium isotopic ratios from South African carbonatites younger than Phalaborwa (Harmer 1985) are similar to the carbonatite data from North America (Bell et al. 1982, Grünenfelder et al. 1986, Tilton et al. 1987) and East Africa (Bell & Blenkinsop 1987) in showing slight depletion to slight enrichment in radiogenic Sr relative to bulk Earth. Phalaborwa is unique among carbonatites because its initial $^{87}Sr:^{86}Sr$ ratios are much higher and initial $^{143}Nd:^{144}Nd$ ratios are lower than the values for bulk Earth, 2050 Ma ago.

10.6.4 U–Pb

The U–Pb isotopic system tells another story. Three carbonatite samples have quite different Pb ratios ($^{206}Pb:^{204}Pb = 19.026 \pm 5$ to 34.655 ± 9; $^{207}Pb:^{204}Pb = 15.935 \pm 5$ to 17.993 ± 4) and the Pb, unlike the Sr, Nd, O, and C isotopic systems, appears to have been in isotopic equilibrium. Figure 10.12 shows a Pb:Pb secondary isochron corresponding to an age of 2050 Ma. Two analyses of one carbonatite sample (P10C) yielded different, highly radiogenic compositions. Partial dissolution showed that the dissolved material (P10C-leach) contained less radiogenic Pb than the residue, a feature that was attributed to the presence of small amounts of uranoan thorianite. The calculated initial ratio for P10C lies at the intersection of the isochron for the three carbonatites with a Pb-growth curve of $\mu = 11.0$ (Stacey & Kramers 1975). The limited data suggest a homogeneous source of common Pb possibly derived from a crustal source, and contrast to the low μ

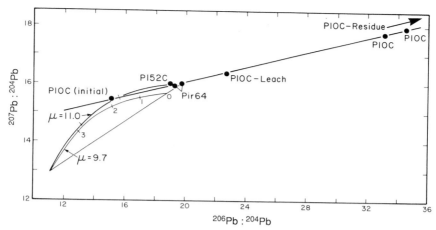

Figure 10.12 Plot of ^{207}Pb:^{204}Pb versus ^{206}Pb:^{204}Pb for three carbonatite samples and leached material and residue from one sample. The data define a secondary isochron corresponding to an age of 2050 Ma.

values for carbonatites from the Superior Province of Canada (Tilton & Grünenfelder 1983) and from Fen, Norway (Andersen & Taylor 1988).

Isotope analyses of S (Mitchell & Krouse 1975, Von Gehlen 1976) show that sulphides from Phalaborwa have mantle values of δ^{34}S of -1 to $+5$‰. The unique abundance of copper sulphides at Phalaborwa might have resulted from the intrusion of a mantle-derived liquid, rich in chalcophile elements, into a crust rich in Cu.

10.7 ORIGIN OF THE COMPLEX

10.7.1 *Conditions of crystallization*

Crystallization conditions are difficult to evaluate in the Phalaborwa Complex. Button (1976) suggested a maximum of 12 km of stratigraphic overburden in the Phalaborwa area at 2100 Ma and Eriksson (1982) estimated a lithostatic pressure of 4.5 Kb for the present level of exposure. This relatively high pressure is in keeping with the lack of metasomatism and brecciation around the main complex.

Estimates of crystallization temperature have been made by several workers. The calcite–dolomite solvus yields temperatures of 600–800 °C (Van Rensburg 1965, Verwoerd 1967, Hennig-Michaeli 1968), but these are recognized as subsolidus equilibration temperatures (Gittins 1979). Magnetite–ilmenite compositions also yield low temperatures of 400–550 °C (Aldous 1980) due to subsolidus equilibration. Partition of F$^-$ and OH$^-$ between phlogopite and apatite (Stormer & Carmichael 1971) suggests temperatures of 150–600 °C (Eriksson 1982), supporting extensive, low-temperature re-equilibration of volatiles with phlogopite.

More complicated calculations (Eriksson 1982) using the magnetite–sanidine–biotite assemblage (Wones & Eugster 1965) along with estimates of f_{O_2}

based on sulphide assemblages (Mitchell & Krouse 1975) and phlogopite compositions, suggest crystallization temperatures > 1000 °C. These temperatures are to be expected in high-potassium, ultrabasic, and carbonatitic liquids (Edgar *et al.* 1976, Wyllie & Huang 1976, Wyllie 1977, Barton & Hamilton 1979, Wendlandt & Eggler 1980).

10.7.2 Nature of the magmas

Mineral and whole rock chemistry, together with isotopic data, point to involvement of multiple magmas at the Phalaborwa Complex. Crystallization of clinopyroxenites in the main complex involved flow differentiation, *in situ* accumulation and gravity separation (Eriksson *et al.* 1985). Cross-cutting relationships and spatial variation of Mg numbers point to multiple surges of magma. The parental magma was peralkaline, ultrabasic, and probably similar to some of the potassium-rich lavas of Uganda and Wyoming. The K:Na ratio of the magma was extremely high. High Mg-numbers coupled with low Zr contents suggest that the silicate magma was undifferentiated, although low Cr concentrations suggest some differentiation. Magnetite may have removed Cr but there is no direct evidence for this.

The clinopyroxenites and the carbonatites and phoscorite can not be the result of magmatic differentiation alone, because of their different isotopic compositions. The mineral chemistry of the phlogopite also argues against immiscibility.

The relationship of the phoscorite to the carbonatite-magmatism is not well understood, although the evidence points to the phoscorite being cogenetic with carbonatite magmatism. Early crystallization and accumulation of apatite, magnetite, phlogopite, and olivine from a silicate-rich carbonatitic magma could have produced the phoscorite. Similarities in Mg number, low Ni content in olivines of phoscorite and carbonatites along with low Cr in the pyroxenes suggest similar crystallization histories for the carbonate and silicate liquids.

The syenites that surround the main complex and appear as pipes have not been studied extensively. Similarities in high K:Na ratios, low Al abundances in silicates, and enriched total REEs and LREEs suggest a similar source to the one that produced the main complex, although the chemistry suggests a greater degree of differentiation in the satellite bodies than in the main complex. The present data suggest that the syenites were not derived directly from the main complex, but may have been derived from different mantle levels by higher degrees of partial melting.

10.8 CONCLUSIONS

Aside from the very interesting, unusual petrogenetic history of the Phalaborwa Complex, the data have some broader implications for carbonatite evolution. Sulphide mineralization at Phalaborwa has been described as late-stage hydrothermal (Heinrich 1970, Palabora Mining Company 1976), implying that the copper sulphides are not related to magmatism but are due to later events. Sulphur isotopic compositions and compositions of fluid inclusions (Aldous 1980) show that copper sulphide liquid was present very early in the crystallization sequence, at least prior to olivine crystallization. Low-temperature sulphide assemblages in phoscorite and carbonatites suggest that sulphide mineralization may be an intrinsic part of carbonatite magmatism. Economic proportions of sulphides at Phalaborwa could be due to the abundance of Cu in the local crust which combined with S from a mantle source.

Metasomatism has been suggested (Borodin & Pavlenko 1974, Gittins *et al.* 1975) as one way of providing alkalis for phlogopite formation in alkaline rocks. However, ample geological (Eriksson *et al.* 1985) and mineralogical evidence at Phalaborwa supports experimental data (Barton & Hamilton 1979, Edgar 1979, Wendlandt & Eggler 1980) that phlogopite is the result of magmatic crystallization. The various clinopyroxenites are not metasomatized dunites, but crystal cumulates. The recognition of magmatic phlogopite may have implications for the origins of other mica-rich rocks associated with carbonatites. In light of the Phalaborwa data, other features, such as high Fe^{3+} in feldspars generally attributed to metasomatism, could also be attributed to a magma crystallizing under high f_{O_2} conditions.

The relationship of silicate magmas to carbonatitic magmas is problematic. At Phalaborwa, isotopic and mineralogical evidence suggests that differentiation, such as crystal fractionation or liquid immiscibility, alone cannot explain the genesis of these rocks. However, other chemical and mineralogical data point to similarities in sources, chemistry, and crystallization history for both silicate and carbonate magmas.

ACKNOWLEDGEMENTS

Thanks are extended to the Palabora Mining Company and Fosker Ltd for logistical support and access to samples, and the Department of Geology and the Bernard Price Institute of Geophysics at the University of Witwatersrand. Help by R. G. Cawthorn and H. L. Allsopp is gratefully acknowledged. R. E. Harmer, L. Hulbert, D. D. Hogarth, and Keith Bell are thanked for critical reviews of an earlier version of this chapter.

REFERENCES

Aldous, R. 1980. *Ore genesis in copper bearing carbonatites: A geochemical, mineralogical and fluid inclusion study.* Ph.D. thesis, Imperial College, University of London, London.
Aldous, R. 1986. Copper-rich fluid inclusions in pyroxenes from the Guide Copper Mine, a satellite intrusion of the Palabora Igneous Complex, South Africa. *Economic Geology* **81**, 143–55.
Allsopp, H. L. & S. C. Eriksson 1986. The Phalaborwa Complex: Isotopic evidence for ancient lithospheric enrichment. *Geological Association of Canada–Mineralogical Association of Canada–Canadian Geophysical Union, Joint Annual Meeting, Ottawa. Program with Abstracts*, 40.
Andersen, T. & P. N. Taylor 1988. Pb isotope geochemistry of the Fen carbonatite complex, S.E. Norway: Age and petrogenetic implications. *Geochimica et Cosmochimica Acta* **52**, 209–15.
Aoki, K. 1974. Phlogopite and potassic richterites from mica nodules in South Africa kimberlites. *Contributions to Mineralogy and Petrology* **48**, 1–7.

Barton, M. & D. L. Hamilton 1978. Water-saturated melting relations to 5 kb of three Leucite Hills lavas. *Contributions to Mineralogy and Petrology* **66**, 41–9.

Barton, M. & D. L. Hamilton 1979. The melting relationships of a madupite from the Leucite Hills, Wyoming, to 30 kb. *Contributions to Mineralogy and Petrology* **69**, 133–42.

Bass, I. T., V. A. Boronikhin & S. M. Kraochenko 1974. Trends in the variation of magnesium, iron, calcium and sodium concentrations in different zones of crystals of monoclinic pyroxenes and phlogopite from carbonatite complexes as criteria of their origin. *Doklady Akademia Nauk SSSR* **291**, 447–50. (in Russian)

Bell, K. & J. Blenkinsop 1987. Nd and Sr isotopic compositions of East African carbonatites: Implications for mantle heterogeneity. *Geology* **15**, 99–102.

Bell, K. & J. L. Powell 1969. Strontium isotopic studies of alkalic rocks: The potassium-rich lavas of the Birunga and Toro-Ankole regions, east and central equatorial Africa. *Journal of Petrology* **10**, 536–72.

Bell, K., J. Blenkinsop, T. J. S. Cole & D. P. Menagh 1982. Evidence from Sr isotopes for long-lived heterogeneities in the upper mantle. *Nature* **298**, 251–3.

Boettcher, A. L. 1966. Vermiculite, hydrobiotite, and biotite in the Rainy Creek igneous complex near Libby, Montana. *Clay Minerals* **6**, 283–96.

Boettcher, A. L. 1967. The Rainy Creek alkaline-ultramafic igneous complex near Libby, Montana, I: Ultramafic rocks and fenite. *Journal of Geology* **75**, 526–53.

Borodin, L. S. & A. S. Pavlenko 1974. The role of metasomatic processes in the formation of alkaline rocks. In *The alkaline rocks*, H. Sørenson (ed.), 515–34. London: Wiley.

Brandt, J. W. 1948. *Die geologie van'n gebied in Noord-oos Transvaal met spesiale verwysing na die verspreiding en petrografie van die rotssoorte van die Palabora-stollings-kompleks*. D.Sc. thesis, University of Stellenbosch, Stellenbosch.

Briden, J. C. 1976. Application of palaeomagnetism to Proterozoic tectonics. *Philosophical Transactions of the Royal Society of London* **280**, 405–16.

Button A. 1976. Stratigraphy and relations of the Bushveld floor in the eastern Transvaal. *Transactions of the Geological Society of South Africa* **79**, 3–12.

Carmichael, I. S. E. 1967. The mineralogy and petrology of the volcanic rocks from the Leucite Hills, Wyoming. *Contributions to Mineralogy and Petrology* **15**, 24–66.

Carmichael. I. S. E., F. J. Turner & J. Verhoogen 1974. *Igneous petrology*. New York: McGraw-Hill.

Cox, K. G., J. D. Bell & R. J. Pankhurst 1979. *The interpretation of igneous rocks*. London: Allen & Unwin.

Dawson, J. B. & J. V. Smith 1973. Alkalic pyroxenite xenoliths from the Lashaine Volcano, Northern Tanzania. *Journal of Petrology* **14**, 113–32.

Dawson, J. B. & J. V. Smith 1977. The MARID (mica–amphibole–rutile–ilmenite–diopside) suite of xenoliths in kimberlite. *Geochimica et Cosmochimica Acta* **41**, 309–23.

Dawson, J. B., D. G. Powell & A. M. Reid 1970. Ultrabasic lavas and xenoliths from the Lashaine volcano, Tanzania. *Journal of Petrology* **11**, 519–48.

Dawson, J. B., J. V. Smith & R. L. Hervig 1977. Late-stage diopside in kimberlite groundmass. *Neues Jahrbuch für Mineralogie Monatshefte* **2**, 529–47.

Deans, T. 1966. Economic mineralogy of African carbonatites. In *Carbonatites*, O. F. Tuttle and J. Gittins (eds), 385–416. London: Wiley.

Duke, J. M. & A. J. Naldrett 1978. A numerical model of the fractionation of olivine and molten sulfide from komatiite magma. *Earth and Planetary Science Letters* **39**, 255–66.

Edgar, A. D. 1979. Mineral chemistry and petrogenesis of an ultrapotassic-ultramafic volcanic rock. *Contributions to Mineralogy and Petrology* **71**, 171–5.

Edgar, A. D., D. H. Green & W. O. Hibberson 1976. Experimental petrology of a highly potassic magma. *Journal of Petrology* **17**, 339–56.

Eggler, D. H. & C. W. Burnham 1973. Crystallization and fractionation trends in the system andesite–H_2O–CO_2–O_2 at pressures to 10 kb. *Bulletin of the Geological Society of America* **84**, 2517–32.

Emeleus, C. H. & J. R. Andrews, 1975. Mineralogy and petrology of a kimberlite dyke and sheet intrusions and included peridotite xenoliths from southeast Greenland. *Physics and Chemistry of the Earth* **9**, 179–98.

Eriksson, S. C. 1982. *Aspects of the petrochemistry of the Phalaborwa Complex, northeastern Transvaal, South Africa*. Ph.D. thesis, University of the Witwatersrand, Johannesburg.

Eriksson, S. C. 1984. Age of carbonatite and phoscorite magmatism of the Phalaborwa complex (South Africa). *Isotope Geoscience* **2**, 291–9.

Eriksson, S. C. 1985. Oscillatory zoning in clinopyroxenes from the Guide Copper Mine, Phalaborwa, South Africa. *American Mineralogist* **70**, 74–9.

Eriksson S. C., P. J. Fourie & D. H. De Jager 1985. A cumulate origin for the minerals in clinopyroxenites of the Phalaborwa complex. *Transactions of the Geological Society of South Africa* **88**, 207–14.

Erlank, A. J., H. L. Allsopp, A. R. Duncan & J. W. Bristow 1980. Mantle heterogeneity beneath southern Africa: evidence from the volcanic record. *Philosophical Transactions of the Royal Society of London* **A297**, 295–307.

Ferrara G., M. A. Laurenzi, H. P. Taylor, Jr., S. Tonarini & B. Turi 1985. Oxygen and strontium isotope studies of K-rich volcanic rocks from the Alban Hills, Italy. *Earth and Planetary Science Letters* **75**, 13–28.

Fourie, P. J. 1981. *Die piroksenitiese gesteentes van die Phalaborwa-kompleks met verwysing na die verspreiding van fosfaat*. M.Sc. thesis, Randse Afrikaanse Universitaat, Johannesburg.

Frick, C. 1975. The Phalaborwa syenite intrusion. *Transactions of the Geological Society of South Africa* **78**, 201–14.

Gehlen, K. Von 1976. Sulphur isotopes from the sulphide-bearing carbonatite of Palabora, South Africa. *Transactions of the Institute of Mineralogy and Metallurgy* **76**, B223. (abstract).

Gevers, T. W. 1948. Vermiculite at Loolekop, Palabora, northeastern Transvaal. *Transactions of the Geological Society of South Africa* **51**, 133–78.

Gibb, F. G. F. 1973. The zoned clinopyroxenes on the Shiant Isles Sill, Scotland. *Journal of Petrology* **14**, 201–30.

Gittins, J. 1979. Problems inherent in the application of calcite–dolomite geothermometry to carbonatites. *Contributions to Mineralogy and Petrology* **69**, 1–4.

Gittins, J., H. R. Hewins & A. F. Laurin 1975. Kimberlitic-carbonatitic dikes of the Saguenay River Valley, Quebec, Canada. *Physics and Chemistry of the Earth* **9**, 137–48.

Grünenfelder, M. H., G. R. Tilton, K. Bell & J. Blenkinsop 1986. Lead and strontium isotope relationships in the Oka carbonatite complex, Quebec. *Geochimica et Cosmochimica Acta* **50**, 461–8.

Haggerty, S. E. 1976. Opaque mineral oxides in terrestrial igneous rocks. In *Oxide minerals*, Mineralogical Society of America Short Course Notes 3, 101–75.

Hall, A. L. 1912*a*. The Palabora plutonic complex of the low country and its relationship to the pegmatites of the Leydsdorp mica-fields. *Transactions of the Geological Society of South Africa* **15**, 4–17.

Hall, A. L. 1912*b*. The crystalline metamorphic limestone of Lulukop and its relationship to the Palabora Plutonic Complex. *Transactions of the Geological Society of South Africa* **15**, 18–25.

Hanekom, H. J., C. M. van Staden, P. J. Smit & D. R. Pike 1965. *The geology of the Palabora Igneous Complex*. South Africa Geological Survey Handbook, Memoir 54.

Harmer, R. E. 1985. A Sr isotope study of Transvaal carbonatites. *Transactions of the Geological Society of South Africa* **88**, 471–2.

Hart, S. R. & K. E. Davis 1978, Nickel partitioning between olivine and silicate melt. *Earth and Planetary Science Letters* **40**, 203–19.

Hawkesworth, C. J. & R. Vollmer 1979. Crustal contamination versus enriched mantle: ^{143}Nd/^{144}Nd and ^{87}Sr/^{86}Sr evidence from the Italian volcanics. *Contributions to Mineralogy and Petrology* **69**, 151–65.

Heinrich, E. W. 1970. The Palabora carbonatitic complex – a unique copper deposit. *Canadian Mineralogist* **10**, 585–98.

Heinrich, E. W. & D. G. Moore 1970. Metasomatic potash feldspar rocks associated with igneous alkalic complexes. *Canadian Mineralogist* **10**, 571–84.

Hennig-Michaeli, C. 1968. *Mikroskopische untersuchungen an Hanstücken der Kupfererzlagerstätte am Loolekop im Phalaborwa Komplex, NE-Transvaal, Sudafrikanische Union*. Mineralogische Diplomarbeit Institut für Mineralogie und Lagerstättenlehre der Rheinisch-Westfälische Technische Hochschule, Aachen.

Hiemstra, S. A. 1955. Baddeleyite from Phalaborwa, Eastern Transvaal. *American Mineralogist* **40**, 275–82.

Holmes, A. & L. Cahen 1956. *Geochronologie Africaine*. Académie royale de Sciences coloniales, Mem-8′, TV, fosc. I.

Irving, A. J. 1978. A review of experimental studies of crystal/liquid trace element partitioning. *Geochimica et Cosmochimica Acta* **42**, 743–70.

Koster van Groos, A. F. 1975. The distribution of strontium between coexisting silicate and carbonate liquids at elevated pressures and temperatures. *Geochimica et Cosmochimica Acta* **39**, 27–34.

Kyser, T. K., J. R. O'Neil & I. S. E. Carmichael 1982. Genetic relations among basic lavas and ultramafic nodules: Evidence from oxygen isotope compositions. *Contributions to Mineralogy and Petrology* **81**, 88–102.

Larsen, L. M. 1976. Clinopyroxenes and coexisting mafic minerals from the alkaline Ilimaussaq intrusion, South Greenland. *Journal of Petrology* **17**, 258–90.

Lindstrom, D. J. 1976. *Experimental study of the partitioning of the transition metals between clinopyroxene and coexisting silicate liquids*. Ph.D. thesis, University of Oregon, Eugene.

Lombaard, A. F., N. M. Ward-Able & R. W. Bruce 1964. The exploration and main geological features of the copper deposit in carbonatite at Loolekop, Palabora Complex. In *The geology of some ore deposits in southern Africa*, S. H. Haughton (ed.), 315–37. Johannesburg: Geological Society of South Africa.

Mariano, A. N. 1978. The application of cathodoluminescence for carbonatite exploration and characterization. In *Proceedings of the First International Symposium on Carbonatites, Poços de Caldas, Brazil*, June 1976, 39–60. Brasilia: Brasil Departamento Nacional da Produção Mineral.

Mellor, E. J. 1906. *The geology of the district about Haenertsburg, Leydsdorp, and the Murchison Range*. Annual Report of the Geological Survey, 21–52.

Mitchell, R. H. & H. R. Krouse 1975. Sulfur isotope geochemistry of carbonatites. *Geochimica et Cosmochimica Acta* **39**, 1505–15.

Modreski, P. J. & A. L. Boettcher 1973. Phase relationships of phlogopite in the system K_2O–MgO–CaO–Al_2O_3–SiO_2–H_2O to 35 kilobars: A better model for micas in the interior of the earth. *American Journal of Science* **273**, 385–414.

Morgan, G. E. & J. C. Briden 1981. Aspects of Precambrian palaeomagnetism, with new data from the Limpopo mobile belt and Kaapvaal craton in southern Africa. *Physics of the Earth and Planetary Interiors* **24**, 1442–68.

Nash, W. P. 1972. Mineralogy and petrology of the Iron Hill carbonatite complex, Colorado. *Bulletin of the Geological Society of America* **83**, 1361–82.

Nash, W. P. & J. F. G. Wilkinson 1970. Shonkin Sag Laccolith, Montana, pt. I, Mafic minerals and estimates of temperature, pressure, oxygen fugacity and silica activity. *Contributions to Mineralogy and Petrology* **25**, 241–69.

Nelson, D. R., M. T. McCulloch & S. Sun 1986. The origins of ultrapotassic rocks as inferred from Sr, Nd and Pb isotopes. *Geochimica et Cosmochimica Acta* **50**, 231–45.

Nelson, D. R., A. R. Chivas, B. W. Chappell & M. T. McCulloch 1988. Geochemical and isotopic systematics in carbonatites and implications for the evolution of ocean-island sources. *Geochimica et Cosmochimica Acta* **52**, 1–17.

Palabora Mining Company Limited, Mine Geological and Mineralogical Staff 1976. The geology and the economic deposits of copper, iron, and vermiculite in the Palabora Igneous Complex: A brief review. *Economic Geology* **71**, 177–92.

Pearce, J. A. & M. J. Norry 1979. Petrogenetic implications of Ti, Zr, Y, and Nb variations in volcanic rocks. *Contributions to Mineralogy and Petrology* **69**, 33–47.

Presnall, D. R. 1966. The join forsterite–diopside–iron oxide and its bearing on the crystallization of basaltic and ultramafic magmas. *American Journal of Science* **264**, 753–809.

Rajamani, V. & A. J. Naldrett 1978. Partitioning of Fe, Co, Ni, and Cu between sulfide liquid and basaltic melts and the composition of Ni–Cu sulfide deposits. *Economic Geology* **73**, 82–93.

Rankin, A. H. 1975. Fluid inclusion studies in apatite from carbonatites of the Wasaki area of Western Kenya. *Lithos* **8**, 123–36.

Rankin, A. H. 1977. Fluid inclusion evidence for the formation conditions of apatite from the Tororo carbonatite complex of Eastern Uganda. *Mineralogical Magazine* **41**, 155–65.

Rensburg, W. C. Van 1965. *Copper mineralization in the carbonate members and phoscorite, Phalaborwa, South Africa*. Ph.D. thesis, University of Wisconsin, Madison.

Rimsaite, J. 1969. Evolution of zoned micas and associated silicates in the Oka carbonatite. *Contributions to Mineralogy and Petrology* **23**, 340–60.

Russell, H. D., S. A. Hiemstra & D. Groeneveld 1954. The mineralogy and petrology of the carbonatite at Loolekop, Eastern Transvaal. *Transactions of the Geological Society of South Africa* **57**, 197–208.

Schreiber, H. D. 1976. *The experimental determination of redox states, properties, and distribution of chromium in synthetic silicate phases and application to basalt petrogenesis.* Ph.D. thesis, University of Wisconsin, Madison.

Shand, S. J. 1931. The granite–syenite–limestone complex of Palabora, eastern Transvaal, and the associated apatite deposits. *Transactions of the Geological Society of South Africa* **34**, 81–105.

Shimizu, N. 1975. Geochemistry of ultramafic inclusions from Salt Lake Crater, Hawaii. *Physics and Chemistry of the Earth* **9**, 655–69.

Smith, C. B. 1983. Pb, Sr, and Nd isotopic evidence for sources of southern African Cretaceous kimberlites. *Nature* **304**, 51–3.

Smith, C. & D. H. Lindsley 1971. Chemical variations in pyroxene and olivine from Picture Gorge basalt. *Carnegie Institute of Washington Yearbook* **69**, 269–74.

Smith J. V., R. Brennesholtz & J. B. Dawson 1978. Chemistry of micas from kimberlites and xenoliths – I. Micaceous kimberlites. *Geochimica et Cosmochimica Acta* **42**, 959–71.

Stacey, J. S. & J. D. Kramers 1975. Approximation of terrestrial lead isotope evolution by a two-stage model. *Earth and Planetary Science Letters* **26**, 207–21.

Steiger, R. H. & E. Jäger 1977. Subcommission of geochronology: convention on the use of decay constants in geo- and cosmochronology. *Earth and Planetary Science Letters* **36**, 359–62.

Stormer, J. C. & I. S. E. Carmichael 1971. Fluorine-hydroxyl exchange in apatite and biotite: A potential igneous geothermometer. *Contributions to Mineralogy and Petrology* **31**, 121–31.

Strauss, C. A. & F. A. Truter 1950. The alkali complex at Spitzkop, Sekukuniland, eastern Transvaal. *Transactions of the Geological Society of South Africa* **53**, 81–125.

Sun, C. C., R. J. Williams & S. S. Sun 1974. Distribution coefficients of Eu and Sr for plagioclase-liquid and clinopyroxene-liquid equilibria in oceanic ridge basalt: an experimental study. *Geochimica et Cosmochimica Acta* **38**, 1415–33.

Suwa, K. & K. Aoki 1975. *Reverse pleochroism of phlogopites in kimberlites and their related rocks from South Africa.* 1st Preliminary Report of African Studies, Nagoya University, 60–4.

Taljaard, M. S. 1936. South African melilite basalts and their relations. *Transactions of the Geological Society of South Africa* **39**, 281–316.

Taylor, H. P., Jr. 1980. The effects of assimilation of country rocks by magmas on $^{18}O/^{16}O$ and $^{87}Sr/^{86}Sr$ systematics in igneous rocks. *Earth and Planetary Science Letters* **47**, 243–54.

Taylor, H. P. Jr., G. Bernardino & B. Turi 1979. Oxygen isotope geochemistry of the potassic igneous rocks from the Roccamonfina volcano, Roman comagmatic region, Italy. *Earth and Planetary Science Letters* **46**, 81–106.

Taylor, H. P., Jr., J. Frechen & E. T. Degens 1967. Oxygen and carbon isotope studies of carbonatites from the Laacher See District, West Germany and the Alnö District, Sweden. *Geochimica et Cosmochimica Acta* **31**, 407–30.

Taylor, H. P., Jr., B. Turi & A. Cundari 1984. $^{18}O/^{16}O$ and chemical relationships in K-rich volcanic rocks from Australia, East Africa, Antarctica, and San Venanzo-Cupaello, Italy. *Earth and Planetary Science Letters* **69**, 263–76.

Tilton, G. R. & M. J. Grünenfelder 1983. Lead isotope relationships in billion-year-old carbonatite complexes, Superior Province, Canadian Shield. *Geological Society of America, Abstracts with Programs* **15**, 707.

Tilton, G. R., S. T. Kwon & D. M. Frost 1987. Isotopic relationships in Arkansas Cretaceous alkalic complexes. In *Mantle metasomatism and alkaline magmatism*, E. M. Morris & J. D. Pasteris (eds), Geological Society of America Special Paper 215, 241–8.

Tyler, R. C. & B. C. King 1967. The pyroxenes of the alkaline igneous complexes of Eastern Uganda. *Mineralogy Magazine* **280**, 5–22.

Verwoerd, W. J. 1967. *The carbonatites of South Africa and South West Africa.* South Africa Geological Survey Handbook 6.

Vollmer R. & M. J. Norry 1983. Possible origin of K-rich volcanic rocks from Virunga, East Africa, by metasomatism of continental crustal material: Pb, Nd and Sr isotopic evidence. *Earth and Planetary Science Letters* **64**, 374–86.

Watkinson, D. H. 1970. Experimental studies bearing on the origin of the alkali rock-carbonatite complex and niobium mineralization at Oka, Quebec. *Canadian Mineralogist* **10**, 350–61.

Wendlandt, R. F. & D. H. Eggler 1980. The origins of potassic magmas: 1. Melting relations in the systems $KAlSiO_4$–$MgSiO_2$ and $KAlSiO_4$–MgO–SiO_2–CO_2 to 30 kb. *American Journal of Science* **280**, 385–420.

Wilkinson, J. F. G. 1966. Clinopyroxenes from the Square Top intrusion, Nundle, New South Wales. *Mineralogical Magazine* **35**, 1061–70.

Wones, D. R. & H. P. Eugster 1965. Stability of biotite: Experiment, theory, and application. *American Mineralogist* **50**, 1228–72.

Wyllie, P. J. 1977. Peridotite–CO_2–H_2O and carbonatitic liquids in the upper asthenosphere. *Nature* **266**, 45–57.

Wyllie, P. J. & W. L. Huang 1976. Carbonation and melting reactions in the system CaO–MgO–SiO_2–CO_2 at mantle pressures with geophysical and petrological applications. *Contributions to Mineralogy and Petrology* **54**, 79–107.

11
SODIUM CARBONATITE EXTRUSIONS FROM OLDOINYO LENGAI, TANZANIA: IMPLICATIONS FOR CARBONATITE COMPLEX GENESIS

J. B. DAWSON

ABSTRACT

Oldoinyo Lengai is the only active carbonatite volcano. The modern natrocarbonatite lava and ash eruptions follow earlier phases of phonolitic and nephelinitic eruptions. Plutonic blocks within the tuffs and agglomerates indicate the presence of a peralkaline igneous-metasomatic complex beneath the volcano. Neodymium and Sr isotope evidence suggests that the natrocarbonatite lavas are genetically linked with the peralkaline silicate rocks, perhaps having separated from them by liquid immiscibility. The existence of alkali carbonate liquids explains the source of the alkalis in cases of fenitization around carbonatite intrusions when peralkaline silicate rocks are absent. The modern natrocarbonatite lavas and ashes are degrading to alkali-poor carbonate rocks. Former natrocarbonatite tuffs, widely distributed around the volcano, have degraded to form 'calcrete' deposits.

11.1 INTRODUCTION

The African rift valleys, cutting across the African Plate, are the sites of some of the classical carbonatite complexes. The spectacular cone of Oldoinyo Lengai, the world's only active carbonatite volcano, rises to nearly 3000 m above Tanzania's Eastern Rift Valley, 15 km south of the southern end of Lake Natron. The volcano owes its place in carbonatite petrology mainly due to the extrusions of high-alkali carbonatite lavas first recognized as such in 1960 (Dawson 1962a, b); the proven existence of high-alkali carbonate liquids has helped to clarify some aspects of carbonatite-complex petrology that had previously been enigmatic. In the quarter century that has passed since the first major description of the volcano (Dawson 1962b), numerous studies have been carried out on the silicate rocks comprising the bulk of the volcano (both lavas and ejected plutonic blocks) and the carbonatite lavas. It is the purpose of this chapter to review these contributions.

11.2 REGIONAL GEOLOGICAL SETTING

The Neogene volcanic province of northern Tanzania, standing astride the Gregory Rift Valley, is a southerly extension of the more extensive volcanic areas of Kenya. The earlier major shield volcanoes, mainly of the continental alkali basalt–phonolite association, were extruded within a major, broad, southerly bifurcating tectonic depression that resulted from up-doming and fracturing of the continental crust in the Middle Tertiary; major volcanoes belonging to this earlier episode include those of the Crater Highlands (including Ngorongoro), Kitumbeine, Gelai (Fig. 11.1) and the Mawenzi and Shira centres of Kilimanjaro.

At approximately 1.2 Ma BP, the volcanic province was subjected to a further major phase of faulting, which produced a major north–south fault that forms the western boundary fault of the present-day Rift Valley in northern Tanzania. Unlike the classical rift valley in Kenya to the north, there is no major fault on the eastern side of the structure in northern Tanzania; its place is taken by a major flexure with associated minor faults. Major inland drainage basins, including those now occupied by Lakes Natron and Manyara, resulted from the faulting, and the sedimentation in previously established sedimentation basins was affected; one such earlier basin is the Olduvai Basin, lying to the west of the main volcanic area, which

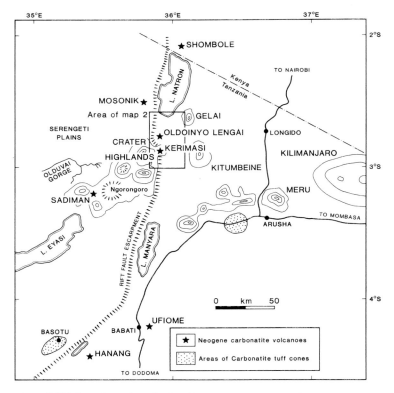

Figure 11.1 Location of Oldoinyo Lengai, and other Neogene carbonatite volcanoes and carbonatite, tuff-cone areas in northern Tanzania.

is not only justifiably famous for its hominid and other vertebrate remains, but also for its record of both alluvial and wind-borne volcanic detritus derived from volcanoes to the east, including Oldoinyo Lengai (Hay 1976). Following this phase of faulting, further volcanoes erupted. However, unlike the earlier phase of volcanic activity, dominated by quiet extrusions of alkali basalts, the products of this second phase of activity were small in volume compared with the earlier basaltic activity, and were dominated by highly explosive nephelinite–phonolite–carbonatite rocks which have given rise to steep, dominantly pyroclastic cones. These volcanoes include Meru and, in the Rift Valley, the carbonatite volcanoes of Hanang, Ufiome, Kerimasi, and Oldoinyo Lengai. Also, in the Basotu and Arusha areas, numerous minor cones have ejected carbonatite ash. These ashes, together with ash from the other bigger centres such as Kerimasi and Oldoinyo Lengai, have covered large parts of the volcanic province, including the Serengeti Plains to the west. Previously interpreted as calcretes, their recognition as carbonatite ashes has been appreciated only recently (Dawson 1964a, b, Hay & Reeder 1978).

Oldoinyo Lengai, and its slightly older twin volcano Kerimasi, erupted close to the

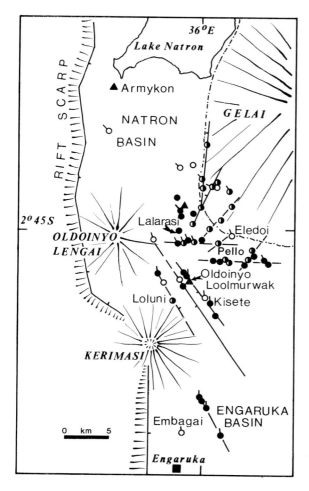

Figure 11.2 The Oldoinyo Lengai–Kerimasi area, showing localities mentioned in the text. Also shown are minor faults, small lava extrusions (▲), explosion craters (○), tuff cones (●), and cratered tuff cones (◐); the tick shows the side of maximum pyroclast deposition on asymmetric tuff rings.

Rift Escarpment some 15–25 km south of Lake Natron (Fig. 11.2). Following the last major movement on this fault at 1.2 Ma (MacIntyre *et al.* 1974), Kerimasi erupted at about 0.4–0.6 Ma (Hay 1976), first erupting nephelinitic–ijolitic fragmental rocks, followed by carbonatite blocks and ashes that have consolidated to form a carbonate carapace over the earlier silicate pile (Dawson 1964*a*). The tephra consist of both sövite and former natrocarbonatite, now replaced by calcite (Hay 1983), and it is apparent that the recent natrocarbonatite ejecta from Oldoinyo Lengai, currently degrading to alkali-poor, residual calcite deposits, and their earlier analogues at Kerimasi. Subsequent explosive eruptions associated with minor faulting have produced many minor tuff cones and explosion craters (Fig. 11.2). Minor flows associated with this phase of activity are of olivine melilitite at Armykon Hill and Lalarasi, and olivine-melilite nephelinite at Oldoinyo Loolmurwak (Dawson & Powell 1969); the chemistry of the Loolmurwak nephelinite flow indicates that it is a primitive mantle melt (Dawson *et al.* 1985), contrasting with the highly evolved nephelinites extruded from Oldoinyo Lengai. Metasomatized peridotite xenoliths occur in the scoria at Pello Hill and Eledoi Crater to the east of Lengai (Dawson & Smith 1988). Tuffs containing large crystals of kaersutite, augite and phlogopite surround the explosion centres at Kisete, Loluni, and Loolmurwak, and another mica-rich tuff cone protrudes through blanketing Lengai tuffs and agglomerates on the lower eastern slopes of Lengai. Phlogopite from the Kisete tuff ring to the south of Lengai has been dated at 0.37 Ma (MacIntyre *et al.* 1974). The fact that all those minor volcanic features are covered by yellow Lengai tuff gives an upper date (post 0.37 Ma) for the onset of the Lengai activity, and early yellow Lengai tuffs have been correlated with the Naisiusiu Beds of the Olduvai Gorge succession which have been dated at 22 000–15 000 y BP (Hay 1976).

11.3 OLDOINYO LENGAI, GENERAL GEOLOGY

Oldoinyo Lengai stands close to the Rift escarpment, against which ejecta from the volcano are piled (Fig. 11.2). The volcano is a steep, almost perfect, dominantly pyroclastic cone, standing 2000 m above the floor of the Rift Valley. It has a basal diameter of approximately 12 km, and an approximate volume of 60 km^3. The cone is cut by numerous, deep, radial, erosion gullies, and there is also a deep landslip scar on the upper eastern slopes, resulting from a major landslide that has deposited extensive lahar deposits on the plains to the east and north-east of the volcano. The summit comprises two craters – the older southern crater, now infilled with sparse vegetation and wind-blown grey ash, which is separated from the northern, active crater by a transverse E–W summit ridge.

The stratigraphy of Oldoinyo Lengai comprises six units, from oldest to youngest, as follows:

(1) Yellow palagonitized tuffs and agglomerates of phonolitic and nephelinitic composition, with rare interbedded nephelinite and phonolite lavas. This unit,

comprising >90% of the volume of the volcano was erupted from the now-extinct southern crater.

(2) Mica and pyroxene tuffs, forming parasitic cones and craters on the lower western and eastern slopes.
(3) Black nephelinitic tuffs that are thickest (up to 200 m) on the western and northern slopes.
(4) Minor flows of melanephelinite, in the summit area and from parasitic cones on the lower northern slopes.
(5) Variegated carbonatite ashes in the summit area and the active crater, but also on the western and northern slopes.
(6) Modern (post-1950) natrocarbonatite lavas and carbonatite–silicate ashes.

With the exception of some minor flows of unit 4 that result from a minor fissure eruption on the lower northern slopes, the rocks of units 3–6 were erupted from the currently active northern crater. This crater has been continuously modified since it was first observed in 1913 to be a shallow tuff-filled depression (Reck 1914). The volcano was next climbed in 1915 by Schulze (Reck & Schulze 1921) who reported that the crater as observed by Reck had been modified by flows of 'soda-mud'. With the benefit of hindsight it appears that this was the first unwitting observation of the now-famous natrocarbonatite lavas. During the 1917 eruption (Richard 1942) the

Figure 11.3 The northern crater of Oldoinyo Lengai, June 1960. Black recent spatter surrounds the vents and the crater floor is covered with older, whitened natrocarbonatite flows. The height of the crater wall is approximately 150 m and the crater floor is approximately 300 m wide. (Photograph by C. M. Bristow.)

Figure 11.4 New, blocky, natrocarbonatite flows erupted from the vent during 8–9 October 1960. By contrast, the slightly older whitened flows (erupted between 23 September and 8 October 1960) were highly mobile. (Photograph by J. B. Dawson.)

shallow northern crater collapsed to form a 150 m-deep steep-sided crater (Fig. 11.3) that existed until the 1966 eruption when it was mainly infilled by ash (Dawson *et al.* 1968).

Carbonatite eruptions took place sporadically throughout the following four decades (reviewed by Dawson 1962*b*); most were inadequately recorded. The 1960 eruption resulted in the extrusion of numerous flows of natrocarbonatite lava within the deep, northern crater (Figs 11.3 & 11.4).

11.3.1 *Post-1960 activity*

Following the extrusions of lava in 1960 and 1961, further flows took place sporadically until August 1966 when the volcano was subject to several months of sporadic, violent ash eruptions (Dawson *et al.* 1968). During the August eruption the former deep, steep-sided crater was infilled by an ash cone (Fig. 11.5). The ash

Figure 11.5 The active crater following the 1966 ash eruption. The former deep crater is filled in by the new ash cone. Note the abundant whitened ash on the outer slopes. The crater, from rim to rim, is 400 m wide. (Photography by J. B. Dawson.)

covered most of the volcano and drifted as far as the soda works at Lake Magadi, 110 km to the north, and to Seronera Lodge on the Serengeti Plains 120 km to the west. Ash sampled in the summit area was found to contain fragmental natrocarbonatite together with crystals of schorlomite, nepheline, wollastonite, aegirine, and pyrrhotite. The garnet and nepheline crystals are euhedral, wollastonite occurs as cleavage splinters, and the pyroxene and pyrrhotite are anhedral. Ejected blocks comprised wollastonite- and schorlomite-bearing ijolite and melteigite (Dawson *et al.* 1968). A recent study of the ash (Dawson *et al.* in press) has shown pyroxene and wollastonite clasts to be replaced by combeite-melilite rims, and melilite and euhedral combeite to be important constituents in the < 100 μm size fraction of the ash.

By December 1966, the activity had ceased. During an overflight of the volcano by J. B. Dawson in November 1969, it was observed that the 1966 ash cone had collapsed to form a deep pit-crater; this was virtually unchanged when the volcano was visited by M. S. Garson and J. B. Dawson in October, 1981.

New activity began in early January 1983, with ash eruptions. The activity changed to extrusion of lavas by April 1983, and lava extrusions have persisted until November 1988 (Nyamweru, 1988; SEAN 1988; J. B. Dawson, pers. obs., 1988). The pit crater formed after the 1966 ash eruption has been steadily infilled by lava. The activity in late November 1988 took the form of extrusions of clinkery and

Figure 11.6 View of the active crater of Oldoinyo Lengai, 23 November 1988 looking north. The most prominent feature is the breached cone containing the dark lava pool from which the dark lava flows in the bottom left of the picture were extruded on 21/23 November. The northern part of the crater floor is covered with older flows from the smaller, less active cones. The crater is much shallower than in 1960 (see Figure 11.3), the maximum height of the northern crater wall is 45 metres.

pahoehoe lava from minor hornitos and larger lava pools. The crater is now considerably shallower (Fig. 11.6) and easier of access than in 1960. Analyses of the November 1988 lavas show them to be natrocarbonatites compositionally similar to those extruded in 1960.

11.3.2 The silicate lavas

A new study of the silicate lavas (Donaldson *et al.* 1987) has supplemented the earlier accounts by Dawson (1962b, 1966); the new investigation covers lava flows from units 1 and 4 and lava blocks in tuffs of units 1 and 3. The lavas are a suite of peralkaline olivine-free nephelinites and phonolites, that display regular changes in their mineral chemistry and major- and trace-element whole-rock chemistry throughout the suite. Phenocrysts consist mainly of nepheline and subordinate barium-bearing sanidine (average $Or_{45.5}$ $Ab_{52.5}$ $Ce_{2.0}$), clinopyroxene (commonly showing strong oscillatory zoning between salite and aegirine), garnet (both high-Ti schorlomite and low-Ti melanite), vishnevite, and cancrinite. Wollastonite is present as both xenocrysts and primary microlites. Microphenocrysts are of nepheline, aegirine, titanomagnetite, apatite and, in the phonolites, alkali feldspar and titanite.

In one specimen are globular patches (segregation vesicles?) mainly infilled by a mineral with approximate stoichiometry $Na_2CaSi_2O_6$, which is suggestive of combeite; it occurs together with microcrystals of loparite, and an unidentified KNaBaSrCaTi silicate. Another NaCa silicate has been found in a further Lengai specimen by Peterson & Marsh (1986) who suggest it to be combeite, $(Na_4(Ca, Al, Fe)_3 Si_6O_{16}(O, OH, F)_2)$, a phase first recognized in nephelinite from Mt. Shaheru, Zaire (Sahama & Hytonen 1957). Residual glasses in the lavas are mainly devitrified. Although variable in composition within the same specimen, the glass compositions are those of peralkaline phonolites, containing appreciable sulphate (0.28–0.44 wt%), chlorine (0.14–0.22 wt%), and (inferred) H_2O or CO_2. Vesicles are filled with analcite, apophyllite, carbonate, and zeolites.

The lava suite displays regular changes in major- and trace-element bulk-rock composition from nephelinite to phonolite. Despite this regular change, the bulk-rock compositions cannot be related to each other by computations modelling fractionation of the phenocrysts, and it is suspected that additional processes, including palagonitization and interaction of magmas with wall rocks, have changed the rock compositions. Overall, the suite is silica-undersaturated (SiO_2 44–53 wt%) and is peralkaline [$(Na + K):Al = 1.1–2.1$]. Although Na_2O is the dominant alkali (7.2–13.5 wt%), the lavas are also rich in K_2O (4.6–5.8 wt%) (Table 11.1). The other major oxide feature of note is the low MgO content which is generally < 1 wt%, and achieves 2 wt% in only one nephelinite specimen. Trace element characteristics are low Cr and Ni (< 10 and < 15 p.p.m. respectively) contrasting with high Zn (130–320 p.p.m.), Rb (96–155 p.p.m.), Sr (1400–3400 p.p.m.), Ba (1400–2600 p.p.m.), Zr (200–850 p.p.m.), Pb (17–51 p.p.m.), and Li (17–65 p.p.m.); both Cl and F show wide variations but can be in high concentrations (up to 7800 and 2180 p.p.m. respectively). All samples, whether nephelinites or phonolites, have very similar chondrite-normalized rare earth patterns, characterized by substantial LREE enrichment, with a steep negative slope between La and Dy (Fig. 11.7). Values relative to chondrite are La $\times 500$, Yb $\times 10$. The shape of the REE patterns is very similar to those for melilitite and nephelinite from minor flows on the floor of the Rift Valley to the east of Oldoinyo Lengai (Fig. 11.7). Only Al_2O_3 shows positive correlation with differentiation index; negative correlation is the case for CaO, FeO, MnO, MgO, P_2O_5, and SO_3 and Cu, Zn, and V. Other oxides (including Na_2O and K_2O) and trace elements (including Zr and Rb) show a random scatter on variation diagrams.

Comparison of the Lengai suite with other peralkaline suites in western Kenya and eastern Uganda (Le Bas 1977, 1978 a, b) shows it to be poorer in MgO for a given CaO content, and to have a lower MgO:(MgO + total Fe) ratio for a given Al_2O_3 content. The low MgO values of the nephelinites combined with their low Ni concentrations (< 15 p.p.m.) and high concentrations of incompatible elements together indicate they are highly evolved rocks rather than primitive, mantle-derived liquids. These low MgO and Ni values (< 2 wt% and < 15 p.p.m.) contrast strongly with those in the small, contemporaneous extrusives of olivine nephelinite and olivine melilitite that occur at Lalarasi and Oldoinyo Loolmurwak in the Rift Valley to the east and south-east of Oldoinyo Lengai (Dawson et al. 1985).

Table 11.1 Major- and trace-element contents and $^{87}Sr:^{86}Sr$ ratios of selected Oldoinyo Lengai silicate lavas, and a coeval olivine nephelinite from Oldoinyo Loolmurwak.

	1	2	3	4	5	6	7
Wt%							
SiO_2	43.07	44.13	46.47	47.94	49.36	52.92	37.52
TiO_2	0.99	1.05	0.99	0.95	0.77	0.90	3.91
Al_2O_3	13.96	13.05	17.05	19.19	18.22	19.83	6.93
Fe_2O_3	5.80	5.59	5.93	5.82	4.10	3.66	6.89
FeO	2.63	3.83	0.85	0.82	0.74	1.78	7.12
MnO	0.34	0.38	0.22	0.18	0.18	0.17	0.23
MgO	1.01	0.80	0.86	0.46	0.53	0.66	15.07
CaO	11.08	7.65	6.96	3.53	4.43	2.88	12.97
Na_2O	9.70	13.53	9.29	10.01	8.54	10.21	2.52
K_2O	4.90	5.43	5.17	5.86	5.59	4.65	1.58
P_2O_5	0.66	0.54	0.27	0.12	0.35	0.18	0.92
SO_3	0.24	0.92	0.14	0.25	0.03	0.05	0
H_2O^+	1.20	1.01	2.85	2.91	4.19	1.48	1.32
CO_2	2.95	0.19	2.15	1.65	2.64	0.09	1.71
Total	98.53	98.10	99.20	99.69	99.67	99.46	98.69
Trace elements (p.p.m.)							
V	211	220	130	144	76	77	276
Cr	5	5	5	5	5	5	698
Ni	14	13	10	9	11	9	356
Cu	47	46	17	18	12	14	114
Zn	274	322	168	181	177	160	117
Rb	119	134	116	140	158	110	48
Sr	2288	2735	2424	2187	2354	1438	845
Y	48	31	40	32	36	28	28
Zr	869	852	516	484	579	640	326
Nb	324	330	140	143	169	183	97
Ba	1720	2634	1605	1575	2085	1448	617
Pb	43	43	26	32	48	34	8
Li	29	46	36	32	53	22	n.a.
Mo	0	7	1	0	0	0	n.a.
Cl	1600	7800	1450	1750	950	1850	n.a.
F	1740	2180	1880	1940	1260	740	n.a.
U	10	9	5	4	7	7	n.a.
Th	20	10	n.d.	31	29	32	n.a.
$^{87}Sr:^{86}Sr$	n.a.	n.a.	0.70507	n.a.	0.70458	0.70461	n.a.

1 (BD 119) and 2 (BD 70), nephelinites; 3 (BD 64) and 4 (BD 49), phonolitic nephelinites; 5 (BD 67) and 6 (BD 74), phonolites; 7 (BD 105), olivine–melilite–nephelinite, Oldoinyo Loolmurwak. n.a., Not analysed; n.d., not detected.
   Major and trace element data for Oldoinyo Lengai rocks (except for U and Th) from Donaldson et al. (1987). U and Th from Dawson & Gale (1970). Sr isotope data from Bell and Dawson (unpublished data). Data in Column 7 from Dawson et al. (1985).

11.3.3 Plutonic blocks

The agglomerates contain numerous blocks of a wide variety of peralkaline and, mainly, silica-undersaturated rock types that together suggest the presence of a lithologically-variable igneous-metasomatic complex beneath the volcano.

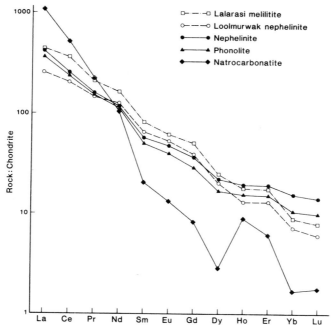

Figure 11.7 Chondrite-normalized REE patterns for natrocarbonatite and representative phonolite and nephelinite from Oldoinyo Lengai. Also shown are patterns for a primitive olivine nephelinite flow from Oldoinyo Loolmurwak and a melilitite from Lalarasi (localities on Figure 11.2).

The blocks belong to two groups, (a) the magmatic series and (b) metasomites. Rocks of both groups are found in units 1 and 3, whereas only blocks of the magmatic series were found as ejecta during the 1966 eruption. The following is a precis of a continuing study on the plutonic blocks in collaboration with J. V. Smith of the University of Chicago.

The magmatic series blocks This group consists of jacupirangite, perovskite pyroxenite, melteigite, ijolite, eucolite nepheline syenite, nepheline syenite, nepheline wollastonite rock, and sövite. Cumulate textures are common in the jacupirangite and pyroxenites, and some wollastonite–nepheline and nepheline–syenite blocks, apparently originating by fragmentation of veins, exhibit comb layering. Many blocks are vesicular and, particularly amongst the ijolites, vesiculated glass is a common component both interstitially and lining gas cavities. The most abundant minerals are clinopyroxene, nepheline, Ti-rich andradite and, in the syenites, sanidine. Perovskite can be an important phase in the jacupirangite and melteigites, together with titanomagnetite and apatite, and titanite occurs in the syenites. In the ijolites, pyrrhotite is a common phase, and its presence with Ti-andradites suggests low f_{O_2} compared with the conditions necessary to stabilize the perovskite–titanomagnetite combination of the jacupirangites. Wollastonite is a common igneous phase, occurring as quench crystals in interstitial glass or in nepheline–wollastonite–glass veins; in one specimen the wollastonite has proved to be para-wollastonite (Dawson & Sahama 1963). Eucolite, only found in nepheline syenite, is the most calcic member of the eudialyte–eucolite series yet recorded (Dawson & Frisch 1971). Skeletal vishnevite and melilite occur in the interstitial glass in a cellular pyroxenite (Donaldson & Dawson 1978).

Within the magmatic series, the pyroxenes, nephelines and Ti-andradites are zoned, usually towards iron-rich rims. The pyroxenes, in particular, are complexly zoned in the melteigites and ijolites, with overgrowths of euhedral, oscillatory zones, varying from hedenbergite to aegirine, on strongly resorbed cores of augite. Ti-augite is the pyroxene in the jacupirangites.

Variants of the normal ijolites are those containing olivine and phlogopite. In these rocks megacrysts of olivine, phlogopite, and augite show reaction rel-

Table 11.2 Representative analyses of magmatic series plutonic blocks.

	1	2	3	4
Wt%				
SiO_2	33.55	41.89	44.28	47.71
TiO_2	6.88	3.34	2.48	1.49
Al_2O_3	3.48	10.69	9.87	19.59
Fe_2O_3	9.02	6.97	6.45	3.52
FeO	7.76	3.12	4.24	2.36
MnO	0.19	0.22	0.27	0.09
MgO	9.71	8.21	6.61	1.08
CaO	23.61	15.02	16.76	5.71
Na_2O	0.65	5.28	5.55	10.53
K_2O	0.21	1.99	2.07	4.80
H_2O^+	0.14	0.09	0.17	0.33
CO_2	0.46	0.09	0.20	0.45
SO_3	1.01	<0.02	0.64	0.54
P_2O_5	3.97	0.54	0.12	0.73
Cl	0.07	n.a.	n.a.	0.18
F	0.16	n.a.	n.a.	0.07
	100.87	97.47	99.71	99.18
Trace elements (p.p.m.)				
V	317	218	264	229
Cr	<20	149	122	<20
Ni	19	75	58	<10
Cu	626	10	98	6
Zn	89	99	99	62
Rb	<5	33	31	80
Sr	686	602	637	834
Y	47	47	65	16
Zr	369	492	638	330
Nb	301	226	119	125
Ba	116	113	165	1439

1, BD 875, jacupirangite; 2, BD 36, melteigite; 3, BD 52, ijolite; 4, BD 71, nepheline syenite.
Analyses by XRF (analyst R. Kanaris-Sotiriou) except FeO, H_2O^+, CO_2, Cl, and F by wet chemical methods (analyst A. Saxby). n.a., Not analysed.

ationships with an ijolite magma. Corroded cores of augite are overgrown by titanium-rich Na-diopside, phlogopite is overgrown by rims richer in Ti and Fe, and the olivines are surrounded by biotite–diopside–titanomagnetite coronas. These rocks are obviously hybrids.

A further ijolite variant is sodalite ijolite in which original nepheline is partially replaced by sodalite.

The magmatic series is not merely the plutonic equivalent of the extrusive nephelinites and phonolites. They show a spectrum of compositions from the jacupirangite (relatively low in SiO_2, Al_2O_3, Na_2O, K_2O, but high in FeO_{Total}, TiO_2, CaO, and P_2O_5) to the ijolites and nepheline syenites that show the reverse (Table 11.2). The plutonic rocks are overall higher in CaO, TiO_2, MgO, FeO_{Total}, but lower in Al_2O_3, Na_2O, and K_2O than the lavas, reflecting their higher content of pyroxene and Ti-andradite relative to nepheline and sanidine. Niobium (correlating with high Ti) and Cu are high in the jacupirangites, whereas Sr and particularly Ba (>1400 p.p.m.) are high in the potassium-rich nepheline syenites.

Metasomites *Fenites* of both granitic and basic original composition, are present (Morogan 1982, Morogan & Martin 1985). In some, the characteristic network of intersecting planar veinlets is present. The ultimate products of the fenitization are nepheline, aegirine–augite, and sanidine, and the highest temperatures achieved during metasomatism were about 800 °C. Disequilibrium melting along grain boundaries took place in the highest-grade assemblages followed by quenching; variable glass compositions, ranging from SiO_2-undersaturated-peralkaline glass to SiO_2-oversaturated peraluminous glass in the same specimen, indicate lack of homogenization.

Olivine–mica pyroxenites are metasomatized ultrabasic rocks. Olivines, some showing strain and kink bands, are relicts from the earlier ultrabasic assemblage. Olivine is replaced by diopside, which is itself replaced by biotite. Compared with other blocks from Lengai, these rocks are relatively high in Cr (up to 1600 p.p.m.) and Ni (900 p.p.m.). Biotite pyroxenites represent the culmination in the replacement of olivine.

Some mica pyroxenites have suffered a second stage of metasomatism from peralkaline magmatic series rocks that have transported them from depth. Reaction rims extending into the blocks, from contacts with mantling ijolite or nephelinite, consist of pyroxene more sodic than the original, and a higher proportion of Ti-magnetite. Biotite is eliminated within the reaction zone. The same effects are seen along nepheline-rich veinlets cutting the mica pyroxenite blocks.

11.3.4 *The natrocarbonate lavas*

The first record of carbonatite activity was the observation by Schulze in 1915 (Reck & Schulze 1921) that the northern crater was filled with 'soda-mud'. Subsequent lava flows have been confined to the northern crater; although white layers, interbedded within unit 1 yellow tuffs high in the cliffs of the landslip-scar on the upper eastern slopes of the volcano, may be the residue of earlier carbonatite eruptions from the older southern crater.

11.3.5 *General petrography and mineralogy*

The extrusion of natrocarbonate lavas in September–October 1960 was observed over a period of approximately two weeks. Prior to the observation period, highly mobile lavas had been extruded from the central vent on the crater floor and had travelled the 200–250 m distance to the foot of the crater walls to lap over fallen debris. These flows were highly vesicular and were, in some cases, no more than 30 mm thick. At the time of observation the vent occupied the centre of a scoria mound approximately 5 m above the general level of the crater floor. The 20 m-wide vent was filled with a heaving pool of black lava that erupted sporadically to throw lava into the air for heights estimated at 15–25 m. Twice during the observation period, the lava breached the rim around the vent to flow slowly for distances of no more than 50 m across the crater floor (Fig. 11.4). These flows were highly scoriaceous, very viscous, and approximately 1.5 m thick, in these respects contrasting strongly with the earlier very mobile lavas. Nonetheless, the two lava types, aa and pahoehoe, have the same composition. Observations of the 1960 lava pool at night showed that the liquid lava did not glow. The lava is jet-black in colour when first extruded but, being hygroscopic, rapidly absorbs atmospheric moisture to form a white, outer layer; X-ray diffraction analysis of the whitened surface layer of one of the 1960 lavas (BD115) has identified the presence of trona, pirrsonite, nyerereite, and sylvite.

The lavas erupted in November 1988 are morphologically similar to the 1960 lavas. Clinkery, gas-rich, more viscous lavas are richer in phenocrysts than the degassed, more mobile pahoehoe lavas that have a higher proportion of glass and nyerereite microphenocrysts. Temperatures of the lavas, measured in flow channels and a lava pool, are consistently in the range 580–590 °C, and maximum viscosities are 100 and 10 poise for the more viscous and pahoehoe lavas respectively (J. B Dawson, H. Pinkerton, D. Pyle – new unpublished data). The lavas were seen at night to be incandescent and, when escaping explosively from the lava pool, to be accompanied by yellow flaring.

The lava is porphyritic containing large, clear plates of nyerereite and rounded brown-coloured grains of gregoryite set in a matrix consisting of small clear plates of nyerereite (aligned to give a trachytic texture), equigranular fine-grained gregoryite, and a fine-grained opaque phase rich in Mn and S (alabandite?). First recognized in the Lengai lavas, nyerereite and gregoryite are alkali carbonates whose compositions lie within the system Na_2CO_3–K_2CO_3–$CaCO_3$; they have not, to date, been found elsewhere. The composition of nyerereite is $(Na_{0.82}K_{0.18})_2Ca(CO_3)_2$ (McKie & Frankis 1977) and that of gregoryite is $(Na_{0.78}K_{0.05})_2Ca_{0.17}(CO_3)$ (Gittins & McKie 1980); however, in both minerals there is substitution of Ba^{2+} and Sr^{2+} for Ca^{2+}, and of SO_4^{2-}, PO_4^{3-}, F^-, and Cl^- for CO_3^{2-} (Table 11.3).

11.3.6 *Bulk chemistry*

The chemistry of the natrocarbonatite lavas is given in Table 11.3. In addition to the virtual absence of two of the commonest rock-forming oxides – SiO_2 and Al_2O_3, the

Table 11.3 Composition of Oldoinyo Lengai carbonatite lavas of the 1960 eruption, a ?pre-1917 altered natro-carbonatite lava, and the phases and matrix of the 1960 lavas.

	1	2	3	4	5	6	7
Wt%							
SiO_2	Tr	Tr	0.10	0.73	—	—	—
TiO_2	0.11	0.08	0.01	0.03	—	—	—
Al_2O_3	0.09	0.09	0.03	0.25	—	—	—
Fe_2O_3T	0.28	0.35	0.29	1.10	—	—	0.61
MnO	0.04	0.04	0.25	0.55	0.07	—	0.81
MgO	0.53	0.58	0.26	0.49	0.12	—	0.25
CaO	13.90	13.96	13.93	25.29	24.27	9.10	14.86
SrO	1.53	1.31	1.41	1.49	2.20	0.66	1.58
BaO	1.04	1.14	0.98	0.91	0.59	0.24	2.2
Na_2O	32.22	32.35	30.30	24.68	23.94	44.87	34.00
K_2O	8.27	7.17	6.55	0.68	7.90	3.95	6.78
CO_2	34.65	35.29	33.90	32.71	38.10*	36.22*	27.61*
SO_3	2.18	2.37	3.23	0.20	1.50	4.28	3.88
P_2O_5	0.90	0.98	1.13	0.64	0.36	1.92	0.08
F	2.93	3.19	1.20	1.06	0.51	0.4	7.7
Cl	4.21	2.87	2.45	0.20	0.26	0.60	2.91
H_2O^+	*	*	4.51	9.33	0.26	n.a.	n.a.
	102.88	101.77	100.53	100.34	100.08	102.24	103.27
Less O ≡ Cl, F	2.18	1.48	1.05	0.44	0.28	0.31	3.86
	100.70	100.29	99.48	99.90	99.80	101.93	99.41
$^{87}Sr:^{86}Sr$	0.70442	0.70445	n.a.	n.a.			
$\delta^{18}O‰$	7.55	7.80	n.a.	n.a.			
$\delta^3C‰$	−6.60	−6.70	n.a.	n.a.			
Trace elements (p.p.m.)							
V	n.a.	n.a.	123	84			
Cr	n.a.	n.a.	10	10			
Ni	n.a.	n.a.	11	12			
Cu	n.a.	n.a.	18	18			
Zn	n.a.	n.a.	100	354			
Rb	n.a.	n.a.	184	55			
Y	n.a.	n.a.	5	75			
Nb	n.a.	n.a.	23	157			
Pb	120	n.a.	89	136			
U^A	6.73	7.31	n.a.	n.a.			
U^B	7.20						
Th^A	2.9	2.8	n.a.	n.a.			
Th^B	1.92						

1. BD 114, pahoehoe lava, and 2. BD 118, aa lava, both extruded October 1960 (Dawson 1962b). Analysed by a combination of optical spectroscopy (CaO, BaO, SrO) and 'wet' methods. Strontium isotope data from Bell & Blenkinsop (1987); oxygen and carbon isotope data from Sheppard & Dawson (1973). Lead value for BD 114 from Williams et al. (1986). Uranium and thorium A and B values from Dawson & Gale (1970) and Williams et al. (1986) respectively.

3. BD 118 (Dawson et al. 1987). Analysed by X-ray fluorescence except for H_2O^+, CO_2, F, and Cl by wet methods.

4. Altered, older (?pre-1917) carbonatite (Dawson et al. 1987).

5, 6, and 7. Nyerereite, gregoryite, and matrix respectively (Gittins & McKie 1980).

*CO_2 calculated; Tr, trace; *, recalculated from original analyses on H_2O-free basis; n.a., not analysed.

lavas are characterized by extremely high abundances of Na_2O, CaO, and K_2O, together with high amounts of CO_2, P_2O_5, and SO_3; they are also enhanced in Sr, Ba, and the halogens – elements usually regarded as minor elements in most igneous rocks. In fact, the lavas can be regarded as a composite of alkali and alkali earth elements combined with volatile elements. Bulk analyses of the November 1988 lavas are very similar to those of the 1960 lavas (see Table 11.3). In addition, they contain high contents of the incompatible elements Rb, Zn, Pb, U, and Th. The U : Th ratio of the lavas is > 1, which is an unusual feature for igneous rocks; high ratios such as these are known in the Kaiserstuhl carbonatite (Verfaillie 1964) and in calcite-rich carbonate segregations in kimberlite from Benfontein, South Africa (Dawson & Hawthorne 1973). Total REE abundances exceed 1000 p.p.m. (Table 11.4) and the chondrite-normalized pattern is steep with highly enhanced LREE values (Fig. 11.7) similar to the pattern in the silicate lavas; although in enhancement of the LREEs, relative depletion of HREEs, and slope variation, the carbonate lavas are more extreme. The complete absence of primary hydrous minerals in the lavas suggests that the magma is anhydrous.

Earlier isotope analyses of the carbonate lavas gave values for initial $^{87}Sr : ^{86}Sr$ which fell within the range for the Lengai silicate lavas (Bell *et al.* 1973), and DePaolo and Wasserburg (1976) found $\varepsilon_{Nd} = 0.1 \pm 0.9$. More recent high-precision analyses give $^{87}Sr : ^{86}Sr$ values of 0.70442 and 0.70445 ($\varepsilon_{Sr} = -1.1$ and -0.7), and $^{143}Nd : ^{144}Nd$ of 0.51261 and 0.51292 ($\varepsilon_{Nd} = -0.5$ and -0.9) (Bell & Blenkinsop 1987). New Sr isotope analyses of the silicate lavas have a range of 0.70417–0.70512, bracketing the carbonate lavas (Bell & Dawson, in preparation). A number of authors (Denaeyer 1970, Vinogradov *et al.* 1970, 1971, O'Neil & Hay 1973, Sheppard & Dawson 1973, Suwa *et al.* 1975) found carbon and oxygen isotope values within the range $\delta^{18}O = 6.2$ to 7.8 ‰ and $\delta^{13}C = -6.6$ to -10.2 ‰, which is

Table 11.4 REE abundances of Oldoinyo Lengai lavas.

	1	2	3
La	170	155	425
Ce	259	261	498
Pr	22.0	22.8	30.9
Nd	75.3	83.1	77.5
Sm	10.3	11.9	4.85
Eu	3.08	3.59	1.15
Gd	8.06	9.10	2.57
Dy	5.69	6.88	1.13
Ho	1.21	1.38	0.77
Er	3.44	4.13	1.60
Yb	2.78	3.01	0.43
Lu	0.40	0.54	0.07

1, BD 70, nephelinite; 2, BD 65, phonolite; 3, BD 118, natrocarbonatite. Values in p.p.m. Analyses by ICP.

mainly within the 'primary carbonatite box' of Taylor *et al.* (1967); however, Suwa *et al.* (1975) do note significant differences in ratios of both carbon and oxygen in powdered and altered specimens relative to fresh specimens. Lead isotope values for lava BD 114 are: ^{206}Pb:^{204}Pb = 19.19, ^{207}Pb:^{204}Pb = 15.55, and ^{208}Pb:^{204}Pb = 39.14 (Williams *et al.* 1986).

The carbonate lavas also have the most extreme disequilibria between the U and Th series nuclides yet measured in volcanic rocks (Williams *et al.* 1986).

11.3.7 Origin of the natrocarbonatite lavas

The carbonate lavas and ashes are the youngest manifestation of the peralkaline igneous activity, preceded by phonolites and nephelinites. Like the earlier silicate rocks, they contain high K_2O in addition to the high Na_2O, and they are likewise high in Ba, Sr, and REEs, and they have similar ^{87}Sr:^{86}Sr ratios. On major- and trace-element grounds, and on the evidence of the strontium isotope ratios, the carbonatite lavas appear to be genetically linked with the silicate rocks. However, in terms of SiO_2 and Al_2O_3 there is a major gap between the least-siliceous silicate lava (a melanephelinite containing 37% SiO and 12% Al_2O_3, Dawson 1966) and the natrocarbonatite lavas. Of known petrogenetic processes to explain such a compositional break, liquid immiscibility is perhaps the most plausible. Immiscibility has been convincingly demonstrated by high P–T experimentation on alkaline carbonate–silicate synthetic systems (e.g. Koster van Groos & Wyllie 1973, Verwoerd 1978) and also on mixtures of natrocarbonatite, phonolite, and nephelinite from Oldoinyo Lengai (Freestone & Hamilton 1980). Additional evidence for liquid immiscibility from fluid inclusions in carbonatite apatites has been reviewed by Le Bas (1987). A review of liquid immiscibility is given in this volume by Kjarsgaard and Hamilton.

The relationship of natrocarbonatite magmas to other calcitic carbonatites is a question of some debate. The direct origin of natrocarbonatite magma by immiscibility from a parental CO_2-rich nephelinite parent (e.g. Le Bas 1987) has been challenged by Twyman & Gittins (1987) who contend that alkali carbonatite liquid is derived by alkali concentration in a fractionating, low alkali (about 8% combined $Na_2O + K_2O$) olivine sövite magma, this magma itself originating by immiscible separation from a carbonated olivine nephelinite magma deep in the mantle. They suggest that, in most circumstances, the olivine sövite magma would be hydrous and the derived supercritical alkali fluids would give rise to fenitization in plutonic settings; by contrast an anhydrous olivine sövite could fractionate to liquid compositions similar to those of Oldoinyo Lengai, since there would be no loss of alkalies in aqueous fluids. This view of derivation of alkali carbonates from a parental sövite magma, is diametrically opposite to those of other petrologists (e.g. von Eckermann 1948, Woolley 1982, Le Bas 1987) who believe that a mixed-cation carbonatite would fractionate to calcite, dolomite, and iron-rich types after alkali loss during fenitization. The Twyman–Gittins hypothesis receives some support from the Kerimasi volcano where the tephra consist of both sövite and (former) natrocarbonatites with primary calcite phenocrysts (Hay 1983, Mariano & Roeder

1983). Furthermore, early calcite enveloped within later nyerereite has now been recognized in natrocarbonatite from Oldoinyo Lengai (Dawson *et al.* 1987); however, olivine–biotite–magnetite sövite, Twyman and Gittins' postulated cumulate, has not been found in the Oldinyo Lengai ejecta.

11.3.8 Oldoinyo Lengai – wider aspects

Origin of the silicate lavas The nephelinites and phonolites are highly evolved and cannot be regarded as parental upper-mantle melts. Parental magma for carbonatite complexes in general is believed to be a highly carbonated olivine nephelinite (e.g. King 1965, Le Bas, 1987, this volume) which fractionates by olivine and pyroxene precipitation. The olivine nephelinite flow at Oldoinyo Loolmurwak represents locally available primitive upper-mantle liquid (Table 11.1) which might be used as a starting point. Comparison of the Lengai and Loolmurwak nephelinite bulk analyses (Table 11.1) show the former to be enhanced in Al_2O_3, Na_2O, and K_2O but depleted in TiO_2, total Fe, MgO, and P_2O_5. These features are consistent with fractionation of diopside, perovskite, magnetite, and apatite – a feature noted in a cumulate pyroxenite from the volcano in which the residual glass is highly enhanced in alkalies and Al_2O_3 (Donaldson & Dawson 1978). In addition, the relatively low MgO, TiO_2, and CaO of the Lengai rocks is consistent with pyroxene and andradite

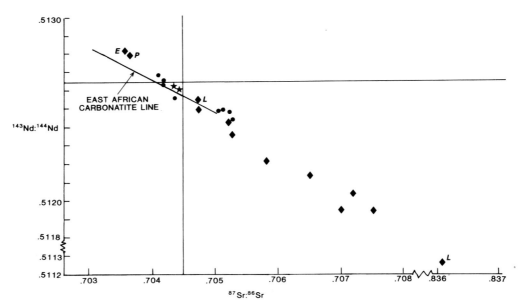

Figure 11.8 Nd v. Sr isotope plot for Oldoinyo Lengai natrocarbonatites (★) (Bell & Blenkinsop 1987), and for Lengai silicate rocks (●) (Bell & Dawson, in preparation). (♦) Diopsides in upper mantle peridotite zenoliths from Eledoi (E), Pello (P), and Lashaine (L), northern Tanzania (Cohen *et al.* 1983), and from southern African kimberlites (ornament unannotated) (Menzies & Murthy 1980). Also shown is the best-fit line for data from 12 East African carbonatites (Bell & Blenkinsop 1987). Note the scale break between $^{143}Nd:^{144}Nd = 0.5118$ and 0.5113, and between $^{87}Sr:^{86}Sr = 0.708$ and 0.836.

fractionation. Accurate modelling must await detailed analysis of the plutonic blocks.

In attempting to assess the type of mantle from which the Lengai rocks derive, it is apparent that they plot in a linear zone on a Nd versus Sr isotope diagram (Fig. 11.8). This zone parallels and overlaps a linear trend found for East African carbonatite lavas by Bell & Blenkinsop (1987), who suggest that the line represents mixtures of two different mantle sources – one with depleted characteristics, the other enriched relative to bulk earth. In northern Tanzania, upper-mantle samples are represented by peridotite blocks that occur in the Pello and Eledoi tuff rings just to the east of Oldoinyo Lengai (Fig. 11.2) and at Lashaine in the carbonatite tuff-cone area to the west of Arusha (Fig. 11.1). Nd and Sr isotope studies on these peridotites (Cohen *et al.* 1983) show the diopsides from Pello and Eledoi peridotites to fall in the depleted quadrant, whereas those from Lashaine fall within the enriched quadrant (Fig. 11.8), in this respect resembling diopsides from mica-bearing lherzolites in kimberlites. If the Lengai lavas represent mixtures of these two putative parents, the depleted asthenosphere component must predominate over the ancient, enriched lithosphere component. The role of metasomatized peridotite, in the form of the blocks of olivine–mica pyroxenite in the Oldoinyo Lengai ejecta, has yet to be evaluated.

Fenitization Perhaps the most important aspect of the Oldoinyo Lengai natrocarbonatites is the proven existence of high-alkali carbonate liquids. Although invoked as a parental magma for the Alnö carbonatite complex (von Eckermann 1948), such a possibility was not entertained seriously by most petrologists until the implications of the natrocarbonatite for the processes of fenitization were recognized (Dawson 1964*c*). This has now been recognized by Gittins and his co-workers (Cooper *et al.* 1975, Gittins *et al.* 1975, Gittins & McKie, 1980) and Le Bas (1981, 1987). Essentially, it is now recognized that the alkali ions required for the metasomatism around plugs and dykes of carbonatites *that are now calcite or dolomite*, must derive from the carbonatite that was alkali-rich when it was intruded. Nonetheless, because some fenites are potassium-rich, it must be acknowledged that the original carbonate liquids could show variations on the Lengai theme, particularly with respect to Na:K ratios. Another variation is that of the carbonatites that lack fenitization aureoles.

Calcite lavas and tuffs Carbonatite lavas and tuffs, although scarce when compared with silicate extrusive rocks, are by no means as rare as was once believed. Although occurring most commonly in the Neogene volcanics associated with the African rift valleys, others, ranging in age from Precambrian to Holocene, are reported from the Kola Peninsula (USSR) (Pyatenko & Saprykina 1976), Greenland (Stewart 1970), southern Mongolia (Kovalenko *et al.* 1977), Kaiserstuhl (FRG) (Keller 1981, this volume), central Tanzania (Sampson 1953), the Cape Verde Islands (Silva *et al.* 1981), and Afghanistan (Yeremenko *et al.* 1975). Most carbonatite extrusives are calcitic with only those from Oldoinyo Lengai containing a high-alkali carbonate component. However, in the paper reporting the 1960 Lengai eruptives,

Dawson (1962b) proposed that alkali carbonate may have been formerly present at other extrusive carbonatite centres but, being water soluble, had been removed during weathering to give a calcite-rich residue. More recently, Hay (1983) has interpreted lath-shaped calcite in carbonatite tephra from Kerimasi as being after original nyerereite and gregoryite. Similar conclusions have been drawn from textural evidence in carbonatite extrusives at Tinderet and Homa Bay, Kenya (Deans & Roberts 1984, Clarke & Roberts 1986). Of direct relevance to this problem is the recent discovery at Oldoinyo Lengai of a pre-1960 lava (possibly from the 1913–15 eruption) that is a 'missing link' between the modern natrocarbonatite lavas and the calcite carbonatites containing pseudomorphed nyerereite (Dawson *et al.* 1987); in this lava, nyerereite and gregoryite are wholly or partially replaced by pirssonite, and the bulk composition of the rock (Table 11.3, analysis 4) is intermediate between natrocarbonatite and calcite carbonatite (Dawson *et al.* 1987). Further degradation and alkali-removal can theoretically lead to formation of a calcite rock. Like the natrocarbonatite lavas themselves, the fortunate preservation of this missing link is due to its relatively young geologic age.

Carbonatite tuffs and 'calcretes' During both the historic and recent explosive eruptions of Oldoinyo Lengai, carbonate ash has been widely dispersed around the volcano (Dawson 1962b, Dawson *et al.* 1968). Analogies between Oldoinyo Lengai and Kerimasi, together with evidence from other small tuff cones, caused Dawson (1964a) to propose that the so-called 'calcretes' of the Olduvai Gorge succession and on the Serengeti Plains to the west are, in fact, carbonatite tuffs from Kerimasi and Oldoinyo Lengai; he also suggested that beds of 'limestone' interbedded with dipping agglomerates and lavas at several other East African carbonatite volcanoes (e.g. Napak, Mt. Elgon, Rangwa, and Hanang) may have originated as carbonatite ashes. Recent detailed work by R. L. Hay and his co-workers has resulted in the firm recognition of carbonatite tuffs at Olduvai and in the nearby Laetolil Beds and also at the Kaiserstuhl volcano, W. Germany (Hay 1976, Hay & Reeder 1978, Hay 1978, Hay & O'Neil 1983).

ACKNOWLEDGEMENTS

C. Bristow provided the photograph for Figure 11.3 and C. Nyamweru kindly gave details of the post-1966 activity. R. Kanaris-Sotiriou and A. Saxby carried out the bulk-rock analyses, and the REE analyses were made at the NERC-funded ICP Unit at Royal Holloway and Bedford New College. B. M. Wilson and P. Mellor typed the manuscript, and M. Cooper drew the figures. The original manuscript was improved by reviews from M. B. Lambert, K. Bell, and an anonymous reviewer. I extend my warmest thanks to them all.

REFERENCES

Bell, K. & J. Blenkinsop 1987. Nd and Sr isotopic compositions of East African carbonatites: implications for mantle heterogeneity. *Geology* **15**, 99–102.

Bell, K., J. B. Dawson & R. M. Farquhar 1973. Strontium isotope studies of alkalic rocks: the active carbonatite volcano Oldoinyo Lengai. *Bulletin of the Geological Society of America* **84**, 1019–30.

Clarke, M. G. C. & B. Roberts 1986. Carbonated melilitites and calcitized alkali carbonatites from Homa Mountain, western Kenya: a reintrepretation. *Geological Magazine* **123**, 683–92.

Cohen, R. S., R. K. O'Nions & J. B. Dawson 1983. Isotope geochemistry of xenoliths from East Africa: implications for the development of mantle reservoirs and their interaction. *Earth and Planetary Science Letters* **68**, 209–20.

Cooper, A. F., J. Gittins & O. F. Tuttle 1975. The system Na_2CO_3–K_2CO_3–$CaCO_3$ at 1 kilobar and its significance in carbonatite petrogenesis. *American Journal of Science* **275**, 534–60.

Dawson, J. B. 1962a. Sodium carbonate lavas from Oldoinyo Lengai, Tanganyika. *Nature* **195**, 1075–6.

Dawson, J. B. 1962b. The geology of Oldoinyo Lengai. *Bulletin Volcanologique* **24**, 349–87.

Dawson, J. B. 1964a. Carbonatitic volcanic ashes in northern Tanganyika. *Bulletin Volcanologique* **27**, 81–92.

Dawson, J. B. 1964b. Carbonate tuff cones in northern Tanganyika. *Geological Magazine* **101**, 129–37.

Dawson, J. B. 1964c. Reactivity of the cations in carbonate magmas. *Proceedings of the Geological Association of Canada* **15**, 103–13.

Dawson, J. B. 1966. Oldoinyo Lengai – an active volcano with sodium carbonatite flows. In *Carbonatites*, O. F. Tuttle & J. Gittins (eds), 155–68. New York: Wiley.

Dawson, J. B. & T. Frisch 1971. Eucolite from Oldoinyo Lengai, Tanzania. *Lithos* **4**, 297–303.

Dawson, J. B. & N. H. Gale 1970. Uranium and thorium in alkalic rocks from the active volcano Oldoinyo Lengai (Tanzania). *Chemical Geology* **6**, 221–31.

Dawson, J. B. & J. B. Hawthorne 1973. Magmatic sedimentation and carbonatitic differentiation in kimberlite sills at Benfontein, South Africa. *Journal of the Geological Society* **129**, 61–85.

Dawson, J. B. & D. G. Powell 1969. The Natron–Engaruka explosion crater area, northern Tanzania. *Bulletin Volcanologique* **33**, 791–817.

Dawson, J. B. & Th. G. Sahama 1963. A note on para-wollastonite from Oldoinyo Lengai, Tanganyika. *Schweizer Mineralogische und Petrologische Mitteilungen* **43**, 131–3.

Dawson, J. B., P. Bowden & G. C. Clark 1968. Activity of the carbonatite volcano Oldoinyo Lengai, 1966. *Geologische Rundschau* **57**, 865–79.

Dawson, J. B., M. S. Garson & B. Roberts 1987. Altered former alkalic carbonatite lava from Oldoinyo Lengai, Tanzania: inferences for calcite carbonatite lavas. *Geology* **15**, 765–8.

Dawson, J. B., J. V. Smith & A. P. Jones 1985. A comparative study of bulk rock and mineral chemistry of olivine melilitites and associated rocks from East and South Africa. *Neues Jahrbuch für Mineralogie Abhandlungen* **152**, 143–75.

Dawson, J. B., J. V. Smith and I. M. Steele. Combeite from Oldoinyo Lengai, Tanzania. *Journal of Geology*, in press.

Deans, T. & B. Roberts 1984. Carbonatite tuffs and lava clasts of the Tinderet foothills, western Kenya: a study of calcified natrocarbonatites. *Journal of the Geological Society* **141**, 563–80.

Denaeyer, M. E. 1970. Rapports isotopiques δO et δC et conditions d'affleurement des carbonatites de l'Afrique central. *Comptes Rendues de l'Académie des Sciences* **270D**, 2155–8.

DePaolo, D. J. & G. J. Wasserburg 1976. Inferences about magma sources and mantle structure from variations of $^{143}Nd/^{144}Nd$. *Geophysical Research Letters* **3**, 743–6.

Donaldson, C. H. & J. B. Dawson 1978. Skeletal crystallization and residual glass compositions in a cellular alkalic pyroxenite nodule from Oldoinyo Lengai. *Contributions to Mineralogy and Petrology* **67**, 139–49.

Donaldson, C. H., J. B. Dawson, R. Kanaris-Sotiriou, R. A. Batchelor & N. J. Walsh 1987. The silicate lavas of Oldoinyo Lengai, Tanzania. *Neues Jahrbuch für Mineralogie Abhandlungen* **156**, 247–79.

Eckermann, H. von 1948. The alkaline district of Alnö Island. *Sveriges Geologiska Undersöking*, Series Ca, No. 36.

Freestone, I. C. & D. L. Hamilton 1980. The role of immiscibility in the genesis of carbonatites – an experimental study. *Contributions to Mineralogy and Petrology* **73**, 105–17.

Gittins, J. & D. Mckie 1980. Alkalic carbonatite magmas: Oldoinyo Lengai and its wider applicability. *Lithos* **13**, 213–15.

Gittins, J., C. R. Allen, & A. F. Cooper 1975. Phlogopitization of pyroxenite; its bearing on the composition of carbonatite magmas. *Geological Magazine* **112**, 503–7.

Hay, R. L. 1976. *Geology of the Olduvai Gorge*. Berkeley: University of California Press.

Hay, R. L. 1978. Melilitite–carbonatite tuffs in the Laetolil Beds of Tanzania. *Contributions to Mineralogy and Petrology* **67**, 357–67.

Hay, R. L. 1983. Natrocarbonatite tephra of Kerimasi volcano, Tanzania. *Geology* **11**, 599–602.

Hay, R. L. & J. R. O'Neil 1983. Carbonatite tuffs in the Laetolil Beds of Tanzania and the Kaiserstuhl in Germany. *Contributions to Mineralogy and Petrology* **82**, 403–6.

Hay, R. L. & R. J. Reeder 1978. Calcretes of Olduvai Gorge and the Ndolanya Beds of northern Tanzania. *Sedimentology* **25**, 649–73.

Keller, J. 1981. Carbonatite volcanism in the Kaiserstuhl alkaline complex. Evidence for highly fluid carbonatite melts at the earth's surface. *Journal of Volcanology and Geothermal Research* **9**, 423–31.

King, B. C. 1965. Petrogenesis of the alkaline igneous rock suites of the volcanic and intrusive centres of eastern Uganda. *Journal of Petrology* **6**, 67–100.

Koster van Groos, A. F. & P. J. Wyllie 1973. Liquid immiscibility in the join $NaAlSi_2O_6$–$CaAlSi_2O_6$–Na_2CO_3–H_2O. *American Journal of Science* **273**, 465–87.

Kovalenko, V. I., V. S. Samoylov, Ye. V. Smirnova, N. V. Vladykin & A. V. Goreglyad 1977. Rare earths in near-surface carbonatite complexes in Mongolia. *Geochemistry International* **1977**, 148–58.

Le Bas, M. J. 1977. *Carbonatite–nephelinite volcanism*. Chichester: Wiley.

Le Bas, M. J. 1978a. Nephelinite volcanism at plate interiors. *Bulletin Volcanologique* **41**, 459–62.

Le Bas, M. J. 1978b. Are olivine-poor nephelinites a primary melt product from the upper mantle? *Bulletin Volcanologique* **41**, 463–5.

Le Bas, M. J. 1981. Carbonatite magmas. *Mineralogical Magazine* **44**, 133–40.

Le Bas, M. J. 1987. Nephelinites and carbonatites. In *Alkaline igneous rocks*, J. G. Fitton & B. G. J. Upton (eds), 53–83. Geological Society, Special Publication 30.

MacIntyre, R. M., J. Mitchell & J. B. Dawson 1974. Age of fault movements in the Tanzanian sector of the East African rift system. *Nature* **247**, 354–6.

McKie, D. & E. J. Frankis 1977. Nyerereite: a new volcanic carbonate mineral from Oldoinyo Lengai, Tanzania. *Zeitschrift für Kristallographie* **145**, 73–95.

Mariano, A. N. & P. L. Roeder 1983. Kerimasi: a neglected carbonatite volcano. *Journal of Geology* **91**, 449–55.

Menzies, M. A. & V. R. Murthy 1980. Enriched mantle Nd and Sr isotopes in diopsides from kimberlite nodules. *Nature* **283**, 634–6.

Morogan, V. 1982. *Fenitization and ultimate rheomorphism of xenoliths from the Oldoinyo Lengai carbonatite volcano, Tanzania*. M.Sc. thesis, McGill University, Montréal.

Morogan, V. & R. F. Martin 1985. Mineralogy and partial melting of fenitized crustal xenoliths in the Oldoinyo Lengai carbonatite volcano, Tanzania. *American Mineralogist* **70**, 1114–26.

Nyamweru, C. 1988. Activity of Oldoinyo Lengai, Tanzania, 1983–1987. *Journal of African Earth Sciences* **7**, 603–10.

O'Neil, J. R. & R. L. Hay 1973. $^{18}O/^{16}O$ ratios in cherts associated with saline lake deposits of East Africa. *Earth and Planetary Science Letters* **19**, 257–66.

Peterson, T. D. & B. D. Marsh 1986. Sodium metasomatism and mineral stabilities in alkaline ultramafic rocks: implications for the origin of the sodic lavas of Oldoinyo L'engai. *EOS Transactions, American Geophysical Union* **67**, 389–90.

Pyatenko, I. K. & L. G. Saprykina 1976. Carbonatite lavas and pyroclastics in the Palaeozoic sedimentary sequence of the Kontozero district, Kola Peninsula. *Doklady, Akademiya Nauk SSSR: Earth Sciences Section* **229**, 185–7.

Reck, H. 1914. Oldoinyo Lengai, ein tätiger Vulkan in Gebiete der Deutsch-Ostafrikanischen Bruchstufe. *Branca Festschrift* **7**, 373–409.

Reck, H. & G. Schulze 1921. Ein Beitrage zur Kenntnis des Baues und der jungsten Veranderung des L'Engai-Vulkanes in nordlichen Deutsch-Ostafrika. *Zeitschrift für Vulkanologie* **6**, 47–71.

Richard, J. J. 1942. Volcanological observations in East Africa. 1 Oldoinyo Lengai. The 1940–41 eruption. *Journal of the East Africa and Uganda Natural History Society* **16**, 89–108.

Sahama, Th. G. & K. Hytonen 1957. Götzenite and combeite, two new silicates from the Belgian Congo. *Mineralogical Magazine* **31**, 503–10.

Sampson, D. N. 1953. The volcanic hills at Igwisi. *Records of the Geological Survey of Tanganyika* **3**, 48–53.

SEAN 1988. Ol Doinyo Lengai (Tanzania): Crater morphology and vent dynamics. *Seismic Event Alert Network Bulletin* **13**, (7), 5–8.

Sheppard, S. M. F. & J. B. Dawson 1973. $^{13}C/^{12}C$ and D/H isotope variations in 'primary igneous carbonatites'. *Fortschritte der Mineralogie* **50**, 128–9.

Silva, L. C., M. J. Le Bas & A. H. F. Robertson 1981. An oceanic carbonatite volcano on Santiago, Cape Verde Islands. *Nature* **294**, 644–5.

Stewart, J. W. 1970. Precambrian alkaline-ultramafic/carbonatite volcanism at Qaqssiarssuk, south Greenland. *Meddelelser om Grønland* **186**, 1–70.

Suwa, K., S. Oana, H. Wada & S. Osaki 1975. Isotope geochemistry and petrology of African carbonatites. *Physics and Chemistry of the Earth* **9**, 735–45.

Taylor, H. P., J. Frechen & E. T. Degens 1967. Oxygen and carbon isotope studies of carbonatites from the Laacher See district, West Germany, and the Alnö district, Sweden. *Geochimica et Cosmochimica Acta* **31**, 407–30.

Twyman, J. D. & J. Gittins 1987. Alkalic carbonatite magmas: parental or derivative? In *Alkaline igneous rocks*, J. G. Fitton & B. G. J. Upton (eds), 85–94. Geological Society, Special Publication 30.

Verfaillie, G. 1964. Dosages par spectrometrie gamma de l'uranium et thorium dans les roches de Kaiserstuhl. In *Les roches alcalines et les carbonatites de Kaiserstuhl*. Euratom Report-1827 d,e,f.

Verwoerd, W. J. 1978. Liquid immiscibility and the carbonatite–ijolite relationship: preliminary data on the join $NaFe^{3+}Si_2O_6-CaCO_3$ and related compositions. *Carnegie Institution of Washington Yearbook* **77**, 767–74.

Vinogradov, A. P., O. I. Kropotova & V. I. Gerasimovski 1970. Carbon isotope composition for carbonatites from East Africa. *Geokhimiya* **1970**, 643–6.

Vinogradov, A. P., E. I. Dontsova, V. I. Gerasimovski & L. D. Kuznetsova 1971. Oxygen isotope composition of carbonatites from the rift of East Africa. *Geokhimiya* **1971**, 507–14.

Williams, R. W., J. B. Gill & K. W. Bruland 1986. Ra-Th disequilibria systematics: Timescale of carbonatite magma formation at Oldoinyo Lengai volcano, Tanzania. *Geochimica et Cosmochimica Acta* **50**, 1249–59.

Woolley, A. R. 1982. A discussion of carbonatite evolution and nomenclature, and the generation of sodic and potassic fenites. *Mineralogical Magazine* **46**, 13–17.

Yeremenko, G. K., B. Ya. Vikhter, V. M. Chmyrev & N. A. Abdullah 1975. A Quaternary volcanic carbonatite complex in Afghanistan. *Doklady, Akademiya Nauk SSSR* **223**, 427–30.

12
NEODYMIUM AND STRONTIUM ISOTOPE GEOCHEMISTRY OF CARBONATITES

K. BELL & J. BLENKINSOP

ABSTRACT

Neodymium and Sr isotopic data show that carbonatites are of mantle origin, that the subcontinental upper mantle is inhomogeneous and that such inhomogeneities are long lived. A depleted mantle source at least 3 Ga old is indicated by most of the data from Canadian carbonatites, while a more complex picture, involving at least two sources, emerges from carbonatite data from Africa and South America, and some Canadian carbonatites. Isotopic signatures of carbonatites and many oceanic island basalts suggest either similar sources or sources that have undergone similar differentiation histories. The Nd and Sr isotopic data from carbonatites cannot, at present, unequivocally distinguish between a lithospheric or asthenospheric source for the parental magma.

12.1 INTRODUCTION

Although carbonatites are relatively rare and small in volume, their unusual geochemistry, deep-seated origin, and widespread occurrence in most continental areas make them particularly suitable for monitoring the chemical evolution of the subcontinental upper mantle. Studies of melt extraction (McKenzie 1985, Richter 1985) suggest that melts of low viscosity, such as carbonatitic or nephelinitic magmas, can sample large mantle volumes. This feature, along with the high Sr (average 7000 p.p.m.) and Nd (average 250 p.p.m.) contents, well in excess of crustal abundances (see Table 12.1), means that the isotope composition of carbonatites should reflect the average isotopic signature of their mantle source.

Surprisingly few papers have been published on the isotope chemistry of carbonatites since the pioneering Sr work by Powell (e.g. Powell 1965*a, b, c*). A resurgence of interest coupled with more sophisticated mass spectrometry have led to publication of a number of papers about the Nd and Sr isotope chemistry of carbonatites. Included among them are papers by Basu & Tatsumoto (1980), Basu & Puustinen (1982), Bell *et al.* (1982), Nielsen & Buchardt (1985), Roden *et al.* (1985), Allsopp & Eriksson (1986), Grünenfelder *et al.* (1986), Bell & Blenkinsop (1987*a, b*), Bell *et al.* (1987), Tilton *et al.* (1987), and Nelson *et al.* (1988).

Table 12.1 Nd and Sr abundances in large-scale reservoirs.

	Nd (p.p.m.)	Sr (p.p.m.)	Ef_{Nd}	Ef_{Sr}	Reference
Continental crust					
Upper crust	26	350	10	20	1
Canadian Shield	26	316	10	22	2
Lower crust	13	230	19	30	1
Scourian granulites	19	570	13	12	3
Bulk continental crust	16	260	23	27	1
Oceanic crust					
MORB	10	134	25	52	4, 5, 6
Primitive mantle (present mantle + crust)	1.0	17.8	250	393	1
Bulk Earth	1.0	21	250	333	7

Ef (enrichment factor) = p.p.m. in carbonatites : p.p.m. in reference reservoir.
References: 1, Taylor & McLennan (1985); 2, Shaw (1976); 3, Weaver & Tarney (1980); 4, Faure (1986); 5, Sun (1980); 6, Basaltic Volcanism Study Project; 7, O'Nions (1987).

12.2 CHEMICAL BACKGROUND

Before the isotopic work is discussed in detail, some of the concepts of isotope chemistry are reviewed, with particular reference to Sr and Nd. Lead isotopes are discussed elsewhere in this volume (Kwon *et al.*).

Both Sr and Nd have radiogenic isotopes produced by the radioactive decay of a long-lived parent isotope. 'Long-lived' means that the half-life of the parent is roughly comparable to the age of the Earth. For Sr, the decay of ^{87}Rb by emission of a β-particle, with a half-life of 49 Ga, produces ^{87}Sr; for Nd, ^{143}Nd is produced by α-decay of ^{147}Sm which has a half-life of 106 Ga. These processes result in variations of the isotopic ratios ^{87}Sr : ^{86}Sr and ^{143}Nd : ^{144}Nd in rocks and minerals. The value of the ratio in a rock or mineral is a function of the parent : daughter ratio, the age of the rock or mineral, and the initial isotopic ratio (the value of the ratio at the time of crystallization).

All of the elements that make up the parent–daughter pairs (Rb, Sr, Sm, Nd) are large-ion lithophile or incompatible elements. Sm and Nd are also rare earth elements (REEs); therefore, both parent and daughter behave chemically in a very similar manner. For postulated mantle mineralogies, partial melting produces a melt enriched in the large-ion lithophile elements (LILEs) relative to the starting material (Gast 1968), the degree of enrichment being greater for small degrees of partial melting. Rubidium is enriched in the melt relative to Sr, and Nd relative to Sm; i.e. Rb is more incompatible than Sr, and Nd more incompatible than Sm. Mass balance considerations require that the resulting residue has a lower Rb:Sr and higher Sm:Nd ratio than the starting material.

These concepts can be applied to the isotopic evolution of Sr and Nd in the Earth. The Earth is believed to have accreted from material very similar in composition to chondritic meteorites. Heating shortly after accretion led to differentiation of the primitive Earth into core and mantle, a process that would not affect isotopic ratios. Rubidium, a volatile element, most probably was lost from the Earth at the time of heating, but refractory elements, such as Sm and Nd, were retained. Therefore, the silicate portion of the Earth (crust plus mantle) has a Sm:Nd ratio identical to that of chondrites; this reservoir was termed the 'chondritic uniform reservoir' or CHUR by DePaolo & Wasserburg (1976). The Rb:Sr ratio of the silicate portion of the Earth is much lower than that of chondrites because of Rb loss during the early history of the Earth, and this reservoir is usually referred to as 'bulk Earth'. The Sr and Nd isotopic ratios of the Earth are determined indirectly from measurements on chondrites, and because the Rb:Sr ratio of bulk Earth and Sm:Nd ratio of CHUR are known (Faure 1986), the isotopic evolution lines showing the variation in Nd and Sr isotopic ratios as a function of time can be determined for the silicate portion of the Earth. These lines are shown on Figures 12.1 and 12.2.

Strontium or Nd initial ratios can be expressed relative to the values for the silicate portion of the Earth using the epsilon notation of DePaolo & Wasserburg (1976), which is somewhat similar to the delta notation used in stable isotopes:

$$\varepsilon_k(T) = \left\{\frac{R_i}{R_r} - 1\right\} \times 10^4$$

where R_i is the initial ratio at time T, R_r is the ratio in the appropriate reservoir at the same time, and k is either Sr or Nd. The reservoirs are bulk Earth or CHUR, as noted above.

The ε values can be used to characterize the source region of the Nd and Sr. Two main sources are continental crust and mantle, but even these can be subdivided into still smaller reservoirs, each characterized by specific Rb:Sr and Sm:Nd ratios. Positive ε_{Nd} and negative ε_{Sr} values indicate that the source region was depleted in the LIL elements for a considerable period of time because its Sm:Nd ratio must be higher than the CHUR value, and the Rb:Sr ratio lower than that of bulk Earth, in order to produce the observed values. Conversely, negative ε_{Nd} and positive ε_{Sr} values indicate a source region with a time-integrated enrichment in the LILEs. The time required for the ε values of a particular reservoir to show a measurable deviation from the bulk Earth values is related to the difference between the parent:daughter ratios and those of the silicate Earth reservoir. Rubidium and Sr, respectively an alkali metal and an alkaline earth, show greater differences than Sm and Nd.

The isotopic variation of a reservoir can be plotted as a function of time, using either the ratios themselves or the epsilon values. However, Sr and Nd initial ratios can be shown together on an anti-correlation plot or ACP (e.g. DePaolo & Wasserburg 1979, Allègre et al. 1979). In a more general form it is a plot of initial

^{143}Nd : ^{144}Nd versus initial ^{87}Sr : ^{86}Sr ratios expressed in terms of the deviation from CHUR or bulk Earth, respectively, at a given time; that is, a plot of $\varepsilon_{Nd}(T)$ against $\varepsilon_{Sr}(T)$. A reservoir is represented by a point on the graph, and the evolution of a reservoir, closed with respect to exchange of Rb, Sr, Sm, and Nd throughout time, is described by a straight line on an ACP. For a reservoir formed from bulk Earth-type material, the distance of the point from the origin is determined by the time since separation, and by the difference between the Rb:Sr and Sm:Nd ratios of the reservoir, and those of bulk Earth. A more restricted form of the ACP, used for rocks of approximately zero age such as oceanic basalts, plots $\varepsilon_{Nd}(0)$ against $\varepsilon_{Sr}(0)$.

Most data from igneous rocks plot in either the upper-left or lower-right quadrants of the ACP that correspond respectively to isotopic evolution of Sr and Nd in a depleted or enriched source region. Most oceanic basalts, for example, plot in the upper-left portion of the ACP, implying an ancient source region depleted in the LILEs. They define the so-called 'mantle array'. Most data from volcanic rocks are represented by a band on the ACP plot that extends from the upper-left quadrant into the lower-right. Data from the continental crust, on the other hand, generally plot in the lower-right quadrant, consistent with the observation that the continental crust is enriched in the incompatible elements. Very few data plot in either the upper-right or lower-left portions of the ACP; these regions correspond to those cases where both the Sm:Nd and Rb:Sr ratios of the sources are either high or low. For example, granulites of the lower crust may represent a reservoir with low Sm:Nd and Rb:Sr ratios.

12.3 STRONTIUM AND Nd ISOTOPE CHEMISTRY OF CARBONATITES

Listed in Table 12.2 are Nd and Sr data from carbonatites from several different continents. The data listed are only those for which Nd and Sr analyses have been obtained from the same sample. The ^{143}Nd : ^{144}Nd ratios have been normalized where necessary to the same ^{146}Nd : ^{144}Nd ratio (0.7219), and epsilons calculated using consistent values for CHUR and bulk Earth. These values are given in Table 12.2. In most cases, the ages of the individual complexes are fairly well constrained.

12.3.1 Sr chemistry

Early work on the Sr isotopic geochemistry of carbonatites (Powell *et al.* 1962, Hamilton & Deans 1963, Powell 1965 *a, b, c*, Powell 1966, Powell *et al.* 1966, Deans & Powell 1968, Pineau & Allègre 1972, Bell *et al.* 1973) showed that most carbonatites have low initial ratios, considerably lower than average continental crust. Faure & Powell (1972) quoted a value of 0.7034 ± 0.0006 for typical carbonatite, pointed out its similarity to values for oceanic basalts, and interpreted the ratio as being consistent with a mantle or deep crustal origin for carbonatitic melts. Exceptions included the high ^{87}Sr : ^{86}Sr ratios of vein carbonatites, considered of

Table 12.2 Summary of Sm, Nd, and Sr data from carbonatites.

Locale	Age (Ma)	Sm (p.p.m.)	Nd (p.p.m.)	Sr (p.p.m.)	^{143}Nd:^{144}Nd* (atomic)	^{87}Sr:^{86}Sr* (atomic)	$\varepsilon_{Nd}(T)$	$\varepsilon_{Sr}(T)$	Reference
AFRICA									
Kalyango	0				0.51257	0.70417	−1.3	−4.7	1
Oldoinyo Lengai	0				0.51261	0.70442	−0.5	−1.1	1
	0				0.51259	0.70445	−0.9	−0.7	1
Rusekere	0				0.51258	0.70417	−1.1	−4.7	1
Homa Bay	13				0.51243	0.70502	−3.7	7.6	1
Napak	20	197.2	1453	1226	0.51276	0.7032	2.9	−18.1	2
Bukusu	40				0.51277	0.70315	3.5	−18.5	1
	40				0.51281	0.70307	4.3	−19.6	1
Sukulu	40				0.51279	0.70312	3.9	−18.9	1
	40	16.83	101.1	2783	0.51276	0.7031	3.4	−19.2	2
Tororo	40				0.51271	0.70360	2.3	−12.1	1
	40	60.31	354.2	7100	0.51264	0.7036	1.0	−12.1	2
Panda Hill	116				0.51242	0.70442	−1.3	0.8	1
Sengeri	116(?)				0.51242	0.70423	−1.3	−1.9	1
Kangankunde	126	720	8620	>20000	0.51263	0.7031	3.0	−17.8	2
Nachendezwaya	655	24.03	130	3754	0.51191	0.7025	2.3	−17.4	2
	1200	34.98	187.4	1637	0.51106	0.7029	−0.6	−2.5	2
Goudini	1200	30.32	94.06	3705	0.51105	0.7027	−0.7	−5.4	2
AUSTRALIA and NEW ZEALAND									
Westland and	25	26.4	354.1	14940	0.51286	0.70348	4.9	−14.1	3
Otago	25	56.7	216.0	12630	0.51279	0.70346	3.6	−14.4	3
	25	94.1	704	18400	0.51281	0.70299	4.0	−21.0	3
Walloway	170	16.61	126.8	999.4	0.51245	0.7054	0.6	15.6	2
Strangways Range	732	18.23	116.3	2731	0.51185	0.7032	3.0	−6.2	2
Mt. Weld	2060	40.88	244.5	6033	0.50999	0.7020	0.4	−0.6	2
	2060	50.61	297.7	7410	0.51000	0.7020	0.6	−0.6	2
EUROPE									
Kaiserstuhl	17	11.63	108.5	17989	0.51278	0.7036	3.2	−12.5	2
	17	14.07	131.5	15012	0.51280	0.7037	3.2	−11.4	2
Fen	550	22.82	127	6457	0.51207	0.7029	2.8	−13.5	2

NORTH AMERICA

Location									Ref
Magnet Cove	97	5.35	33.53	5597	0.51271	0.7036	3.8	−11.2	2
	97	4.1	36.84	12079	0.51271	0.70373	3.8	−9.3	4
	97	3.72	33.63	9899	0.51267	0.70378	3.1	−8.6	4
	97	2.06	12.73	5161	0.51278	0.70367	5.2	−10.2	4
	97	1.75	12.14	5405	0.51278	0.70355	5.2	−11.9	4
	97	2.26	15.21	5186	0.51275	0.70359	4.6	−11.3	4
	97	1.94	12.56	5671	0.51275	0.70370	4.6	−9.7	4
Oka	110	25.8	231	13740	0.51277	0.70331	5.4	−15.1	5
	110	58.1	478	7180	0.51284	0.70327	6.7	−15.6	6
(a)	110	541	5410	6340	0.51281	0.70323	6.1	−16.2	6
St Honoré	600	50.8	314.0	6450	0.51206	0.70289	3.7	−12.8	5
Burritt Island	600	44.2	279.0	7130	0.51198	0.70309	2.3	−10.0	5
Iron Island	600	10.7	138.7	20500	0.51197	0.70287	2.0	−13.1	5
Nemegosenda Lake	1015	31.4	230.1	12000	0.51122	0.70371	−2.1	5.8	unpublished data
Big Beaver House	1060	65.6	413.0	11000	0.51144	0.70244	3.4	−11.5	5
Firesand River	1060	19.6	116.3	7490	0.51139	0.70242	2.4	−11.7	5
Clay-Howells	1075	67.4	456.4	8500	0.51128	0.70372	0.6	7.0	unpublished data
Seabrook Lake	1100	78.3	592.0	2310	0.51145	0.70256	4.6	−9.1	5
Schryburt Lake	1140	22.4	140.8	8870	0.51140	0.70234	4.6	−11.5	5
Prairie Lake	1170	41.1	269.0	9580	0.51128	0.70256	3.0	−7.9	5
	1170	144.7	824.0	4240	0.51133	0.70272	3.9	−5.6	5
(a)	1170	426	2662	11000	0.51132	0.70251	3.8	−8.6	7
Spanish River	1840	14.3	92.9	4300	0.51033	0.70205	1.3	−3.7	5
Borden (c)	1870	22.4	109.9	4530	0.51031	0.70181	1.7	−6.6	5
Cargill	1900	49.2	283.0	2770	0.51027	0.70183	1.7	−5.8	5

SOUTH AMERICA

Location									Ref
Jacupiranga (c)	130	15.8	91.7	6110	0.51251	0.7051	0.8	10.7	8
(c)	130	13.3	72.7	6510	0.51256	0.7051	1.7	10.7	8
(c)	130	15.9	89.4	6960	0.51251	0.7049	0.8	7.8	8
(c)	130	4.97	26.2	3420	0.51248	0.7054	0.2	14.9	8
	130	26.3	152.5	5370	0.51248	0.7050	0.2	9.3	2
	130	15.79	88.27	6161	0.51244	0.7052	−0.6	12.1	2

References: 1, Bell & Blenkinsop (1987b); 2, Nelson *et al.* (1988); 3, Barreiro & Cooper (1987); 4, Tilton *et al.* (1987); 5, Bell & Blenkinsop (1987a); 6, Wen *et al.* (1987); 7, Pollock (1987); 8, Roden *et al.* (1985).

All data from whole rocks other than (a) apatite and (c) carbonate. Epsilon values calculated using the following values for bulk Earth:
$^{87}Sr:^{86}Sr = 0.7045$, $^{87}Rb:^{86}Sr = 0.0827$, $^{143}Nd:^{144}Nd = 0.512638$, $^{147}Sm:^{144}Nd = 0.1967$.

b contents of all samples are <10 p.p.m.
* Initial ratios.

hydrothermal origin (Powell 1965*a*), and the relatively high value of 0.706 for the Amba Dongar carbonatite from India (Deans & Powell 1968).

Hand-in-hand with these studies went the controversy involving the significance of initial $^{87}Sr:^{86}Sr$ ratios in distinguishing carbonatites from limestones and marbles (Gittins *et al.* 1970), a discussion that continued until 1978 (Iyer *et al.* 1978). As the number of analyses increased it became clear that there was some degree of overlap between the measurements from these two groups of rocks, although it is still true that most carbonatites have lower initial $^{87}Sr:^{86}Sr$ ratios than other carbonate-rich rocks. Pineau & Allègre (1972) also pointed out that the involvement of older carbonates in the genesis of carbonatites cannot be excluded because they, too, have low $^{87}Sr:^{86}Sr$ ratios, and also presented evidence for interaction between carbonatitic magma and radiogenic Sr from continental crust for some carbonatites from Angola.

Within the last few years high-precision isotope measurements have helped constrain the source of carbonatites and unravel the chemical characteristics of the subcontinental mantle. Order of magnitude better precision in ratio measurements showed subtle variations in the isotopic signatures of carbonatites that clearly separated them from limestones, and with this followed a more detailed understanding of carbonatite evolution and genesis.

An example of one such study is our investigation of the Sr and Nd initial ratios of carbonatites and some silicate rocks from a major alkalic province that occurs over an area of at least 1 million km² (Erdosh 1979). Samples of different age from the Grenville and Superior Provinces of the Canadian Shield (Bell *et al.* 1982, Bell &

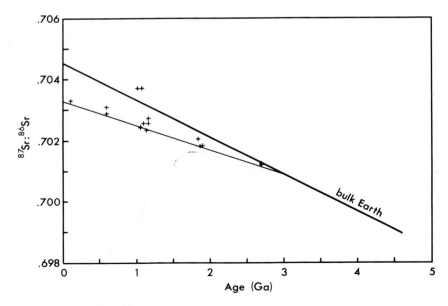

Figure 12.1 Plot of initial $^{87}Sr:^{86}Sr$ ratios v. time for Canadian carbonatites. Data from Bell & Blenkinsop (1987*a*), Bell *et al.* (1987), Pollock (1987), Sharpe (1987), Wen *et al.* (1987). The development line for carbonatites has been drawn through the lowest initial $^{87}Sr:^{86}Sr$ ratios for each group of data. Shown for comparison is the development line for bulk Earth. Data from syenite from the 2700 Ma Poohbah complex are included.

Blenkinsop 1987a) have Sr initial ratios that define a reasonable straight line when plotted as a function of age. All but one of the samples, a syenite from the Poohbah complex, were carbonatites. The data, given in Figure 12.1, are from complexes that range in age from 2700 Ma (Poohbah complex; Mitchell 1976) to 110 Ma (Oka; Wen et al. 1987). The simplest interpretation of the linear relationship defined by those samples with the lowest $^{87}Sr:^{86}Sr$ ratios involves a source for the Sr contained in the carbonatites that has behaved as a closed system with respect to exchange of Rb and Sr over the range in ages of the complexes from 2700 to 110 Ma. From the slope of the line an estimate of the Rb:Sr ratio of the source region was made; the value of 0.020 ± 0.002 obtained is considerably lower than that of bulk Earth. The intersection of the carbonatite line with the bulk Earth development line, at about 3 Ga, was taken to represent the mean age of formation of the carbonatite source from bulk Earth-type material by a single-stage process. Because the same reservoir has been the source of Sr in complexes of such different ages, we suggested that the depleted reservoir probably lay within the subcontinental lithosphere, the complementary enriched reservoir being the continental crust of the Superior Province. Recent seismic studies (Silver & Chan 1988) are consistent with this model.

Recent work by Nelson et al. (1988) has shown that the ε_{Sr} values for carbonatites of different ages from other continents are generally negative, with the younger carbonatites having more negative values, a pattern similar to that observed for the Canadian carbonatites. Lack of an adequate data base for carbonatites from other continents prevented Nelson et al. (1988) from making any estimates of the Rb:Sr ratios of the carbonatite sources.

12.3.2 Nd chemistry

The first measurements of the Nd isotopic composition of carbonatites (DePaolo 1978, Basu & Tatsumoto 1980) yielded epsilon values similar to those found in oceanic island basalts, indicating a source with a time-averaged Sm:Nd ratio greater than bulk Earth. On the basis of this information, Basu & Tatsumoto (1980) suggested the presence of sub-oceanic mantle below the continents and noted that the sources of carbonatites are unrelated to the mantle source of kimberlites.

Further investigations (Bell & Blenkinsop 1985, 1986, 1987a, Bell et al. 1987) of the Canadian carbonatites demonstrated that the Nd initial ratios support the findings from the Sr data. The Nd in the complexes has been derived from a depleted source characterized by a Sm:Nd ratio of 0.36 ± 0.01, higher by about 10% than the value of CHUR (DePaolo & Wasserburg 1976, Jacobsen & Wasserburg 1980). The pattern in Figure 12.2 is not as definitive as that for the Rb–Sr data because both parent and daughter elements are geochemically similar and hence do not fractionate relative to their source to the same extent as Rb and Sr. Furthermore, initial $^{143}Nd:^{144}Nd$ ratios cannot be measured directly because mineral phases with very low Sm:Nd ratios are not common in igneous rocks, particularly carbonatites, and hence corrections have to be made for in situ growth of radiogenic ^{143}Nd.

Most carbonatites have initial $^{143}Nd:^{144}Nd$ ratios higher than the corresponding values for CHUR, a phenomenon now shown to be of global extent (Nelson et al.

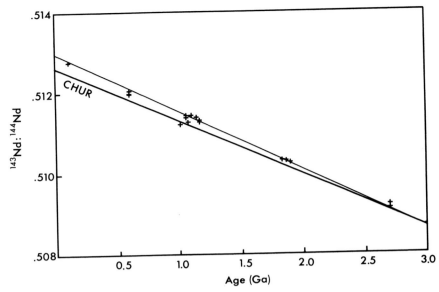

Figure 12.2 Plot of initial ^{143}Nd : ^{144}Nd ratios v. time for Canadian carbonatites. Data from Bell & Blenkinsop (1987a), Bell *et al.* (1987), Pollock (1987), Sharpe (1987), Wen *et al.* (1987). Shown for comparison is the development line for CHUR. Data from syenite from the 2700 Ma Poohbah complex are included.

1988). The estimated Sm : Nd ratio for the source of the Nd contained within Canadian carbonatites is higher than the value for bulk Earth, indicating a source depleted in the LIL-elements relative to bulk Earth. Both the Rb : Sr and Sm : Nd ratios for the carbonatite source are consistent with a residue resulting from extraction of the LIL elements during partial melting.

12.3.3 The anti-correlation plot

Figure 12.3 summarizes the isotope data from young carbonatites (< 40 Ma old), mid-ocean ridge basalts (MORBs) and oceanic island basalts (OIBs). The relatively tight field in the upper left-hand corner of the plot shows that MORBs were derived from an ancient depleted mantle source, while the spread of the OIB data from the depleted into the enriched quadrant of the ACP suggests a more heterogeneous, but still long-lived, mantle reservoir. Although the carbonatite data closely follow the oceanic island trend, in particular that based on data from the Atlantic Ocean, they do tend to lie below and to the left of the main mantle array, a feature that seems to characterize most carbonatites. It is interesting to note that Menzies & Wass (1983) argued that mantle metasomatized by CO_2-rich fluids will generate, with time, isotopic ratios that lie to the left of the mantle array.

High-precision data for Nd and Sr from carbonatites of different ages and from different continents are shown on an $\varepsilon_{Nd}(T)$ v. $\varepsilon_{Sr}(T)$ diagram (Fig. 12.4). This type of plot, however, is limited because data from rocks of different ages cannot be directly compared. The $\varepsilon(T)$ values reflect the difference between the isotopic ratios of a sample and the reference reservoir at one point in time, and under closed-system

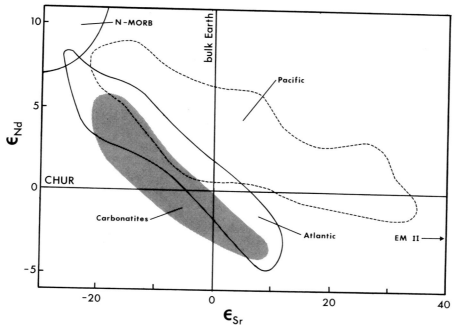

Figure 12.3 Plot of $\varepsilon_{Nd}(T)$ v. $\varepsilon_{Sr}(T)$ summarizing data from MORBs, OIBs, and young carbonatites. Based on data from DePaolo (1978), Hart *et al.* (1986), Barreiro & Cooper (1987), Bell & Blenkinsop (1987*b*), Nelson *et al.* (1988).

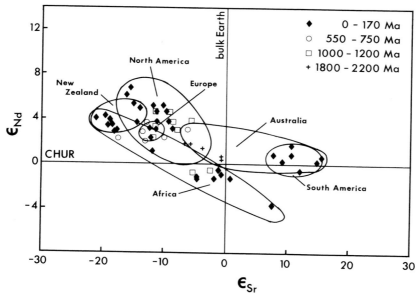

Figure 12.4 Plot of $\varepsilon_{Nd}(T)$ v. $\varepsilon_{Sr}(T)$ showing carbonatite data from different continents. Based on data from Roden *et al.* (1985), Barreiro & Cooper (1987), Bell & Blenkinsop (1987*a, b*), Bell *et al.* (1987), Pollock (1987), Sharpe (1987), Tilton *et al.* (1987), Wen *et al.* (1987), and Nelson *et al.* (1988). Not plotted are the data from Clay-Howells, Nemegosenda, and Poohbah (Canada) and the carbonatite data from Phalaborwa (South Africa).

conditions the ε values for a given reservoir will increase as a function of time. In spite of this limitation, Figure 12.4 is useful because it does show that almost all of the data fall within the depleted quadrant, indicating that most carbonatites evolved from a melt that was generated from a reservoir depleted in LILEs relative to bulk Earth. No clear-cut distinction can be made between the carbonatite data from the Northern and Southern Hemispheres that reflect a mantle enriched in radiogenic daughters characteristic of the widespread Dupal anomaly found in the Southern Hemisphere (Hart 1984).

Data from young African carbonatites are presented in Figure 12.5, along with data from two nephelinites from Kisingiri, a volcano that lies close to the Homa Bay carbonatite. The linear array extends from the depleted into the enriched quadrant, and of significance is the fact that most data fall well within the depleted quadrant, and closely follow the trend shown by some of the OIB data (see Fig. 12.3). A generally similar trend is shown by data from the 1100 Ma complexes of the Canadian Shield. The linear array in Figure 12.5 can result from either mixing of two reservoirs (one depleted, the other slightly enriched or similar to bulk Earth) or sampling of several ancient reservoirs formed at the same time from the same parent material.

Figure 12.6 compares data from different parts of the African plate. Excluded are data from the Ugandan lamproites (Vollmer & Norry 1983a, b), and the Pha-

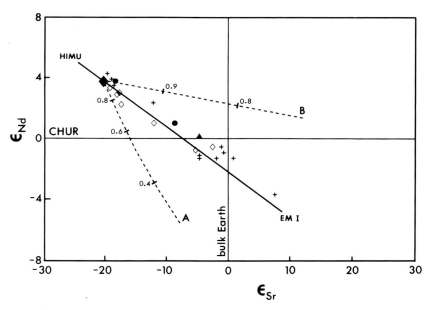

Figure 12.5 Plot of $\varepsilon_{Nd}(T)$ v. $\varepsilon_{Sr}(T)$ for data from East African carbonatites. Data from: (+) Bell & Blenkinsop (1987b); (◇) Nelson et al. (1988); (▲) DePaolo (1978). Nephelinite data (●) from Norry et al. (1980). Also shown is the 'LoNd' line of Hart et al. (1986). Two mixing curves (---) are shown. A parent magma (◆) (Nd = 130 p.p.m., $^{143}Nd:^{144}Nd = 0.51283$, Sr = 2800 p.p.m., $^{87}Sr:^{86}Sr = 0.70307$) is mixed with lower crust, curve A (Nd = 20 p.p.m., $^{143}Nd:^{144}Nd = 0.5110$, Sr = 115 p.p.m., $^{87}Sr:^{86}Sr = 0.713$) and with upper crust, curve B (Nd = 32 p.p.m., $^{143}Nd:^{144}Nd = 0.5113$, Sr = 405 p.p.m., $^{87}Sr:^{86}Sr = 0.747$). Numbers on mixing curves refer to fraction of parent magma in the mixture.

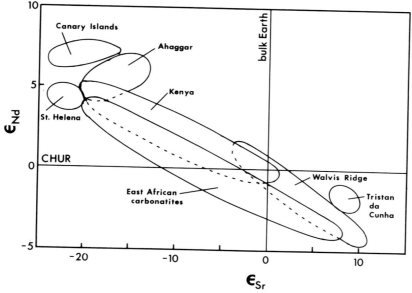

Figure 12.6 Comparison of data from East African carbonatites with data from basalts and nephelinites from other parts of the African plate. Data from Norry et al. (1980), Allègre et al. (1981), Richardson et al. (1982), White & Hofmann (1982), and Staudigel et al. (1984). Data from lamproites and the Phalaborwa complex are excluded.

laborwa complex of South Africa (see Eriksson, this volume) that lie well within the enriched quadrant. The Phalaborwa data from three samples (Eriksson, pers. comm.), with ε_{Nd} values of -4.2 to -7.3 and ε_{Sr} of 26.7 to 113.1, appear to be unique for carbonatites. Data fields from some young oceanic and continental volcanic rocks from other parts of the African plate, shown on Figure 12.6, overlap the carbonatite data.

The relationship between carbonatites and kimberlites is still the source of much controversy, and although both Haggerty and Wyllie (this volume) argue for a genetic link between the two, neither feels it is direct. The Nd and Sr isotopic compositions can distinguish between basaltic and micaceous kimberlites (Smith 1983). Data from both types are plotted in Figure 12.7. Data from Group I (basaltic) kimberlites lie to the right of the East African carbonatite line, a feature that is supported by additional data from Weis & Demaiffe (1985). Most data from basaltic kimberlites overlap those from oceanic islands and most are characterized by constant $^{143}Nd:^{144}Nd$ and variable $^{87}Sr:^{86}Sr$ ratios. On the other hand, the high $^{87}Sr:^{86}Sr$ and low $^{143}Nd:^{144}Nd$ ratios for the Group II (micaceous) kimberlites fall well within the enriched quadrant, and show that the source regions for basaltic and micaceous kimberlites are isotopically quite distinct, one depleted and the other enriched, and that they have been separated from one another for some considerable period of time.

Group I kimberlite and East African carbonatite data have some similarities. The $^{143}Nd:^{144}Nd$ ratios from both suggest a depleted source, both have similar spreads in $^{87}Sr:^{86}Sr$ ratios, and two of the kimberlite points fall directly on the East African

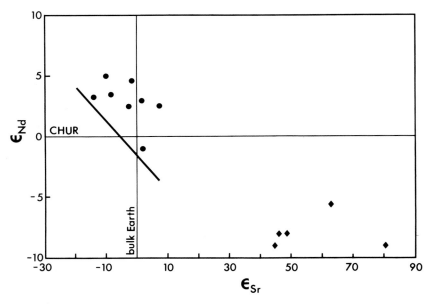

Figure 12.7 Data from kimberlites (Smith 1983, Weis & Demaiffe 1985) and carbonatites. The Group I (basaltic) kimberlite points (●) lie close to the East African carbonatite line, while the Group II (micaceous) kimberlite points (♦) fall well within the enriched quadrant. Data from kimberlite nodules have not been included.

line. Smith (1983) considers the Group I kimberlites to be derived from an undifferentiated to slightly depleted source in terms of Sr and Nd, but with high U : Pb ratios. The Nd and Sr data and the few Pb analyses presently available from East African carbonatites (Tilton & Bell, unpublished data) show many features in common with basaltic kimberlites, but resolution of any genetic relationships awaits further study.

12.4 ISOTOPIC CONSTRAINTS ON THE EVOLUTION OF CARBONATITE MAGMAS

Models proposed for the generation of carbonatitic magmas include:

(a) direct partial melting of the upper mantle;
(b) fractional crystallization of a basic silicate magma; and
(c) separation of an immiscible fraction from a silicate magma.

These models are discussed in some detail in other papers of this volume, and we do not propose to elaborate on them in this chapter. It is unfortunate that the isotopic data do not enable us to favour any one of these models. The isotopic data can, however, be used to evaluate the behaviour of the REEs during magma generation and evolution, the relationship between carbonatites and associated silicate rocks, and the cause of the isotopic variation in some carbonatites from the same complex.

12.4.1 Behaviour of the REEs

The average Sm:Nd ratio directly measured from Canadian carbonatites is about 0.15 (Bell & Blenkinsop 1987a, Pollock 1987, Wen et al. 1987), a value considerably lower than the ratio of 0.36 calculated by Bell & Blenkinsop (1987a) for the source of Nd in the carbonatites. Whatever process produced the carbonate melts, it had the overall effect of enriching the light REEs in the melt relative to its source. Extreme light REE enrichment of carbonatites has been noted by numerous other workers (e.g. Eby 1975, Cullers & Medaris 1977, Bowden 1985, Nelson et al. 1988, van Straaten, this volume).

The near-linear relationship between the ^{143}Nd:^{144}Nd ratio and time, suggested in Figure 12.2, indicates that such enrichments in the LREEs, i.e. a reduction in the Sm:Nd ratio, must have occurred close to the time of magma genesis, otherwise the linear relationship would not be maintained. The simplest way of producing the LREE enrichment observed in all carbonatites appears to be preferential partitioning of the LREEs into a melt during either magma formation or magma differentiation. Immiscibility (Hamilton et al., this volume), vapour phase transfer (Wendlandt & Harrison 1979), or small degrees of partial melting of a garnet-rich source (Nelson et al. 1988) are all possible ways of producing enrichment in the LREEs, a feature that characterizes all carbonatites.

12.4.2 Relationship between carbonatites and associated silicate rock

Many workers have drawn attention to the high abundance of silicate rocks such as pyroxenites, ijolites, and syenites that are associated with many carbonatites, and have used this as an argument in favour of producing carbonatites by liquid immiscibility or crystal fractionation. Although the number of studies that include isotope ratio measurements from both carbonatite and related silicate rocks from the same intrusion is comparatively small, those available suggest that the silicate rocks have undergone a complex history.

Liquid immiscibility and crystal fractionation should have no effect on Nd and Sr isotopic compositions, and if these processes were the only ones operating during magma evolution then the initial ^{87}Sr:^{86}Sr and ^{143}Nd:^{144}Nd ratios should be the same in both carbonatite and silicate melts.

Carbonate and silicate rocks from the Oka complex, Canada (Wen et al. 1987) have isotopic ratios that are fairly similar, although some variation occurs that lies just outside the limits of experimental uncertainties. These differences, of the order of 0.02%, for the initial ^{87}Sr:^{86}Sr and ^{143}Nd:^{144}Nd ratios, are similar to those observed in other complexes (Pollock 1987, Sharpe 1987). Although small, these differences suggest lower initial ^{87}Sr:^{86}Sr and higher initial ^{143}Nd:^{144}Nd ratios for carbonatites than their silicate counterparts.

A more definitive story emerges from Oldoinyo Lengai (Bell & Dawson, in preparation), where Sr and Nd isotopic data from both carbonate and silicate rocks define a trend identical to that found for other East African carbonatites on an ACP (Bell & Blenkinsop 1987b, Dawson, this volume). Data from the natrocarbonatites lie close to bulk Earth on the ACP (see Fig. 12.5), and are bracketed by those from the silicate rocks. The ε values, however, from trachytes, phonolites, and nephelinites fall well within the enriched quadrant (ε_{Nd} values of about -4, ε_{Sr} values of about $+11$), indicating that processes other than simple magmatic differentiation were involved in the generation of some of the Oldoinyo Lengai rocks. Simple crystal fractionation or liquid immiscibility cannot, by themselves, produce the observed variations in the Nd and Sr isotopic data. Additional processes such as magma mixing, crustal contamination, or metasomatism have to be invoked.

12.4.3 Isotopic variations within carbonatites

Most studies to date are based on a limited number of samples from any one carbonatite, and few have addressed the question of the isotopic variation of carbonatites within the same complex. In some cases the isotope ratios within a single complex are relatively uniform (Wen et al. 1987), while in others significant Sr isotopic variation have been documented (Pineau & Allègre 1972, Basu & Puustinen 1982, Nielsen & Buchart 1985, Roden et al. 1985).

Isotopic variation within some carbonatites has been attributed to crustal contamination, promoted by (a) the high reactivity of carbonatitic melts with crust, and (b) heat released during crystallization (Roden et al. 1985). An inverse correlation between initial $^{87}Sr:^{86}Sr$ and Sr contents from several carbonatite samples, mostly from the Tchivira complex of Angola (Pineau & Allègre 1972), was attributed to contamination of a magma by a fluid phase enriched in radiogenic Sr. Simple mixing of a carbonatite magma with granite was ruled out because the elevated Sr contents of the carbonatites require enormous amounts of contaminant in order to change the Sr isotopic composition of the carbonatitic melt. Similar conclusions can be drawn using models that involve both Nd and Sr. Binary mixing between a carbonatite magma (4000 p.p.m. Sr, 330 p.p.m. Nd) and continental crust (upper or lower) either fail to generate the Nd and Sr patterns observed in the East African carbonatites or require unrealistically large amounts of crustal assimilation (Bell & Blenkinsop 1987b). If carbonatite liquids are derivative, then an alternative model can be invoked that involves contamination of a mafic magma, such as nephelinite, at crustal levels followed by differentiation to form a carbonatite liquid. Even assuming a parent nephelinite magma with 2800 p.p.m. Sr and 130 p.p.m. Nd (e.g. Fadaie 1987), values considerably lower than those used in the example given earlier, large amounts of bulk assimilation of continental crust are still required to produce some of the isotope variation observed in Figure 12.5. The amount of continental crust that has to be assimilated in order to produce the isotope variations can be seen from some examples of binary mixing curves plotted on Figure 12.5.

Isotopic variation observed in four carbonatite samples from the Jacupiranga complex of Brazil were duplicated by Roden et al. (1985) using the AFC (assimilation fractional crystallization) model (DePaolo 1981). In this model both fractional crystallization and assimilation take place simultaneously, which is probably more realistic in terms of what actually happens in an evolving magma. By assuming a parent nephelinite or basaltic magma, Roden *et al.* (1985) were able to model the data by using distribution coefficients for the precipitating minerals of either 0.4 or 0.5 for Sr, and either 0.5 or 0.7 for Nd. Few measurements have been made of the distribution coefficients of minerals such as calcite, dolomite, and apatite associated with carbonatites. Differentiation at crustal levels, as well as interaction between carbonate melt and country rock, no doubt occur during emplacement, but the degree and extent are as yet unknown and more detailed studies of individual complexes should offer valuable insights into the late-stage evolution of carbonatite melts.

12.5 THE SOURCE REGION OF CARBONATITES – LITHOSPHERE OR ASTHENOSPHERE?

12.5.1 The models

Although the number of isotopic analyses from carbonatites is relatively small compared to that from other mantle-derived rocks such as basalts, attempts have been made to compare the carbonatite data to those from known, depleted mantle reservoirs. Three mantle sources have so far been proposed for carbonatitic melts or their parents, based on isotopic data:

(a) a lithosphere depleted by crustal extraction (Bell & Blenkinsop 1987*a*);
(b) a depleted lithosphere modified by metasomatism (Barreiro & Cooper 1987, Meen 1987, Meen *et al.*, this volume); and
(c) asthenosphere (Nelson *et al.* 1988, Kwon *et al.*, this volume).

We chose to interpret our earlier data from Canadian carbonatites in terms of a lithospheric origin for the carbonatitic liquids, basing our argument on the close agreement between the age of the Superior crustal segment and the intersection age calculated from the development lines for bulk Earth and the source of the Nd and Sr contained in carbonatites. Although most Canadian carbonatites seem to have been derived from one source, some members of the 1100 Ma group of complexes have initial $^{87}Sr:^{86}Sr$ ratios that are similar to those of bulk Earth (Bell *et al.* 1982) and point to the tapping of more than one mantle reservoir. The involvement of more than one mantle source in carbonatite genesis is also shown by the East African carbonatite data. We originally thought of the isotopic data as reflecting large-scale reservoirs, one corresponding to a lithospheric keel favoured by some geophysicists (see Lay 1988), the other asthenosphere.

Metasomatism is another way of generating significant isotopic changes in the upper mantle. The model proposed by Barreiro & Cooper (1987) for lamprophyre and carbonatite dykes from New Zealand also involved a depleted mantle source, but one that was enriched during the Palaeozoic by the addition of CO_2, Sr, LREEs, and U by carbonated hyperalkaline magma. A second event triggered melting of the metasomatized mantle, with resulting dyke swarm emplacement. Barreiro & Cooper

(1987), Meen (1987) and Meen *et al.* (this volume) have elaborated on the phase relations of a carbonated, hydrated peridotite, the generation of a carbonated hyperalkaline magma and its truncation with the peridotite solidus, and the overall effect this will have on the isotopic composition of any melt produced by melting of the metasomatized mantle. Meen *et al.* (this volume) attribute the isotopic variations in carbonatites to melting of a metasomatized lithosphere produced by incursions of carbonated alkalic melt. Water- and CO_2-bearing magmas from depths in excess of 70 km react with peridotite to produce a variety of mineral assemblages as they cross the peridotite solidus at temperatures $< 1100°C$ and at pressures between 17 and 22 Kb. At pressures < 17 Kb the metasomatized mantle produced is a carbonated lherzolite (with or without amphibole), a potential carbonatite source rock. The products of high temperature metasomatism are envisaged as veins of different widths enriched in clinopyroxene relative to the surrounding depleted mantle, and the metasomatized mantle will have Rb:Sr, Sm:Nd, and U:Pb ratios controlled by partitioning of each element between mantle minerals and melt or fluid. Later melting of the metasomatized mantle can yield carbonated nephelinite or basanite magmas with relatively low $^{143}Nd:^{144}Nd$, moderate $^{87}Sr:^{86}Sr$, and high Pb ratios similar to some of the isotope data obtained from carbonatites.

Other workers (Nelson *et al.* 1988) also considered that carbonatite melts are generated from a depleted mantle source, but attribute their origin to asthenospheric mantle plumes produced at depths >350 km. Similarity between the trace element and isotopic data from OIB and carbonatites, two quite dissimilar rock types, was considered by them to be the result of the re-fertilization of a depleted source by low viscosity carbonatite melts and the extreme LREE enrichment, typical of carbonatites, was attributed to small degrees of partial melting of a garnet-rich eclogitic source, considered to be volatile-rich, subducted oceanic lithosphere.

The model implied, but not spelled out, by Nelson *et al.* (1988) certainly explains many of the chemical characteristics of carbonatites, but some features of their model require further clarification. First, the array of carbonatite and OIB data, such as that shown in Figure 12.3, has to be interpreted by Nelson *et al.* as the result of mixing between a MORB source and a carbonatitic melt. The latter end member must have Nd and Sr isotopic compositions characteristic of an enriched source, assuming simple binary mixing. Although carbonatites from Phalaborwa, a clearly anomalous case, have such characteristics, we have yet to find such a carbonatite from North America, or Europe. Secondly, the absence of any young subduction zones associated with the African plate means that any subducted material has to be trapped and stored within the mantle for substantial periods of time (Hofmann & White 1982). In addition, carbonatites have yet to be documented from any active, consuming plate margin. Thirdly, there is the absence of carbonatites from most oceanic islands. The only known examples are from the Canary and Cape Verde Islands (Silva *et al.* 1981, Le Bas 1984) and although these islands are commonly cited as typically oceanic, there is still considerable debate about whether they are underlain by continental crust.

Although the isotopic evidence presented so far is insufficient to make a simple choice between lithosphere or asthenosphere as the source for carbonatitic melts or their parental magmas, some generalizations can be made that help constrain some of the models. Any model for carbonatite genesis must explain:

(a) the similarity of the Nd and Sr isotopic data from OIBs, carbonatites, and some basaltic kimberlites;
(b) the lack of carbonatites that have Nd and Sr isotopic ratios that fall within the MORB field; and
(c) the involvement of a long-lived depleted mantle source.

12.5.2 The nature of the mantle source

Initially the isotopic evolution of the upper mantle was based on a rather simple box model, in which the mantle was divided into three large-scale reservoirs (Allègre 1982). As more isotopic data became available, it became clear that the mantle was more heterogeneous than previously thought, and the box model is now being replaced by a more complex scenario involving a number of mantle components, each with its own isotopic signature. Zindler & Hart (1986), in their review of mantle evolution, recognized at least six components based on isotopic data from oceanic basalts, and some of these have isotopic characteristics similar to those observed in carbonatites. These components can be considered as mineral phases, larger-scale reservoirs, or even fluids trapped within the mantle.

The linear array of the African data (Fig. 12.5) is very similar to the 'LoNd' line of Hart *et al.* (1986) defined by isotopic data from basalts from Tubuaii, St. Helena, the New England seamounts, the Comores, San Felix, and the Walvis Ridge. This array lies below the oceanic island field on the ACP (hence the name, low Nd), and is also marked by an inverse correlation between $^{87}Sr:^{86}Sr$ and $^{206}Pb:^{204}Pb$. The array is considered the result of mixing one end member termed HIMU (high U:Pb ratio, Tubuaii) and a second, EM-I (slightly enriched mantle I, Walvis). Because of its ubiquitous presence HIMU is considered by Hart *et al.* (1986) to be a dispersed component in the mantle. The EM-I type has a signature that is very similar to that of bulk Earth and may be produced by only minor modification of bulk Earth-type material. Many mantle xenoliths from rocks of continental setting yield data that fall on the 'LoNd' line and this led Hart *et al.* (1986) to speculate that the continental lithosphere may carry the LoNd signature. These sources would be available for convective disruption and recycling to produce OIBs. The data from African carbonatites suggested to us that the depleted source was lithospheric and the other, perhaps asthenospheric. Both sources, however, have isotopic signatures remarkably similar to the HIMU and EM-I end members proposed by Hart *et al.* (1986).

The EM-I end member can be attributed to metasomatized mantle originating either at the lithosphere–asthenosphere boundary, or at even shallower depths within the lithosphere, such as the proposed $Na + CO_2$ metasomatized layer at < 60 km (see Haggerty, this volume). The depleted source (HIMU) must be fairly widespread because most carbonatites are derived in part from this reservoir, and relative to bulk Earth, such a source requires high U:Pb and Sm:Nd ratios, coupled with low Rb:Sr ratios. Subducted old oceanic or continental crust, or mantle that has lost Pb to the core (Allègre 1982, Hofmann & White 1982, Nelson *et al.* 1988) have been suggested as possible sources, but Hart *et al.* (1986) argue on chemical grounds that this is unlikely and that the HIMU end member is metasomatically produced within the lithosphere. Unlike Hart *et al.* (1986) we prefer, at present, to attribute the isotopic signature of the HIMU component to large-scale differentiation during crustal formation that produced a depleted mantle source.

We initially thought of the sources for carbonatite melts or their parents as being large-scale reservoirs, but did not rule out the possibility that mineralogical control may be an alternative way of explaining the isotopic data (Bell & Blenkinsop 1987*b*). Jones, Kwon *et al.*, and Meen *et al.* (this volume) show that small amounts of certain mantle phases can profoundly affect melt compositions, including their isotopic signatures. Emphasizing the inverse correlation between Sr and Pb isotope ratios, Kwon *et al.* (this volume) favour garnet and clinopyroxene control, while Meen *et al.* (this volume) suggest that a U-bearing accessory mineral, such as whitlockite, may also play an important role in controlling the Pb isotopic signature of carbonatite magmas. No matter what mantle components are responsible for the African carbonatites, the linear arrays shown by ^{87}Sr:^{86}Sr v. ^{206}Pb:^{204}Pb (Kwon *et al.*, this volume), and ^{87}Sr:^{86}Sr v. ^{143}Nd:^{144}Nd plots require that the Sr:Pb and Sr:Nd ratios be the same in both end members if these can be attributed to mixing.

It is difficult to assess the scale of mantle heterogeneity and extent of mantle mixing in carbonatite genesis, but the spatial relationships among carbonatites within the same petrographic province and their related silicate rocks present evidence for localized isotopic differences. In the Kapuskasing Structure in the Superior Province of the Canadian Shield, alkalic complexes with depleted mantle signatures are located only a few kilometres away from carbonatites of the same age with bulk Earth values. The same is true in East Africa. Samples from Kisingiri, less than 50 km from Homa Bay, have depleted signatures while those from the Homa Bay carbonatite suggest a slightly enriched source. Together these observations suggest that, whatever process produced the carbonatites and nephelinites, it had to have occurred on a very local scale within the mantle.

12.6 CONCLUSIONS

Several conclusions can be drawn from the Nd and Sr initial isotopic compositions of carbonatites:

(a) The close similarity between the isotopic data from most carbonatites and some oceanic islands suggest that the parental magmas to carbonatites are produced within the mantle.
(b) Most isotopic data from carbonatites indicate the involvement of a long-lived, depleted mantle source. In the case of the Canadian carbonatites this source is about 3 Ga old.

(c) The linear array displayed by the Nd and Sr isotopic data from young African carbonatites, and from some of the Proterozoic Canadian carbonatites suggests more than one mantle source. If such an array is attributed to simple binary mixing, one end member is depleted and the other is similar to, or slightly more enriched than, bulk Earth. Both end members have similar Nd:Sr ratios. These two sources are probably contiguous, and are thoroughly mixed to provide discrete magma batches.

The similarity between the isotopic data from OIBs and carbonatites suggests that both originated from the same sources or from source regions that underwent similar differentiation histories. Either carbonatitic melts (or their parents) are produced from the asthenosphere, in which case a mantle plume model is favoured, or alternatively from ancient lithosphere, in which case oceanic islands have to be considered the products of lithospheric delamination (McKenzie & O'Nions 1983, Hart *et al.* 1986).

In summary, any model for carbonatite genesis must explain the virtual absence of carbonatites in most oceanic areas; the similar isotopic signatures in carbonatites, OIBs, as well as some basaltic kimberlites; and the involvement of a long-lived, depleted mantle source. Although it is difficult, at present, to choose between a lithospheric or asthenospheric source, we tend to favour the former. Carbonatitic activity over millions of years in restricted parts of the Earth's crust (e.g. Kapuskasing Structural Zone, Canada; southern Greenland) as well as the near agreement in age between the crustal rocks of the Superior Province of Canada and the estimated age of chemical differentiation of the subcontinental upper mantle suggest strong lithospheric control in the generation of carbonatites. If, on the other hand, the melts are from the asthenosphere, either the convecting asthenosphere was unable to purge itself of a Precambrian memory or the asthenosphere stagnated for about 3 Ga in both continental and oceanic environments.

ACKNOWLEDGEMENTS

The work at Carleton was initiated by a grant from the Ontario Geological Survey. Subsequent support was from NSERC operating grants A-7813 and A-7960, and equipment grants E-4167 and E-5761. R. P. Sage (Ontario Geological Survey) is especially thanked for providing samples and leading us to others. J. W. Card is thanked for his invaluable contribution to the manuscript at all stages. Thoughtful reviews by G. Faure and M. F. Roden significantly improved the paper.

REFERENCES

Allègre, C. J. 1982. Chemical geodynamics. *Tectonophysics* **81**, 109–32.
Allègre, C. J., D. Ben Othman, M. Polvé & P. Richard 1979. The Nd-Sr isotopic correlation in mantle materials and geodynamic consequences. *Physics of the Earth and Planetary Interiors* **19**, 293–306.

Allègre, C. J., B. Dupré, B. Lambret & P. Richard 1981. The subcontinental versus suboceanic debate, I. Lead–neodymium–strontium isotopes in primary alkali basalts from a shield area: The Ahaggar volcanic suite. *Earth and Planetary Science Letters* **52**, 85–92.

Allsopp, H. L. & S. C. Eriksson 1986. The Phalaborwa complex; isotopic evidence for ancient lithospheric enrichment. *Geological Association of Canada–Mineralogical Association of Canada–Canadian Geophysical Union, Joint Annual Meeting, Ottawa. Program with Abstracts* **11**, 40.

Barreiro, B. A. & A. F. Cooper 1987. A Sr, Nd, and Pb isotope study of alkaline lamprophyres and related rocks from Westland and Otago, South Island, New Zealand. In *Mantle metasomatism and alkaline magmatism*, E. M. Morris & J. D. Pasteris (eds), 115–25. Geological Society of America Special Paper 215.

Basaltic Volcanism Study Project 1981. *Basaltic Volcanism on the terrestrial planets*. New York: Pergamon.

Basu, A. R. & K. Puustinen 1982. Nd-isotopic study of the Siilinjarvi carbonatite complex, eastern Finland and evidence of early Proterozoic mantle enrichment. *Geological Society of America, Abstracts with Programs* **14**, 440. (abstract)

Basu, A. R. & M. Tatsumoto 1980. Nd-isotopes in selected mantle-derived rocks and minerals and their implications for mantle evolution. *Contributions to Mineralogy and Petrology* **75**, 43–54.

Bell, K. & J. Blenkinsop 1985. Carbonatites – clues to mantle evolution. *Geological Society of America, Abstracts with Programs* **17**, 151. (abstract)

Bell, K. & J. Blenkinsop 1986. Evolution of carbonatites. *Geological Association of Canada–Mineralogical Association of Canada–Canadian Geophysical Union, Joint Annual Meeting, Ottawa. Program with Abstracts* **11**, 44.

Bell, K. & J. Blenkinsop 1987a. Archean depleted mantle – evidence from Nd and Sr initial isotopic ratios of carbonatites. *Geochimica et Cosmochimica Acta* **51**, 291–8.

Bell, K. & J. Blenkinsop 1987b. Nd and Sr isotopic compositions of East African carbonatites: Implications for mantle heterogeneity. *Geology* **15**, 99–102.

Bell, K., J. B. Dawson & R. M. Farquhar 1973. Strontium isotope studies of alkalic rocks: The active carbonatite volcano Oldoinyo Lengai. *Geological Society of America* **84**, 1019–30.

Bell, K., J. Blenkinsop, T. J. S. Cole & D. P. Menagh 1982. Evidence from Sr isotopes for long-lived heterogeneities in the upper mantle. *Nature* **298**, 251–3.

Bell, K., J. Blenkinsop, S. T. Kwon, G. R. Tilton & R. P. Sage 1987. Age and radiogenic isotopic systematics of the Borden carbonatite complex, Ontario, Canada. *Canadian Journal of Earth Sciences* **24**, 24–30.

Bowden, P. 1985. The geochemistry and mineralization of alkaline ring complexes in Africa (a review). *Journal of African Earth Sciences* **3**, 17–39.

Cullers, R. L. & G. Medaris, Jr. 1977. Rare-earth elements in carbonatite and cogenetic alkaline rocks: examples from Seabrook Lake and Callandar Bay, Ontario. *Contributions to Mineralogy and Petrology* **65**, 143–53.

Deans, T. & J. L. Powell 1968. Trace elements and strontium isotopes in carbonatites, fluorites, and limestones from India and Pakistan. *Nature* **218**, 750–2.

DePaolo, D. J. 1978. Nd and Sr isotope systematics of young continental igneous rocks. In *Short Papers of the Fourth International Conference, Geochronology, Cosmochronology, Isotope Geology*, R. E. Zartman (ed.), 91–3. United States Geological Survey Open-File Report 78–701.

DePaolo, D. J. 1981. Trace element and isotopic effects of combined wallrock assimilation and fractional crystallization. *Earth and Planetary Science Letters* **53**, 189–202.

DePaolo, D. J. & G. J. Wasserburg 1976. Nd isotopic variations and petrogenetic models. *Geophysical Research Letters* **3**, 249–52.

DePaolo, D. J. & G. J. Wasserburg 1979. Petrogenetic mixing models and Nd-Sr isotopic patterns. *Geochimica et Cosmochimica Acta* **43**, 615–27.

Eby, G. N. 1975. Abundance and distribution of the rare earth elements and yttrium in the rocks and minerals of the Oka carbonatite complex, Quebec. *Geochimica et Cosmochimica Acta* **39**, 597–620.

Erdosh, G. 1979. The Ontario carbonatite province and its phosphate potential. *Economic Geology* **74**, 331–8.

Fadaie, K. 1987. *Geophysical and isotopic constraints on the East African Rift System*. Ph.D. thesis, Carleton University, Ottawa.

Faure, G. 1986. *Principles of isotope geology*. New York: Wiley.

Faure, G. & J. L. Powell 1972. *Strontium isotope geology*. Berlin: Springer.

Gast, P. W. 1968. Trace element fractionation and the origin of tholeiitic and alkaline magma types. *Geochimica et Cosmochimica Acta* **32**, 1057–86.

Gittins, J., A. Hayatsu & D. York 1970. A strontium isotope study of metamorphosed limestones. *Lithos* **3**, 51–8.

Grünenfelder, M. H., G. R. Tilton, K. Bell & J. Blenkinsop 1986. Lead and strontium isotope relationships in the Oka carbonatite complex, Quebec. *Geochimica et Cosmochimica Acta* **50**, 461–8.

Hamilton, E. I. & T. Deans 1963. Isotopic composition of strontium in some African carbonatites and limestones and in strontium minerals. *Nature* **198**, 776–7.

Hart, S. R. 1984. A large-scale isotope anomaly in the Southern Hemisphere mantle. *Nature* **309**, 753–7.

Hart, S. R., D. C. Gerlach & W. M. White 1986. A possible new Sr–Nd–Pb mantle array and consequences for mantle mixing. *Geochimica et Cosmochimica Acta* **50**, 1551–7.

Hofmann, A. W. & W. M. White 1982. Mantle plumes from ancient oceanic crust. *Earth and Planetary Science Letters* **57**, 421–36.

Iyer, S. S. S., P. J. Woodford & A. F. Wilson 1978. Carbonatites and granulite facies marbles, Strangways Range, Central Australia: Sr isotopic investigation. In *Proceedings of the First International Symposium on Carbonatites, Poços de Caldas, Brazil*, June 1976, 119–20. Brasilia: Brasil Departamento Nacional da Produção Mineral.

Jacobsen, S. B. & G. J. Wasserburg 1980. Sm–Nd isotopic evolution of chondrites. *Earth and Planetary Science Letters* **50**, 139–55.

Lay, T. 1988. The deep roots of continents. *Nature* **333**, 209–10.

Le Bas, M. J. 1984. Oceanic carbonatites. In *Kimberlites I. Kimberlites and related rocks*. Proceedings of the Third International Kimberlite Conference, J. Kornprobst (ed.), 169–78. New York: Elsevier.

McKenzie, D. 1985. The extraction of magma from the crust and mantle. *Earth and Planetary Science Letters* **74**, 81–91.

McKenzie, D. & R. K. O'Nions 1983. Mantle reservoirs and ocean island basalts. *Nature* **301**, 229–31.

Meen, J. K. 1987. Mantle metasomatism and carbonatites; An experimental study of a complex relationship. In *Mantle metasomatism and alkaline magmatism*, E. M. Morris & J. D. Pasteris (eds), 91–100. Geological Society of America Special Paper 215.

Menzies, M. A. & S. Y. Wass 1983. CO_2- and LREE-rich mantle below eastern Australia; a REE and isotopic study of alkaline magmas and apatite-rich mantle xenoliths from Southern Highlands Province, Australia. *Earth and Planetary Science Letters* **65**, 287–302.

Mitchell, R. H. 1976. Potassium–argon geochronology of the Poohbah Lake alkaline complex, northwestern Ontario. *Canadian Journal of Earth Sciences* **13**, 1456–9.

Nelson, D. R., A. R. Chivas, B. W. Chappell & M. T. McCulloch 1988. Geochemical and isotopic systematics in carbonatites and implications for the evolution of ocean-island sources. *Geochimica et Cosmochimica Acta* **52**, 1–17.

Nielsen, T. F. D. & B. Buchardt 1985. Sr–C–O isotopes in nephelinitic rocks and carbonatites, Gardiner Complex, Tertiary of east Greenland. *Chemical Geology* **53**, 207–17.

Norry, M. J., P. H. Truckle, S. J. Lippard, C. J. Hawkesworth, S. D. Weaver & G. F. Marriner 1980. Isotopic and trace element evidence from lavas bearing on mantle heterogeneity beneath Kenya. *Philosophical Transactions of the Royal Society of London* **A297**, 259–71.

O'Nions, R. K. 1987. Relationships between chemical and convective layering in the Earth. *Journal of the Geological Society, London* **144**, 259–74.

Pineau, F. & C. J. Allègre 1972. Étude et signification des rapports isotopiques $^{87}Sr : ^{86}Sr$ dans les carbonatites. *Comptes Rendus Hebdomadaires des Séances de l'Académie des Sciences Paris* **D274**, 2620–3.

Pollock, S. J. 1987. *The isotopic geochemistry of the Prairie Lake carbonatite complex, Ontario.* M.Sc. thesis, Carleton University, Ottawa.

Powell, J. L. 1965a. Isotopic composition of strontium in four carbonatite vein-dikes. *American Mineralogist* **50**, 1921–8.

Powell, J. L. 1965b. Low abundance of ^{87}Sr in Ontario carbonatites. *American Mineralogist* **50**, 1075–9.

Powell, J. L. 1965c. Isotopic composition of strontium in carbonate rocks from Keshya and Mkwisis, Zambia. *Nature* **206**, 288–9.

Powell, J. L. 1966. Isotopic composition of strontium in carbonatites and kimberlites. In *International*

Mineralogical Association Fourth General Meeting Papers and Proceedings, P. R. J. Naidu & M. N. Viswanathiah (eds), 58–66. Mysore: Mineralogical Society of India.

Powell, J. L., P. M. Hurley & H. W. Fairbairn 1962. Isotopic composition of strontium in carbonatites. *Nature* **196**, 1085–6.

Powell, J. L., P. M. Hurley & H. W. Fairbairn 1966. The strontium isotopic composition and origin of carbonatites. In *Carbonatites*, O. F. Tuttle & J. Gittins (eds), 365–78. London: Wiley.

Richardson, S. H., A. J. Erlank, A. R. Duncan & D. L. Reid 1982. Correlated Nd, Sr and Pb isotope variation in Walvis Ridge basalts and implications for the evolution of their mantle source. *Earth and Planetary Science Letters* **59**, 327–42.

Richter, F. M. 1985. Simple models for trace element fractionation during melt segregation. *Earth and Planetary Science Letters* **77**, 333–44.

Roden, M. F., V. R. Murthy & J. C. Gaspar 1985. Sr and Nd isotopic composition of the Jacupiranga carbonatite. *Journal of Geology* **93**, 212–20.

Sharpe, J. L. 1987. Geochemistry of the Cargill carbonatite complex, Kapuskasing, Ontario. M.Sc. thesis, Carleton University, Ottawa.

Shaw, D. M. 1976. Additional estimates of continental surface Precambrian Shield composition in Canada. *Geochimica et Cosmochimica Acta* **40**, 73–83.

Silva, L. C., M. J. Le Bas & A. H. F. Robertson 1981. An oceanic carbonatite volcano on Santiago, Cape Verde Islands. *Nature* **304**, 51–4.

Silver, P. G. & W. W. Chan 1988. Implications for continental structure and evolution from seismic anisotropy. *Nature* **335**, 34–9.

Smith, C. B. 1983. Pb, Sr and Nd isotopic evidence for sources of southern African Cretaceous kimberlite. *Nature* **304**, 51–4.

Staudigel, H., A. Zindler, S. R. Hart, T. Leslie, C.-Y. Chen & D. Clague 1984. The isotope systematics of a juvenile intraplate volcano: Pb, Nd and Sr isotope ratios of basalts from Loihi Seamount. *Earth and Planetary Science Letters* **69**, 13–29.

Sun, S. S. 1980. Lead isotopic study of young volcanic rocks from mid-ocean ridges, ocean islands and island arcs. *Philosophical Transactions of the Royal Society of London, Series A* **297**, 409–45.

Taylor, S. R. & S. M. McLennan 1985. *The continental crust; its composition and evolution*. Oxford: Blackwell.

Tilton, G. R., S. T. Kwon & D. M. Frost 1987. Isotopic relationships in Arkansas Cretaceous alkalic complexes. In *Mantle metasomatism and alkaline magmatism*, E. M. Morris & J. D. Pasteris (eds), Geological Society of America Special Paper 215, 241–8.

Vollmer, R. & M. J. Norry 1983a. Unusual isotopic variations in Nyiragongo nephelinites. *Nature* **301**, 141–3.

Vollmer, R. & M. J. Norry 1983b. Possible origin of K-rich volcanic rocks from Virunga, East Africa, by metasomatism of continental crustal material: Pb, Nd and Sr isotopic evidence. *Earth and Planetary Science Letters* **64**, 374–86.

Weaver, B. L. & J. Tarney 1980. Continental crust composition and nature of the lower crust: constraints from mantle Nd–Sr isotope correlation. *Nature* **386**, 342–6.

Weis, D. & D. Demaiffe 1985. A depleted mantle source for kimberlites from Zaire: Nd, Sr and Pb isotopic evidence. *Earth and Planetary Science Letters* **73**, 269–77.

Wen, J., K. Bell & J. Blenkinsop 1987. Nd and Sr isotope systematics of the Oka complex, Quebec, and their bearing on the evolution of the sub-continental upper mantle. *Contributions to Mineralogy and Petrology* **97**, 433–7.

White, W. M. & A. W. Hofmann 1982. Sr and Nd isotope geochemistry of oceanic basalts and mantle evolution. *Nature*, **296**, 821–5.

Zindler, A. & S. Hart 1986. Chemical geodynamics. *Annual Review of Earth and Planetary Sciences* **14**, 493–571.

13
STABLE ISOTOPE VARIATIONS IN CARBONATITES

P. DEINES

ABSTRACT

The $^{13}C : {}^{12}C$, $^{18}O : {}^{16}O$, and $^{34}S : {}^{32}S$ ratios of carbonatite minerals vary as a result of differences in (a) the isotopic composition of the magma sources, (b) isotope fractionation processes during magma formation, and (c) isotope effects in the crystallization of the magmas. The latter lead to an increase in $\delta^{13}C$ of the later carbonatite phases and correlated variations in $\delta^{13}C$ and $\delta^{18}O$ in 90% of the carbonatites studied. Some of the isotopic variation is produced by post-magmatic processes. Closed-system autometamorphic changes are also documented. Hydrothermal solutions may be important in establishing the isotope record of some carbonatites, although at Oka, the only complex for which sufficiently detailed data are available, no evidence is found for large-scale, hydrothermal circulating systems. Detailed combined studies of δD, $\delta^{13}C$, $\delta^{18}O$, and $\delta^{34}S$ variations in carbonatites have yet to be carried out and would be a valuable adjunct to radiogenic isotope studies.

13.1 INTRODUCTION

Variations in the stable isotope ratios of carbonatite minerals may be attributed to (a) isotopic composition differences in the sources of magmas, (b) fractionation processes during carbonatite evolution, and (c) post-magmatic alteration. It is the aim of this review to identify those processes responsible for the stable isotope distribution in carbonatites. For this purpose the isotopic composition variations of the elements C, O, and S in carbonatites reported in the literature, and some unpublished results, will be used (Table 13.1).

The isotope analyses are reported in the conventional delta notation as per mil deviations (‰) from a reference sample:

$$\delta = \left(\frac{R_{sample}}{R_{reference}} - 1\right) \times 10^3$$

where R stands for the appropriate isotope ratio, and the reference samples are: Standard Mean Ocean Water (SMOW) for $^{18}O : {}^{16}O$, the Pee Dee Belemnite (PDB) for $^{13}C : {}^{12}C$, and Canyon Diablo Troilite (CDT) for $^{34}S : {}^{32}S$ ratios.

The equilibrium isotope fractionation between two phases A and B is expressed as

Table 13.1 List of carbonatites and related rocks used in the review.

Name and location	Reference
Afrikanda, USSR	Kukharenko & Dontsova (1964)
Akjoujt, Mauritania	Pineau & Javoy (1969)
Albany Forks, Canada	Deines & Gold (1973)
Alnö, Sweden	von Eckermann et al. (1952)
	Wickman (1956)
	Baertschi (1957)
	Taylor et al. (1967)
	Pineau et al. (1973)
Bailundo, Angola	Pineau et al. (1973)
Bear Paw, Montana, USA	Mitchell & Krouse (1975)
	this volume
Belaya Zima, East Sayan, USSR	Korzhinskiy & Mamchur (1978, 1980)
Beloziminskiy, East Sayan, USSR	Plyusnin et al. (1980, 1982)
Big Beaver House, Canada	this volume
Bol'shaya Tagna, East Sayan, USSR	Korzhinskiy & Mamchur (1978, 1980)
Bol'shetagninskiy, East Sayan, USSR	Plyusnin et al. (1980, 1982)
Bol'shezhidoy, East Sayan, USSR	Plyusnin et al. (1980, 1982)
Bol'shoy Sayan, USSR	Grinenko et al. (1970a, b)
	Dontsova et al. (1977a, b)
	Kononova & Yashina (1984)
Bukusu, Uganda	Deines & Gold (1973)
	Sheppard & Dawson (1973)
Busumbu, Uganda	Denaeyer (1970)
	Pineau et al. (1973)
Cargill, Canada	Dontsova et al. (1977a, b)
	Kononova & Yashina (1984)
	this volume
Chadobets Uplift, Siberia, USSR	Bagdasarov et al. (1969a, b)
Changit, Maimecha-Kotuy, USSR	Plyusnin et al. (1980, 1982)
Chernigov, Azov, USSR	Bagdasarov & Grinenko (1981)
	Bagdasarov et al. (1982)
Chigwakwalu-Hill, Malawi	Baertschi (1957)
Coola, Angola	Pineau et al. (1973)
Dalbykha, Maimecha-Kotui, USSR	Bagdasarov & Grinenko (1983)
	Bagdasarov & Buyakayte (1985a, b)
Fen, Norway	von Eckermann et al. (1952)
	Wickman (1956)
	Friedrichsen (1968)
	Pineau & Javoy (1969)
	Pineau et al. (1973)
	this volume
Firesand, Canada	Vinogradov et al. (1971a, b)
Fort Portal, Uganda	Plyusnin et al. (1980, 1982)
Gardiner Complex, Greenland	Nielsen & Buchardt (1985)
Gek, Siberia, USSR	Vinogradov et al. (1967a, b)
	Kropotova (1969)
Gornoye Ozero, Sette Deban, USSR	Plyusnin et al. (1980, 1982)
Goudini, Transvaal, South Africa	Sheppard & Dawson (1973)
Greenland	Sheppard & Dawson (1973)
Gula, Maimecha-Kotui, USSR	Korzhinskiy & Mamchur (1978, 1980)
Gula-North, Maimecha-Kotui, USSR	Plyusnin et al. (1980, 1982)
Gula-South, Maimecha-Kotui, USSR	Plyusnin et al. (1980, 1982)
Haast River, New Zealand	Blattner & Cooper (1974)

Table 13.1 (*contd*).

Name and location	Reference
Homa Mountain, Kenya	Deines & Gold (1973)
	Suwa *et al.* (1975)
Ile de Fogo, Cap Vert	Pineau *et al.* (1973)
Ile de Fuerte Ventura, Canary Isl.	Pineau *et al.* (1973
Il'men, USSR	Kononova (1980)
Iron Hill, Colorado, USA	Baertschi (1957)
	Taylor *et al.* (1967)
Ishkulusk, USSR	Kononova (1980)
Kaiserstuhl, Germany	Gonfiantini & Tongiorgi (1964)
	Hay & O'Neil (1983)
Kalyango, Fort Portal, Uganda	Deines & Gold (1973)
	Vinogradov *et al.* (1978, 1980)
Kerimasi, Tanzania	Vinogradov *et al.* (1970*a, b*)
	Sheppard & Dawson (1973)
	Kuleshov (1986)
Kharla, Sanilen Highlands, Tuva, USSR	Kuleshov (1986)
Kibuye, Rwanda	Pineau & Javoy (1969)
Kirumba, Zaire	Denaeyer (1970)
	Pineau *et al.* (1973)
Kiya-Shaltysk, USSR	Kononova (1980)
Kolyano, Uganda	Baertschi (1957)
Kotamu, Malawi	Baertschi (1957)
Kovdor, Kola-Karelia, USSR	Kukharenko & Dontsova (1964)
	Grinenko *et al.* (1970*a, b*)
	Dontsova *et al.* (1977*a, b*)
	Korzhinskiy & Mamchur (1978, 1980)
	Plyusnin *et al.* (1980, 1982)
	Kononova & Yashina (1984)
	Kuleshov (1985, 1986)
Krasnyye Obryvy, Maimecha-Kotui	Korzhinskiy & Mamchur (1978, 1980)
Kugda, Maimecha-Kotui, USSR	Galimov *et al.* (1974)
	Plyusnin *et al.* (1980, 1982)
	Kononova & Yashina (1984)
Kwaraha, Tanzania	Vinogradov *et al.* (1970*a, b*, 1971*a, b*)
Laacher See, Germany	Taylor *et al.* (1967)
Lackner Lake, Canada	Deines & Gold (1973)
Laetoli, Tanzania	Hay & O'Neil (1983)
Lelyaki, Dnieper-Donets, USSR	Lyashkevich *et al.* (1978)
	Panov *et al.* (1981)
Lesnaya Varaka, USSR	Kukharenko & Dontsova (1964)
Lueshe, Zaire	Pineau & Javoy (1969)
	Denaeyer (1970)
	Pineau *et al.* (1973)
Magnet Cove, USA	Conway & Taylor (1969)
	Mitchell & Krouse (1975)
Makome, Uganda	Kuleshov (1986)
Maly Sayanski, East Sayan, USSR	Grinenko *et al.* (1970*a, b*)
	Dontsova *et al.* (1977*a, b*)
	Kononova & Yashina (1984)
Marongwe-Hill, Malawi	Baertschi (1957)
Mbeya, Tanzania	Pineau & Javoy (1969)
	Suwa *et al.* (1969)
	Pineau *et al.* (1973)
	Sheppard & Dawson (1973)

(*continued*)

Table 13.1 (*contd*).

Name and location	Reference
Mbuga crater, Uganda	Baertschi (1957)
Monte Verde, Angola	Pineau *et al.* (1973)
Mountain Pass, California, USA	Mitchell & Krouse (1971)
	Mitchell & Krouse (1975)
Mrima, Kenya	Vinogradov *et al.* (1970*a*, *b*, 1971*a*, *b*)
Mud Tank, Australia	Wilson (1979)
Murun Syenite Block, USSR	Plyusnin *et al.* (1984*a*, *b*)
Napak, Fort Portal, Uganda	Denaeyer (1970)
	Pineau *et al.* (1973)
	Kuleshov (1986)
Ndeke, Uganda	Deines & Gold (1973)
Nemegosenda Lake, Canada	Deines & Gold (1973)
North Ruri, Nyanza, Kenya	Deines & Gold (1973)
Novaya Poltava, Ukraine, USSR	Dontsova *et al.* (1977*a*, *b*)
	Galimov *et al.* (1974)
Odikhincha, Maimecha-Kotui, USSR	Dontsova *et al.* (1977*a*, *b*)
	Plyusnin *et al.* (1980, 1982)
	Kononova & Yashina (1984)
Oka, Canada	Deines (1967, 1970)
	Conway & Taylor (1969)
	Mitchell & Krouse (1975)
	Dontsova *et al.* (1977*a*, *b*)
	Plyusnin *et al.* (1980, 1982)
	Kononova & Yashina (1984)
Oldoinyo Lengai, Tanzania	Denaeyer (1970)
	Vinogradov *et al.* (1970*a*, *b*, 1971*a*, *b*)
	O'Neil & Hay (1973)
	Suwa *et al.* (1975)
	Kuleshov (1986)
Ozernaya Varaka, Kola-Karelia, USSR	Kukharenko & Dontsova (1964)
	Plyusnin *et al.* (1980, 1982)
Ozernyi, Siberia, USSR	Vinogradov *et al.* (1967*a*, *b*)
	Kropotova (1969)
Pesochnaya Varaka, USSR	Kukharenko & Dontsova (1964)
Phalaborwa, South Africa	von Gehlen (1967)
	Hoefs *et al.* (1968)
	Pineau & Javoy (1969)
	Pineau *et al.* (1973)
	Sheppard & Dawson (1973)
	Mitchell & Krouse (1975)
	Suwa *et al.* (1975)
Povorotnyy, Siberia, USSR	Vinogradov *et al.* (1967*a*, *b*)
	Kropotova (1969)
Prairie Lake, Canada	this volume
Rangwe, Nyanza, Kenya	Deines & Gold (1973)
St. André, Canada	Deines & Gold (1973)
St. Honoré, Canada	this work
Saka, Fort Portal, Uganda	Deines & Gold (1973)
Sallanlatva, USSR	Kukharenko & Dontsova (1964)
Seabrook Lake, Canada	Deines & Gold (1973)
	this volume
Sebl-Jarvi, USSR	Kukharenko & Dontsova (1964)
Seiland, Norway	von Eckermann *et al.* (1952)
Sevathur	Sheppard & Dawson (1973)

Table 13.1 (*contd*).

Name and location	Reference
Sokli, Finland	Dontsova *et al.* (1977*a,b*)
	Mäkelä & Vartiainen (1978)
	Kononova & Yashina (1984)
Southern Siberia, USSR	Bolonin & Zhukov (1983)
South Ruri, Kenya	Deines & Gold (1973)
Spanish River, Canada	this work
Spitskop, South Africa	Holmes (1958)
	Baertschi (1957)
	Suwa *et al.* (1975)
Sredneziminskiy, East Sayan, USSR	Plyusnin *et al.* (1980, 1982)
Stjernøy, Norway	von Eckermann *et al.* (1952)
	Wickman (1956)
Sukulu, Uganda	Baertschi (1957)
	Denaeyer (1970)
	Deines & Gold (1973)
	Pineau *et al.* (1973)
	Sheppard & Dawson (1973)
Syanskiy, USSR	Vinogradov *et al.* (1967*a,b*)
Tamazert, Morocco	Pineau & Javoy (1969)
	Pineau *et al.* (1973)
Tanga, East Sayan, USSR	Dontsova *et al.* (1977*a,b*)
Tanginski, Sayan, USSR	Kononova & Yashina (1984)
Tatarsky Fault, Yenisei Mts., USSR	Lapin *et al.* (1986)
Tchivira-Bonga, Angola	Pineau *et al.* (1973)
Tororo, Uganda	Denaeyer (1970)
	Vinogradov *et al.* (1971*a,b*)
	Deines & Gold (1973)
	Pineau *et al.* (1973)
	Sheppard & Dawson (1973)
Tundulu, Malawi	Sheppard & Dawson (1973)
Turii Peninsula, Kola Karelia, USSR	Plyusnin *et al.* (1980, 1982)
Vishnev Hills, USSR	Dontsova *et al.* (1977*a,b*)
Vuorijarvi, Kola Karelia, USSR	Kukharenko & Dontsova (1964)
	Grinenko *et al.* (1970*a,b*)
	Korzhinskiy & Mamchur (1978, 1980)
	Plyusnin *et al.* (1980, 1982)
Vyshenevogorsk, USSR	Kononova (1980)
Wet Mountains, Colorado, USA	Armbrustmacher (1979)
Yessey, Maimecha-Kotui, USSR	Bagdasarov & Grinenko (1983)
	Bagdasarov & Buyakayte (1985*a,b*)
Yraas, Maimecha-Kotui, USSR	Bagdasarov & Grinenko (1983)
	Plyusnin *et al.* (1980, 1982)

the fractionation factor:

$$\alpha_{A-B} = R_{\text{phase A}}/R_{\text{phase B}}$$

which also equals:

$$\alpha_{A-B} = (\delta_A + 1000)/(\delta_B + 1000).$$

13.2 OXYGEN

13.2.1 Overview of available data

Figure 13.1 shows about 400 $\delta^{18}O$ analyses for intrusive carbonates from 75 localities. In order not to bias the distribution unduly, results from the Oka and St. Honoré carbonatites, for which a larger number of analyses are available, have been excluded.

The $\delta^{18}O$ values of carbonatite carbonates cover a wide range from about 5 to 25‰ v. SMOW. About 50% of the analyses fall, however, in a narrow interval between 6 and 9‰. Deines & Gold (1973) have pointed out that the isotopic composition range of subvolcanic carbonatites (Verwoerd 1966) is more restricted than that of the volcanic–subvolcanic association.

The $\delta^{18}O$ sample frequency distribution of carbonates from carbonatite dykes and veins differs from that of carbonatite intrusions. The clear peak at 6–9‰ is absent (Fig. 13.2). Differences also exist between the isotopic compositions of dyke and vein carbonates: about 26% of the dyke samples have $\delta^{18}O$ values between 6 and 9‰, while only 4% of the vein samples fall in this range.

Isotope analyses of carbonate lavas and tuffs (Fig. 13.3) cover essentially the same total $\delta^{18}O$ range as intrusive carbonatites. It is interesting to note, however, that slags and tuffs have higher ^{18}O contents than the lavas; of the latter 45% of the analyses fall in the narrow range from 8 to 11.5‰.

For comparison the $\delta^{18}O$ distribution of kimberlite carbonates is shown in Figure

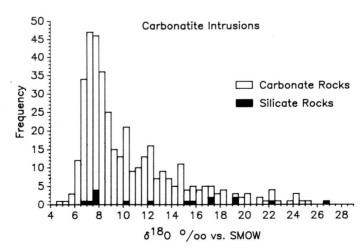

Figure 13.1 Oxygen isotopic composition of carbonatite intrusions. Sources of data: von Eckermann *et al.* (1952), Wickman (1956), Baertschi (1957), Gonfiantini & Tongiorgi (1964), Taylor *et al.* (1967), Vinogradov *et al.* (1967a, b, 1970a, b, 1971a, b), Friedrichsen (1968), Bagdasarov *et al.* (1969a, b), Conway & Taylor (1969), Suwa *et al.* (1969, 1975), Denaeyer (1970), Deines & Gold (1973), Pineau *et al.* (1973), Dontsova *et al.* (1977a, b), Lyashkevich *et al.* (1978), Armbrustmacher (1979), Wilson (1979), Bolonin & Zhukov (1983), Hay & O'Neil (1983), Kononova & Yashina (1984), Plyusnin *et al.* (1984a, b), Nielsen & Buchardt (1985), Kuleshov (1986), Lapin *et al.* (1986), this chapter.

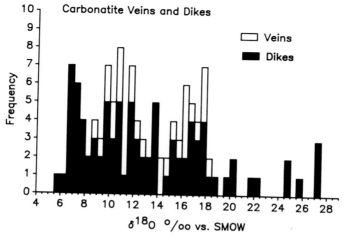

Figure 13.2 Oxygen isotopic composition of carbonatite veins and dikes. Sources of data: von Eckermann *et al.* (1952), Wickman (1956), Baertschi (1957), Gonfiantini & Tongiorgi (1964), Taylor *et al.* (1967), Suwa *et al.* (1969, 1975), Deines & Gold (1973), Pineau *et al.* (1973), Blattner & Cooper (1974), Dontsova *et al.* (1977a, b), Vinogradov *et al.* (1980), Bolonin & Zhukov (1983), Nielsen & Buchardt (1985), Kuleshov (1986).

13.4. Only 7‰ of the kimberlite analyses fall in the range from 6 to 9‰, and there is a strong maximum at about 12‰, similar to the values shown by many extrusive carbonatites. A more recent study of kimberlite carbonates by Kirkley (pers. comm.) confirms this observation.

The $\delta^{18}O$ sample frequency distribution of carbonates from the St. Honoré carbonatite indicates the existence of two modes (Figs 13.5, 13.6). Samples containing only calcite show a much more restricted range (Fig. 13.5) than those containing only dolomite or both carbonates (Fig. 13.6).

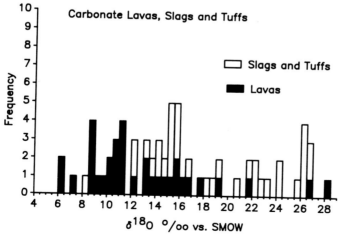

Figure 13.3 Oxygen isotopic composition of carbonate lavas, slags, and tuffs. Sources of data: Baertschi (1957), Vinogradov *et al.* (1970a, b, 1971a, b, 1980), Deines & Gold (1973), O'Neil & Hay (1973), Sheppard & Dawson (1975), Suwa *et al.* (1975), Lyashkevich *et al.* (1978), Wilson (1979), Hay & O'Neil (1983), Kuleshov (1986).

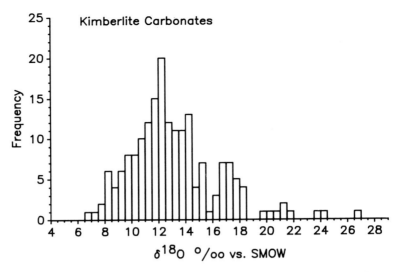

Figure 13.4 Oxygen isotopic composition of kimberlite carbonates. Sources of data: Deines & Gold (1973), Kobelski (1977), Kobelski *et al.* (1979).

In the Oka complex (Fig. 13.7), in contrast to the St. Honoré carbonatite, the number of samples with $\delta^{18}O$ values larger than 15‰ is very small. No significant differences are observed among the mean $\delta^{18}O$ values of the various carbonate rock types (Fig. 13.8). Rauhaugites have essentially the same O isotopic composition as sövites and there are no major differences in the ^{18}O content that can be attributed to late-stage magmatic or hydrothermal processes. The very coarse-grained pegmatitic sövites, niobium-rich sövites classified as niobium ore, and carbonate veins that cut the associated silicate rocks have, on average, $\delta^{18}O$ values essentially identical to massive sövites.

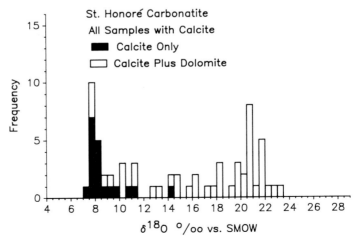

Figure 13.5 Oxygen isotopic composition of calcite from dolomite-free and dolomite-containing sövite, St. Honoré carbonatite. Source of data: this chapter.

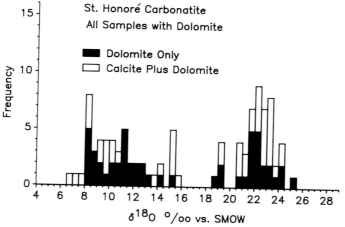

Figure 13.6 Oxygen isotopic composition of dolomite in calcite-free and calcite-containing carbonate rocks, St. Honoré carbonatite. Source of data: this chapter.

Carbonates from silicate rocks have slightly higher ^{18}O contents than the sövites and rauhaugites (Fig. 13.7); in addition, their mean $\delta^{18}O$ varies slightly between various silicate rock types. On average, the ^{18}O content increases in the following sequence: carbonate rocks < ijolites < okaites < fenites and rocks of possible metasomatic origin (Fig. 13.9). It is interesting to note that the carbonate concentrations in these rocks decrease in the same sequence.

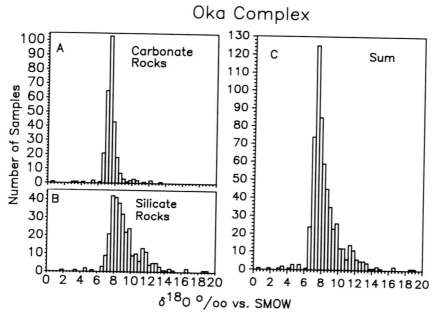

Figure 13.7 Oxygen isotopic composition of calcite and dolomite from carbonate and silicate rocks of the Oka complex. (A) Carbonate rocks; (B) silicate rocks; (C) summary distribution. Source of data: Deines (1970).

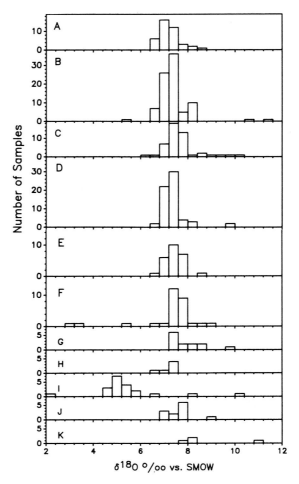

Figure 13.8 Oxygen isotopic composition of different carbonate rocks from the Oka complex. (A) Banded calcite rocks with accessory pyroxene, biotite, and magnetite; (B) banded sodapyroxene–biotite–magnetite–calcite rocks; (C) monticellite calcite rocks; (D) ore phase; (E) calcite rocks; (F) dolomite rocks; (G) melilite–calcite rocks; (H) niocalite–calcite rocks; (I) carbonate veins in silicate rocks; (J) pegmatite phase; (K) carbonate dikes in gneiss. Source of data: Deines (1970).

13.2.2 Expected oxygen isotopic composition of a carbonate magma derived from the mantle

The $\delta^{18}O$ range of a mantle-derived carbonate magma can be estimated from the isotope record of meteorites, ultramafic xenoliths, and basalts, in conjunction with the isotope fractionation expected at upper-mantle temperatures. Data from stony meteorites are summarized in Figure 13.10. Excluded are the results for carbonaceous chondrites, whose unusual isotopic compositions have been discussed by Clayton *et al.* (1973, 1977) and Clayton & Mayeda (1984). The mean $\delta^{18}O$ value of chondrites (5.3‰) is slightly higher than that of achondrites (4.5‰). The grand

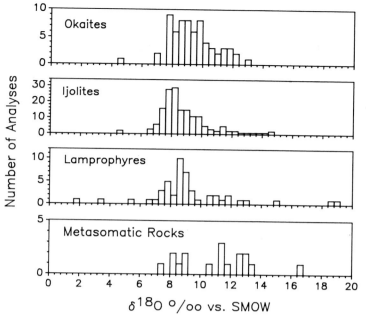

Figure 13.9 Oxygen isotopic composition of different silicate rock types from the Oka carbonatite. Source of data: Deines (1970).

mean for all stony meteorites is 4.9‰; 70% of the analyses fall in the range from 4 to 6‰.

The $\delta^{18}O$ measurements on mafic and ultramafic xenoliths from basalts and kimberlites have been compiled in Figure 13.11. If the data from eclogite xenoliths from the Roberts Victor kimberlite are excluded, as they may reflect subduction of

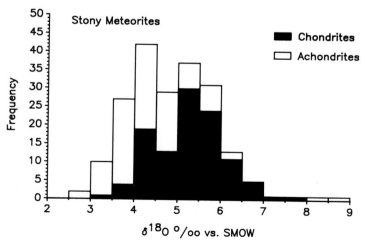

Figure 13.10 Oxygen isotopic composition of stony meteorites, exclusive of carbonaceous chondrites. Sources of data: Reuter *et al.* (1965), Taylor *et al.* (1965), Onuma *et al.* (1972), Clayton *et al.* (1973, 1976), Clayton & Mayeda (1978, 1983).

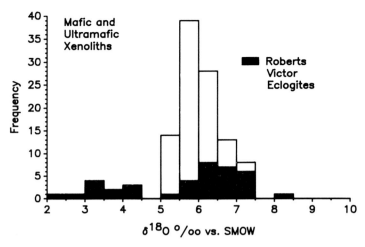

Figure 13.11 Oxygen isotopic composition of mafic and ultramafic xenoliths. Sources of data: Taylor & Epstein (1962a, b), Reuter et al. (1965), Garlick (1966), Sheppard & Epstein (1970), Garlick et al. (1971), Pinus & Dontsova (1971), Kyser & O'Neil (1978), Kyser et al. (1981, 1982), Jagoutz et al. (1984), MacGregor & Manton (1986), Ongley et al. (1987).

ancient oceanic crust (Jagoutz et al. 1984, MacGregor & Manton 1986, Ongley et al. 1987), the analyses group tightly between 5 and 7‰ (mean = 5.9 ± 0.4‰); 92% fall in the range from 5 to 6.5‰.

The summary of O isotope analyses of unaltered basalts (Fig. 13.12) indicates a mean isotopic composition of 6.1‰. The distribution has a strong mode between 5.5 and 6‰ and 56% of the analyses fall between 5 and 6.5‰.

Data from meteorites, mafic and ultramafic xenoliths, and basalts indicate that

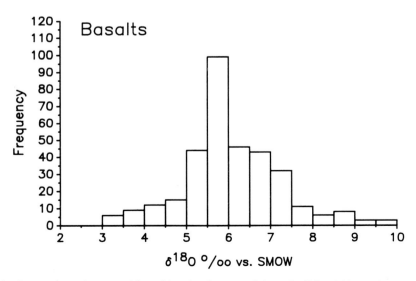

Figure 13.12 Oxygen isotopic composition of basalts. Sources of data: Garlick (1966), Taylor (1968), Muehlenbachs & Clayton (1971, 1972), Matsuhisa et al. (1973), Muehlenbachs et al. (1974), Muehlenbachs (1976, 1977a, b), Pineau et al. (1976b), Gray & Cumming (1977), Matsuhisa (1979), Kyser et al. (1982).

the O isotopic composition of the mantle is fairly restricted and lies between 5 and 6‰, although the existence of mantle reservoirs of isotopic compositions outside this range has been considered (Kyser *et al.* 1982, Kyser 1986, Taylor & Sheppard 1986).

In order to estimate the isotopic composition of a mantle-derived, carbonate magma the isotope fractionation between such a melt and an average peridotite must be established. The ^{18}O fractionation of peridotite was computed by summing the fractionation factors for its component minerals. The modal data for common peridotites (Cox *et al.* 1973) were used to weight the fractionations for olivine, ortho- and clinopyroxene, and garnet (Kieffer 1982). The ^{18}O fractionation of garnet was estimated from those of andradite, grossular, and pyrope, using the chemical data for peridotite garnets from Cox *et al.* (1973) to weight the fractionations of the three components. It was assumed that to a first approximation there is no difference in δ^{18}O between almandine and andradite; no significant errors result from this assumption. The fractionations among calcite, the silicate minerals, and peridotite are shown in Figure 13.13.

At 1000 °C the isotopic composition difference between peridotite and calcite amounts to about 2‰. Hence, one would expect that if the mantle silicates had average δ^{18}O values between 5 and 6‰, and carbonate melts fractionated δ^{18}O to the same degree as calcite, the δ^{18}O value of carbonate melts in equilibrium with mantle minerals should be between 7 and 8‰. Figure 13.1 clearly indicates that the mode of the δ^{18}O distribution occurs in this range; about 23% of the analyses fall

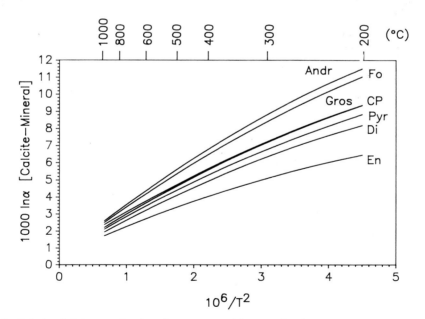

Figure 13.13 Calculated O isotope fractionations among calcite, mantle minerals, and common peridotite. The fractionation has been expressed as $1000 \ln \alpha_{\text{(calcite-mineral)}}$ which equals to a first approximation $\delta^{18}O_{\text{calcite}} - \delta^{18}O_{\text{mineral}}$. Andr, andradite; Di, diopside; En, enstatite; Fo, forsterite; Gros, grossular; Pyr, pyrope; CP, common peridotite. Sources of data: isotope fractionation factors, Kieffer (1982); chemical data, Cox *et al.* (1973).

between 7 and 8‰. Assuming that most mantle silicates have a $\delta^{18}O$ value of 5.7‰ and that the mode of carbonatite carbonates at 7.5‰ represents isotopic equilibrium between carbonatite melt and mantle silicates, the estimated equilibration temperature lies between 1100 and 1200 °C.

Considering the range of O isotopic composition in meteorites, ultramafic xenoliths, and basalts it can be postulated that the original O isotopic composition of carbonatite magmas would probably lie between 6.5 and 9.5‰ v. SMOW. Whether within this range there are systematic differences between complexes is not clear at this time. The question of O isotope heterogeneity in the mantle has been discussed by Kyser (1986).

13.2.3 Oxygen isotope fractionation between carbonatite minerals

Differences in ^{18}O content between minerals formed in isotopic equilibrium are temperature dependent, and they can be used, if appropriate conditions are met, to estimate crystallization temperatures (Clayton 1981, O'Neil 1986). The size of the anticipated isotope fractionation between carbonatite minerals at magmatic temperatures is illustrated in Figure 13.14. Few $\delta^{18}O$ analyses of silicates from carbonatites exist (Taylor *et al.* 1967, Friedrichsen 1968, Conway & Taylor 1969, Friedrichsen 1973, Blattner & Cooper 1974). Isotopic compositions of silicate minerals are consistent with isotopic equilibrium with carbonatite carbonates at magmatic temperatures. Using the fractionation curves shown in Figure 13.14, equilibration temperatures between 600 and 800 °C are obtained. Based on differences in $\delta^{18}O$ between carbonates and coexisting silicates and iron oxides Friedrichsen

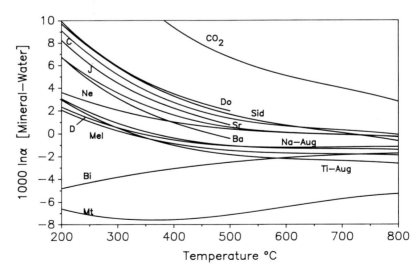

Figure 13.14 Oxygen isotope fractionation among carbonatite minerals and water. Ba, witherite; Bi, biotite; C, calcite; D, diopside; Do, dolomite; J, jadeite; Mel, melilite; Mt, magnetite; Na-Aug, sodium-augite; Ne, nepheline; Sid, siderite; Sr, strontianite; Ti-Aug, titanaugite. Sources of data: Northrop & Clayton (1966), O'Neil *et al.* (1969), Becker (1971), Matthews & Katz (1977), Hoernes & Friedrichsen (1978), Matsuhisa *et al.* (1979), Downs *et al.* (1981), Matthews *et al.* (1983). The fractionations for augite, melilite, and nepheline have been estimated.

(1973) deduced for samples from Alnö, the Fen area, and the Kaiserstuhl that:

(a) rocks containing calcite equilibrated at 660–680 °C;
(b) with increasing dolomite content the equilibration temperature was reduced (655–630 °C); and
(c) the lowest temperatures were recorded in rocks consisting mainly of dolomite.

Without doubt, part of the O isotope record of the carbonatites was established at high temperatures and was retained.

More analyses are available for coexisting dolomite–calcite pairs. The isotope fractionation is, however, highly variable as can be seen from Figures 13.15 and 13.16. The bimodal distribution of $\delta^{18}O$ of the St. Honoré carbonatite (Figs 13.5, 13.6) may indicate the involvement of two distinct magmatic or hydrothermal phases in the carbonatite formation. Figures 13.5 and 13.6 demonstrate that the high $\delta^{18}O$ samples either are dolomite or contain dolomite. Hence, results from the St. Honoré carbonatite may indicate the introduction of secondary dolomite with a distinct isotopic composition.

If magnesium-bearing solutions caused dolomite formation at low temperatures, large fractionations between dolomite and calcite would be observed in samples in which dolomite is enriched in ^{18}O. Therefore, the ^{18}O fractionation between dolomite and calcite and the $\delta^{18}O$ of dolomite have been considered jointly for those carbonatites for which more than one mineral pair has been analysed (Figs 13.15, 13.16). It is apparent that the relationships between the two variables are very different for the four carbonatites.

Unfortunately, the equilibrium isotope fractionation factor between dolomite and calcite is not well established. The available data, summarized in Figure 13.14

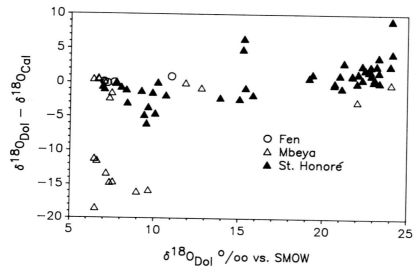

Figure 13.15 Oxygen isotope fractionation between coexisting dolomite and calcite for Fen, Mbeya, and St. Honoré carbonatites. Sources of data: Friedrichsen (1968), Suwa *et al.* (1969, 1975), this chapter.

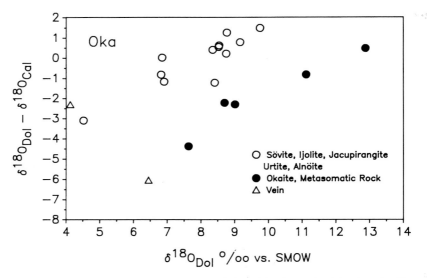

Figure 13.16 Oxygen isotope fractionation between coexisting dolomite and calcite for the Oka carbonatite. Source of data: Deines (1970).

together with information about the fractionation of other carbonates, lead to the expectation that at high temperatures there should be a small enrichment in ^{18}O in dolomite with respect to calcite. As seen in Figures 13.15 and 13.16 this is not the case for most of the calcite–dolomite pairs that have been analysed.

The data from Fen come closest to what might be expected for a high-temperature formation of dolomite in a carbonatite. For the St. Honoré complex most of the positive fractionations are observed for carbonate pairs enriched in ^{18}O, and most negative fractionations occur when $\delta^{18}O$ of the dolomite is < 11‰. A correlation between the ^{18}O fractionation between dolomite and calcite and dolomite isotopic composition is suggested; a similar trend is indicated by the Oka data (Fig. 13.16). Part of these correlations may be attributed to the lower temperatures of formation of the ^{18}O enriched dolomite. Thirty per cent of the St. Honoré samples with $\delta^{18}O > 20$‰ have, however, fractionations that are small enough to be compatible with magmatic equilibration temperatures. There is not enough information available to come to a firm conclusion about the temperature of formation of the ^{18}O enriched carbonate rocks of this carbonatite.

Negative fractionations, indicative of isotopic disequilibrium between the two carbonates, are also observed. The most extreme cases occur in the Mbeya carbonatite, where the dolomite, by and large, retained the primary igneous isotopic composition while calcite with higher ^{18}O contents, was probably introduced by secondary processes.

13.2.4 Causes of ^{18}O enrichment in carbonatite carbonates

Massive carbonates Many carbonatites contain carbonates that have $\delta^{18}O$ values larger than the estimated upper limit of 9.5‰. This is particularly true of carbonatites intruded at shallower levels. The ^{18}O enrichment may be due to

(a) the loss of fluids during pressure reduction at the time of emplacement;
(b) exchange with magmatic fluids of high ^{18}O content;
(c) retrograde exchange with magmatic waters;
(d) exchange with ^{18}O-rich hydrothermal fluids;
(e) influx of meteoric water and isotope exchange at low temperatures.

Loss of fluids as a result of pressure release during the emplacement of carbonatites can result in ^{18}O enrichment if ^{18}O-depleted water escapes preferentially or the $CO_2:H_2O$ ratio in the remaining fluid is reduced. As indicated in Figure 13.17, the fractionation between calcite and fluid decreases as the mole fraction of CO_2 decreases. This would lead to an increase in the ^{18}O content of a carbonate in equilibrium with the residual fluid. The magnitude of the isotopic composition changes that could be produced would depend on the size of the relevant isotope fractionation factors, the fraction of O of the system carried by the fluid, the degree of volatile loss, and the compositional changes in the fluid resulting from the loss.

Carbonatitic fluids of high $CO_2:H_2O$ ratios can attain $\delta^{18}O$ values of about 10‰ in equilibrium with mantle minerals; the largest enrichments are achieved for fluids which are rich in CO_2 (Fig. 13.18). However, calcite precipitated from a fluid will be enriched in ^{18}O with respect to it only if the CO_2 content of the fluid is low (Fig. 13.17). Hence, an increase in $\delta^{18}O$ of carbonatitic carbonates involving mantle-derived ^{18}O-rich fluids seems unlikely.

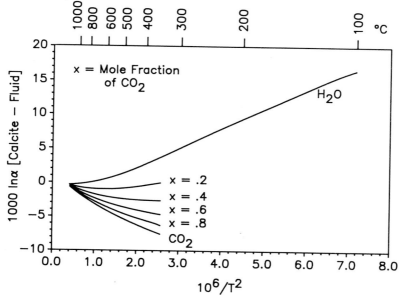

Figure 13.17 Oxygen isotope fractionation between calcite and a $H_2O:CO_2$ fluid as a function of fluid composition and temperature. Sources of data: Epstein *et al.* (1953), Clayton (1959), O'Neil *et al.* (1969), Tarutani *et al.* (1969), Richet *et al.* (1977).

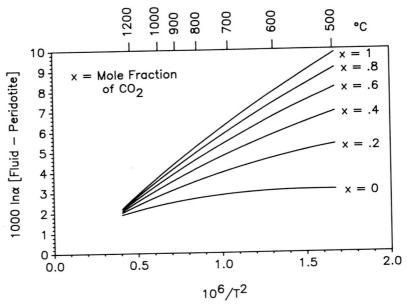

Figure 13.18 Oxygen isotope fractionation between peridotite and a $H_2O:CO_2$ fluid as a function of fluid composition and temperature. Sources of data: Richet et al. (1977), see also Figure 13.13.

Exchange between a primary magmatic carbonate and a magmatic fluid of $\delta^{18}O = 8‰$ can raise the ^{18}O content of the carbonate. It would have to occur at temperatures of about 100 °C in order to change the primary carbonate isotopic composition to $\delta^{18}O$ values of 25‰ (generally the upper limit of ^{18}O enrichment of carbonatite carbonates). This estimate assumes the presence of an infinite reservoir (at least locally) of magmatic water; smaller shifts in $\delta^{18}O$ would require correspondingly smaller amounts of magmatic fluid.

If it is assumed that the skewness observed in Figure 13.1 is the result of an autometasomatic process involving a magmatic fluid, the frequency of a particular $\delta^{18}O$ value in conjunction with the water:carbonate ratio required to achieve it, can be used to compute a weighted mean water:carbonate ratio for the system. A molar ratio of 1.5 has been obtained using Figure 13.1, assuming an initial carbonate and magmatic fluid isotopic composition of $\delta^{18}O = 7.5‰$ and a fractionation factor of 1.017 between the carbonate and the solution at the end of the exchange process. This high ratio indicates that the very large ^{18}O enrichments cannot be attributed to autometamorphic processes alone.

The very high $\delta^{18}O$ values observed in some massive carbonatites are more readily explained by invoking solutions that undergo isotope exchange with an ^{18}O-rich reservoir such as sediments or metasediments before they interact with the carbonatite. In the St. Honoré carbonatite such a process may be responsible for the bimodal O isotope distribution, and most of the dolomite may be secondary. Comparison of Figures 13.1, 13.6, and 13.7 indicates that the St. Honoré complex is unusual.

Study of the isotopic composition of carbonates along several traverses towards the contact of the Oka carbonatite with country rock (Deines 1967) did not reveal significant ^{18}O enrichments from the interior of the complex towards the contacts. An example is shown in Figure 13.19. No evidence was found for any large-scale circulating hydrothermal systems in and around this intrusion. The carbonates in the fenitized gneiss have $\delta^{18}O$ values indistinguishable from those of the sövites in the interior of the complex. It should also be noted that $\delta^{18}O$ values of carbonate > 16‰ are exceedingly rare in this carbonatite (Fig. 13.7).

Meteoric waters generally have negative $\delta^{18}O$ values, and if such waters were involved in secondary alteration the isotope exchange would have to occur below 200 °C to 250 °C in order to raise the ^{18}O content of the carbonates. At higher temperatures an exchange with ordinary surface- or groundwaters would lead to an ^{18}O depletion in the carbonate. If water as heavy as 0‰ was available for secondary alteration, isotope exchange would have to take place at temperatures as low as 40 °C in order to achieve a carbonate isotopic composition of 25‰. The conditions under which isotope equilibration might occur at such low temperatures are not understood. It is apparent that large volumes of carbonatites have not changed their

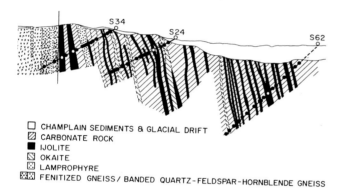

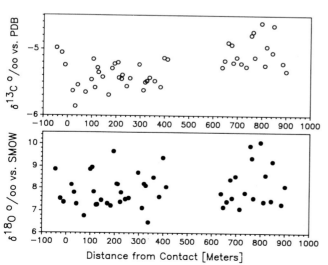

Figure 13.19 $\delta^{13}C$ and $\delta^{18}O$ variations in a vertical section through the Bond Zone of the Oka complex. Source of Data: Deines (1967).

isotopic composition significantly, in spite of the fact they have been exposed to meteoric waters for long periods of time. It has also been observed that during the weathering of carbonatites no significant changes in the O isotopic composition occur (Deines 1967). Assuming a more realistic $\delta^{18}O$ value of $-6‰$ for average groundwater for the globe, equilibration to 40 °C would lead to carbonate $\delta^{18}O$ values only as heavy as 19‰, while exchange to 100 °C would result in values of about 11‰. If the ^{18}O enrichment observed in some carbonatite carbonates was mainly the result of exchange with meteoric waters, one might expect some relationship between the O isotope shift and geographic location, because the isotopic composition of meteoric waters varies with latitude, altitude, distance from the sea, mean annual precipitation, temperature, etc. (see Yurtsever & Gat 1981). Insufficient data are available to assess whether such a correlation exists.

Silicate rocks The carbonate content of the ultramafic and feldspathoidal rocks that occur in carbonatite complexes can vary widely. In these rocks the ^{18}O content of the carbonates may be determined by processes which have not been discussed so far. Few studies have reported $\delta^{18}O$ values of carbonates from silicate rocks associated with massive sövites. A large body of data exists only for the Oka carbonatite and the following discussion is based exclusively on it.

The ^{18}O content of silicate rocks that are of possible metasomatic origin (Fig. 13.9) is partly determined by the composition of the source rocks (Precambrian paragneisses). As these generally have $\delta^{18}O$ values higher than 6‰, the slightly elevated ^{18}O content of the carbonates from fenites and metasomatites can be expected.

Closed-system re-equilibration between carbonate and silicates to low temperatures can bring about major changes in the carbonate isotopic composition because the carbonate is only a minor phase in the silicate rocks. The process can be elucidated through the study of small-scale isotopic composition variations in silicate rocks. Figures 13.20 and 13.21 indicate the variation found in two hand-specimens each about 15×15 cm in size.

The data displayed in Figure 13.20 indicate that while $\delta^{18}O$ of the calcite varies widely, 98% of the $\delta^{13}C$ values fall in the range -4.8 to $-5.8‰$ v. PDB. It is important to note that the $\delta^{18}O$ range of the vein carbonate is limited (7.5 to 9‰) and distinct from that of the calcite in the ijolite which is enriched in ^{18}O. It is unlikely that fluids moving through the veins could be responsible for the ^{18}O enrichment and the range in isotopic composition of the ijolite carbonate in this hand-specimen.

The complexities of the isotope variations on a small scale are further exemplified by 75 analyses on the second hand-specimen summarized in Figure 13.21. A mean isotopic composition was computed for volumes of sövite (S) ijolite (I), biotitized ijolite (B) and contact zones (C) which have similar isotopic compositions. The sövite has the $\delta^{18}O$ value (7.3‰) expected for massive sövite, as does the carbonate from the ijolite I_1 (6.9‰). Carbonate from ijolite of the central part of the specimen shows ^{18}O enrichment, and on the basis of the measured $\delta^{18}O$ values the ijolite can be separated into two units: I_2 ($\delta^{18}O = 7.5‰$), and I_3 ($\delta^{18}O = 10.2‰$). The largest

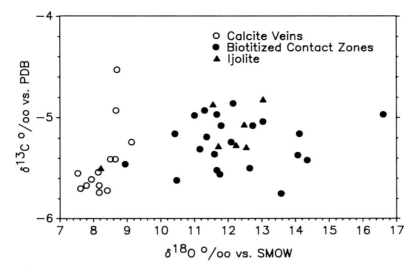

Figure 13.20 Variation of C and O isotopic composition on hand-specimen scale in the Oka carbonatite. Source of data: Deines (1967).

enrichment in ^{18}O is observed in the central part of the ijolite where $\delta^{18}O$ values as high as 14.2‰ were determined.

Two major carbonate veins cut the specimen. While the $\delta^{18}O$ values of vein V_2 vary little over its whole length, those of V_1 increase more-or-less systematically along the vein. The heaviest isotopic compositions in the latter vein are encountered at its narrowest part; as it widens the ^{18}O content decreases. The distance between the ijolite of highest ^{18}O enrichment (14.2‰) and the vein carbonate of highest $\delta^{18}O$ (15.3‰) is about 10 cm, and much lower $\delta^{18}O$ values are found in the carbonate of

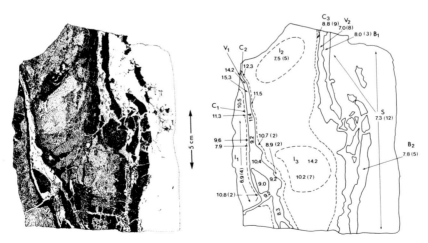

Figure 13.21 Variation of O isotopic composition on hand-specimen scale in the Oka carbonatite. B, biotitized ijolite; C, contact zone; I, ijolite; S, sövite; V, vein; if averages are given, the number of analyses entering the mean are given in parentheses, otherwise the isotopic compositions represent individual analyses. The standard deviations are in all cases less than 0.5‰, the mean standard deviation is 0.3‰. Source of data: Deines (1967).

veins and ijolite located between them. The geometric distribution of the $\delta^{18}O$ values in vein carbonate and ijolite makes it unlikely that the elevated $\delta^{18}O$ values in the silicate portion of the specimen were caused by fluids associated with the vein V_1.

Calcite from the biotitized ijolite B_1 (8.0 ± 0.17‰, $n = 3$) and B_2 (7.8 ± 0.50‰, $n = 5$) bounded by V_2 (7.0 ± 0.20‰, $n = 8$) and the sövite (7.3 ± 0.20‰, $n = 12$) have distinctly higher ^{18}O contents than the surrounding carbonate; calcite from the contact zone C_3 is significantly enriched in ^{18}O compared to S, V_2, and B_1 and B_2. These observations indicate that the increased ^{18}O content of the calcite is not caused by a secondary introduction of an ^{18}O-rich fluid. Over much of the specimen the $\delta^{18}O$ of the carbonate appears to be related to the local proportion of silicate to carbonate. This observation, and the manner in which the ^{18}O-enriched carbonate is distributed across the specimen suggests that the isotope record was established in a closed system.

The observed large- and small-scale $\delta^{18}O$ variations can be evaluated by examination of the fractionations expected between the major minerals coexisting in carbonatites (Fig. 13.14). At high temperatures calcite is generally enriched in ^{18}O by about 1–2‰ with respect to melilite and the pyroxenes; with respect to nepheline it is slightly depleted in ^{18}O.

If carbonate crystallizes from a carbonate or carbonated silicate melt, mass balance requires that, for a given total magma O isotopic composition, the $\delta^{18}O$ of the carbonate will depend on the proportion of silicate to carbonate and water. One would expect that in silicate rocks which contain only minor amounts of carbonate, the ^{18}O content of the carbonate would be slightly higher than in sövites, and that the ^{18}O enrichment of the carbonate would increase with decreasing carbonate content of the rock. The slight increase in the ^{18}O content and decrease of carbonate concentration in the sequence from sövite to ijolite to okaite discussed earlier is consistent with this concept.

If re-equilibration between carbonates and silicates occurred at lower temperatures, the carbonate in the silicate rocks, as a minor phase, would change its isotopic composition most noticeably towards higher ^{18}O concentrations, while the silicate isotopic composition would remain largely unaffected. The change in the carbonate isotopic composition could be several per mil towards more positive values, depending on the temperature at which the exchange ceased. In the carbonate-rich rocks, on the other hand, little change in the isotope record of the carbonates is expected, a feature noted in Figure 13.21.

The O isotope fractionation data indicate also that the difference in $\delta^{18}O$ between calcite and silicates would be considerably increased if biotite formed and participated in the exchange. The isotope distribution observed in Figure 13.21 could then be related to temperature, and silicate : carbonate : water ratio as well as the growth of biotite.

Dykes, veins, lavas, and tuffs The $\delta^{18}O$ range in carbonatite dykes and veins (Fig. 13.2) can be, in part, attributed to secondary isotope exchange. As carbonate is less massive in dykes and veins than in plutons, the degree of exchange would likely be larger in them and they would be less likely to retain their original isotopic composition.

Secondary alteration may have also affected the $\delta^{18}O$ record of lavas, tuffs, and slags. In view of the $\delta^{18}O$ distribution of kimberlite carbonates, in particular the existence of a fairly sharp peak in the frequency distribution at about 12‰ (Fig. 13.4), the question arises whether there is a, yet unrecognized, process that leads to systematic ^{18}O enrichment in the eruption of carbonate or carbonated lavas. Primary kimberlite carbonates should have isotopic compositions similar to those of unaltered carbonatites, but the data collected so far do not bear this out. Although it can be argued that very few of the analysed kimberlite carbonates represent primary carbonate, an explanation is still required for the maximum observed in the frequency distribution diagram. Because similar ^{18}O enrichments are found in the kimberlites of southern Africa and Colorado, a set of very special and generally prevailing circumstances would be required if secondary exchange processes were responsible for the observed frequency distribution. Of the chemical species involved in both the carbonatite and kimberlite formation, CO_2 has the largest ^{18}O enrichment and it might be responsible for introduced ^{18}O in the system; however, the details of the chemical and isotopic equilibria that might be important remain open to speculation.

13.3 CARBON

13.3.1 Overview of available data

The variation of $\delta^{13}C$ in carbonatite intrusions (Fig. 13.22) is more restricted than that of $\delta^{18}O$; 91% of the $\delta^{13}C$ values fall in the narrow range between -2 and -8‰ v. PDB. Because significant differences in C isotopic composition exist among

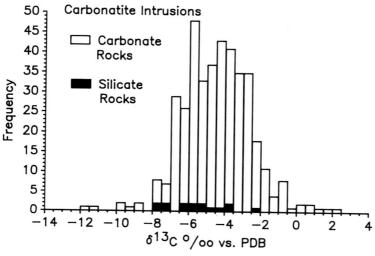

Figure 13.22 Carbon isotopic composition of carbonatite intrusions. Sources of data: see Figure 13.1.

carbonatite complexes (Deines & Gold 1973), and many more analyses are available for the Oka and St. Honoré complexes than for any other carbonatite, the analyses for these two localities are shown separately (Figs 13.23 & 13.24).

In general, carbonates from silicate rocks and sövites have very similar $\delta^{13}C$ values (Figs 13.22, 13.24). In the case of the Oka carbonatite, Deines (1970) notes that the ^{13}C content of calcite increases in the sequence:

(1) sövites ($-5.08‰$);
(2) syn-intrusive igneous rocks (ijolites, okaites) ($-4.98‰$); and
(3) post-intrusive rocks (lamprophyres) ($-4.4‰$).

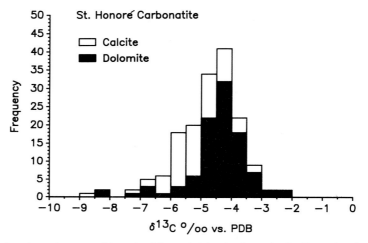

Figure 13.23 Carbon isotopic composition of calcite and dolomite from the St. Honoré carbonatite. Source of data: this volume.

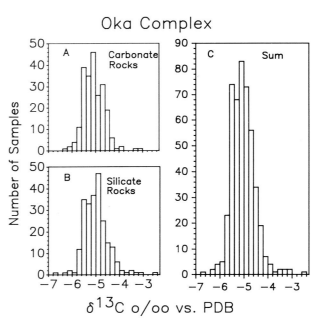

Figure 13.24 Carbon isotopic composition of calcite from carbonate and silicate rocks from the Oka carbonatite. (A) Carbonate rocks; (B) silicate rocks; (C) summary distribution. Source of data: Deines (1970).

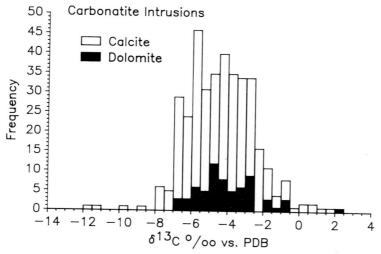

Figure 13.25 Comparison of calcite and dolomite C isotopic composition for carbonatite intrusions. Sources of data: see Figure 13.1

Dolomite may be slightly enriched in ^{13}C compared to calcite (Fig. 13.25); the isotopic composition difference between these two minerals becomes particularly noticeable if $\delta^{13}C$ values are compared within a given complex (e.g. Fig. 13.23).

There are no significant differences among the ranges of $\delta^{13}C$ values in carbonatite intrusions, dykes, and veins (compare Fig. 13.22 & Fig. 13.26). The carbonate lavas, slags and tuffs have commonly lower ^{13}C contents (Fig. 13.27) than sövites. Figure 13.27 includes lava samples from only seven volcanoes, and tuff and slag samples from nine eruptive centres, mostly located in east Africa; thus it is difficult to draw general conclusions. In those few cases where lavas and intrusive

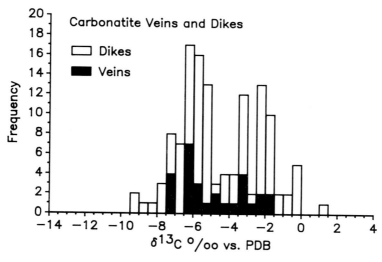

Figure 13.26 Carbon isotopic composition of carbonatite veins and dikes. Sources of data: see Figure 13.2.

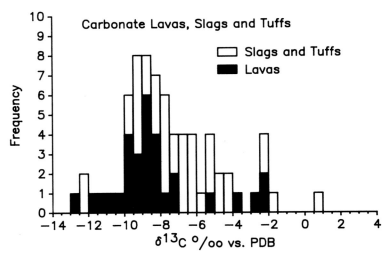

Figure 13.27 Carbon isotopic composition of carbonate lavas, slags, and tuffs. Sources of data: see Figure 13.3.

carbonates from the same centre have been analysed, the lavas generally show lower ^{13}C contents than the intrusive carbonatite (Table 13.2). Only in the case of the Kaiserstuhl is the difference between plutonic and extrusive carbonate (carbonatite globules from tuff) statistically significant. Although the data may be suggestive of a systematic depletion in ^{13}C in carbonate extrusives, they are inadequate to establish this with any degree of certainty.

13.3.2 Expected carbon isotope composition of a mantle-derived magma and ^{13}C fractionation between coexisting magmatic carbonates

Carbon, unlike O, constitutes a trace element in the mantle. Its modes of occurrence are not established with certainty, and the isotope fractionation properties of the carbon-bearing phases that might be important for the formation of carbonate and carbonated silicate melts are only partially known. It is hence difficult to carry out the type of extrapolations made for O; yet, the following observations may be made. During the formation of carbonate magma, C is highly concentrated. One might

Table 13.2 Mean carbon isotopic composition (X, ‰ v. PDB), standard deviation (s), and number of analyses (n) of carbonatites and associated carbonate lavas.

Location	Carbonatite			Lava		
	X	s	n	X	s	n
Kaiserstuhl	−5.7	1.3	21	−8.6	0.5	3
Kalyango	−7.8	1.2	5	−9.1	1.3	15
Kerimasi	−5.6	1.9	3	−2.4	0.4	3
Oldoinyo Lengai	−5.2	0.9	2	−9.8	1.9	6

expect, therefore, a fairly complete extraction of the element from its reservoir and an inheritance of the isotopic composition of its source. Because the average C content of the mantle is low, the concentration of the element into a carbonatite melt requires extraction over a volume some 1000–10 000 times the volume of the carbonatite magma. Consequently, the mean $\delta^{13}C$ value of a carbonatite represents an average over such a volume, and can be expected to be less variable than the C isotopic composition of other mantle C samples, such as diamonds or mantle-derived magmas of low C content. For these, such a homogenization of $\delta^{13}C$ over a large volume has not necessarily occurred.

One might suggest that, as in the case of O, meteorite data could be used to establish the expected range of C isotopic composition for the mantle, and consequently for carbonatites. In contrast to O, meteoritic C (Fig. 13.28) shows, however, a wide range in isotopic composition. The data given are for bulk reduced meteoritic C and do not include the highly ^{13}C-enriched minor carbonate that has been found in some carbonaceous chondrites (e.g. Clayton 1963, Krouse & Modzelski 1970, Kvenvolden et al. 1970, Smith & Kaplan 1970, Chang et al. 1978, Hirner 1979a, b) nor the extremely ^{13}C-enriched trace C fraction that has been observed in a few meteorites during stepwise combustion (e.g. Chang et al. 1978, Becker & Epstein 1982, Kerridge 1982, 1983, Robert & Epstein 1982, Swart et al. 1982, 1983, Yuen et al. 1984). The mean $\delta^{13}C$ value for iron meteorites lies between −10 and −11‰, while for the stony meteorites the value is about −14‰. The average value of −12‰ corresponds to the value suggested by Craig (1963) for juvenile C and is similar to the value for carbonaceous material incorporated in olivine crystals from Hualalai, Hawaii (Watanabe et al. 1983). Allègre et al. (1983) have suggested, based on the isotopic compositions of He, Ar, and Xe, that

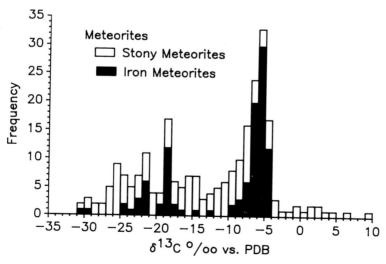

Figure 13.28 Carbon isotopic composition of stony and iron meteorites. Sources of data: Boato (1954), Vinogradov et al. (1967c), Begemann & Heinzinger (1969), Flory (1969), Belsky & Kaplan (1970), Krouse & Modzeleski (1970), Kvenvolden et al. (1970), Smith & Kaplan (1970), Deines & Wickman (1973, 1975, 1985), Chang et al. (1978), Grady et al. (1982, 1986), Robert & Epstein (1982).

Hawaiian basalts are derived from an undegassed mantle reservoir, which might record the original C isotopic composition of the Earth.

The isotopic composition of diamonds might also be considered to constrain the $\delta^{13}C$ value of the carbonatite source region. Figure 13.29 summarizes $\delta^{13}C$ values of diamonds from Russian kimberlites and nine kimberlites in southern Africa. Because the frequency distributions for diamond suites from individual kimberlites are quite distinct, and different numbers of analyses were available for different pipes, a weighted frequency distribution for the diamonds from southern Africa is also given.

The range in C isotopic composition in diamonds from Russia and southern Africa is surprisingly similar and matches fairly well that of meteorites; the major mode at about $-5‰$ has been well established. Based on the work of Galimov (1984) the mean $\delta^{13}C$ of diamonds from Russian kimberlites is $-7.1‰$, identical to the mean of the weighted $\delta^{13}C$ frequency distribution of southern African diamonds (Fig. 13.29C).

The C isotopic composition of basalts and associated carbon-bearing gases, a third potential source of information, has been reviewed by Kyser (1986) and Allard (1983).

The results of $\delta^{13}C$ determinations reported for the total C of basalts vary greatly

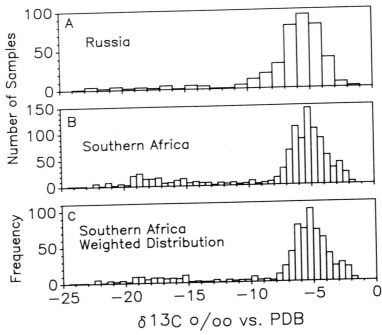

Figure 13.29 Carbon isotopic composition of diamonds from kimberlites. (A) Russia, (B) southern Africa, (C) weighted C isotopic composition frequency distribution of diamonds from southern Africa. Southern African kimberlites represented include: Dan Carl, Finsch, Jägersfontein, Jwaneng, Koffiefontein, Newlands, Orapa, Premier, Roberts Victor. Sources of data: Russian diamonds, Galimov (1984); southern African diamonds, Deines (1980), Deines et al. (1984, 1987); Deines, unpublished results.

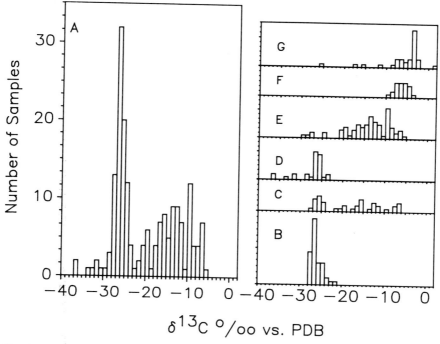

Figure 13.30 Carbon isotopic composition of basalts and high-temperature volcanic CO_2. (A) Summary distribution of basalts, (B) continental basalts, (C) Hawaiian basalts, (D) Icelandic basalts, (E) ocean-floor basalts, (F) CO_2 released by crushing – sources of data: Craig (1953), Naughton & Terada (1954), Friedman & O'Neil (1971), Fuex & Baker (1973), Hoefs (1973), Pineau *et al.* (1976a), Galimov & Gerasimovskiy (1978), Rison & Craig (1983), Pineau & Javoy (1983), Watanabe *et al.* (1983), Des Marais & Moore (1984), Mattey *et al.* (1984), Sakai *et al.* (1984), Exley *et al.* (1986). (G) High-temperature CO_2 (430–1130 °C) – sources of data: Naughton & Terada (1954), Wasserburg *et al.* (1963), Friedman & O'Neil (1971), Matsubaya *et al.* (1975), Kratsov *et al.* (1976, 1979), Allard (1979, 1980a, b 1981a, b, 1983), Allard *et al.* (1981).

(Fig. 13.30A) and the causes of the diversity in isotopic composition are not well understood. Part of the variability may be due to differences in the sample-preparation techniques used by different authors.

In the earlier studies samples were cleaned carefully, either exclusively outside or outside and inside the vacuum combustion line, and subsequently reacted in an atmosphere of O_2 (Fuex & Baker 1973, Hoefs 1973, Galimov & Gerasimovskiy 1978). More recently, the combustion of samples was carried out by stepwise heating under O_2 (Des Marais & Moore 1984), or a low-temperature combustion step was followed by heating in vacuum without the presence of O_2 (Mattey *et al.* 1984).

The stepwise combustion procedures have shown (Fig. 13.31) that the CO_2 formed below 600 °C is generally depleted in ^{13}C compared to CO_2 formed above 600 °C (e.g. Pineau *et al.* 1976a, Des Marais & Moore 1984, Mattey *et al.* 1984). This has led some authors (e.g. Mattey *et al.* 1984) to suggest that the low-temperature CO_2 results from surficial organic contamination, and that C contamination prior to and during the analysis may be an important factor.

The following four observations lead, however, to the conclusions that not all of

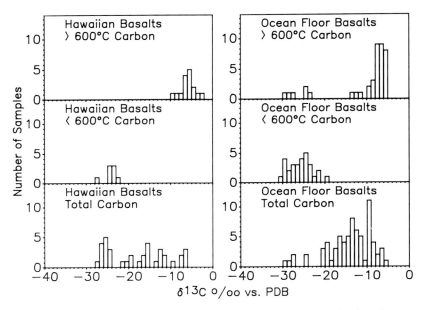

Figure 13.31 Carbon isotopic composition of basalts determined by stepwise combustion. Sources of data: Des Marais & Moore (1984), Mattey *et al.* (1984), Sakai *et al.* (1984), Exley *et al.* (1986).

the variability seen in Figure 13.30 results from secondary contamination:

(a) The CO_2 released above 600 °C (Fig. 13.31) still shows a large range in $\delta^{13}C$ and includes values as low as -29 to -30‰.

(b) Watanabe *et al.* (1983) have isolated, in a vacuum dissolution experiment, C of $\delta^{13}C = -26.9$‰ from olivines from Hawaiian basalts.

(c) The sample-cleaning procedures used by Galimov & Gerasimovskiy (1978) should have reduced any surface contamination by at least 70%, with a consequent shift towards more positive $\delta^{13}C$ values. Nevertheless, these authors' data for Icelandic basalts (Fig. 13.30D) include the lowest $\delta^{13}C$ values measured for basalts.

(d) In Figure 13.32 the $\delta^{13}C$ values of the total C of basalts have been plotted v. their C content; also indicated are the concentration and isotopic composition range of the high-temperature combustion results. If this range were characteristic for indigenous C in basalts and the variation in $\delta^{13}C$ and total C concentration were the result of the introduction of secondary contaminants of $\delta^{13}C = -25$‰, the mixtures should lie in the region between the mixing hyperbolae indicated in Figure 13.32. Very high levels of relative and absolute amounts of contamination would have to be postulated to explain a large fraction of the data set. This seems very unlikely in view of the care that was taken by all investigators to minimize contamination of the samples that were analysed. Secondary contamination may well be the explanation for some of the measured low $\delta^{13}C$ values; however, it is clearly ruled out for many, in particular for the suite of samples from Iceland. Hence, there are basalts with indigenous $\delta^{13}C$ values considerably lower than -10‰.

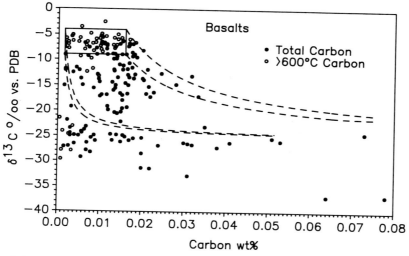

Figure 13.32 Relationship between C isotopic composition and C content of basalts. Sources of data: see Figure 13.30.

The question of how to interpret the light indigenous C isotopic compositions has also been debated. One school attributes the low $\delta^{13}C$ to organic material either subducted or caught up in the eruptive process, while another, mainly represented by Javoy and Pineau, explains the low ^{13}C content of some of the C in basalts as the result of a degassing process which can be characterized by Rayleigh fractionation. In such a process, CO_2 enriched in ^{13}C is removed from the magma, trapped in vesicles and liberated by crushing of the samples (Fig. 13.30F). The residual C is depleted in ^{13}C, and is released upon high-temperature pyrolysis or combustion. Arguments have been raised both in favour of and in opposition to the two hypotheses, and the discussions concerning the measurement techniques and the interpretation of the results are ongoing (e.g. Des Marais 1986, Craig 1987, Exley *et al.* 1987). At this time it is therefore not possible to use the $\delta^{13}C$ data for basalts to limit the $\delta^{13}C$ range expected for carbonatites.

Allard (1983) has argued that the high-temperature CO_2 from central volcanic vents provides a reliable sample of the C initially in the magma. Other authors (e.g. Gerlach & Thomas 1986) have interpreted the C isotopic composition of low-temperature fumarolic gases in a similar manner. In Figure 13.30G the $\delta^{13}C$ measurements of CO_2 from high-temperature volcanic gases have been summarized. Compositions enriched in ^{13}C have been interpreted as the result of thermal decomposition of limestones, while ^{13}C depleted compositions have been attributed either to contamination by organic material caught up in the volcanic eruption process or to isotope separation as a result of magma degassing.

Contamination would be minimal in areas of ocean-floor spreading for which Allard (1983) reports $\delta^{13}C$ values of -6.4 and $-6.0‰$. Gerlach & Thomas (1986) suggest that the $\delta^{13}C$ values of $-3.2‰$ of CO_2 from Kilauean low-temperature fumaroles indicate ^{13}C enrichment in the source region of Hawaiian basalts compared to mid-ocean ridge basalts. The authors propose that there might be a significant difference in $\delta^{13}C$ between hot-spot and mid-ocean ridge magmas. It should be noted, however, that all of the high temperature CO_2 samples with $\delta^{13}C$ less than $-12‰$ come from Mauna Loa and Kilauea (Naughton & Terada 1954, Wasserburg *et al.* 1963, Friedman & O'Neil 1971). Because the number of isotopic composition measurements of high-temperature CO_2 are so few and their interpretation is not unique, they are of little help in estimating the expected $\delta^{13}C$ range of carbonatites.

In view of the variability of the C isotopic composition in meteorites, diamonds, basalts, and high-temperature volcanic gases, the restricted $\delta^{13}C$ range of carbonatites is surprising. The more limited variability may, in part, be due to averaging of the source region isotopic composition during carbonatite magma formation. Although meteorite, diamond, and carbonatite $\delta^{13}C$ sampling frequency distributions show a mode at about $-5‰$, carbonatite carbonates tend to have, on the average, higher $\delta^{13}C$ values than diamonds or meteoritic C, suggesting that there may be an isotope effect in the formation of the carbonatite magmas which enriched ^{13}C in the melt. Figure 13.33 summarizes some relevant C isotope fractionation factors. Assuming the fractionation behaviour of a carbonate melt is similar to that of calcite and dolomite, one would expect ^{13}C concentration in a carbonate rather than a silicate melt.

Figure 13.33 indicates that dolomite concentrates ^{13}C slightly, compared to calcite. From the isotopic composition differences between coexisting carbonates (Fig. 13.34) it can be seen that only about one-third of the mineral pairs could be in isotopic equilibrium at high temperatures. Many mineral pairs have reversals in fractionation indicative of isotopic disequilibrium. The most extreme examples are found in mineral pairs from the Mbeya carbonatite, Tanzania. The large, coupled

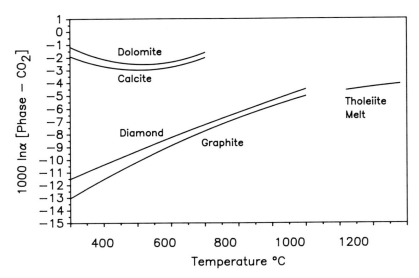

Figure 13.33 Carbon isotope fractionation between C-bearing minerals, silicate melt, and carbon dioxide. Sources of data: Bottinga (1969*a, b*), Sheppard & Schwarcz (1970), Javoy *et al.* (1978).

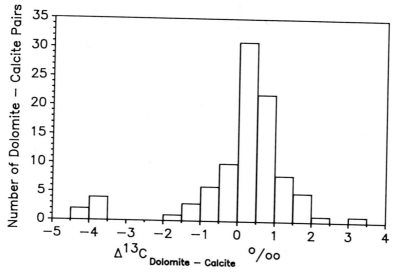

Figure 13.34 Carbon isotope fractionation between coexisting dolomite–calcite pairs. Sources of data: see Figures 13.15, 13.16.

negative C and O isotope fractionations are probably the result of secondary processes.

13.3.3 Regional differences in the carbon isotopic composition of the source of carbonatite magmas

In order to establish whether significant differences exist among the mean $\delta^{13}C$ values of carbonatite complexes, only those samples whose C isotope record was not disturbed by secondary processes are considered. Therefore $\delta^{13}C$ data were used only for carbonates whose $\delta^{18}O$ values fall between 5.5 and 8.5‰. There were 30 localities for which two or more samples fulfilled the criterion. In Figure 13.35 the mean for each locality and its 95% confidence interval are shown. It is obvious that the mean $\delta^{13}C$ of the lavas from Oldoinyo Lengai differs from those of the rest of the carbonatites. In order to test for smaller $\delta^{13}C$ differences among the remaining 29 localities, the grand mean of their means (-5.4‰) and its 95% confidence interval (± 0.2‰) are also indicated in Figure 13.35.

In spite of the large uncertainty of some of the means, which is largely the result of insufficient data, Figure 13.35 indicates that there are significant differences in C isotopic composition among carbonatites. These may be related either to the processes involved in the extraction of C or the nature of the primary reservoir within the mantle.

Regional isotopic composition variations can be evaluated in East Africa where data from a sufficiently large number of carbonatites are available. Figure 13.36 represents an update of the original compilation made by Deines & Gold (1973). The data suggest a north-south trending zone of carbonatites characterized by average

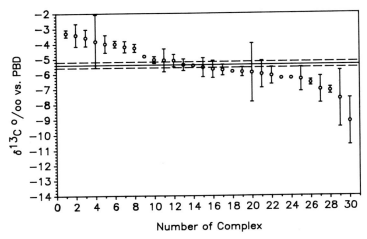

Figure 13.35 Comparison of the C isotopic composition of carbonatite complexes. Mean $\delta^{13}C$ and 95% confidence intervals are given for complexes for which several carbonate samples were analysed and whose $\delta^{18}O$ values fell in the range 5–8‰. The solid line indicates the mean C isotopic composition of all complexes, the dashed lines the 95% confidence interval around this mean. Carbonatites: 1, Tororo; 2, Mbeya; 3, Sukulu; 4, Homa Mountain; 5, Phalaborwa; 6, Cargill; 7, Kovdor; 8, Mud Tank; 9, Wet Mountains; 10, St. Honoré; 11, Alnö; 12, Tatarsky; 13, Sokli; 14, Magnet Cove; 15, Iron Hill; 16, Fen; 17, Oka; 18, Lueshe; 19, Kaiserstuhl; 20, Prairie Lake; 21, Chernigov; 22, Bailundo; 23, Maly Sayanski; 24, Bol'shoy; 25, Monte Verde; 26, Nemegosenda Lake; 27, Ile de Fuerte Ventura; 28, Laacher See; 29, Ile de Fogo; 30, Oldoinyo Lengai. Sources of data: see Table 13.1.

$\delta^{13}C$ values between -2.8 and -3.9‰. In contrast, carbonate lavas and tuffs, as well as intrusive carbonatites from the eastern and western rift areas, have $\delta^{13}C$ values that are significantly lower (-6 to -7‰). Regional isotopic composition differences are hence indicated, suggesting a relationship between the C isotopic composition of carbonatites and tectonic setting. These conclusions were confirmed by the new results of Nelson *et al.* (1988) published recently. Mantle heterogeneity in the C isotope distribution has also been deduced from the study of diamonds (Deines *et al.* 1987).

13.3.4 Carbon isotope fractionation in carbonatite emplacement

Fractionation processes during the emplacement and crystallization of a carbonatite will modify the isotope distribution of the parent magma. The observed $\delta^{13}C$ frequency distribution is an expression of all isotope fractionation processes involved in the formation of the carbonatite. If the number of such processes is very large, the resulting distribution pattern should tend toward normality. The observed, well-defined skewness demonstrated in Figure 13.24 suggests that only one or a few processes fractionating isotopes in the same direction are responsible for most of the observed variability.

To interpret the sampling frequency distribution in terms of a process, it must be assumed that the observed frequency of a certain δ value is proportional to the volume of carbonatite of that isotopic composition. Such an assumption can be supported only if the sampling of the carbonatite has been truly representative. In

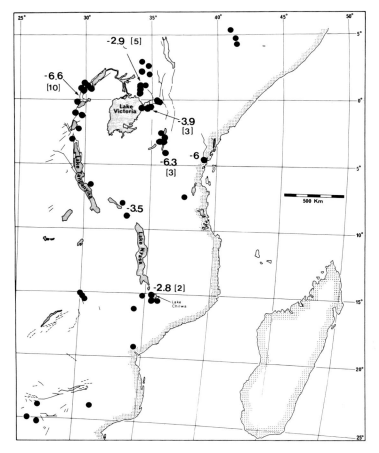

Figure 13.36 Location and C isotopic composition of some East African carbonatites. Mean values are given for several areas, the numbers in brackets represent the number of locations represented by the mean. Sources of data: see Table 13.1.

the case of the Oka carbonatite, where a large number of samples from all representative rock types and a wide area and depth distribution have been analysed, such an assumption has some validity.

Fractionation by diffusive separation or by crystallization mechanisms which could play a role in carbonatite formation, can be characterized by Rayleigh fractionation if it takes place in a reservoir of given volume. The following relationship exists between the isotopic composition, δ, of the element in the product formed, the fraction, F, of the reservoir left, the isotopic composition, δ_0, of the element in the reservoir at the beginning of the process, and the fractionation factor between product and reservoir, α:

$$\delta = \{\alpha \times (\delta_0 + 1000) F^{(\alpha - 1)}\} - 1000.$$

A plot of log F v. log $(1000 + \delta)$ should be linear with a slope of $(\alpha - 1)$. This relationship is shown for the Oka data in Figure 13.37, and a fractionation factor of

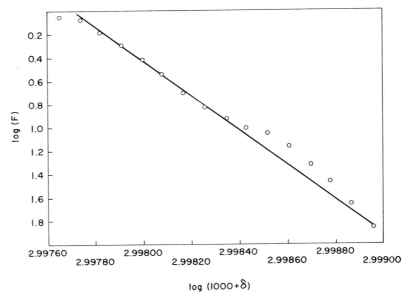

Figure 13.37 Correlation between the amount of accumulated carbonate (%) and C isotopic composition for the Oka carbonatite. Source: Deines (1967).

0.9993 is calculated. Figure 13.38 shows the remarkable fit of the data (dots) to the Rayleigh fractionation model (solid line). It was assumed that the initial reservoir had a range of isotopic compositions characterized by a normal distribution with a standard deviation of 0.3‰.

The Rayleigh fractionation model also explains the greater than equilibrium ^{13}C enrichment in some of the Oka rauhaugites. Formation of the rauhaugites after a

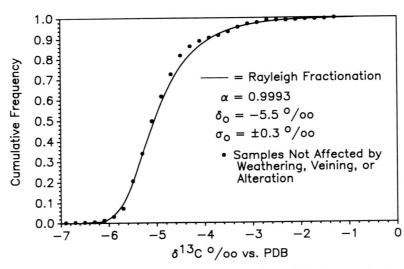

Figure 13.38 Observed (circles) and calculated Rayleigh fractionation (solid line) cumulative frequency distribution for Oka carbonatite data. Source of data: Deines (1970).

substantial removal of calcite from the reservoir (geologic observations, experimental data, and $\delta^{18}O$ data indicate that this is a reasonable assumption) would result in ^{13}C enrichment in rauhaugites.

Observations which agree with the $\delta^{13}C$ trends identified at Oka have been made by other authors. Korzhinskiy & Mamchur (1978, 1980) found that in the carbonatites of the Vuorijarvi alkaline massif the $\delta^{13}C$ values increase from the earliest calcite carbonatite ($-4.1‰$) through later phases (-3.3 to $-3.2‰$) to the lowest-temperature fine-grained ankerite–dolomite rock ($-2.4‰$). For the Belaya Zima carbonatite, East Sayan, the authors report that $\delta^{13}C$ increases with increasing dolomite content. An increase in $\delta^{13}C$ in the later carbonatite stages was also observed by Lyashkevich & Panov (1983) for complexes in the Dnieper-Donets basin and by Stepanenko & Sukhanov (1980) for those in central Timan.

While it is interesting to note that the Rayleigh fractionation model successfully explains the C isotope variations in carbonatite carbonates, many other factors can influence the distribution of ^{13}C. The Rayleigh model may not explain some of the local variations in C isotopic composition.

13.4 THE RELATION BETWEEN CARBON AND OXYGEN ISOTOPE COMPOSITION

The relation between $\delta^{13}C$ and $\delta^{18}O$ of carbonatite carbonates is displayed in Figure 13.39; the large range in $\delta^{18}O$ is accompanied by only minor changes in $\delta^{13}C$. The more extreme $\delta^{18}O$ values can be attributed to secondary processes in which $\delta^{13}C$ and $\delta^{18}O$ may or may not be correlated. Closer inspection of the data shown in Figure 13.39 suggests that in the $\delta^{18}O$ range from 5 to 15‰ the isotopic compositions of these two elements may be correlated.

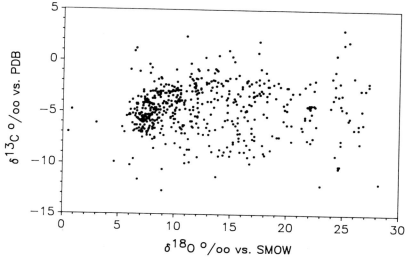

Figure 13.39 Carbon and O isotopic composition variations in carbonatite minerals. Sources of data: see Table 13.1. Not included were the analyses from Oka and St. Honoré.

Data for each locality within the $\delta^{18}O$ range 5.5–14.5‰ were divided into three groups:

(a) $\delta^{18}O = 5.5$–8.5‰;
(b) $\delta^{18}O = 8.5$–11.5‰;
(c) $\delta^{18}O = 11.5$–14.5‰.

For each group a mean $\delta^{13}C$ and $\delta^{18}O$ value was computed. For 25 localities analyses fell in at least two of these groups, and for 21 of the 25 localities the mean $\delta^{13}C$ showed a systematic increase from group (a) through group (c). Analyses from Kaiserstuhl, Mt. Homa, and Odoinyo Lengai indicated no real change in $\delta^{13}C$ over this $\delta^{18}O$ range, while samples from Phalaborwa showed a 1‰ decline in $\delta^{13}C$. In Figure 13.40 lines connecting the means of the groups have been drawn for the 21 localities.

The trends shown by the data in Figure 13.40 are too systematic to be coincidental. Regression analyses indicate that the correlations between $\delta^{13}C$ and $\delta^{18}O$ have, on the average, a slope of about 0.4. Through a detailed examination of the Oka data, Deines (1970) calculated a slope of 0.3, and the data for the St. Honoré carbonatite (Fig. 13.41) suggest a slope of about 0.2 for samples with $\delta^{18}O$ values between 5 and 16‰. Pineau *et al.* (1973) also identified a group of carbonatites showing a correlation between $\delta^{13}C$ and $\delta^{18}O$, with a slope of 0.4. It seems unlikely that secondary alteration is responsible for such a commonly observed relationship, and it can be concluded that the positive correlation between $\delta^{13}C$ and $\delta^{18}O$ of carbonates in the $\delta^{18}O$ range from about 5 to 15‰ reflects a process fundamental to carbonatite formation.

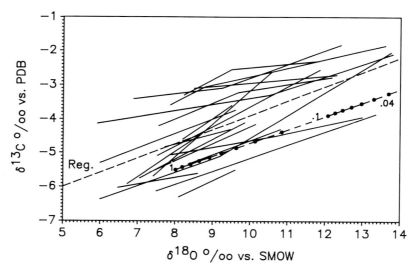

Figure 13.40 Correlation between C and O isotopic composition in the O isotopic composition range 5.5–14.5‰ v. SMOW. Solid lines connect C isotopic composition means for different O isotopic composition ranges; see text. --- = regression for whole data set; ●-●-●-● = Rayleigh fractionation model for the simultaneous separation of two phases from a common reservoir; see text. The numbers along the fractionation line represent the fraction of the reservoir left. Between 1 and 0.1 each dot represents a decrease by 0.1, between 0.1 and 0.01 a decrease by 0.01.

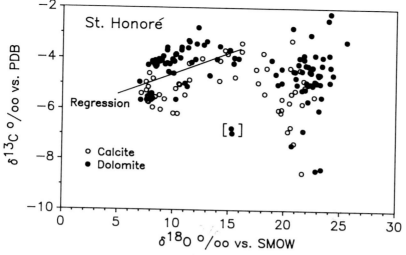

Figure 13.41 Associated C and O isotopic composition variations for St. Honoré carbonates. The regression was computed for a range in O isotopic composition from 5 to 16‰; two outliers (in brackets) were not included in the analysis. Source of data: this chapter.

Fractionation of ^{13}C and ^{18}O between dolomite and calcite might help in interpreting the positive correlation. In Figure 13.42 the difference in isotopic composition between coexisting calcite and dolomite is shown, as well as the relationship which would be expected if coexisting calcite and dolomite were formed in isotopic equilibrium. Although a rough positive correlation between the variables is apparent, much of the data is not consistent with the existence of isotopic

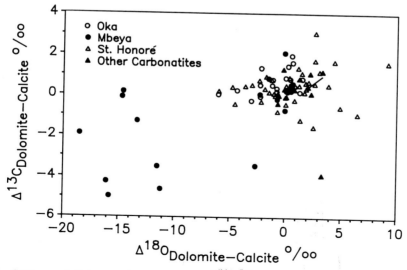

Figure 13.42 Carbon and O isotope fractionations between associated calcite and dolomite. The short line indicates the equilibrium fractionation relationship; see Figure 13.43. Sources of data: see Figures 13.15 and 13.16.

equilibrium between coexisting calcite and dolomite. Even if samples that might have been affected by secondary processes are eliminated, and only those carbonates are considered with $\delta^{18}O$ values in the range from 5 to 10‰, the correlation is still not improved (Fig. 13.43). It is interesting to note that for many of the samples whose O isotope fractionation and $\delta^{18}O$ values suggest a primary magmatic carbonate pair, the C isotope fractionation is much larger than would be expected at magmatic temperatures.

Alternative interpretations can be proposed for the observed correlation shown in Figure 13.40. Pineau *et al.* (1973) suggested that it could be the result of a Rayleigh fractionation process during the precipitation of calcite at 700 °C from a phase in which CO_2 is the dominant O carrier and < 30% of the O is present in the form of water or silicates. The C and O isotope fractionation is determined by the carbonate–CO_2 equilibrium. Deines (1970) interpreted the $\delta^{13}C$ variations of the Oka complex as the result of reservoir effects in the crystallization of carbonate from a magma. The variations in $\delta^{18}O$ were thought to be governed not only by reservoir effects but also by the proportions of silicates and carbonate precipitating from the melt and, in some rocks, by metasomatic changes. Because the $\delta^{13}C$ and $\delta^{18}O$ variations were not attributed to a single chemical process, a strict correlation between the two isotopic compositions was not expected.

The consistent slope between $\delta^{13}C$ and $\delta^{18}O$ observed for many carbonatites suggests that fractionation of the two elements is linked to a process which is an integral part of the formation of carbonatites.

Carbonatites are closely associated with magmatic silicate rocks containing carbonates. For such rocks the assumption of Pineau *et al.* (1973) that < 30% of

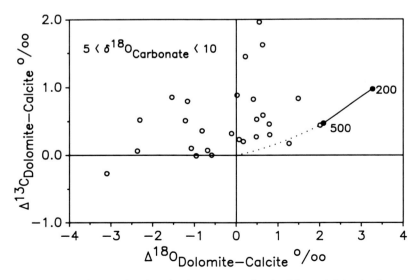

Figure 13.43 Carbon and O isotope fractionation between associated calcite and dolomite for samples whose O isotopic composition falls in the range 5–10‰. The expected fractionation based on the data of Figures 13.14 and 13.33 is indicated; the fractionations for 200 and 500 °C are marked; the dotted curve represents an extrapolation for high-temperature fractionations.

the O is present in the form of water and silicates is not valid. The fact that the isotopic compositions of the carbonates from both silicate and carbonate rocks are closely related, suggests that the model of Pineau *et al.* (1973) is not a suitable explanation of their C and O isotopic composition variability.

Under some circumstances the simultaneous formation of a highly carbonaceous and a siliceous phase might be described by a Rayleigh fractionation model in which two products are removed simultaneously from a common reservoir. Assumed are:

(a) that there are C and O isotopic composition differences between the two products formed and the reservoir;
(b) that the proportions of the separating products throughout the process are fixed; and
(c) that after the formation of the products no exchange between them or the remaining reservoir occurred.

The isotopic composition of C in product P_1 when the fraction, F, of the reservoir is left is given by:

$$\delta^{13}C_{P_1} = \{\alpha_1^c \times (1000 + \delta^{13}C_R) \times F^{ec1}\} - 1000,$$

where α_1^c stands for the C isotope fractionation factor between P_1 and the reservoir and $\delta^{13}C_R$ denotes the original isotopic composition of the reservoir; ec1 is given by:

$$ec1 = \frac{[(B^c \times \alpha_1^c) + \alpha_2^c]}{(B^c + 1)} - 1.$$

The fractionation factor α_2^c characterizes the ^{13}C separation between the second product and the reservoir, and B^c represents the molar ratio of C in product 1 to product 2. A similar expression can be written for the $\delta^{13}C$ value of the second product.

The O isotopic composition of product P_1, $\delta^{18}O_{P_1}$ and product P_2, $\delta^{18}O_{P_2}$, are given by corresponding expressions. The common variable among all four equations is the fraction of the reservoir remaining (F). Through elimination of F between the relationship for $\delta^{13}C_{P_1}$ and that for $\delta^{18}O_{P_1}$, the process-introduced correlation between $\delta^{13}C$ and $\delta^{18}O$ for product P_1 can be found. To a first approximation this relationship is linear with a slope:

$$S_1 = \frac{[B^c(\alpha_1^c - 1) + \alpha_2^c - 1][B^o + 1]}{[B^o(\alpha_1^o - 1) + \alpha_2^o - 1][B^c + 1]}$$

where B^o is the molar ratio of O in product P_1 to product P_2, and α_1^o and α_2^o are the O isotope fractionation factors between product P_1 and the reservoir and product P_2 and the reservoir, respectively. Because the fractionation factors change with temperature and B^c and B^o could change during the process, the relationship between $\delta^{13}C$ and $\delta^{18}O$ does not necessarily remain linear.

As one example, it can be assumed that product P_1 represents a carbonatite magma which carries most of the C and product P_2 represents a silicate magma or precipitating silicates. One can assume B^c to be in the order of 1000, because the amount of carbonate in product P_2 is small, and B^o to be in the order of 0.1, because the volume of carbonatite magma is generally small compared to product P_2. On the basis of the Oka data, α_1^c and α_2^c are considered of similar size and equal to 0.9993. From the fractionation factors compiled in Figure 13.14, α_1^o is assumed to be 1.0025. Using these values we find $\alpha_2^o = 0.9978$ for a value of $S_1 = 0.4$. Although the value of α_2^o may appear to be somewhat high, it could be easily reduced by using different values for B^o, B^c, and different fractionation factors. The formation of a third product, e.g. a CO_2-containing fluid phase, likewise reduces the size that this fractionation factor would have to have in order to attain a slope of $S_1 = 0.4$.

Using these assumptions one can determine the $\delta^{13}C$ and $\delta^{18}O$ variations in carbonatite carbonates that result if two products are formed from a reservoir which initially had a $\delta^{13}C_R = -4.8‰$ (extrapolated from Oka data) and a $\delta^{18}O_R = 5.7‰$ (average mantle). The heavy dotted line in Figure 13.40 shows the results of such a model computation. Correlated C and O isotopic composition variations covering the range from 8 to 14‰ can be produced in this way, assuming only small degrees of C and O isotope fractionations. Obviously, the model computations could readily be made to coincide with the regression relationship by choosing slightly different values for $\delta^{13}C_R$ and $\delta^{18}O_R$.

Coupled silicate-carbonate formation can hence explain the correlation between $\delta^{13}C$ and $\delta^{18}O$ of carbonatite carbonates. In view of the complete lack of detailed knowledge of the distribution of isotopes during the separation of carbonate and silicate magmas or their differentiation from a common parent and the development of a fluid phase (loss of CO_2), further testing of this type of model must await the results of experimental studies.

13.5 SULPHUR

13.5.1 Overview of available data

Sulphur isotope data are available for relatively few carbonatites. In Figure 13.44, 462 analyses from 14 complexes have been compiled. The sulphides include chalcopyrite, galena, pyrite, and pyrrhotite; all of the sulphates analysed were barite except for one celestite sample. Schneider (1970) reported 12 analyses for nosean and hauyne from Laacher See ejecta with an average isotopic composition of 4.4‰; these results are not included in Figure 13.44. The ^{34}S enrichment in the sulphate compared to the sulphides is apparent. The sampling is too limited, however, to establish whether there are systematic differences in isotopic composition between

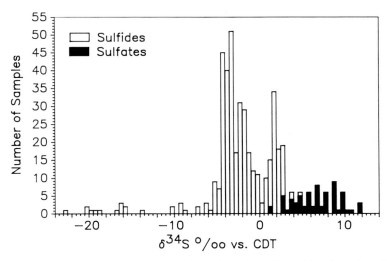

Figure 13.44 Sulphur isotopic composition of sulphides and sulphates from 14 carbonatites. Sources of data: Hoefs *et al.* (1968), Grinenko *et al.* (1970a, b), Mitchell & Krouse (1971, 1975), Mäkelä & Vartiainen (1978), Panov *et al.* (1981), Bagdasarov & Grinenko (1983), Bolonin & Zhukov (1983).

different sulphides; the apparent bimodality in the $\delta^{34}S$ distribution is not linked to the sulphide mineralogy.

13.5.2 *The sulphur isotope composition of the source of carbonatite magmas*

Figure 13.45 shows $\delta^{34}S$ data from meteorites, basic sills, and ocean-floor basalts. Most analyses fall in a fairly narrow range between -1 and $+1‰$ v. CDT and this range is generally considered to be characteristic of pristine mantle. It should be noted, however, that Ueda & Sakai (1984) have reported $\delta^{34}S$-enriched sulphides for some submarine basalts that fall well outside this range. Significant ^{34}S enrichments compared to the mean meteoritic value are also found in some basic intrusions (e.g. Muskox, $+5‰$, Sasaki 1969; Noril'sk, $+9‰$, Godlevskii & Grinenko 1963; Sudbury, $+2.1$, Schwarcz 1973). Basalts from continental volcanos (Fig. 13.46) can also have increased ^{34}S contents. Crustal contamination, changing O fugacity conditions and degassing processes are all possible explanations for these elevated $\delta^{34}S$ values. It is also noteworthy that ejecta from the Laacher See volcano have $\delta^{13}C$ values of $-7‰$, $\delta^{18}O$ values compatible with an isotopic equilibrium with mantle silicates, and $\delta^{34}S$ values of 3–4‰. If the C and O isotopic compositions of the ejecta reflect those of the mantle source, then the same should be true for S.

The evidence suggests that some parts of the mantle may be enriched in ^{34}S compared to meteoritic S. A similar conclusion was also reached on the basis of studies of volcanic rocks from Germany (Schneider 1970) and Japan (Ueda & Sakai 1984). At this time it would be imprudent to rule out the possible existence of heterogeneities in the S isotopic composition of the mantle, and one can expect differences in the mean $\delta^{34}S$ of carbonatite magmas of the order of several per mil.

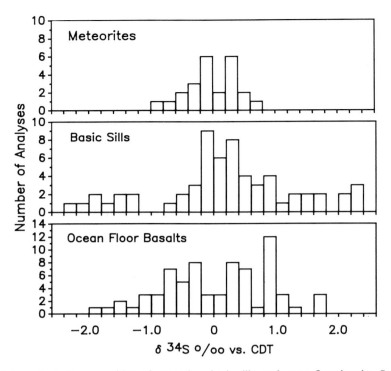

Figure 13.45 Sulphur isotopic composition of meteorites, basic sills, and ocean-floor basalts. Sources of data: McNamara & Thode (1950), Ault & Kulp (1959), Shima & Thode (1961), Thode *et al.* (1961), Jensen & Nakai (1962), Shima *et al.* (1963), Smitheringale & Jensen (1963), Krouse & Folinsbee (1964), Hulston & Thode (1965), Monster *et al.* (1965), Kaplan & Hulston (1966), Kanehira *et al.* (1973), Grinenko *et al.* (1975), Sakai *et al.* (1978, 1980, 1982, 1984), Puchelt & Hubberten (1980), Ueda & Sakai (1984).

13.5.3 Sulphur isotope composition difference between carbonatites

In Figure 13.47 the mean isotopic compositions of 12 complexes are compared. The figure includes only sulphide analyses. Data from the Mountain Pass carbonatite, for which sulphides and sulphates show a large difference in $\delta^{34}S$ (see Fig. 13.50) were not used. It is apparent that carbonatite complexes differ in their mean S isotopic composition. To what extent the mean $\delta^{34}S$ values are related to isotope variations in the source of the carbonatites or to isotope effects in the formation or crystallization of the carbonatite magmas is not known. If the mantle had a truly uniform S isotopic composition of $\delta^{34}S = 0‰$, and isotope fractionation during carbonatite emplacement did not significantly affect the mean isotopic composition of the complex, the isotope effects in the carbonatite magma generation would have to differ by at least 7‰ among the complexes shown in Figure 13.47, excluding the data from Yrass (11‰ if these data are included).

Considering the size of the S isotope effects at the appropriate temperatures (Fig. 13.48) the following points may be made:

(a) no equilibrium fractionation factors are presently known which would produce a melt $\delta^{34}S$ value of $-5‰$ from a source with $\delta^{34}S = 0‰$;

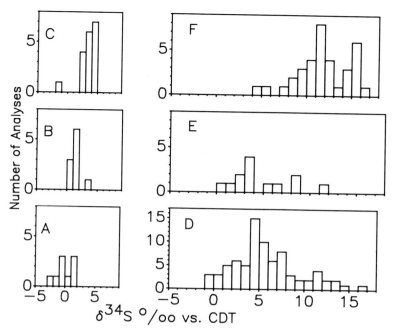

Figure 13.46 Sulphur isotopic composition of volcanic rocks. (A) Tholeiites, Germany; (B) olivine alkali basalts, Germany; (C) alkali-rich undersaturated rocks, Germany; (D) Japanese arc; (E) Greece; (F) Saudi Arabia. Sources of data: Schneider (1970), Hubberten *et al.* (1977), Ueda & Sakai (1984).

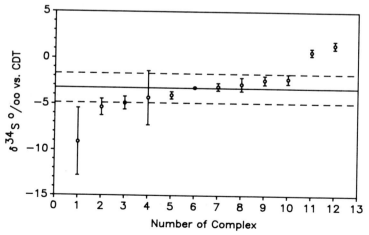

Figure 13.47 Comparison of the S isotopic composition of carbonatite complexes. Mean isotopic compositions and 95% confidence intervals are given. The solid line indicates the mean S isotopic composition of all complexes, the dashed lines the 95% confidence interval around this mean. 1, Yraas, Maymecha-Kotuy; 2, Magnet Cove; 3, Oka; 4, Yessey, Maymecha-Kotuy; 5, Bear Paw; 6, Kovdor; 7, Vuorijarvi; 8, Dalbykha, Maymecha-Kotuy; 9, Sokli; 10, Maly; 11, Bol'shoy; 12, Phalaborwa. Sources of data: see Table 13.1.

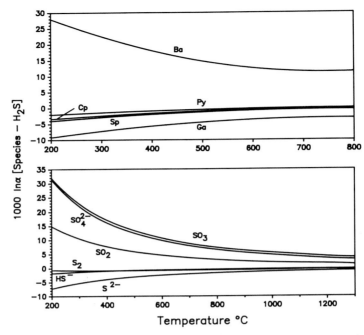

Figure 13.48 Measured and computed S isotope fractionation factors. Sources of data: Sakai (1968), Kajiwara & Krouse (1971), Kiyosu (1973), Robinson (1973), Kusakabe & Robinson (1977), Richet *et al.* (1977).

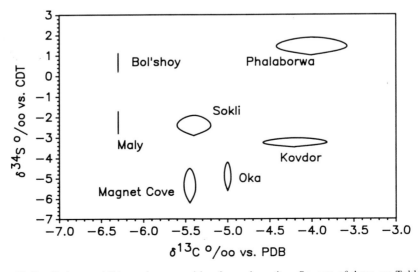

Figure 13.49 Carbon and S isotopic composition for carbonatites. Sources of data: see Table 13.1.

(b) melt $\delta^{34}S$ values as heavy as $+5‰$ may be produced only if fractionations are governed by oxidized S species in the melt, and if only a small fraction of the S were extracted from the reservoir; and

(c) substantial amounts of S would have to be lost during magma emplacement to produce large deviations of the carbonatite mean $\delta^{34}S$ value from a magma value of 0‰.

In view of these observations, some of the isotopic composition differences between carbonatite complexes are attributed to isotope heterogeneity of S within the mantle.

Because the sources of carbonatite magmas may differ in their $\delta^{13}C$ and $\delta^{34}S$ values, it is possible that there may be some relationship between them. Unfortunately, no detailed joint $\delta^{13}C$ and $\delta^{34}S$ studies of carbonatites have been undertaken. In Figure 13.49 $\delta^{13}C$ and $\delta^{34}S$ from some complexes are compared for which several analyses are available. Different carbonatite complexes are characterized by distinct mean $\delta^{13}C$ and $\delta^{34}S$ values. Only two C isotope analyses were included from the Bol'shoy and Maly complex, hence their mean C isotopic compositions are very uncertain. If only Phalaborwa, Sokli, Magnet Cove, Oka, and Kovdor are considered, one might suggest that there is a correlation between $\delta^{13}C$ and $\delta^{34}S$.

13.5.4 Variations in sulphur isotope composition among different stages of carbonatite formation

There are few detailed studies of $\delta^{34}S$ variations within carbonatites. In some cases systematic trends have been observed, and these have been related to the different stages of carbonatite formation. Results are summarized in Table 13.3.

Figure 13.48 summarizes the available data relevant to the fractionation behaviour of the different S species potentially present during carbonatite formation. Notably lacking is information on the partitioning of S isotopes during the formation of a carbonate melt. It is apparent from Figure 13.48 that the relative proportions of the oxidized and reduced S species in the melt will determine the isotope fractionation between sulphide minerals and the total S in the melt.

The trend in the carbonatite sulphides towards more negative $\delta^{34}S$ values in later stages can be interpreted readily as resulting from an increase in oxidizing conditions

Table 13.3 Sulphur isotopic composition of different carbonatite stages.

| Location | Stage | | | | | Mineralogy |
	1	2	3	4	5	
Kovdor	−1.9	−3.6	−3.6	−6.1		Sulphides
Vuorijarvi	−1.2	−3.9				Sulphides
Bol'shoy	+2.0		+1.6	−1.0		Sulphides
Maly Sayanski		+1.9 to −2.2		−4.3		Sulphides
Mountain Pass	+5.4	+4.8	+6.9	+8.7	+9.4	Barite

NB The stage number has different connotations for different complexes, the original publications should be consulted for details.

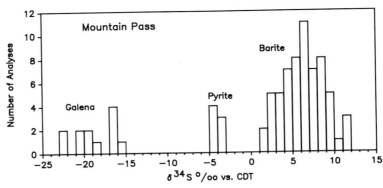

Figure 13.50 Sulphur isotope variations in the Mountain Pass carbonatite. Sources of data: Mitchell & Krouse (1971, 1975).

in consecutive carbonatite stages. This is not true for barite data from the Mountain Pass carbonatite (Fig. 13.50). The fractionation between barite and galena has been used by Mitchell & Krouse (1971, 1975) to estimate a formation temperature of 300 °C, and the authors have suggested that a continuous decrease in the S content of the hydrothermal solutions would lead to an increase in the relative abundances of the reduced S species. This, in turn, would enlarge the fractionation between the hydrothermal solution and the precipitating sulphate and consequently lead to an increased ^{34}S content of the latter. Assuming that the solutions had a $\delta^{34}S = 0‰$ for total S, a large fraction of the S has still to be accounted for. The total S isotope budget for the carbonatite remains to be established before the sulphide and sulphate precipitation processes can be constrained on the basis of δ^{34}S measurements.

13.6 SUMMARY

The stable isotopic composition of carbonatite minerals is governed by several factors. Some can be identified on the basis of the available data. For C and S the source regions of carbonatite magmas are probably isotopically heterogeneous. In the case of δ^{13}C differences, these may be of regional extent and related to tectonic setting. The limited data available also suggest that some source regions may possess distinct combinations of δ^{13}C and δ^{34}S values.

Isotope effects may occur in the formation of the carbonatite magma. In the case of C the higher mean ^{13}C content of all carbonatites compared to those of meteorites and diamonds may be the result of such a fractionation. An estimate of the δ^{18}O of carbonatite melts indicates that they could have formed in isotopic equilibrium with average mantle peridotite at temperatures between 1000 and 1200 °C.

Reservoir effects which can be simulated by a Rayleigh fractionation process may be important during the separation of immiscible carbonate–silicate melts and fractional crystallization processes. The C isotope distribution of the Oka carbonatite can be interpreted readily on the basis of such a model. A correlation between $\delta^{13}C$ and $\delta^{18}O$ in the $\delta^{18}O$ range from 5 to 15‰ for 90% of the carbonatites studied probably reflects fundamental processes involved in carbonatite formation. The slope of the linear array, as well as the range of observed isotopic compositions, can be predicted assuming coupled element fractionation during a Rayleigh fractionation process involving the simultaneous separation of two products from a common reservoir.

Part of the stable isotope record of carbonatites is established by equilibrium fractionation between cogenetic minerals. This is demonstrated by the O isotope fractionation between some of the coexisting silicates, oxides, and carbonates. It also explains, on the basis of mass balance considerations, the slight systematic enrichment in ^{18}O in the carbonates from silicate rocks. In rocks of metasomatic origin the ^{18}O content of the carbonates is in part governed by the O isotopic composition of the source rocks. Systematic changes in the S isotopic composition between different carbonatite stages may be related to changes in O fugacity.

Formation at hydrothermal temperatures has been suggested for the Mountain Pass carbonatite. Whether the high ^{18}O content of some parts of the St. Honoré carbonatite is related to such a process remains to be established. The absence of significant ^{18}O gradients towards the contact of the Oka carbonatite can be taken as evidence that the interaction between the magma of this carbonatite and the surrounding rocks was minimal and that no significant hydrothermal circulating systems were set up at the margin of the intrusion.

Post-magmatic processes may have little effect on the C isotopic composition but can produce major changes in the O isotope record. Studies of the isotope variations on hand-specimen scale demonstrate that closed-system re-equilibration in the presence of a magmatic fluid is, in part, responsible for the large range in $\delta^{18}O$ values in carbonates from silicate rocks. Biotite formation also influences the O isotope distribution significantly.

Exchange with heated groundwaters may be the cause of the elevated ^{18}O content of some of the carbonates, but it would have to occur below 200–250 °C in order to raise the $\delta^{18}O$ of the carbonate above 6‰. The changes in isotopic composition of the carbonates during weathering have been evaluated for the Oka carbonatite and are found to be very small.

There are only a few *detailed* stable isotope studies of carbonatites, and investigations that combine δD, $\delta^{13}C$, $\delta^{18}O$, and $\delta^{34}S$ measurements are absent. Such investigations would constitute a fruitful approach to evaluate the relative importance of the various processes that have been involved in the generation and evolution of carbonatite melts. Combined chemical, radiogenic, and stable isotope studies will help to characterize mantle source differences and identify problems of crustal contamination.

ACKNOWLEDGEMENTS

The previously unpublished analytical work used in this review was in part supported by NSF Grants GP-3949, GA-1272, and EAR-8416386.

REFERENCES

Allard, P. 1979. $^{13}C/^{12}C$ and $^{34}S/^{32}S$ ratios in magmatic gases from ridge volcanism in Afar. *Nature* **282**, 56–8.
Allard, P. 1980a. Composition isotopique du carbone dans les gaz d'un volcan d'arc Momotombo (Nicaragua). *Competes Rendus des Séances de l'Académie des Sciences, Série D: Sciences Naturelles* **290**, 1525–8.
Allard, P. 1980b. Proportions des isotopes ^{13}C et ^{12}C du carbone émis à haute température par un dome andésitique en cours de croissance: Le Mérap (Indonesié). *Comptes Rendus des Séances de l'Académie des Sciences, Série D: Sciences Naturelles* **291**, 613–16.
Allard, P. 1981a. Composition chimique et isotopique des gaz émis par le nouveau dome andésitique de la Soufrière de Saint-Vincent (Antilles). *Comptes Rendus des Séances de l'Académie des Sciences, Série 2: Mécanique-Physique, Chimie, Sciences de l'Univers, Sciences de la Terre* **293**, 721–4.
Allard, P. 1981b. Composition isotopique du carbone dans les gaz d'un magma dacitique (volcan Usu, Japon). Relation entre le rapport $^{13}C/^{12}C$ des volatiles et le rapport $^{87}Sr/^{86}Sr$ de la phase silicatée dans le volcanisme d'arc. *Comptes Rendus des Séances de l'Académie des Sciences, Série 2: Mécanique-Physique, Chimie, Sciences de l'Univers, Sciences de la Terre* **293**, 583–6.
Allard, P. 1983. The origin of hydrogen, carbon, sulfur, nitrogen and rare gases in volcanic exhalation: Evidence from isotope geochemistry. In *Forecasting volcanic events*, H. Tazieff & J.-C. Sabroux (eds), 337–86. Amsterdam: Elsevier.
Allard, P., C. Jehanno & J.-C. Sabrous 1981. Composition chimique et isotopique de produits gazeux et solides de l'activité éruptive du Krakatau (Indonésie) pendant la période 1978–80. *Comptes Rendus des Séances de l'Académie des Sciences, Série 2: Mécanique-Physique, Chimie, Sciences de l'Univers, Sciences de la Terre* **293**, 1095–8.
Allard, P., F. Le Guern & J.-C. Sabroux 1977. Thermodynamic and isotopic studies in eruptive gases. *Geothermics* **5**, 37–40.
Allègre, C. J., F. Pineau, M. Bernat & M. Javoy 1971. Evidence for the occurrence of carbonatites on the Cape Verde and Canary Islands. *Nature* **233**, 103–4.
Allègre, C. J., T. Staudacher, P. Sarda & M. Kurz 1983. Constraints on the evolution of Earth's mantle from rare gas systematics. *Nature* **303**, 762–6.
Armbrustmacher, T. J. 1979. Replacement and primary magmatic carbonatites from the Wet Mountains area, Fremont and Custer Counties, Colorado. *Economic Geology* **74**, 888–901.
Ault, W. U. & J. L. Kulp 1959. Isotopic geochemistry of sulphur. *Geochimica et Cosmochimica Acta* **16**, 201–35.
Baertschi, P. 1957. Messung und Deutung relativer Häufigkeitsvariationen von ^{18}O und ^{13}C in Karbonatgesteinen und Mineralien. *Schweizerische Mineralogische und Petrographische Mitteilungen* **37**, 73–152.
Bagdasarov, Yu. A. & M. Buyakayte 1985a. Features of carbonatite formation in carbonate sedimentary rocks according to isotope-geochemical data. *Geokhimiya* **1985**, 559–68.
Bagdasarov, Yu. A. & M. I. Buyakayte 1985b. Isotopic data on carbonatite formation in carbonate sediments. *Geochemistry International* **22**, (8), 30–8.
Bagdasarov, Yu. A. & L. N. Grinenko 1981. Isotopic composition of sulfur in carbonatites of the Chernigovsk zone Azov Sea coastal region and reasons for its variations in carbonatite complexes. *Doklady Akademii Nauk SSSR, Seriya Geologiya* **258**, 1192–5.
Bagdasarov, Yu. A. & L. N. Grinenko 1983. Isotopic composition of sulfur in sulfides from carbonatite massifs of the Maimecha-Kotui province and some conditions of their formation. *Doklady Akademii Nauk SSSR, Seriya Geologiya* **271**, 1484–8.
Bagdasarov, Yu. A., L. A. Bannikova & I. N. Ivanoskaya 1982. Isotopic composition of carbon of coexisting graphites and carbonates from carbonatites of the Chernigov zone (Sea of Azov region) and some features of their genesis. *Doklady Akademii Nauk SSSR, Seriya Geologiya* **262**, 967–70.
Bagdasarov, Yu. A., E. M. Galimov & V. S. Prokhorov 1969a. Isotopic composition of the carbon in ankerite carbonatites and sources of carbonatite material in sedimentary rocks. *Doklady Akademii Nauk SSSR, Seriya Geologiya* **188**, 1327–75.
Bagdasarov, Yu. A., E. M. Galimov & V. S. Prokhorov 1969b. Isotopic composition of carbon in ankerite carbonatite and the source of carbonatite material present in sedimentary rocks. *Doklady Akademii Nauk SSSR, Seriya Geologiya* **188**, 201–4.
Becker, R. H. 1971. *Carbon and oxygen isotope ratios in iron-formation and associated rocks from the*

Hamersley Range of Western Australia and their implications. Ph.D. thesis, University of Chicago, Chicago.

Becker, R. H. & S. Epstein 1982. Carbon, hydrogen and nitrogen isotopes in solvent-extractable organic matter from carbonaceous chrondrites. *Geochimica et Cosmochimica Acta* **46**, 97–103.

Begemann, F. & K. Heinzinger 1969. Content and isotopic composition of carbon in the light and dark portion of gas-rich chondrites. In *Meteorite research*, P. M. Millman (ed.), 87–92. Dordrecht: Reidel.

Belsky, T. & I. R. Kaplan 1970. Light hydrocarbon gases, ^{13}C, and origin of organic matter in carbonaceous chondrites. *Geochimica et Cosmochimica Acta* **34**, 257–78.

Blattner, P. & A. F. Cooper 1974. Carbon and oxygen isotopic composition of carbonatite dikes and metamorphic country rock of the Haast Shist Terrain, New Zealand. *Contributions to Mineralogy and Petrology* **44**, 17–27.

Boato, G. 1954. The isotopic composition of hydrogen and carbon in carbonaceous chondrites. *Geochimica et Cosmochimica Acta* **6**, 209–20.

Bolonin, A. V. & F. I. Zhukov 1983. Isotopic composition of carbon, oxygen and sulfur of carbonatites from deposits in southern Siberia. *Izvestiya Vysshikh Uchebnykh Zavedenii, Geologiya i Razvedka* **26**, 67–72.

Bottinga, Y. 1969a. Carbon isotope fractionation between graphite, diamond and carbon dioxide. *Earth and Planetary Science Letters* **5**, 301–7.

Bottinga, Y. 1969b. Calculated fractionation factors for carbon and hydrogen isotope exchange in the system calcite–carbon dioxide–graphite–methane–hydrogen–water vapor. *Geochimica et Cosmochimica Acta* **33**, 49–64.

Chang S., R. Mack & K. Lennon 1978. *Carbon chemistry of separated phases of Murchison and Allende meteorites.* Lunar and Planetary Science Abstracts and Papers submitted to the Lunar and Planetary Science Conference 9, 157–9.

Clayton, R. N. 1959. Oxygen isotope fractionation in the system calcium carbonate–water. *Journal of Chemical Physics* **30**, 1246–50.

Clayton, R. N. 1963. Carbon isotope abundance in meteoritic carbonates. *Science* **140**, 192–3.

Clayton, R. N. 1981. Isotopic thermometry. In *Advances in physical geochemistry*, Vol. 1: *Thermodynamics of minerals and melts*. R. C. Newton, A. Navrotsky & B. J. Wood (eds), 85–109. New York: Springer-Verlag.

Clayton, R. N. & T. K. Mayeda 1978. Genetic relations between iron and stony meteorites. *Earth and Planetary Science Letters* **40**, 168–74.

Clayton, R. N. & T. K. Mayeda 1983. Oxygen isotopes in eucrites, shergottites, nakhlites, and chassignites. *Earth and Planetary Science Letters* **62**, 1–6.

Clayton, R. N. & T. K. Mayeda 1984. The oxygen isotope record in Murchison and other carbonaceous chondrites. *Earth and Planetary Science Letters* **67**, 151–61.

Clayton, R. N., L. Grossman & T. K. Mayeda 1973. A component of primitive nuclear composition in carbonaceous meteorites. *Science* **182**, 485–8.

Clayton, R. N., N. Onuma & T. K. Mayeda 1976. A classification of meteorites based on oxygen isotopes. *Earth and Planetary Science Letters* **30**, 10–18.

Clayton, R. N., N. Onuma, L. Grossman & T. K. Mayeda 1977. Distribution of the pre-solar component in Allende and other carbonaceous chondrites. *Earth and Planetary Science Letters* **34**, 209–24.

Conway, C. M. & H. P. Taylor 1969. $^{18}O/^{16}O$ and $^{13}C/^{12}C$ ratios of coexisting minerals in the Oka and Magnet Cove carbonatite bodies. *Journal of Geology* **77**, 618–26.

Cox, K. G., J. J. Gurney & B. Harte 1973. Xenoliths from the Matsoku Pipe. In *Lesotho kimberlites*, P. H. Nixon (ed.), 76–100. Lesotho National Development Corporation, Maseru. Cape Town: Cape and Transvaal Printers Ltd.

Craig, H. 1953. The geochemistry of stable carbon isotopes. *Geochimica et Cosmochimica Acta* **3**, 53–92.

Craig, H. 1963. The isotopic geochemistry of water and carbon in geothermal areas. In *Nuclear geology in geothermal areas*, E. Tongiorgi (ed.), 17–53. Pisa: Consiglio Nazionale Delle Ricerche, Laboratorio Di Geologia Nucleare.

Craig, H. 1987. Comment on "Carbon isotope systematics of a mantle hotspot: A comparison of Loihi Seamount and MORB glasses" by R. A. Exley, D. P. Mattey, D. A. Clague & C. T. Pillinger. *Earth and Planetary Science Letters* **82**, 384–6.

Deines, P. 1967. *Stable carbon and oxygen isotopes of carbonatite carbonates and their interpretation*. Ph.D. thesis, The Pennsylvania State University, University Park.

Deines, P. 1970. The carbon and oxygen isotopic composition of carbonates from the Oka carbonatite, Quebec, Canada. *Geochimica et Cosmochimica Acta* **34**, 1199–225.

Deines, P. 1980. The carbon isotope composition of diamonds: relationship to diamond shape, color, occurrence and vapor composition. *Geochimica et Cosmochimica Acta* **44**, 943–61.

Deines, P. & D. P. Gold 1973. The isotopic composition of carbonatites and kimberlite carbonates and their bearing on the isotopic composition of deep-seated carbon. *Geochimica et Cosmochimica Acta* **37**, 1709–33.

Deines, P. & F. E. Wickman 1973. The isotopic composition of "graphitic" carbon from iron meteorites and some remarks on the troilitic sulfur in iron meteorites. *Geochimica et Cosmochimica Acta* **37**, 1295–319.

Deines, P. & F. E. Wickman 1975. A contribution to the stable carbon isotope geochemistry of iron meteorites. *Geochimica et Cosmochimica Acta* **39**, 547–57.

Deines, P. & F. E. Wickman 1985. Stable carbon isotopes in enstatite chondrites and Cumberland Falls. *Geochimica et Cosmochimica Acta* **49**, 89–95.

Deines, P., J. J. Gurney & J. W. Harris 1984. Associated chemical and carbon isotopic composition variations in diamonds from Finsch and Premier kimberlite, South Africa. *Geochimica et Cosmochimica Acta* **48**, 325–42.

Deines, P., J. W. Harris & J. J. Gurney 1987. Carbon isotopic composition, nitrogen content and inclusion composition of diamonds from the Roberts Victor kimberlite, South Africa: Evidence for ^{13}C depletion in the mantle. *Geochimica et Cosmochimica Acta* **51**, 1227–43.

Denaeyer, M. E. 1970. Rapports isotopiques δO et δC et conditions d'affleurement des carbonatites de l'Afrique centrale. *Comptes Rendus des Séances de l'Académie des Sciences, Série D: Sciences Naturelles* **270**, 2155–8.

Des Marais, D. J. 1986. Carbon abundance measurements in oceanic basalts: the need for a consensus. *Earth and Planetary Science Letters* **79**, 21–6.

Des Marais, D. J. & J. G. Moore 1984. Carbon and its isotopes in mid-ocean basaltic glasses. *Earth and Planetary Science Letters* **69**, 43–57.

Dontsova, Ye. I., V. A. Kononova & L. D. Kuznetsova 1977a. Isotopic composition of oxygen of carbonatites and carbonatite-like rocks in connection with the sources of their material and the problems of ore content. *Geokhimiya* **1977**, 963–75.

Dontsova, Ye. I., V. A. Kononova & L. D. Kuznetsova 1977b. The oxygen–isotope composition of carbonatites and similar rocks in relation to their sources and mineralization. *Geochemistry International* **14**, (4), 1–11.

Downs, W. F., Y. Touysinhthiphonexay & P. Deines 1981. A direct determination of the oxygen isotope fractionation between quartz and magnetite at 600 and 800 °C and 5 kb. *Geochimica et Cosmochimica Acta* **45**, 2065–72.

Eckerman, H. von, H. von Ubisch, & F. E. Wickman 1952. A preliminary investigation into the isotopic composition of carbonates from some alkaline intrusions. *Geochimica et Cosmochimica Acta* **2**, 207–10.

Epstein, S., R. Buchsbaum, H. A. Lowenstam & H. C. Urey 1953. Revised carbonate–water isotopic temperature scale. *Bulletin of the Geological Society of America* **64**, 1315–26.

Exley, R. A., D. P. Mattey & C. T. Pillinger 1987. Low temperature carbon components in basaltic glasses – reply to comment by H. Craig. *Earth and Planetary Science Letters* **82**, 387–90.

Exley, R. A., D. P. Mattey, D. A. Claque & C. T. Pillinger 1986. Carbon isotope systematics of a mantle 'hotspot': A comparison of Loihi Seamount and MORB glasses. *Earth and Planetary Science Letters* **78**, 189–99.

Flory, D. A. 1969. *Stable isotope measurements of meteoritic carbonaceous material*. Ph.D. thesis, University of Houston, Houston.

Friedman, I. & J. O'Neil 1971. C^{13}–C^{12} *ratios of CO_2 from Hawaiian lava lake*. United States Geological Survey Professional Paper 750, A216.

Friedrichsen, H. 1968. Sauerstoffisotope einiger Minerale der Karbonatite des Fengebietes, Süd Norwegen. *Lithos* **1**, 70–5.

Friedrichsen, H. 1973. Die Verteilung der Sauerstoffisotope bei der Differentiation karbonatitischer Magmen. *Fortschritte der Mineralogie* **50**, (3), 34–5.

Fuex, A. N. & D. R. Baker 1973. Stable carbon isotopes in selected granitic, mafic and ultramafic igneous rocks. *Geochimica et Cosmochimica Acta* **37**, 2509–21.

Galimov, E. M. 1984. The relationship between formation conditions and variations in isotopic composition of diamonds. *Geokhimiya* **1984**, 1091–118.

Galimov, E. M. & V. I. Gerasimovskiy 1978. Isotopic composition in Icelandic magmatic rocks. *Geochemistry International* **15**, (6), 1–6.

Galimov, E. M., V. A. Kononova & V. S. Prokhorov 1974. Isotopic composition of carbonatites and carbonatite-like rocks (in connection with the problem of matter source). *Geokhimiya* **1974**, 708–16.

Garlick, G. D. 1966. Oxygen isotope fractionation in igneous rocks. *Earth and Planetary Science Letters* **1**, 361–8.

Garlick, G. D., I. D. MacGregor & D. E. Vogel 1971. Oxygen isotope ratios in eclogites from kimberlites. *Science* **172**, 1025–7.

Gehlen, K. von 1967. Sulphur isotopes from the sulfide-bearing carbonatite of Phalabora, South Africa. *Transactions, Institution of Mining and Metallurgy*, **76**(B), 223. (Abstract).

Gerlach, T. M. & D. M. Thomas 1986. Carbon and sulfur isotopic composition of Kilauea parental magma. *Nature* **319**, 480–3.

Godlevskii, M. N. & L. N. Grinenko 1963. Some data on the isotopic composition of sulphur in sulfides of the Noril'sk deposit. *Geokhimiya* **1963**, 35–40.

Gonfianitini, R. & E. Tongiorgi 1964. La composition isotopique des carbonatites du Kaiserstuhl. In *Les roches alcalines et les carbonatites du Kaiserstuhl*, L. V. Wambeke, J. W. Brinck, W. Duetzmann, R. Gonfianitini, A. Hubaux, D. Métais, P. Omenetto, E. Tongiorgi, G. Verfaillie, K. Weber & W. Wimmenauer (eds), 139–9. Report EUR 1827.d.f.e. of the European Atomic Energy Commission. Luxembourg: European Atomic Energy Commission.

Grady, M. M., P. K. Swart & C. T. Pillinger 1982. Variable carbon isotope composition of Type 3 ordinary chondrites. *Proceedings of the 13th Lunar and Planetary Science Conference, Journal of Geophysical Research Section* **A87**, Supplement Part 1, A289–96.

Grady, M. M., I. P. Wright, L. P. Carr & C. T. Pillinger 1986. Compositional differences in enstatite chondrites based on carbon and nitrogen stable isotope measurements. *Geochimica et Cosmochimica Acta* **50**, 2799–813.

Gray, J. & G. L. Cumming 1977. *Oxygen and strontium isotopic compositions and thorium and uranium contents of basalts from DSDP 37 cores*. Initial Reports of the Deep Sea Drilling Project, 1969, **37**, 607–11.

Grinenko, L. N., V. A. Kononova & V. A. Grinenko 1970a. Isotopic composition of sulfide sulphur in carbonatites. *Geochemistry International* **7**, (1), 45–53.

Grinenko, L. N., V. A. Kononova & V. A. Grinenko 1970b. Isotopic composition of sulfur in sulfides from carbonatites. *Geokhimiya* **970**, 66–75.

Grinenko, V. A., L. V. Dimitriev, A. A. Migdisov & A. Y. Sharas'kin 1975. Sulfur contents and isotope compositions for igneous and metamorphic rocks from Mid-Ocean Ridges. *Geochemistry International* **12**, (1), 132–7.

Hay, R. L. & J. R. O'Neil 1983. Carbonatite tuffs in the Laetolil Beds of Tanzania and the Kaiserstuhl in Germany. *Contributions to Mineralogy and Petrology* **82**, 403–6.

Hirner, A. 1979a. $^{13}C/^{12}C$-Bestimmungen am Murchison-Meteoriten. *Bulletin der Vereinigung Schweizerischer Petroleum-Geologen und -Ingenieure* **45**, 47–55.

Hirner, A. 1979b. Isotopenverhältnisse des stabilen Kohlenstoffs in C-haltigen Phasen des Murchison-Meteoriten. *Fortschritte der Mineralogie* **57**, 191.

Hoefs, J. 1973. Ein Beitrag zur Isotopengeochemie des Kohlenstoffs in magmatischen Gesteinen. *Contributions to Mineralogy and Petrology* **41**, 277–330.

Hoefs, J., H. Nielsen & M. Schidlowski 1968. Sulphur isotope abundances in pyrite from the Witwatersrand conglomerates. *Economic Geology* **63**, 975–7.

Hoernes, S. & H. Friedrichsen 1978. Oxygen and hydrogen isotope study of the polymetamorphic area of the northern Ötztal-Stubai Alps (Tyrol). *Contributions to Mineralogy and Petrology* **67**, 305–15.

Holmes, A. 1958. Spitzkop carbonatite, Eastern Transvaal. *Bulletin of the Geological Society of America* **69**, 1525–6.

Hubberten, H. W., H. Nielsen & H. Puchelt 1977. Sulfur isotope investigations on rocks and ores of the Santorini Archipelago, Greece. *Annales Géologiques des Pays Helléniques 1e Séries* **XXVIII-1976**, 334–48.

Hulston, J. R. & H. G. Thode 1965. Variations in the ^{33}S, ^{34}S, and ^{36}S contents of meteorites and their relation to chemical and nuclear effects. *Journal of Geophysical Research* **70**, 3475–84.

Jagoutz, E., J. B. Dawson, S. Hoernes, B. Spettel & H. Wänke 1984. *Anorthositic oceanic crust in the*

Archean Earth. Lunar and Planetary Science. Abstracts and Papers submitted to the Lunar and Planetary Science Conference 15, 395–6.

Javoy, M., F. Pineau & I. Iiyama 1978. Experimental determination of the isotopic fractionation between gaseous CO_2 and carbon dissolved in tholeiitic magma. *Contributions to Mineralogy and Petrology* 67, 35–9.

Jensen, M. L. & N. Nakai 1962. Sulfur isotope meteorite standards: Results and recommendations. In *Biochemistry of Sulfur Isotopes*, M. L. Jensen (ed.), 30–41. Proceedings of a National Science Foundation Symposium 12–14 April 1962, Yale University.

Kajiwara, Y. & H. R. Krouse 1971. Sulfur isotope partitioning in metallic sulfide systems. *Canadian Journal of Earth Sciences* 8, 1397–408.

Kanehira, K., S. Yiu, H. Sakai & A. Sasaki 1973. Sulphide globules and sulfur isotope ratios in the abyssal tholeiite from the Mid-Atlantic Ridge near 30°N latitude. *Geochemical Journal* 7, 89–96.

Kaplan, I. R. & J. R. Hulston 1966. The isotopic abundance and content of sulfur in meteorites. *Geochimica et Cosmochimica Acta* 30, 479–96.

Kerridge, J. F. 1982. *Isotopic composition of C, H, N in carbonaceous chondrite polymer using stepwise combustion*. Lunar and Planetary Science. Abstracts and Papers submitted to the Lunar and Planetary Science Conference 13, 381–2.

Kerridge, J. F. 1983. Isotopic composition of carbonaceous-chondrite kerogen: evidence for an interstellar origin of organic matter in meteorites. *Earth and Planetary Science Letters* 64, 186–200.

Kieffer, S. W. 1982. Thermodynamics and lattice vibrations of minerals: 5. Applications to phase equilibria, isotopic fractionation, and high-pressure thermodynamic properties. *Reviews of Geophysics and Space Physics* 20, 827–49.

Kiyosu, Y. 1973. Sulfur isotopic fractionation among sphalerite, galena and sulfide ions. *Geochemical Journal* 7, 191–9.

Kobelski, B. J. 1977. *South African and Lesothan kimberlites: With emphasis on the stable carbon and oxygen isotopic composition of kimberlite carbonates*. M.Sc. thesis, The Pennsylvania State University, University Park.

Kobelski, B. J., D. P. Gold & P. Deines 1979. Variations in stable isotope compositions for carbon and oxygen in some south African and Lesothan kimberlites. In *Kimberlites, diatremes, and diamonds: Their geology, petrology and geochemistry*. Proceedings of the Second International Kimberlite Conference, Vol. 1. F. R. Boyd & H. O. A. Meyer (eds), 252–71. Washington: American Geophysical Union.

Kononova, V. A. 1980. Alkaline magmatic rocks and sources of matter in their composition based on data of strontium, oxygen and carbon isotopic ratios. *Petrologiya* 1980, 30–40.

Kononova, V. A. & R. M. Yashina 1984. Geochemical criteria for differentiating between rare-metallic carbonatites and barren carbonatite-like rocks. *The Indian Mineralogist* 1985, Sukhneswala Volume, 136–50.

Korzhinskiy, A. F. & G. P. Mamchur 1978. Isotopic composition of carbon from carbonatites and probable sources of their matter. *Zapiski Vsesoyuznogo Mineralogicheskogo Obschestva* 107, 129–36.

Korzhinskiy, A. F. & G. P. Mamchur 1980. Carbon isotope composition of the carbonatites and a probable source for their material. *International Geology Review* 22, 1390–6.

Kratsov, A. I., O. I. Kropotova, S. K. Onikiyenko & B. V. Federenko 1976. Isotopic composition of carbon gases of active volcanoes and hot springs of the Kuril-Kamchatka volcanic arc. *International Geology Review* 18, 773–6.

Kratsov, A. I., V. A. Bobrov, O. S. Kropotova, Y. K. Markhinin & N. Y. Podkletnov 1979. $\delta^{13}C$ for gases and organic compounds in juvenile material from volcanoes in the Kurile-Kamchatka volcanic arc. *Geochemistry International* 16, (5), 152–6.

Kropotova, O. I. 1969. Die Isotopenzusammensetzung des Kohlenstoffs in Kalziten von Karbonatitmassiven und das Problem der Genese der Karbonatite. *Zeitschrift für angewandte Geologie* 15, 135–40.

Krouse, H. R. & R. E. Folinsbee 1964. The $^{32}S/^{34}S$ ratio in troilite from the Bruderheim and Peace River meteorites. *Journal of Geophysical Research* 69, 4192–3.

Krouse, H. R. & V. E. Modzeleski 1970. $^{13}C/^{12}C$ abundances in components of carbonaceous chondrites and terrestrial samples. *Geochimica et Cosmochimica Acta* 34, 459–74.

Kukharenko, A. A. & E. I. Dontsova 1964. A contribution to the problem of the genesis of carbonatites. *Economic Geology USSR* (English transl.) 1, 31–46.

Kuleshov, V. N. 1985. Isotopic composition of high-temperature carbonates of the Kovdor Massif USSR. *Soviet Geology* **1985**, 107–13.

Kuleshov, V. N. 1986. Isotope composition and origin of deep-seated carbonates. *Trudy Akademiya Nauk SSSR, Geologicheskii Institut* **405**, 1–123.

Kusakabe, M. & B. W. Robinson 1977. Oxygen and sulfur isotope equilibria in the $BaSO_4$–HSO_4^-–H_2O system from 110 to 350 °C and applications. *Geochimica et Cosmochimica Acta* **41**, 1033–40.

Kvenvolden, K., J. Lawless, K. Pering, E. Peterson, J. Flores, C. Ponnamperuma, I. R. Kaplan & C. Moore 1970. Evidence for extraterrestrial amino-acids and hydrocarbons in Murchison meteorite. *Nature* **228**, 923–6.

Kyser, K. T. 1986. Stable isotope variations in the mantle. In Stable isotopes in high temperature geological processes, J. W. Valley, H. P. Taylor & J. R. O'Neil (eds). *Reviews in Mineralogy* **16**, 141–64.

Kyser, K. T. & J. R. O'Neil 1978. Oxygen isotope relations among oceanic tholeiites, alkali basalts, and ultramafic nodules. In Short Papers of the Fourth International Conference, Geochronology, Cosmochronology, Isotope Geology, R. E. Zartman (ed.), 237–40, United States Geological Survey Open-File Report 78–701.

Kyser, K. T., J. R. O'Neil & I. S. E. Carmichael 1981. Oxygen isotope thermometry of basic lavas and mantle nodules. *Contributions to Mineralogy and Petrology* **77**, 11–23.

Kyser, K. T., J. R. O'Neil & I. S. E. Carmichael 1982. Genetic relations among basic lavas and ultramafic nodules: Evidence from oxygen isotope compositions. *Contributions to Mineralogy and Petrology* **81**, 88–102.

Lapin, A. V., V. N. Gushchin & I. P. Lugovaya 1986. Isotopic composition of carbonatites and metamorphosed carbonate sedimentary rocks. *Geokhimiya* **1986**, 979–86.

Lyashkevich, Z. M. & B. S. Panov 1983. Distribution of carbon and oxygen isotopes in carbonates of the Dnieper-Donets basin. *Geologiya i Geokhimiya Goryuchikh Iskopaemykh* **60**, 74–7.

Lyashkevich, Z. M., A. I. Marushkin, G. P. Mamchur & O. D. Yarnych 1978. Explosive carbonatites in the Dnieper-Donets basin. *Geologiya i Geokhimiya Goryuchikh Iskopaemykh* **51**, 68–72.

MacGregor, I. D. & W. I. Manton 1986. Roberts Victor Eclogites: Ancient oceanic crust. *Journal of Geophysical Research* **91**, 14063–79.

MacNamara, J. & H. G. Thode 1950. Comparison of the isotopic constitution of terrestrial and meteoritic sulfur. *Physical Review* **78**, 307–8.

Mäkelä, M. & H. Vartiainen 1978. A study of sulfur isotopes in the Sokli multi-stage carbonatite (Finland). *Chemical Geology* **21**, 257–65.

Matsubaya, O., A. Ueda, M. Kusakabe, Y. Matsuhisa, H. Sakai & A. Sakai 1975. An isotopic study of the volcanoes and the hot springs in Satsuma Iwojima and some areas in Kyushu. *Bulletin of the Geological Survey of Japan* **26**, 375–92.

Matsuhisa, Y. 1979. Oxygen isotopic composition of volcanic rocks from the East Japan island arcs and their bearing on petrogenesis. *Journal of Volcanology and Geothermal Research* **5**, 271–96.

Matsuhisa, Y., J. R. Goldsmith & R. N. Clayton 1979. Oxygen isotopic fractionation in the system quartz–albite–anorthite–water. *Geochimica et Cosmochimica Acta* **43**, 1131–40.

Matsuhisa, Y., O. Matsubaya & H. Sakai 1973. Oxygen isotope variations in magmatic differentiation processes of the volcanic rocks in Japan. *Contributions to Mineralogy and Petrology* **39**, 277–88.

Mattey, D. P., R. H. Carr, I. P. Wright & C. T. Pillinger 1984. Carbon isotopes in submarine basalts. *Earth and Planetary Science Letters* **70**, 196–206.

Matthews, A. & A. Katz 1977. Oxygen isotope fractionation during the dolomitization of calcium carbonate. *Geochimica et Cosmochimica Acta* **41**, 1431–8.

Matthews, A., J. R. Goldsmith & R. N. Clayton 1983. Oxygen isotope fractionations involving pyroxenes: the calibration of mineral-pair geothermometers. *Geochimica et Cosmochimica Acta* **47**, 631–44.

Mitchell, R. H. & H. R. Krouse 1971. Isotopic composition of sulfur in carbonatite at Mountain Pass, California. *Nature* **231**, 182.

Mitchell, R. H. & H. R. Krouse 1975. Sulphur isotope geochemistry of carbonatites. *Geochimica et Cosmochimica Acta* **39**, 1505–13.

Monster, J., E. Anders & H. G. Thode 1965. $^{34}S/^{32}S$ ratios for the different forms of sulphur in the Orgueil meteorite and their mode of formation. *Geochimica et Cosmochimica Acta* **29**, 773–9.

Muehlenbachs, K. 1976. *Oxygen isotope geochemistry of DSDP Leg 34 basalts*. Initial Reports of the Deep Sea Drilling Project 1969, **34**, 337–9.

Muehlenbachs, K. 1977a. *Oxygen isotope geochemistry of DSDP Leg 37 rocks*. Initial Reports of the Deep Sea Drilling Project 1969, **37**, 617–19.

Muehlenbachs, K. 1977b. Oxygen isotope geochemistry of rocks from DSDP Leg 37. *Canadian Journal of Earth Sciences* **14**, 771–6.

Muehlenbachs, K. & R. N. Clayton 1971. Oxygen isotope ratios of submarine diorites and their constituent minerals. *Canadian Journal of Earth Sciences* **8**, 1591–5.

Muehlenbachs, K. & R. N. Clayton 1972. Oxygen isotope studies of fresh and weathered submarine basalts. *Canadian Journal of Earth Sciences* **9**, 172–84.

Muehlenbachs, K., A. T. Anderson & G. E. Sigvaldason 1974. Low-^{18}O basalts from Iceland. *Geochimica et Cosmochimica Acta* **38**, 577–88.

Naughton, J. J. & K. Terada 1954. Effect of Hawaiian volcanoes on the composition and carbon isotope content of associated volcanic and fumarolic gases. *Science* **120**, 580–1.

Nelson, D. R., A. R. Chivas, B. W. Chappell & B. W. McCulloch 1988. Geochemical and isotopic systematics in carbonatites and implications for the evolution of ocean island sources. *Geochimica et Cosmochimica Acta* **52**, 1–17.

Nielsen, T. F. D. & B. Buchardt 1985. Sr–C–O isotopes in nephelinitic rocks and carbonatites, Gardiner complex, Tertiary of East Greenland. *Chemical Geology* **53**, 207–17.

Northrop, D. A. & R. N. Clayton 1966. Oxygen-isotope fractionations in systems containing dolomite. *Journal of Geology* **74**, 174–96.

O'Neil, J. R. 1986. Theoretical and experimental aspects of isotope fractionation. In Stable isotopes in high temperature geological processes, J. W. Valley, H. P. Taylor & J. R. O'Neil (eds). *Reviews in Mineralogy* **16**, 1–40.

O'Neil, J. R. & R. L. Hay 1973. ^{18}O/^{16}O ratios in cherts associated with the saline lake deposits of East Africa. *Earth and Planetary Science Letters* **19**, 257–66.

O'Neil, J. R., R. N. Clayton & T. K. Mayeda 1969. Oxygen isotope fractionation in divalent metal carbonates. *Journal of Chemical Physics* **51**, 5547–58.

Ongley, J. S., A. R. Basu & K. T. Kyser 1987. Oxygen isotopes in coexisting garnets, clinopyroxenes and phlogopites of Roberts Victor eclogites: Implications for petrogenesis and mantle metasomatism. *Earth and Planetary Science Letters* **83**, 80–4.

Onuma, N., R. N. Clayton & T. K. Mayeda 1972. Oxygen isotope temperatures of "equilibrated" ordinary chondrites. *Geochimica et Cosmochimica Acta* **36**, 157–68.

Panov, B. S., Z. M. Lyashkevich & I. Pilot 1981. Isotopic sulfur, oxygen and carbon content of Devonian mineral formations of the Dnieper Donetsk basin. *Dopovidi Akademii Nauk Ukrains'koi RSR, Seriya B: Geologichni, Khimichni ta Biologichni Nauki* **1981**, 21–3.

Pineau, F. & M. Javoy 1969. Détermination des rapports isotopiques ^{18}O/^{16}O et ^{13}C/^{12}C dans diverses carbonatites, implications génétiques. *Comptes Rendus des Séances de l'Académie des Sciences, Série D: Sciences Naturelles* **269**, 1930–3.

Pineau, F. & M. Javoy 1983. Carbon isotopes and concentrations in mid-oceanic ridge basalts. *Earth and Planetary Science Letters* **62**, 239–57.

Pineau, F., M. Javoy & C. J. Allègre 1973. Etude systématique des isotopes de l'oxygène, du carbone et du strontium dans les carbonatites. *Geochimica et Cosmochimica Acta* **37**, 2363–77.

Pineau, F., M. Javoy & Y. Bottinga 1976a. ^{13}C/^{12}C ratios of rocks and inclusions in popping rocks of the Mid-Atlantic ridge and their bearing on the problem of isotopic composition of deep-seated carbon. *Earth and Planetary Science Letters* **29**, 413–21.

Pineau, F., M. Javoy, J. W. Hawkins & H. Craig 1976b. Oxygen isotope variations in marginal basins and ocean ridge basalts. *Earth and Planetary Science Letters* **28**, 299–307.

Pinus, G. V. & Ye. I. Dontsova 1971. Oxygen isotope ratios in olivine from ultrabasic rocks of varied genesis. *Geologiya i Geofizika* **1971**, 3–8.

Plyusnin, G. S., V. S. Samoylov & S. I. Gol'shev 1980. The δ^{13}C, δ^{18}O isotope pair method and temperature facies of carbonatites. *Doklady Akademii Nauk SSSR, Seriya Geologiya* **254**, 1241–5.

Plyusnin, G. S., V. S. Samoylov & S. I. Gol'shev 1982. The relationship between isotopic pairs δ^{13}C and δ^{18}O and the temperature facies of the carbonatites. *Doklady of the Academy of Sciences of the USSR, Earth Sciences Section* (English transl.) **254**, 227–31.

Plyusnin, G. S., Ye. I. Vorob'yev & A. V. Perminov 1984a. Isotopic composition of (δ^{18}O, δ^{13}C) carbonatites of the Murun alkaline massif. *Doklady Akademii Nauk SSSR, Seriya Geologiya* **275**, 999–1033.

Plyusnin, G. S., Ye. I. Vorob'yev & A. V. Perminov 1984b. Isotopic composition (δ^{18}O, δ^{13}C) of carbonatites in the Murun alkalic rock block. *Doklady of the Academy of Sciences of the USSR, Earth Sciences Section* (English transl.) **275**, 156–9.

Puchelt, H. & H. W. Hubberten 1980. *Preliminary results on sulfur isotope investigation on deep sea drilling project cores from Legs 52 and 53.* Initial Reports of the Deep Sea Drilling Projects 51, 52, 53, Part 2, 1145–8.

Reuter, J. H., S. Epstein & H. P. Taylor 1965. $^{18}O/^{16}O$ ratios of some chondritic meteorites and terrestrial ultramafic rocks. *Geochimica et Cosmochimica Acta* **29**, 481–8.

Richet, P., Y. Bottinga & M. Javoy 1977. A review of hydrogen, carbon, nitrogen, oxygen, sulfur, and chlorine stable isotope fractionation among gaseous molecules. *Annual Review of Earth and Planetary Sciences* **5**, 65–110.

Rison, W. & H. Craig 1983. Helium isotopes and mantle volatiles in Lohi Seamount and Hawaiian Island basalts and xenoliths. *Earth and Planetary Science Letters* **66**, 407–26.

Robert, F. & S. Epstein 1982. The concentration and isotopic composition of hydrogen, carbon and nitrogen in carbonaceous meteorites. *Geochimica et Cosmochimica Acta* **46**, 81–95.

Robinson, B. W. 1973. Sulfur isotope equilibrium during sulphur hydrolysis at high temperatures. *Earth and Planetary Science Letters* **18**, 443–50.

Sakai, H. 1968. Isotopic properties of sulfur compounds in hydrothermal processes. *Geochemical Journal* **2**, 29–49.

Sakai, H., T. J. Casadevall & J. G. Moore 1982. Chemistry and isotope ratios of sulfur in basalts and volcanic gases at Kilauea Volcano, Hawaii. *Geochimica et Cosmochimica Acta* **46**, 729–38.

Sakai, H., A. Ueda & C. W. Field 1978. $\delta^{34}S$ and concentration of sulfide and sulfate sulfur in some ocean-floor basalts and serpentinites. In *Short Papers of the Fourth International Conference, Geochronology, Cosmochronology, Isotope Geology*, R. E. Zartman (ed.), 372–4, United States Geological Survey Open-File Report 78-701.

Sakai, H., D. J. Des Marais, A. Ueda & J. G. Moore 1984. Concentrations and isotope ratios of carbon, nitrogen and sulfur in ocean-floor basalts. *Geochimica et Cosmochimica Acta* **48**, 2433–41.

Sakai, H., E. Gunnlaugsson, J. Tomasson & J. E. Rouse 1980. Sulfur isotope systematics in Icelandic geothermal systems and influence of seawater circulation at Reykjanes. *Geochimica et Cosmochimica Acta* **44**, 1223–31.

Sasaki, A. 1969. *Sulphur isotope study of the Muskox intrusion, district of Mackenzie.* Geological Survey of Canada, Paper 68-46, 1–68.

Schneider, A. 1970. The sulfur isotope composition of basaltic rocks. *Contributions to Mineralogy and Petrology* **25**, 95–124.

Schwarcz, H. P. 1973. Sulfur isotope analyses of some Sudbury, Ontario, ores. *Canadian Journal of Earth Sciences* **10**, 1444–59.

Sheppard, S. M. F. & J. B. Dawson 1973. $^{13}C/^{12}C$ and D/H isotope variations in "primary igneous carbonatites". *Fortschritte der Mineralogie* **50**, 128–9.

Sheppard, S. M. F. & J. B. Dawson 1975. Hydrogen, carbon and oxygen isotope studies of megacrysts and matrix minerals from Lesothan and South African kimberlites. *Physics and Chemistry of the Earth* **9**, 747–63.

Sheppard, S. M. F. & S. Epstein 1970. D/H and $^{18}O/^{16}O$ ratios of minerals of possible mantle or lower crustal origin. *Earth and Planetary Science Letters* **9**, 232–9.

Sheppard, S. M. F. & H. P. Schwarcz 1970. Fractionation of carbon and oxygen isotopes and magnesium between coexisting metamorphic calcite and dolomite. *Contributions to Mineralogy and Petrology* **26**, 161–98.

Shima, M. & H. G. Thode 1961. The sulfur isotope abundances in Abee and Bruderheim meteorites. *Journal of Geophysical Research* **66**, 3580.

Shima, M., W. H. Gross & H. G. Thode 1963. Sulfur isotope abundances in basic sills, differentiated granites, and meteorites. *Journal of Geophysical Research* **68**, 2835–47.

Smith, J. W. & I. R. Kaplan 1970. Endogenous carbon in carbonaceous meteorites. *Science* **167**, 1367–70.

Smith, J. W., D. Doolan & E. F. McFarlane 1977. A sulfur isotope geothermometer for the trisulfide system galena–sphalerite–pyrite. *Chemical Geology* **19**, 83–90.

Smitheringale, W. G. & M. L. Jensen 1963. Sulfur isotopic composition of the Triassic igneous rocks of eastern United States. *Geochimica et Cosmochimica Acta* **27**, 1183–207.

Stepanenko, V. I. & N. V. Sukhanov 1980. Isotopic composition of carbon and oxygen of carbonatites from Central Timan. *Doklady Akademii Nauk SSSR, Seriya Geologiya* **251**, 699–702.

Suwa, K., S. Osaki, I. Shiida & K. Miyakawa 1969. Isotope geochemistry and petrology of the Mbeya carbonatite, south-western Tanzania, East Africa. *Journal of Earth Sciences, Nagoya University* **17**, 125–67.

Suwa, K., S. Oana, H. Wada & S. Osaki 1975. Isotope geochemistry and petrology of African carbonatites. *Physics and Chemistry of the Earth* **9**, 735–45.

Swart, P. K., M. M. Grady & C. T. Pillinger 1982. Isotopically distinguishable carbon phases in the Allende meteorite. *Nature* **297**, 381–3.

Swart, P. K., M. M. Grady, C. T. Pillinger, R. S. Lewis & E. Anders 1983. Interstellar carbon in meteorites. *Science* **220**, 406–10.

Tarutani, T., R. N. Clayton & T. K. Mayeda 1969. The effect of polymorphism and magnesium substitution on oxygen isotope fractionation between calcium carbonate and water. *Geochimica et Cosmochimica Acta* **33**, 987–96.

Taylor, H. P. 1968. The oxygen isotope geochemistry of igneous rocks. *Contributions to Mineralogy and Petrology* **19**, 1–71.

Taylor, H. P. & S. Epstein 1962a. Relationship between $^{18}O/^{16}O$ ratios in coexisting minerals of igneous and metamorphic rocks Part 1. Principles and experimental results. *Bulletin of the Geological Society of America* **73**, 461–80.

Taylor, H. P. & S. Epstein 1962b. Relationship between $^{18}O/^{16}O$ ratios in coexisting minerals of igneous and metamorphic rocks Part 2. Applications to petrologic problems. *Bulletin of the Geological Society of America* **73**, 675–94.

Taylor, H. P. & S. M. F. Sheppard 1986. Igneous rocks: I. Processes of isotopic fractionation and isotope systematics. In Stable isotopes in high temperature geological processes, J. W. Valley, H. P. Taylor & J. R. O'Neil (eds). *Reviews in Mineralogy* **16**, 227–68.

Taylor, H. P., J. Frechen & E. T. Degens 1967. Oxygen and carbon isotope studies of carbonatites from the Laacher See district, West Germany and the Alnö district, Sweden. *Geochimica et Cosmochimica Acta* **31**, 407–30.

Taylor, H. P., M. B. Duke, L. T. Silver & S. Epstein 1965. Oxygen isotope studies of minerals in stony meteorites. *Geochimica et Cosmochimica Acta* **29**, 489–512.

Thode, H. G., J. Monster & H. B. Dunford 1961. Sulphur isotope geochemistry. *Geochimica et Cosmochimica Acta* **25**, 159–74.

Ueda, A. & H. Sakai 1984. Sulfur isotope study of Quaternary volcanic rocks from the Japanese Island Arc. *Geochimica et Cosmochimica Acta* **48**, 1837–48.

Verwoerd, W. J. 1966. South African carbonatites and their probable mode of origin. *Annale Universiteit van Stellenbosch, Serie A* **41**, (2), 233.

Vinogradov, A. P., O. I. Kropotova & V. I. Gerasimovskiy 1970a. Carbon isotopic composition for carbonatites from East Africa. *Geokhimiya* **1970**, 643–6.

Vinogradov, A. P., O. I. Kropotova & V. I. Gerasimovskiy 1970b. Carbon isotopic composition for carbonatites from East Africa. *Geochemistry International* **7**, 562.

Vinogradov, A. P., E. I. Dontsova, V. I. Gerasimovskiy & L. D. Kuznetsova 1971a. Isotopic composition of oxygen in carbonatites of continental rift zones in Eastern Africa. *Geokhimiya* **1971**, 507–14.

Vinogradov, A. P., E. I. Dontsova, V. I. Gerasimovskiy & L. D. Kuznetsova 1971b. Oxygen isotopic composition of carbonatites from the rift zones of East Africa. *Geochemistry International* **8**, (5), 307–13.

Vinogradov, A. P., O. I. Kropotova, E. M. Epshtein & V. A. Grinenko 1967a. Isotopic carbon composition of calcites of different stages of the carbonatite process in connection with carbonatite genesis. *Geokhimiya* **1967**, 499–509.

Vinogradov, A. P., O. I. Kropotova, E. M. Epshtein & V. A. Grinenko 1967b. Isotopic composition of carbon in calcites representative of different temperatures of carbonatite formation and the problem of genesis of carbonatites. *Geochemistry International* **4**, 431–41.

Vinogradov, A. P., O. I. Kropotova, G. P. Vdovykin & V. A. Grinenko 1967c. Isotopic composition of different phases of carbon in carbonaceous meteorites. *Geokhimiya* **1967**, 267–73.

Vinogradov, V. I., A. A. Krasnov, V. N. Kuleshov & L. D. Sulerzhitskiy 1978. Carbon-13/carbon-12 and oxygen-18/oxygen-16 ratios and carbon-14 concentration in carbonatites of the Kaliango volcano (eastern Africa). *Izvestiya Akademii Nauk SSSR, Seriya Geologicheskaya* **1978**, (6), 33–41.

Vinogradov, V. I., A. A. Krasnov, V. N. Kuleshov & L. D. Sulerzhitskiy 1980. $^{13}C/^{12}C$ and $^{18}O/^{16}O$ ratio and ^{14}C concentration in carbonatites of the Kaliango volcano East Africa. *International Geology Review* **22**, 51–7.

Wasserburg, G. J., E. Mazor & R. E. Zartman 1963. Isotopic and chemical composition of some terrestrial natural gases. In *Earth Science and Meteoritics*, J. Geiss & E. D. Goldberg (eds), 219–40. Amsterdam: North-Holland Publishing Company.

Watanabe, S., K. Mishima & S. Matsuo 1983. Isotopic ratios of carbonaceous materials incorporated in olivine crystals from the Hualalai Volcano, Hawaii – An approach to mantle carbon. *Geochemical Journal* **17**, 95–104.

Wickman, F. E. 1956. The cycle of carbon and the stable carbon isotopes. *Geochimica et Cosmochimica Acta* **9**, 136–53.

Wilson, A. F. 1979. Contrast in the isotopic composition of oxygen and carbon between the Mud Tank carbonatite and the marbles in the granulite terrane of the Strangways Range, central Australia. *Journal of the Geological Society of Australia* **26**, 39–44.

Yuen, G., N. Blair, D. J. Des Marais & S. Chang 1984. Carbon isotope composition of low molecular weight hydrocarbons and monocarboxylic acids from Murchison meteorite. *Nature* **307**, 252–4.

Yurtsever, Y. & J. R. Gat 1981. Atmospheric waters. In *Stable isotope hydrology deuterium and oxygen-18 in the water cycle*, I.A.E. Technical Reports Series – International Atomic Energy Agency **210**, 103–42.

14
LEAD ISOTOPE RELATIONSHIPS IN CARBONATITES AND ALKALIC COMPLEXES: AN OVERVIEW

S.-T. KWON, G. R. TILTON & M. H. GRÜNENFELDER

ABSTRACT

Lead isotope data from carbonatites and related silicate rocks spanning an age-range from 100 to 2700 Ma, mainly from North America, are reviewed. The Pb data from 100 Ma-old complexes plot well within the fields of present-day oceanic volcanic rocks as defined by MORB and OIB in $^{207}Pb:^{204}Pb-^{206}Pb:^{204}Pb$ and $^{208}Pb:^{204}Pb-^{206}Pb:^{204}Pb$ diagrams, indicating mantle sources. The initial Pb ratios in the 100–1900 Ma age-range plot distinctly below crustal Pb evolution curves in the $^{207}Pb:^{204}Pb-^{206}Pb:^{204}Pb$ diagram, providing evidence for large-ion lithophile element (LILE) depletion in the magma sources over that time-span. $^{206}Pb:^{204}Pb$ ratios are negatively correlated with $^{87}Sr:^{86}Sr$ ratios in all age groups except at 2700 Ma, which has a slight positive correlation. The Sr–Pb isotope relationship is used to correlate $^{206}Pb:^{204}Pb$ with bulk Earth $^{87}Sr:^{86}Sr$ ratios as a function of age. The derived ratios fit isotopic evolution in a closed-system source that possibly characterizes the U:Pb ratio for mean silicate Earth. The bulk Earth Pb evolution data further suggest that the depleted mantle source of the Canadian Shield alkalic complex magmas originated c. 3000 Ma ago, in agreement with Sr and Nd studies by earlier investigators. The Pb data further indicate that the source was a dynamic, open, rather than a closed, system. Least radiogenic Pb isotope data from three galena deposits covering an age-range of 3210–3770 Ma are consistent with Pb isotope evolution in a source having the same U:Pb ratio as that derived for bulk Earth from the alkalic complex data, further supporting the differentiation model. The parameters found for bulk silicate Earth are: age of the source = 4520 Ma, with initial Pb isotope ratios identical to those from Canyon Diablo troilite; $^{238}U:^{204}Pb$ (μ) = 8.38 ± 0.04; $^{232}Th:^{238}U$ ratio ($\varkappa$) = 4.21 ± 0.06. The LILE depletion isotope signature for carbonatite and oceanic island magmas results from decrease in Rb:Sr, and increase in Sm:Nd, U:Pb, and Th:Pb ratios in the source. The resemblance between isotopic data from ocean island basalts and carbonatites argues for production of carbonatite magmas in the asthenosphere rather than in subcontinental lithosphere. In some cases the magmas may be modified by interaction with the overlying lithosphere.

14.1 INTRODUCTION

Many isotopic studies, both radiogenic and stable, of carbonatites and associated alkalic rocks suggest that those rocks are derived from the mantle (e.g. Deines & Gold 1973, Bell *et al.* 1982, Grünenfelder *et al.* 1986, Bell & Blenkinsop 1987). In particular, the well-characterized radiogenic isotopic signatures of the mantle-derived oceanic island basalts (OIBs) and young ($\leqslant 100$ Ma old) carbonatites indicate a close affinity between these two groups. In contrast to the younger ages of oceanic basalts, which are limited to the Phanerozoic, alkalic complexes encompass a time-span reaching back to the late Archaean. Thus, in principle, carbonatites provide a unique opportunity to study the geochemical evolution of the mantle from Precambrian time up to the present.

The Canadian Shield contains numerous intrusions of carbonatites and related alkalic rocks ranging in age from 100 to 2700 Ma (Fig. 14.1). Bell *et al.* (1982), in a study of initial Sr ratios in Canadian Shield carbonatite complexes, showed that the magmas appear to have sampled mantle sources that were depleted in large-ion lithophile elements (LILEs). They furthermore postulated that the depletion occurred *c.* 3000 Ma ago, and that the mantle source has been coupled to the overlying crust since that time. Newer Nd studies (Bell & Blenkinsop 1987) yield results consistent with the Sr data.

While there have been several studies of the secular isotopic evolution of the mantle by the Rb–Sr and Sm–Nd systems, there are fewer U–Th–Pb studies. Brevart *et al.* (1986) made a detailed evaluation of isotopic Pb data from 1.9 to 3.5-Ga-old komatiites from different continents. They concluded that the mantle 2700 Ma ago was very different from the primitive mantle in terms of the Th:U ratio, and ascribed the difference to formation of continental crust. We have re-examined the mantle evolution model using Pb isotope data from alkalic complexes to obtain an alternative evaluation that parallels the Sr–Nd work. The interpretation of the Pb isotopic systematics is more complex than those for Sr and

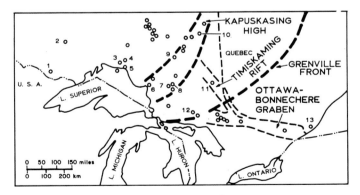

Figure 14.1 Locations of alkalic complexes of Ontario and Quebec, Canadian Shield. The complexes studied are, from west to east: (1) Poohbah Lake, (2) Sturgeon Lake, (3) Prairie Lake, (4) Killala Lake, (5) Port Coldwell, (6) Firesand River, (7) Borden, (8) Lake Nemegosenda, (9) Cargill, (10) Goldray, (11) Otto Stock, (12) Spanish River, (13) Oka.

Nd, owing to the coupled decay of the two U isotopes with their greatly different half-lives. However, the Pb isotopes potentially contain more information than the Sr and Nd systems since the short half-life (0.7 Ga) of ^{235}U makes ^{207}Pb : ^{204}Pb ratios sensitive to U : Pb differentiation processes that occurred prior to $c.$ 2500 Ma ago, while the long half-life (4.5 Ga) of ^{238}U makes ^{206}Pb : ^{204}Pb ratios sensitive to U : Pb differentiation that occurred at any time over Earth history. In addition, the Th–Pb system provides still another index for Pb isotope evolution.

Carbonatites and alkaline complexes have a disadvantage *vis-à-vis* oceanic basalts in that they have been intruded through continental crust, which raises the possibility of contamination from the older crustal materials. This is particularly serious for Pb, which has a much lower concentration in carbonatites compared to granitoid crustal rocks (2 v. 15–20 p.p.m.), but may also cause problems in some cases for Nd and Sr (Cavell & Baadsgaard 1986b, Andersen 1987). Therefore careful examination of data is necessary in estimating initial isotopic ratios for a given complex (e.g. Tilton *et al.* 1987).

In this chapter we summarize Pb isotopic data for selected carbonatites and alkaline complexes, mainly from North America, and compare them with available Sr and Nd data. We emphasize North American complexes because that is where most of our work has been done, and because the most data exist for that continent. The data are next used to evaluate the geochemical evolution of the mantle from the standpoint of the U–Th–Pb system. Finally, the results will be compared with Pb isotope data of carbonatite complexes from other continents. From these data we discuss and derive estimates of the U : Pb and Th : Pb ratios for mean, or bulk, silicate Earth material.

14.2 NORTH AMERICAN ALKALIC COMPLEXES

14.2.1 Pb isotope evolution framework

Before discussing the Pb isotope results a few remarks concerning the framework used in the interpretation are in order. We will use two models in the interpretations that follow. One is the single-stage, or closed-system evolution condition. This assumes that a mantle source formed 4.55 Ga ago and remained closed to gain or loss of U, Th or Pb except for radioactive decay until production of a magma or Pb ore. The initial Pb ratios in the source are assumed to be those measured in Canyon Diablo troilite (Tatsumoto *et al.* 1973), i.e. primordial lead. A second model is that for average continental upper crustal lead. Here we will use the evolution curve of Stacey & Kramers (1975), although the orogene evolution curve of Zartman & Doe (1981) would serve just as well. The 0 Ma isochron for single-stage evolution (Geochron) is shown in Figure 14.2, along with the Stacey & Kramers (1975) evolution curve. The main justification for using the Stacey–Kramers curve is that the leads they used to define the curve have originated in island arc environments where the isotopic composition of Pb is in most cases controlled by the Pb in the pelagic sediments (e.g. Doe & Stacey 1974, Sun 1980, Barreiro 1983), which in turn

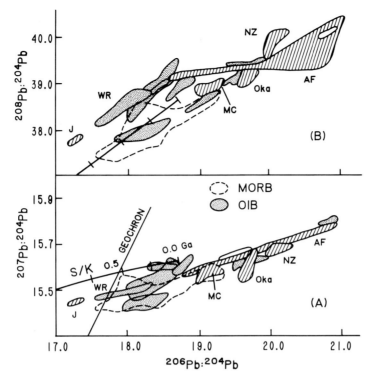

Figure 14.2 Lead isotope data for the Oka, Quebec, and Magnet Cove, Arkansas alkalic complexes. Other young complexes: J, Jacupiranga, Brazil; NZ, New Zealand; Af, East Africa. See text for sources of data. Fields of oceanic island basalt (OIB) and mid-ocean ridge basalt (MORB) are shown for comparison. WR is Walvis Ridge. Geochron: 0 Ma isochron for closed-system Pb isotope evolution using 4550 Ma for the age of the Earth and Canyon Diablo troilite Pb isotopic composition for primordial Pb. S/K: Pb isotope evolution curve of Stacey & Kramers (1975).

sample large areas of continental surface rocks. Note that the fields of MORB and OIB in Figure 14.2A tend to plot below the Stacey–Kramers curve. This will be taken as an indication of derivation from a LILE-depleted mantle source in subsequent discussions. The relationship for the Th decay system in Figure 14.2B is varied.

14.2.2 Analytical methods

Chemical processing Details for the analytical procedures have been described previously in other publications (Grünenfelder *et al.* 1986, Bell *et al.* 1987, Tilton *et al.* 1987). All samples were initially washed in dilute HCl to expose fresh surfaces (calcites), or hot 7N HNO_3 to remove any loosely held elements (feldspars). All analyses were made with a mixed $^{235}U-^{230}Th-^{205}Pb$ spike in order to insure proper corrections for *in situ* radiogenic Pb produced by decay of U and Th in the samples. We do not consider the determination of initial ratios to be reliable when the correction for radiogenic Pb exceeds *c.* 4% of the observed ratios, since experience has shown that samples requiring large corrections generally yield erratic initial ratios. Lead blanks for processing of 50 mg samples were typically in the range of 0.2–0.5 ng.

Mass spectrometry Most of the data reported here were obtained on an AVCO single collector mass spectrometer, although the Arkansas and Borden complex data were obtained on a multiple collector Finnigan MAT 261 machine. For the AVCO data the precisions of ratio measurements are: $^{207}Pb:^{206}Pb$ and $^{208}Pb:^{206}Pb$, 0.04%; $^{206}Pb:^{204}Pb$, 0.08% (2σ mean). For multiple collector runs the precisions are improved by approximately a factor of 2. The ratios were corrected to absolute values by comparison with multiple runs on the NBS 981 Pb reference standard, using the values reported by Todt *et al.* (1984). The resulting corrections were of the order of $0.12 \pm 0.03\%$ per mass unit.

14.2.3 Young carbonatites (c. 100 Ma)

Figure 14.2 shows the Pb isotopic data for carbonatites from Oka, Quebec (Grünenfelder *et al.* 1986), and Magnet Cove, Arkansas (Tilton *et al.* 1987) in the conventional Pb isotope correlation diagrams, along with the fields of MORB and OIB. A recent estimate of the age of the Oka intrusion is 110 Ma (Wen 1985); Magnet Cove is 97 Ma (Zartman & Howard 1987). Both studies showed that carbonatites and associated alkalic rocks have rather homogeneous isotopic signatures, suggesting that rocks from alkalic complexes without carbonatites can often be used instead of carbonatites, although Tilton *et al.* (1987) cited evidence for crustal Pb contamination in some of the Arkansas syenites.

The Pb isotopic data of the two complexes plot within the field of MORB and OIB in the $^{207}Pb:^{204}Pb-^{206}Pb:^{204}Pb$ correlation diagram, indicating mantle origins. Significant departures from the single-stage isochron (plotting to the right of the single-stage isochron) indicate that the time-averaged U:Pb ratios in the sources are greater than the bulk Earth U:Pb ratio. Note that the 100 Ma age difference between the complexes and present-day oceanic basalts does not significantly affect the isotope relationships between the two suites. For reasonable estimates of the source U:Pb ratios (e.g. $^{238}U:^{204}Pb = 8-9$) any correction for age differences will cause the carbonatite leads to appear more radiogenic relative to the oceanic basalt fields.

The increase in time-averaged U:Pb ratio in the source was accompanied by decrease in Rb:Sr and Nd:Sm relative to bulk Earth values, as indicated by the initial $^{87}Sr:^{86}Sr$ and $^{143}Nd:^{144}Nd$ ratios (Bell & Blenkinsop 1987, Tilton *et al.* 1987), although the timing of the fractionation events is unknown. In contrast to the $^{207}Pb:^{204}Pb-^{206}Pb:^{204}Pb$ diagram, the $^{208}Pb:^{204}Pb-^{206}Pb:^{204}Pb$ diagram does not serve to distinguish between mantle and crustal rocks.

The initial Pb isotope ratios, especially the $^{206}Pb:^{204}Pb$ ratios, correlate with initial Sr and Nd. That is, the Oka complex has lower $^{87}Sr:^{86}Sr$, but higher $^{206}Pb:^{204}Pb$ and $^{143}Nd:^{144}Nd$ ratios than Magnet Cove. These correlations are also evident in many other young carbonatites, including Jacupiranga, Brazil (Roden *et al.* 1985, Nelson *et al.* 1988), New Zealand (Barreiro & Cooper 1987), and East

Africa (Bell and Tilton, unpublished data). We believe these correlations are significant, and discuss them further below.

Other young carbonatite data shown in Figure 14.2 are those from New Zealand with an age of c. 24 Ma (Barreiro & Cooper 1987); Jacupiranga, Brazil, 130 Ma (Nelson *et al.* 1988); and East Africa, 0–10 Ma (Lancelot & Allègre 1974). The Walvis Ridge tholeiites, with ages of 70 Ma (Richardson *et al.* 1982), are included for comparison with the carbonatite data. The East African and Jacupiranga data provide the extremes in the carbonatite fields in Figure 14.2. Note that the Jacupiranga ratios plot approximately along an extension of the Walvis Ridge data in Figure 14.2.

14.2.4 1000–1100 Ma carbonatites

At least ten complexes belonging to this age-group are identified in the Canadian Shield (Bell *et al.* 1982). Here we present new data for the Coldwell, Nemegosenda,

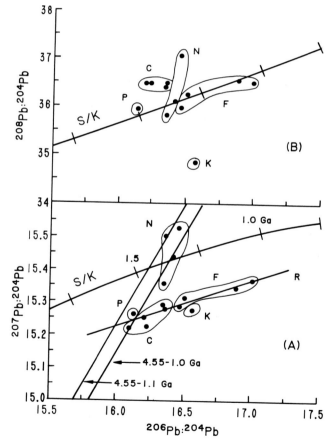

Figure 14.3 Lead isotope data for 1100-Ma-old alkalic complexes from Ontario. P, Prairie Lake; K, Killala Lake; C, Port Coldwell; F, Firesand River; N, Nemegosenda Lake. The two isochrons indicate closed-system Pb isotope evolution from 4550 to 1100 and 1000 Ma. R: regression line fitted to all data in A except the three Nemegosenda Lake samples with high $^{207}Pb:^{204}Pb$ ratios. Other symbols as in Figure 14.2.

Table 14.1 U, Th, and Pb isotopic data for c. 1100-Ma-old alkalic complexes in the Canadian Shield.

	Observed ratios			Concentrations			Atomic ratios			in situ decay corrected[2]		
Sample[1]	$^{206}Pb:^{204}Pb$	$^{207}Pb:^{204}Pb$	$^{208}Pb:^{204}Pb$	Pb (p.p.m.)	U (p.p.m.)	Th (p.p.m.)	$^{238}U:^{204}Pb$	$^{232}Th:^{238}U$		$^{206}Pb:^{204}Pb$	$^{207}Pb:^{204}Pb$	$^{208}Pb:^{204}Pb$
Nemegosenda												
NL-024C	17.715	15.618	38.758	17.81	1.997	8.80	7.09	4.55		16.461	15.525	37.037
1-4C-1	16.421	15.361	36.677	10.36	0.0557	0.263	0.323	4.88		16.361	15.356	35.796
-2	16.477	15.444	36.946	10.44	0.0539	0.259	0.312	4.96		16.419	15.440	36.080
1-6F-1	19.755	15.608	37.760	2.53	0.950	0.665	24.06	0.723				
-2	39.789	17.244	47.084	1.877	1.862	1.883	85.57	1.044				
1-7C	16.413	15.508	37.177	12.24	0.0563		0.279			16.361	15.504	
Ap	134.86	23.954	118.14	16.05								
Coldwell												
6-3-2F	17.325	15.308	38.424	9.35	0.889	5.13	5.93	5.96		16.235	15.225	36.471
6-4F	17.323	15.337	36.927	8.11	0.809	1.09	6.09	1.40		16.204	15.252	36.457
6-7F	16.568	15.297	36.632	14.64	0.280	0.734	1.15	2.67		16.357	15.281	36.462
6-10F	18.780	15.339	39.787	16.33	4.015	16.28	15.9	4.19				
6-11F	18.340	15.302	39.008	13.11	1.893	7.04	9.19	3.85				
Kf[3]										16.098	15.220	
Galena[4]										16.342	15.290	36.377
Prairie Lake												
7-2C	27.965	16.220	41.910	0.572	0.462	0.801	60.8	1.79				
7-5C-1	76.207	20.001	43.105	3.851	0.774	3.20	24.3	4.27				
-2	81.184	20.338	42.555	3.511								
7-6C-1	16.614	15.298	36.340	3.509	0.153	0.421	2.61	2.84		16.128	15.261	35.925
-2	22.173	15.683	38.244	0.405	0.213	0.247	35.0	1.20				
p-17C	37.489	16.888	54.360	4.535	0.423		8.84					
16-33C	17.472	15.404	36.455	3.445	0.009		0.16			17.442	15.402	36.455
Firesand River												
5-1C	16.895	15.343	36.385	19.46	0.666	0.834	2.06	1.31		16.512	15.314	36.234
5-4C	17.702	15.418	36.576	6.12	0.373	0.177	3.73	0.488		17.008	15.365	36.474
5-6C	16.892	15.343	36.546	2.81	0.001		0.021			16.888	15.343	36.546
5-8C	16.532	15.288	37.503	3.71	0.020	1.650	0.334	83.7		16.470	15.283	35.940
Killala												
pyrite	16.576	15.274	36.312	1.15	0.0007	0.578	0.037	743		16.569	15.274	34.830

1 C; carbonate; F, feldspar; Ap, apatite.
2 Assumed age = 1100 Ma.
3 From Turek et al. (1985).
4 From Thorpe (1986).

Killala, Prairie Lake, and Firesand complexes. The initial Pb isotope ratios are plotted in Figure 14.3. The numerical data, together with U, Th, and Pb concentrations are given in Table 14.1.

Coldwell complex In contrast to most other complexes of this age-group, the Coldwell complex does not have carbonatite. We therefore analysed five K-feldspar separates, of which two show weathering effects under the binocular microscope. The other three samples define an age of 1094 ± 10 Ma in a $^{208}Pb:^{204}Pb-^{232}Th:^{204}Pb$ isochron diagram (not shown), in good agreement with the zircon U–Pb age of 1088 ± 10 Ma given in Thorpe (1986). However, the isochron age of 900 ± 10 Ma obtained in a $^{206}Pb:^{204}Pb-^{238}U:^{204}Pb$ plot is lower, and could be related to the relative mobilities of U and Th in near-surface conditions. The initial Pb ratios of the complex have been estimated previously by Turek et al. (1985) and Thorpe (1986). The initial Pb ratios (corrected for in situ decay of U) in this study are slightly higher than those from the feldspar leach experiment by Turek et al. (1985), but are lower than those of 'galena within a large augite crystal from a pegmatitic ferroaugite syenite' analysed by Thorpe (1986). Although the galena data by Thorpe are probably the best estimate of the initial Pb ratios, we cannot exclude a possibility of initial isotopic heterogeneity. We will, therefore, use an average of the three ratios as representative of the Coldwell complex. The estimated initial ratios plot well below the Stacey–Kramers crustal evolution curve in Figure 14.3, indicating a uranium-depleted source relative to the ore sources. The Pb data plot close to the 1100 Ma single-stage isochron, which agrees with the initial Sr (Bell et al. 1982, Platt & Mitchell 1982) and Nd (Kwon 1986) ratios that are very close to the bulk Earth values. In the $^{208}Pb:^{204}Pb-^{206}Pb:^{204}Pb$ diagram the data plot above the Stacey–Kramers evolution curve. This mirrors the relationship found in many young OIBs with lower $^{206}Pb:^{204}Pb$ ratios (Fig. 14.2).

Prairie Lake complex Five calcite separates were analysed (two in duplicate) from the Prairie Lake complex. The measured Pb isotopic ratios are variable, from extremely radiogenic to non-radiogenic, and define a Pb:Pb errorchron age of 1155 ± 36 Ma (Fig. 14.4). Bell & Blenkinsop (1980) reported a Rb–Sr age of 1023 ± 74 Ma. Although disturbances are apparent in the U:Pb and Th:Pb systems, three samples define errorchron U:Pb and isochron Th:Pb ages of 1094 ± 160 and 1080 ± 6 Ma, respectively, with initial $^{206}Pb:^{204}Pb = 16.12 ± 0.13$ and $^{208}Pb:^{204}Pb = 35.94 ± 1.0$. The estimated initial $^{206}Pb:^{204}Pb$ and $^{207}Pb:^{204}Pb$ ratios of Prairie Lake are very similar to those of the Coldwell complex, while the $^{208}Pb:^{204}Pb$ ratio is slightly lower. In general, the Pb isotope data from Prairie Lake indicate a history similar to that for the Pb in the Coldwell complex. In contrast to the Coldwell results, the initial Sr ratios for Prairie Lake appear to be significantly lower ($\varepsilon_{Sr} = -10$ to -8), while initial Nd ratios are slightly higher ($\varepsilon_{Nd} = +1$ to $+2$) (Bell & Blenkinsop 1987). Sr and Nd therefore indicate sources depleted in LILEs relative to bulk Earth. It is not clear whether the inconsistency of isotope ratios relative to the bulk Earth parameters results from contamination or intrinsic variation in the mantle sources.

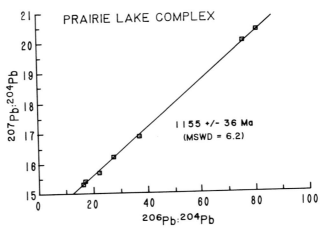

Figure 14.4 $^{207}Pb:^{204}Pb-^{206}Pb:^{204}Pb$ isochron diagram for Prairie Lake carbonatite samples.

Firesand River complex The Pb isotopic data for calcite separates from four carbonatites are relatively unradiogenic, allowing good estimates of the initial Pb ratios. The calculated *in situ* decay-corrected initial Pb ratios, using an age of 1000 Ma (Bell & Blenkinsop 1980) are not uniform. This can be attributed to open-system behaviour of the samples, or to initial isotopic heterogeneity. We return to this question below. In any case, the initial Pb ratios for the Firesand River complex share the same characteristics as those of the other complexes in that they plot significantly below the Stacey–Kramers curve in Figure 14.3A. Firesand calcite has the highest initial $^{206}Pb:^{204}Pb$ ratios found in our study.

Killala complex One sulphide (pyrite) was analysed from the Killala complex. the U:Pb ratio was very low, yielding small *in situ* decay corrections for the initial $^{206}Pb:^{204}Pb$ and $^{207}Pb:^{204}Pb$ ratios. These ratios plot with Pb from the other complexes described above in Figure 14.3A. However, the $^{208}Pb:^{204}Pb$ ratio, when corrected for *in situ* decay of Th, is lower than the data from the other complexes, suggesting an overcorrection. In fact, the measured $^{208}Pb:^{204}Pb$ ratio is conformable with the other data.

Nemegosenda complex Three calcites and one apatite from carbonatites, and one K-feldspar from syenite were analysed from Nemegosenda Lake. The calcite–apatite data from sample 1–7 yield a Pb–Pb isochron age of 966 ± 20 Ma. This age is slightly younger than the Rb–Sr isochron age of 1015 ± 63 Ma, but the values agree within error limits. The feldspar Pb is rather radiogenic and does not provide useful information about initial ratios, whereas the calcites have relatively low U:Pb and Th:Pb ratios, and therefore provide good estimates of initial ratios. Again, the calculated initial ratios are not uniform. Three samples plot on or close to the Stacey–Kramers curve and make a nearly vertical trend. This suggests crustal contamination, although we have not studied the country rocks at this locality. If we interpret the vertical trend as binary mixing, extrapolation of the mixing line to the mantle trend defined by the other alkalic complexes indicates a mantle component for Nemegosenda similar to that for the Coldwell complex.

Summary The initial Pb isotopic data from the five alkalic complexes generally show the same characteristics observed in the younger complexes. Several relationships are noteworthy. First, all ratios, except for the Nemegosenda leads, plot below the Stacey–Kramers evolution curve in Figure 14.3A. Secondly, most of the initial Pb ratios plot to the right of the single-stage isochron for 1100 Ma in Figure 14.3A, indicating single-stage model ages that are younger than the geological age. Thirdly, except for the Nemegosenda samples, all Pb ratios define an approximately linear trend in Figure 14.3A that is similar to the mantle regression field for young oceanic volcanic rocks (e.g. Church & Tatsumoto 1975, Tatsumoto 1978). Tilton & Grünenfelder (1983) found that samples from Firesand, Killala Lake, Coldwell, and Prairie Lake, for which the correction for *in situ* radiogenic $^{206}Pb:^{204}Pb$ was less than 3%, fit a regression line having a slope of 0.128 and an intercept of 13.186, with a correlation coefficient of 0.96. Adding the Coldwell galena does not change the slope significantly. The slope is steeper than that defined by modern oceanic volcanic rocks, which is $c.\ 0.102 \pm 0.012$ (Church & Tatsumoto 1975, Tatsumoto 1978), and thus qualitatively fits the criteria for the mean slope of an oceanic regression field at 1100 Ma ago, analogous to that defined by MORB and OIB at present. From these observations we conclude that the processes resulting in the present-day Pb isotopic signatures of sub-oceanic LILE-depleted mantle have been operating since at least 1100 Ma ago.

14.2.5 *1900 Ma carbonatites*

We have studied four carbonatite complexes belonging to this age-group in the Canadian Shield, the Borden, Cargill, Goldray and Spanish River plutons. The first three complexes occur along the Kapuskasing Structural Zone defined by linear aeromagnetic and gravity anomalies (Innes 1960). The Borden data have already been reported in a separate publication (Bell *et al.* 1987); the remaining data will be published separately (Kwon, in preparation). The initial Pb isotopic data from calcite separates for the 1870-Ma-old Borden and 1900-Ma-old Cargill carbonatites are plotted in Figure 14.5. The significant difference in the initial Pb ratios (especially $^{206}Pb:^{204}Pb$) between the two complexes indicates isotopic heterogeneity in the sources (the 30 Ma difference in age is not great enough to account for the difference in $^{206}Pb:^{204}Pb$ ratios). This is consistent with a small, but significant, difference in the $^{87}Sr:^{86}Sr$ ratios for the two complexes (Bell & Blenkinsop 1987).

Calcite separates from three Goldray carbonatites were analysed for study of the Pb isotope relationships. All samples have very low U:Pb and Th:Pb ratios, so that the measured ratios are close to the calculated initial ratios. The Pb data for the complex generally show the same features as those for the Borden and Cargill complexes, with slightly, but significantly lower $^{206}Pb:^{204}Pb$ ratios at Goldray. These isotope differences add to the evidence for Pb isotopic heterogeneity in the mantle $c.$ 1900 Ma ago, as already suggested by Kwon & Tilton (1986).

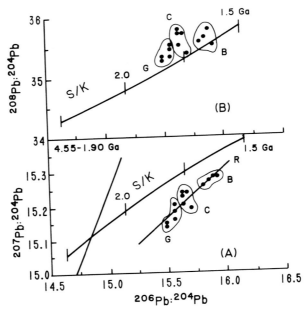

Figure 14.5 Lead isotope data for 1900 Ma alkalic complexes, Ontario. G, Goldray; C, Cargill; B, Borden. The isochron is drawn for closed-system Pb isotope evolution from 4550 to 1900 Ma. S/K is the Stacey–Kramers (1975) Pb evolution curve. R: regression line fitted to all data.

The initial Pb isotopic data of calcite separates from the Spanish River complex are not homogeneous, but define a linear array corresponding to radiogenic Pb produced between 1.9 and 1.5 Ga in a $^{207}Pb:^{204}Pb-^{206}Pb:^{204}Pb$ diagram (not shown). The linear array appears to be a relict isochron after a thermal event at c. 1.5 Ga, which is supported by the biotite K–Ar age of 1.56 Ga reported by Gittins et al. (1967).

In summary, the Pb data for the Goldray, Cargill, and Borden complexes plot below the Stacey–Kramers evolution curve in Figure 14.5A, in a manner analogous to the data for the two younger carbonatite groups cited above. We take this to indicate derivation from LILE-depleted mantle sources. If true, this extends to 1900 Ma the time over which the processes which produce sub-oceanic mantle isotope phenomenon have existed. A regression line through all of the data in Figure 14.5A has a slope of 0.3094 ± 0.03 (1σ) with MSWD = 0.34. As in the case of the 1100 Ma data, this may qualitatively approximate the slope of the field of oceanic volcanic rock Pb data at 1900 Ma. An inverse correlation between initial $^{206}Pb:^{204}Pb$ and $^{87}Sr:^{86}Sr$ ratios of the Borden, Cargill and Goldray complexes also resembles the data from the younger carbonatites.

14.2.6 2200 Ma alkalic complex

The oldest carbonatite-bearing alkalic complex known in the Canadian Shield is the Big Spruce Lake complex of the Yellowknife district (Cavell & Baadsgaard

1986a, b). Zircon from syenite gave an age of 2188 ± 15 Ma, in agreement with less precisely defined Pb:Pb and Sm:Nd whole rock ages (Cavell & Baadsgaard 1986a). Cavell & Baadsgaard (1986a) list Pb isotopic compositions and concentrations for eight carbonatite samples. Although no U concentration data were reported, one carbonatite contained the least radiogenic Pb out of 43 samples analysed from the complex. Since the sample contains 94 p.p.m. Pb, correction for radiogenic Pb produced by *in situ* U decay is likely small, and we will assume that the observed isotope ratios approximate the initial ratios. These are ^{206}Pb:^{204}Pb = 14.478; ^{207}Pb:^{204}Pb = 14.931; ^{208}Pb:^{204}Pb = 35.091. The ratios are plotted in Figure 14.7. In Figure 14.7A the carbonatite Pb plots below the Stacey & Kramers (1975) evolution curve and to the right of the single-stage isochron for 2190 Ma. In this way, the Big Spruce Lake Pb resembles the 1100 and 1900 Ma alkalic complex Pb data. The ^{208}Pb:^{204}Pb ratio seems anomalously high, and probably contains radiogenic Pb from decay of Th in the sample (not determined).

Cavell & Baadsgaard (1986a) also list Sm:Nd analyses for seven carbonatites. Taking the age as 2188 Ma, and assuming closed-system evolution of ^{143}Nd, the initial ratios yield initial ε_{Nd} = + 0.9 to + 2.8, with an average of + 2.2. This again indicates a LILE-depleted mantle origin. Eight syenite samples yield initial ε_{Nd} = + 0.2 to + 2.0, with an average of + 0.9. The syenite values, on the whole, are somewhat lower than the carbonatite values, possibly an indication of some degree of crustal interaction during intrusion.

Cavell & Baadsgaard (1986b) give data from which an initial ^{87}Sr:^{86}Sr ratio of 0.7016 can be calculated for the carbonatite. The ratio plots on the depleted mantle trend of Bell & Blenkinsop (1987) for Sr.

14.2.7 2700 Ma alkalic complexes

Because no alkalic complexes with associated carbonatites are known in the Canadian Shield for this age group, we use data from three syenitic complexes, the Poohbah Lake, Sturgeon Lake, and Otto Stock plutons (Fig. 14.1). The data will be published separately (Kwon & Tilton, in preparation). Kwon (1986) determined isotopic Pb ages for each of the complexes, using apatite, titanite, and feldspar acid leach experiments. The Pb:Pb ages are: Poohbah, 2667 ± 5 Ma; Sturgeon, 2659 ± 19 Ma; Otto, 2671 ± 14 Ma, all of which agree within error limits. The initial Pb ratios were obtained in a series of leaching experiments on K-feldspars (Kwon 1986).

The initial Pb ratios are plotted in Figure 14.6. Figure 14.6A shows that they plot close to the single-stage isochron for 2670 Ma; also close to the Stacey–Kramers crustal evolution curves. These are substantial departures from the data of the younger complexes, which plot below the crustal evolution curve in the ^{207}Pb:^{204}Pb–^{206}Pb:^{204}Pb diagram, and which yield single-stage model ages that are markedly younger than the isotopic ages. Significantly, only small differences between isotopic signatures in mantle and crustal rocks in Archaean time have been predicted by numerical modelling of radiogenic isotopic evolution in terrestrial reservoirs (O'Nions *et al.* 1979, Zartman & Doe 1981) in which mixing between the

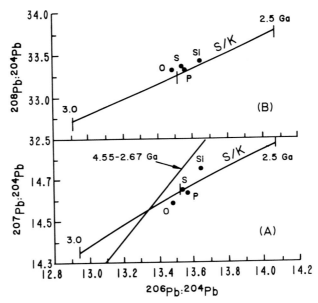

Figure 14.6 Lead isotope data for 2700 Ma alkalic complexes. Ontario: O, Otto Stock; P, Poohbah Lake; S, Sturgeon Lake. Si: Siilinjarvi, Finland. Isochron is for closed-system Pb isotope evolution from 4550 to 2670 Ma. S/K: Stacey–Kramers (1975) Pb evolution curve.

mantle and crust is one of the main processes that has produced mantle isotopic heterogeneity.

Another major difference between the Late Archaean alkalic and the younger complexes is the greater homogeneity in the Archaean data. This is still true even if the Pb isotope data for Late Archaean komatiites (Tilton 1983, Dupré et al. 1984; Brevart et al. 1986) and granitoid rocks (Gariepy & Allègre 1985) are included. The single-stage ^{238}U : ^{204}Pb (μ) ratios for these Late Archaean rocks vary only from 7.5 to 7.8, with slightly lower ratios for komatiites. Although more Pb data are needed to confirm this observation, it suggests that there has been no major U:Pb differentiation in the mantle prior to Late Archaean time (c. 3 Ga ago), presumably due to efficient mixing processes, at least on a regional scale.

Data for the Siilinjarvi, Finland, carbonatite are also given in Figure 14.6. Zircon from the complex gives nearly concordant U:Pb ages, with ^{207}Pb : ^{206}Pb age = 2605 ± 6 Ma (O. Kouvo, pers. comm.). An analysis by Grünenfelder for the initial Pb isotope ratios is listed in Tilton (1983). The Siilinjarvi data cannot be compared directly with Canadian data because ^{207}Pb : ^{204}Pb ratios (i.e. source μ values) for Finnish rocks appear to be systematically higher than those from Canadian Shield rocks of equal age (Tilton 1983), suggesting regional differences between sources. Nevertheless, the Siilinjarvi Pb ratios resemble the Canadian data in giving a single-stage model Pb age that is close to the measured isotopic age.

Figure 14.7 is a summary of the initial ratio data from all of the age-groups we have studied. A striking feature is the change in the apparent source μ values between the 1900 and 1100 Ma data, as evidenced by the discontinuity in

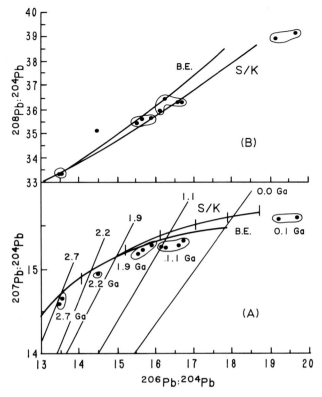

Figure 14.7 Summary of Pb isotope data from North American alkalic complexes cited in Figures 14.2–14.6. S/K is the Stacey–Kramers (1975) evolution curve. BE denotes Pb isotope evolution for estimated silicate bulk Earth, with parameters derived in the text. Initial Pb ratios are taken from Canyon Diablo troilite values (Tatsumoto *et al.* 1973); $^{238}U:^{204}Pb = 8.38$; $^{232}Th:^{238}U = 4.21$. Solid lines are isochrons for closed-system evolution from primordial Pb from 4.52 Ga to the designated times.

$^{207}Pb:^{204}Pb$ ratios. (Note the difference in $^{207}Pb:^{204}Pb$ ratios relative to the S/K curve, which represents closed-system Pb evolution in this age-range.) This indicates some kind of a fundamental change in the Pb sources. Two possible explanations are:

(a) the mantle source for the alkaline rocks is layered, with different depths sampled at different times; or
(b) it has changed isotopic composition for Pb over time by some kind of mixing process.

Contamination of the 1100 Ma plutons with lower crustal, granulitic material would tend to lower the $^{206}Pb:^{204}Pb$ and $^{207}Pb:^{204}Pb$ ratios, but would not be expected to preserve the negative correlation between $^{87}Sr:^{86}Sr$ and $^{206}Pb:^{204}Pb$ discussed above. The Pb data therefore contrast with the Sr data, which are consistent with a single, closed-system, depleted mantle source that has been coupled to the continental crust over the past 3000 Ma (Bell *et al.* 1982). No particular trends are evident in the ^{208}Pb : ^{204}Pb data in Figure 14.7B.

14.3 BULK EARTH Pb ISOTOPIC EVOLUTION

The U : Pb and Th : U ratios for bulk Earth (controlled by the silicate phase) are not well defined. The Sm : Nd ratio for bulk Earth can confidently be taken from the ratio observed in chondrites (DePaolo & Wasserburg 1976). The Rb : Sr ratio is not directly observed because of the relative volatility of Rb, but must be inferred from correlation with Sm : Nd data (DePaolo & Wasserburg 1976). The U : Pb ratio likewise cannot be directly observed, because Pb, like Rb, shows differential losses due to volatilization processes in early solar material (Larimer 1967). Thus Pb isotope evolution, like Sr evolution, must be inferred from comparison with another known system. In this section we first constrain the U : Pb ratio from present-day crustal and mantle rocks, and then use the ^{206}Pb : ^{204}Pb–^{87}Sr : ^{86}Sr correlation to refine the limits further. Finally, we will use data from Pb ores that are older than 3000 Ma to evaluate the ratio from samples that originated early in Earth history when, as postulated above, mixing in the mantle may have been more efficient, and the volume of crustal material small.

From a theoretical standpoint, present-day bulk Earth Pb isotope ratios must plot along the 0 Ma single-stage isochron in the ^{207}Pb : ^{204}Pb–^{206}Pb : ^{204}Pb diagram. Thus, if we know the age of the Earth, which we may take as 4550 Ma, we need constrain only one of the ratios in order to determine both present-day bulk Earth ratios. If we follow the commonly accepted idea that crustal materials have formed by partial melting of the mantle, then crust and mantle are complementary reservoirs with regard to mass balance, as appears to be true for the Rb–Sr and Sm–Nd systems. For Pb, the original undifferentiated mantle should have had a U : Pb ratio intermediate between crust and residual mantle. Because the differences in ^{207}Pb : ^{204}Pb ratios between crust and mantle rocks are mainly controlled by early differentiation events (i.e. older than *c.* 2000 Ma) and are not much affected by younger ones due to the low abundance of ^{235}U in younger rocks, the ^{207}Pb : ^{204}Pb ratio is probably a better indicator of the mean Earth U : Pb ratio than is the ^{206}Pb : ^{204}Pb ratio in young rocks. In fact, most young crustal and mantle-derived rocks plot substantially to the right of the 0 Ma single-stage isochron (Fig. 14.2), indicative of disturbances in the U : Pb ratio in relatively recent times after evolution of ^{207}Pb : ^{204}Pb had nearly ceased (cf. Fig. 14.2). At present MORB has, in general, the least radiogenic Pb isotope signature among mantle-derived rocks, and will be used as one end member. For average crust, we can use the Stacey–Kramers (1975) Pb evolution curve or the Zartman & Doe (1981) orogene Pb as the other end member. Taking average Pb isotopic compositions for present-day mantle and orogene as given by Zartman & Doe (1981), we obtain 8.4 and 8.6 for the time-averaged ^{238}U : ^{204}Pb ratios (μ values) for mantle and crust, based on the ^{235}U–^{207}Pb system. With the ^{238}U–^{206}Pb system the time-averaged values are 8.5 for mantle and 9.3 for crust. Because the recent U : Pb fractionations, cited above, quite obviously make the second data set less reliable, we will use the ^{235}U–^{207}Pb system to fix the limits on the U:Pb ratio of bulk silicate Earth (BSE).

It should be remembered that the above values are a function of the assumed age of the Earth. Ages observed for meteorites vary from c. 4.55 to 4.53 Ga. Changing the age of the Earth from 4550 to 4530 Ma increases the values of the μ limits by c. 0.15 for both the mantle and crust reservoirs (^{235}U : ^{207}Pb system).

14.4 ^{206}Pb : ^{204}Pb–^{87}Sr : ^{86}Sr CORRELATION

Best estimates for the age and initial Sr and Pb ages of the Canadian Shield and Arkansas complexes, from Kwon (1986), are given in Table 14.2. Figure 14.8 shows the correlation between initial ^{206}Pb : ^{204}Pb and ^{87}Sr : ^{86}Sr ratios for the various age-groups of alkaline rocks based on these data. The parameter ^{206}Pb : ^{204}Pb is chosen because it is a better indicator than ^{207}Pb : ^{204}Pb of the averaged variation of U:Pb ratios in the sources over the entire age of the Earth owing to the longer half-life of ^{238}U. Strontium, rather than Nd, isotope ratios are used because Sr isotope variations are larger than Nd variations, and because the initial Sr ratios are directly observed, while Nd initial ratios generally involve correlations for the *in situ* decay of ^{147}Sm. The age groups exhibit strong negative correlation except for the 2700 Ma suite, which shows a slight positive correlation. Such correlations are common among oceanic island basalts, both on regional and local scales. For example, Vidal & Dosso (1978) have reviewed the negative correlations between ^{206}Pb : ^{204}Pb and ^{87}Sr : ^{86}Sr on regional scales, and suggested a possible mechanism to explain the data. Basalts from the Walvis Ridge (Richardson *et al.* 1982) (shown in Fig. 14.8) and early-stage rocks (mainly tholeiitic) of Hawaiian volcanoes, as compiled by Staudigal *et al.* (1984), are good examples of the correlation on local

Table 14.2 Summary of initial ratios for the alkalic complexes.[1]

Complex	Age (Ma)	Atomic ratio ^{87}Sr : ^{86}Sr [2]	$\varepsilon_{Sr}(T)$ [3]	Atomic ratio ^{206}Pb : ^{204}Pb	^{207}Pb : ^{204}Pb	^{208}Pb : ^{204}Pb	^{232}Th : ^{283}U [4]
Arkansas	100	0.70365	−13.2	19.159	15.577	38.901	3.91
Oka	110	0.70331	−17.8	19.675	15.588	39.091	3.79
Firesand	1060	0.70242	−14.2	16.721	15.326	36.299	3.93
Coldwell	1090	0.7035	+1.6	16.247	15.254	36.442	4.29
Borden	1870	0.70184	−8.4	15.865	15.275	35.613	4.15
Cargill	1900	0.70198	−5.9	15.656	15.221	35.572	4.26
Otto	2670	0.70120	−3.4	13.476	14.590	33.322	4.25
Poohbah	2670	0.70129	−2.1	13.551	14.632	33.317	4.17
Sturgeon	2670	0.70164	+2.8	13.547	14.645	33.349	4.21

1 Data are from Kwon (1986), Bell & Blenkinsop (1987), Bell *et al.* (1987), and Tilton *et al.* (1987).
2 Normalized to ^{86}Sr : ^{88}Sr = 0.1194; ^{87}Sr : ^{86}Sr = 0.71025 for NBS983 standard.
3 Present-day B.E. ^{87}Sr : ^{86}Sr = 0.7047, ^{87}Rb : ^{86}Sr = 0.085.
4 Time-averaged ^{232}Th : ^{238}U ratio estimated from 4.55-Ga-old primordial Pb isotopic composition (Tatsumoto *et al.* 1973).

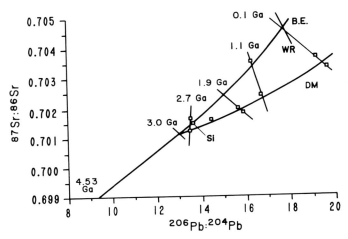

Figure 14.8 Correlation between ^{206}Pb : ^{204}Pb and ^{87}Sr : ^{86}Sr for North American alkalic complexes and Walvis Ridge tholeiitic basalts (WR). BE is a bulk Earth evolution curve determined by the intersection of lines drawn through the data points for each age-group with the bulk Earth ^{87}Sr : ^{86}Sr for the age of the Pb samples. DM: depleted mantle; Si: Siilinjarvi carbonatite.

scales. The slopes of the correlation lines in Figure 14.8 are not constant, but that is expected because the bulk Earth Sr ratios are fixed by the model for any given age, while ^{206}Pb : ^{204}Pb ratios vary widely in oceanic rocks and in the rocks from the alkalic complexes 1900 Ma and younger. The contrast between the Walvis Ridge and Oka–Magnet Cove data illustrate the variation in patterns (Fig. 14.8).

Extrapolation of the correlation lines to the bulk Earth Sr values, using average ^{206}Pb : ^{204}Pb ratios from each complex, yields apparent bulk Earth ^{206}Pb : ^{204}Pb ratios as a function of age. We assume a linear correlation between the two ratios, although the actual trend is linear only for the special case in which the ratios, Sr : (^{206}Pb + ^{204}Pb), are equal in the two end members. Otherwise the trend is hyperbolic. For the range of Sr and Pb ratios involved, the departure from a linear dependence is likely to be small, as found in the Walvis Ridge data (Richardson *et al.* 1982), for example. The estimated bulk Earth Pb ratios are obviously dependent on the choice of 0.7047 for the present ^{87}Sr : ^{86}Sr ratio, although that is probably not the main uncertainty in the derived Pb ratios.

By definition, the bulk Earth values for each of the ages should exhibit internal consistency, i.e. one value for μ should fit all of the data. In order to test for that requirement, we use the method of Albarede & Juteau (1984) in which the ^{206}Pb : ^{204}Pb ratios are plotted against the parameter $e^{\lambda t} - 1$, where t is the age of the Pb sample. This has the advantage of yielding a linear regression for closed-system isotope evolution when the ages of the Pb samples are correctly known. Such a regression analysis of the four bulk Earth Pb ratios in Figure 14.8 as a function of age yields a correlation coefficient of 0.9999, indicating a good linear correlation. For ^{206}Pb : ^{204}Pb = 9.307 (i.e. the Canyon Diablo Pb ratio) the bulk Earth ^{206}Pb : ^{204}Pb ratio derived from this treatment is given by ^{206}Pb : ^{204}Pb (t) = 9.307 + (8.35 ± 0.10) $(e^{\lambda T} - e^{\lambda t})$, where 8.35 is the present ^{238}U : ^{204}Pb (μ) of the source, T is the age of the source (4520 Ma), t is the time at which the ratio is evaluated, and λ is the decay constant for ^{238}U. It is not clear why the value for T is slightly younger than the one often assumed for the Earth, 4550 Ma, but the difference may be related to U:Pb differentiation processes in the early Earth. We derive a similar age for the source in a second calculation given below. In addition, the age is identical to the 4530 Ma ^{207}Pb:^{206}Pb ages calculated from the data from several basaltic achondrites (Tatsumoto et al. 1973). The calculated present ^{206}Pb:^{204}Pb ratio is 17.82 ± 0.17 (1σ), corresponding to 15.48 for ^{207}Pb:^{204}Pb. The μ value is close to the lower limit derived above from the ^{235}U:^{207}Pb system for young volcanic rocks. The present-day bulk Earth ^{206}Pb:^{204}Pb ratio is also similar to those that can be estimated from the Pb–Sr isotope correlations for the Hawaiian (Staudigal et al. 1984) and Walvis Ridge (Richardson et al. 1982) basalts.

A good correlation, which is insensitive to the age of the Earth, is found when model time-integrated Th:U ratios derived from the Pb isotope data are plotted against the ε_{Sr} values for the various suites (Kwon 1986). Kwon derived a value of 4.28 ± 0.08 for ^{232}Th:^{238}U ($\varkappa$) of bulk silicate Earth from such a diagram.

It is interesting to compare the Pb evolution curve just derived with the fit of data from samples older than 3000 Ma. According to the model suggested by earlier discussions, Pb isotope ratios should be much more uniform for those samples, and it should make little difference whether samples are derived from mantle or crust. Because of the extreme age, the data should be based on galena samples in order to minimize the effects of post-crystallization contamination or lack of closed-system conditions for U and Pb. We consider data from three galenas, which are the least radiogenic ones reported in the literature for each locality: Isua, with an age of 3770 Ma (Appel et al. 1978); Pilbara, 3450 Ma (Richards et al. 1981); and Barberton, 3230 Ma (Sinha 1972, Stacey & Kramers 1975, Cumming & Richards 1975). The isotopic data are listed in Table 14.3. The Barberton galena is associated with the Onverwacht lavas for which the age is somewhat problematic. Sinha (1972) gave a Pb:Pb isochron age of 3190 ± 30 Ma and a concordia intercept age of

Table 14.3 Comparison of observed and calculated[1] Pb ratios for Archaean galenas.

Locality	Age (Ma)	^{206}Pb:^{204}Pb	^{207}Pb:^{204}Pb	^{208}Pb:^{204}Pb
Barberton[2]	3210	12.461 (12.414)	14.077 (14.072)	32.285 (32.245)
Pilbara	3450	11.892 (11.891)	13.690 (13.689)	31.786 (31.751)
Isua	3770	11.146 (11.163)	13.025 (13.016)	31.148 (31.083)

1. Ratios in parentheses are calculated using $T^0 = 4520$ Ma, $\mu = 8.38$, $\varkappa = 4.21$. See text for derivation of these parameters.
2. The Barberton galena isotope composition is the average of French Bob's (Stacey & Kramers 1975) and Daylight (Cumming & Richards 1975) mines.

3250 ± 45 Ma for mafic members of the lava series. However, Hamilton et al. (1979) reported a Sm–Nd age of 3540 ± 30 Ma for Onverwacht lavas. Since Sinha's data include the Barberton galenas as an end member, his data apply best to the U : Pb system. The weighted mean of his ages yields a value of 3210 Ma, which we shall use here.

Using those results, the galena data yield values of 8.30 ± 0.02 for μ (from the ^{235}U : ^{207}Pb system) and 4.13 ± 0.06 for $\varkappa$, which agree within error limits with the bulk silicate Earth Pb evolution parameters derived from the Canadian Shield alkali complex data. A regression for the combined galena plus alkali complex data yields a slope of 8.38 ± 0.04 (1σ), i.e. μ for the source, and an intercept of 17.83 ± 0.08 (1σ), the present-day ^{206}Pb : ^{204}Pb. We will use that solution for the bulk Earth μ value. For $\varkappa$ we take the average of the two values, 4.21 ± 0.06. Although these values are somewhat higher than the value of 3.8–4.0 often quoted for terrestrial ^{232}Th : ^{238}U, they agree well with the ratio of 4.2 that Allègre et al. (1986) deduced from komatiite data. The age of the source is 4520 Ma. Tera (1980), using an approach similar to ours, found a source age of 4.53 Ga and a value for μ of 8.05 using galena samples covering an age range of 2.64–3.5 Ga. The agreement between observed and calculated ratios for the galenas based upon our bulk silicate Earth model is shown in Table 14.3. A listing of the bulk Earth model Pb ratios as a function of age is given in Table 14.4.

An important relationship is shown by the line marked DM in Figure 14.8, which approximates a curve drawn through the lower ^{87}Sr : ^{86}Sr data for each age group. The Sr data are the ones which define the depleted mantle trend in Bell et al. (1982) and Bell & Blenkinsop (1987). Kwon (1986) has obtained additional Sr data that give the same pattern as those of Bell et al. (1982) and Bell & Blenkinsop (1987). The corresponding ^{206}Pb : ^{204}Pb ratios yield a smooth correlation with the ^{87}Sr : ^{86}Sr ratios, providing strong evidence for a very regular coherence between the U : Pb and Rb : Sr systems in the carbonatite sources. The line through the data intersects the bulk Earth trend at c. 3000 Ma, the same age found by Sr. This is supporting

Table 14.4 BE Pb isotopic compositions.

Age (Ga)	^{206}Pb : ^{204}Pb	^{207}Pb : ^{204}Pb	^{208}Pb : ^{204}Pb
0.00	17.822	15.445	38.317
0.50	17.146	15.407	37.434
1.00	16.416	15.343	36.528
1.50	15.626	15.240	35.599
2.00	14.774	15.070	34.648
2.50	13.852	14.793	33.672
3.00	12.856	14.340	32.672
3.50	11.779	13.597	31.647
4.00	10.616	12.383	30.597
4.52	9.307	10.294	29.476

Calculated for ^{238}U : ^{204}Pb = 8.38 and ^{232}Th : ^{238}U = 4.21 (see text for discussion).

evidence from the U:Pb system that the apparent source of the Canadian Shield alkaline complexes experienced major chemical differentiation c. 3000 Ma ago. It remains to be seen how well this scheme works on a world-wide basis. A ^{208}Pb:^{204}Pb–^{87}Sr:^{86}Sr diagram based on the data from Table 14.2 (not shown) yields a correlation between the Th:Pb and Rb:Sr systems that closely resembles U:Pb–Rb:Sr correlation.

14.5 EVOLUTION MODELS FOR CANADIAN COMPLEXES

A plot of ^{206}Pb:^{204}Pb v. $e^{\lambda t}-1$ for all of the Canadian Shield data is shown in Figure 14.9. It is clear from the scatter of data that the time-averaged U:Pb ratios in the sources are variable and that two diverging trends are seen in the data. We will call plutons whose data plot above the crustal evolution line, COL (conformable ore lead), 'enriched' and those below the line 'depleted' for identification purposes. (COL is similar to S/K used earlier, but is based on more data.) It is important to note that the convention here is the reverse of the terminology for the groups based on Sr or Nd data. Reference to Figure 14.7 shows, in fact, that the depleted mantle

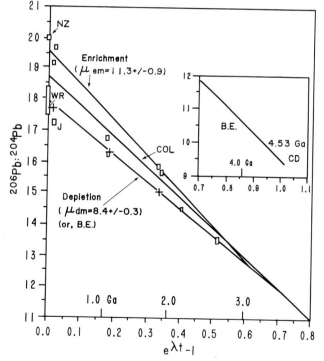

Figure 14.9 ^{206}Pb:^{204}Pb ratios plotted against $e^{\lambda t}-1$ for alkalic complex samples. The conformable ore evolution curve (COL) and calculated bulk Earth Pb isotope ratios (+) derived in this work are shown for comparison. COL is similar to S/K, but is based on more data. NZ: New Zealand; WR: Walvis Ridge; J: Jacupiranga. Symbols: λ is the decay constant for ^{238}U, t is the age of the sample, μ is ^{238}U:^{204}Pb for the sources, CD represents ratios for Canyon Diablo troilite Pb (Tatsumoto et al. 1973). Note that BE and DM evolution curves essentially coincide.

signature as defined by the Pb isotope system is due more to an excess of ^{206}Pb than to a deficiency of ^{207}Pb. This is an expression of the negative correlation between ^{87}Sr:^{86}Sr and ^{206}Pb:^{204}Pb for the alkalic complex data as discussed above. Each age-group tends to be characterized by a specific trend, with the 1100, 2200, and 2700 Ma samples belonging to the depleted trend and the 100 and 1900 Ma groups belonging to the enriched one. This grouping of Pb isotopic signatures might be attributed to a sampling bias or, alternatively, it could be due primarily to tapping of two different mantle sources at different times in the magma generation processes.

The two trends were fitted into line by using the least square regression method of York (1969). Errors for the Pb isotopic data were estimated from the scatter of the data for each alkalic complex, and by assuming 5% errors for each of the ages. With those limits the depleted trend has a present-day μ value of 8.4 ± 0.3 and the enriched trend 11.3 ± 0.9. The depleted and enriched mantle trends converge at c. 3100 Ma, in close agreement with the intersection age found above in Figure 14.8. The intersections can be taken as the approximate age of the differentiation of depleted mantle from primitive mantle for the rock suite, or the beginning of the stable craton system.

The enriched trend appears to involve some kind of mantle differentiation process. Although the higher ^{206}Pb:^{204}Pb ratios might be ascribed to contamination with aged granitic crust, the low ^{87}Sr:^{86}Sr and ^{207}Pb:^{204}Pb ratios in the enriched samples are not consistent with such a model. Instead, the isotope pattern resembles one commonly seen in oceanic volcanic rocks. The values of the above intersection ages are not precisely determined, but the important fact is that they are all much younger than 4550 Ma, suggesting that depleted mantle, at least as known in the Canadian Shield, is a later feature in Earth history. We consider below a possible mechanism to account for the enriched trend.

The calculated μ value of 11.3 for the enriched Pb isotope trend does not give a satisfactory fit to the ^{207}Pb data, and cannot therefore be taken literally. Lead that has evolved from 4530 to 3000 Ma with the bulk Earth μ of 8.35, then with $\mu = 11.3$ to the present, would have a ^{207}Pb:^{204}Pb ratio of 15.82, while the observed values for the young complexes (Oka, Magnet Cove and New Zealand) are 15.58–15.68. The mechanism that has increased ^{206}Pb:^{204}Pb ratios must be one in which μ has increased towards more recent time in order to avoid excessive increases in ^{207}Pb:^{204}Pb ratios. Detailed discussion of this process is beyond the scope of this report. Continuous mixing models, such as those discussed by O'Nions *et al.* (1979) provide possible examples of solutions.

In a ^{207}Pb:^{204}Pb v. $e^{\lambda t} - 1$ plot (not shown), the scatter in the data is limited but the enrichment and depletion trends follow the pattern in Figure 14.9. A value of 8.4 ± 0.4 is derived for μ for 'depleted mantle'.

Figure 14.10 is the corresponding plot for ^{208}Pb:^{204}Pb. Because the half-life of ^{232}Th, like that of ^{238}U, is substantially longer than 3000 Ma, the ^{208}Pb:^{204}Pb data provide a test of the regularities seen in the ^{206}Pb:^{204}Pb data. The ^{208}Pb data do, in fact, agree fairly well with the ^{206}Pb observations. The enriched and depleted trends for the carbonatites can be distinguished in Figure 14.10. Moreover, the trends are defined by the same age-groups as in Figure 14.10, indicating that increase in U:Pb

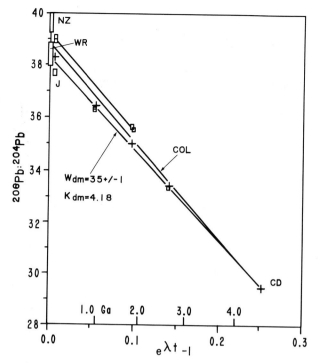

Figure 14.10 Diagram for ^{208}Pb:^{204}Pb and ^{232}Th. The symbols are the same as in Figure 14.9, except λ = decay constant for ^{232}Th.

is accompanied by a parallel increase in Th:Pb in the mantle differentiation processes. Again, the depleted mantle isotope signature for ^{87}Sr:^{86}Sr in the carbonatites argues against crustal contamination as the cause of the higher ^{208}Pb:^{204}Pb ratio samples belonging to the enriched trend. The depleted trend defines a ^{232}Th:^{204}Pb ratio of 35.1 ± 1.0 which, coupled with the derived μ value, yields 4.18 for $\varkappa$ of 'depleted mantle'. This value agrees within limits of uncertainty with the ratio of 4.21 we derived above for $\varkappa$ in BSE from the Sr–Pb correlation in carbonatites, and from the pre-3.0 Ga galena data. The enriched trend corresponds to a value of 36.6 ± 1.3 for ^{232}Th:^{204}Pb. In this diagram the enriched and depleted trends for Canadian Shield data do not intersect, but the trend is not very accurately defined since it is based upon only two age-groups. If the New Zealand carbonatite data (Barreiro & Cooper 1987) are used as a limit, then an intersection is obtained at c. 3000 Ma. As in the case of the ^{206}Pb:^{204}Pb–^{87}Sr:^{86}Sr relationship, the depleted mantle trend corresponds closely to the value deduced for bulk Earth.

14.6 DIFFERENTIATION OF U, Th AND Pb IN MANTLE PROCESSES

Data from the alkalic complexes demonstrate that enrichment of U over Pb was accompanied by depletion of Rb and Nd over Sr and Sm, respectively, in a manner similar to that observed in OIB. Several models have been proposed as explanations for such a differentiation pattern, for example, subduction of sediments, and oceanic crust which was enriched in U during seawater alteration, into the deeper mantle (Hofmann & White 1982), or pumping of Pb into the Earth's core (Allègre et al. 1980). Noting that Pb isotopic data of carbonatites are generally radiogenic, Nelson et al. (1988) suggested CO_2- and volatile-rich subducted oceanic crust as the source of carbonatites, following the Hofmann & White (1982) model. These models assume rather large amounts of transfer of specific elements between the mantle and other reservoirs, which is not proven. Alternatively, it seems more reasonable to consider mantle differentiation mechanisms through partial melting controlled by specific minerals. Kramers et al. (1983) have proposed one such process, emphasizing the role of phlogopite and K-richterite in the subcontinental mantle to explain the negative correlation between Pb and Sr. Meijer (1985) suggested that U and Pb budgets in the mantle may be controlled by accessory phases which possess high U contents, and that in partial melting processes the U:Pb ratio was higher in the residual phase than in the melt. In a formal way this hypothesis explains many otherwise enigmatic Pb isotope data. In any case, we note that the Pb–Sr correlations among OIB and alkalic complexes from continents are global in scale and are probably related to some fundamental mantle mineralogy.

Assuming a garnet lherzolite mantle, only clinopyroxene and garnet are important carriers of LILEs. Although there have been several attempts to determine the distribution coefficients of U and Pb between coexisting clinopyroxene and melt, the results are inconclusive (cf. review in Meijer 1985). Recently, the experiments of Watson et al. (1986) indicated that U is more incompatible than Pb in clinopyroxene coexisting with basaltic melts, suggesting that if clinopyroxene is the only residual phase controlling LILE differentiation during partial melting, the residual material would have a lower U:Pb ratio than undifferentiated mantle. The reverse trend is needed to explain the Pb isotopic data of the depleted trend. Watson et al. (1986) also noted that distribution coefficients for Hf and Lu do not yield the proper correlation between Hf and Sr or Nd for control of fractionation by residual clinopyroxene. They therefore postulated that garnet plays an important role as an additional phase in the residual mineralogy needed to explain the Lu–Hf fractionation. Garnet may satisfy the Pb isotope data as well. Although the distribution coefficients of U and Pb for garnet are not known, estimates can be made on the basis of Pb isotope data in garnets. Kwon (1986) and Mezger et al. (1987) have found highly radiogenic Pb in garnets from Archaean high-grade metamorphic rocks in the Canadian Shield. In addition, Mattinson (1986) reports radiogenic Pb in garnets from ampibolites from the Franciscan complex. We suggest that isotopic variations in the source of OIB and alkalic complexes can be explained as the result of interplay between garnet and clinopyroxene during partial melting in the mantle.

14.7 THE SOURCE OF CARBONATITES

To discuss the origin of the radiogenic isotope patterns in the alkaline complexes, we consider three types of mantle reservoirs: (a) strongly depleted (source of MORB),

(b) less depleted to slightly enriched (source of OIB), and (c) variably enriched and depleted (subcontinental lithospheric mantle: SCLM). Significantly, the degree of isotopic heterogeneity increases in that order (see Zindler & Hart 1986 for compilation of data). Because the majority of carbonatites occur in continental settings, we need to test whether there is any relationship between SCLM and carbonatite sources. The isotopic signatures of SCLM have been best identified by continental flood basalts and associated mantle xenoliths (Menzies et al. 1983, Hart 1985, McDonough et al. 1985, Carlson & Hart 1987, Dudas et al. 1987). Particularly good examples are found in the Columbia River (Carlson 1984, Church 1985) and Snake River (Leeman 1975) basalt flows. In some cases (Saddle Mountains member of the Columbia River, and the olivine tholeiite (SROT) member of the Snake River flows) the data reflect the c. 2.6 Ga age of the basement in ^{207}Pb : ^{206}Pb isochron diagrams. The authors ascribe the relationship to 2.6 Ga old SCLM rather than to crustal contamination. The isotope signatures are variably enriched and depleted, commonly scattered outside of, but converging to, the OIB and MORB fields in isotope correlation diagrams. The Nd and Sr isotope data for the saddle Mountains basalts are also anomalous, yielding high ^{87}Sr : ^{86}Sr ratios of 0.70755 to 0.71317 and ε_{Nd} values of -5.8 to -16.9 (Carlson 1984); SROT basalts yield similarly high ^{87}Sr : ^{86}Sr ratios of c. 0.708 (Leeman 1975). Thus SCLM appears to be a geochemically complex system infiltrated by magmas from below and modified during subduction processes (Hart 1985). Also, the presumed long-term separation of SCLM from the rest of the convecting mantle (e.g. Leeman 1975, Carlson & Hart 1987) appears to enhance isotopic heterogeneities, perhaps in part through metasomatic processes. The rather restricted isotopic variations and good correlation between isotope pairs exhibited by carbonatite data strongly resemble the patterns seen in OIB and suggest that the magma sources of the two rock types are related. This argues for production of carbonatite magmas in the asthenosphere rather than SCLM; however, the isotopic signatures of carbonatites may, in some cases, be influenced by the SCLM through which the magmas must pass.

14.8 CONCLUDING STATEMENT

It is evident that alkalic complexes, especially carbonatites, provide highly promising mantle probes for tracing crust–mantle geochemical evolution back into the Precambrian Era. In some cases problems arise due to crustal contamination or lack of closed chemical system behaviour in minerals since crystallization. However, it appears that those complications can be overcome most of the time. The study of radiogenic isotope data from alkalic complexes is still in an elementary stage, and much more work must be done to put the conclusions reached in this survey on a firm basis. In particular, the database, which is presently weighted strongly towards North America, needs to be expanded to encompass a world-wide scale.

ACKNOWLEDGEMENTS

Work described in this report has been supported by National Science Foundation Grants EAR82 12931 and EAR86 07163 to G.R.T. We especially thank Keith Bell and John Blenkinsop, Carleton University, Ottawa, for providing samples and for helpful discussions throughout the course of the project. R. P. Sage, Ontario Geological Survey, expertly guided much of the field work, in several cases enabling us to obtain samples in poorly exposed areas where we would otherwise have failed to locate suitable material. R. E. Zartman provided a thoughtful review of the manuscript. Finally, we thank David Crouch for drafting of the figures.

REFERENCES

Albarede, F. & M. Juteau 1984. Unscrambling the lead model ages. *Geochimica et Cosmochimica Acta* **48**, 207–12.

Allègre, C. J., B. Dupré & E. Lewin 1986. Thorium/uranium ratio of the earth. *Chemical Geology* **56**, 219–27.

Allègre, C. J., O. Brevart, B. Dupré & J.-F. Minster 1980. Isotopic and chemical effects produced in a continuously differentiating convecting earth mantle. *Philosophical Transactions of the Royal Society of London* **A297**, 447–77.

Andersen, T. 1987. Mantle and crustal components in a carbonatite complex, and the evolution of carbonatite magma: REE and isotopic evidence from the Fen complex, southeast Norway. *Chemical Geology* **65**, 147–66.

Appel, P. W. U., S. Moorbath & P. N. Taylor 1978. Least radiogenic terrestrial lead from Isua, West Greenland. *Nature* **272**, 524–6.

Barreiro, B. 1983. Lead isotopic composition of South Sandwich Island volcanic rocks and their bearing on magma genesis in intra-oceanic island arcs. *Geochimica et Cosmochimica Acta* **47**, 817–22.

Barreiro, B. & A. F. Cooper 1987. A Sr, Nd, and Pb isotope study of alkaline lamprophyres and related rocks from Westland and Otago, South Island, New Zealand. In *Mantle metasomatism and alkaline magmatism*, E. M. Morris & J. D. Pasteris (eds), 115–27. Geological Society of America Special Paper 215.

Bell, K. & J. Blenkinsop 1980. Ages and initial $^{87}Sr:^{86}Sr$ ratios from alkalic complexes of Ontario. In *Geoscience research grant program, summary of research 1979–1980*, E. G. Pye (ed.), 16–23. Ontario Geological Survey Miscellaneous Paper 93.

Bell, K. & J. Blenkinsop 1987. Archean depleted mantle: Evidence from Nd and Sr initial isotopic ratios of carbonatites. *Geochimica et Cosmochimica Acta* **51**, 291–8.

Bell, K., J. Blenkinsop, T. J. S. Cole & D. P. Menagh 1982. Evidence from Sr isotopes for long-lived heterogeneities in the upper mantle. *Nature* **298**, 251–3.

Bell, K., J. Blenkinsop, S.-T. Kwon, G. R. Tilton & R. P. Sage 1987. Age and radiogenic isotopic systematics of the Borden carbonatite complex, Ontario, Canada. *Canadian Journal of Earth Sciences* **24**, 24–30.

Brevart, O., B. Dupré & C. J. Allègre 1986. Lead-lead age of komatiitic lavas and limitations on the structure and evolution of the Precambrian mantle. *Earth and Planetary Science Letters* **77**, 293–302.

Carlson, R. J. 1984. Isotopic constraints on Columbia River basalt genesis and the nature of the subcontinental mantle. *Geochimica et Cosmochimica Acta*, **48**, 2357–72.

Carlson, R. J. & W. K. Hart 1987. Crustal genesis on the Oregon plateau. *Journal of Geophysical Research* **92**, 6191–206.

Cavell P. A. & H. Baadsgaard 1986a. Geochronology of the Big Spruce Lake alkaline intrusion. *Canadian Journal of Earth Sciences* **23**, 1–10.

Cavell, P. A. & H. Baadsgaard 1986b. Sm–Nd, Rb–Sr and U–Pb systematics of the Big Spruce Lake alkaline carbonatite complex. *Geological Association of Canada–Mineralogical Association of Canada–Canadian Geophysical Union, Joint Annual Meeting, Ottawa. Program with Abstracts* **11**, 53.

Church, S. E. 1985. Genetic interpretation of lead-isotopic data from the Columbia River Basalt Group, Oregon, Washington and Idaho. *Bulletin of the Geological Society of America* **96**, 676–90.

Church, S. E. & M. Tatsumoto 1975. Lead isotope relations in oceanic ridge basalts from the Juan de Fuca-Gorda Ridge area, N.E. Pacific Ocean. *Contributions to Mineralogy and Petrology* **53**, 253–79.

Cumming, G. L. & J. R. Richards 1975. Ore lead isotope ratios in a continuously changing earth. *Earth and Planetary Science Letters* **28**, 155–71.

Deines, P. & D. P. Gold 1973. The isotopic composition of carbonatite and kimberlite carbonates and their bearing on the isotopic composition of deep-seated carbon. *Geochimica et Cosmochimica Acta* **37**, 1709–33.

DePaolo, D. J. & G. J. Wasserburg 1976. Inferences about magma sources and mantle structure from variations of $^{143}Nd/^{144}Nd$. *Geophysical Research Letters* **3**, 743–6.

Doe, B. R. & J. S. Stacey 1974. The application of lead isotopes to the problems of ore genesis and ore prospect evolution: A review. *Economic Geology* **69**, 757–76.

Dudas F. O., R. W. Carlson & D. H. Eggler 1987. Regional Middle Proterozoic enrichment of the subcontinental mantle source of igneous rocks from Central Montana. *Geology* **15**, 22–5.

Dupré, B., C. Chauvel & N. T. Arndt 1984. Pb and Nd isotopic study of two Archean komatiitic flows from Alexo, Ontario. *Geochimica et Cosmochimica Acta* **48**, 1965–72.

Gariepy, C. & C. J. Allègre 1985. The lead isotope geochemistry of late-kinematic intrusives from the Abitibi greenstone belt, and implications for late Archaean crustal evolution. *Geochimica et Cosmochimica Acta* **49**, 2371–83.

Gittins, J., R. M. MacIntyre & D. York 1967. The age of carbonatite complexes in Eastern Canada. *Canadian Journal of Earth Sciences* **4**, 651–5.

Grünenfelder, M. H., G. R. Tilton, K. Bell & J. Blenkinsop 1986. Lead and strontium isotope relationships in the Oka carbonatite complex, Quebec. *Geochimica et Cosmochimica Acta* **50**, 461–8.

Hamilton, P. J., N. M. Evensen, R. K. O'Nions, H. S. Smith & A. J. Erlank 1979. Sm–Nd dating of Onverwacht group volcanics, Southern Africa. *Nature* **279**, 298–300.

Hart, W. K. 1985. Chemical and isotopic evidence for mixing between depleted and enriched mantle, northwestern U.S.A. *Geochimica et Cosmochimica Acta* **49**, 131–44.

Hofmann, A. W. & W. M. White 1982. Mantle plumes from ancient oceanic crust. *Earth and Planetary Science Letters* **57**, 421–36.

Innes, M. J. J. 1960. *Gravity and isostasy in northern Ontario and Manitoba, Cochrane District.* Geological Survey of Canada Paper 61-16.

Kramers, J. D., J. C. M. Roddick & J. B. Dawson 1983. Trace element and isotopic studies on veined, metasomatic and "MARID" xenoliths from Bultfontein, South Africa. *Earth and Planetary Science Letters* **65**, 90–106.

Kwon, S.-T. 1986. *Pb–Sr–Nd isotope study of the 100 to 2700 Ma old alkalic rock–carbonatite complexes in the Canadian Shield: Inferences on the geochemical and structural evolution of the mantle.* Ph.D. thesis, University of California, Santa Barbara.

Kwon, S.-T. & G. R. Tilton 1986. Comparative isotopic studies of Cargill and Borden carbonatite complexes from the Kapuskasing gravity high zone, Ontario. *Geological Association of Canada–Mineralogical Association of Canada–Canadian Geophysical Union, Joint Annual Meeting, Ottawa. Program with Abstracts* **11**, 92.

Lancelot, J. R. & C. J. Allègre 1974. Origin of carbonatitic magma in the light of the Pb–U–Th isotope system. *Earth and Planetary Science Letters* **22**, 233–8.

Larimer, J. W. 1967. Chemical fractionation in meteorites – I. Condensation of the elements. *Geochimica et Cosmochimica Acta* **31**, 1215–38.

Leeman, W. P. 1975. Radiogenic tracers applied to basalt genesis in the Snake River Plain–Yellowstone National Park region – Evidence for a 2.7 B.Y. old upper mantle keel. *Geological Society of America Abstracts with Programs* **7**, 1165.

McDonough, W. F., M. T. McCulloch & S. S. Sun 1985. Isotopic and geochemical systematics in Tertiary–Recent basalts from southeastern Australia and implications for the evolution of the sub-continental lithosphere. *Geochimica et Cosmochimica Acta* **49**, 2051–67.

Mattinson, J. M. 1986. *Geochronology of high pressure–low-temperature Franciscan metabasites: A new approach using the U–Pb system.* Geological Society of America Memoir 164, 95–105.

Meijer, A. 1985. Pb isotope evolution in the earth: A proposal. *Geophysical Research Letters* **12**, 741–4.

Menzies, M. A., W. P. Leeman & C. J. Hawkesworth 1983. Isotope geochemistry of Cenozoic volcanic rocks reveals mantle heterogeneity below western USA. *Nature* **303**, 205–9.

Mezger, K., G. N. Hanson & S. R. Bohlen 1987. U–Pb garnet ages used to date multiple metamorphic events and partial melting, Pikwitonei domain, Manitoba. *EOS. Transactions, American Geophysical Union* **68**, 453. (abstract)

Nelson, D. R., A. R. Chivas, B. W. Chapell & M. T. McCulloch 1988. Geochemical and isotopic systematics in carbonatites and implications for the evolution of ocean-island sources. *Geochimica et Cosmochimica Acta* **52**, 1–17.

O'Nions, R. K., N. M. Evensen & P. J. Hamilton 1979. Geochemical modeling of mantle differentiation and crustal growth. *Journal of Geophysical Research* **84**, 6091–101.

Platt, R. G. & R. H. Mitchell 1982. Rb–Sr geochronology of the Coldwell complex, northwestern Ontario, Canada. *Canadian Journal of Earth Sciences* **19**, 1796–801.

Richards, J. R., I. R. Fletcher & J. G. Blockley 1981. Pilbara galenas: Precise assay of the oldest Australian leads; model ages and growth-curve implications. *Mineralium Deposita* **16**, 7–30.

Richardson, S. H., A. J. Erlank, A. R. Duncan & D. L. Reid 1982. Correlated Nd, Sr and Pb isotope variation in Walvis Ridge basalts and implications for the evolution of their mantle source. *Earth and Planetary Science Letters* **59**, 327–42.

Roden, M. F., R. V. Murthy & J. C. Gaspar 1985. Sr and Nd isotopic composition of the Jacupiranga carbonatite. *Journal of Geology* **93**, 212–20.

Sinha, A. K. 1972. U–Th–Pb systematics and age of the Onverwacht series, South Africa. *Earth and Planetary Science Letters* **16**, 219–27.

Stacey, J. S. & J. D. Kramers 1975. Approximation of terrestrial lead isotope evolution by a two-stage model. *Earth and Planetary Science Letters* **26**, 207–21.

Staudigal, H., A. Zindler, S. R. Hart, T. Leslie, C. Y. Chen & D. Clague 1984. The isotope systematics of a juvenile intraplate volcano: Pb, Nd and Sr isotope ratios of basalts from Loihi seamount, Hawaii. *Earth and Planetary Science Letters* **69**, 13–29.

Sun, S-S. 1980. Lead isotopic studies of young volcanic rocks from mid-ocean ridges, oceanic islands and island arcs. *Philosophical Transactions of the Royal Society of London* **A297**, 409–45.

Tatsumoto, M. 1978. Isotopic composition of lead in oceanic basalt and its implication to mantle evolution. *Earth and Planetary Science Letters* **38**, 63–87.

Tatsumoto, M., R. J. Knight & C. J. Allègre 1973. Time differences in the formation of meteorites as determined from the ratio of lead-207 to lead-206. *Science* **180**, 1279–83.

Tera, F. 1980. Reassessment of the "Age of the Earth". *Carnegie Institution of Washington Year Book* **79**, 524–31.

Thorpe, R. I. 1986. U–Pb geochronology of the Coldwell complex, northwestern Canada: Discussion. *Canadian Journal of Earth Sciences* **23**, 125–7.

Tilton, G. R. 1983. Evolution of depleted mantle: The lead perspective. *Geochimica et Cosmochimica Acta* **47**, 1191–7.

Tilton, G. R. & M. H. Grünenfelder 1983. Lead isotope relationships in billion-year-old carbonatite complexes, Superior Province, Canadian Shield. *Geological Society of America Abstracts with Programs* **15**, 707.

Tilton, G. R., S.-T. Kwon & D. M. Frost 1987. Isotopic relationships in Arkansas Cretaceous alkalic complexes. In *Mantle metasomatism and alkaline magmatism*, E. M. Morris & J. D. Pasteris (eds), Geological Society of America Special Paper 215, 241–8.

Todt, W., R. A. Cliff, A. Hanser & A. W. Hofmann 1984. ^{202}Pb + ^{205}Pb spike for lead isotopic analysis. *Terra Cognita* **4**, 209. (abstract)

Turek, A., P. E. Smith & D. T. A. Symons 1985. U–Pb geochronology of the Coldwell complex, northwestern Ontario, Canada. *Canadian Journal of Earth Sciences* **22**, 621–6.

Vidal, Ph. & L. Dosso 1978. Core formation: Catastrophic or continuous? Sr and Pb isotopic constraints. *Geophysical Research Letters* **5**, 169–72.

Watson, E. B., D. Ben Othman, J.-M. Luck & A. W. Hofmann 1986. Partitioning of U, Pb, Cs, Yb, Hf, Re and Os between chromian diopsidic pyroxene and haplobasaltic liquid. *Chemical Geology* **62**, 191–208.

Wen, J.-P. 1985. *Isotope geochemistry of the Oka Carbonatite Complex, Quebec*. M.Sc. thesis, Carleton University, Ottawa.

York, D. 1969. Least squares fitting of a straight line with correlated errors. *Earth and Planetary Science Letters* **5**, 320–4.

Zartman, R. E. & B. R. Doe 1981. Plumbotectonics – the model. *Tectonophysics* **75**, 135–62.

Zartman, R. E. & J. M. Howard 1987. U–Th–Pb ages of large zircon crystals from the Potash Sulfur Springs igneous complex, Garland County, Arkansas. In *Mantle metasomatism and alkaline magmatism*, E. M. Morris & J. D. Pasteris (eds), 235–41. Geological Society of America Special Paper 215.

Zindler, A. & S. R. Hart 1986. Chemical dynamics. *Annual Review of Earth and Planetary Sciences* **14**, 493–571.

15
THE GENESIS OF CARBONATITES BY IMMISCIBILITY

B. A. KJARSGAARD & D. L. HAMILTON

ABSTRACT

New experimental data clearly show that immiscibility between carbonate and silicate liquids can occur in alkali-poor systems at pressures at least to 15 Kb. A model is proposed in which calcite-rich liquids separate from carbonated silicate magma. The separation of carbonatite liquids by immiscibility from a silicate melt is envisaged as a continuous process occurring over a wide compositional range. Subsequent fractional crystallization of these alkali-poor carbonatite melts generates the range of compositions observed in carbonatites. Cumulate calcite (and/or dolomite) forms the coarse-grained intrusive carbonatites. Residual fluids are considered responsible for lubrication during carbonatite emplacement and fenitization. The sequence melilitite–nephelinite–phonolite is thought to reflect a parent magma undergoing both carbonate-liquid and crystal fractionation.

15.1 INTRODUCTION

The idea that liquid immiscibility plays an important role in the genesis of carbonatites has increased in popularity over the past decade. Until recently the main stumbling block for its acceptance as a major process or even *the* major process in the production of carbonatite melts from parent carbonated silicate magmas was the absence of supportive experimental data for silicate–carbonate immiscibility in alkali-poor systems. Several groups of experimental workers demonstrated immiscibility in alkali-rich systems (e.g. phonolite–natrocarbonatite), but attempts to show immiscibility in calcium-rich systems (e.g. nepheline–calcite) largely failed. The present work was initiated because we were not convinced that a sufficient difference exists between the melt structures of Na_2CO_3 and $CaCO_3$ such that immiscibility occurs only in the alkali-rich system.

This chapter has three main divisions, viz.

(a) a review of relevant experimental studies;
(b) evidence from rocks for liquid immiscibility; and
(c) the proposal of a model that explains the diverse nature of carbonatites, based primarily on immiscibility, but also fractional crystallization.

15.2 EVIDENCE FOR SILICATE–CARBONATE IMMISCIBILITY

15.2.1 Pre-1988 experimental data

One of the earliest modern papers bearing on silicate–carbonate liquid immiscibility is that of Koster van Groos & Wyllie (1966). In this pioneering paper it was shown that at 1 Kb a large miscibility gap separates albite-rich and sodium carbonate-rich liquids, which exist at temperatures down to about 870 °C (tie line 1, Fig. 15.1). The same authors (Koster van Groos & Wyllie 1968) investigated the effect of adding H_2O to the albite–sodium carbonate system, and concluded that:

(a) H_2O does not destroy the immiscibility;
(b) compositions of the conjugate liquids are little changed by the addition of

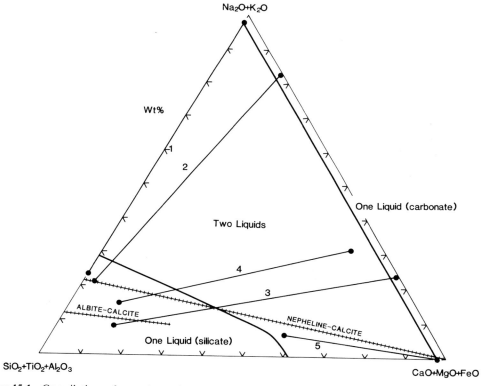

Figure 15.1 Compilation of experimental data relevant to silicate–carbonate immiscibility plotted on the Freestone & Hamilton (1980) pseudoternary diagram projected from CO_2. The two heavy lines are the silicate and carbonate limbs of the two-liquid field determined at 2 Kb and 1250 °C (this study). The thin lines numbered 1–5 are conjugation lines between silicate–carbonate liquid pairs from the following sources: (1) albite–sodium carbonate at 1 Kb and 900 °C (Koster van Groos & Wyllie 1966), (2) plagioclase–sodium carbonate at 1 Kb and 925 °C (Koster van Groos & Wyllie 1973), (3) carbonated pantellerite enriched in CaO and MgO (composition P5) at 10 Kb and 725 °C (Koster van Groos 1975), (4) nephelinite–carbonatite (sample FH24) at 7.6 Kb and 1100 °C (Freestone & Hamilton 1980), (5) albite–calcite at 2 Kb and 1250 °C (this study). The two hachured lines are the joins nepheline–calcite and albite–calcite (Watkinson & Wyllie 1969, 1971). No immiscibility was detected in either of these systems at 1 Kb. Data are plotted as wt%.

small amounts of H_2O (approx. 10 wt%) – however, the two-liquid field tends to narrow with > 20 wt% H_2O; and
(c) the two liquids are stable down to 725 °C.

Because the main carbonate in carbonatites is calcite, the results of the high-sodium systems studied by Koster van Groos & Wyllie are not directly applicable to the genesis of most carbonatites. Watkinson & Wyllie (1969, 1971) studied the systems albite–calcite + 25 wt% H_2O, and nepheline–calcite + 25 wt% H_2O. The calcite content of the charges of both studies makes the results more relevant to natural compositions. In neither of these systems was liquid immiscibility detected (see hachured lines, Fig. 15.1). However, the 25 wt% H_2O used in the experiments is much higher than amounts observed in most magmas, and the liquidus data for the albite–calcite system extended only to 40 wt% $CaCO_3$, which falls outside our two-liquid field. These two shortcomings probably resulted from the limited temperatures that could be attained by the equipment used by Watkinson & Wyllie.

The next major paper (Koster van Groos & Wyllie 1973) reported on the system plagioclase–sodium carbonate plus 10 wt% H_2O. Details of two joins ($Ab_{50}An_{50}$–Na_2CO_3 and $Ab_{80}An_{20}$–Na_2CO_3) are published. Both of these systems show a miscibility gap (e.g. line 2 on Fig. 15.1), that narrows in the An-rich system. Results from the two joins, plotted on a ternary Ab–An–Na_2CO_3 diagram, illustrated the closure of the two-liquid field before it intersected the An–Na_2CO_3 sideline, consistent with earlier published data. This paper, for the first time, provided electron microprobe analyses of the silicate glass.

Koster van Groos (1975) published data for a multi-component system that indicated the presence of a miscibility gap. These results are not in agreement with the nepheline–calcite study of Watkinson & Wyllie (1971). Such differences are likely due to the effects of pressure (10 Kb versus 1 Kb). Of particular interest is Koster van Groos' composition, P5, described as an 'artificial pantellerite with additional CaO and MgO' which was run at 10 Kb and 725 °C with 16 wt% added H_2O; the starting bulk composition contained 16.6 wt% CO_2. The run products (in modal %) consisted of silicate glass (80%) which contained carbonate spherules (15%) and pyroxene crystals (5%). Unfortunately, a silicate glass separate was not chemically analysed. Because the P5 sample (run at 10 Kb and 725 °C) contains only 5% pyroxene crystals, the glass composition will therefore be close to that of the total silicate. The resulting silicate–carbonate melt tie line, shown in Figure 15.1 as line 3 is significant for two reasons:

(a) the carbonate liquid composition has the highest $CaCO_3$ content (approx. 75 wt%) of any experimental compositions published prior to 1988; and
(b) this silicate–carbonate tie line cuts across the nepheline–calcite and albite–calcite joins.

Freestone & Hamilton (1980) published the results of an immiscibility study using natural lavas as starting materials. The highlights of this work were:

(a) coarse-grained, two-liquid textures were produced;
(b) carbonate melts with up to 60 wt% $CaCO_3$ were immiscible with silicate liquids, e.g. line 4 on Figure 15.1;
(c) with pressure increase from 0.7 to 7.6 Kb (at constant TX) the two-liquid field increased in width; and
(d) with temperature decrease from 1250 to 1050 °C (at constant PX) the two-liquid field increased in width.

The use of a triangular diagram $(SiO_2 + Al_2O_3) - (Na_2O + K_2O) - (CaO)$, projected from CO_2, was general enough so that all the previous silicate–carbonate experimental data could be plotted and compared.

15.2.2 1988 Results

Kjarsgaard & Hamilton (1988), working with synthetic compositions at 5 Kb and in the temperature range 1100–1250 °C, showed that the miscibility gap between silicate and carbonate melts, contrary to widely held beliefs, does not close off at alkali-poor compositions, but continues to the alkali-free base line. Consequently a pure $CaCO_3$ melt is immiscible with a silicate melt. New data presented here extend the scope of the Kjarsgaard & Hamilton study to include a determination of two-liquid phase relations at 2 Kb, plus reconnaissance runs at 15 Kb. These new results are given in Tables 15.1 and 15.2, together with data gathered at 5 Kb by Kjarsgaard & Hamilton (1988). The 2 Kb data are shown on Figure 15.2, and the 5 Kb results along with one 15 Kb tie line are shown on Figure 15.3. It should be

Table 15.1 Run data from immiscibility studies.

Run No.	Composition (wt%)	Added components (wt%)	P (Kb)	T (°C)	Time (h)	Assemblage
BK5	$Ab_{25} CC_{75}$	10% SO	2.0	1225	2	LS + LC + V
BK18	$Ab_{25} CC_{75}$	14.70% NC	2.1	1250	2	LS + LC + V
BK23	$Ab_{50} CC_{50}$	15.13% NC	2.0	1250	1	LS + LC + V
BK34	$Ab_{38} CC_{62}$	37.05% NC	2.1	1250	2	LS + LC + V
BK125	$El_{35} CC_{65}$		2.0	1100	24	LS + LC + W
BK129	$Ab_{50} CC_{50}$		2.0	1250	30	LS + LC + V
BK10	$Ab_{38} CC_{62}$	10% SO	4.9	1250	3	LS + LC + V
BK21	$Ab_{38} CC_{62}$	14.12% NC	5.0	1250	4	LS + LC + V
BK25	$Ab_{38} CC_{62}$	9.00% NC	5.1	1250	3	LS + LC + V
BK27	$Ab_{38} CC_{62}$	22.54% NC	5.1	1250	3	LS + LC + V
BK42	$Ab_{25} CC_{75}$	37.08% NC	5.1	1250	26	LS + LC + V
BK58	$Ab_{50} NC_{50}$		5.1	1250	26	LS + LC + V
BK70	$An_{15} CC_{85}$		5.0	1250	12	LS + LC + V
BK73	$Ab_{38} CC_{62}$	45.69% NC	5.0	1250	12	LS + LC + V
BK119	$El_{25} CC_{75}$		5.2	1175	12	LS + LC + V
RB7	$Ab_{38} CC_{62}$		15.0	1300	3	LS + LC

Ab, Albite; An, anorthite; El, composition of the quartz–anorthite–wollastonite eutectic with Si:Al ratio similar to albite; CC, calcium carbonate; NC, sodium carbonate; SO, silver oxalate; LS, silicate melt; LC, carbonate melt; V, vapour; W, wollastonite crystals.

Table 15.2 Electron microprobe data of quenched liquids from experimental runs (in wt%).

Run data	Phase	SiO$_2$	Al$_2$O$_3$	Na$_2$O	CaO	Total
BK5	Bulk	25.64	7.25	4.41	62.70	100.00
	LS	28.53	3.78	5.38	49.64	87.33
	LS*	32.67	4.33	6.16	56.84	100.00
	LC	0.60	0.12	0.15	55.94	56.81
	LC*	1.06	0.21	0.26	98.47	100.00
BK18	Bulk	22.22	6.29	17.00	54.49	100.00
	LS	35.60	10.03	11.44	30.41	87.48
	LS*	40.70	11.47	13.08	34.75	100.00
	LC	2.12	0.82	10.47	42.38	55.79
	LC*	3.80	1.47	18.77	75.96	100.00
BK23*	Bulk	38.80	10.97	18.60	31.63	100.00
	LS	38.42	11.13	14.66	25.56	89.77
	LS*	42.83	12.41	16.34	28.42	100.00
	LC	0.52	0.12	23.63	34.58	58.85
	LC*	0.88	0.20	40.16	58.76	100.00
BK34	Bulk	24.35	6.89	36.36	32.40	100.00
	LS	48.52	13.76	20.66	13.46	96.40
	LS*	50.32	14.27	21.43	13.98	100.00
	LC	0.42	0.12	28.54	28.01	57.09
	LC*	0.74	0.21	49.99	49.06	100.00
BK125	Bulk	31.36	6.29		62.35	100.00
	LS	20.49	17.57		56.61	94.67
	LS*	21.64	18.56		59.80	100.00
	LC	0.29	0.04		56.60	56.93
	LC*	0.51	0.07		99.42	100.00
	W	51.12	0.12		48.81	100.05
BK129	Bulk	43.37	12.27	44.36		100.00
	LS	50.81	14.61	27.01		92.43
	LS*	54.97	15.81	29.22		100.00
	LC	0.51	0.17	53.29		53.97
	LC*	0.94	0.32	98.74		100.00
BK10	Bulk	35.91	10.16	6.17	47.76	100.00
	LS	35.52	9.76	6.72	41.32	93.32
	LS*	38.00	10.50	7.20	44.30	100.00
	LC	1.35	0.36	0.74	55.38	57.83
	LC*	2.30	0.60	1.30	95.80	100.00
BK21*	Bulk	31.47	8.90	17.45	42.18	100.00
	LS	44.16	11.25	12.26	23.16	90.83
	LS*	48.60	12.40	13.50	25.50	100.00
	LC	2.00	0.15	14.47	41.10	57.72
	LC*	3.50	0.30	25.00	71.20	100.00
BK25*	Bulk	33.27	9.41	13.08	44.24	100.00
	LS	39.01	10.70	11.05	28.67	89.43
	LS*	43.60	12.00	12.40	32.00	100.00
	LC	1.25	0.23	8.86	45.99	56.33
	LC*	2.20	0.40	15.70	81.70	100.00

Table 15.2 (contd).

Run data	Phase	SiO$_2$	Al$_2$O$_3$	Na$_2$O	CaO	Total
BK27*	Bulk	29.08	8.22	24.00	38.70	100.00
	LS	46.88	11.54	14.87	19.35	92.64
	LS*	50.60	12.50	16.00	20.90	100.00
	LC	1.11	0.11	21.35	36.10	58.67
	LC*	1.90	0.20	36.40	61.50	100.00
BK42*	Bulk	16.93	4.79	36.88	41.40	100.00
	LS	45.10	17.91	17.58	14.40	94.99
	LS*	47.40	18.90	18.50	15.20	100.00
	LC	0.83	0.15	25.75	30.93	57.66
	LC*	1.40	0.30	44.70	53.60	100.00
BK58	Bulk	43.37	12.27	44.36		100.00
	LS	53.35	15.68	25.47		94.50
	LS*	56.50	16.60	26.90		100.00
	LC	0.31	0.31	52.73		53.35
	LC*	0.60	0.60	98.80		100.00
BK70	Bulk	10.68	8.52		80.80	100.00
	LS	26.93	27.22		42.46	96.61
	LS*	27.90	28.20		43.90	100.00
	LC	0.36	0.20		54.43	54.99
	LC*	0.60	0.40		99.00	100.00
BK73	Bulk	21.42	6.06	44.03	28.49	100.00
	LS	50.94	17.56	21.60	7.20	97.30
	LS*	52.40	18.00	22.20	7.40	100.00
	LC	1.14	0.26	32.65	22.16	56.21
	LC*	2.00	0.50	58.10	39.40	100.00
BK119	Bulk	23.53	5.13		71.34	100.00
	LS	36.05	9.90		48.44	94.39
	LS*	38.20	10.50		51.30	100.00
	LC	0.34	0.04		56.51	56.89
	LC*	0.60	0.10		99.30	100.00

Abbreviations: Bulk, composition of the starting material calculated free of CO_2; LS, silicate melt; LS*, silicate melt recalculated to 100%; LC, carbonate melt; LC*, carbonate melt recalculated to 100%; W, wollastonite crystals. Run numbers with an asterisk indicate that the carbonate melt produced was an intergrowth of two quenched carbonates, and LC has been calculated utilizing a tie line construction method (see Kjarsgaard & Hamilton 1988).

noted that the apparent rotation of the tie lines in Figures 15.2 and 15.3 is an artefact resulting from the projection technique used. This can be clearly seen from Figure 15.4 (a quaternary 5 Kb plot of the data used to construct Fig. 15.3), where the tie lines are parallel.

The starting materials for all these runs were finely ground mixtures of silicate glass (e.g. albite, anorthite, E1) and one or two carbonates ($CaCO_3$, Na_2CO_3). Run products were analysed by electron microprobe. Further analytical and experimental techniques are given in Kjarsgaard & Hamilton (1988). The liquid compositions are

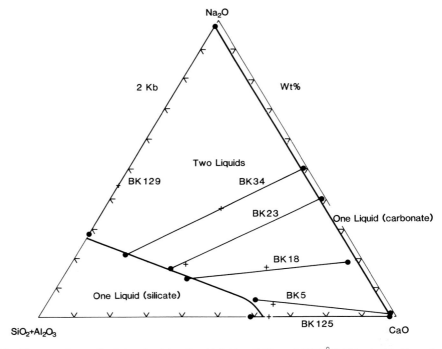

Figure 15.2 The experimentally determined two-liquid field at 2 Kb and 1250 °C (this study). Quenched silicate and carbonate liquid compositions, determined by electron microprobe analyses are shown as dots, conjugate liquid pairs are joined by tie lines. Bulk starting compositions are shown as crosses. Corresponding run numbers are indicated by each line. Data are plotted as wt%.

considered to be equilibrium values and not merely residual from the starting materials. The reasons for this are:

(a) 2- and 26-hour run results are essentially identical;
(b) the silicate glass in the run products is not the same as in the starting materials, nor is it a simple mixture of the starting materials;
(c) textural features of the charge are consistent with a pair of immiscible liquids quenched from run conditions (see Fig. 15.5); and
(d) in any one sample, the composition of variously sized, carbonate spheres within silicate glass is the same as larger masses of carbonate from the same run. Similarly, the silicate spheres surrounded by carbonate also have identical compositions to the massive silicate glass.

The versatile nature of the Freestone & Hamilton pseudoternary diagram has enabled us to plot data, on Figure 15.1, from all previous experimental studies relevant to carbonatites. Most other data are reasonably consistent with our 2 Kb data (Fig. 15.2). One explanation for the absence of immiscibility on the albite–calcite join of Watkinson & Wyllie is that they studied that part of the join which stopped short of the two-liquid field. Their nepheline–calcite join, however, cuts across our two-liquid field and their results should have indicated immiscibility.

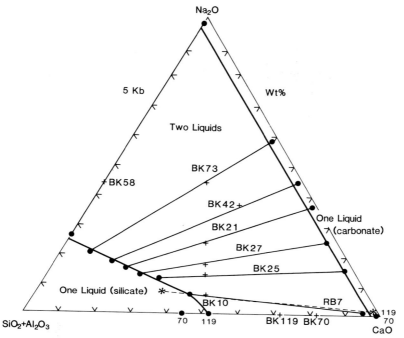

Figure 15.3 The experimentally determined two-liquid field at 5 Kb and 1250 °C (modified from Kjarsgaard & Hamilton 1988). All symbols are the same as those shown in Figure 15.2. RB7 is an albite–calcite (15 Kb and 1300 °C) two-liquid tie line; compositions of the silicate and carbonate liquids are marked by asterisks. Data are plotted as wt%.

The main differences between the two studies are that Watkinson & Wyllie carried out their experiments at temperatures 200–300 °C below ours, and their runs were water-saturated whereas ours were mostly dry. Water is probably responsible for the differences in the experimental results. Koster van Groos & Wyllie (1968), and Freestone & Hamilton (1980) both illustrated that the miscibility gap decreases in width with increasing H_2O content at constant PT conditions. Temperature differences are unlikely to destroy immiscibility; in fact, the two-liquid field should widen with falling temperature.

15.2.3 *Natural evidence in support of immiscibility*

Textural and chemical criteria should be used when examining for evidence of immiscibility between silicate and carbonate melts. Textural evidence includes ocelli (Philpotts 1982) and the diapir-like structures documented from the Benfontein Sill, South Africa (Dawson & Hawthorne 1973). Chemical compositions of coexisting carbonate and silicate fractions should give tie lines that are similar to those determined experimentally. Using these criteria, few examples of immiscibility in natural systems can be found in the literature. In fact, we have located only four:

(a) diabase dyke with dolomite ocelli, Sinai, Egypt (Bogoch & Magaritz 1983);

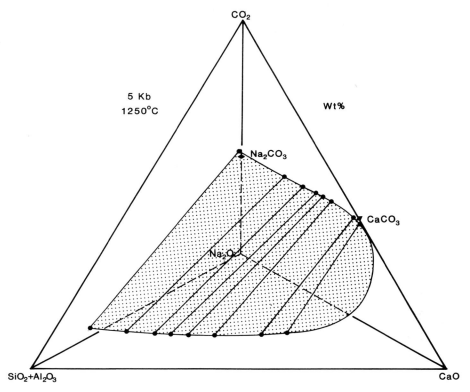

Figure 15.4 Quaternary $(SiO_2 + Al_2O_3)-(Na_2O)-(CaO)-(CO_2)$ diagram of the two-liquid field at 5 Kb and 1250 °C. Symbols are the same as Figure 15.2. A comparison of Figures 15.3 and 15.4 (both constructed from the same data) shows that the rotation of the tie lines in Figure 15.3 is an artefact of the projection from CO_2. Data are plotted as wt%.

(b) ultramafic glass with carbonate spherules, Spitzbergen, Norway (Amundsen 1987);
(c) kimberlite sill containing centimetre-sized carbonate diapirs, Benfontein, South Africa (Dawson & Hawthorne 1973); and
(d) nephelinite lava with carbonate globules (see Fig. 15.6), Shombole, Kenya (Peterson, in press).

Chemical compositions from these studies are shown in Figure 15.7, and it can be seen that the slopes of the silicate–carbonate tie lines are similar to those experimentally determined (see Figs 15.2 & 15.3). Data from the Benfontein kimberlite sill, point 3 on Figure 15.7, are in good agreement with the experimentally determined 2 Kb silicate limb of the solvus. Points 1, 2, and 4 also show good agreement with experimental results, but they lie outside the solvus. A possible explanation is that these samples represent liquid immiscibility at higher pressures and/or lower temperatures than those used during the experimental work.

This scarcity of evidence to support silicate–carbonate immiscibility might, of course, be due to the disparate physical properties of the silicate and carbonate

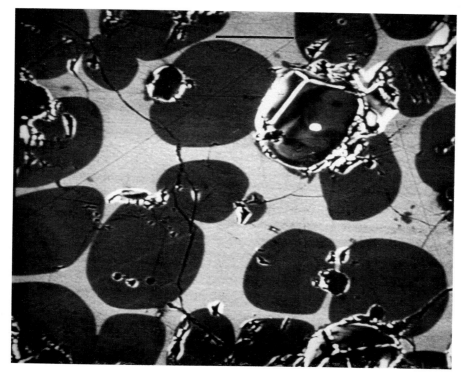

Figure 15.5 Back-scattered electron image. Run BK119, El_{25}–CC_{75} at 5 Kb and 1175 °C, shows coalescing calcium carbonate spheres (dark grey) in a calcium aluminosilicate glass (light grey). Note the tendency of spheres to flatten on the coalescing side. Several irregular-shaped gas vesicles (black) are also present. Scale bar is 100 μm long.

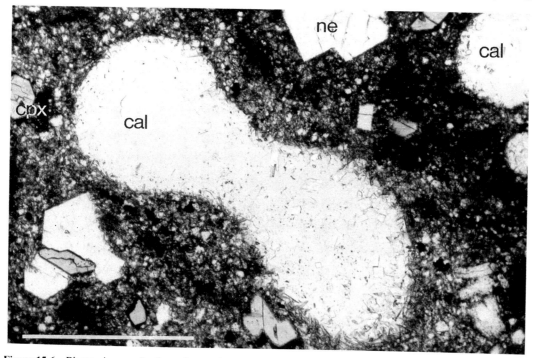

Figure 15.6 Photomicrograph of a carbonated nephelinite from Shombole, Kenya. Spherical and dumbell-shaped calcite (cal) makes up approximately 9% of the sample. The phenocryst assemblage is nepheline (ne), alkali clinopyroxene (cpx), and perovskite. Scale bar is 0.5 mm long. Plane-polarized light.

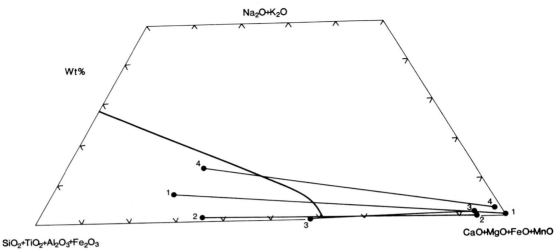

Figure 15.7 Plot of data from natural rocks compared to the experimentally determined silicate limb of the solvus at 2 Kb and 1250 °C. The four labelled tie lines are silicate–carbonate pairs from natural samples: (1) diabase dyke with dolomite ocelli, Sinai, Egypt (Bogoch & Magaritz 1983), (2) ultramafic glass with carbonate spherules, Spitzbergen, Norway (Amundsen 1987), (3) kimberlite sill with carbonate diapirs, Benfontein, South Africa (Dawson & Hawthorne 1973), (4) nephelinite with carbonate dumbells and spheres, Shombole, Kenya (Peterson, in press). Data are plotted as wt%.

melts. Mutual separation of these two liquids is probably much more efficient than between a pair of silicate–silicate liquids (e.g. the Fe–alkali pair seen in many basalts and the basic–alkali pairs described by Philpotts 1982). Efficient separation of carbonate liquids exsolved from silicate melts will produce two independently evolving systems that are unlikely to interact.

15.3 DISCUSSION

15.3.1 Opening statement

The new experimental results indicate that the separation of a carbonatite magma by immiscibility from a CO_2-rich silicate parent magma can take place over a much wider compositional range than was previously thought. Kjarsgaard & Hamilton (1988, this chapter) using a system containing SiO_2, Al_2O_3, CaO, Na_2O, and CO_2 demonstrated that there is a continuous miscibility gap between a silicate melt and a carbonate melt at 2 and 5 Kb. Freestone & Hamilton (1980) showed that K can substitute for Na with little effect on the miscibility gap. Recent, unpublished work of Kjarsgaard & Brooker demonstrates that immiscibility persists when MgO is added to the five-component system, and that the miscibility gap persists to 15 Kb.

We have demonstrated experimentally that carbonate liquids, very rich in $CaCO_3$, are immiscible with silicate liquids. Is there any supportive evidence in natural systems that $CaCO_3$ liquids can exist? A review of the literature shows several examples. Included among these are the fine-grained vein and dyke rock, alvikite, very common in carbonatite complexes (e.g. Le Bas 1977), the comb-layered carbonatite dykes of the Kaiserstuhl (Katz & Keller 1981), the Pele's tear-drop lapilli tuffs of the Kaiserstuhl (Keller 1981, this volume), the lavas of the Fort Portal area, Uganda (von Knorring & DuBois 1961), and several examples that appear on Figure 15.7, to name but a few. While an origin for $CaCO_3$-rich magmas by alkali loss has several supportive and well-documented examples (e.g. the description of Deans & Roberts (1984) of a largely calcite lava which contains pseudomorphs after nyerereite), our prejudice is that in the absence of evidence supporting metasomatic activity (e.g. nyereite pseudomorphs, or close proximity to a fenitization zone) the more likely origin for the rock is one directly linked to a $CaCO_3$-rich magma.

15.3.2 Evolution of immiscible magmas

The observation that silicate–carbonate immiscibility is not restricted to alkali-rich melts but is possible over a wide range of chemical compositions raises some interesting consequences for carbonatite genesis. Most important is the fact that immiscibility will occur when the carbonate concentration in the host silicate magma reaches some critical level (a function of PTX). If the parent silicate magma has a high carbonate content, then only a small degree of cooling will be required before the onset of immiscibility. The composition of the carbonate melt will be rich in $CaCO_3$. As an example, if the parent magma is close to a carbonated melilitite in composition, then the carbonatite magma exsolved will be virtually alkali-free, as shown by tie line BK10 in Figure 15.8. However, if the parent magma carbonate content is quite low, then the onset of immiscibility would only be reached after extensive cooling had occurred, with crystal fractionation building up the carbonate concentration. Under such conditions, both the host silicate melt and the exsolved carbonate melt will be enriched in alkalies relative to the parental magma. One example is shown by the phonolite–natrocarbonatite tie line BK73 in Figure 15.8.

Volatiles and superheat are significant to any discussion of carbonatite magma evolution. The temperature of the exsolving carbonate melt is fixed to that of the host silicate melt, and both are controlled by the confining pressure and the composition of the system; the amount of H_2O will be particularly relevant. In the early stages we suggest that the dominant vapour phase is CO_2 and the temperature of the system will be appropriate for a dry primitive magma, i.e. rather high. The silicate melt is unlikely to have any superheat but the exsolving carbonatite melt might have. The extent of the superheating will largely depend on how the total water content is distributed between the two liquids and vapour phases of the system. Preliminary data from mixed $CO_2 : H_2O$ immiscibility runs indicate that the vapour phase is predominantly CO_2.

Evidence from rock compositions in carbonatite complexes suggests that ultimately the silicate melt composition approaches that of a phonolite. The conjugate carbonate melt composition, at this stage, will be a natrocarbonate (Freestone & Hamilton 1980). The temperature at this phonolitic stage will depend on the PX conditions. In the absence of the precipitation of hydrous phases, the water content of the fluid phases will probably increase with fractionation, causing the $CO_2 : H_2O$

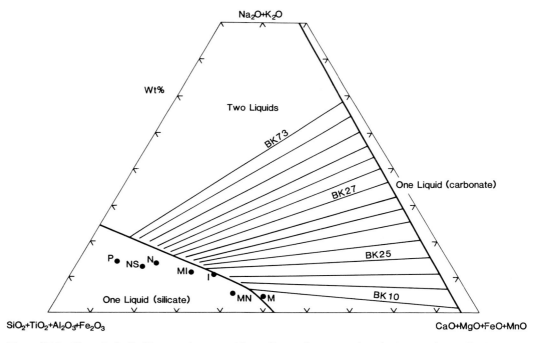

Figure 15.8 Plot of alkali silicate rock compositions. Shown for comparison is the experimentally determined silicate limb of the solvus at 5 Kb and 1250 °C. Compositions are taken from Le Bas (1977); samples are all from carbonatite complexes. M, melilitite; MN, melanephelinite; I, ijolite; MI, micro-ijolite; N, nephelinite; NS, nepheline syenite; P, phonolite. BK73, 27, 25, and 10 are silicate–carbonate liquid tie lines determined at 5 Kb. Also shown is an interpolated family of two-liquid tie lines. Data are plotted in wt%.

ratio in the vapour phase to decrease. The temperature of the system will fall towards that of a water-saturated phonolite (e.g. 750 °C at 1 Kb, Hamilton & MacKenzie 1965); how closely the temperature approaches this minimum temperature will depend on the late-stage $CO_2:H_2O$ ratio of the vapour phase. Erupting natrocarbonate lava has been seen at Oldoinyo Lengai several times (e.g. Dawson 1962, this volume) and has been described as not 'incandescent' which probably means $T < 750$ °C. Thus it seems possible for a natrocarbonate melt produced by immiscibility to arrive at the low temperature required by the observed non-incandescent nature of the Oldoinyo Lengai carbonate lavas. Certainly an origin by immiscibility cannot be ruled out because of these low observed temperatures.

The compositions of silicate rocks found in carbonatite complexes indicate the changing chemical composition of the silicate magma during both liquid and crystal fractionation. Plotted on Figure 15.8 are a number of chemical compositions of alkali silicate rocks from carbonatite complexes. The evolutionary trend of the alkali silicate rocks lies approximately parallel to the silicate limb of the experimentally determined two-liquid field. While these alkali silicate rocks are thought to be fairly representative of magmatic liquids, their bulk chemistry will have definitely been influenced by cumulate minerals. Chemical compositions of the basic alkali silicate rocks (melanephelinites, ijolites, and melilitites) plot fairly close to the experimental solvus. The more evolved alkali silicates (nephelinites, nepheline syenites, and phonolites) lie slightly off the experimental solvus, and this can be attributed to their lower liquidus temperatures compared to those used to construct the solvus.

It should be noted that the evolutionary trend of alkali silicate rock types found in carbonatite complexes cannot be produced by crystal fractionation. Figure 15.9 shows that this trend could be attained if exsolved carbonate liquid plus minerals were fractionated together out of the melt. We therefore feel that alkali silicate magma composition and evolution are mainly controlled by the solvus at the existing *PTX* conditions. This implies continuous liquid immiscibility (or nearly so) and subsequent carbonate-liquid fractionation is possible over a large temperature

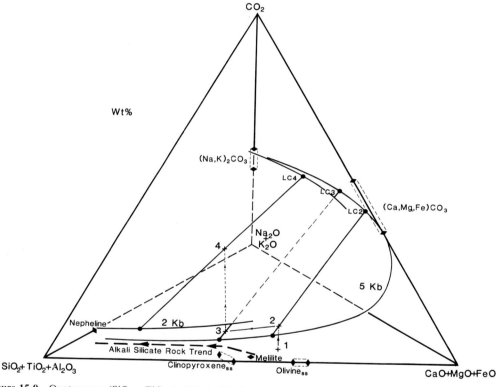

Figure 15.9 Quaternary $(SiO_2 + TiO_2 + Al_2O_3)$–$(Na_2O + K_2O)$–$(CaO + MgO + FeO)$–$(CO_2)$ diagram showing residual liquid trends in a hypothetical alkali silicate–carbonatite system. Solid curved lines labelled 2 Kb and 5 Kb are theoretical isobaric, polythermal solvi based on the experimental work of Freestone & Hamilton (1980) and Kjarsgaard & Hamilton (1988, this chapter and unpublished data). Only part of the 2 Kb solvus has been shown for clarity. The apparent evolutionary trend of alkali silicate rock types found in carbonatite complexes is shown as a heavy dashed line. Point 1 (denoted by a cross) is a carbonated basic alkali silicate parent magma at 5 Kb in the one-liquid field. With cooling and fractionation of olivine ± clinopyroxene ± melilite the liquid becomes enriched in carbonate and moves into the two-liquid field (Point 2), where carbonatite melt of composition LC2 is exsolved. Subsequent cooling and fractionation of exsolved carbonatite melt (LC2–LC3) plus ferromagnesian silicates ± nepheline keep the bulk composition of the system inside the solvus (i.e. thin solid line between Points 2 and 3), with continuous liquid immiscibility occurring. If the *PTX* conditions of the magma change (in this example depressurization from 5 to 2 Kb), the bulk composition is no longer inside the solvus but is in the one-liquid field. However, with further cooling and mineral fractionation, carbonate becomes enriched in the liquid progressively (dashed line between Points 3 and 4) until the solvus is met. At point 4 carbonatite melt of composition LC4 is exsolved. Oxides are plotted in wt%.

interval, with the alkali silicate melt composition maintaining close proximity to the solvus. If, on cooling, the carbonated silicate magma remains within the immiscibility dome, then a carbonate melt can be continuously exsolved. However, if the magma's *PTX* conditions change (e.g. due to depressurization, contamination, or magma mixing), this will move the melt into the one-liquid field. Subsequent fractionation of nepheline and ferromagnesian silicates will increase the carbonate content of the liquid until the solvus is met, at which point a carbonate melt would again be exsolved. In this manner, a series of carbonate melts could be produced in a discontinuous fashion. If a carbonate liquid exsolves, in either a continuous or discontinuous way, these liquids will become richer in alkalis and poorer in CaO and MgO as the temperature falls.

15.3.3 *Crystal fractionation within the carbonatite melt*

An exsolved carbonatite liquid rich in $CaCO_3$ can be considered as an independently evolving system. The initial temperature of the carbonatite magma is not controlled by the phase relations in the carbonate system, but is inherited from the silicate magma. Initially, the carbonatite magma just after separation from the silicate melt probably has a small amount of H_2O, and a small amount of superheat. This magma will crystallize over a wide temperature interval, thus allowing ample opportunity for fractionation. As the H_2O content of the magma increases during differentiation, the effect will be to depress the solidus to lower temperatures.

Fractionation of crystals from this melt should be very efficient because of the low viscosity of carbonatite melts (see Treiman, this volume). Crystallization will be dominated by calcite (and/or dolomite). Kjarsgaard & Hamilton (1988) determined that > 50%crystals could be fractionated before an alkali-bearing carbonate crystal phase appears on the liquidus. Other crystalline phases such as apatite and magnetite, with a range of exotic minerals such as bastnaesite, baddeleyite, and pyrochlore might also precipitate from the magma at this stage and join the calcite (and/or dolomite) as crystal cumulates. The variety and abundance of the rarer minerals will depend on the trace element content of the parent silicate magma, and also the *PTX* conditions during the immiscibility event (Hamilton *et al.*, this volume).

The origins of the rarer ferrocarbonatites pose some problems and we can only speculate on these. We propose that these rocks reflect residual liquids that remain after crystal fractionation has taken place at low oxygen fugacities. By this process Fe should concentrate in the melt with increasing fractionation.

The residual carbonate melt formed during any fractionation process will be enriched in alkalis, F, Cl, incompatible elements, and H_2O. One possible residual liquid may well be an alkali-rich carbonatite, possibly similar in composition to the Oldoinyo Lengai natrocarbonatite. Thus a natrocarbonatite magma may be derived in two ways:

(a) as a late-stage immiscible liquid from a host phonolite; or

(b) as a late-stage liquid produced by fractional crystallization of an alkali-poor carbonatite melt.

Although the exact nature and origin of fluids responsible for fenitization are unknown, evolved carbonatite liquids with high concentrations of halogens, incompatible elements, and to a lesser degree H_2O are potential candidates.

15.4 SUMMARY

The acceptance of liquid immiscibility as a means of producing the carbonatite suite of rocks has increased slowly over the past ten years. The extent to which immiscibility can be applied has been severely restricted, however, because it was thought to be limited to very alkali-rich melts (e.g. the phonolite and natrocarbonatite of Oldoinyo Lengai). The results of Kjarsgaard & Hamilton (1988) and new results given here show that there is a wide miscibility gap which persists even in alkali-free systems (e.g. anorthite–calcite) in the pressure range of 1–15 Kb. We now propose that immiscibility is a major process in the development of both the carbonatites and their attendant silicate rocks. Immiscibility can account for the origin of the many types of carbonatites more easily than any other process. There are still many outstanding problems connected with the generation of carbonatites. These include the effect of H_2O, P, F, and Mg:Fe on the miscibility gap, and the effect of f_{O_2} and $H_2O:CO_2$:halogen ratios on fractional crystallization within carbonatite magmas.

ACKNOWLEDGEMENTS

The authors thank W. S. MacKenzie and two reviewers for making improvements to the manuscript, and the editor, Keith Bell, for his encouragement and help with the revisions. Tony Peterson of the Geological Survey of Canada kindly provided chemical data and a sample from the Shombole carbonated nephelinite. The technical and secretarial staff of the Department of Geology, University of Manchester, are also thanked for their help. Financial assistance from the NERC (grant GR3/6477) to the experimental petrology laboratory, and financial assistance from the University of Manchester and the CVCP to B.A.K. are gratefully acknowledged.

REFERENCES

Amundsen, H. E. F. 1987. Evidence for liquid immiscibility in the upper mantle. *Nature* **327**, 692–5.
Bogoch, R. & M. Magaritz 1983. Immiscible silicate–carbonate liquids as evidenced from ocellar diabase dykes, Southeast Sinai. *Contributions to Mineralogy and Petrology* **83**, 227–30.
Dawson, J. B. 1962. The geology of Oldoinyo Lengai. *Bulletin Volcanologique* **24**, 349–87.
Dawson, J. B. & J. B. Hawthorne 1973. Magmatic sedimentation and carbonatitic differentiation in

kimberlite sills at Benfontein, South Africa. *Journal of the Geological Society of London* **129**, 61–85.

Deans, T. & B. Roberts 1984. Carbonate tuffs and lava clasts of the Tinderet foothills, western Kenya: a study of calcified natrocarbonatites. *Journal of the Geological Society of London* **141**, 563–80.

Freestone, I. C. & D. L. Hamilton 1980. The role of liquid immiscibility in the genesis of carbonatites – an experimental study. *Contributions to Mineralogy and Petrology* **73**, 105–17.

Hamilton, D. L. & W. S. MacKenzie 1965. Phase-equilibrium studies in the system nepheline–kalsilite–silica–water. *Mineralogical Magazine* **34**, 214–31.

Katz, K. & J. Keller 1981. Comb-layering in carbonatite dykes. *Nature* **294**, 350–2.

Keller, J. 1981. Carbonatitic volcanism in the Kaiserstuhl Alkaline Complex: Evidence for highly fluid carbonatitic melts at the earth's surface. *Journal of Volcanology and Geothermal Research* **9**, 423–31.

Kjarsgaard, B. A. & D. L. Hamilton 1988. Liquid immiscibility and the origin of alkali-poor carbonatites. *Mineralogical Magazine* **52**, 43–55.

Knorring, O. von & C. G. B. DuBois 1961. Carbonatitic lava from Fort Portal area in western Uganda. *Nature* **192**, 1064–5.

Koster van Groos, A. F. 1975. The effect of high CO_2 pressures on alkaline rocks and its bearing on the formation of alkalic ultrabasic rocks and the associated carbonatites. *American Journal of Science* **275**, 163–85.

Koster van Groos, A. F. & P. J. Wyllie 1966. Liquid immiscibility in the system Na_2O–Al_2O_3–SiO_2–CO_2 at pressures to 1 kilobar. *American Journal of Science* **264**, 234–55.

Koster van Groos, A. F. & P. J. Wyllie 1968. Liquid immiscibility in the join $NaAlSi_3O_8$–Na_2CO_3–H_2O and its bearing on the genesis of carbonatites. *American Journal of Science* **266**, 932–67.

Koster van Groos, A. F. & P. J. Wyllie 1973. Liquid immiscibility in the join $NaAlSi_3O_8$–$CaAl_2Si_2O_8$–Na_2CO_3–H_2O. *American Journal of Science* **273**, 465–87.

Le Bas, M. J. 1977. *Carbonatite–nephelinite volcanism*. Chichester: Wiley.

Peterson, T. D. Peralkaline nephelinites I. Comparative petrology of Shombole and Oldoinyo Lengai, East Africa. *Contributions to Mineralogy and Petrology* (in press).

Philpotts, A. R. 1982. Compositions of immiscible liquids in volcanic rocks. *Contributions to Mineralogy and Petrology* **80**, 201–18.

Watkinson, D. H. & P. J. Wyllie 1969. Phase equilibria studies bearing on the limestone-assimilation hypothesis. *Geological Society of America Bulletin* **80**, 1565–76.

Watkinson, D. H. & P. J. Wyllie 1971. Experimental study of the composition join $NaAlSiO_4$–$CaCO_3$–H_2O and the genesis of the alkalic rock–carbonatite complexes. *Journal of Petrology* **12**, 357–78.

16
THE BEHAVIOUR OF TRACE ELEMENTS IN THE EVOLUTION OF CARBONATITES

D. L. HAMILTON, P. BEDSON & J. ESSON

ABSTRACT

Distribution coefficients (K_D) of the trace elements Ba, Ce, Cr, Cu, Eu, Gd, Hf, La, Lu, Sm, Ta, Yb, and Zr between phonolite/sodium-rich carbonate (phonolite system) and nephelinite/calcium-rich carbonate (nephelinite system) liquid pairs have been determined in the pressure range 1–6 Kb and the temperature range 1050–1250 °C. The enrichment of most of these elements into the carbonate liquid is favoured by high pressure, low temperature, and increased polymerization of the silicate conjugate melt. At pressures and temperatures experienced by a carbonated alkali-rich magma in the upper mantle, most of the above elements would be concentrated into the carbonate melt of any carbonate/silicate immiscible pair of liquids by a factor of about ten or greater.

16.1 INTRODUCTION

Carbonatites are common in complexes associated with alkali-rich silicate rocks. Many, but not all, of these complexes have unusually high concentrations of incompatible elements which are normally found within the carbonatite itself or in adjacent fenitized zones. Some of these elements are exploited commercially, and carbonatites are the chief source of REEs (rare earth elements) and are also mined for P, Cu, Zr, and Nb. This present study was undertaken to provide more data on the behaviour of selected incompatible elements in an igneous environment and K_{DS} were determined for selected trace elements between immiscible silicate and carbonatite melts over a range of P and T (K_D in this chapter is the concentration of an element in a silicate melt divided by its concentration in a conjugate immiscible carbonate melt). The trace elements studied were chosen bearing in mind their importance in nature and the ability to determine their amounts in the run products. The inability to analyse for Nb using neutron activation analysis (NAA) explains the absence of this very important element from our list. We thought it unnecessary to include all the REEs but chose seven, spread throughout the lanthanide group, thus allowing us at least to infer REE trends without making the experimental and analytical techniques too complicated.

No comparison of the merits of fractional crystallization and liquid immiscibility in the genesis of carbonatite complexes is included: we merely present one side of the argument. The possibility of extensive fractional crystallization within a calcium carbonate-rich carbonatite magma (see Kjarsgaard & Hamilton, this volume) and its effect on incompatible element concentrations in carbonatite rocks are explored.

16.2 METHODS

16.2.1 Starting materials

Four compositions were used in this study (Table 16.1):

(a) natural phonolite from Oldoinyo Lengai;
(b) natural nephelinite from Oldoinyo Lengai;
(c) synthetic carbonatite similar in composition to the sodium-rich carbonatite lavas of Oldoinyo Lengai (Dawson 1962);
(d) synthetic carbonatite richer in calcium than (c). This corresponds to the carbonate liquid found by Freestone & Hamilton (1980) to be in equilibrium with a liquid of similar composition to the natural nephelinite (b).

Professor J. B. Dawson kindly provided rock powders of the Oldoinyo Lengai

Table 16.1 Composition of starting materials for trace-element studies.

	Nephelinite BD119	Phonolite BD50	Synthetic sodic carbonatite SC	Synthetic calcic carbonatite CC
Wt%				
SiO_2	41.93	52.88	n.d.	n.d.
TiO_2	0.92	0.93	n.d.	n.d.
Al_2O_3	15.55	19.89	n.d.	n.d.
Fe_2O_3	6.41	4.07	n.d.	n.d.
FeO	2.28	1.58	n.d.	n.d.
MnO	0.24	0.03	n.d.	n.d.
MgO	1.28	0.45	n.d.	1.00
CaO	10.89	2.67	16.69	33.32
Na_2O	10.26	10.63	33.38	16.66
K_2O	4.95	4.91	8.34	5.55
P_2O_5	0.54	0.11	1.00	1.00
H_2O^+	0.80	0.80	n.d.	n.d.
H_2O^-	0.85	0.02	2.00	2.00
CO_2	2.39	0.15	33.59	35.47
F	n.a.	n.a.	n.d.	n.d.
Cl	n.a.	n.a.	3.00	3.00
Others	0.39	0.95	n.d.	n.d.
Total	99.68	100.07	98.00	98.00

BD119, Nephelinite, Oldoinyo Lengai, analyst D. G. Powell; BD50, phonolite, Oldoinyo Lengai, analyst D. G. Powell; SC, synthetic sodic carbonatite, mean of Oldoinyo Lengai carbonatite lavas; CC, synthetic calcium-rich alkali carbonatite. n.a., Not analysed; n.d., not detected.

Table 16.2 Trace-element content of charges for immiscibility studies.

Element	Source	Approximate content in experimental charges (p.p.m.)
Ba	$BaCO_3$	1000
Zr	ZrO_2	1100
Ce	CeO_2	100
Gd	Gd_2O_3	90
Hf	HfO_2	25
Cr	Cr_2O_3	15
Cu	CuO	500
Yb	Yb_2O_3	15
Lu	Lu_2O_3	5
Ta	Ta_2O_5	5
La	Rock powder	50
Sm	Rock powder	15
Eu	Rock powder	2
Mn	Rock powder	200–1000

material, while the synthetic carbonatites (c) and (d) were prepared from high purity Na_2CO_3, K_2CO_3, $CaCO_3$, NaF, NaCl, $Ca_3(PO_4)_2$, and MgO. The synthetic carbonatites were then mixed with a blend of trace elements, containing Ba, Ce, Cr, Cu, Gd, Hf, Lu, Ni, Ta, Yb, and Zr, so that trace-element concentrations in the final starting materials (see Table 16.2) were suitable for NAA. The amounts of Eu, La, and Sm contained in the natural phonolite and nephelinite were sufficiently high for NAA, so no more of these elements were added. The trace-element doped, synthetic carbonatites were mixed with equal weights of silicate rock powders, the sodium-rich carbonatite with the phonolite and the calcium-rich carbonatite with the nephelinite. These starting materials were stored at 110 °C; thus, the only water present in experimental charges was structurally bound in the lavas.

16.2.2 Experimental and analytical methods

Approximately 0.5 g batches of powder were sealed in 5.0 mm-diameter silver–palladium alloy ($Ag_{60}Pd_{40}$) tubes. Experiments were performed in internally heated, argon, medium-pressure vessels, with the long axis of the furnace horizontal. At the end of an experiment the furnace was turned off and the vessel tilted so that its long axis was vertical. The tilting caused cool gas from the end of the furnace to displace the hot gas surrounding the sample, resulting in a rapid quench of the sample. The temperature fell by 600 °C in the first 60 s.

Pressures were measured using Bourdon tube pressure gauges and manganin cells, and are believed accurate to ± 0.25 Kb. Temperatures were measured using Pt–$Pt_{87}Rh_{13}$ thermocouples and are believed accurate to ± 15 °C. Run times were from 16 to 24 h.

Experiments were performed on phonolite/sodium-rich carbonatite mixtures at 1150 °C at pressures of 0.8, 0.86, 1, 2, 3, 4, and 6 Kb, and also at 1050 °C and

Table 16.3 Major element analyses, wt%, of run products determined by electron microprobe.

Mixture	Ph		Ph		Ph		Ph		Ph		Ph		Ph	
$T\,^\circ C$	1150		1150		1150		1150		1150		1150		1150	
P Kb	0.8		0.86		1		2		3		4		6	
Experiment	PB95		PB97		PB25		PB24		PB77		PB29		PB76	
Phase	Sil	Car	Sil	Car	Sil	Car	Sil	Car	Sil	Car	Sil	Car	Sil	Car
SiO_2	40.26	1.68	42.14	0.63	45.63	1.08	47.69	0.40	46.44	n.a.	48.51	0.78	49.55	1.53
TiO_2	0.71	0.20	0.73	0.18	0.70	0.16	0.76	0.19	0.65	n.a.	0.73	0.23	0.70	0.31
Al_2O_3	14.34	0.12	15.18	0.09	16.37	0.06	17.09	0.06	17.03	n.a.	17.37	0.03	18.60	0.21
MgO	0.03	0.09	0.13	0.24	0.21	n.d.	0.10	0.01	0.10	n.a.	0.03	0.01	0.10	0.01
FeO	2.17	0.37	2.03	0.43	3.32	0.29	1.89	0.18	1.29	n.a.	1.73	0.03	1.56	0.77
CaO	10.07	20.38	9.07	24.18	8.73	24.10	6.83	23.13	4.06	n.a.	4.14	23.15	3.36	22.81
Na_2O	20.65	20.25	19.62	13.42	12.53	26.13	13.06	28.37	16.59	n.a.	16.26	29.02	15.25	22.73
K_2O	5.14	7.11	5.25	4.19	5.88	7.85	6.21	7.23	6.10	n.a.	6.06	6.16	6.02	3.47
P_2O_5	0.28	0.57	0.36	1.09	0.34	0.75	0.29	0.78	0.32	n.a.	0.28	0.76	0.07	0.47
S	0.05	0.88	0.02	0.95	0.01	0.19	0.02	0.06	0.02	n.a.	0.02	0.07	0.05	0.21
Total	93.70	51.65	94.53	45.40	93.72	60.61	93.94	60.41	92.60	–	95.13	60.24	95.26	52.52
n	20	12	39	9	20	12	20	15	14	–	20	9	20	10

Mixture	Ne		Ne		Ne		Ne		Ne	
$T\,^\circ C$	1150		1150		1150		1150		1150	
P Kb	1		2		3		3.6		6	
Experiment	PB59		PB66		PB105		PB171		PB81	
Phase	Sil	Car	Sil	Car	Sil	Car	Sil	Car	Sil	Car
SiO_2	31.92	1.78	38.88	1.85	43.93	2.98	45.49	0.37	42.94	6.54
TiO_2	0.70	0.07	0.81	0.16	0.68	0.33	0.78	0.04	0.77	0.35
Al_2O_3	10.63	0.08	12.59	0.12	16.07	0.20	15.91	0.73	14.63	0.76
MgO	1.14	0.01	0.95	0.01	0.08	0.46	0.22	0.06	0.73	0.94

FeO	4.04	0.34	4.37	0.54	1.52	0.68	2.03	0.32	2.71	1.49
CaO	24.65	25.50	18.46	29.81	7.43	33.37	8.61	29.90	12.36	33.06
Na$_2$O	12.54	21.71	12.00	19.85	18.50	17.25	14.80	22.92	11.83	15.12
K$_2$O	4.10	8.04	4.38	6.51	5.34	4.83	5.98	5.68	5.10	4.93
P$_2$O$_5$	0.59	1.56	0.49	1.56	0.79	1.30	0.32	1.14	0.35	1.14
S	0.04	0.17	0.01	0.02	0.04	0.01	0.02	0.03	0.01	0.12
Total	90.35	59.26	92.94	69.43	94.38	61.41	94.16	61.19	91.43	64.45
n	20	10	20	13	20	12	15	10	19	13

Mixture	Ph		Ph		Ne		Ne	
T °C	1050		1250		1050		1250	
P Kb	3		3		3		3	
Experiment	PB93		PB90		PB69		PB118	
Phase	Sil	Car	Sil	Car	Sil	Car	Sil	Car
SiO$_2$	50.54	0.53	38.10	1.11	44.31	6.75	42.56	2.20
TiO$_2$	0.76	0.12	0.81	0.10	0.85	0.32	0.78	0.17
Al$_2$O$_3$	18.07	0.03	12.63	0.07	14.77	1.20	14.78	0.13
MgO	0.06	n.d.	0.98	0.06	0.88	0.54	0.40	n.d.
FeO	2.66	0.56	2.72	0.27	3.88	1.38	2.03	0.27
CaO	3.05	25.64	19.26	29.80	13.40	27.39	11.76	27.84
Na$_2$O	15.21	27.01	11.31	21.09	9.76	18.31	14.11	22.32
K$_2$O	5.72	4.70	4.45	5.96	5.25	5.46	5.32	6.85
P$_2$O$_5$	0.20	0.71	0.30	1.20	0.39	1.44	0.12	1.07
S	0.03	0.07	0.04	0.03	0.05	0.11	0.04	0.04
Total	96.30	59.37	90.60	59.62	93.54	62.90	91.90	60.89
n	35	20	20	10	21	13	20	11

Ph, phonolitic/sodium carbonate-rich carbonate system; Ne, nephelinitic/calcium carbonate-rich carbonate system; Sil, silicate phase; Car, carbonate phase. n, Number of points analysed; n.a., not analysed; n.d., not detected.

Table 16.4 Measured K_D, silicate/carbonate, values for trace elements.

Mixture	Ph	Ph	Ph	Ph	Ph	Ph	Ph	Ph	Ph
$T\,°C$	1150	1150	1150	1150	1150	1150	1150	1050	1250
P Kb	0.8	0.86	1	2	3	4	6	3	3
Experiment	PB95	PB86	PB85	PB170	PB77	PB29	PB76	PB93	PB90
Cr	5.57	3.19	1.77	4.64	1.00	1.55	0.60	1.15	2.68
Mn	2.19	2.17	1.97	1.68	0.96	0.84	2.20	0.79	2.89
Cu	1.33	0.83	0.91	0.52	0.85	0.35	0.63	0.72	0.16
Zr	7.14	1.41	0.85	0.69	1.61	2.15	0.81	1.12	1.66
Ba	0.63	0.61	0.56	0.37	0.32	0.16	0.17	0.17	0.59
La	1.84	1.76	1.64	1.04	0.54	0.33	0.31	0.26	1.68
Ce	2.23	2.05	1.78	1.03	0.62	0.39	0.25	0.40	1.45
Sm	2.50	2.47	2.28	1.63	0.75	0.51	0.45	0.42	2.31
Eu	2.46	2.43	2.18	1.55	0.84	0.51	0.65	0.46	2.25
Gd	2.59	2.85	2.30	1.65	0.93	0.60	1.60	0.48	1.96
Yb	4.84	4.22	4.08	2.59	1.54	1.14	0.73	1.25	2.38
Lu	4.22	5.52	4.67	2.97	1.70	1.37	0.73	1.55	2.99
Hf	4770	59.6	408	366	50.5	278	34.7	192	93.4
Ta	1.69	3.26	3.67	4.56	2.85	3.99	2.78	5.56	1.89
Mixture	Ne	Ne	Ne	Ne	Ne	Ne	Ne	Ne	
$T\,°C$	1150	1150	1150	1150	1150	1150	1050	1250	
P Kb	1	2	3	3.6	4	6	3	3	
Experiment	PB59	PB66	PB105	PB171	PB35	PB81	PB67	PB118	SD
Cr	0.37	3.76	9.94	0.98	5.92	96.74	1.07	1.37	0.1–0.5
Mn	3.22	n.a.	0.67	1.10	2.15	1.04	1.33	1.43	0.01–0.02
Cu	1.31	0.93	0.84	1.01	1.33	1.97	0.85	1.13	0.05–0.1
Zr	2.83	1.92	0.71	4.48	3.71	1.92	1.97	2.20	0.01–0.1
Ba	4.48	0.80	5.47	0.35	0.81	1.69	0.50	0.44	0.01–0.02
La	3.18	1.68	0.70	0.72	2.02	0.62	0.90	1.21	0.01–0.02
Ce	2.84	1.87	0.81	1.24	1.16	0.77	0.97	1.33	0.01
Sm	4.56	2.24	0.93	1.02	2.62	0.77	1.25	n.a.	0.01–0.05
Eu	3.78	2.27	0.89	0.99	2.23	0.73	1.29	1.53	0.01–0.05
Gd	3.73	2.32	1.58	1.21	3.12	0.81	1.39	1.71	0.05–0.2
Yb	4.48	3.56	n.a.	1.84	3.43	1.35	2.27	2.35	0.03–0.3
Lu	4.80	3.19	0.88	1.93	3.56	0.76	2.40	2.40	0.05–0.4
Hf	116	115	135	300	11.2	396	64.1	5370	5–200
Ta	2.61	4.32	2.64	3.33	2.13	2.00	3.06	2.44	0.01–0.07

Analyses by neutron activation. Abbreviations as in Table 16.3. SD, typical values of standard deviation based on counting statistics. n.a., Not analysed.

1250 °C at 3 Kb; experiments were also performed on nephelinite–calcium-rich carbonatite mixtures at 1150 °C at 1, 2, 3, 3.6, and 6 Kb and at 1050 °C and 1250 °C at 3 Kb pressure.

Run products consisted of two phases, a silicate glass representing quenched silicate liquid and a fine-grained mixture of carbonate crystals which represent quenched carbonate liquid. These two phases were separated by hand-picking beneath a binocular microscope; the fractions were ground under acetone and dried for 24 h at 110 °C before irradiation.

The hand-picking operation was done with diligence and we are aware of the fact that small spheres of one phase sometimes occur in the other phase well away from the meniscus. The possibility that a few of the very small (say < 10 µm) spheres escaped our attention exists, and the presence of this contamination will tend to make the K_D numbers less extreme. Without further experimental work it is not possible for us to quantify the diminution in K_D values caused by this small amount of contamination; we feel the effect should be minor on the majority of the K_{DS} but could be significant on the more extreme ones, i.e. Hf.

Run products and standard solutions were irradiated together for 7.5 h in a thermal neutron flux of 10^{12} n cm^{-2} s^{-1} at Manchester University's Research Reactor, Risley. After a 16-hour decay period, the samples were decomposed in a mixture of HF and HClO$_4$ and, after evaporation to dryness, dissolved in 6 M HCl. These solutions were passed though hydrated antimony peroxide (HAP) columns (Girardi & Sabbioni 1968) to remove Na, because the relatively high Na contents of untreated samples resulted in the γ-ray spectra being dominated by ^{24}Na activity, which made it impossible to measure the activities of the isotopes of interest. Gamma-ray spectra were recorded, using a Ge–Li detector, for sample and standard solutions after 1, 3, 14, and 28 days. Because Ta is also retained by HAP, fresh HAP columns were used for each sample and standard and the HAP portions were retained to record the Ta activity after 28 days when the ^{24}Na activity had decayed.

Small portions of the silicate and carbonate phases were also analysed for major elements using a Cameca Camebax electron probe microanalyser fitted with a Link Systems Model 860 energy dispersive spectrometer/computerized analysis system. In all but two samples (PB97 and PB24), the pair of quenched melts analysed for major elements were the same as the pair analysed for minor elements; in the two exceptions, PB97 was substituted by PB86 and PB24 by PB170 for the NAA but the *PT* conditions of the substitute pair were the same as the originals. An electron beam with a 15 kV accelerating potential was used with an incident beam current of approximately 3 nA and a take-off angle of 40°. Various beam diameters were used for the analysis of different materials. Beams of diameter 1–30 μm were used to analyse silicates, and up to 50 μm diameter beams were used for the carbonates. Counting times were commonly 100 s while shorter periods (typically 60 s) were used for carbonate analyses. Analyses were corrected for the sodium volatilization by correcting back to zero live-time. Results are shown in Table 16.3 for the major elements and Table 16.4 for the trace elements.

16.3 EFFECT OF COMPOSITION ON K_D

The effect of melt composition on K_{DS} may be illustrated by comparing the K_D values of trace elements in the phonolite system to those of the nephelinite system at the same *P* and *T*. To show this effect graphically we need a simple parameter which is able to represent the major element chemistry of the system and which is also informative about the melt structure. The ratio of the number of non-bridging oxygens (NBO) to the number of tetrahedral sites (T) per formula unit seems to be suitable on both counts (Mysen 1986). This ratio can be calculated for a melt from its composition and is indicative of the degree of polymerization of the melt, e.g. the ratio is 0 for a feldspar or SiO$_2$ melt (highly polymerized) and 4 for an olivine melt

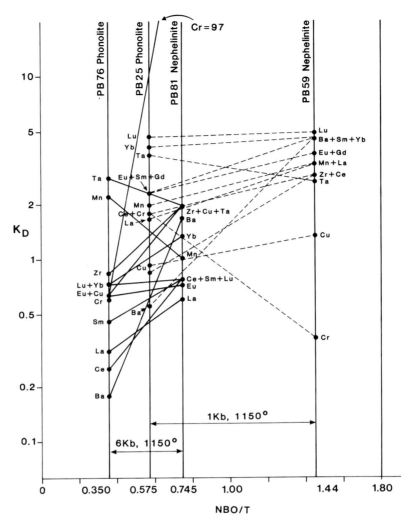

Figure 16.1 The diagram shows the K_Ds for Ba, Cu, Cr, Mn, Ta, Zr, La, Ce, Sm, Eu, Gd, Yb, and Lu between pairs of silicate/carbonate conjugate liquids (PB76, 25, 59, and 81). The compositions of the four silicate melts are shown by appropriate NBO:T values. The K_Ds of the elements for the four liquid pairs are plotted on the relevant NBO:T vertical line. The K_Ds for corresponding elements have been joined by tie lines in 2 pairs: (a) the values for PB76 (phonolite, 6 Kb, 1150 °C) have been joined by solid lines to those for PB81 (nephelinite, 6 Kb, 1150 °C); and (b) the values for PB25 (phonolite, 1 Kb, 1150 °C) have been joined to those for PB59 (nephelinite, 1 Kb, 1150 °C). NBO, number of non-bridging oxygens; T, number of cations in tetrahedral co-ordination; K_D, concentration of element in the silicate melt divided by concentration of the element in the carbonate melt. K_D values are presented in Table 16.4.

(highly depolymerized). There is no equivalent parameter to NBO:T for carbonate melts, so on Figure 16.1 we have only shown the value for the silicate member of a silicate/carbonate pair of immiscible liquids. For example, the nephelinitic melt of run PB59 has an NBO:T value of 1.44 and is in equilibrium with a carbonatite melt at 1 Kb and 1150 °C. On Figure 16.1 we have plotted the K_Ds (silicate/carbonate) of 13 elements along the vertical line with a NBO:T value of 1.44. Similarly, the silicate melt (PB25) of a phonolite–carbonatite melt pair, also equilibrated at 1 Kb and 1150 °C, has its NBO:T value (0.575) shown on Figure 16.1 and the K_{DS} for the same 13 elements are plotted along its vertical line. The 13 corresponding trace-element points have been connected by dashed lines on Figure 16.1. The slopes of these dashed lines indicate the effect of silicate-melt composition on the K_D; a positive slope on this diagram indicates that increasing polymerization of the silicate melt enhances enrichment of the element into the carbonate melt. All elements but Cr and Ta have tie lines with a positive slope. In the cases of Ba, Zr, and Cu the dashed tie lines cross the $K_D = 1$ line so that the enrichment of these three, at 1 Kb and 1150 °C, is into the silicate melt in the depolymerized system but into the carbonate-melt in the polymerized system.

A second pair of melt compositions is shown in Figure 16.1; this pair is analogous to the previous pair but have been run at 6 Kb; PB76 represents the relatively polymerized phonolite system, whereas PB81 represents the nephelinite system. The following points emerge from an examination of the tie lines of Figure 16.1:

(a) most tie lines have positive slopes;
(b) the lines for Ta at both 1 and 6 Kb have negative slopes;
(c) the tie-line slopes for Cr and Mn are opposite at 1 and 6 Kb. We would like to repeat the determination for Cr and Mn before drawing any conclusions about the reversals but one possible explanation is valency differences at 1 and 6 Kb.

16.4 EFFECT OF T ON K_D

All results discussed in this section were obtained at 3 Kb.

16.4.1 REEs

K_D values for seven REEs in the phonolite system have been determined at 1050, 1150, and 1250 °C and the results are plotted against atomic number in Figure 16.2. At each temperature the K_D values tend to increase with rising atomic number, and for each REE the K_D value increases with increasing temperature. At 1250 °C the K_D values are all greater than 1, so that all REEs will be relatively enriched in the silicate melt. At the two lower temperatures only the heavier REEs have K_D values greater than 1, so that the LREEs are more concentrated in the carbonate melt. These results are shown in an alternative way in Figures 16.3a and b, where K_D values are plotted against temperature. Figures 16.2 and 16.3 also show clearly that the K_D values of different REEs tend to converge as temperature increases. Temperature is a crucial factor in determining which phase is the preferred host for REEs.

Figures 16.4a and b show the REE K_D values plotted against temperature for the nephelinite system. The K_{DS} are mostly close to 1, and generally show only slight variation with temperature, in contrast to the findings in the phonolite system.

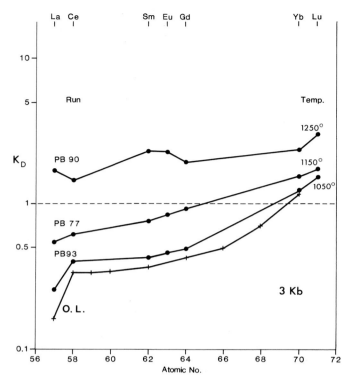

Figure 16.2 K_Ds of seven REEs in a phonolite/sodium-rich carbonate pair of conjugate liquids (i.e. phonolite system, PS) as a function of their atomic number and at three temperatures. $P = 3$ Kb. Graph marked O.L. is for Oldoinyo Lengai rocks (see text).

16.4.2 Ba, Cr, Cu, Mn, Ta, Zr

Figures 16.5a and b show the results for these elements as plots of K_D against temperature for the phonolite system. Chromium and Mn show similar behaviour, with K_D values close to 1 at 1050 and 1150 °C but increasing sharply at 1250 °C. Zirconium shows a slight preference for the silicate phase and its behaviour is not very sensitive to temperature change. Tantalum shows a strong preference for the silicate phase, and its K_D values fall as temperature increases. In contrast, Ba is always strongly enriched in the carbonate liquid under the PT conditions investigated, and its K_D values increase as temperature increases. Copper has K_D values close to 1 at 1050 and 1150 °C but, unlike Cr, there is a sharp decrease at 1250 °C.

The results for the nephelinite-system are shown in Figures 16.6a and b. Apart from the Cr result at 1150 °C, the K_D values for Cr and Cu are close to 1 and not temperature dependent. The extreme value for Cr at 1150 °C may be anomalous but we have no way of judging this at present. Tantalum exhibits a trend which is very similar to its behaviour in the phonolitic system. Zirconium and Mn, as in the phonolitic system, show similar patterns of behaviour, but there are apparent minima at 1150 °C. Barium appears to favour the carbonate phase, as in the

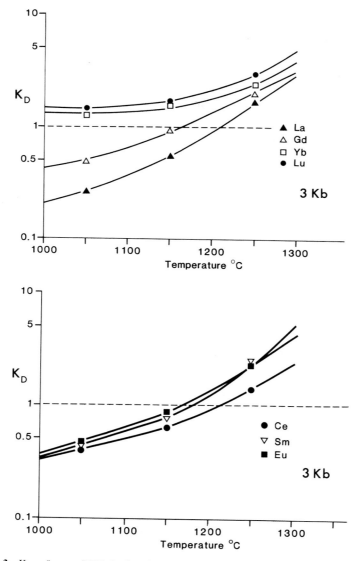

Figure 16.3 K_Ds of seven REEs in the phonolite system plotted against temperature. $P = 3$ Kb.

phonolitic system, but it has a rather high maximum at 1150 °C which, in common with Cr, could be an anomalous value.

16.4.3 Summary of effects of T

In the phonolite system, the K_D values for most of the trace elements investigated show a considerable temperature dependence. The REE K_D values increase smoothly as temperature increases, with the temperature dependence being greatest for the LREEs and least for the HREEs. Above 1200 °C the K_D values for all REEs are

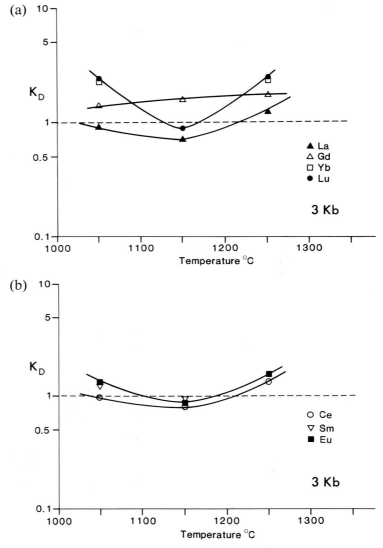

Figure 16.4 As for Figure 16.3, but in the nephelinite/calcium-rich carbonate pair of conjugate liquids (i.e. nephelinite system, NS). $P = 3$ Kb.

greater than 1, while below 1100 °C only the heaviest REEs have K_Ds greater than 1. In the nephelinite system the REE, Cr, Cu, Mn, and Zr K_D values are all quite close to 1 and do not vary significantly with temperature.

The K_D values for Ta are considerably greater than unity (2–9) at all temperatures in both systems, and decrease with increasing temperature. Barium shows a marked preference for the carbonate phase at all temperatures but its partition is clearly temperature dependent in the phonolite system. Of the remaining elements studied in the phonolite system, Zr has K_D values slightly greater than 1; Cr and Mn K_Ds only depart significantly from unity by increasing above 1200 °C; Cu behaves

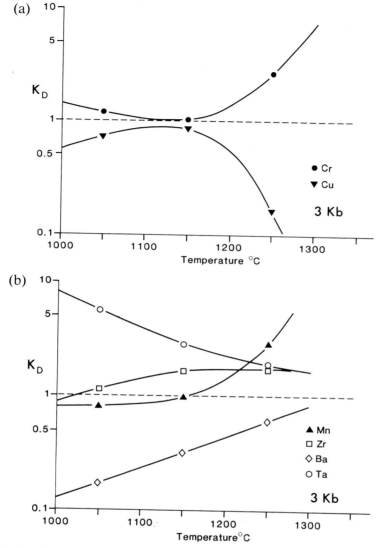

Figure 16.5 K_Ds for Ba, Cr, Cu, Mn, Ta, and Zr in the phonolite system plotted against temperature. $P = 3$ Kb.

similarly to Cr and Mn up to 1200 °C and then shows a strong preference for the carbonate phase.

16.5 EFFECT OF P ON K_D

16.5.1 REEs

Figure 16.7 shows the experimentally determined distribution coefficients for seven REEs in the phonolite system plotted as a function of atomic number and at six

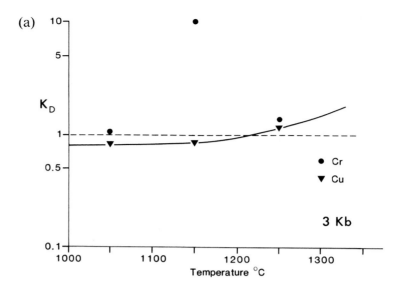

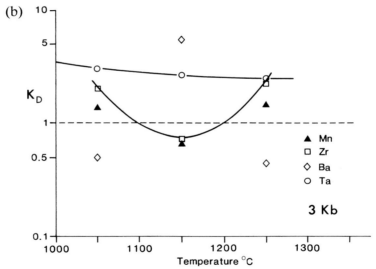

Figure 16.6 K_Ds for Ba, Cr, Cu, Mn, Ta, and Zr in the nephelinite system plotted against temperature. $P = 3$ Kb.

different pressures. The temperature in all cases was 1150 °C. On Figure 16.7 there are six isobaric, isothermal curves which all have a positive slope, showing that the LREEs are relatively more abundant in the carbonate melt than the HREEs at a chosen P and T. Also the isobaric curves are depressed to lower values of K_D with increasing pressure (ignoring the slight overlap of the 4 and 6 Kb curves), meaning that increasing pressure favours the enrichment of the REEs in the carbonate melt.

Figures 16.8a–g show individual plots of K_D against P for each of the seven REEs, all at 1150 °C. With the exception of one point for Gd, the data all lie on smooth curves. These diagrams show very clearly that higher pressures are needed to

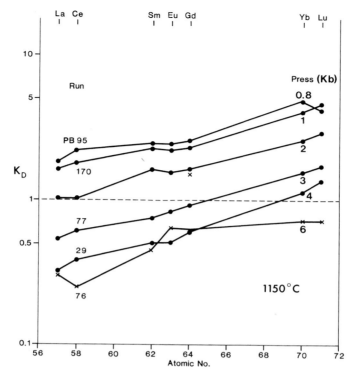

Figure 16.7 K_{DS} for seven REEs as a function of their atomic number at six different pressures. $T = 1150\ °C$. Phonolite system.

depress the value of K_D below 1 and achieve enrichment of these elements in the carbonate melt.

The variation of K_D with pressure for seven REEs in the nephelinite system is shown on Figures 16.9a–g. Smooth curves have been drawn through the data points, but compared to Figure 16.8 there is considerably more scatter of the points about these curves; however, the curves have similar negative slopes on both figures. Therefore, the conclusion is once more that increasing pressure decreases the K_{DS} and favours enrichment in the carbonate melt.

16.5.2 Ba, Cr, Cu, Mn, Ta, Zr

Figures 16.10a–f show the variation of K_{DS} for these elements in the phonolite system as a function of pressure, all at 1150 °C. Where feasible, a smooth curve has been drawn through the data points. The curve for Ba (Fig. 16.10a) has a negative slope, and is entirely below the $K_D = 1$ line, so for the TX conditions of these runs Ba is always enriched in the carbonate melt. Figure 16.10b shows that the data points for Cr are somewhat scattered but Cr seems to be concentrated into the silicate melt until at least 4 Kb. Similarly, the data points for Cu on Figure 16.10c show some scatter but there seems to be a definite trend to lower K_D values at higher pressures. The points on this figure are all below $K_D = 1$ at all pressures greater than

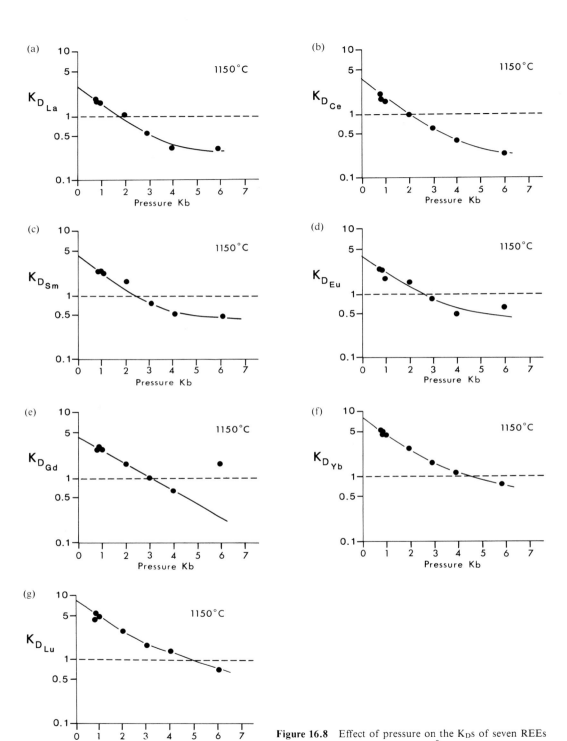

Figure 16.8 Effect of pressure on the K_Ds of seven REEs in the phonolite system. $T = 1150\,°C$.

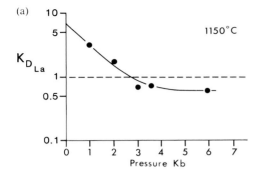
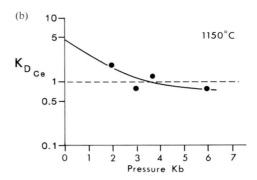
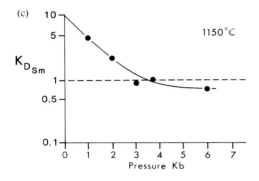
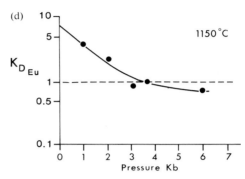
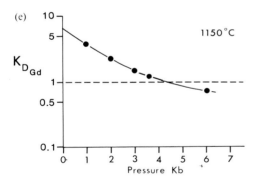
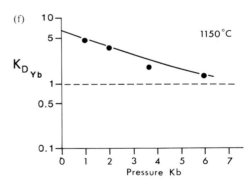
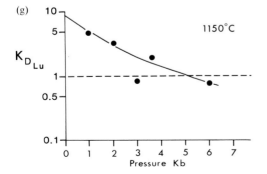

Figure 16.9 Effect of pressure on the K_Ds of seven REEs in the nephelinite system. $T = 1150\,°C$.

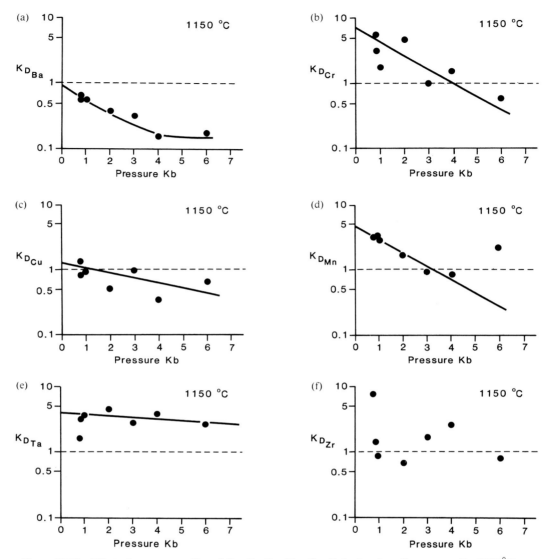

Figure 16.10 Effect of pressure on K_Ds of Ba, Cr, Cu, Mn, Ta, Zr in the phonolite system. $T = 1150\,°C$.

about 1 Kb so that Cu under the TX conditions of these runs is concentrated into the carbonate above 1 Kb. The data points for Mn shown on Figure 16.10d, other than one point, mostly fall very close to a straight line with a negative slope; up to about 3.5 Kb the line is above the $K_D = 1$ line but at greater pressures Mn will be enriched into the carbonate liquid at the TX conditions shown. Figure 16.10e illustrates the situation for Ta, the straight line through the data points has a gentle slope and is well above the $K_D = 1$ line for all pressures. This is an interesting result because Ta might be expected to be more concentrated in the carbonate melt under fractional crystallization conditions, and therefore the relative Ta content of associated silicate and carbonate rocks might prove indicative of the mode of genesis. Finally, the K_D data for Zr are shown on Figure 16.10f; the data points do not lie on any simple curve so no attempt has been made to draw a line through them.

Figures 16.11a–f correspond to Figures 16.10a–f respectively, but for the nephelinite system: again the temperature for all runs was 1150 °C. Figure 16.11a for Ba shows wide scatter of the data points. The Cr data (Fig. 16.11b) also show scattered points but there seems to be a positive correlation of K_D and P, the

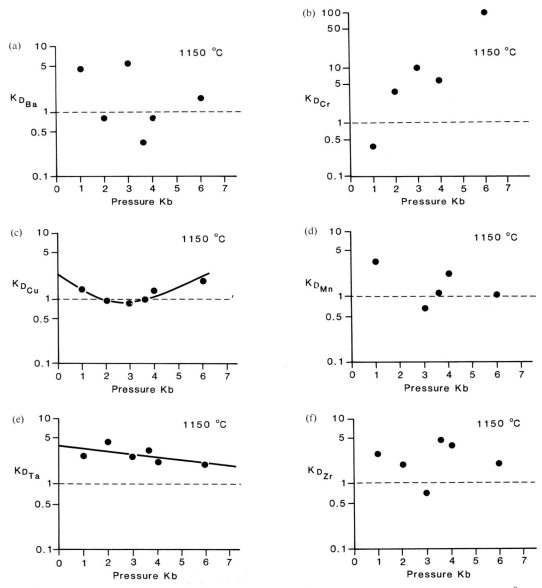

Figure 16.11 Effect of pressure on K_Ds of Ba, Cr, Cu, Mn, Ta, Zr in the nephelinite system. $T = 1150$ °C.

opposite for Cr in the phonolite–carbonate pair (Fig. 16.10b). The results for Cu (Fig. 16.11c) fall on a smooth curve which is concave upwards and which grazes the $K_D = 1$ line at about 3 Kb. This is in contrast to the phonolite system which shows a shallow negative slope. The results shown for Mn in Figure 16.11d are scattered. There seems to be a general negative relation of K_D to P and, as the majority of points are above the $K_D = 1$ line, Mn is enriched into the silicate melt at the TX of this diagram. The data points for Ta (Fig. 16.11e) fall fairly close to a straight line with a shallow negative slope. All K_Ds are > 1 up to at least 6 Kb; this graph is very similar to that for the phonolite system. Finally, the data for Zr are shown on Figure 16.11f and a great deal of scatter is apparent.

16.5.3 Summary of effects of pressure

The REE K_D values decrease as P increases and are only slightly less than 1 at 6 Kb, which means that if immiscibility is controlling carbonatite genesis then the high level of REE enrichment commonly found in carbonatites must have taken place at > 6 Kb. Results for Ta show that it would be preferentially partitioned into the silicate melt at all pressures. In the phonolite system, Ba is strongly concentrated in the carbonate melt, which is consistent with the high Ba contents of many carbonatites; however, the Ba K_D values for the nephelinite system are scattered and do not show any pattern. Zirconium, although the results are somewhat scattered, has K_D values between 0.7 and 7 (phonolite system) and 0.7–5 (nephelinite system), suggesting that Zr would tend to be concentrated in the silicate melt. At pressures below 3–4 Kb Mn would be concentrated in the silicate melt, but the results suggest that this would not be the case at > 6 Kb. Copper shows no strong preference for either phase in the pressure range investigated. In the phonolite–carbonate system Cr behaves like Mn but in the nephelinite–carbonate system the results are too scattered to define a trend. The K_D values for Hf show a very wide scatter, mainly due to the large analytical errors involved in determining the very low Hf contents of the carbonate fraction, and the results were not plotted. However, Hf was invariably strongly enriched (K_D values all > 10 and commonly > 100) in the silicate melt.

At present we are unable to explain the cause of scattered results for some elements. Apart from the occasional errant point, the various plots for the REEs, Ta, Mn, and Cu show systematic trends in K_D with changing P and T which indicate that the experimental methods and mechanical separation of the phases are satisfactory. Other possible causes of scatter are (a) analytical errors, (b) disequilibrium, and (c) trace-element heterogeneity in the run products. The analytical errors for Hf have already been mentioned. If equilibrium was not achieved, it is likely that scattered results would have been obtained for more elements. No barium- zirconium-, or chromium-rich accessory phases were detected but their presence could result in these elements being heterogeneously distributed. In future work the analytical errors associated with Hf and, possibly, Cr could be reduced by increasing their concentrations in the starting materials.

16.6 SUMMARY AND APPLICATIONS

As pointed out in the introduction, carbonatites are well known for, or even because of, their high concentrations of uncommon elements. Not all carbonatites show these enrichments. Any hypothesis for carbonatite genesis should be able to explain at least the mechanisms of this enrichment and why some carbonatites are enriched in, for instance, the REEs while others are not noticeably so. We have presented data on the effect of P, T, and X on K_D for a number of elements and have concluded that high P, low T, and a relatively polymerized silicate melt all favour the enrichment into the carbonatite melt of most of the minor and trace elements studied here. Our results show that over a wide range of PTX conditions many of the minor elements will actually be concentrated into the silicate melt and a barren carbonatite rock will result. Alternatively, concentrations up to a maximum of about eight times ($K_D = 0.125$) those of the conjugate silicate melt can be achieved in the carbonate melt under the optimum PTX-conditions covered by this study. Many carbonatites contain abundances of the REEs (e.g. Mountain Pass) or P and Cu (e.g. Phalaborwa) which are far in excess of abundances that could be produced by an eightfold increase from a silicate parent magma. This massive concentration of incompatible elements in some carbonatite rocks may be met, to some extent, if the PTX conditions in nature are more favourable to low K_Ds, e.g. lower liquidus temperatures due to the presence of other components, especially F and H_2O, and higher pressures. In our opinion, however, carbonatites very enriched in incompatible elements probably owe their origin to a multi-stage process involving an initial moderate concentration due to immiscibility, followed by an increasing concentration in the carbonate melt due to the removal of large amounts of calcite (or dolomite) crystals upon cooling, and culminating in the precipitation and accumulation of phases which have high contents of the incompatible elements, e.g. bastnaesite, pyrochlore, and baddeleyite. This hypothesis is discussed in more detail by Kjarsgaard and Hamilton (this volume).

16.6.1 K_Ds from rocks

If we had a silicate rock and a carbonate rock which represented quenched liquids originally related by immiscibility and their chemistry was known in detail, then the results presented here could lead to an estimation of the PT conditions at the time of separation of the two melts. Alternatively, if the PT conditions of separation could be independently estimated, then by comparing the K_Ds of the two rocks to those determined in this study we could test the validity of the immiscibility model. In the real world hardly any of the prerequisites are met, e.g. there is uncertainty whether the rocks represent quenched liquids and which silicate rock is conjugate with which carbonatite; and, in addition, the chemistry is usually not fully known. Despite this pessimism, however, we would like to present two natural examples which we think are relevant to this discussion.

(a) *Gardar Province, South Greenland.* Pearce (1988) has described a number of dyke rocks (both silicate and carbonatite) from this area and has presented REE abundances. Field evidence suggests that all these dykes are contemporaneous, so the possibility exists that the carbonatites are related to the silicates by immiscibility. This notion would be supported if the K_Ds observed in the rocks corresponded to those of this study. To simplify matters we will only consider Ce. Its concentration (in p.p.m.) varies from 1000 to 2000 in the silicate rocks and from 2500 to 20 000 in the carbonatites. These values yield extreme K_Ds of 0.05 and 0.8 with an average of 0.2. This latter value compares well with a K_D of 0.25 determined in run PB76 (Table 16.4). Also there are five other values for the K_D of Ce in Table 16.4 which fall within the above range. These observations are compatible with the hypothesis that the silicate and carbonate dyke rocks are related by immiscibility and that the liquid separation took place at $P \geqslant 6$ Kb and at temperatures $\leqslant 1050°C$.

(b) *Oldoinyo Lengai.* Gerasimovskiy *et al.* (1972) have published the REE contents of the carbonate, phonolite, and nephelinite rocks from Oldoinyo Lengai and, from their results, we have calculated the K_Ds for the phonolite–carbonatite and nephelinite–carbonatite pairs. The phonolite K_Ds are shown on Figure 16.2. The nephelinite points are not too different from the phonolite ones but for clarity have been left off the figure. There is a striking parallelism and proximity between the points for the natural phonolite system and those for the experimentally-derived results for a phonolite system at 3 Kb and 1050 °C. We suggest that this very close correspondence is strong evidence supporting the hypothesis of an immiscibility relationship between the carbonate and silicate magmas of this volcano. Twyman & Gittins (1987) used Gerasimovskiy's data and suggested that it shows that the carbonate lavas of Oldoinyo Lengai are not the products of immiscibility. We are of the opinion that our conclusion on the origin of the Oldoinyo Lengai carbonatite by liquid immiscibility should be the preferred one as it is based on many more K_D values than were at the disposal of Twyman & Gittins.

ACKNOWLEDGEMENTS

We acknowledge the financial support of the NERC without which the experimental petrology laboratory at Manchester could not function. The activation analyses were done at the Research Reactor at Risley and we thank Dr G. R. Gilmore and his staff for valuable assistance. Many staff provided invaluable support during the research work and the preparation of the manuscript and we would like to record our thanks to them. The manuscript was improved by suggestions from W. S. Mackenzie, C. H. Donaldson, and two reviewers.

REFERENCES

Dawson, J. B. 1962. The geology of Oldoinyo Lengai. *Bulletin Volcanologique* **24**, 349–87.

Freestone, I. C. & D. L. Hamilton 1980. The role of liquid immiscibility in the genesis of carbonatites. *Contributions to Mineralogy and Petrology* **73**, 105–17.

Gerasimovskiy, V. I., Yu. A. Balashov & V. A. Karpushina 1972. Geochemistry of the rare earth elements in the extrusive rocks of East African rift zones. *Geochemistry International* **9**, 305–19.

Girardi, F. & E. Sabbioni 1968. Selective removal of radio-sodium from neutron-activated materials. *Journal of Radioanalytical Chemistry* **1**, 106.

Mysen, B. O. 1986. Magmatic silicate melts: Relations between bulk composition, structure and properties. In *Magmatic processes: physicochemical principles*. Geochemical Society Special Publication 1, 375–99.

Pearce, N. J. G. 1988. *The petrology and geochemistry of the Igaliko dyke swarm, South Greenland*. Ph.D. thesis, University of Durham, Durham.

Twyman, J. D. & J. Gittins 1987. Alkalic carbonatite magmas: parental or derivative? In *Alkaline igneous rocks*, J. G. Fitton and B. G. Upton (eds), 85–94. Geological Society of London Special Publication 30.

17
DIVERSIFICATION OF CARBONATITE

M. J. LE BAS

ABSTRACT

Carbonatites vary: some are calcite-rich, others are rich in dolomite or ankerite; some are apatite-rich, others contain abundant magnetite, pyrochlore, or fluorite.

Several parameters can contribute to the chemical diversity of carbonatite:

(a) nature of the upper mantle source;
(b) chemical composition of the parental magma;
(c) pressure and temperature at which liquid immiscibility may take place;
(d) crystal fractionation of carbonate minerals and the early precipitation of apatite, magnetite, and olivine;
(e) loss of alkalis by fenitization; and
(f) contamination by adjacent country rocks.

Each can drastically affect the composition of carbonatite magma.

17.1 INTRODUCTION

Only 25 years ago few carbonatites were recognized and the idea of an average and typical chemical composition for carbonatite was useful. However, the great increase in the number of carbonatites studied since then and their great diversity of mineralogical composition show that the term 'average carbonatite' has limited meaning. Different carbonatite magmas result from different *PTX* conditions at the time of their formation, and each carbonatite may crystallize along a particular path governed by the bulk composition of the carbonatite magma, the composition and pressure of the volatile phase, the total pressure, the degree of loss of components by alkali metasomatism, and the degree of gain of components by contamination with wall-rock. The reactivity of carbonatite magma with wall-rock, particularly at low total pressures, renders these processes very important in determining the carbonatite that is finally produced. The diversification seen in carbonatites is greater than that observed for most silicate igneous rocks.

Experiments by Wyllie (summarized in Wyllie 1978, this volume) and by Kjarsgaard & Hamilton (this volume) have demonstrated that carbonate–silicate liquid immiscibility can occur at magmatic temperatures and over a range of pressures. The experiments also revealed that:

(a) silicate melts can dissolve 10–20 wt % of carbonate;
(b) the coexisting carbonate melt contains only about 5% dissolved silicate; and
(c) the proportion of these dissolved components decreases with decrease in temperature.

These results explain why calcite can occur as a minor late-stage constituent in ijolites, urtites, and nepheline syenites and in fluid inclusions trapped in apatite in ijolite. Verwoerd's experiments (1978) showed pyroxene on the silicate liquidus surface of the two-liquid silicate–carbonate system, thus bringing the experiments nearer to actual rock compositions. Freestone & Hamilton (1980), using natural nephelinite and phonolite from Oldoinyo Lengai as starting materials, together with a simplified synthetic natrocarbonatite, showed that melts of both these silicate compositions are immiscible with alkali-rich carbonate melts.

Field evidence also favours liquid immiscibility. Most carbonatitic complexes show syenites and ijolites as discrete intrusions distinct from carbonatite intrusions. Furthermore, carbonatites commonly occur without associated syenites or ijolites, and vice versa. Gradations between carbonatitic and ijolitic rocks are rare, and when seen can be ascribed to intrusion and brecciation of one into the other. Likewise, the extrusive equivalents of these rocks occur as discrete interbedded units. Recent work by the author reveals examples of extrusive carbonatite interbedded with nephelinite and phonolite lavas at Shombole and Tinderet in Kenya and on the Cape Verde Islands, particularly the island of Brava, which is currently under investigation (Turbeville *et al.* 1987). The discreteness of occurrence favours liquid immiscibility rather than a relationship involving fractional crystallization for which a continuous series from silicate, probably phonolitic, to carbonate would be expected, but is not found. Additional evidence supporting immiscibility comes from fluid inclusions where actual liquid immiscibility of silicate–carbonate components has been recorded by time-lapse photography (Rankin & Le Bas 1974), and from globules of carbonatitic bulk composition found in apatite from ijolite (Le Bas & Aspden 1981). The similarity in chemical composition of the apatite crystallizing from both carbonatite and nephelinitic rocks in west Kenyan alkaline complexes also points to carbonate and silicate melts being in equilibrium with each other, i.e. related by immiscibility, rather than one being the crystallization fractionation product of the other (Le Bas & Handley 1979).

It is still debatable whether primitive carbonatite magma is dry or hydrous, but the fresh, hot natrocarbonatite collected by Guest during the 1960 eruption of Oldoinyo Lengai contained $< 2.0\%$ H_2O (Du Bois *et al.* 1963). In addition, the feldspathic fenites surrounding high-level carbonatite complexes more often than not contain no hydrous minerals. Fluid inclusions in carbonatites are aqueous but have relatively low $H_2O:CO_2$ ratios because the main minerals in the fluid inclusions are bicarbonates, carbonates, halides, and sulphates. No precise analytical figure for H_2O can be given for the fluids because the original fluid content of many fluid inclusions has changed by 'necking off' of the fluids, but a figure of $< 0.5\%$ is likely (Le Bas & Aspden 1981).

Carbonatites, although small in volumetric terms, are widespread and this poses the problem of which carbonatites are nearer in composition to the magmas from which they crystallized, and which are modified, i.e. altered by secondary processes, such as fenitization and contamination.

17.2 THE CHOICE OF CARBONATITES

Four approaches might be utilized in selecting carbonatites for study. First, in petrographic provinces that contain numerous igneous complexes, it is possible to recognize features that are common to several carbonatites and thus are genetically significant. This approach was employed in studies of the East African Cenozoic alkaline province (Le Bas 1977) and of the carbonatitic and related rocks of the Cape Verde Islands in the Central Atlantic, where one or more carbonatitic complexes occur on at least seven of the ten islands of the archipelago (Le Bas 1984).

Secondly, carbonatitic centres can be chosen from areas where contamination by continental crust is minimal. The only two provinces known where such carbonatitic complexes exist, are the Cape Verde Islands and the Canary Islands, both in the Atlantic Ocean. Evidence for contamination of the carbonatite magmas has been found for the occurrences in these islands, but not by granitic or other strongly fractionated rocks.

The third desirable factor is to find rocks which best represent carbonatitic magma (Le Bas 1981); this is discussed by Keller (this volume). Chilled margins of small intrusions which show no fenitization could serve because carbonatite lavas are rare. But since the first description of carbonatite lavas by Dawson (1962) based on observations at Oldoinyo Lengai in Tanzania during 1960 (see Dawson, this volume), many other carbonatite lavas have now been recognized, e.g. in Kenya (Deans & Roberts 1984); in Uganda (Nixon & Hornung 1973); the Cape Verde

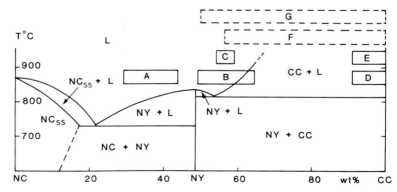

Figure 17.1 Estimated positions of carbonatitic lavas (liquid + phenocrysts) on the join NC–CC (wt%) at 1 Kb. NC_{ss}, sodium carbonate solid solution; NY, nyerereite ($Na_2Ca(CO_3)_2$); CC, $CaCO_3$; L, liquid. (After Cooper *et al.* 1975). A, Oldoinyo Lengai, Tanzania (Gittins & McKie 1980); B, Kerimasi, Tanzania (Mariano & Roeder 1983, Le Bas 1987); C, Tinderet, central Kenya (Deans & Roberts 1984, Le Bas 1987); D, Fort Portal, SW Uganda (Nixon & Hornung 1973); E, Kaiserstuhl, SW Germany (Keller 1981); F, Kruidfontein, South Africa (new data); G, Cape Verde Islands (new data).

Islands (Silva *et al.* 1981, Silva & Ubaldo 1985, Turbeville *et al.* 1987); Kaiserstuhl, West Germany (Keller 1981); in Kola, USSR (Pyatenko & Saprykina 1976); in Kamchatka, USSR (Rass & Frikh-Khar 1987); and in the Kruidfontein carbonatitic complex in South Africa originally described by Verwoerd (1967). All carry groundmass calcite and are variously studded with phenocrysts of calcite, biotite, or pseudomorphs after nyererite. Apatite is rare, and olivine and magnetite are even rarer.

The binary system Na_2CO_3–$CaCO_3$ at 1 Kb is appropriate for evaluating carbonatitic lavas (Fig. 17.1). Apart from the natrocarbonatite from Tanzania, all plots are based on modal data (see Le Bas 1987). Although Figure 17.1 suffices for the interpretation of the data from carbonatite lavas, it is a simplification because carbonatite magma can also contain Fe, K, Mg, Ti, Sr, Ba, REEs, PO_4, SO_4, and (F, Cl, OH). The presence of minor amounts of these elements or oxides would reduce the liquidus temperature and might alter the position of the minima, but the thermal divide at nyererite would remain. The binary system can thus be useful in interpreting the composition of carbonatite magma. The Oldoinyo Lengai natrocarbonatite lavas plot at A and contain nyererite and gregoryite (NC_{ss}) but no calcite. Data from rocks of the other six localities (B–G) plotted are based on the interpreted original mineralogy of the lavas, and all give reasonable liquidus phase relations. For instance, rocks from Tinderet (C on Fig. 17.1) show some nyererite phenocrysts, now pseudomorphed by secondary calcite, which enclose unaltered quench calcite.

Fourthly, it is useful to distinguish those carbonatites with abnormal or unique features. The Phalaborwa complex in South Africa is one such example (see Erikkson, this volume), because it is associated with copper mineralization, and hence is atypical of most carbonatites. The mineralized and metasomatized nature of rauhaugite at Fen in Norway, the type locality, sets that rock apart from other Mg-carbonatites in spite of the fact that the petrology of most of the complex is similar to that shown by many other carbonatites. That carbonatites in the Cape Verde Islands show considerable trace-element and O, C, and Sr isotopic exchange by water–rock reaction, suggests that many of these carbonatites cannot be considered as retaining normal geochemical characteristics (Hodgson 1985). Hay & O'Neil (1983) describe similar isotopic and trace-element exchange with meteoric waters from the carbonatitic tuffs at Laetolil, Tanzania and at Kaiserstuhl, West Germany.

Eggler (1978) and others have shown that at pressures >20 Kb it is possible to produce a direct partial melt of carbonated peridotite from which carbonatite magma could evolve. Such carbonatites would be magnesium-rich, and would have low alkali contents and Na:K ratios, and could make up the carbonatitic material which is associated with kimberlites. These carbonatites show no associated fenitization. Bedson (1983) noted that no liquid immiscibility was observed for mixtures of kimberlites and carbonatites melted at pressures up to 25 Kb. Taking the further evidence that the normal field association of carbonatites is with nephelinitic rather than kimberlitic rocks, it is assumed here that the rare kimberlitic carbonatites are abnormal carbonatites.

17.3 BASIC ASSUMPTIONS

Three basic assumptions are made in this discussion on the possible lines of evolution of carbonatite magma:

(a) That carbonatites are igneous rocks with their ultimate source in the mantle (Bell & Blenkinsop, this volume, Kwon et al., this volume).
(b) That most primitive carbonatite magmas are generally rich in alkalis. The natrocarbonatite of Oldoinyo Lengai is one such example. Arguments for the assumption that pristine carbonatites are normally rich in alkalis are detailed in Le Bas (1987). Natrocarbonatite is the only known occurrence of fresh carbonatite in the world. Pseudomorphs after nyerereite have been identified in many altered carbonatite lavas but rarely among carbonatite intrusive bodies. Early carbonatites, usually sövites, in shallow-seated carbonatitic complexes characteristically produce potassic and sodic fenites, whereas later carbonatites within the same complex rarely show any alkali fenitization, the capacity to fenitize having been apparently exhausted as a result of the loss of alkalis from the earlier carbonatites.
(c) That most carbonatites are produced by liquid immiscibility from a silicate magma. This is in keeping with the premise that early carbonatite magmas are alkali-rich, and with the experiments conducted by Verwoerd (1978), Wyllie (1978), Freestone & Hamilton (1980) and Kjarsgaard and Hamilton (this volume). The arguments in favour of this latter assumption are reviewed in Le Bas (1987).

17.4 THE NEPHELINITE–CARBONATITE ASSOCIATION

Carbonatites are typically associated with nephelinitic and phonolitic magmatism. This has been shown by King & Sutherland (1960), and accounts of these nepheline-bearing rocks are given by Le Bas (1987) and by Donaldson et al. (1987).

17.4.1 Liquid immiscibility

Figure 17.2 shows the liquid immiscibility relations that can take place between nephelinites and carbonatites at crustal pressures. It demonstrates that carbonatites separate immiscibly from phonolite as well as from nephelinite magma (Hamilton et al. 1979). Fields II and III in Figure 17.2 are interpreted (Le Bas 1987) to abut the solvus defining the two-liquid field at c. 15 Kb. Fractionation of nephelinite magma will result in liquids whose compositions migrate along the solvus towards phonolitic compositions continuously exsolving successively different compositions of carbonate melt. Thus at the very birth of carbonatite magmas, the potential exists for producing a continuous spread of diverse carbonatite compositions with variable $(Na_2O + K_2O):CaO$ ratios.

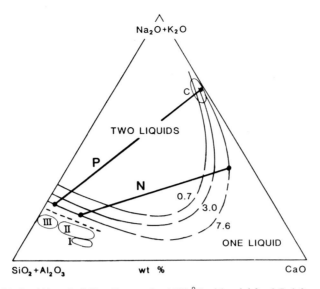

Figure 17.2 Major oxide liquid immiscibility diagram for 1150 °C with solvi for 0.7, 3.0, and 7.6 Kb with tie lines P (phonolite–natrocarbonatite) and N (nephelinite–calcic natrocarbonatite), after Freestone & Hamilton (1980). The short dashed line in the lower-left corner is the estimated c. 15 Kb solvus (incorporating data from Bedson 1983). Field I is the plot of olivine nephelinite, melanephelinite, and melilite nephelinite compositions. Field II is nephelinite + ijolite. Field III is phonolite + nepheline syenite. Field C is natrocarbonatite. All other silicate rocks, including kimberlite, plot nearer the $SiO_2 + Al_2O_3$ apex, except melilitite which plots immediately to the right of Field I (see Le Bas 1987). Except perhaps at much higher pressures, melilitites do not lie on the solvus as proposed by Nielsen & Buchardt for the production of the Gardiner carbonatite in E. Greenland (1985).

The range for liquid immiscibility taking place at 7.6 Kb is shown by the tie lines on Figure 17.2 and yields $(Na_2O + K_2O):CaO$ weight ratios from 0.6 to 2.3. This range of $(Na_2O + K_2O):CaO$ ratios spans the NY and CC fields in Figure 17.1 (the minor K_2O is ignored) and, with the exception of the wholly calcic compositions, corresponds to all actual systems. Recent work by Kjaarsgaard and Hamilton (this volume) shows that at high pressures (5 Kb) and temperatures (1250 °C) liquid immiscibility can occur between even more calcic carbonatite compositions and silicate liquids.

The present knowledge of the phase relations represented on Figure 17.2 suggests that the natrocarbonatite at Oldoinyo Lengai was derived from phonolite. At Kerimasi, only 5 km away, and at Tinderet in central Kenya, primary calcite occurs in lavas. If their mineralogical compositions (given in Le Bas 1987) are plotted on Figures 17.1 and 17.2, it appears that these carbonatite magmas must have been in equilibrium with nephelinitic compositions. The interpretation and identification of the phenocryst phases from Kruidfontein, South Africa, and from Santiago in the Cape Verde Islands are less certain because the carbonates in the droplets of lava preserved in tuffs found at both localities are strongly altered, commonly to dolomite. However, the original morphology of the quench crystals, similar to that seen in the lavas at Tinderet, leaves little doubt that both calcite and nyerereite were originally present and in variable proportions, indicating cumulate processes. The Cape Verde carbonatite lapilli on Santiago are associated in the field with nephelinites and ijolites, while the carbonatite tephra and lavas on Brava occur more often with phonolitic pyroclastic rocks, and an origin by liquid immiscibility appears likely in both cases. Von Eckermann's description (in Tuttle & Gittins 1966, p. 12) of globular carbonate structures containing melilite pseudomorphs in fine-grained dyke rocks near Alnö in Sweden, similarly supports an origin by liquid immiscibility.

Although the carbonatites at Fort Portal, Uganda, and Kaiserstuhl, West Germany, are reasonably fresh and carry abundant prismatic quench calcite crystals, they show no evidence of nyerereite. The Fort Portal lavas commonly contain phenocrysts of biotite as well as magnetite, olivine, and perovskite. Although carbonatite lapilli at Kaiserstuhl carry magnetite and apatite, they are otherwise almost pure $CaCO_3$ (Keller 1981). The possible occurrence of calcic carbonatite lava at Kaiserstuhl is confirmed by the presence of carbonatite dykes composed of rhythmically layered, comb-textured, quench calcite needles interpreted to have been derived from an almost pure calcium carbonate liquid (Katz & Keller 1981). However, Hay & O'Neil (1983) did show, on the basis of C and O isotopic evidence, that the groundmass of the Kaiserstuhl tuff could have reacted with groundwater which may have removed any alkalis. These calcium-rich carbonate liquids might be related to a parental liquid of natrocarbonatite composition liquid by loss of alkalis as a result of wall-rock reaction, but it is more likely that they are primary carbonatite magmas and the natural analogues of the calcic carbonatite immiscible melts (Kjarsgaard & Hamilton, this volume).

17.4.2 *Chemical variation*

Variation in the bulk composition of a silicate magma undergoing liquid immiscibility would affect the composition of the conjugate carbonate melt being formed, and, more significantly, the effects of variable pressure and temperature would also affect the silicate–carbonate melt distribution coefficients. Both Ca and Na would fractionate strongly into the carbonate melt, and as the pressure increases (experiments were run from 1 to 6 Kb) the Ca:Na ratio increases at $1150°C$ in carbonate melts coexisting with phonolitic, and to some extent with nephelinitic, compositions.

Bedson's experiments (1983, 1984) confirm those of Freestone and Hamilton. Not only does the distribution coefficient K_D (ratio of concentration in silicate : concentration in carbonate) of K fall below unity for pressures less than *c.* 4 Kb, but also the Na:K ratio changes. Figure 17.3 shows the $Na_2O:K_2O$ ratios in nephelinites and phonolites coexisting with carbonatite. Although the data points are few, it shows that the ratio for the nephelinite–carbonate conjugate-melt pairs is relatively constant at different pressures, unlike the phonolite–carbonate pairs. At the pressures of deeper crustal levels, the conjugate carbonate melts have much higher Na:K ratios. The question has frequently been asked (e.g. Woolley 1969) whether different carbonatites can have different Na:K ratios and hence generate various sodium-rich and potassium-rich fenites. While the answer could lie in the chemical compositions of the source from which the partial melts are derived, these experimental data show that formation of carbonatites with different Na and K

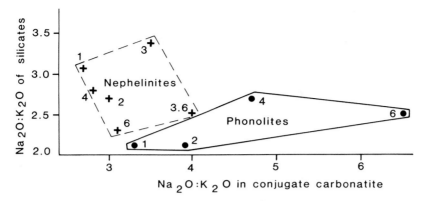

Figure 17.3 $Na_2O : K_2O$ ratios in nephelinitic and phonolitic melts coexisting with synthetic carbonatites, showing the relative enrichment of Na in carbonate melts conjugate with phonolite, which increases as pressure increases. Figures in the fields are pressures in Kb. Nephelinites show no systematic effect. (Data after Bedson 1983). The $Na_2O : K_2O$ ratio of Oldoinyo Lengai natrocarbonatite is 4.7 ± 1 and, presuming that it formed immiscibly with phonolite as proposed by Freestone & Hamilton (1980), the plot suggests that this natrocarbonatite magma formed at pressures between 2 and 5 Kb.

contents could depend on the pressure at which the carbonate melt separated from a conjugate phonolitic melt.

Magnesium also varies considerably in its distribution between conjugate nephelinite and carbonatite melts. Over the pressure range of 1–5 Kb, Mg partitions at c. 1100 °C into the silicate liquid ($K_D > 1$). Experiments at 6 Kb (Bedson 1983) and 7.6 Kb (Freestone & Hamilton 1980) give K_{DS} of 0.8 and 0.7 respectively. This could suggest that Mg-carbonatite magmas can be generated only in the lower portions of the crust or within the mantle. It has been argued in the past that Mg-carbonatites are fractionated from calcic carbonatite melts (Woolley 1982), but now it is seen that, within one magmatic system, primary Mg-carbonatites could be generated by liquid immiscibility at greater depths than those proposed for the generation of calcic carbonatites. Thus Mg-carbonatites might be expected to be found as the early formed members of a carbonatite complex. Kapustin (1980) reports a few cases where Mg-carbonatites precede calcic carbonatites. However, in many cases it is not always possible to determine the ages of associated Mg- and Ca-carbonatites, e.g. at Dorowa and Shawa in Zimbabwe (Johnson 1961). The problem becomes greater because some carbonatites become dolomitized during or after emplacement.

At all pressures employed experimentally by Bedson (1983), Fe partitioned into the silicate magma, and this is consistent with the field evidence that early stage iron-rich carbonatites do not occur, but that they are normally the product of fractional crystallization from earlier Ca-carbonatites. Manganese also partitions preferably into the conjugate silicate melt, but less than Fe. Hence, carbonatite melts produced by liquid immiscibility, whether from nephelinitic or phonolitic melts, should have relatively high Mn : Fe ratios. Carbonatites do have high Mn : Fe ratios (Le Bas 1981).

Carbonatites rarely show more than 0.5 wt% Al_2O_3, and SiO_2 commonly reaches several wt%. The low Al content might result from loss during fenitization involving growth of alkali feldspar. However, the natrocarbonatite lava from Oldoinyo Lengai has negligible Al, along with most carbonatites, and this closely reflects the partition observed in Bedson's experiments where the $SiO_2:Al_2O_3$ ratio in most conjugate carbonate melts is between 5 and 10.

Phosphorus is strongly partitioned into the conjugate carbonate melts at all experimental temperatures and pressures, and this is reflected in high modal apatite contents in early Ca- and Mg-carbonatites.

In an immiscible relationship, not only are the pair of conjugate liquids in equilibrium with each other but also with any contained crystalline phase. The compositional similarity of phenocryst apatite found in both nephelinites and early carbonatites thus supports liquid immiscibility (Le Bas & Handley 1979). Olivine also occurs in some early carbonatites and has the same composition range Fo_{85-95} as that in associated nephelinites and ijolites (Le Bas 1987). Thus olivine gives further evidence of liquid immiscibility. Perovskite is another phase which occurs in both carbonatites and nephelinites.

Experiments reveal differences in the partitioning of trace elements between silicate and carbonate melts, particularly Zr, Hf, Ta, and the REEs (Hamilton *et al.*, this volume). The Zr:Hf ratio differs considerably in the two conjugate melts as a result of the strong preference of Hf for silicate melts; Zr shows no particular preference. Thus, the carbonatites on the Cape Verde island of San Vicente have Zr:Hf ratios that range from 50 to 200 at Zr contents of 0–400 p.p.m., but the

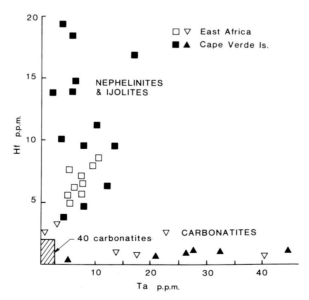

Figure 17.4 Hf v. Ta plot of ijolitic and nephelinitic (boxes) and carbonatitic (triangles) rocks from W. Kenya (open symbols) and from the Cape Verde islands (solid symbols), showing the strong partitioning of Hf into the silicate melt as predicted by Bedson (1983). In the shaded area near the origin, there are 23 carbonatites from W. Kenya and 17 carbonatites from the Cape Verde Islands, but no silicate rocks. INAA analytical data courtesy of J. Hertogen, University of Leuven.

silicate rocks, including phonolites, have Zr:Hf ratios usually < 50 for Zr contents ranging up to 1000 p.p.m. (Hodgson 1985). Tantalum also shows a mild preference for the silicate melts ($K_D = 2-5$) in contrast to Hf. Figure 17.4 shows a plot of Hf v. Ta for silicate and carbonatitic rocks taken from three Cape Verde islands, San Vicente, Santiago, and Maio, and from four West Kenyan complexes. The relationship supports liquid immiscibility. Likewise, the strong preference of the LREEs for conjugate carbonatite melts at lower temperatures and higher pressures (Bedson 1984) would correspond to that observed in carbonatites.

Thus, because the element abundance data from carbonatites supports liquid immiscibility, a continuous range of conjugate carbonate melts can be created, each with their own chemical composition depending on the composition of the parent silicate magma and depth of formation. Variation in the temperature at which separation takes place can also affect the composition of the conjugate phases, but since the onset of liquid immiscibility occurs first during the fractionation of melanephelinite to nephelinite and can continue through to phonolite, it must take place at liquidus temperatures. Piotrowski & Edgar (1970) showed that the liquidus temperatures of East African nephelinite and phonolite lavas differed only slightly and lay between 1100 and 1200 °C. Figure 17.5 shows diagrammatically the several paths that nephelinitic magma could take to produce carbonatite magmas by liquid immiscibility at various pressures.

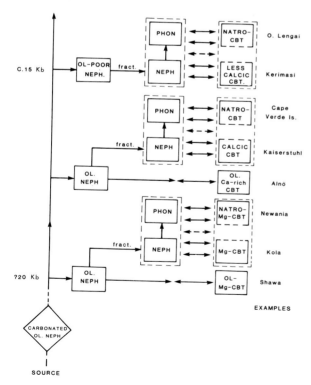

Figure 17.5 Simplified schematic flow chart representing the main flow lines of silicate fractionation followed by liquid immiscible separation of carbonate magmas in reservoirs at different levels in the lithosphere. No liquid immiscibility takes place if the nephelinitic magmas are low in depolymerizing agents (e.g. P, Ti, CO_2, F). Boxes outlined by long-dashed lines enclose a range of compositions of fractionating silicate magmas as they coast along the two-liquid solvus surface. Boxes outlined by short-dashed lines enclose carbonate magmas individually produced from the fractionating silicate magmas. Double-ended arrows indicate conjugate melts. Some natural examples are listed on the right.

17.5 PRIMITIVE CARBONATITES

Once generated, primitive carbonatite magmas can differentiate by fenitization and crystal fractionation. The early-stage carbonatites are Na-carbonatites, Ca-carbonatites, and Mg-carbonatites. Most plutonic carbonatites are either calcium- or magnesium-rich.

17.5.1 Fenitization

Calcium- and magnesium-rich carbonatites, particularly those at higher levels within intrusive complexes are characterized by metasomatic aureoles marking expulsion of alkali-rich fluids from carbonatite into the country rocks.

Most carbonatite-related fenites consist of low-temperature K-feldspar (Or content averages 90%), and may be accompanied by aegirine and Na-amphibole and/or pure albite. Woolley (1969, 1982) and others have demonstrated that the K and Na in fenites must come from carbonatite. Although some carbonatite complexes include both Na- and K-fenites, others are associated only with Na-fenites. Examples include the arfvedsonite-rich fenites developed in the granite-gneiss at Newania in northern India (Viladkar 1980), and in shales at Loe Shilman in northwest Pakistan (Mian & Le Bas 1986), and albite-rich fenites in the quartzites at Kruidfontein, South Africa (L. Clarke, pers. comm.). Examples of potassium-rich fenites are seen in the schists and gneisses at Songwe Scarp, Tanzania (Brown 1964); in the Tertiary volcanic rocks of the Kaiserstuhl (Sutherland 1967); in the basalts, andesites, and dacites of Homa Mountain, western Kenya (Le Bas 1977); in the granites at Bayan-Obo, N. China (Liu Tiegen 1985); and in the ijolites on Brava, Cape Verde islands. These fenites are considered to be quite distinct from the similar-looking syenitic fenites formed around ijolites, which show a much higher temperature mineralogy.

Both sodium-rich and potassium-rich fenites can occur around a single intrusion of carbonatite, and it has been suggested that these fenites are depth zone-related, with the potassic fenitization occurring towards the top and capping carbonatite intrusions, and the sodic fenitization developing only deeper. Examples are seen at Amba Dongar in India (Viladkar 1981) and in the Wasaki complex of western Kenya (Le Bas 1977). In contrast to the above mechanism, whereby one carbonatite apparently produces both sodic and potassic fenitizations, is the possibility that separate sodic and potassic carbonatite magmas could individually produce the sodic and potassic fenites respectively. It has already been shown (Fig. 17.3) that carbonatites of variable Na:K ratio might be produced by varying the pressure at which phonolite and carbonatite melts immiscibly separate, and it also appears possible that potassium-rich carbonatite magma might be generated from sodium-rich carbonatite magma by fractional crystallization (discussed below, see Fig. 17.6).

Thus there are three independent factors which might lead to the formation of sodic and potassic fenitizations:

(a) the depth zone-relation within a carbonatite intrusion at which fenitization takes place;

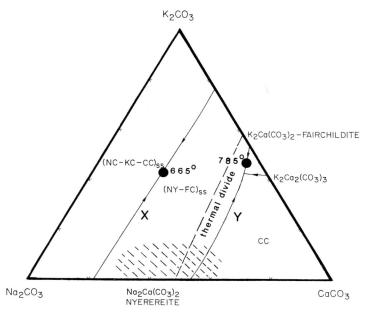

Figure 17.6 The two potassium-enrichment trends, along cotectics X and Y, possible during fractionation of natrocarbonatite. Trend X, appropriate for natrocarbonatite magma separating immiscibly from phonolitic melts, would achieve a $K_2O:(Na_2O + K_2O)$ ratio (wt) at the liquidus temperature minimum (filled circle) of 0.6; but trend Y, appropriate for natrocarbonatite magma separating immiscibly from nephelinitic melts, would achieve a ratio of 0.9 at the liquidus temperature minimum (filled circle). Potassium-rich feldspathization or phlogopitization could result from fenitization by either of these high-potassium carbonatite melts. The shaded portion is the area of immiscible natrocarbonatite melts from Figure 17.3. Gregoryite plots at the low-potassium end of the $(NC-KC-CC)_{ss}$ field. NC is Na_2CO_3, KC is K_2CO_3, CC is $CaCO_3$, NY is nyerereite, and FC is fairchildite. (After Cooper et al. 1975.)

(b) the depth at which phonolitic and carbonatitic melts immiscibly separate;
(c) the degree of fractional crystallization of sodium-rich carbonatite magma.

That some carbonatites seemingly lose Na and others K is not specific to either calcium-rich or magnesium-rich carbonatites. Either fenite can occur around either carbonatite. Of the examples cited, most are associated with calcium-rich carbonatites, and although the Newania and Bayan Obo complexes are dolomite carbonatite, the former produced Na-fenites and the latter K-fenites.

Fenitization demonstrates that sodium- and potassium-free carbonatites seen at outcrop once contained major amounts of Na and K early in their history. The natrocarbonatite lavas of Oldoinyo Lengai are considered to be representatives of these early carbonatite magmas.

17.5.2 Wall-rock contamination

Primitive alkali-bearing carbonatite magmas are prone to crustal contamination because they are out of equilibrium with most crustal rocks. Fenitization around carbonatite complexes shows that the magma had reacted strongly with the surrounding country rocks and had frequently incorporated fenitized wall-rock material into the carbonatite magma.

The wall-rock thus incorporated occurs near the intrusive contact, sometimes as angular xenolithic fenite blocks. More commonly, however, the outlines of the xenoliths are indistinct because of reaction with, and disaggregation of, the minerals making the xenolith. All degrees of breakup can be seen, flow deformation is common and, in extreme cases, the fenite xenoliths are reduced to no more than long wisps of mafic minerals. At Smedsgården in Alnö, angular masses of aegirine silicocarbonatite and biotite silicocarbonatite, sometimes with strong flow deformation, occur scattered in massive sövite. In other examples, the contaminating material is so concentrated in the outer portions of the intrusion that the carbonatite becomes massive silicocarbonatite. A Na-amphibole silicocarbonatite zone $c.$ 1 km wide, which surrounds a sövite core in the large (20 km^2) carbonatite at Sokli in northern Finland is one such example (Vartiainen & Paarma 1979).

The larger the carbonatite intrusion, the greater the contamination observed, and the many clusters and schlieren rich in sodic pyroxenes, sodic amphiboles, micas, feldspars, or other fenitic minerals, may be the relicts of incorporated fenitized wall-rock. The material incorporated into the carbonatite magma may not be completely assimilated; the reaction may be restricted to the margins producing reaction rims. Aegirine augite is commonly seen enclosing diopsidic aegirine augite crystals, and amphiboles are commonly rimmed by Mg-arfvedsonite or richterite, and biotites and phlogopites by tetraferriphlogopite. These are fenitic reactions and indicate that the core compositions were out of equilibrium with the carbonatite magma.

17.5.3 Crystal cumulates in carbonatites

The principal minerals precipitated early in the crystallization history of carbonatites are apatite, magnetite, and olivine.

Most intrusive carbonatites carry about 1–2 wt% of P_2O_5 (Woolley & Kempe, this volume), although the amount is variable. Assuming a K_D for P_2O_5 of 0.3, a former conjugate nephelinite would have had < 1 wt% P_2O_5, which is similar to that observed in many East African nephelinites (Le Bas 1977). Natrocarbonatite has less P_2O_5 (0.7–1.1 wt%) than the intrusive carbonatites, and this is consistent for a carbonate melt conjugate with phonolites which have P_2O_5 0.2–0.3 wt% as is commonly the case in East Africa. These P_2O_5 contents approximate to the amount of phosphate required to co-precipitate calcite and apatite in a carbonatite magma (Wyllie & Jones 1984).

Petrographic observations show that apatite is one of the first minerals to crystallize on the carbonatite liquidus. Apatites occur both as dense clusters of prismatic crystals with adcumulate texture, or, more typically, as isolated, large, ovoid crystals. The former usually occur as schlieren, in high-level, subvolcanic complexes such as those found in East Africa, and, more commonly, in deeply eroded complexes such as those in the Kola Peninsula. Very early quench, acicular apatite crystals are recorded by Sommerauer & Katz-Lehnert (1985) in the comb-layered alvikite dykes at Kaiserstuhl, which quenched from a phosphate-rich carbonate melt.

Ovoid apatite crystals occur in most sövites, commonly in quantities sufficient to indicate that apatite may appear early in the crystallization history of carbonatite magma (Biggar 1969). In other sövites, apatite occurs only rarely, although as phenocrysts indicative of early precipitation. It is evident that considerable apatite fractionation can take place within carbonatites, a process that is favoured by the high specific gravity of apatite (3.2) in contrast to that of carbonatite magma (2.2), and the very low viscosity of the carbonate melt (5×10^{-2} poise; Treiman & Schedl 1983). The latter authors argue that carbonatite magma in chambers would convect turbulently, inhibiting crystal settling. The petrographic evidence suggests that precipitation of early apatite produces prismatic crystals because of rapid cooling (Wyllie & Biggar 1966), and these settle to form cumulate apatite rocks such as those observed at Vuorijarvi in Kola (Kukharenko et al. 1965). With the transition from an early conductive to a convective heat-flow regime in a carbonatite magma chamber, the later apatites are more liable to be swept around the chamber. Thus the apatites seen dispersed in the carbonatites are large and well-rounded ovoids.

Apatite fractionation clearly bears on the evolution of carbonatitic magmas, and is particularly evident in the REE pattern of many sövites. The REE pattern of apatite shows relative enrichment in the middle REEs compared with the light and heavy REEs. Hence fractionation of apatite from carbonatite magma which has a straight, but steep, REE-enriched pattern would produce a REE pattern in the residual carbonate melt with relative depletion of the middle REEs. Such middle REE-depleted patterns are seen in many early stage carbonatites, including the natrocarbonatite lavas of Oldoinyo Lengai.

Olivine (Fo_{85}–Fo_{90}) precipitates from many early stage carbonatite magmas, but in lesser amounts than apatite. There is a close parallel between the properties and crystallizing behaviour of forsterite and apatite; both have similar densities and similar sizes (c. 0.1 mm). Olivinites with cumulate textures occur in the Kola Peninsula and in other carbonatites, commonly associated with apatite-rich rock, and are believed to have accumulated under the same physical conditions. Olivine also occurs as isolated crystals in many sövites and Mg-carbonatites, particularly those from the more deeply eroded complexes such as Alnö, Sweden; Oka, Canada; and Shawa, Zimbabwe.

If the oxygen fugacity is high enough, titanium-poor magnetite is another phase precipitated early in carbonatite evolution. Vuorijarvi, like other Kola carbonatitic complexes, has economic cumulate apatite–forsterite–magnetite deposits, unlike the East African high-level complexes where magnetite occurs in schlieren and as isolated euhedral crystals. The low Ti content of the magnetite in most carbonatites correlates well with the high K_D (> 2) for Ti in laboratory experiments (Freestone & Hamilton 1980; Bedson 1983). The small amount of Ti normally present in carbonatite magmas is usually contained in perovskite, which can also form cumulates, e.g. Kovdor in Kola (Kukharenko et al. 1965).

The cumulate processes involving some or all of apatite, olivine, magnetite, and perovskite explain some of the diversity seen in early carbonatites. The precipitation

CARBONATITE DIVERSIFICATION

of other early cumulate phases in carbonatites, such as zircon and pyrrhotite, also increases the possible diversification of carbonatite magmas.

17.5.4 Fractional crystallization of carbonate minerals

The main phases precipitating in carbonatites are carbonate minerals. Natrocarbonatite magma can precipitate either gregoryite and nyerereite, or nyereite and calcite along cotectics X and Y, depending on which side of the thermal divide the melt lies in the simple carbonatite system at 1 Kb (Fig. 17.6). Fractionation along either trend will produce potassium-rich carbonatite melts at the temperature minima. Woolley (1982) proposed alkali fractionation of carbonatite magma, with increase in K:Na ratios in an attempt to explain the early sodic and later potassic fenitizations associated with carbonatites. The explanation of the K:Na increase is provided by Figure 17.6. Particularly high K:Na ratios of 0.9 can be achieved by melts fractionating along path Y, and such a path could be taken by carbonatite melts separating immiscibly from nephelinitic melts.

Apart from the crystallization of any alkali carbonates from a natrocarbonatite melt, the earliest crystals to precipitate from a carbonate melt are usually strontium-rich calcites, which may then be followed by dolomite or ankerite, depending on the bulk composition and oxygen fugacity of the carbonatite magma. Occurrences of dolomite being followed by more ankeritic compositions are well known (Barber 1974, Kapustin 1980, Woolley 1982). However, the carbonatites of northern

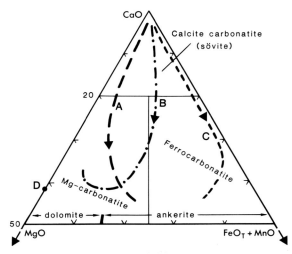

Figure 17.7 Plot, on the carbonatite classification diagram (see Woolley & Kempe, this volume), of three evolutionary paths of carbonatitic crystal fractionation: A, Chilwa, Malawi; B, Loe Shilman, Pakistan; C, Tundulu, Malawi. All begin in the field of calcite carbonatite but diverge depending on the oxygen fugacity, which is relatively high for A, intermediate for B, and low for C. These oxidation values have been modelled from a modified carbonatite norm programme written by Hodgson (1985). A shows Mg:Fe ratios increasing at first and then decreasing, B and C show Mg:Fe ratios decreasing at first and then increasing. Mg-carbonatite = magnesio-carbonatite. D is position of pure dolomite. The boundary between dolomite solid–solution and ankerite solid–solution is taken at Mg:Fe = 4:1 (atomic). Data for A and C after Woolley (1982); data for B from Mian (1987).

Pakistan show a well-defined trend beginning with sövite and progressing to ankerite-bearing ferrocarbonatite and then to dolomite-bearing magnesiocarbonatite (Mian 1987). At low $Fe^{3+}:Fe^{2+}$ ratios, Fe partitions preferentially into ankerite rather than magnetite. At higher $Fe^{3+}:Fe^{2+}$ ratios more magnetite forms and the carbonate mineral that precipitates is more dolomitic, providing that the carbonatite magma contains sufficient Mg (the Ca:Mg ratio of carbonatite separating immiscibly from phonolitic conjugate metals is greater than that separating from nephelinitic melts). These trends are shown in Figure 17.7.

Fractional crystallization of calcic carbonatites, whether to late-stage ferrocarbonatite or Mg-carbonatite, leads to final strong enrichment in the incompatible elements and the formation of economic mineral deposits.

17.6 LATE-STAGE CARBONATITES

The commonest late-stage carbonatites are ankerite or ankeritic dolomite-bearing ferrocarbonatites in which REEs, fluorite, U–Th, and other mineralizations develop. These iron-enriched, dark-brown carbonatites contrast markedly with the more common white carbonatites such as sövite.

17.6.1 Ferrocarbonatite

Ferrocarbonatites make up only a few per cent of carbonatite complexes, and most occur as thin dykes cutting other carbonatites. Some have chilled margins and are clearly magmatic, but not all form discrete bodies. Some occur as anastomosing fractures marked by iron oxide-staining of the original calcite, or dolomite, or sometimes ankerite of the carbonatite; and it can be seen that oxidizing iron-rich fluids have penetrated along cleavages, progressively introducing Fe^{3+} into the earlier carbonate minerals. Some of the Fe^{3+} is undoubtedly produced by oxidation of Fe in the carbonate lattice. Further introduction of ferriferous fluids can lead to merging of the iron-enriched areas and the formation of dyke-like masses of ferrocarbonatite commonly rich in hematite.

One such example of this end-stage of the carbonatite fractionation process is the development of rödberg, an iron-rich low-temperature hematite–calcite–dolomite metasomatic rock that forms a body within the Fen carbonatitic complex. Andersen (1984) describes rödberg as the product of post-magmatic pervasive alteration of ferrocarbonatite caused by the influx of groundwater-derived hydrothermal fluids.

17.6.2 Mineralization

The four main stages of mineralization associated with ferrocarbonatite in many complexes are: (1) REEs, (2) fluorite, (3) baryte, and (4) U–Th commonly accompanied by silicification. Discrete veins and pockets of REE-rich carbonatite, some within the dark-brown ferrocarbonatites, typically contain bastnaesite $((Ce,La)FCO_3)$, and show strong enrichment in the LREEs (La:Yb ratios of

10^2–10^3). REE mineralization occurs within both oceanic and continental carbonatites (Jones & Wyllie 1983, Le Bas 1984), and the experiments of Jones & Wyllie suggest temperatures below 500 °C for the precipitation of most bastnaesite.

Fluorite mineralization can be fairly extensive and may be associated with, but more commonly is independent of, REE mineralization. Fluorite-rich carbonatite occurs in the roof zone of the Amba Dongar carbonatite in India, and within the carbonatitic tuffs filling the caldera above the Proterozoic carbonatitic complex at Kruidfontein, South Africa. Natrocarbonatite contains 1–2% F which can behave as an incompatible element. Some F is concentrated early in apatite and biotite/phlogopite, and F is an important constituent of fenitizing fluids where it is recorded in the fluorine-rich fenitic amphiboles and micas. Fluorite, however, is finally precipitated in quantity only at a late stage in the mineralization, and is commonly followed by baryte.

Barium and SO_4^{2-} similarly occur in natrocarbonatite (1–2 wt%) and likewise behave as incompatible components. Baryte can occur disseminated within ferrocarbonatite, and some barian ferrocarbonatites have 3–4% Ba. At Araxá in Brazil, the barian ferrocarbonatite also contains bariopyrochlore which is mined for Nb (Deans 1978).

The ultimate stage of carbonatite fractionation is reflected in U–Th mineralization, seen in such carbonatites as the Loe Shilman complex in northwest Pakistan (Jan et al. 1981) and in the Kola Peninsula (Kapustin 1980). Chalcedonic silicification, sometimes pervasive, usually follows the U–Th mineralization, with the later development of hydrothermal minor calcite veining as the final event.

17.7 SUMMARY

Fractional crystallization and related differentiation processes produce a wide range of different carbonatites. But even before these processes begin to operate, the course of liquid immiscibility produces a wide spectrum of carbonatitic compositions, with $Na_2O + K_2O : CaO$ wt ratios ranging from less than 0.6 to 2.3. MgO can substitute for CaO up to 50 mole%, and each of these carbonatites can additionally have a range of Na:K ratios. Sodium-carbonate fractionation within the natrocarbonatite system can lead to potassium-rich carbonatites and thus, depending on the degree of fractionation of natrocarbonatite magma, sodium-dominated or potassium-dominated fenitization can take place. The Ca:Mg ratio of primitive carbonatite is pressure-dependent; with increase in pressure the Mg appears to be preferentially partitioned into the carbonatite magma.

An early stage of fractional crystallization of carbonatite involves precipitation of apatite and other phases, variously producing apatite-rich, forsterite-rich, and magnetite-rich carbonatites. These carbonatites can be either sövitic or dolomitic. The commonest is sövite, and usually forms 75% or more of carbonatitic complexes. Apatite carbonatite occurs in most complexes but rarely exceeds 5% of the bulk. Interaction between early carbonatites and their fenitic or conjugate silicate wall-rocks produces silicocarbonatite, arfvedsonite sövite, biotite sövite, phlogopite sövite, aegirine sövite, feldsparphyric sövite, or dolomitic carbonatites.

Fractional crystallization of sövite can produce magnesiocarbonatite or ferrocarbonatite, depending on bulk composition and oxygen fugacity. With low oxygen fugacity, ferrocarbonatite can fractionate to magnesiocarbonatite, but at high oxygen fugacity the sequence of fractionation is from magnesiocarbonatite to ferrocarbonatite. The later-stage carbonatitic compositions are the mineralized carbonatites such as the REE-rich, fluorite-rich, and barium-rich carbonatites. The final U–Th mineralization of carbonatite is more disseminated and rarely of economic value.

Thus Na-, K-, Ca-, Mg-, and Fe-carbonatites occur, and most can carry one or more of the following principal minerals: apatite, biotite, fluorite, magnetite, Na-amphibole, Na-pyroxene, phlogopite, pyrochlore, and REE-fluorcarbonates, each indicative of different stages and diverse paths of development.

REFERENCES

Andersen, T. 1984. Secondary processes in carbonatites: petrology of "rodberg" (hematite–calcite–dolomite carbonatite) in the Fen central complex, Telemark (South Norway). *Lithos* **17**, 227–45.

Barber, C. 1974. The geochemistry of carbonatites and related rocks from two carbonatite complexes, south Nyanza, Kenya. *Lithos* **7**, 53–63.

Bedson, P. 1983. *The origin of the carbonatites and their relation to other rocks by liquid immiscibility.* Ph.D. thesis, University of Manchester, Manchester.

Bedson, P. 1984. Rare earth element distribution between immiscible silicate and carbonate liquids. *Progress in Experimental Petrology 1981–1984 (NERC)* **6**, 12–19.

Biggar, G. M. 1969. Phase relationships in the join $Ca(OH)_2$–$CaCO_3$–$Ca_3(PO_4)_2$–H_2O at 1000 bars. *Mineralogical Magazine* **37**, 75–82.

Brown, P. E. 1964. The Songwe scarp carbonatite and associated feldspathization in the Mbeya Range, Tanganyika. *Quarterly Journal of the Geological Society of London* **120**, 223–40.

Cooper, A. F., J. Gittins & O. F. Tuttle 1975. The system Na_2CO_3–K_2CO_3–$CaCO_3$ at 1 kilobar and its significance in carbonatite petrogenesis. *American Journal of Science* **275**, 534–60.

Dawson, J. B. 1962. Sodium carbonate lavas from Oldoinyo Lengai, Tanganyika. *Nature* **195**, 1075–6.

Deans, T. 1978. Mineral production from carbonatite complexes: a world review. *In Proceedings of the First International Symposium on Carbonatites. Poços de Caldas, Brazil*, June 1976, 123–33. Brasilia: Brazil Departamento Nacional da Produção Mineral.

Deans, T. & B. Roberts 1984. Carbonatite tuffs and lava clasts of the Tinderet foothills, western Kenya: a study of calcified natrocarbonatites. *Journal of the Geological Society, London* **141**, 563–80.

Donaldson, C. H., J. B. Dawson, R. Kanaris-Sotiriou, R. A. Batchelor & J. N. Walsh 1987. The silicate lavas of Oldoinyo Lengai, Tanzania. *Neues Jahrbuch für Mineralogie* **156**, 247–79.

Du Bois, C. G. B., J. Furst, N. J. Guest & D. J. Jennings 1963. Fresh natro-carbonatite lava from Oldoinyo Lengai. *Nature* **197**, 445–6.

Eggler, D. H. 1978. The effect of CO_2 upon partial melting of peridotite in the system Na_2O–CaO–Al_2O_3–MgO–SiO_2–CO_2 to 35 kbar with an analysis of melting in a peridotite–H_2O–CO_2 system. *American Journal of Science* **278**, 305–43.

Freestone, I. C. & D. L. Hamilton 1980. The role of liquid immiscibility in the genesis of carbonatites – an experimental study. *Contributions to Mineralogy and Petrology* **73**, 105–17.

Gittins, J. & D. McKie 1980. Alkalic carbonatite magmas: Oldoinyo Lengai and its wider applicability. *Lithos* **13**, 213–15.

Hamilton, D. L., I. Freestone, J. B. Dawson & C. H. Donaldson 1979. Origin of carbonatites by liquid immiscibility. *Nature* **279**, 52–4.

Hay, R. L. & J. R. O'Neil 1983. Carbonatite tuffs in the Laetolil Beds of Tanzania and the Kaiserstuhl in Germany. *Contributions to Mineralogy and Petrology* **82**, 403–6.

Hodgson, N. A. 1985. *Carbonatites and associated rocks from the Cape Verde Islands*. Ph.D. thesis, University of Leicester, Leicester.

Jan, M. Q., M. Kamal & A. A. Qureshi 1981. Petrography of the Loe Shilman carbonatite complex, Khyber Agency. *Geological Bulletin, University of Peshawar* **14**, 29–43.

Johnson, R. L. 1961. The geology of the Dorowa and Shawa carbonatite complexes, Southern Rhodesia. *Proceedings of the Geological Society of South Africa* **64**, 101–45.

Jones, A. P. & P. J. Wyllie 1983. Low-temperature glass quenched from a synthetic, rare earth carbonatite: implications for the origin of the Mountain Pass deposit, California. *Economic Geology* **78**, 1721–3.

Kapustin, Y. L. 1980. *Mineralogy of carbonatites*. New Delhi: Amerind Publishing.

Katz, K. & J. Keller 1981. Comb-layering in carbonatite dykes. *Nature* **294**, 350–2.

Keller, J. 1981. Carbonatitic volcanism in the Kaiserstuhl alkaline complex: evidence for highly fluid carbonatitic melts at the earth's surface. *Journal of Volcanology and Geothermal Research* **9**, 423–31.

King, B. C. & D. S. Sutherland 1960. Alkaline rocks of eastern and southern Africa. *Science Progress* **48**, 298–321, 504–24, 709–20.

Kukharenko, A. A., M. P. Orlova, A. G. Bulakh, E. A. Bagidasarov, O. M. Rimskaya-Korsakova, E. I. Nefedov, G. A. Ilinski, A. B. Sergeev, & N. B. Abakumova 1965. *The Caledonian complex of ultrabasic and alkaline rocks and carbonatites of the Kola Peninsula and northern Karelia*. Moscow: Nedra.

Le Bas, M. J. 1977. *Carbonatite-nephelinite volcanism*. Chichester: Wiley.

Le Bas, M. J. 1981. Carbonatite magmas. *Mineralogical Magazine* **44**, 133–40.

Le Bas, M. J. 1984. Oceanic carbonatites. In *Kimberlites I: kimberlites and related rocks*. Proceedings of the Third International Kimberlite Conference, J. Kornprobst (ed.), 169–78. New York: Elsevier.

Le Bas, M. J. 1987. Nephelinites and carbonatites. In *Alkaline igneous rocks*, J. G. Fitton & B. G. J. Upton (eds), 53–83. Geological Society of London Special Publication 30.

Le Bas, M. J. & J. A. Aspden 1981. The comparability of carbonatitic fluid inclusions in ijolites with natrocarbonatite lava. *Bulletin Volcanologique* **44**, 429–38.

Le Bas, M. J. & C. Handley 1979. Variation in apatite composition in ijolitic and carbonatitic igneous rocks. *Nature* **279**, 54–6.

Liu Tiegen 1985. Geological and geochemical character of Baiyon Ebo rauhaugite. *Acta Petrologica Sinica* **1**, (3), 15–28.

Mariano, A. N. & P. L. Roeder 1983. Kerimasi: a neglected carbonatite volcano. *Journal of Geology* **91**, 449–55.

Mian, I. 1987. *Mineralogy and petrology of the carbonatites, syenites and fenites of the North West Frontier Province, Pakistan*. Ph.D. thesis, University of Leicester, Leicester.

Mian, I. & M. J. Le Bas 1986. Sodic amphiboles from the Loe Shilman carbonatitic complex, NW Pakistan. *Mineralogical Magazine* **50**, 187–97.

Nielsen, T. F. D & B. Buchardt 1985. Sr–C–O isotopes in nephelinitic rocks and carbonatites, Gardiner complex, Tertiary of East Greenland. *Chemical Geology* **53**, 207–17.

Nixon, P. H. & G. Hornung 1973. The carbonatite lavas and tuffs near Fort Portal, western Uganda. *Overseas Geology and Mineral Resources* **41**, 168–79.

Piotrowski, J. M. & A. D. Edgar 1970. Melting relations of undersaturated alkaline rocks from S. Greenland compared with those from Africa and Canada. *Meddelelser om Grønland* **181**, No. 9.

Pyatenko, I. K. & L. G. Saprykina 1976. Carbonatite lavas and pyroclastics in the Palaeozoic sedimentary volcanic sequence of the Kontozero District, Kola Peninsula. *Doklady Akademii Nauk SSSR* **229**, 185–7.

Rankin, A. H. & M. J. Le Bas 1974. Liquid imiscibility between silicate and carbonate melts in naturally-occurring ijolite magma. *Nature* **250**, 206–9.

Rass I. T. & D. I. Frikh-Khar 1987. On the discovery of carbonatites in the Berkhnemelov ultrabasic volcanoes of Kamchatka. *Doklady Akademii Nauk SSSR* **294**, 182–6.

Silva, L. C., M. J. Le Bas & A. H. F. Robertson 1981. An oceanic carbonatite volcano on Santiago, Cape Verde Islands. *Nature* **294**, 644–5.

Silva, L. C. & M. L. Ubaldo 1985. Consideracoes geologicas e petrogeneticas sobre os tufos carbonatiticos globulares da estrutura alcalino-carbonatitica do Norte de Santiago, arquipelago de Cabo Verde. *Garcia de Orta, Série de geologia, Lisboa* **8**, 1–6.

Sommerauer, J. & K. Katz-Lehnert 1985. Trapped phosphate melt inclusions in silicate-carbonate-hydroxyapatite from comb-layer alvikites from the Kaiserstuhl carbonatite complex (SW-Germany). *Contributions to Mineralogy and Petrology* **91**, 354–9.

Sutherland, D. S. 1967. A note on the occurrence of potassium-rich trachytes in the Kaiserstuhl carbonatite complex, west Germany. *Mineralogical Magazine* **36**, 334–41.

Treiman, A. H. & A. Schedl 1983. Properties of carbonatite magma and processes in carbonatite magma chambers. *Journal of Geology* **91**, 437–47.

Turbeville, B. N., J. A. Wolff, D. J. Miller & M. J. Le Bas 1987. An oceanic nephelinite–phonolite–carbonatite association, Brava, Cape Verde Islands. *EOS. Transactions, American Geophysical Union* **68**, (44), 1522.

Tuttle, O. F. & J. Gittins 1966. *Carbonatites*. New York: Wiley.

Vartiainen, H. & H. Paarma 1979. Geological characteristics of the Sokli carbonatite complex, Finland. *Economic Geology* **74**, 1296–306.

Verwoerd, W. J. 1967. *The carbonatites of South Africa and South West Africa*. Geological Survey of South Africa, Handbook 6.

Verwoerd, W. J. 1978. Liquid immiscibility and the carbonatite–ijolite relationship: preliminary data on the join $NaFe^{3+}Si_2O_6$–$CaCO_3$ and related compositions. *Annual Report of the Geophysical Laboratory, Washington* **77**, 767–74.

Viladkar, S. G. 1980. The fenitized aureole of the Newania carbonatite, Rajasthan. *Geological Magazine* **117**, 285–92.

Viladkar, S. G. 1981. The carbonatites of Amba Dongar, Gujarat, India. *Bulletin of the Geological Society of Finland* **53**, 17–28.

Woolley, A. R. 1969. Some aspects of fenitization with particular reference to Chilwa Island and Kangankunde, Malawi. *Bulletin of the British Museum (Natural History), Mineralogy* **2**, 189–219.

Woolley, A. R. 1982. A discussion of carbonatite evolution and nomenclature, and the generation of sodic and potassic fenites. *Mineralogical Magazine* **46**, 13–17.

Wyllie, P. J. 1978. Silicate–carbonate systems with bearing on the origin and crystallization of carbonatites. *Proceedings of the First International Symposium on Carbonatites. Poços de Caldas, Brazil*, June 1976, 61–78. Brasilia: Brasil Departamento Nacional da Produção Mineral.

Wyllie, P. J. & G. M. Biggar 1966. Fractional crystallization in the "carbonatite systems" CaO–MgO–CO_2 and CaO–CaF_2–P_2O_5–CO_2–H_2O. *Mineralogical Society of India*, IMA volume, 92–105.

Wyllie, P. J. & A. P. Jones 1984. Experimental data bearing on the origin of carbonatites, with particular reference to the Mountain Pass rare earth deposit. In *Proceedings of the Second International Congress on Applied Mineralogy, Los Angeles, California*. W. C. Park, D. M. Hausen & R. D. Hagni (eds), 937–49.

18
UPPER-MANTLE ENRICHMENT BY KIMBERLITIC OR CARBONATITIC MAGMATISM

A. P. JONES

ABSTRACT

Recognized mantle enrichment processes beneath South Africa include crystallization of MARID pegmatites from kimberlitic magma trapped in the subcontinental lithosphere. Residual melts or fluids enriched in Nb, REEs (rare earth elements), Ti, U, and Zr, and many other LIL elements, migrated to form distal metasomites. These elements are embodied in mantle titanate minerals, whose chemistries resemble enriched kimberlite. The same process throughout geological time could have metasomatized large parts of the lithosphere. Availability of accessory REE-bearing titanates during magma genesis in the mantle would provide a highly enriched kimberlitic component capable of being scavenged by subsequent melts/fluids. The unusual physical and chemical properties of carbonatitic melts suggest that they may be related to the titanates in two ways:

(a) they could be efficient scavengers, and obtain the major proportion of their incompatible elements and REEs from the titanates; and
(b) small-volume, hydrous, carbonatitic melts could be the agents responsible for the metasomatism.

18.1 INTRODUCTION

Carbonatites and kimberlites are crustal expressions of mantle magmatism involving large concentrations of C and H, as carbonate and hydrous minerals, from depths close to the low-velocity zone. The sources of these volatile species are unknown but may include ancient subducted lithosphere, primordial mantle, or even the outer core (Smith 1979, Williams *et al.* 1987). The volatiles cause magmatism when the local geotherm exceeds the mantle solidus. The petrology of silicate-poor, carbonate melts (i.e. carbonatites) and carbonate-rich kimberlites is similar at high pressures (e.g. Wyllie 1987*a*). Therefore, the term 'kimberlite' in this paper is used in a broad sense, and repeated similarities between kimberlites, carbonatites, and metasomites emerge. A variety of kimberlitic magmas are likely to have crystallized sporadically near the lithosphere–asthenosphere boundary over geological time, where released volatiles may have caused mantle metasomatism and triggered further partial melting in the overlying lithosphere. Upwelling of mantle plumes can lead to magma genesis in the depth interval 90–65 km (Wyllie 1980, 1987a, b), with production of a metasomatic layer that contains accessory REE-bearing titanates. Below thickened continents these enrichments may be driven by mantle plumes rising from greater depths in the sub-lithospheric mantle, similar to oceanic hot-spots in origin except that they are trapped by cold, insulating lithosphere (e.g. Le Roex 1986). Xenolith suites characterize the metasomatic and igneous products of the mantle, and several petrogenetic schemes have been proposed (Bailey 1982, Erlank et al. 1987, Menzies et al. 1987) that are based on an assumed peridotitic bulk composition for the mantle. This paper reviews the important characteristics of a kimberlitic component in the differentiation of the Earth, by examining the ability of accessory titanate minerals in the mantle to pass on a kimberlite signature during metasomatism. There are repeated geochemical similarities between carbonatites, kimberlites, and metasomatized mantle, and it is suggested that the agent responsible for metasomatism is likely to have been a small-volume, hydrous carbonatitic melt.

18.2 MANTLE METASOMATISM

Studies of mantle metasomatism have progressed rapidly in recent years, and it has been recognized that such a process can explain trace-element and isotopic enrichment or depletion in the upper mantle, particularly in relation to basaltic volcanism (e.g. Hawkesworth et al. 1984, Menzies et al. 1987). Two different types of mantle metasomatism can be distinguished (Dawson 1980). Patent or modal metasomatism is believed to be caused by migrating fluids resulting in secondary replacement textures and veining. The term 'fluid' is used in the general sense to include, for example, volatile-rich melt or supercritical fluid (see Eggler 1987, Wyllie 1987a). Modal metasomatism (Harte 1983) is more specific, but is taken here to be synonymous with patent metasomatism for reasons stated elsewhere (Erlank et al. 1987). By contrast, cryptic metasomatism enriches xenoliths in certain trace elements, and the xenolith shows no petrographic evidence of mineralogical change. Some xenoliths offer evidence of more than one generation of metasomatism (e.g. Jones et al. 1982). There is a useful analogy between metasomatism in the mantle, and fenitization (e.g. Le Bas 1977) where the metasomatized products are available for study, but the fugitive parental fluids or melts are not. Different types of metasomatism have been perceived as either (a) iron-, titanium-rich, related to alkali basaltic melts; or (b) potassium-rich, related to kimberlitic and lamproitic hydrous fluids (Erlank et al. 1987, Harte 1987, Menzies et al. 1987). Identification of the likely mantle metasomatic agent can be attempted by assessing the various geochemical data relative to presumed unaffected mantle such as average mantle, or bulk Earth. Recent high-pressure experiments have provided important new constraints. Solubility determinations clearly suggest that CO_2–H_2O-rich fluids are very ineffective metasomatic agents in the mantle compared with melts (Eggler 1987). Similar fluids and brines develop universally high wetting angles in upper-mantle conditions, which led Watson & Brenan (1987) to conclude that fluids are unable to penetrate mantle rocks by infiltration. These experiments suggest that mantle metasomatism is very unlikely to be caused by migration of hydrous fluids, and is caused instead by melts.

Much information about mantle metasomatism has been accumulated from studies of xenolith suites entrained in kimberlites from South Africa, where a petrogenetic link between kimberlite magmatism in the broadest sense, and potassium-rich mantle metasomatism may exist. The principal metasomatic minerals are amphibole, mica, and, to a lesser extent, diopside. There is thus a first-order correlation between the enriching agent and hydrous mineral assemblages. At their most developed, these minerals form anastomosing networks, or linear veinlets on a millimetre scale in peridotite xenoliths from the Bultfontein kimberlite (Jones *et al.* 1982, Erlank *et al.* 1987). In addition, minor but ubiquitous titanium-rich minerals may include Mg-ilmenite, rutile, and a small group of LREE-rich titanates. Also, discrete carbonate minerals are present, with low initial $^{87}Sr:^{86}Sr$ ratios near 0.704 (Jones *et al.* 1982, Exley 1983). The metasomatized and veined peridotites, collectively called metasomites (Dawson 1987*a*), have mineralogical similarities to a distinctive suite of South African xenoliths called the MARID suite (mica–amphibole–rutile–ilmenite–diopside: Dawson & Smith 1977). Differences between mineral chemistries of the metasomites and the MARID suite involve variations in Cr, Fe, and Ti contents. Thus, coexisting mica and amphibole have higher TiO_2 (Fig. 18.1) in the MARID suite compared with metasomites. The origin of the metasomites is compatible with interaction between a kimberlite and wall-rock, chromium-bearing mantle peridotites (Jones *et al.* 1982, Kramers *et al.* 1983). The MARID suite has been interpreted as an igneous association of mantle pegmatites crystallized from kimberlitic or lamproitic magmas (Dawson & Smith 1977, Kramers *et al.* 1983, Waters 1987). The evidence is consistent with a model whereby MARID rocks crystallized from kimberlite magma trapped in the subcontinental lithosphere, from which metasomatic melt or fluid enriched, in particular,

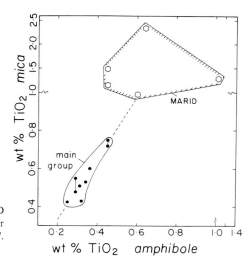

Figure 18.1 Mineral chemistry distinction between MARID suite and metasomites with veins ('main group') illustrated for wt% TiO_2 in phlogopite and coexisting richterite (Jones *et al.* 1982).

in Nb, REEs, Ti, U, Zr, and many other LIL elements, migrated to form distal metasomites (Jones et al. 1982, Kramers et al. 1983, Dawson 1987b, Waters 1987). These elements are incorporated to varying degrees in a small group of mantle titanate minerals, dominated by members of the crichtonite series. The euhedral to subhedral titanate grains occur in the metasomatic veins, and may be poikilitically enclosed by amphibole, diopside, or mica. They are opaque and easily mistaken for spinel in transmitted light. The presence of a titanate mineral as an accessory phase in the mantle would provide a low-melting component capable of being scavenged by melts or fluids (Jones et al. 1982). The same LREE-rich titanates can occur in kimberlites themselves (Haggerty 1983), though LREE-bearing perovskite is far more common (Jones & Wyllie 1984).

Detailed studies of mantle xenoliths from the Kimberley area (Erlank et al. 1987) have confirmed pervasive metasomatism in the sub-crustal lithosphere. Of the peridotites studied, relatively few (approximately 10%) of the garnet peridotites (GP) are phlogopite free, and most bear primary phlogopite (garnet–phlogopite peridotites or GPP). Garnet-free peridotites are either phlogopite peridotites (PP) with higher phlogopite ($> 1\%$), or phlogopite–K-richterite peridotites (PKP), which also contain titanate minerals of the crichtonite series. Erlank et al. (1987) have interpreted their geochemical and mineralogical data in terms of an overall metasomatic process resulting in increasing trace-element enrichment in the series GP–GPP–PP–PKP (see also Dawson 1987b). Discrete veins in the Kimberley peridotites (Jones et al. 1982) only occur in the PP and PKP nodules.

18.3 MANTLE TITANATE MINERALS

The mantle titanate minerals have the general formula $AM_{21}O_{38}$ and show extensive solid solution between several end members named according to the dominant large radius cation in the A-site. These are principally lindsleyite (Ba), mathiasite (K), crichtonite (Sr), and loveringite (Ca), and, to a lesser extent, davidite (REE) (Smyth et al. 1978, Jones et al. 1982, Haggerty 1983, Haggerty et al. 1983, Haggerty et al. 1986). The compositional range of mantle titanates is shown in Figure 18.2. The position of the mantle titanate analysed in detail and referred to throughout this chapter is indicated by point 'M' (Jones & Ekambaram 1985). Because the acronym 'LIMA', based on a predominance of lindsleyite and mathiasite (Haggerty 1983), precludes the existence of a considerable crichtonite (sensu stricto) and loveringite component, the term titanate is preferred in this paper. This also simplifies further complexities associated with the mantle titanates, including additional new minerals (Haggerty et al. 1986). A large number of potential cations, including Cr, Fe, Mn, Nb, Ti, Zn, Zr, can be accommodated in the M-site, and the titanate series can be considered an efficient scavenger of a large number of incompatible elements. Jones & Ekambaram (1985) presented multi-elemental analyses of one mantle titanate grain. The INAA data demonstrated a strong LREE enrichment in the natural titanate (Fig. 18.3). High La:Y ratios in the cores of zoned titanates (Jones et al. 1982) might reflect changes in the REE character of the metasomatic melt. In detail,

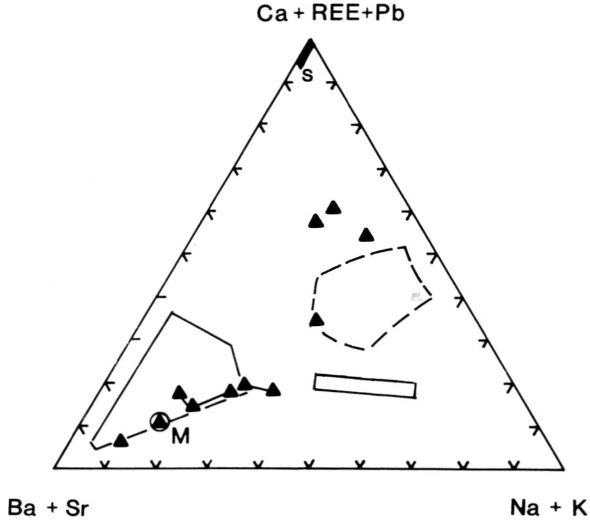

Figure 18.2 Compositions of mantle titanates of the crichtonite series. Natural compositions in atomic per cent plotted for Bultfontein metasomites (▲) (Jones *et al.* 1982). Best analysed titanate, M (Jones & Ekambaram 1985); dashed field outlines, titanates from Jägersfontein (Haggerty 1983); and solid fields, additional mantle and kimberlite titanates (Haggerty *et al.* 1983). Solid inset near Ca + REE + Pb apex (S) indicates range of synthetic loveringite–davidites (Green & Pearson 1987).

the REE behaviour in the crichtonite series can be quite variable, because natural davidites show LREE and HREE enrichment (Hayton 1960). The results of recent experiments (Green & Pearson 1987) to synthesize loveringite–davidite from HREE-rich compositions (chondrite-normalized) indicate depletion in the middle REEs. It should be noted, however, that these experiments also yielded coexisting titanite (sphene) known to be enriched in the middle REEs. Thus, the crichtonite minerals also exhibit an extraordinary flexibility in REE composition, both in terms of absolute and relative abundance. It is this variability of REE composition that enables the mantle titanates to mimic the geochemical signatures of their parent

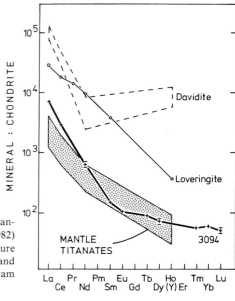

Figure 18.3 Chondrite-normalized REE concentrations in titanates showing the range of electron-probe data (Jones *et al.* 1982) for mantle titanates compared with INAA data for M in Figure 18.2. Note the steep LREE-enriched pattern. Loveringite and davidite data are plotted for comparison (Jones & Ekambaram 1985).

melts so successfully. The textural association of titanate and Cr-spinel (Jones et al. 1982) and the high concentration of Cr_2O_3 (14–17 wt%) in titanates indicate that crichtonites may be stabilized by Cr. Chromium is probably derived from Cr-spinel in host peridotites (Jones et al. 1982); Cr may also have played a similar role in the crystallization of loveringite in crystal cumulates (Campbell & Kelly 1978).

Additional end members of the crichtonite series are rare in the mantle titanates, but these are important because they contain Pb and U. If the low Pb levels determined by electron probe ($PbO_2 < 0.06$ wt%) are entirely the result of radiogenic decay of U and Th (98 p.p.m., 14.7 p.p.m. respectively), then it might be possible to calculate a U–Pb age for the titanates (Campbell & Kelly 1978). The low Th : U ratio (0.15) in the titanate is particularly interesting because it is very low in comparison to the value of 5.2 found in kimberlites (Wedepohl & Muramatsu 1979). The presence of U should also enable these phases to be located by fission-track methods.

18.4 KIMBERLITE SIGNATURE

Although kimberlites are variable in composition they are characterized by unusually high abundances of many incompatible trace elements, including the LREEs, and compatible elements, such as Cr, Mg, and Ni, features attributed to an origin by small degrees of partial melting of a garnet peridotite mantle source. Kimberlites are also depleted in Al, Ca, and Na relative to other alkaline rocks believed to originate by small degrees of partial melting, e.g. nephelinites. Trace-element differences may reflect variations in the diopside : phlogopite and $CO_2 : H_2O$ ratios in their source regions (Dawson et al. 1985, Fraser et al. 1985). The enriched geochemistries of carbonatites and kimberlites (see Table 18.1) make them important in the budget for bulk Earth calculations. Thus, Anderson (1984) compared the origin of terrestrial kimberlites to extraction of enriched residual KREEP from a crystallizing magma ocean on the Moon. The chondrite-normalized trace-element abundances for KREEP and kimberlite are surprisingly similar (see Fig. 18.4), except that kimberlite is relatively depleted in eclogite-compatible elements and Na, reflecting garnet and pyroxene crystallization. In his model, kimberlite crystallized from a residual fluid remaining after eclogite cumulates formed from a terrestial magma ocean. The mantle titanate has a chondrite-normalized chemistry similar to that for kimberlite but at greatly enriched levels (Fig. 18.4). Compositional variations among titanates from different metasomites in terms of Ba, Ca, K, and REEs are small in comparison to kimberlite. The enrichment levels in titanate are considered relative to average kimberlite in Figure 18.5. Normalization to specific types of kimberlite, such as the Group I basaltic kimberlites or Group II micaceous kimberlites (Smith 1983), hardly affects the overall pattern. Most kimberlites, therefore, fall in a band between +0.5 and −0.5 of the 1.0 line on Figure 18.5. The titanate is depleted relative to kimberlite in Ca, K, Na, and Rb, and other elements contained in associated metasomite minerals, such as amphibole, diopside, and phlogopite. The titanate is even more LREE-enriched than kimberlite and contains

Table 18.1 Titanate mineral compared with average rock, compositions (p.p.m.).

	Titanate	Kimberlite	Carbonatite
Na	190	3580	2151
K	4650	13300	2241
Rb	500	58	21
P	1000	4890	8902
Nb	1000	110	1017
Th	14.7	16	71
U	98.0	3.1	11.1
Sr	11050	632	6825
Ba	30360	808	3898
La	2279	150	667
Ce	2595	200	1904
Nd	404	85	776
Sm	30.4	13	66
Eu	8.0	3.0	18.8
Tb	4.2	1.0	5.6
Dy	24.4	7.3	34
Y	100	13	93
Yb	13.1	1.2	55
Lu	2.0	0.2	0.4
Ti	374000	9110	1220
Zr	24200	240	182
Hf	692	7	3.2
Ca	3359	67500	308970
Fe	67700	61400	36889
Mn	800	1240	5090
Ni	500	1160	23
V	3475	98	82
Mg	20500	153000	58684
Al	900	17840	5495
Cr	103000	1380	33
Pb	<3300	8.5	74
Sc	168	15	91

Titanate data from Jones *et al.* (1982), Jones & Ekambaram (1985); average kimberlite from Wedepohl & Muramatsu (1979); carbonatite, average of calciocarbonatite and magnesiocarbonatite from Woolley & Kempe (this volume).

some HREE, also slightly elevated relative to kimberlite. This suggests that the parent melt was not in equilibrium with a residual HREE phase, such as mantle garnet or zircon associated with MARID rocks (Exley & Smith 1982). The Th:U ratios are quite different in titanates relative to kimberlites (Fig. 18.5) and the titanate contains anomalously high U (98 p.p.m.). Zirconium and Hf are also enriched in the titanate, with a Zr:Hf ratio of 35, similar to average kimberlite (Wedepohl & Muramatsu 1979).

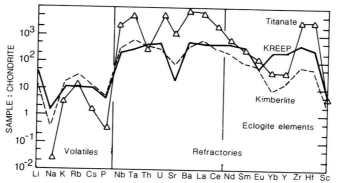

Figure 18.4 Chondrite-normalized trace-element composition for average kimberlite, mantle titanate, and lunar KREEP (Anderson 1984). The relative depletion of eclogite elements (HREEs, Y, Zr, Hf, Sc) and Na in kimberlite reflects garnet and pyroxene crystallization, with kimberlite being the composition of the residual liquid. The mantle titanate could represent an enriched kimberlite component with a significantly fractionated Th : U ratio.

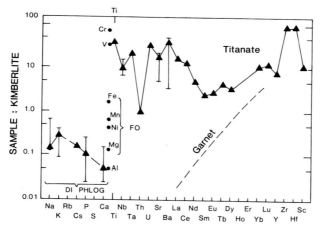

Figure 18.5 Plot of selected element-abundance data in mantle titanate minerals relative to average kimberlite. Mineral composition shows average (▲ or ●) or analysed range (bars) where significantly variable (Jones *et al.* 1982, Jones & Ekambaram 1985, Jones, unpublished data). Mantle titanate is enriched relative to average kimberlite (Wedepohl & Muramatsu 1979) in most trace elements considered incompatible in silicate melts, and depleted in olivine (FO) and MARID elements (DI, diopside; PHLOG, phlogopite, amphibole). Notable is the strong enrichment in Cr and large fractionation of U relative to Th. The titanate is enriched in both LREEs (La–Sm) and HREEs (Dy–Lu) *relative to kimberlite*; approximate HREE control for garnet is shown for comparison.

18.5 SOUTH AFRICAN MODEL

The integrated results from mantle xenoliths in the Kimberley pipes, one of the best known and deepest sections of upper-mantle lithologies, clearly indicate that the mantle was extensively metasomatized. Erlank *et al.* (1987) have shown that the majority of the peridotites have been affected by the same metasomatic process. Unmetasomatized garnet peridotites without phlogopite (GP) are surprisingly few, and form about 10% of the nodules sampled. In the model proposed by Erlank *et al.* (1987), a progressive increase in hydrous minerals leads to phlogopite–K-richterite peridotites (PKP), the most metasomatized mantle rocks. There is a coherent increase in many incompatible trace elements in the sequence GP–GPP–PP–PKP and Erlank *et al.* draw analogy to stockwork-veined, metasomatized mantle linked by channelways rich in incompatible elements (Bailey 1982). Even harzburgites and dunites have been enriched in the same incompatible trace-element suite (especially Ba, K, Nb, Rb, Sr, Zr) relative to primitive or primordial mantle (Dawson 1987*a*).

When the MARID rocks are considered together with the metasomites and kimberlite, it is clear, on the basis of chemistry and mineralogy, that there could be a broad genetic relationship among all three. Some effects of minerals controlling the whole-rock chemistry can be shown using plots of Sr v. CaO (Fig. 18.6) and TiO_2 v. Nb (Fig. 18.7). Both diagrams show the influence of Sr- and Nb-bearing titanate on the chemistry of the PKP metasomites. Additionally, MARID rocks may be crude mixtures between peridotites and kimberlite. Other important minerals which control the Sr : Ca ratios and Ca and Sr abundances include diopside and apatite in metasomites, MARID and kimberlite, and carbonate in metasomites and kimberlite. Minerals controlling Nb : Ti ratios and Nb and Ti abundances include ilmenite and rutile in metasomites and MARID and perovskite in kimberlite.

Erlank *et al.* (1987, Figs 28, 29) show that there are no simple inter-element

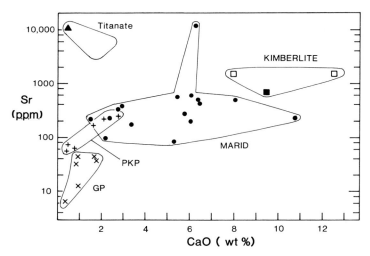

Figure 18.6 Sr (p.p.m.) plotted on a log scale versus CaO (wt%) for Kimberley peridotites, kimberlite, and titanates. This diagram shows relative enrichment in Sr of the most metasomatized phlogopite–K-richterite peridotites (PKP) compared to unmetasomatized (?) garnet peridotites (GP). MARID rocks (Waters 1987) may represent mixtures between the peridotites and kimberlite (□, Bultfontein, Erlank *et al.* 1987; ■, average kimberlite, data from Table 18.1). PKP rocks, including veined peridotites (Jones *et al.* 1982), are displaced towards the field for the titanate (approximate range from data in Figure 18.2). These rocks contain titanate as an important metasomatic mineral. The rock chemistries of the MARID suite, and to a lesser extent the metasomatized xenoliths, are strongly influenced by the variability in the modes of metasomatic minerals. Other important minerals which control Sr : Ca rock ratios and contents are diopside and apatite in PKP, MARID, and kimberlite; carbonate in PKP and kimberlite; and perovskite in kimberlite.

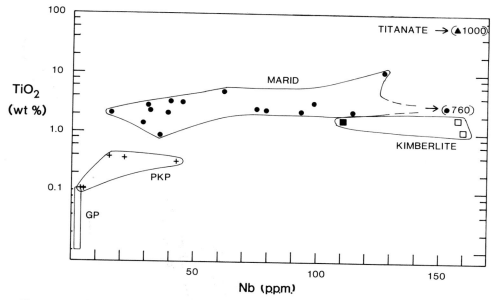

Figure 18.7 TiO$_2$ (wt%) plotted on a log scale v. Nb (p.p.m.) for whole-rock Kimberley peridotites, kimberlites, and titanates. Note, in comparison with Figure 18.6, the consistent relative displacements between GP, PKP, MARID, and kimberlite. Data from Nb-bearing titanate samples and one Nb-rich MARID sample are plotted on the diagram, to indicate Nb levels of 1000 and 760 p.p.m. respectively. Other minerals controlling rock Ti : Nb ratios and contents include rutile and ilmenite (in PKP, MARID, and kimberlite) and perovskite (in kimberlite). Symbols are as in Figure 18.6.

relationships among metasomites, MARID, and kimberlite. However, the metasomites and MARID rocks show wide variations in proportions of modal metasomatic minerals that largely control trace-element chemistry of the mantle rocks. The total range in incompatible trace elements for the mantle lithologies shows a remarkable coherence, and overlap between the rock groups. To a first approximation, the enriched mantle rock chemistries can be explained as mixtures between peridotites and kimberlite or titanate.

Average MARID rock chemistry (Waters 1987) and titanate mineral chemistry, both normalized to kimberlite, are plotted in Figure 18.8. Of particular note is the parallel behaviour of the volatile elements (Na–P). These elements are uniformly enriched in MARID rocks because of increased diopside, phlogopite, and richterite compared with kimberlite, and depleted in titanate. Most notable is the enrichment in the elements Nb to Ba and the striking mirror-image between LREEs (La–Nd), which are enriched in titanate and depleted in MARID relative to kimberlite. Elevated Zr in metasomite titanate provides a ready explanation for relatively low Zr levels in MARID rocks (Waters 1987), if both are kimberlite-related, and the titanate crystallized from fugitive MARID-derived melts or fluids. The high Zr and Hf levels in titanate may partly reflect partitioning of the elements between melt and solid, but presumably these elements were present at high levels in the metasomatic melts or fluids. Significant LREE enrichment in the titanate associated with carbonates in the metasomites but not in MARID rocks, again suggests a carbon

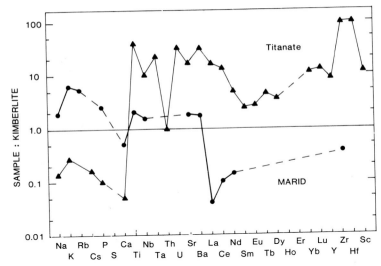

Figure 18.8 Kimberlite-normalized trace-element abundances for MARID xenoliths and titanate. The data are compatible with a simple model in which kimberlite crystallized to form MARID rocks and titanate minerals formed in distal metasomites (essentially the PKP rocks of Erlank *et al.* 1987). In particular, note the parallel behaviour of volatile elements (Na–P) and the mirror-image between LREEs (La–Nd) which are enriched in titanate and depleted in MARID relative to kimberlite. MARID data are averages of 16 samples from Waters (1987). Other data are taken from references given in Figure 18.2.

dioxide-rich metasomatic agent. Insufficient elements have been analysed in MARID rocks to make further detailed comparisons.

The simplest explanation of the chemical data is to form the igneous MARID suite by crystallization of a kimberlitic melt, and to precipitate titanate minerals from fugitive residual melts or fluids to form distal metasomites (Jones *et al.* 1982, Kramers *et al.* 1983). An alternative proposal of an ultrapotassic lamproitic parent for the MARID rocks (Waters 1987) more closely matches the major element chemistry (especially K), but assumes mostly closed-system crystallization, which is somewhat unrealistic (Dawson 1987*a, b*).

18.6 DISCUSSION

There is an apparent chemical continuity in titanium- and niobium-bearing minerals between the LREE-bearing titanates in metasomatized mantle, LREE-rich perovskites in kimberlites (Jones & Wyllie 1984, 1985), and pyrochlore in carbonatites that may be of genetic importance. Although the modal percentages differ greatly, these minerals can account for most Ti, Nb, and other kimberlite-enriched elements in each of the host rocks, while the REEs can be readily accommodated in carbonatites. The series metasomite–kimberlite–carbonatite also represents rocks with radically increasing $CO_2:H_2O$ ratios, and genetic links between kimberlites and carbonatites have been widely proposed (e.g. Dawson & Hawthorne 1973, Wyllie 1980, 1987*b*, Wyllie & Jones 1985, Wyllie, this volume). Average whole-rock compositions for kimberlite and carbonatite are compared with titanate in Table 18.1. It is widely accepted that carbonatites may have more than one origin (e.g. Wyllie & Jones 1985) which is ultimately in the mantle. The following discussion is focused towards primary carbonatitic magmas in the mantle, and does not directly evaluate those derived by prolonged fractionation from nephelinitic magmas or by liquid immiscibility (Le Bas 1977, this volume, Kjarsgaard & Hamilton, this volume).

The petrogenesis of carbonatites must bear a first-order relationship with mantle peridotites suitably enriched in carbonate components and incompatible elements. Primary carbonates, though rarely preserved in mantle xenoliths for a variety of reasons (e.g. Boyd & Gurney 1982, Boettcher 1984, Haggerty 1986) clearly do exist (Jones *et al.* 1982, Berg 1986, Smith 1987). The petrological effects of carbonate in mantle peridotites have been studied experimentally. A spectrum of carbonatitic to kimberlitic melts can be produced by small degrees of partial melting of normal mantle peridotites bearing C–O–H species at a range of depths below 75 km (Wyllie 1980, 1987*a*, Eggler & Baker 1982). However, though carbonatitic in terms of major elements, these melts would lack much of the incompatible trace-element enrichment characteristic of carbonatite magmas. Geochemical and isotopic evidence indicates that carbonatites, and perhaps most incompatible-element enriched primary magmas, originate from previously enriched, or metasomatized, mantle source rocks. It is suggested that the link between primary carbonatites and metasomatized mantle is fundamental, and that trace-element enrichment in carbonatites is largely controlled by LREE-enriched titanate minerals.

Arguments have been made to suggest that mantle titanate minerals represent enriched kimberlite. The same titanate was discovered in seven of 14 metasomites from Bultfontein (Jones *et al.* 1982) and has been synthesized at upper-mantle temperatures (900–1100 °C) and pressures (20–30 Kb) (Podpora & Lindsley 1984). On this basis it is a suitable contender for a more widespread mantle accessory mineral, perhaps the fifth phase of classic four-phase peridotite, and the KIMB or Q component of Anderson (1983). As an accessory phase with a kimberlitic signature, normal mantle would be enriched in an important array of incompatible trace elements, including LREEs, Ba, and Sr. The presence of titanates in the mantle would provide a low melting phase ideal for scavenging by successive melts or fluids. Addition of 1 part in 1000 titanate to normal mantle would approximately double the LREEs (La, Ce), Ba, Sr, Rb, and U abundances in average mantle. This would be approximately equivalent to one 0.3 mm grain in a typical thin section. Petrographic textures of titanate are complex, but subhedral grains in veined peridotites from Bultfontein (Jones *et al.* 1982) are commonly enclosed in phlogopite, diopside, and amphibole. This suggests that the titanate crystallized early in the metasomites and may be stable to slightly higher temperatures than the associated hydrous minerals and diopside. If titanate in a mantle source region could survive independently of phlogopite, this would provide a mechanism for obtaining the relatively high Ti:K ratios characteristic of most oceanic and continental within-plate basalts (Hawkesworth *et al.* 1984).

Carbonatites may be related to the metasomatic titanates in two ways:

(a) primary carbonatitic melts may obtain a large proportion of their incompatible elements and REEs by scavenging of existing titanates; and
(b) the metasomatizing melts which crystallized titanates may also have been carbonatitic.

In the first instance (a), carbonate-rich fluids and carbonatite melts can both accommodate large amounts of the LREEs (Wendlandt & Harrison 1979, Jones & Wyllie 1986). The mantle titanate would be a major source of REEs and incompatible elements in carbonatites, as shown by comparative chemistries in Table 18.1. Major element discrepancies, such as Ti, make simple mixing of titanate into carbonatite melt unlikely, unless in equilibrium with a residual chromium-bearing Ti-mineral such as rutile. The transfer mechanism is unknown and need not be restricted to carbonatites, but scavenging of titanate from the lithosphere (Jones *et al.* 1982) would be aided by the extraordinary fluidity of carbonatitic melts (see Treiman, this volume). Furthermore, compaction calculations suggest that very small amounts of partial melt appropriate for carbonatites may be interconnected in the mantle matrix (McKenzie 1984), which would assist in scavenging. In the second case (b), the above arguments are equally important since they point to the likely mobility of carbonatitic melts in the mantle. There is also the observation that in addition to hydrous minerals, occasional carbonate is observed in the metasomites (see §18.5). Experimental data also favour melts as metasomatic agents rather than fluids (see §18.2), which would be buffered towards increasing $CO_2:H_2O$ ratios during removal of hydrous minerals. The titanate-bearing metasomites may therefore have been caused by infiltration of hydrous carbonatitic melts, whose exact nature and timing remain uncertain.

New techniques have increased experimental pressures by almost an order of magnitude (e.g. Irifune & Ringwood 1987), which may alter established concepts of carbonatite and kimberlite genesis. For example, subsolidus phase relations between mantle garnet and pyroxenes show increasing solid solution from pressures as low as 50 Kb. At approximately 100 Kb, 20 mole% pyroxene is soluble in garnet (Akaogi & Akimoto 1977, 1979), and by 190 Kb pyroxene is completely replaced by majorite garnet (Irifune & Ringwood 1987). This must surely affect REE partitioning and is immediately relevant to the genesis of kimberlites and dynamic upper-mantle models (e.g. Menzies & Hawkesworth 1987). At the very least, the routine assumption of HREE enrichment in residual garnet must become suspect at pressures beyond about 50 Kb.

18.7 CONCLUSIONS

(a) Studies of xenoliths from South African kimberlites suggest that MARID pegmatites crystallized from failed kimberlitic magma trapped in the subcontinental lithosphere. Residual melts migrated and reacted with host peridotites to form distal metasomites containing titanate minerals.

(b) Availability of an accessory REE-bearing titanate mineral in the mantle would provide a 'kimberlitic component' capable of being scavenged through absorption by subsequent melts or fluids.

(c) Approximately 1 part in 1000 titanate, occurring as discrete grains or grain boundary phases, would double the content of Ba, LREE (La, Ce), Rb, Sr, and U in average mantle.

(d) Because of their special physical properties and chemistries, carbonatite melts would be efficient scavengers, and could derive most of their incompatible elements, including the REEs, from the titanates.

(e) It is suggested that the agent responsible for metasomatism and crystallization of the titanates in the upper mantle could have been a small-volume hydrous carbonatitic melt.

ACKNOWLEDGEMENTS

C. J. Hawkesworth and S. E. Haggerty kindly provided useful comments on an earlier version of the manuscript. J. V. Smith and J. B. Dawson are thanked for providing the specimens and initial introduction to the subject. P. J. Wyllie is thanked for supplying further incentives through preprints and fruitful discussions.

REFERENCES

Akaogi, M. & S. Akimoto 1977. Pyroxene-garnet solid solution equilibria in the system $Mg_4Si_4O_{12}-Mg_3Al_2Si_3O_{12}$ and $Fe_4Si_4O_{12}-Fe_3Al_2Si_3O_{12}$ at high pressures and temperatures. *Physics of the Earth and Planetary Interiors* **15**, 90–106.

Akaogi, M. & S. Akimoto 1979. High pressure phase equilibria in a garnet lherzolite, with special reference to $Mg^{2+}-Fe^{2+}$ partitioning among constituent minerals. *Physics of the Earth and Planetary Interiors* **19**, 31–51.

Anderson, D. L. 1982. Chemical composition and evolution of the mantle. In *Advances in Earth and planetary sciences. High pressure research in geophysics*, S. Akimoto & M. H. Manghnani (eds), 301–18. Dordrecht: Reidel.

Anderson, D. L. 1983. Chemical composition of the mantle. *14th Proceedings of the Lunar and Planetary Science Conference Part I. Journal of Geophysical Research* **88**, supplement B41–52.

Anderson, D. L. 1984. Kimberlite and the evolution of the mantle. In *Kimberlites I: Kimberlites and related rocks*. Proceedings of the Third International Kimberlite Conference, J. Kornprobst (ed.), 395–403. New York: Elsevier.

Bailey, D. K. 1982. Kimberlite: "The mantle sample" formed by ultrametasomatism. *Terra Cognita* **2**, 232.

Berg, G. W. 1986. Evidence for carbonate in the mantle. *Nature* **324**, 50–1.

Boettcher, A. L. 1984. The source regions of alkaline volcanoes. In *Explosive volcanism: inception, evolution, and hazards*, 13–22. Washington, DC: National Academy Press.

Boyd, F. R. & J. J. Gurney 1982. Low-calcium garnets: keys to craton structure and diamond crystallisation. *Carnegie Institute of Washington Year Book* **81**, 261–7.

Campbell, I. H. & P. R. Kelly 1978. The geochemistry of loveringite, a uranium-rare-earth-bearing accessory phase from the Jimberlana Intrusion of Western Australia. *Mineralogical Magazine* **42**, 187–93.

Dawson, J. B. 1980. *Kimberlites and their xenoliths*. Berlin: Springer.

Dawson, J. B. 1987a. Metasomatised harzburgites in kimberlite and alkaline magmas: enriched restites and "flushed" lherzolites. In *Mantle metasomatism*, M. A. Menzies & C. J. Hawkesworth (eds), 125–44. London: Academic Press.

Dawson, J. B. 1987b. The MARID suite of xenoliths in kimberlite: relationship to veined and metasomatised peridotite xenoliths. In *Mantle xenoliths*, P. H. Nixon (ed.), 465–73. Chichester: Wiley.

Dawson, J. B. & J. B. Hawthorne 1973. Magmatic sedimentation and carbonatite differentiation in kimberlite sills at Benfontein, South Africa. *Journal of the Geological Society of London* **129**, 61–85.

Dawson, J. B. & J. V. Smith 1977. The MARID (mica–amphibole–rutile–ilmenite–diopside) suite of xenoliths in kimberlite. *Geochimica et Cosmochimica Acta* **41**, 309–23.

Dawson, J. B., J. V. Smith & A. P. Jones 1985. A comparative study of bulk rock and mineral chemistry of olivine melilitites and associated rocks from East and South Africa. *Neues Jahrbuch für Mineralogie, Abhandlungen* **152**, 143–75.

Eggler, D. H. 1987. Solubility of major and trace elements in mantle metasomatic fluids: experimental constraints. In *Mantle metasomatism*, M. A. Menzies & C. J. Hawkesworth (eds), 21–41. London: Academic Press.

Eggler, D. H. & D. R. Baker 1982. Reduced volatiles in the system C–O–H; implications to mantle melting, fluid formation, and diamond genesis. In *Advances in Earth and planetary sciences. High pressure research in geophysics*, S. Akimoto & M. H. Manghnani (eds), 237–50. Dordrecht: Reidel.

Erlank, A. J., F. G. Waters, C. J. Hawkesworth, S. E. Haggerty, H. L. Allsopp, R. S. Rickard & M. A. Menzies 1987. Evidence for mantle metasomatism in peridotite nodules from the Kimberley Pipes, South Africa. In *Mantle metasomatism*, M. A. Menzies & C. J. Hawkesworth (eds), 221–311. London: Academic Press.

Exley, R. A. 1983. Evaluation and application of the ion microprobe in the strontium isotope geochemistry of carbonates. *Earth and Planetary Science Letters* **65**, 303–10.

Exley, R. A. & J. V. Smith 1982. The role of apatite in mantle enrichment processes and in the petrogenesis of some alkaline basalt suites. *Geochimica et Cosmochimica Acta* **46**, 1375–84.

Fraser, K. J., C. J. Hawkesworth, A. J. Erlank, R. H. Mitchell & B. H. Scott-Smith 1985. Sr, Nd, and Pb isotope and minor element geochemistry of lamproites and kimberlites. *Earth and Planetary Science Letters* **76**, 57–70.

Green, T. H. & N. J. Pearson 1987. High-pressure, synthetic loveringite-davidite and its rare earth element geochemistry. *Mineralogical Magazine* **51**, 145–9.

Haggerty, S. E. 1983. The mineral chemistry of new titanates from the Jägersfontein kimberlite, South Africa: implications for metasomatism in the upper mantle. *Geochimica et Cosmochimica Acta* **47**, 1833–54.

Haggerty, S. E. 1986. Diamond genesis in a multiply constrained mantle. *Nature* **320**, 34–8.

Haggerty, S. E., A. J. Erlank & I. E. Grey 1986. Metasomatic mineral titanate complexing in the upper mantle. *Nature* **319**, 761–3.

Haggerty, S. E., J. R. Smyth, A. J. Erlank, R. S. Rickard & R. V. Danchin 1983. Lindsleyite (Ba) and mathiasite (K): two new chromium titanates in the crichtonite series from the upper mantle. *American Mineralogist* **68**, 494–505.

Harte, B. 1983. Mantle peridotites and processes – the kimberlite sample. In *Continental basalts and mantle xenoliths*, C. J. Hawkesworth & M. J. Norry (eds), 46–91. Cheshire: Shiva Publishing.

Harte, B. 1987. Metasomatic events recorded in mantle xenoliths: an overview. In *Mantle xenoliths*, P. H. Nixon (ed.), 625–40. Chichester: Wiley.

Hawkesworth, C. J., N. W. Rogers, P. W. C. van Calsteren & M. A. Menzies 1984. Mantle enrichment processes. *Nature* **311**, 331–5.

Hayton, J. D. 1960. The constitution of davidite. *Economic Geology* **55**, 1030–8.

Irifune, T. & A. E. Ringwood 1987. Phase transformations in a harzburgite composition to 26 GPa: implications for dynamical behaviour of the subducting slab. *Earth and Planetary Science Letters* **86**, 365–76.

Jones, A. P. & V. Ekambaram 1985. New INAA analysis of a mantle-derived titanate mineral of the crichtonite series, with particular reference to the rare earth elements. *American Mineralogist* **70**, 414–18.

Jones, A. P. & P. J. Wyllie 1984. Minor elements in perovskite from kimberlites and distribution of the rare-earth elements: an electron probe study. *Earth and Planetary Science Letters* **69**, 128–40.

Jones, A. P. & P. J. Wyllie 1985. Paragenetic trends of oxide minerals in carbonate-rich kimberlites, with new analyses from the Benfontein Sill, South Africa. *Journal of Petrology* **26**, 210–22.

Jones, A. P. & P. J. Wyllie 1986. Solubility of rare earth elements in carbonatite magmas, indicated by the liquidus surface in $CaCO_3$–$Ca(OH)_2$–$La(OH)_3$ at 1 kbar pressure. *Applied Geochemistry* **1**, 95–102.

Jones, A. P., J. V. Smith & J. B. Dawson 1982. Mantle metasomatism in 14 veined peridotite xenoliths from Bultfontein Mine, South Africa. *Journal of Geology* **90**, 435–53.

Kramers, J. D., J. C. M. Roddick & J. B. Dawson 1983. Trace element and isotope studies on veined, metasomatic and "MARID" xenoliths from Bultfontein, South Africa. *Earth and Planetary Science Letters* **65**, 90–106.

Le Bas, M. J. 1977. *Carbonatite–nephelinite volcanism*. New York: Wiley.

Le Roex, A. P. 1986. Geochemical correlation between southern African kimberlites and South Atlantic hotspots. *Nature* **324**, 243–5.

McKenzie, D. 1984. The generation and compaction of partially molten rock. *Journal of Petrology* **25**, 713–65.

Menzies, M. A. & C. J. Hawkesworth 1987. Upper mantle processes and composition. In *Mantle xenoliths*, P. H. Nixon (ed.), 725–38. Chichester: Wiley.

Menzies, M. A., N. W. Rogers, A. Tindle & C. J. Hawkesworth 1987. Metasomatic enrichment processes in lithospheric peridotites, an effect of asthenosphere–lithosphere interaction. In *Mantle metasomatism*, M. A. Menzies & C. J. Hawkesworth (eds), 313–61. London: Academic Press.

Podpora, C. & D. H. Lindsley 1984. Lindsleyite and mathiasite: synthesis of chromium-titanates in the crichtonite ($A_1M_{21}O_{38}$) series. *EOS. Transactions, American Geophysical Union* **65**, 293. (abstract)

Smith, C. B. 1983. Pb, Sr and Nd isotopic evidence for sources of southern African Cretaceous kimberlites. *Nature* **304**, 51–4.

Smith, D. 1987. Genesis of carbonate in pyrope from ultramafic diatremes on the Colorado Plateau, southwestern United States. *Contributions to Mineralogy and Petrology* **97**, 389–96.

Smith, J. V. 1979. Mineralogy of the planets: a voyage in space and time. *Mineralogical Magazine* **325**, 1–89.

Smyth, J. R., A. J. Erlank & R. S. Rickard 1978. A new Ba–Sr–Cr–Fe titanate mineral from a kimberlite nodule. *EOS. Transactions, American Geophysical Union* **59**, 394. (abstract)

Waters, F. G. 1987. A suggested origin of MARID xenoliths in kimberlites by high pressure crystallisation of an ultrapotassic rock such as lamproite. *Contributions to Mineralogy and Petrology* **95**, 523–33.

Watson, E. B. & J. M. Brenan 1987. Fluids in the lithosphere, 1. Experimentally-determined wetting characteristics of CO_2–H_2O fluids and their implications for fluid transport, host-rock physical properties, and fluid inclusion formation. *Earth and Planetary Science Letters* **85**, 497–515.

Wedepohl, K. H. & Y. Muramatsu 1979. The chemical composition of kimberlites compared with the average composition of three basaltic magma types. In *Kimberlites, diatremes and diamonds: their geology, petrology, and geochemistry*. Proceedings of the Second International Kimberlite Conference, Vol. 1, F. R. Boyd & H. O. A. Meyer (eds), 300–12. Washington: American Geophysical Union.

Wendlandt, R. F. & W. J. Harrison 1979. Rare earth partitioning between immiscible carbonate and silicate liquids and CO_2 vapor: results and implications for the formation of light rare earth-enriched rocks. *Contributions to Mineralogy and Petrology* **69**, 409–19.

Williams, Q., R. Jeanloz, J. Bass, B. Svendsen & T. J. Ahrens 1987. The melting curve of iron to 250 gigapascals: a constraint on the temperature at the Earth's center. *Science* **236**, 181–2.

Wyllie, P. J. 1980. The origin of kimberlites. *Journal of Geophysical Research* **85**, 6902–10.

Wyllie, P. J. 1987a. Metasomatism and fluid generation in mantle xenoliths. In *Mantle xenoliths*, P. H. Nixon (ed.), 465–73. Chichester: Wiley.

Wyllie P. J. 1987b. Transfer of subcratonic carbon into kimberlites and rare earth carbonatites. In *Magmatic processes: physicochemical principles*, B. O. Mysen (ed.), 107–20. University Park: The Geochemical Society.

Wyllie, P. J. & A. P. Jones 1985. Experimental data bearing on the origin of carbonatites, with particular reference to the Mountain Pass rare earth deposit. In *Applied mineralogy*, W. C. Park, D. M. Hausen & R. D. Hagni (eds), 935–49. New York: American Institute of Mining, Metallurgical, and Petroleum Engineers.

19
A MODEL OF MANTLE METASOMATISM BY CARBONATED ALKALINE MELTS: TRACE-ELEMENT AND ISOTOPIC COMPOSITIONS OF MANTLE SOURCE REGIONS OF CARBONATITE AND OTHER CONTINENTAL IGNEOUS ROCKS

J. K. MEEN, J. C. AYERS, E. J. FREGEAU

ABSTRACT

Experimental evidence suggests that carbonated alkaline melts may metasomatize depleted upper mantle to produce vein networks enriched in low-temperature melting components. This metasomatism is a necessary result of phase relations of peridotite–H_2O–CO_2 and occurs at depths < 70 km. Mineral assemblages in such metasomatic veins depend on the temperature and pressure of reaction and the effective melt:solid ratio. Veins are generally enriched in calcium-rich clinopyroxene, carbonate (at P > ~17 Kb), and, at subsolidus temperatures, amphibole. The potential stability of mineral phases that have Zr, Ti, or P as essential structural constituents has been determined. Zirconium-bearing minerals are stable at temperatures above the solidus only if melts are extremely enriched in this element. Ilmenite is a stable Ti-bearing phase at all conditions studied. Melts with 'normal' amounts of P may crystallize whitlockite at super-solidus temperatures.

Metasomatic veins formed in equilibrium with an alkaline melt have Rb:Sr ratios similar to those of bulk Earth, and Sm:Nd ratios that are significantly lower and develop, with time, low ^{87}Sr:^{86}Sr and ^{143}Nd:^{144}Nd ratios that lie to the left of the mantle array. Metasomatic veins formed by complete crystallization of melt have higher Rb:Sr and lower Sm:Nd ratios than the super-solidus veins, and will develop high ^{87}Sr:^{86}Sr and low ^{143}Nd:^{144}Nd ratios.

Most metasomatic veins formed at temperatures above the melt solidus have low U:Pb and Th:Pb ratios, and develop relatively non-radiogenic Pb contents. Formation of whitlockite concentrates U relative to Pb, and Th relative to U, so that high ^{206}Pb:^{204}Pb and, especially, ^{208}Pb:^{204}Pb ratios are produced. Metasomatic veins formed at temperatures below the solidus have high U:Pb and Th:Pb ratios and also produce radiogenic Pb compositions, although no Th–U fractionation is expected.

These metasomatic veins may subsequently melt and the melts may infiltrate and interact with the wall-rocks. Such interaction of melt and matrix, possibly a natural requirement of formation of a large melt body, may result in effective cryptic metasomatism of the matrix. Although the effects of the interaction on the major and trace element compositions of the newly-formed melts are likely to be slight, their isotopic compositions may be perturbed greatly. Variable interaction between melt and wall-rock can result in the melts having isotopic compositions describing linear arrays between compositions of veins and of depleted mantle. Major- and trace-element compositions of various magmas along such 'mixing' trends vary only slightly.

19.1 INTRODUCTION

Clear evidence that parts of the crust have been underlain by lithospheric mantle for significant portions of their history has been provided by a number of sources. Jordan (1975) argued from seismic and thermal evidence that there is a profound difference between the subcontinental and sub-oceanic mantle to depths as great as 400 km and, possibly, deeper. He referred to the region beneath the continents, in which heat transfer was conductive as opposed to advective, as the continental tectosphere.

Leeman (1975) suggested that the cratonic northwestern United States is underlain by an Archaean mantle keel. More recent isotopic studies of volcanic rocks from that area (Doe *et al.* 1982, Carlson 1984, Vollmer *et al.* 1984, Carlson *et al.* 1985, Fraser *et al.* 1985, Hart 1985, Salters & Barton 1985, Dudas *et al.* 1987, Meen & Eggler 1987, Scambos, 1987, O'Brien *et al.* 1988) support this view and suggest that the subcontinental upper mantle (SCUM) is very homogeneous in age and isotopic composition. Meen & Eggler (1987) interpreted Pb isotopic compositions of rocks from Independence volcano, Montana, as defining 3.80 and 1.98 Ga isochrons reflecting enrichment events in the mantle. Carlson *et al.* (1985) showed that mantle ages defined by Pb isochrons from volcanics in the northwestern United States decrease westward from Independence volcano, although all Pb isotopic compositions lie close to a single isochron with an age of ~2.5 Ga, similar to the age defined by most crustal rocks in the Wyoming Province (Peterman 1979). O'Brien *et al.* (1988) present a Pb–Pb pseudoisochron age of 1.85 Ga for potassic rocks that may lie north of the Wyoming Province.

Much of the Wyoming Province SCUM has low $^{87}Sr:^{86}Sr$, $^{143}Nd:^{144}Nd$, and $^{206}Pb:^{204}Pb$, suggesting low time-integrated Rb:Sr and μ, but high Nd:Sm (Vollmer *et al.* 1984, Fraser *et al.* 1985, Salters & Barton 1985, Dudas *et al.* 1987, Meen & Eggler 1987). This is consistent with trace-element enrichments of volcanics from this area (Dudas *et al.* 1987, Meen & Eggler 1987) insofar as these rocks have low Rb:Sr, U:Pb, and Th:Pb, and high Nd:Sm, and are extremely enriched in their contents of Ba, Sr, LREEs, U, and Th.

Bell *et al.* (1982) and Bell & Blenkinsop (1987*a*) have shown that source regions of many carbonatites and hyperalkaline rocks within the Canadian Shield separated from the asthenosphere ~2.9 Ga ago. They showed that part of the mantle beneath the Canadian Shield had time-integrated Rb:Sr = 0.020 ± 0.002 (Bell *et al.* 1982) compared to 0.03 of bulk Earth (DePaolo & Wasserburg 1976, O'Nions et al. 1977), and time-integrated Sm:Nd = 0.358 compared to 0.32 of CHUR (chondritic uniform reservoir, Jacobsen & Wasserburg 1980). If this source region still exists today, it has ^{87}Sr:^{86}Sr = 0.70342 and ^{143}Nd:^{144}Nd = 0.51292. Pb isotopic compositions of alkaline rocks of the Canadian Shield (Tilton & Grünenfelder 1983, Grünenfelder et al. 1986) indicate that sub-Canadian Shield mantle was enriched in U over Pb. The Cretaceous Oka complex has Pb isotopic compositions that lie to the right of the geochron (Grünenfelder et al. 1986).

Barreiro & Cooper (1987) argued that the crust and lithospheric mantle beneath New Zealand were stabilized during the Palaeozoic. The mantle has high LREE:HREE and μ but low Rb:Sr, so that it developed high ^{206}Pb:^{204}Pb and relatively low ^{143}Nd:^{144}Nd and ^{87}Sr:^{86}Sr ratios. Remelting of the mantle during Miocene time produced lamprophyric and carbonatitic magmas intruded as the Westland dyke swarm.

Direct evidence that some portions of the upper mantle have undergone modifications of primary mineralogy and chemistry by interaction with a fugitive agent was provided by petrographic studies of mantle xenoliths (Lloyd & Bailey 1975, Wilshire & Shervais 1975). They showed that some xenoliths exhibit evidence of replacement of the original mineral assemblages by ones enriched in basaltic components and in large ion lithophile elements (LILEs). They believed the replacements to be due to the influx of a hydrous fluid or a silicate melt, respectively.

Other evidence supporting LILE enrichment was provided by combined trace-element and isotopic studies of alkaline igneous rocks and their entrained mantle xenoliths (Basu & Tatsumoto 1980, Menzies & Murthy 1980a). Menzies & Murthy (1980b) showed that such rocks at Ataq (South Yemen) and Nunivak (Alaska) are enriched in light rare earth elements (LREEs) but have Nd isotopic compositions that indicate a time-integrated LREE-depleted source. Addition of the LREEs to the mantle source regions must have occurred no more than a few hundred million years before melting or fragmentation and entrainment of the xenoliths occurred. Roden & Murthy (1985) summarized evidence for the relative timing of mantle metasomatism based on the available Sm–Nd isotopic data for mantle-derived xenoliths and igneous rocks.

Garnets in a peridotite xenolith from the Thumb minette of the Colorado Plateau demonstrate significant chemical zoning, whereas coexisting olivine and clinopyroxene are unzoned or only slightly zoned (Smith & Ehrenberg 1984). Zonation of the garnets apparently resulted from Fe–Ti metasomatism of a pre-existing peridotite. Solid-state diffusion homogenized the pyroxenes and olivine but not the garnets. Compositional zoning of garnets can survive only about 1000 years in the mantle, indicating that the nodules were brought to the surface shortly after mantle enrichment occurred. A much greater difference in time between enrichment of subcontinental upper mantle (SCUM) and its fragmentation was found by Richardson et al. (1984) who determined model ages of 3.2–3.4 Ga for garnets included in diamonds in Cretaceous kimberlites from South Africa, suggesting an extremely ancient lithospheric mantle.

Metasomatism also affected many mantle nodules from kimberlites in the

Kimberley area of South Africa (Erlank *et al.* 1987). The SCUM in this area experienced LILE enrichment prior to 1 Ga, at some time before fragmentation and entrainment in the kimberlites. This event was sufficiently close to time of kimberlite emplacement that the Sm–Nd systematics were not affected; the exact time of enrichment is unknown but may have been penecontemporaneous with fragmentation or may have been some tens of Ma earlier.

Calculation by Menzies *et al.* (1987*a*) of Sm–Nd model ages for cryptically metasomatized clinopyroxenes relative to the MORB evolution line, yielded dates from 0.5–3.0 Ga, whereas most clinopyroxenes from rift valleys and oceanic areas produced very young model ages (<100 Ma).

Studies of mantle-derived materials from North America, South Africa, and New Zealand indicate that mantle metasomatism accompanied crustal stabilization in a number of areas and at various times, from the Archaean to the Phanerozoic. Furthermore, trace-element characteristics of a number of the mantle source regions appear to have generally similar patterns, in particular, high ratios of alkaline-earth elements to alkali elements (Rb–Sr ratios are somewhat below estimated bulk Earth values) and high LREE:HREE, although values of U:Pb vary. This paper presents a model that explains the combination of low Rb:Sr and low Sm:Nd ratios by metasomatism of lithospheric mantle by carbonated hyperalkaline melts derived from the asthenosphere. Variable U:Pb and Th:Pb ratios may be explained by mineral assemblages formed during metasomatism. In particular the potential stability of a phosphate at super-solidus conditions may concentrate U and, especially, Th in the mantle relative to Pb.

19.2 MODELS OF MANTLE METASOMATISM

Mantle metasomatism may be defined as the process(es) whereby materials are added to the mantle either by diffusion or by infiltration of a fugitive phase. Two different, general kinds of metasomatism may be distinguished. In one case, the major element chemistry or mineralogy of peridotite is modified. This is modal metasomatism or patent metasomatism (in the sense of Dawson 1984). In the other case, the mineralogy and major element chemistry are not affected but certain trace elements are enriched in the mantle. This is cryptic metasomatism (latent metasomatism in Dawson's terminology).

Solid-state diffusion is a slow process and cannot operate over significant distances in geologically reasonable periods of time (Hofmann & Hart 1978). Consequently, most general models of mantle metasomatism call upon carrier phases that fall into three general categories, a) supercritical CO_2–H_2O fluid, b) hydrous silicate melt, and c) carbonated alkaline melt.

Lloyd & Bailey (1975) invoked influx of a hydrous fluid in their study of West Eifel and Ugandan xenoliths. The attractions of a hydrous fluid are numerous. Schneider & Eggler (1986) demonstrated that H_2O can dissolve substantial amounts of silicate material (10–12 wt% at 13–20 Kb and 1080–1100 °C). Eggler (1987) investigated the solubility relations of silicate melt and H_2O and CO_2 under mantle conditions and showed that the miscibility gap between H_2O and felsic silicate melt diminishes greatly in size as pressure and temperature increase. Furthermore, metasomatism of peridotite by hydrous fluid is required by stability of amphibole in peridotitic compositions at pressures < 22 Kb. Thus, rising fluids react with peridotite to create amphibole at such pressures (Schneider & Eggler 1986).

The stability of amphibole at depths shallower than ~70 km restricts the existence of hydrous fluids to levels deeper than this but shallower than ~200 km, where brucite becomes stable (Ellis & Wyllie 1979). Schneider & Eggler (1986) emphasized that interaction of fluids and peridotite adds much more water than SiO_2 or alkalis to the mantle. The dominant effect is thus hydration, not enrichment in LILE or basaltic components. The added silicate material is, therefore, disseminated throughout the hydrated mantle. If this mantle was originally of 'typical' peridotitic mineralogy and chemistry, the added material would produce tschermakitic pargasite (as synthesized by Mysen & Boettcher 1975 in hydrated peridotite) with minor amounts of Na-, K-, and Ti-components rather than the pargasitic, richteritic, or kaersutitic amphiboles characteristic of metasomatized xenoliths.

Fluids rich in CO_2 have been suggested as important metasomatic agents. Experimental studies of Fregeau & Eggler (1985) and Schneider & Eggler (1986) strongly suggest that addition of relatively small amounts of CO_2 to hydrous fluids severely depresses the solubility of major-, and many trace-elements in the fluid. On the other hand, Wendlandt & Harrison (1979) and Mysen (1983) suggested that LREEs may be partitioned into a CO_2-fluid from diopside and from melts, respectively. Although carbonic fluids are unable to metasomatize modally peridotite, they may cryptically metasomatize existing minerals.

Frey & Green (1974) suggested that a silicate melt was the carrier phase ('Component B') and Wilshire et al. (1980) and Wilshire (1987) strongly advocated the importance of hydrous basaltic melts in metasomatizing the mantle, particularly beneath the southwestern United States. Basaltic magmas may crystallize as pyroxenites and simultaneously enrich wall-rocks in Fe, Ti, and other elements. Volatile-rich differentiated magmas metasomatize surrounding mantle by infiltration-metasomatism. These evolved magmas may become saturated in volatiles and so release a fluid phase that can further penetrate the wall-rocks and enrich them in LREEs cryptically.

Erlank et al. (1987) described metasomatic alterations in nodules from South African kimberlites. They argue that deeper fugitive agents (perhaps at 100 km depth) produced modal diopside and phlogopite; shallower metasomatism was more intense and produced great enrichments at 75–100 km depth in Fe, Ti, K, Na, Rb, Sr, Ba, Zr, Nb, and LREEs. They suggest that the latter metasomatism is produced by introduction of H_2O-rich fluids containing halogens and sulphur. Menzies et al. (1987a) distinguish two different types of metasomatism. One is the highly potassic enrichment discussed by Erlank et al., and this they ascribe to either a hydrous fluid or a hydrous and potassic melt such as kimberlite or lamproite. The other metasomatism is due to interaction of peridotite and basanitic melts and is characterized by Fe–Ti enrichments.

Menzies & Wass (1983) and Meen (1987) have suggested carbonated alkaline melts as potential metasomatic agents. Meen stressed that, at $P < 22$ Kb, the interaction of hydrated–carbonated alkaline melt and peridotitic wall-rocks is a necessary consequence of the phase relations of peridotite–H_2O–CO_2 (Olafsson & Eggler 1983). This model is developed further below.

19.3 PHASE RELATIONS OF PERIDOTITE–H_2O–CO_2

Figure 19.1 (Olafsson & Eggler 1983, Meen 1987) shows phase relations of peridotite in the presence of small amounts of H_2O and CO_2. Subsolidus mineral assemblages may be divided into three general categories:

(a) amphibole peridotite ($P < 17$ Kb);
(b) amphibole–carbonate peridotite ($17 < P < 22$ Kb); and
(c) phlogopite–carbonate peridotite ($P > 22$ Kb).

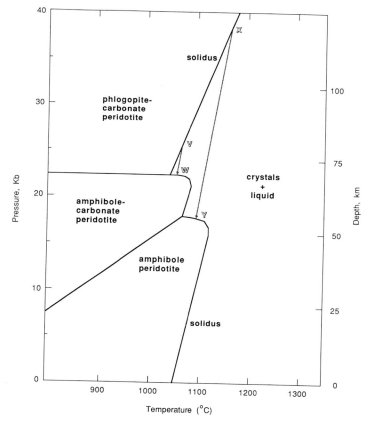

Figure 19.1 Phase relations of peridotite in the presence of small amounts of H_2O and CO_2 (from Meen 1987, after Olafsson & Eggler 1983). Low-degree partial melts formed at depths in excess of 70 km cross the peridotite–H_2O–CO_2 solidus at lower pressure. A melt formed at V will intersect the solidus at W; one formed at X will cross it at Y. After crossing the solidus, the melt will metasomatize the wall-rocks.

Fluids at $P < 17$ Kb are buffered to CO_2-rich compositions; those at $P > 22$ Kb are hydrous; at intermediate pressures fluids can exist only locally (Schneider & Eggler 1986).

The appearance of amphibole and the disappearance of carbonate in peridotite both correspond to almost isobaric changes in solidus temperature. Consequently, low-degree partial melts of hydrated carbonated peridotite at depths > 70 km must cross the peridotite–H_2O–CO_2 solidus at either ~22 Kb or ~17 Kb. The melt travels along either path V–W or X–Y in Figure 19.1. After crossing the solidus, melt and peridotitic wall-rock may react to produce mineral assemblages more fertile than the original mantle materials.

Meen (1987) modelled the interaction by considering phase relations on the join ijolite–harzburgite at 14 Kb. He found that, although olivine is the liquidus phase at all compositions studied, olivine and orthopyroxene have reaction relationships with alkaline melts at sub-liquidus temperatures, and crystal assemblages correspond to lherzolite, olivine clinopyroxenite, or websterite. Melt composition changes with decreasing temperature, from ijolite containing 5% CO_2 to highly carbonated urtite (enriched in feldspathoidal components). Alkali-rich minerals do not occur above the solidus. Subsolidus assemblages are feldspathoid lherzolites and CO_2-rich fluid. By analogy with Olafsson & Eggler's (1983) experiments, amphibole would be a stable subsolidus phase if $X^{H_2O}_{H_2O+CO_2} > \sim 0.3$, but would occur above the solidus only if $X^{H_2O}_{H_2O+CO_2} > \sim 0.9$. Consequently, CO_2- and H_2O-rich low-temperature melts react with peridotite at $P < 17$ Kb to form olivine clinopyroxenite (in equilibrium with melt at 1100 °C) or amphibole lherzolite (in equilibrium with CO_2 at $T < 1100$°C). Results of experimental work presented below extend the study of ijolite–peridotite phase relations to 20 Kb and permit consideration of the super-solidus stability of minor phases in mantle peridotite and pyroxenite.

19.4 EXPERIMENTAL TECHNIQUES

Experiments were performed in solid-media, high-pressure apparatus similar to that of Boyd & England (1960), using the piston-out technique. Sodium chloride-pyrex assemblies with a diameter of 25.4 mm were used. A pressure correction of -3% relative to the reaction of enstatite + magnesite $\rightleftharpoons$ forsterite + CO_2 (Newton & Sharp 1975) was applied. Pressures are considered accurate to ± 0.5 Kb. Temperatures, measured with Pt–$Pt_{90}Rh_{10}$ thermocouples, were automatically controlled. No correction for pressure effect on emf output of thermocouples was made. Temperatures are considered accurate to ± 10 °C.

Graphite was present in all run products. We consider it unlikely that a free fluid phase was present in any super-solidus runs and cannot, therefore, define the oxygen fugacity of any experiment in which carbonate was not stable. All carbonate-bearing runs also contain graphite, olivine, and a pyroxene. Oxygen fugacity was, therefore, presumably buffered to conditions close to those of EMOG (Eggler & Baker 1982), consistent with the f_{O_2} of fertile mantle (Eggler 1983).

19.4.1 Phase relations on the ijolite–harzburgite join

Starting materials The ijolite used was P-061 of Deines (1967) from the Oka complex, Quebec. It was ground to $< 10\ \mu m$ (size determined by microscopic examination), and dried at 700 °C for 1 hour. Analysis of the powder after drying (Table 19.1, column 1) showed it to be nominally free of H_2O and CO_2. The harzburgite was artificially prepared from MgO, Fe_2O_3, and SiO_2, equivalent to a ratio of 3:1, olivine:orthopyroxene (Table 19.1, column 5). The chemicals were ground together under acetone and fired at 1 atm, 1400 °C, $f_{O_2} = 10^{-12}$ bar, three separate times.

Four starting compositions were employed in this study (Table 19.1), corresponding to harzburgite:ijolite weight ratios of 0, 1, 2, and 3. An amount of silver oxalate was added to each sample equivalent to CO_2 amounting to 5 wt% of the ijolite. Each charge also contained a small amount of graphite. Graphite was present in all run products and the bearing of this on the f_{O_2} of the runs was noted above.

Table 19.1 Compositions of materials employed in ijolite–harzburgite phase study.

	Chemical compositions				
	1	2	3	4	5
Wt%					
SiO_2	41.2	43.3	43.9	43.3	45.3
TiO_2	0.39	0.20	0.13	0.10	0.00
Al_2O_3	15.1	7.55	5.03	3.78	0.00
FeO^*	7.56	7.88	7.99	8.04	8.20
MnO	1.16	0.58	0.39	0.29	0.00
MgO	6.20	26.4	33.1	36.4	46.5
CaO	16.7	8.35	5.57	4.18	0.00
Na_2O	6.75	3.38	2.25	1.69	0.00
K_2O	3.18	1.59	1.06	0.80	0.00
P_2O_5	0.47	0.24	0.16	0.12	0.00
S	1.20	0.60	0.40	0.30	0.00
	99.9	100.0	100.0	100.0	100.0
CIPW Norms					
Or	–	–	1.00	4.73	–
Ab	–	–	–	2.22	–
An	1.51	0.73	0.49	0.36	–
Lc	14.7	7.37	4.13	–	–
Ne	30.9	15.5	10.3	6.55	–
Di	13.0	19.4	20.7	15.5	–
Hy	–	–	–	–	24.8
Ol	16.1	50.6	62.2	69.7	75.2
Ln	19.4	4.56	–	–	–
Mt	0.50	0.18	0.14	0.12	–
Il	0.74	0.38	0.25	0.19	–
Ap	1.03	0.52	0.35	0.26	–
Py	2.02	0.75	0.46	0.33	–

FeO^* is total iron calculated as FeO. The value of the $FeO:Fe_2O_3$ used for normative compositions was calculated by the equations of Sack *et al.* (1980) using T = 1200 °C and $f_{O_2} = 10^{-12}$ bar.
Ln = larnite.

Table 19.2 Results of experimental study on ijolite–harzburgite join.

Solid composition harzburgite : ijolite (wt.)	wt% CO_2	T, °C	Time, hours	Stable phase assemblage
0	5.0	1400	3.0	ol, L
1	2.5	1400	3.0	ol, L
2	1.67	1400	3.0	ol, L
3	1.25	1400	3.0	ol, L
0	5.0	1300	5.0	ol, cpx, L
1	2.5	1300	5.0	ol, L
2	1.67	1300	5.0	ol, opx, L
3	1.25	1300	5.0	ol, opx, L
0	5.0	1200	9.0	ol, cpx, L
1	2.5	1200	9.0	ol, opx, cpx, L
2	1.67	1200	9.0	ol, opx, cpx, L
3	1.25	1200	9.0	ol, opx, cpx, L
0	5.0	1100	12.0	ol, opx, cpx, carb, L
1	2.5	1100	12.0	ol, opx, cpx, L
2	1.67	1100	12.0	ol, opx, cpx, L
3	1.25	1100	12.0	ol, opx, cpx, L
0	5.0	1000	15.0	ol, opx, cpx, carb, L
1	2.5	1000	15.0	ol, opx, cpx, carb, L
2	1.67	1000	15.0	ol, opx, cpx, L
3	1.25	1000	15.0	ol, opx, cpx, L
0	5.0	950	28.0[1]	ol, opx, cpx, carb, amp
1	2.5	950	28.0[1]	ol, opx, cpx, carb, amp
2	1.67	950	28.0[1]	ol, opx, cpx, carb, amp
3	1.25	950	28.0[1]	ol, opx, cpx, carb, amp

Abbreviations used: ol, olivine; opx, orthopyroxene; cpx, calcium-rich clinopyroxene; carb, carbonate; amp, amphibole; L, melt phase.

[1] Includes 3 hours at 1500 °C and 11 hours decreasing temperature to run conditions (50 °C per hour).

Experimental charges were loaded in iron-preconditioned Pt capsules. Conditioning with Fe was accomplished by loading capsules with wüstite and running them in a 1 atm furnace at 1150 °C and $f_{O_2} = 10^{-12}$ bar for 8 hours. The wüstite was removed and the capsules cleaned ultrasonically. Capsules were sealed for experimental runs and checked for weight loss during loading and welding. Run durations varied from 3 to 28 hours according to the temperature. Run data are presented in Table 19.2. Subsolidus runs were held at 1500 °C for 3 hours to induce melting, and then cooled to the run temperature at a rate of 50 °C per hour.

Material characterization Polished experimental charges were examined by petrographic microscope and by electron microprobe. An accelerating potential of 15 kV, ~1.2 µamp specimen current, and a 2 µm beam diameter were used when analysing crystalline materials. Quenched liquids were analysed using a defocussed beam (10 µm diameter) to minimize alkali volatilization. Points centred at 5–10 µm intervals across pools of glass from one crystal to another were analysed. Analyses on areas centred within 15 μm of crystals did not give consistent results. Analyses performed on glasses farther than that from grains were generally consistent with each other. Lower-temperature runs did not contain large enough pools for reproducible analysis according to this criterion. However, this was not a problem as quenched melt in all samples at temperatures below 1200 °C, and in more alkaline samples at 1200 °C, contained too much quench carbonate for microprobe analysis. Mineralogical compositions of runs at 1200 °C and lower temperatures were determined entirely by X-ray diffraction.

Alkali counts were monitored for ten successive 20 seconds of impingement of the electron beam. No change in the number of counts was observed. Furthermore, mass balance calculations performed using compositions of the crystals, the melt, and the bulk system gave similar results whether or not the content of Na_2O was employed in the calculation. Matrix corrections (Bence & Albee 1968, Albee & Ray 1970) were applied to all microprobe data. Phase compositions used are averages of at least six different analyses (involving at least three grains).

Experimental results Part of the ijolite–harzburgite join was studied at 20 Kb and phase relations are shown in Figure 19.2a. Phase relations for the same compositions at 14 Kb are shown in Figure 19.2b (Meen 1987). The two diagrams are similar in many respects. For all compositions studied at both pressures, olivine is the liquidus phase and pyroxene is the second phase to crystallize. Orthopyroxene crystallizes from harzburgitic compositions, whereas a calcium-rich clinopyroxene forms in ijolitic compositions. At lower temperatures, solid assemblages include olivine and two pyroxenes. One major difference between experiments at 14 Kb and 20 Kb is the appearance of carbonate as a supersolidus phase for more ijolitic compositions at 20 Kb; carbonate is stable only at temperatures significantly below the solidus at 14 Kb (Fig. 19.1). Amphibole joins the crystallizing assemblage at or near the solidus at 20 Kb. The small amount of H_2O in the run charges is concentrated into the melt by crystallization of anhydrous phases and the $H_2O:CO_2$ of the melt increases as a result of carbonate crystallization. During cooling to the solidus, the run at 950 °C precipitated major minerals to leave small pools of residual liquid that crystallized to very fine-grained aggregates of amphibole, clinopyroxene, carbonate, and, possibly, other phases. Subsolidus assemblages at 14 Kb contained trace amounts of nepheline, kalsilite, and sanidine (Meen 1987), and these minerals may be present at 20 Kb in the absence of sufficient H_2O to produce enough amphibole or mica to hold all the alkalis. If present, however, they were not identified by X-ray diffraction.

Table 19.3 gives the melt and solid compositions in each of the ten runs in which analysis of the melt phase was possible. Compositions were determined by electron microprobe, and melt analyses normalized to 100%. Modal proportions of phases were calculated by mass balance of all analysed phases, using all elements determined. A simple petrological mixing calculation was employed to obtain mass balance using a least-squares approach. The CO_2 contents of the melts quoted assume that all CO_2 in the system was dissolved in the melt.

Clinopyroxenes have 0.7–2.3% Na_2O and 4–10% Al_2O_3 and orthopyroxenes

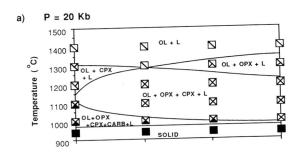

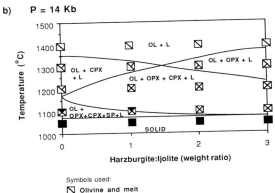

Figure 19.2 Phase relations on the ijolite–harzburgite join at (a) 20 Kb (see Tables 19.2 & 19.3), and (b) at 14 Kb (Meen 1987).

have 0.1–0.3% and 3–6% respectively. Consequently, Na:K and Al:(Na+K) ratios of the melts fall with decreasing temperature.

19.4.2 Saturation of melts with titanium-, zirconium- and phosphorus rich minerals

Starting materials Five bulk compositions were used to determine the Ti, Zr, and P contents needed to crystallize titanium-rich, zirconium-rich, and phosphorus-rich minerals, respectively. Four of these were melt compositions determined by Meen (1987) to be in equilibrium with peridotites or pyroxenites at 14 Kb for harzburgite:ijolite = 2, and temperatures of 1100, 1200, 1300, and 1400 °C (Table 19.4, columns 1–4). The fifth composition is that determined by Wendlandt & Mysen (1980) to be in equilibrium with peridotite at 1415 °C and 15 Kb (Table 19.4, column 5). Three separate charges were made up for each composition, with 5% TiO_2, 5% ZrO_2, and 5% P_2O_5, respectively.

The compositions were prepared by grinding together appropriate amounts of oxides and carbonates under ethanol and firing at 1 atm, 1400 °C, and $f_{O_2} = 10^{-12}$ bar, three separate times. An amount of silver oxalate was added to each sample

Table 19.3 Compositions (in wt%) of melts and crystal assemblages obtained on the ijolite–harzburgite join at 20 Kb.

Run composition (harzburgite:ijolite)	0	1	2	3	0	1	2	3	2	3
T (°C)	1400	1400	1400	1400	1300	1300	1300	1300	1200	1200
Fraction melt	0.978	0.637	0.546	0.478	0.652	0.544	0.393	0.323	0.278	0.194
Compositions of melts										
% CO_2 (melt)	5.11	3.92	3.06	2.62	7.67	4.60	4.25	3.87	6.01	6.44
SiO_2	42.8	45.5	46.2	48.5	39.6	45.9	44.6	44.2	42.6	46.8
TiO_2	0.42	0.29	0.22	0.12	0.24	0.35	0.26	0.20	0.32	0.50
Al_2O_3	16.1	12.8	10.9	8.24	19.6	15.4	12.9	11.2	18.7	21.3
FeO*	7.57	9.41	10.8	12.4	5.82	9.96	10.8	10.1	6.30	3.83
MgO	5.54	11.4	15.9	17.1	2.36	4.68	8.88	14.0	10.5	7.15
CaO	17.4	12.8	9.40	8.18	18.5	14.8	14.6	11.7	9.87	7.73
Na_2O	6.91	5.32	4.36	3.63	9.11	6.10	5.41	5.78	7.85	8.63
K_2O	3.27	2.51	2.15	1.82	4.78	2.72	2.56	2.81	3.81	4.09
CIPW norms										
Or	–	–	4.31	8.05	–	–	–	–	–	0.28
Ab	–	–	–	6.83	–	–	–	–	–	–
An	3.00	3.04	2.97	0.61	–	6.16	2.96	–	3.70	6.78
Le	14.0	9.72	4.38	–	–	19.7	11.8	10.5	10.2	14.4
Ac	–	–	–	–	2.26	–	–	3.46	–	–
Ns	–	–	–	–	–	–	–	0.35	–	–
Wo	1.87	8.86	–	–	4.75	17.9	1.17	–	3.97	–
Di	42.9	40.4	26.8	24.0	22.8	45.5	49.7	34.6	29.0	23.0
Ol	–	15.4	44.1	50.0	–	–	10.9	33.3	17.8	14.0
Cs	6.01	–	–	–	13.3	–	–	0.84	–	–
Mt	2.30	1.80	1.47	1.53	1.02	2.81	2.48	–	1.20	0.92
Il	0.74	0.46	0.33	0.17	0.41	0.62	0.44	0.30	0.50	0.88
Percentages of crystal assemblages										
Ol	100.0	100.0	100.0	100.0	21.1	100.0	89.3	77.2	61.8	63.2
Opx	–	–	–	–	–	–	10.7	22.8	18.1	22.4
Cpx	–	–	–	–	78.9	–	–	–	20.1	14.4

Melt analyses normalized to 100%. Percentage CO_2 in melt calculated by mass balance. Value of FeO:Fe_2O_3 ratio used for normative equations calculated by equations of Sack *et al.* (1980) using run temperatures and f_{O_2} of EMOG buffer (Eggler & Baker 1982). FeO* is total iron calculated as FeO.

equivalent to the CO_2 content required for the melt. Each charge also contained a small amount of graphite which was present in all run products.

Experimental charges were analysed using methods similar to those outlined earlier. Quenched liquids were analysed for 20–180 seconds, depending on the oxide content, using a defocused beam to minimize alkali volatilization. Glasses were analysed from the margins of grains towards the centres of pools of quenched melt. In most cases, a slight increase in content of the oxide in the melt occurred away from grains over a distance < 20 μm. The compositions of the centres of the pools were, however, approximately constant. Quoted contents are averages of at least six average compositions (at least four analyses) of these pool centres. Minerals were identified by microprobe analysis and X-ray diffraction.

Table 19.4 Compositions of materials employed in saturation studies.

	1[1]	2[1]	3[1]	4[1]	5[2]
T (°C)	1400	1300	1200	1100	1415
Wt%					
SiO_2	49.8	45.5	44.2	45.2	56.1
TiO_2[3]	0.42	0.29	0.42	0.33	0.82
Al_2O_3	12.2	13.2	18.2	25.2	11.4
FeO*	7.10	6.01	4.42	1.72	10.7
MgO	9.49	10.2	9.88	2.99	9.89
CaO	12.5	15.2	9.10	2.12	10.1
Na_2O	5.54	6.57	9.32	14.20	0.58
K_2O	2.81	3.03	4.51	8.26	0.40
CO_2[4]	4.14	4.79	7.31	12.3	10.0

FeO* is total iron calculated as FeO.
[1] From Meen (1987); compositions in equilibrium with peridotite or pyroxenite with a bulk composition of harzburgite : ijolite = 2.
[2] From Wendlandt & Mysen (1980); composition in equilibrium with peridotite.
[3] TiO_2 content for Zr- and P-saturation runs.
[4] Content of CO_2 added to charge; composition 5 was CO_2-saturated.

Experimental results Table 19.5 and Fig. 19.3 show the TiO_2, ZrO_2, and P_2O_5 contents in melts of the five compositions and at the five temperatures studied, and indicate the minor-element-concentrating, solid phases present. The titanium-concentrating phase obtained at all temperatures was ilmenite. Titanite, perovskite, and rutile were not observed in any run products. Contents of TiO_2 in melts vary from 1.2% at 1100 °C to 2.9% at 1415 °C. Baddeleyite (monoclinic ZrO_2) was found to be the stable mineral phase in equilibrium with zirconium-saturated melts from 1100–1400 °C, whereas zircon occurred in the run at 1415 °C. The concentration of ZrO_2 in the melts increased from 0.2% at 1100 °C to 3.0% at 1415 °C. The phosphorus-rich phase in all experiments was whitlockite, a volatile-free calcium phosphate [$Ca_3(PO_4)_2$]. Concentrations of P_2O_5 in the melts were 0.2–3.2%. The melt with the highest P_2O_5 content was that at 1400 °C; whitlockite is stabilized at 1415 °C at a substantially lower P_2O_5 content in the melt.

Major element compositions of some melts were modified from those of the initial compositions. Melts crystallizing ilmenite or whitlockite were depleted in FeO or

Table 19.5 Concentrations of TiO_2, ZrO_2, and P_2O_5 required to saturate melts in trace-element concentrating phases at 14 Kb.

T (°C)	TiO_2 (wt%)	Phase	ZrO_2 (wt%)	Phase	P_2O_5 (wt%)	Phase
1100	1.20 ± 0.08	Ilmenite	0.20 ± 0.08	Baddeleyite	0.23 ± 0.06	Whitlockite
1200	1.68 ± 0.08	Ilmenite	1.32 ± 0.07	Baddeleyite	0.54 ± 0.09	Whitlockite
1300	2.37 ± 0.07	Ilmenite	1.79 ± 0.10	Baddeleyite	1.17 ± 0.07	Whitlockite
1400	2.67 ± 0.08	Ilmenite	2.32 ± 0.07	Baddeleyite	3.18 ± 0.09	Whitlockite
1415	2.85 ± 0.10	Ilmenite	3.01 ± 0.08	Zircon	2.25 ± 0.11	Whitlockite

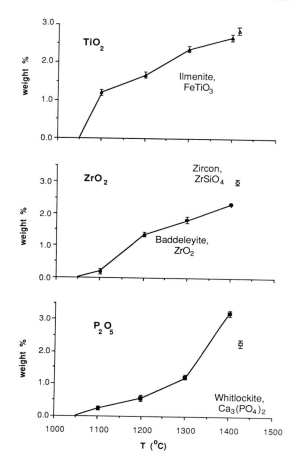

Figure 19.3 Contents of TiO_2, ZrO_2, and P_2O_5 required to saturate various melts in equilibrium with peridotite or pyroxenite at 14 Kb and temperatures of 1100–1415 °C. Solid symbols indicate that the melt compositions used were those determined by Meen (1987), formed from a 2:1 harzburgite–ijolite mixture at the appropriate temperatures. The open symbol is for a melt determined by Wendlandt & Mysen (1980) to be in equilibrium with peridotite at 1415 °C. The mineral phases stabilized in these experiments are indicated on the appropriate diagrams. The phosphate formed in all cases was whitlockite; the titanate was ilmenite. Runs at temperatures $\leqslant 1400$ °C produced baddeleyite, whereas the one at 1415 °C stabilized zircon.

CaO, respectively; increase in MgO:FeO ratio of the melts resulted in crystallization of small amounts of magnesian olivine.

Comparison with other saturation studies Watson & Ryerson (1986) found that rutile crystallized from TiO_2-rich melts of basaltic to rhyodacitic composition at 15–30 Kb and 1000–1300 °C. The amount of TiO_2 in the melts was found to have a positive correlation with temperature and to decrease markedly as the melts became more felsic. Watson & Ryerson note that there is little dependence on the volatile constituents in the melts. They found that 1–3 wt% TiO_2 was sufficient to stabilize rutile in felsic melts but that 7–9 wt% were required for more basaltic melts. The melt compositions used in this study are, presumably, very different from those used by Watson & Ryerson (1986). The TiO_2 contents required to stabilize rutile in rhyodacites and ilmenite in the compositions listed in Table 19.4 are very similar, but these values are much lower than required for basaltic liquids.

The difference in the mineral phases stabilized in the different melts may be a function of the f_{O_2} of the runs as well as the different bulk compositions used. The high Al_2O_3 of the lower temperature melts used in this study may depress the solubility of TiO_2, but similar arguments cannot be used to explain the lower TiO_2 contents in the higher temperature melts. The high contents of CO_2 and alkalis of the melts used in this study may affect the melt structures sufficiently to depress the solubility of TiO_2.

Watson & Harrison (1983) determined the contents of Zr required to saturate granitic liquids in zircon. The liquids considered here are of very different compositions from those studied by Watson & Harrison. The high temperatures and very high ratios of the sum of Na, K, Ca, Mg, and Fe to Al in these melts result in much higher solubilities of Zr than for granitic melts. The Zr contents in the highest temperature and most peralkaline melts of Watson & Harrison are similar to those in the lowest temperature melts studied here.

Watson (1980) found that P_2O_5 contents in basaltic melts in equilibrium with apatite were higher for lower contents of SiO_2 and lower for lower temperatures. His 12–14 Kb experiments at 1300 °C for SiO_2 contents of 45–50% contained 3–4 wt% P_2O_5 compared with the 1.2–3 wt% P_2O_5 obtained here. Furthermore, CaO contents of the two sets of melts are very similar and CaO content is another factor known to affect apatite solubility. Watson's runs were, however, free from CO_2 and, in fact, many of the runs contained significant amounts of depolymerizing fluorine. Addition of CO_2 is expected to polymerize the melt structure, so the presence of high concentrations of CO_2 in the melts (Table 19.4) may depress the solubility of the polymerizing P_2O_5. The suppression of the solubility of phosphorus in these melts is consistent with the fact that their TiO_2 contents are lower than those noted by Watson & Ryerson (1986).

19.5 CONSEQUENCES OF IJOLITE–PERIDOTITE INTERACTION

19.5.1 Metasomatically-produced mineral assemblages

Mineral assemblages produced in mantle rocks formed by ijolite–peridotite interaction at $P < 17$ Kb were discussed by Meen (1987) who argued that interaction at high temperatures would dissolve olivine and orthopyroxene and precipitate clinopyroxene to produce clinopyroxenitic wall-rocks in magma conduits. The ijolite may be completely expended in this process at ~1000 °C to produce amphibole–lherzolite and a CO_2-rich fluid.

Interactions of melt and peridotite at pressures between 17 and 22 Kb are similar to those at lower pressures, except for the presence of carbonate at both super-solidus and subsolidus conditions and consequent lack of a fluid phase. These relations are summarized in Figure 19.4, a modified version of the peridotite–H_2O–CO_2 solidus diagram of Olafsson & Eggler (1983), that illustrates four different mantle mineral assemblages produced by metasomatism:

(I)   olivine clinopyroxenite at $T > ~1100$ °C;
(II) carbonate–olivine clinopyroxenite or lherzolite at super-solidus conditions and $P > 17$ Kb;

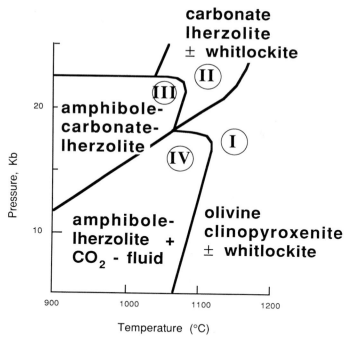

Figure 19.4 Modified version of Figure 19.1 showing the P–T fields in which different metasomatic materials may form, as deduced from Figures 19.1 and 19.2. Regions I and II are both above the solidus; III and IV are subsolidus. II and III are in the carbonate-stability field; I and IV are in fields where carbonate is not a stable mineral phase.

(III)   subsolidus carbonate–amphibole–lherzolite ($22 > P > 17$ Kb);
(IV)   subsolidus amphibole–lherzolite ($+ CO_2$) at $P < 17$ Kb.

In addition to the silicate and carbonate minerals, ilmenite and whitlockite may occur at super-solidus conditions. Baddeleyite or zircon will occur in Type I or II metasomatic veins only if the metasomatizing melts contain, at least, several thousand p.p.m. Zr (Table 19.5). Obviously, subsolidus parageneses may contain minerals other than those noted in Figure 19.4, according to the composition of the metasomatizing melt. Melts that are very strongly LILE-enriched may crystallize mineral phases that have trace-elements as essential structural components (e.g. high Ba concentrations may stabilize lindsleyite-rich crichtonites, Haggerty 1983), but it appears unlikely that such minerals can form at temperatures significantly above that of the solidus.

19.5.2   *Trace-element and isotopic compositions of metasomatized mantle*

The postulated metasomatic agents, carbonated hyperalkaline melts, are assumed to have trace-element enrichments characteristic of alkaline rocks that reach the surface of the Earth. In particular, although they almost certainly have high contents of all LILEs and values of Rb:Sr that are similar to that of bulk Earth, such rocks have values of Nd:Sm, U:Pb, and Th:Pb that are much higher than bulk Earth values (e.g. Loubet *et al.* 1972, Lancelot & Allègre 1974, Eby 1984).

Trace-element compositions of Type I and II metasomatic veins are determined by partitioning between melt and solid phases. In a simple lherzolite, LILEs are dominantly contained in calcium-rich clinopyroxene, and so the trace-element enrichments in the olivine clinopyroxenite or lherzolite are determined by diopside-melt partitioning. Arth (1976) reviewed partition coefficients then available. Values quoted for Rb, Sr, Sm, and Nd between diopside and melt are 0.015, 0.15, 0.26, and 0.18, respectively. Ray *et al.* (1983) determined a rather lower partition coefficient for Sm between diopside and haplobasalt (~ 0.14). In the interests of obtaining a consistent Sm:Nd ratio, the older coefficients were used and indicate that $(Rb:Sr)_{diopside} : (Rb:Sr)_{melt} \sim 0.125$ and that $(Sm:Nd)_{diopside} : (Sm:Nd)_{melt} \sim 1.5$.

Consequently, the first diopside to form from melt with 26 p.p.m. Rb, 600 p.p.m. Sr, 60 p.p.m. Nd, and 11 p.p.m. Sm (Mount Royal gabbro, Eby 1984) has Rb:Sr = 0.00542 and Sm:Nd = 0.275 compared to bulk Earth values of 0.08 and 0.32, respectively (DePaolo & Wasserburg 1976, Jacobsen & Wasserburg 1980). Such a diopside initially with bulk Earth isotopic composition would have $^{87}Sr:^{86}Sr = 0.70243$ ($\varepsilon_{Sr} = -33$) and $^{143}Nd:^{144}Nd = 0.51208$ ($\varepsilon_{Nd} = -11$) after 2.75 Ga, thus lying to the left of the mantle array (Table 19.6, Fig. 19.5). As mentioned above, this will be very close to the composition of a pyroxenitic or peridotitic vein in the absence of exotic phases (Types I and II in the terminology of Fig. 19.4). The additional presence of carbonate in such veins would lower the Rb:Sr ratio of the metasomatic material, while probably leaving the Sm:Nd ratio unaltered.

The model for types I and II veins was calculated using a fairly primitive melt as the metasomatic agent. Precipitation of a mineral assemblage with Rb:Sr ratios less than, and Sm:Nd ratios greater than, the melt values given above raises the Rb:Sr ratio and lowers the Sm:Nd ratio in the remaining liquid. Veins produced from the evolved melt will have progressively higher Rb:Sr and lower Sm:Nd ratios, and thus develop isotopic compositions that lie to the right of the compositions indicated on Figure 19.5. In this way, a spectrum of compositions with similar $^{143}Nd:^{144}Nd$ but very different $^{87}Sr:^{86}Sr$ ratios may be formed.

Partition coefficients for U and Th between clinopyroxene and basaltic melt are similar, at 0.002 each (Benjamin *et al.* 1980). Partition coefficients of Pb and U between calcium-rich clinopyroxene and haplobasaltic melt at 1 atm are 0.01 and 0.0003, respectively (Watson *et al.* 1987). Most clinopyroxenes of mantle xenoliths from kimberlites have low values of μ (< 5, with one value of 10), and low U, Th, and Pb contents (Kramers 1977). Assuming a μ value of 10 for a melt in equilibrium with the peridotitic clinopyroxenes ($\mu = 3.9$) yields a partition coefficient of Pb between clinopyroxene and melt of 0.005. A Pb partition coefficient of 0.01 (Watson *et al.* 1987) is used for these calculations. If U, Th, and Pb concentrations in the metasomatic agent were equivalent to those in the Finsch kimberlite of 1.46, 5.77, and 7.36 p.p.m., respectively (Kramers 1977), then clinopyroxenes of Type I or II metasomatic veins would have $\mu = 2.46$ and Th:U = 3.95 (Table 19.6). Metasomatism 2.75 Ga ago of a mantle, with the average composition of active terrestrial Pb of Stacey & Kramers (1975), would give relatively low present-day Pb isotopic

Table 19.6 Calculated isotopic compositions of 2.75-Ga-old SCUM.

	$^{87}Rb:^{86}Sr$	$^{87}Sr:^{86}Sr^1$	ε_{Sr}	$^{147}Sm:^{144}Nd$	$^{143}Nd:^{144}Nd^2$	ε_{Nd}	μ	Th:U	$^{206}Pb:^{204}Pb^3$	$^{207}Pb:^{204}Pb^3$	$^{208}Pb:^{204}Pb^3$
Type Ia–IIa[4]	0.0157	0.70243	−33.0	0.166	0.51208	−10.9	2.46	3.95	14.83	14.89	34.69
Type Ib–IIb[4]	0.0157	0.70243	−33.0	0.166	0.51208	−10.9	8.68	7.96	18.14	15.52	43.33
Type III[4]	0.125	0.70679	+29.0	0.111	0.51108	−30.4	12.3	3.98	20.06	15.89	40.39
Matrix[5]	0	0.7018	−41.9	0.302	0.51455	+37.3	0	–	13.52	14.64	33.26

[1] Calculated assuming that enrichment occurred 2.75 Ga ago and that pre-existing mantle had a $^{87}Sr:^{86}Sr$ at that time of 0.70180 – equivalent to bulk Earth that has present-day $^{87}Rb:^{86}Sr = 0.074$ and $^{87}Sr:^{86}Sr = 0.70475$.
[2] Calculated assuming that enrichment occurred 2.75 Ga ago and that pre-existing mantle had a $^{143}Nd:^{144}Nd$ at that time of 0.50907 – equivalent to bulk Earth that has present-day $^{147}Sm:^{144}Nd = 0.1967$ and $^{143}Nd:^{144}Nd = 0.512638$.
[3] Calculated assuming that enrichment occurred 2.75 Ga ago and that pre-existing mantle had Pb isotopic compositions equivalent to that of Stacey & Kramers' (1975) active terrestrial lead – $^{206}Pb:^{204}Pb = 13.52$; $^{207}Pb:^{204}Pb = 14.64$; $^{208}Pb:^{204}Pb = 33.26$.
[4] Types I, II, and III metasomatites as described in text and in Figure 19.4. Type Ib and IIb are taken to contain 1% whitlockite and 99% calcium-rich clinopyroxenite.
[5] Matrix is the depleted mantle residual to removal of the crust assumed to have taken place at 2.75 Ga from a mantle with the isotopic compositions defined in footnotes 1–3 above.

Epsilon units are relative deviations of isotopic compositions from bulk Earth (Sr) or CHUR (Nd) expressed in parts in 10^{-4}.

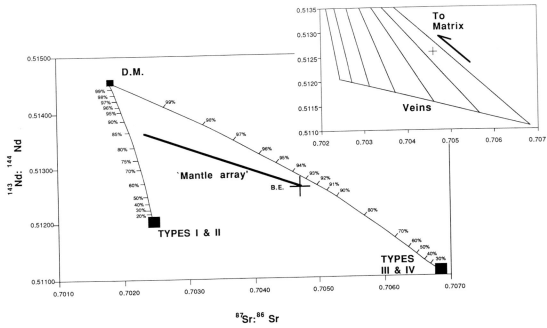

Figure 19.5 Present-day Nd–Sr isotopic compositions of postulated SCUM formed 2.75 Ga ago. Compositions are listed in Table 19.6. 'Types I and II' indicate the position of supersolidus metasomatites crystallized from a model metasomatic fluid; 'Types III and IV' indicate the position of subsolidus metasomatites that crystallized from the same fluid. D.M. is the calculated position of the depleted mantle considered to be the material that comprises the bulk of SCUM, also 2.75-Ga-old. Mixing lines between D.M. and the two types of metasomatized mantle are shown. These assume that the metasomatism involved a 90% dilution of the metasomatic minerals by the depleted wall-rock. Numbers on the curves are the weight fraction of D.M. B.E. is present-day bulk Earth. The line marked 'mantle array' shows the approximate position of the centre of the compositions of those modern igneous rocks postulated to be derived from the asthenosphere. The compositions of the metasomatites formed from a single metasomatizing melt will vary with evolution of the melt (see text for discussion). Consequently, veins will define a trend of shallow-slope on the Nd–Sr diagram. Mixing curves between veins and matrix will, then, define an array towards the matrix (inset figure).

ratios of $^{206}Pb:^{204}Pb = 14.83$ $^{207}Pb:^{204}Pb = 14.89$, and $^{208}Pb:^{204}Pb = 34.69$ (Table 19.6).

Whitlockite in Type I or II metasomatic veins may drastically change the U:Pb and Th:Pb ratios of these rocks. Benjamin *et al.* (1980) reported partition coefficients for U and Th between whitlockite and haplobasaltic melt at 20 Kb and ~1400 °C of 0.5 and 1.2, respectively. Assuming 1% whitlockite and 99% clinopyroxene in the metasomatic material, its μ of 8.68 and Th:U of 7.96, results in high values of $^{206}Pb:^{204}Pb$ and very high values of $^{208}Pb:^{204}Pb$ (Table 19.6). Whitlockite-free, metasomatic materials are referred to as Ia and IIa, whereas whitlockite-bearing ones are Ib and IIb.

Metasomatic type III veins are formed by complete addition of melt to the mantle. Consequently, metasomatized mantle has a LILE enrichment pattern the same as that of the metasomatizing agent. Calculations summarized in Table 19.6 show that Type III veins develop higher ε_{Sr} and lower ε_{Nd} than Type I and II veins. Similarly,

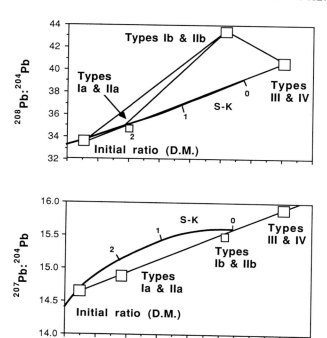

Figure 19.6 Present-day Pb isotopic compositions of postulated SCUM formed 2.75 Ga ago. Compositions are listed in Table 19.6. Positions of different types of metasomatite and depleted mantle matrix are indicated as on Figure 19.5. Ia and IIa have no whitlockite, whereas Ib and IIb contain 1% whitlockite. Curves S-K are Pb evolution curves of Stacey & Kramers (1975). Numbers on curves indicate positions of curves at times before present (in Ga).

the high U and Th contents of the metasomatizing agents are reflected in high Pb isotopic ratios of old metasomatic materials (Table 19.6, Fig. 19.6).

Type IV metasomatic materials are formed in equilibrium with CO_2-rich fluid, that carries away some LILEs. Although fluids rich in CO_2 may transport LREEs and, possibly other LILEs, the total amount of material removed by the limited amount of CO_2 in the fluid is probably quite small. In the absence of massive amounts of CO_2-fluxing, the trace-element characteristics of Types III and IV materials are assumed to be very similar.

Figure 19.6 illustrates the modelled Pb isotopic compositions of the metasomatic regions. Types Ia and IIa (whitlockite-free) are depleted in U and Th, thus retaining relatively non-radiogenic Pb isotopic compositions, whereas Types Ib, IIb, III, and IV have radiogenic Pb isotopic compositions. Types Ib and IIb have higher Th : U ratios than those of Types III and IV, and so develop higher ratios of $^{208}Pb : ^{204}Pb$ to $^{206}Pb : ^{204}Pb$. Obviously, evolution of magmas due to interaction with the mantle may increase their U : Pb ratio, thus resulting in a range of U : Pb ratios in the Type I or II metasomatic materials. In time, these will develop a range in Pb isotopic compositions.

Variation in $^{207}Pb : ^{204}Pb$ and $^{206}Pb : ^{204}Pb$ ratios of different regions is totally a function of variation in μ, so these compositions (and those of magmas formed from them) define a secondary isochron that gives the age of metasomatism and emanates from the initial ratio of the mantle. The $^{208}Pb : ^{204}Pb$ and $^{206}Pb : ^{204}Pb$ compositions can also define a straight line, providing that no mineral that fractionates Th from U forms above the solidus.

We envisage SCUM as being dominated by a matrix composed of dunite, harzburgite, and depleted (i.e. calcium- and iron-poor) lherzolites, material presumably residual to removal of melts of basaltic or more mafic compositions. Enrichment of this depleted SCUM by alkaline melts would produce a system of dykes, veins, and veinlets of more fertile material, as discussed above. Many authors have noted that enrichment by melts is likely to occur by either crystallization or reaction. Crystallization occurs by precipitation of minerals within magma conduits without interaction with wall-rocks (e.g. the solid assemblages at the left side of Fig. 19.2). Reaction involves contributions to mineral assemblages from both melt and wall-rock.

Vein systems in SCUM are undoubtedly as complex as the methods by which they may form. Motion of melts in the lithosphere appears to occur primarily by fracture propagation (Shaw 1980, Spera 1980, 1987) and such melts will interact with their wall-rocks unless their velocity is great. Spera (1987) calculated that an excess pressure of 300 bar would result in an alkalic magma ascending through the lithosphere at a rate of $0.1–10$ m s^{-1} through fractures ~ 1 m across. This rate and the small size of the conduit can result in extensive interaction and cooling. The first magma to move through cold lithosphere will almost certainly suffer 'heat death' (Spera 1984), thus forming a metasomatic vein by reaction or crystallization. Later magmas may exploit the pre-heated pathway to rise to shallower levels. Large numbers of magma bodies of small volume passing through the lithospheric mantle will very efficiently produce extensive networks of metasomatic veins. The chemical properties of such a network will be highly variable, both laterally and vertically.

19.5.3 *Products of melting of metasomatized mantle*

Type I veins melt at the volatile-free solidus at temperatures between 1400–1500 °C (Kushiro *et al.* 1968), conditions unlikely to be attained in continental lithosphere. These rocks occur as xenoliths in magmas that either originated in SCUM or originated in the asthenosphere and rose along the same paths as earlier metasomatizing magmas.

Type II materials melt at the carbonated-lherzolite solidus, 1100–1200 °C (Eggler 1978, Wendlandt & Mysen 1980) to carbonatitic or carbonated alkaline magmas (Wendlandt & Mysen 1980). Characteristic of such magmas is the possession of very high $CO_2:H_2O$ ratios, because no hydrous mineral is involved in their genesis. These magmas inherit the isotopic compositions of Type II veins, and are strongly enriched in the elements concentrated in carbonate and clinopyroxene (and, if present, whitlockite). Thus, they are strongly enriched in alkaline earths over alkali metals and probably have relatively high Sr:Ba ratios, in common with expected enrichments in clinopyroxenes and carbonates.

Type III regions also melt to carbonatite or carbonated alkaline melts, but at the peridotite–H_2O–CO_2 solidus at temperatures of 1000–1100 °C (Olafsson & Eggler 1983). Depending upon whether carbonate is consumed or is residual to melting, the melts have $CO_2 : H_2O$ ratios similar to or less than those of the metasomatic agents. Furthermore, they have trace-element compositions that reflect those of metasomatizing agents.

Type IV veins melt at the amphibole–lherzolite solidus at temperatures of 1000–1100 °C (Green 1973) to non-carbonated alkaline magmas. Such magmas also have trace-element contents similar to those of the metasomatic agent.

19.5.4 Processes of melting in metasomatized mantle

Melts of metasomatic mantle can only be sampled if they approach the surface of the Earth. In order for this to occur, a melt body of sufficient size to penetrate the cold, overlying lithosphere without experiencing heat death must be formed; such a melt is 'viable'. The ability to produce a viable melt depends upon the amount of fertile material available for melting and the degree of partial melting that the mantle experiences, and is highly temperature-dependent. This ability also depends upon the geometry of the vein network.

Figure 19.7 illustrates some of the variation that might be expected in a complex network of veins within SCUM plausibly within a single network. Figure 19.7A represents a single thick dyke in SCUM capable of producing a viable melt. Consequently, the dyke yields a magma that retains the LILE pattern and isotopic

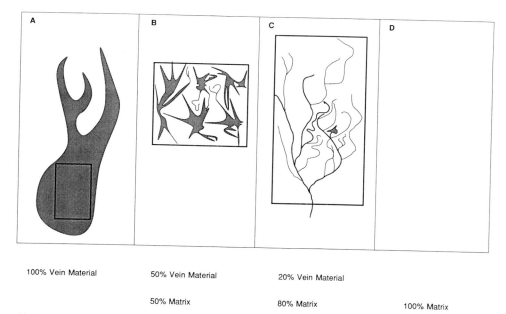

Figure 19.7 Cartoon representing a model of SCUM. The matrix of SCUM is shown unshaded and is considered to be composed of depleted dunite, harzburgite, and some lherzolite. The shaded areas indicate dykes, veins, and veinlets in SCUM. These bodies may be metasomatites that formed by reaction of alkaline melt and wall-rock. Veins in the vein systems are of various sizes, decreasing from left to right. Boxes indicate the amount of material required to melt in order to form a viable melt body. Large dykes may melt without contribution from the wall-rocks (A). Systems of smaller veins (B and C) may require contributions from both veins and wall-rocks as discussed in the text. Vein-free SCUM cannot yield melts at reasonable temperatures.

composition of the dyke, and there is not necessarily any interaction of melt and wall-rock. Smaller veins may not be able to produce viable melt bodies on their own. Any melts produced from individual veins may move only a short distance to react with and enrich neighbouring parts of the mantle. Systems of small veins (e.g. Fig. 19.7B, C) may contain as much fertile material as a large dyke and, on that basis, be just as able to produce viable melt bodies. The fractions of the melt must, however, be able to communicate with each other. If the vein system is not continuous (or if the connecting route is long or passes through regions that are at temperatures below the solidus), melt collection can occur only by passage through the depleted matrix material (unshaded region of Fig. 19.7).

We shall assume that the melt formed from the veins is in thermal, major element, and textural equilibrium with the wall-rocks. If distances involved are small, thermal equilibrium seems likely. Major element equilibrium is likely if the melting is a chemical reversal of the metasomatism. Because veins were formed from depleted mantle similar to that still forming wall-rocks, the material in the vein residual to the melting event will probably be similar to unmetasomatized wall-rocks in major element chemistry. In the absence of external forces, the melt will infiltrate the wall-rocks by surface-energy induced, grain-edge infiltration (Bulau *et al.* 1979, Waff & Bulau 1979, Watson 1982). In effect, the melt infiltrates by extending fingers along grain edges (mutual contacts of three grains) to connect at grain corners and propagates by dissolution of minerals at the head of the tubule and reprecipitation of the same mineral in the magma chamber. Stevenson (1986) undertook a theoretical analysis of the flow of the melt through such tubules and estimated $D \sim 10^{-16}$–10^{-12} m^2 s^{-1}. These values are consistent with the experimental results of Watson (1982) and similar to rates of diffusion in a melt (Hofmann 1980). This may not be the rate of infiltration, however, as dissolution of the head of the melt tubules may be the rate-limiting step in this process.

The process of dissolution–reprecipitation of wall-rock means that a large amount of that material is passed through the melt. Furthermore, wall-rock will communicate with the melt by solid-state diffusion and exchange at the mineral–melt interface. A simple spherical diffusion model indicates that a mineral grain with radius 5 mm will essentially equilibrate in 100 000 years if the diffusion coefficient is 10^{-18} m^2 s^{-1} (e.g. garnet at 1000 °C) or 1000 years if $D \sim 10^{-16}$ m^2 s^{-1} (e.g. olivine at 1000 °C or garnet at 1200 °C). Clearly, the combined effects of infiltration and exchange with the infiltrated wall-rocks can lead to fairly extensive communication of melt and wall-rocks.

Interaction of the melts and their wall-rocks will not affect the major-element compositions of either melt or wall-rock. Similarly, melt contents of compatible trace elements are little affected. Minerals in the residual mantle will, presumably, have LILE contents considerably out of equilibrium with the melt. Minerals dissolved by the melt during infiltration will have low contents of LILE; those precipitated on the walls of the chamber will, presumably, be significantly higher. Similarly, exchange between minerals and melt will result in extraction of LILEs from the liquid. The propensity of these minerals to dissolve LILEs is, however, quite slight, and the effect on the melt contents of these elements is limited.

Assuming a bulk mantle:melt ratio of 1:1, a residual mantle containing 80% olivine, 12% orthopyroxene, and 8% calcium-rich clinopyroxene, originally with a bulk Nd content of 0.63 p.p.m., and a melt with 60 p.p.m. Nd, after equilibration the mantle contains 3 p.p.m. Nd. This constitutes a significant cryptic metasomatism resulting in an increase of 370% in the original content of Nd. The Nd content in the melt, however, has fallen only to 58 p.p.m., a decrease of a mere 4%.

The main effect on the melt of the melt–solid interaction will be seen in its isotopic composition. Although the depleted mantle has low contents of LILEs, it may have extreme $^{87}Sr:^{86}Sr$ and $^{143}Nd:^{144}Nd$ ratios because unusual values of parent:daughter may have been maintained for long periods of time.

A model depleted mantle has been developed in Table 19.6. We assume formation at 2.75 Ga of a mantle with olivine:orthopyroxene:calcium-rich clinopyroxene of 80:12:8 as residue of a basaltic or komatiitic melting event. Such a mantle is likely to represent more enriched portions of the residue. This mantle is taken as having 0.314 p.p.m. Sm, 0.628 p.p.m. Nd, and 0.133 p.p.m. Sr. Rubidium, U, and Th contents are set to zero and the content of Pb is taken as greater than zero. Isotopic ratios calculated for this depleted mantle (D.M.) are given in Table 19.6 and plotted in Figures 19.5 and 19.6.

Mixing curves between depleted mantle and model metasomatic veins are shown on Figures 19.5 and 19.6 using the isotopic compositions in Table 19.6. The veins used are considered formed by extensive reaction of melt with wall-rock. Contents of LILEs in the veins are, therefore, calculated as 10% of the values of the model metasomatic agent employed above. Even with this dilution, isotopic compositions of Nd and Sr of the melts are displaced significantly from those of veins only when large amounts of matrix material are involved in melting. Melts that have interacted with very large amounts of the wall-rocks may not be extracted from the source as the melt will be used up enriching the residual mantle. Furthermore, major melt bodies formed by extraction of very low-degree partial melts require a very large mantle source region. McKenzie (1985) calculated that a nephelinite magma produced by 0.3% partial melting of the mantle could be effectively separated in 10^7 years. In order to form a 20 km^3 melt body, however, some 10 000 km^3 of the mantle must be melted and the melt collected into a single body.

Melting of veins alone (Figure 19.7a) produces a melt with the isotopic composition of the veins. Melting of veins and equilibration of melt and wall-rocks displaces the isotopic composition of the melts along trajectories towards D.M. values. These lines are almost straight and are indistinguishable from bulk mixing lines. The major- and trace-element compositions of the liquids are little altered by melt–solid interaction, however, and consequently, melts that lie along apparent mixing lines may be very similar in major- and trace-element chemistry.

Lead isotopic compositions will also betray involvement of depleted and enriched components. Note that a linear array of compositions on a plot of $^{207}Pb:^{204}Pb$ v. $^{206}Pb:^{204}Pb$ is a mixing line but has a slope that defines the age of metasomatism. Variable amounts of interaction of melt and wall-rock will also result in the melts possessing $^{208}Pb:^{204}Pb-^{206}Pb:^{204}Pb$ ratios that lie on a straight line.

So far the simplest possible case has been adopted, i.e. one in which the wall-rocks to the veins are essentially identical to the material that was enriched to form the veins. Obviously, more complex scenarios may be envisaged, including variation in the age of metasomatism. Formation of SCUM by extraction of crust and its enrichment by melts may have occurred (episodically) from the Archaean to the present day. Figure 19.8 shows the possible Nd–Sr isotopic compositions of SCUM of various ages. Asthenospheric mantle is assumed to have evolved from a bulk Earth composition at 2.75 Ga to somewhat depleted mantle values at the present day. SCUM formed at any given time evolves along growth lines that emanate from the asthenosphere mantle growth curve. SCUM of three separate ages, 2.75 Ga, 1.5 Ga, and 500 Ma, are shown. The three lines that define the end members of SCUM growth for each of these ages are the model Type I and II, model Type III and IV, and the model D.M. of Table 19.6. The 2.75 Ga triangle on Figure 19.8 is, thus, an idealized form of the mixing triangle of Figure 19.5.

Gradual depletion of asthenosphere results in the movement of initial SCUM compositions to gradually higher ^{143}Nd : ^{144}Nd and slightly higher ^{87}Sr : ^{86}Sr values. Old metasomatized SCUM may maintain low Sm : Nd for long periods of time and develop very low ^{143}Nd : ^{144}Nd values, whereas younger material does not have a long enough time to produce such low ^{143}Nd : ^{144}Nd values. Similarly, the possible range of values of ^{87}Sr : ^{86}Sr of older SCUM is considerably greater than that of younger SCUM.

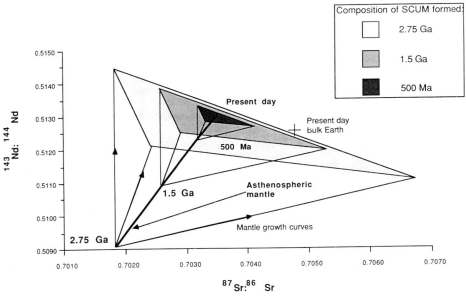

Figure 19.8 Compositions of SCUM formed from asthenospheric mantle at various times in the past. Three component SCUMs are shown, with the three components corresponding to depleted mantle, type I/II metasomatites, and type III/IV metasomatites of Figure 19.5. Asthenospheric mantle is assumed to have separated from bulk Earth composition at 2.75 Ga and to have evolved to more depleted compositions so that present-day asthenospheric mantle has Nd–Sr isotopic compositions similar to those of enriched MORB or more depleted ocean island basalts.

19.6 EXAMPLES OF MATERIALS DERIVED FROM METASOMATIC MANTLE REGIONS

Figure 19.9 shows Nd–Sr isotopic compositions of rocks and mantle minerals that may have been derived from mantle metasomatized by processes outlined in this chapter. All these rocks have ^{143}Nd : ^{144}Nd ratios that are lower at a given ^{87}Sr : ^{86}Sr ratio than the mantle array shown by MORBs, ocean island basalts, and bulk Earth.

Some of the lowest ^{143}Nd : ^{144}Nd and highest ^{87}Sr : ^{86}Sr ratios illustrated on this diagram are possessed by rocks from the Wyoming Province of the cratonic northwestern United States. The Crazy Mountain field (Dudas *et al.* 1987) includes data from alkaline igneous rocks (crosses) and from mantle-derived xenoliths (squares). The xenoliths were probably derived from metasomatically produced veins and have Sm–Nd model ages of 1.3–1.8 Ga relative to depleted mantle. These xenoliths demonstrate considerable variation in ^{87}Sr : ^{86}Sr ratios with much less variation in their ^{143}Nd : ^{144}Nd ratios, and most of the data from the alkaline igneous rocks analysed from the Crazy Mountains lie along this trend. The scatter to higher ^{143}Nd : ^{144}Nd ratios of some samples, most notably the Bear Paw carbonatite, may be due to greater amounts of interaction between melt and wall-rocks or to involvement of a younger lithospheric component.

Samples from Independence volcanic complex, south of the Crazy Mountains, lie in a field displaced even farther from the mantle array (Meen & Eggler 1987). Mantle-derived rocks at Independence volcanic complex fall into two categories, tholeiitic basalts to shoshonites and high-magnesium andesites to banakites (Meen & Eggler 1987). The former group have significantly higher ^{143}Nd : ^{144}Nd and somewhat lower ^{87}Sr : ^{86}Sr ratios than do the latter, and have similar ^{143}Nd : ^{144}Nd ratios to the xenoliths and most alkaline rocks from the Crazy Mountains (Fig. 19.9a). Furthermore, these basalts and derivatives from the Independence volcanic complex possess Pb isotopic compositions that indicate a secondary isochron with an age of 2.0 Ga, similar to model ages of xenoliths from the Crazy Mountains. Basalts from Independence and the alkaline rocks from the Crazy Mountains and Leucite Hills may have been derived from a system of Type II and, possibly, Type III veins that formed 1.3–2.0 Ga ago.

The high-magnesium andesites and derivatives at the Independence volcanic complex apparently had a distinctly different genesis, possessing considerably lower ^{143}Nd : ^{144}Nd ratios than Independence basalts or rocks from the Crazy Mountains. They have ^{143}Nd : ^{144}Nd ratios similar to the rocks from Smoky Butte, but rather lower ^{87}Sr : ^{86}Sr ratios (Fig. 19.9b). They also define a secondary Pb isochron, but show poor correlations of ^{208}Pb : ^{204}Pb and ^{206}Pb : ^{204}Pb. Lead isotopic compositions apparently record the age of a distinct metasomatic event at 3.8 Ga. One suite of high-magnesium andesites has a steep slope on a plot of ^{208}Pb : ^{204}Pb v. ^{206}Pb : ^{204}Pb (Meen & Eggler 1987), indicating that these rocks had a high Th : U ratio. Small and variable amounts of phosphate in the individual source regions of magmas could provide this radiogenic Pb, although the dominant contribution of Pb was the relatively non-radiogenic Pb from the silicate minerals. The possibility of

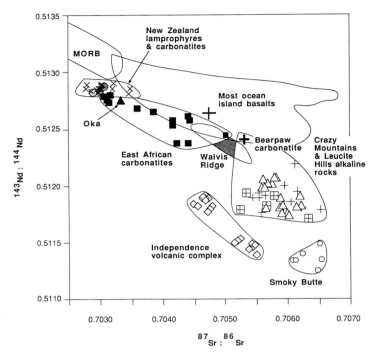
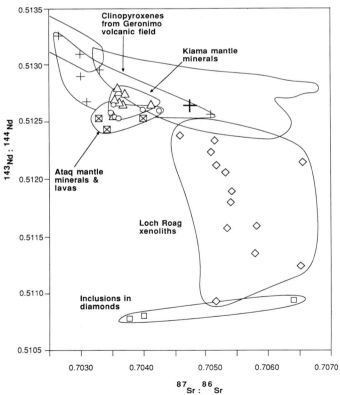

phosphate in the source regions of this suite of high-magnesium andesites is also suggested by high contents of P_2O_5, Ba, and Th. These rocks may have formed from melts of Type II metasomatic veins (containing whitlockite or apatite) and are probably older than the veins that were the sources of Independence basalts and alkaline rocks from the Crazy Mountains.

Some, but not all, of the lavas and mantle-derived minerals from Ataq, South Yemen (Menzies & Murthy 1980b) define a field that lies to the left of the mantle array (Fig. 19.9b). These rocks have a considerable variation in their $^{87}Sr:^{86}Sr$ ratios but a more limited range in $^{143}Nd:^{144}Nd$ ratios. In this regard, they are similar to the xenoliths and most of the alkaline magmatic rocks from the Crazy Mountains. Displacement to much higher $^{143}Nd:^{144}Nd$ than values for the rocks from Montana suggests a more recent age of metasomatism for the sub-Yemen mantle.

Lamprophyres and carbonatites from New Zealand (Barreiro & Cooper 1987) have Nd–Sr isotopic compositions that fall to the left of the mantle array (Fig. 19.9a) and high Pb isotopic ratios. These are considered good candidates for melting of Type III metasomatic veins, with metasomatism having occurred penecontemporaneously with stabilization of New Zealand crust in the Palaeozoic.

Carbonatites and nephelinites from East Africa (Bell & Blenkinsop 1987b) define a very good linear array (Fig. 19.9a) and show a considerable range in isotopic compositions. Figure 19.10, a Nd isotopic growth diagram, demonstrates the extent of this range (ε_{Nd} values of $+4.3$ to -3.7) compared with other carbonatites. The data are mostly from carbonatites (two are nephelinites), so there is no reason to expect that parental magmas possessed ranges in major- or trace-element compositions commensurate with the range of isotopic compositions. We suggest that primary carbonated magmas were formed by melting of Type III veins ($\varepsilon_{Nd} < \sim -4$ and $^{87}Sr:^{86}Sr > \sim 0.705$) in the East African SCUM, and the melts of the veins interacted with the depleted mantle to a variable extent, producing the range of isotopic compositions.

The composition of the Oka carbonatite (Bell & Blenkinsop 1987a) is shown on Figure 19.9a as an example of data from the Canadian carbonatite complexes. Nd isotopic compositions of these complexes are shown on a growth curve (Fig. 19.10).

Figure 19.9 Nd–Sr isotopic compositions of some materials that are possibly melts of metasomatized upper mantle: (a) shows compositions of some Mesozoic and Tertiary igneous rocks; (b) gives the compositions of mantle-derived xenoliths. Fields of MORB and ocean island basalts are shown for reference. The dark cross represents the present-day composition of CHUR and bulk Earth ($^{143}Nd:^{144}Nd = 0.512638$, $^{87}Sr:^{86}Sr = 0.70475$).

Symbols used in (a): + Crazy Mountain alkaline rocks, ⊞ xenoliths (Dudas et al. 1987); ◇ igneous rocks of Independence volcanic complex (Meen & Eggler 1987); ○ Smoky Butte igneous rocks (Fraser et al. 1985); △ Leucite Hills igneous rocks (Vollmer et al. 1984); ■ carbonatites and nephelinites from East Africa (Bell & Blenkinsop 1987b); ▲ Oka complex, Canada (Bell & Blenkinsop 1987a); × lamprophyres and carbonatites from New Zealand (Barreiro & Cooper 1987). Field shaded and marked 'Walvis Ridge' denotes compositions of those rocks from the South Atlantic (Richardson et al. 1982). Three small circles in field of New Zealand alkaline rocks denote compositions of rocks from the islands of St. Helena (Cohen & O'Nions 1982), Tubaii (Vidal et al. 1984), and some Cook–Austral–Samoa Islands (Palacz & Saunders 1986).

Symbols used in (b): + clinopyroxenes from Geronimo volcanic field (Menzies et al. 1985); ◇ xenoliths from Loch Roag (Menzies et al. 1987b); ○ Ataq igneous rocks, ⊠ xenoliths (Menzies & Murthy 1980b); △ Kiama xenoliths (Menzies & Wass 1983); □ Garnet inclusions in diamonds (Richardson et al. 1984).

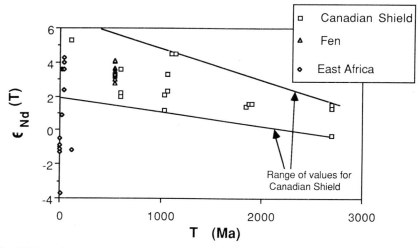

Figure 19.10 Nd isotopic evolution diagram for carbonatites and related rocks from the Canadian Shield (Bell & Blenkinsop 1987a), East Africa (Bell & Blenkinsop 1987b), and Fen complex, Norway (Andersen 1987). The East African rocks show a considerable range of Nd isotopic composition that correlates with Sr isotopic composition (Figure 19.9). The Fen data are similar to the more depleted compositions for the East African rocks. The carbonatites and related rocks from the Canadian Shield define a smaller range in Nd isotopic composition than do the East African rocks, and all have $\varepsilon_{Nd} > 0$, i.e. they were derived from a source that had a time-integrated history of depletion in LREEs relative to HREEs with respect to bulk Earth.

The variation in the Nd isotopic compositions of the Canadian rocks is slight compared with that defined by data from East African carbonatites, but encompasses the range of data from the Fen complex, Norway (Andersen 1987). The spread of compositions of Canadian carbonatites may reflect interaction of melts and matrix during the melting events. If the lower limit of Nd isotopic compositions of Canadian bodies lies close to that of the metasomatic veins, the veins have time-integrated LREE-depletion with respect to bulk Earth and indicate a mild metasomatic event. On the other hand, metasomatism of Canadian SCUM may have produced networks of thin veins and derived melts that may have interacted significantly with SCUM matrix so that their isotopic compositions were displaced towards depleted mantle values. In this regard, it is interesting to note that data from the Bear Paw carbonatite (Dudas et al. 1987) lie on the extension of the linear array that includes the East African carbonatites, Oka, and, if age corrected, Fen, as well as some Canadian complexes.

Pyroxenite and lherzolite xenoliths and megacrysts from a monchiquite dyke at Loch Roag on the Isle of Lewis (Menzies et al. 1987b) have isotopic compositions that define a field extending to very low $^{143}Nd : ^{144}Nd$ ratios at moderate $^{87}Sr : ^{86}Sr$ ratios (Fig. 19.9b). The xenoliths were probably derived from a variably LREE-enriched pyroxene-rich stockwork in the upper mantle. Despite the presence of mica, this stockwork apparently had moderate time-integrated Rb:Sr ratios. Such characteristics are those expected for Type I or II metasomatic veins. We consider that the Loch Roag samples may have been derived from such a vein network in the mantle below Scotland.

Eastern Australia contains a vast province of xenolith-bearing alkaline igneous rocks (O'Reilly & Griffin 1987). Many of the xenoliths are pyroxenites and lherzolites that demonstrate clear evidence of metasomatism. The pressure–temperature path along which these rocks equilibrated (Griffin *et al.* 1984) has a remarkable resemblance to the path VW marked on Figure 19.1 and this may chart the pathway of alkaline magmas rising through the lithosphere. Interaction of melt and peridotite in such a case would be most intense at $P < 22$ Kb. The PT estimates of Griffin *et al.* (1984) on garnet pyroxenites are 900 °C, 11 Kb to 1100 °C, 16 Kb. Estimates from other pipes in eastern Australia extend the pressure range to 25 Kb. The Nd–Sr isotopic compositions of some mantle xenoliths from Kiama, eastern Australia, are shown on Figure 19.9b (Menzies & Wass 1983) and are displaced towards lower $^{143}Nd:^{144}Nd$ ratios at low $^{87}Sr:^{86}Sr$ ratios.

Andersen *et al.* (1984) and O'Reilly (1987) studied the fluid inclusions in wehrlite-series xenoliths from the pipes described by Griffin *et al.* (1984), and noted that these are rich in CO_2. Some xenoliths contain vugs 1 cm across. The amphiboles in the xenoliths are halogen-rich and contain CO_2-rich fluid inclusions. These xenoliths apparently formed by reaction between melt and peridotitic wall-rocks at pressures less than the stability field of carbonate (< 17 Kb) which is consistent with other aspects of their mineral assemblages and reported palaeopressures. We suggest that these xenoliths were derived from a vein network formed by the complete consumption of melt due to reaction with peridotite at $T < 1100$ °C and $P = 14$–17 Kb, i.e. they represent a system of Type IV veins. The higher pressure pyroxenites and lherzolites may reflect interaction of melts at greater pressures and be parts of a Type II and III network.

Also plotted on Figure 19.9a are compositions of some oceanic rocks that lie to the left of the mantle array. Volcanic rocks of St. Helena (Cohen & O'Nions 1982), Tubaii (Vidal *et al.* 1984), some Cook–Austral–Samoa islands (Palacz & Saunders 1986), and New Zealand (Barreiro & Cooper 1987) all lie to the left of the mantle array; also they have Pb isotopic compositions that reflect an enrichment in U:Pb and Th:Pb relative to bulk Earth. Rocks from the Walvis Ridge (Richardson *et al.* 1982) also lie to the left of the mantle array but have $^{206}Pb:^{204}Pb$ ratios that are lower than for the ocean island basalts noted above, although they have $^{208}Pb:^{204}Pb$ ratios much higher than the Northern Hemisphere Regression Line, indicating relatively high, time-integrated Th:U.

Apparently two different types of mantle with low $^{87}Sr:^{86}Sr$ and low $^{143}Nd:^{144}Nd$ ratios are required in the source regions of some oceanic magmas. One of these has Pb isotopic ratios higher than estimated for bulk Earth; the other has lower U:Pb but high Th:U. Both of these components may have their ultimate origins in SCUM (Figs 19.5 & 19.6) as suggested, for example, by McKenzie & O'Nions (1983).

19.7 SUMMARY

Rising low-temperature carbonated alkaline melts originating at depths > 70 km cross the peridotite–H_2O–CO_2 solidus at $T = 1000$–1050 °C, $P = 22$ Kb, or at

$T = 1050-1100\ °C$, $P = 17$ Kb. Under these conditions, melt and wall-rock react. At high temperatures, the main effect is to create clinopyroxene at the expense of olivine and orthopyroxene. Lower temperature interactions produce carbonate at $P > 17$ Kb. Reaction to completion (i.e. with loss of melt) produces amphibole peridotite, with carbonate at $P > 17$ Kb and CO_2-fluid at lower pressures. Experiments conducted using minor-element-doped melts in equilibrium with peridotite show that ilmenite and whitlockite may be super-solidus phases in the mantle, and that other titanium- and zirconium-rich phases are unlikely to form at temperatures above the solidus.

Metasomatized veins formed at temperatures above the solidus have relatively low Rb:Sr and Sm:Nd ratios, so they develop low $^{87}Sr:^{86}Sr$ and $^{143}Nd:^{144}Nd$ ratios and fall to the left of the mantle array. Metasomatic regions produced by complete loss of melt possess somewhat higher Rb:Sr and significantly lower Sm:Nd ratios so that they lie in the enriched quadrant of the $^{143}Nd:^{144}Nd$-$^{87}Sr:^{86}Sr$ diagram.

Super-solidus metasomatic veins have U–Th–Pb characteristics that vary according to their mineral assemblages. Super-solidus olivine clinopyroxenites and lherzolites have low U:Pb and Th:Pb ratios and thus develop non-radiogenic Pb isotopic compositions, whereas subsolidus amphibole peridotites have high U:Pb and Th:Pb and thus develop radiogenic Pb isotopic compositions. The presence of whitlockite in super-solidus mineral assemblages would increase U:Pb and decrease U:Th, so that these assemblages develop relatively high $^{206}Pb:^{204}Pb$ and very high $^{208}Pb:^{204}Pb$ ratios.

Some metasomatic veins may melt to give alkaline or tholeiitic magmas which inherit the isotopic and trace-element characteristics of the source regions. Interaction of the melt derived from the veins and the rocks that form the matrix to these veins is quite likely to occur during melting of the veins and coalescing of the melt fraction. If the area of SCUM undergoing melting contains a network of small metasomatic veins, interaction of melt and wall-rock appears to be a necessary condition for forming large melt volumes. If veins in SCUM are, on the other hand, wide and extensive, no reaction of the melt and the wall-rocks appears to be necessary to form large bodies. Infiltration of the wall-rocks by the melt in the veins occurs by surface-energy-induced infiltration and appears to be a necessary result if the wall-rocks and the melt are in thermal, textural, and major-element equilibrium and in the absence of external forces.

Reaction of wall-rock and melt will almost certainly metasomatize the former. If the minerals in the wall-rocks and those residual to the melt are similar in composition, the metasomatism will be cryptic, not patent. The metasomatism not only raises the LILE content of the mantle but also perturbs their isotopic compositions. Major- and trace-element compositions of the melts are little affected by metasomatism, but isotopic compositions can be severely affected. Furthermore, variable amounts of interaction of melts derived from veins in the SCUM and the wall-rocks can yield melts that have similar major- and trace-element compositions but that lie along an isotopic mixing curve.

ACKNOWLEDGEMENTS

This paper is an outgrowth of conversations that were initiated at the Dept of Terrestrial Magnetism with R. W. Carlson, B. A. Barreiro, and W. K. Hart, and at Pennsylvania State University with D. Baker, M. E. Schneider, and D. H. Eggler. Development of the concepts has required the ears and questions of graduate students whose persistence and inquisitiveness have furthered our understanding of mantle processes.

We are grateful to E. B. Watson and M. A. Menzies for extremely thoughtful reviews that improved the paper immeasurably. K. Bell is applauded for his careful editing, encouragement, and patience. We would like to acknowledge the generosity of D. H. Eggler in this study. Not only was the work largely financed by his National Science Foundation grant EAR-8206769 but Dave Eggler also provided us with the facilities, the discussion, and the large degree of cynicism required for work on such a project.

Additional funding for microprobe analysis was provided by the Dept of Geology, University of North Carolina. P. Fullagar graciously made facilities available and paid the senior author during manuscript preparation.

REFERENCES

Albee A. L. & L. Ray 1970. Correction factors for electron probe microanalysis of silicates, oxides, carbonates, phosphates, and sulfates. *Analytical Chemistry* **42**, 1408–14.

Andersen, T. 1987. Mantle and crustal components in a carbonatite complex, and the evolution of carbonatite magma: REE and isotopic evidence from the Fen complex, southeast Norway. *Chemical Geology (Isotope Geoscience Section)* **65**, 147–66.

Andersen, T., S. Y. O'Reilly & W. L. Griffin 1984. The trapped fluid phase in upper mantle xenoliths from Victoria, Australia: Implications for mantle metasomatism. *Contributions to Mineralogy and Petrology* **88**, 72–85.

Arth, J. G. 1976. Behavior of trace elements during magmatic processes: a summary of theoretical models and their applications. *Journal of Research of the U.S. Geological Survey* **4**, 41–7.

Barreiro, B. A. & A. F. Cooper 1987. A Sr, Nd, and Pb isotope study of alkaline lamprophyres from Westland and Otago, South Island, New Zealand. In *Mantle metasomatism and alkaline magmatism*, E. M. Morris & J. D. Pasteris (eds), 115–25. Geological Society of America Special Paper 215.

Basu, A. R. & M. Tatsumoto 1980. Nd-isotopes in selected mantle-derived rocks and minerals and their implications for mantle evolution. *Contributions to Mineralogy and Petrology* **75**, 43–54.

Bell, K. & J. Blenkinsop 1987a. Archean depleted mantle: Evidence from Nd and Sr initial isotopic ratios of carbonatites. *Geochimica et Cosmochimica Acta* **51**, 291–8.

Bell, K. & J. Blenkinsop 1987b. Nd and Sr isotopic compositions of East African carbonatites: implications for mantle heterogeneity. *Geology* **15**, 99–102.

Bell, K., J. Blenkinsop, T. J. S. Cole & D. P. Menagh 1982. Evidence from Sr isotopes for long-lived heterogeneities in the upper mantle. *Nature* **298**, 251–3.

Bence, A. E. & A. L. Albee 1968. Empirical correction factors for the electron microanalysis of silicates and oxides. *Journal of Geology* **76**, 382–403.

Benjamin, T., W. R. Heuser, D. S. Burnett & M. G. Sietz 1980. Actinide crystal–liquid partitioning for clinopyroxene and $Ca_3(PO_4)_2$. *Geochimica et Cosmochimica Acta* **44**, 1251–64.

Boyd, F. R. & J. L. England 1960. Apparatus for phase equilibrium measurements at pressures up to 50 kilobar and temperatures up to 1750 °C. *Journal of Geophysical Research* **65**, 741–8.

Bulau, J. R., H. S. Waff & J. A. Tyburczy 1979. Mechanical and thermodynamic constraints on fluid distribution in partial melts. *Journal of Geophysical Research* **84**, 6102–8.

Carlson, R. W. 1984. Isotopic constraints on Columbia River flood basalt genesis and the nature of the subcontinental mantle. *Geochimica et Cosmochimica Acta* **48**, 2357–72.

Carlson, R. W., F. O. Dudas, J. K. Meen & D. H. Eggler 1985. Formation and evolution of the Archean subcontinental mantle beneath the northwestern U.S. *EOS. Transactions, American Geophysical Union* **66**, 1109.

Cohen, R. S. & R. K. O'Nions 1982. The lead, neodymium and strontium isotopic structure of ocean ridge basalts. *Journal of Petrology* **23**, 299–324.

Dawson, J. B. 1984. Contrasting types of upper-mantle metasomatism? In *Kimberlites II: The mantle and crust-mantle relationships*, J. Kornprobst (ed.), 289–94. Amsterdam: Elsevier.

Deines, P. 1967. *Stable carbon and oxygen isotopes of carbonatite carbonates and their interpretation*. Ph.D. thesis, The Pennsylvania State University, University Park.

DePaolo, D. J. & G. J. Wasserburg 1976. Inferences about magma sources and mantle structure from variations of $^{143}Nd/^{144}Nd$. *Geophysical Research Letters* **3**, 743–6.

Doe, B. R., W. P. Leeman, R. L. Christiansen & C. E. Hedge 1982. Lead and strontium isotopes and related trace elements as genetic tracers in the Upper Cenozoic rhyolite–basalt association of the Yellowstone plateau volcanic field. *Journal of Geophysical Research* **87**, 4785–806.

Dudas, F. O., R. W. Carlson & D. H. Eggler 1987. Regional mid-Proterozoic enrichment of the subcontinental mantle source of igneous rocks from central Montana. *Geology* **15**, 22–5.

Eby, G. N. 1984. Monteregian Hills: I. Petrography, major and trace element geochemistry, and strontium isotopic chemistry of the western intrusions; Mounts Royal, St. Bruno, and Johnson. *Journal of Petrology* **25**, 421–52.

Eggler, D. H. 1978. The effect of CO_2 upon partial melting of peridotite in the systems $Na_2O–CaO–Al_2O_3–MgO–SiO_2–CO_2$ to 35 kb, with an analysis of melting in a peridotite–$H_2O–CO_2$ system. *American Journal of Science* **278**, 305–43.

Eggler, D. H. 1983. Upper mantle oxidation state: Evidence from olivine–orthopyroxene–ilmenite assemblages. *Geophysical Research Letters* **10**, 365–8.

Eggler, D. H. 1987. Solubility of major and trace elements in mantle metasomatic fluids: Experimental constraints. In *Mantle metasomatism*, M. A. Menzies & C. J. Hawkesworth (eds), 21–41. London: Academic Press.

Eggler, D. H. & D. R. Baker 1982. Reduced volatiles in the system C–O–H: Implications to mantle melting, fluid formation, and diamond genesis. In *Advances in Earth and planetary sciences. High-pressure research in geophysics*, S. Akimoto & M. H. Manghnani (eds), 237–50. Dordrecht: Reidel.

Ellis, D. E. & P. J. Wyllie 1979. Hydration and melting reactions in the system $MgO–SiO_2–H_2O$ at pressures up to 100 kbar. *American Mineralogist* **64**, 41–8.

Erlank, A. J., F. G. Waters, C. J. Hawkesworth, S. E. Haggerty, H. L. Allsopp, R. S. Rickard & M. A. Menzies 1987. Evidence for mantle metasomatism in peridotite nodules from the Kimberley pipes, South Africa. In *Mantle metasomatism*, M. A. Menzies & C. J. Hawkesworth (eds), 221–311. London: Academic Press.

Fraser, K. J., C. J. Hawkesworth, A. J. Erlank, R. H. Mitchell & B. H. Scott-Smith 1985. Sr, Nd, and Pb isotope and minor element geochemistry of lamproites and kimberlites. *Earth and Planetary Science Letters* **76**, 57–70.

Fregeau, E. J. & D. H. Eggler 1985. Partitioning of minor elements between supercritical H_2O fluid and silicate melt at 15–20 kbar pressure: Implications to metasomatism. *Geological Society of America Abstracts with Program* **17**, 586.

Frey, F. A. & D. H. Green 1974. The mineralogy, geochemistry, and origin of lherzolite inclusions in Victorian basanites. *Geochimica et Cosmochimica Acta* **38**, 1023–59.

Green, D. H. 1973. Experimental melting studies on a model upper mantle composition at high water pressure under water-saturated and water-undersaturated conditions. *Earth and Planetary Science Letters* **19**, 37–53.

Griffin, W. L., S. Y. Wass & J. D. Hollis 1984. Ultramafic xenoliths from Bullenmerri and Gnotuk Maars, Victoria, Australia: Petrology of a sub-continental crust–mantle transition. *Journal of Petrology* **25**, 53–87.

Grünenfelder, M. H., G. R. Tilton, K. Bell & J. Blenkinsop 1986. Lead and strontium isotope relationships in the Oka carbonatite complex, Quebec. *Geochimica et Cosmochimica Acta* **50**, 461–8.

Haggerty, S. E. 1983. The mineral chemistry of new titanates from the Jägersfontein kimberlite, South Africa: implications for metasomatism in the upper mantle. *Geochimica et Cosmochimica Acta* **47**, 1833–54.

Hart, W. K. 1985. Chemical and isotopic evidence for mixing between depleted and enriched mantle, northwestern U.S.A. *Geochimica et Cosmochimica Acta* **49**, 131–44.

Hofmann, A. W. 1980. Diffusion in natural silicate melts: a critical review. In *Physics of magmatic processes*, R. B. Hargraves (ed.), 385–417. Princeton, NJ: Princeton University Press.

Hofmann, A. W. & S. R. Hart 1978. An assessment of local and regional isotopic equilibrium in the mantle. *Earth and Planetary Science Letters* **38**, 44–62.

Jacobsen, S. B. & G. J. Wasserburg 1980. Sm–Nd isotopic evolution of chondrites. *Earth and Planetary Science Letters* **50**, 139–55.

Jordan, T. H. 1975. The continental tectosphere. *Reviews of Geophysics and Space Physics* **13**, (3), 1–12.

Kramers, J. D. 1977. Lead and strontium isotopes in Cretaceous kimberlites and mantle-derived xenoliths from southern Africa. *Earth and Planetary Science Letters* **34**, 419–27.

Kushiro, I., Y. Syono & S. Akimoto 1968. Melting of peridotite nodule at high pressure and high water pressure. *Journal of Geophysical Research* **73**, 6023–9.

Lancelot, J. R. & C. J. Allègre 1974. Origin of carbonatitic magma in the light of the Pb–U–Th isotope system. *Earth and Planetary Science Letters* **22**, 233–8.

Leeman, W. P. 1975. Radiogenic tracers applied to basalt genesis in the Snake River Plain-Yellowstone National Park region: evidence for a 2.7-b.y.-old upper mantle keel. *Geological Society of America Abstracts with Program* **7**, 1165.

Lloyd, F. E. & Bailey, D. K. 1975. Light element metasomatism of the continental mantle: The evidence and the consequences. *Physics and Chemistry of the Earth* **9**, 389–416.

Loubet, M., M. Bernat, M. Javoy & C. J. Allègre 1972. Rare earth contents in carbonatites. *Earth and Planetary Science Letters* **14**, 226–32.

McKenzie, D. 1985. The extraction of magma from the crust and mantle. *Earth and Planetary Science Letters* **74**, 81–91.

McKenzie, D. & R. K. O'Nions 1983. Mantle reservoirs and ocean island basalts. *Nature* **301**, 229–31.

Meen, J. K. 1987. Mantle metasomatism and carbonatites; An experimental study of a complex relationship. In *Mantle metasomatism and alkaline magmatism*, E. M. Morris & J. D. Pasteris (eds), 91–100. Geological Society of America Special Paper 215.

Meen, J. K. & D. H. Eggler 1987. Petrology and geochemistry of the Cretaceous Independence volcanic suite, Absaroka Mountains, Montana. *Geological Society of America Bulletin* **98**, 238–47.

Menzies, M. A. & V. R. Murthy 1980a. Enriched mantle: Nd and Sr isotopes in diopsides from kimberlite nodules. *Nature* **283**, 634–6.

Menzies, M. A. & V. R. Murthy 1980b. Nd and Sr isotope geochemistry of hydrous mantle nodules and their host alkali basalts: Implications for local heterogeneities in metasomatically veined mantle. *Earth and Planetary Science Letters* **46**, 323–34.

Menzies, M. & S. Y. Wass 1983. CO_2- and LREE-rich mantle below eastern Australia: A REE and isotopic study of alkaline magmas and apatite-rich mantle xenoliths from the Southern Highlands Province, Australia. *Earth and Planetary Science Letters* **65**, 287–302.

Menzies, M., P. Kempton & M. Dungan 1985. Interaction of continental lithosphere and asthenospheric melts below the Geronimo Volcanic Field, Arizona, U.S.A. *Journal of Petrology* **26**, 663–93.

Menzies, M. A., N. Rogers, A. Tindle & C. J. Hawkesworth 1987a. Metasomatic and enrichment processes in lithospheric peridotites, an effect of asthenosphere–lithosphere interaction. In *Mantle metasomatism*, M. A. Menzies & C. J. Hawkesworth (eds), 313–61. London: Academic Press.

Menzies, M. A., A. N. Halliday, Z. Palacz, R. H. Hunter, B. G. J. Upton, P. Aspen & C. J. Hawkesworth 1987b. Evidence from mantle xenoliths for an enriched lithospheric keel under the Outer Hebrides. *Nature* **325**, 44–7.

Mysen, B. O. 1983. Rare earth element partitioning between ($H_2O + CO_2$) vapor and upper mantle minerals: Experimental data bearing on the conditions of formation of alkali basalt and kimberlite. *Neues Jahrbuch für Mineralogie Abhandlungen* **146**, 41–65.

Mysen, B. O. & A. L. Boettcher 1975. Melting of a hydrous mantle: II. Geochemistry of crystals and liquids formed by anatexis of mantle peridotite at high pressures and temperatures as a function of controlled activities of water, hydrogen, and carbon dioxide. *Journal of Petrology* **16**, 549–93.

Newton, R. C. & W. E. Sharp 1975. Stability of forsterite + CO_2 and its bearing on the role of CO_2 in the mantle. *Earth and Planetary Science Letters* **26**, 239–44.

O'Brien, H. E., A. J. Irving, I. S. McCallum & M. F. Thirlwall 1988. Characterization of source components of potassic mafic magmas of the Highwood Mountains Province, Montana. *EOS. Transactions, American Geophysical Union* **69**, 519.

Olafsson, M. & D. H. Eggler 1983. Phase relations of amphibole-, amphibole-carbonate, and phlogopite-carbonate peridotite: Petrologic constraints on the asthenosphere. *Earth and Planetary Science Letters* **64**, 305–15.

O'Nions, R. K., P. J. Hamilton & N. M. Evensen 1977. Variations in ^{143}Nd/^{144}Nd and ^{87}Sr/^{86}Sr ratios in oceanic basalts. *Earth and Planetary Science Letters* **34**, 13–22.

O'Reilly, S. Y. 1987. Volatile-rich mantle beneath eastern Australia. In *Mantle xenoliths*, P. H. Nixon (ed.), 661–70. Chichester: Wiley.

O'Reilly, S. Y. & W. L. Griffin 1987. Eastern Australia; 4000 kilometers of mantle samples. In *Mantle xenoliths*, P. H. Nixon (ed.), 267–80. Chichester: Wiley.

Palacz, Z. A. & A. D. Saunders 1986. Coupled trace element and isotope enrichment in the Cook–Austral–Samoa islands, southwest Pacific. *Earth and Planetary Science Letters* **79**, 270–80.

Peterman, Z. E. 1979. Geochronology and the Archean of the United States. *Economic Geology* **74**, 1544–62.

Ray, G. L., N. Shimizu & S. R. Hart 1983. An ion microprobe study of the partitioning of trace elements between clinopyroxene and liquid in the system diopside–albite–anorthite. *Geochimica et Cosmochimica Acta* **47**, 2131–40.

Richardson, S. H., A. J. Erlank, A. R. Duncan & D. L. Reid 1982. Correlated Nd, Sr and Pb isotope variations in Walvis Ridge basalts and implications for the evolution of their mantle source. *Earth and Planetary Science Letters* **59**, 327–42.

Richardson, S. H., J. J. Gurney, A. J. Erlank & J. W. Harris 1984. Origin of diamonds in old enriched mantle. *Nature* **310**, 198–202.

Roden, M. F. & V. R. Murthy 1985. Mantle metasomatism. *Annual Reviews of Earth and Planetary Sciences* **13**, 269–96.

Sack, R. O., I. S E. Carmichael, M. Rivers & M. S. Ghiorso 1980. Ferric–ferrous equilibria in natural silicate liquids at 1 bar. *Contributions to Mineralogy and Petrology* **75**, 369–76.

Salters, V. J. M. & M. Barton 1985. The geochemistry of ultrapotassic lavas from the Leucite Hills, Wyoming. *EOS. Transactions, American Geophysical Union* **66**, 1109.

Scambos, T. A. 1987. Sr and Nd isotope ratios for the Missouri Breaks diatremes, central Montana. *Geological Society of America Abstracts with Programs* **19**, 830–1.

Schneider, M. E. & D. H. Eggler 1986. Fluids in equilibrium with peridotite minerals: implications for mantle metasomatism. *Geochimica et Cosmochimica Acta* **50**, 711–24.

Shaw, H. R. 1980. The fracture mechanisms of magma transport from the mantle to the surface. In *Physics of magmatic processes*, R. B. Hargraves (ed.), 201–64. Princeton, NJ: Princeton University Press.

Smith, D. G. & S. N. Ehrenberg 1984. Zoned minerals in garnet peridotite nodules from the Colorado Plateau: implications for mantle metasomatism and kinetics. *Contributions to Mineralogy and Petrology* **86**, 274–85.

Spera, F. J. 1980. Aspects of magma transport. In *Physics of magmatic processes*, R. B. Hargraves (ed.), 265–323. Princeton, NJ: Princeton University Press.

Spera, F. J. 1984. Carbon dioxide in petrogenesis III: Role of volatiles in the ascent of alkaline magmas with special reference to xenolith-bearing mafic lavas. *Contributions to Mineralogy and Petrology* **88**, 217–32.

Spera, F. J. 1987. Dynamics of translithospheric migration of metasomatic fluid and alkaline magma. In *Mantle metasomatism*, M. A. Menzies & C. J. Hawkesworth (eds), 1–20. London: Academic Press.

Stacey, J. S. & J. D. Kramers 1975. Approximation of terrestrial lead evolution by a two-stage model. *Earth and Planetary Science Letters* **26**, 207–21.

Stevenson, D. J. 1986. On the role of surface tension in the migration of melts and fluids. *Geophysical Research Letters* **13**, 1149–52.

Tilton, G. R. & M. H. Grünenfelder 1983. Lead isotope relationships in billion-year old carbonatite complexes, Superior Province, Canadian Shield. *Geological Society of America Abstracts with Programs* **15**, 707.

Vidal, P., C. Chauvel & R. Brousse 1984. Large mantle heterogeneity beneath French Polynesia. *Nature* **307**, 536–8.

Vollmer, R., P. Ogden, J.-G. Schilling, R. H. Kingsley & D. H. Waggoner 1984. Nd and Sr isotopes in ultrapotassic volcanic rocks from the Leucite Hills, Wyoming. *Contributions to Mineralogy and Petrology* **87**, 359–68.

Waff, H. S. & J. R. Bulau 1979. Equilibrium fluid distribution in an ultramafic partial melt under hydrostatic stress conditions. *Journal of Geophysical Research* **84**, 6109–14.

Watson, E. B. 1980. Apatite and phosphorus in mantle source regions: An experimental study of apatite/melt equilibria at pressures to 25 kbar. *Earth and Planetary Science Letters* **51**, 322–35.

Watson, E. B. 1982. Melt infiltration and magma evolution. *Geology* **10**, 236–40.

Watson, E. B. & T. M. Harrison, 1983. Zircon saturation revisited: Temperature and composition effects in a variety of crustal magma types. *Earth and Planetary Science Letters* **64**, 295–304.

Watson, E. B. & F. J. Ryerson 1986. Rutile saturation in magmas: Implications for Nb–Ta–Ti depletion in orogenic magmas. *EOS. Transactions, American Geophysical Union* **67**, 412.

Watson, E. B., D. Ben-Othman, J.-M. Luck & A. W. Hofmann 1987. Partitioning of U, Pb, Cs, Yb, Hf, Re, and Os between chromian diopsidic pyroxene and haplobasaltic liquid. *Chemical Geology* **62**, 191–208.

Wendlandt, R. F. & W. J. Harrison 1979. Rare earth partitioning between immiscible carbonate and silicate liquids and CO_2 vapor: Results and implications for the formation of light rare earth-enriched rocks. *Contributions to Mineralogy and Petrology* **69**, 409–19.

Wendlandt, R. F. & B. O. Mysen 1980. Melting phase relations of natural peridotite + CO_2 as a function of degree of partial melting at 15 and 30 kbar. *American Mineralogist* **65**, 37–44.

Wilshire, H. G. 1987. *A model of mantle metasomatism*. In *Mantle metasomatism and alkaline magmatism*, E. M. Morris & J. D. Pasteris (eds), 47–60. Geological Society of America Special Paper 215.

Wilshire, H. G. & J. W. Shervais 1975. Al-augite and Cr-diopside ultramafic xenoliths in basaltic rocks from western United States. *Physics and Chemistry of the Earth* **9**, 257–72.

Wilshire, H. G., J. E. N. Pike, C. E. Meyer & E. C. Schwarzman 1980. Amphibole-rich veins in lherzolite xenoliths, Dish Hill and Deadman Lake, California. *American Journal of Science* **280-A**, 576–93.

20
ORIGIN OF CARBONATITES: EVIDENCE FROM PHASE EQUILIBRIUM STUDIES

P. J. WYLLIE

ABSTRACT

An integrated model is developed for the generation of crustal carbonatites from deep-mantle carbon, some probably derived through subduction. After generation of silicate melts in the asthenosphere, the melts are then processed in the lithosphere associated with rifting, and parental nephelinitic melts from about 75 km depth then yield the carbon in the form of carbonatite at depths between 75 km and the surface. Experimental phase equilibrium data are reviewed covering:

(a) melting temperature for Ca–Mg carbonates to 30 Kb (100 km depth) and the effect of H_2O and alkalis in lowering liquidus temperatures down to those appropriate for the precipitation of calcite and dolomite in magmatic carbonatites;

(b) conditions for the precipitation with calcite of accessory minerals pyrochlore, apatite, sulphide, bastnaesite, barite, and fluorite;

(c) the relationships between silicate and carbonatite melts, with respect to crystal fractionation, the syntexis hypotheses, and silicate–carbonate liquid immiscibility;

(d) the effect of CO_2 in generating Ca–Na carbonatite melts from peridotite and other rocks at temperatures lower than normal rock melting temperatures;

(e) the effect of CO_2 in the generation of Ca–Mg carbonatite melts from peridotite at depths greater than 75 km;

(f) the generation of kimberlite-like magmas in the asthenosphere and lower lithosphere;

(g) the uprise of those magmas with thinning of the lithosphere beneath a rift, and the formation of magma chambers at about 75 km depth;

(h) the eruption of parental nephelinitic or melilititic magmas from this level; and

(i) the formation of carbonatites by fractionation or immiscibility from these parents, at 75 km, or within the crust.

The real prospect that primary carbonatites could be erupted from the mantle justifies a search for them, but the high ratio of silicate:carbonatite in most alkalic complexes argues against this origin.

20.1 INTRODUCTION

The history of debate about whether or not carbonatite magmas exist has been detailed elsewhere (e.g. Heinrich 1966, Tuttle & Gittins 1966), and the results outlined in this chapter support a magmatic origin. There have been several hypotheses for the origin of carbonatite magmas based on field and petrological studies. It has been proposed that carbonatites may be:

(a) primary, possibly very rich in alkalis (von Eckermann 1948, Dawson 1966, Cooper *et al.* 1975, Koster van Groos 1975);
(b) residual melts derived by fractional crystallization of a parent 'carbonated alkali peridotite magma' (King & Sutherland 1960); or
(c) immiscible liquid fractions separated from kimberlitic, melilititic, or nephelinitic parent magma (Le Bas 1977, this volume, Kjarsgaard & Hamilton, this volume).

My associates and I have been following dendritic paths through the phase relationships of carbonate and carbonate–silicate systems with non-quenchable liquids ever since the late O. F. Tuttle asked me one day: 'When you have time, see if you can make calcite melt with water.' I tried, calcite melted, and this led to the study of synthetic carbonatite magmas in the system $CaO-CO_2-H_2O$ (Wyllie & Tuttle 1960), followed by a series of more complex systems, combining the carbonate melts with silicates, that were summarized in Tuttle & Gittins' (1966) book *Carbonatites* (Wyllie 1966*a*) and in two reviews for the symposium on 'Carbonatites, kimberlites, and their minerals' organized by the International Mineralogical Association (Wyllie & Biggar 1966, Wyllie 1966*b*). My charge from the editor of this volume is to review a selection of the relevant experimental data from my laboratory and gathered by associates since then, without detailed coverage of experimental results from other laboratories.

Following the discovery of synthetic carbonatite magmas in the system $CaO-CO_2-H_2O$, three approaches were used in the investigation of more complex systems at pressures to 4 Kb. The first was to add MgO, P_2O_5, and SiO_2 as a fourth component. The second was to add silicate components such as feldspars, feldspathoids, olivine, and pyroxene. The third was to add excess alkali in the form of carbonate to feldspars and feldspathoids. The composition joins studied were selected in order to test the relationships suggested by field and petrological studies.

Unravelling the history of carbonatite complexes is difficult (Le Bas 1977). Repeated intrusions of fluid, reactive, volatile-charged carbonatite magma may be followed by differentiation, the release of vapours and solutions from the crystallizing carbonatite or magmas at greater depth, metasomatism occurring within as well as around the complex, and the development of explosion breccias with concomitant changes in pressure on the magma at greater depths. Superimposed may be the effects of remobilization of the crystal cumulates, plastic flow, shearing, and recrystallization associated with solid state intrusions, and later metamorphism.

Since the symposium on 'Carbonatites, kimberlites, and their minerals' (Naidu 1966), the existence of petrogenetic links between kimberlites and carbonatites has been examined. The petrogenesis of kimberlites and carbonatites (Daly 1925) has been reviewed in several books (Heinrich 1966, Tuttle & Gittins 1966, Wyllie 1967, Dawson 1980, Mitchell 1986) and the proceedings volumes of the International Kimberlite Conferences (e.g. Kornprobst 1984a, b). Many carbonatite complexes include rocks with affinities to kimberlites, and most kimberlites contain carbonates either as dispersed minerals or in discrete bodies (Nixon 1973, Ahrens et al. 1975, Boyd & Meyer 1979a, b). Petrological studies confirm that crystallization of kimberlite magmas can produce residual magmas that precipitate carbonate (e.g. Watson 1955, Dawson & Hawthorne 1973, Jones & Wyllie 1985). Mitchell (1979) argued against the 'alleged kimberlite–carbonatite relationship', pointing out that carbonate-residues of kimberlites differ in many respects from the intrusions of alkalic complexes, which have probably evolved from nephelinitic or melilititic parent magmas, and that typical carbonatite complexes and kimberlites occur in tectonically different environments. In order to test the proposed links between kimberlites and carbonatites, we have followed the low-temperature melting reactions determined in the silicate–carbonate–H_2O systems up to pressures corresponding to depths of ~100 km.

Information to be obtained from experimental phase equilibria includes the following:

(a) the chemical and physical nature of carbonate-precipitating melts, and the paths of crystallization causing differentiation of carbonatites;
(b) the paths of crystallization of synthetic carbonatite magmas with accessory components and minerals, including phosphate, sulphide, and REE minerals;
(c) the relationships between high-temperature silicate melts with dissolved volatile components and low-temperature synthetic carbonatite magmas, including liquid immiscibility; and
(d) the conditions for the formation of CO_2-rich parent silicate melts (or of primary carbonatites) from mantle peridotite at high pressures.

Experimental results will be reviewed in this sequence. Applications of the experimental data to various aspects of carbonatite petrogenesis will be summarized as the data are reviewed, with special attention devoted to the limestone-syntexis hypothesis and its obverse, the sial-syntexis hypothesis. After consideration of the conditions for the formation and crystallization of carbonatite magmas, I move to the source of the parents of carbonatites and associated alkalic igneous rocks and conclude that these represent late-stage products in a series of events initiated within the mantle, with source materials rising from the asthenosphere and being processed within the lithosphere. The origin of carbonatites cannot be resolved without better understanding of volatile components from deeper than the lithosphere, and of their effects on the generation of kimberlites and kimberlite-like magmas from mantle peridotite.

20.2 CARBONATE SYSTEMS: PRECIPITATION OF CARBONATES FROM MELTS

Working on the thesis that carbonatites are crystal cumulates dominated by carbonates precipitated from carbonatite magmas, the obvious place to start an experimental investigation is to define the conditions for precipitation of carbonates. The melting temperatures of Ca–Mg carbonates are very high, but it has been established that addition of H_2O or alkalis brings the liquidus temperatures down to levels corresponding to those estimated from field studies for the emplacement of carbonatites.

20.2.1 The systems CaO–CO_2 and MgO–CO_2

Figure 20.1 shows the arrays of decarbonation and fusion curves in these two systems emanating from the invariant points (Q_1 and Q_2) for the assemblages oxide + carbonate + L + V. Irving & Wyllie (1973, 1975) measured the fusion curves for calcite and magnesite, and extended the dissociation curve of magnesite above 10 Kb from that of Goldsmith & Heard (1961). Huang & Wyllie (1976) determined the other curves involving carbonate and liquid. In each system, the carbonate dissociation curve extends to higher temperatures with increasing pressure until the assemblage begins to melt, generating the invariant point, from which fusion curves

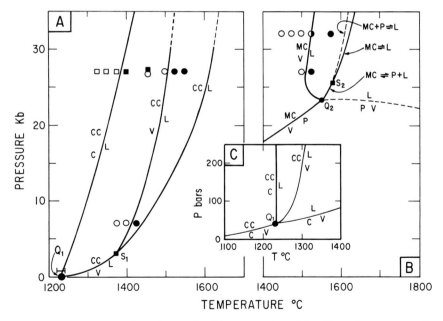

Figure 20.1 (A) Dissociation (decarbonation) and melting curves for calcite; (B) analogous curves for magnesite; (C) details around invariant point Q_1 in A. Abbreviations: calcite, CC; magnesite, MC; lime, C; periclase, P; liquid, L; and CO_2 vapour, V. For compositions of L and V at different pressures see Figures 20.2 and 20.3 (Huang & Wyllie 1976). Q and S are invariant and singular points, respectively.

ORIGIN OF CARBONATITES

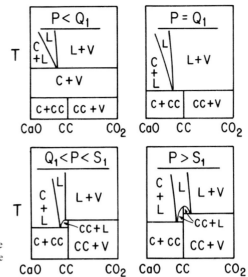

Figure 20.2 Schematic isobaric sections for $CaO-CO_2$. See Figure 20.1A and C; contrast Figure 20.3 (Huang & Wyllie 1976).

extend to higher temperatures and pressures. The invariant point, Q_1, for calcite is at 40 bar, whereas Q_2 for magnesite is at 23 Kb. The congruent or incongruent melting of the carbonates at different pressures, as well as the varying solubility of CO_2 in the liquids as a function of pressure relative to Q, are illustrated in Figures 20.2 and 20.3. Huang & Wyllie (1976) found the solubilities of CO_2 at 27 Kb to be 11.5 wt% in molten $CaCO_3$, and 6.5 wt% in $MgCO_3$. Melting temperatures were lowered by 105 °C and 80 °C, respectively, by the dissolved CO_2.

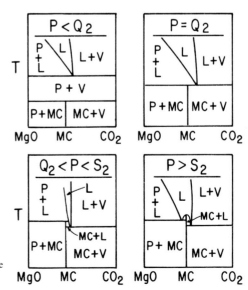

Figure 20.3 Schematic isobaric sections for $MgO-CO_2$. See Figure 20.1B; contrast Figure 20.2 (Huang & Wyllie 1976).

20.2.2 Phase relations of dolomite in the system $CaO-MgO-CO_2$

The phase relationships for dolomite are compared with those for calcite and magnesite, without excess CO_2, in Figure 20.4. Irving & Wyllie (1975) determined the dolomite–magnesite solvus at 27 Kb and its intersection with the carbonate melting loop, with results corresponding to those shown in Figure 20.4B for 30 Kb. These results, incorporated with previously published experimental data on aragonite and the subsolidus decarbonation reactions, provided the complete PT projection on Figure 20.4A. Figures 20.4B and 20.4C are isobaric sections through the composition join $CaCO_3-MgCO_3$ (no excess CO_2, compare Figs 20.1, 20.2 & 20.3) with reaction temperatures taken from Figure 20.4A. Note the solvus crest between calcite and dolomite (Cc = Cd), and the temperature minimum on the solidus and liquidus between calcite and dolomite (Ccd = L). Figure 20.4C is consistent with the 10 Kb isobaric section subsequently determined by Byrnes &

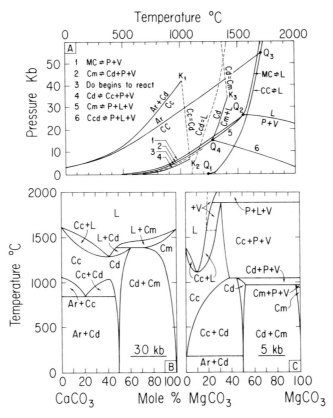

Figure 20.4 (A) Dissociation and melting reactions involving $CaCO_3-MgCO_3$, incorporating Figure 20.1 and experimental data obtained and compiled by Irving & Wyllie (1975). Dashed lines are crests of solvi and a temperature minimum on liquidus. Note critical end-points, K. (B) Isobaric section at 30 Kb, between Q_2 and K_3. (C) Isobaric section at 5 Kb, between Q_1 and K_2. Abbreviations: see Figure 20.1; Ar, aragonite; Do, dolomite; Cc, Ccd, Cd, and Cm, carbonate solid solutions with compositions extending from end members calcite, dolomite, and magnesite. Note that at some pressures there is complete solid solution between Cc and Cd (Ccd), but at others the solvus intervenes.

Wyllie (1981). The results in Figure 20.4C for 6 Kb persist in general form without much change in temperature down to low pressures. According to Figure 20.4, carbonates can be precipitated from melts in the system $CaO-MgO-CO_2$ only at temperatures above 1000 °C, and calcite, without significant $MgCO_3$ in solid solution, can be precipitated only at temperatures in excess of 1240 °C, rising to 1400 °C by 5 Kb. These temperatures are much higher than those estimated from field relationships for carbonatite intrusions. The addition of dissolved CO_2 does not change this conclusion (Figs 20.1–20.3). Additional experiments with $FeCO_3$ are desirable.

20.2.3 Precipitation of calcite in $CaO-CO_2-H_2O$

Wyllie & Tuttle (1960) determined that addition of H_2O extends the liquidus field for calcite down to about 650 °C at 1 Kb until it meets the liquidus field for portlandite. Wyllie & Boettcher (1969) extended the study to 40 Kb, determining the calcite–aragonite transition in the process, and Figure 20.5 is their summary of the phase relationships for the fusion of calcite, portlandite, and the ternary eutectics, with and without excess vapour. Their estimated curves for $CaO-CO_2$ require some modification following subsequent experimental determinations (see Fig. 20.1A). The composition of the vapour-saturated ternary eutectic liquid precipitating calcite and portlandite (with vapour) at 655 °C is illustrated in Figure 20.6. Wyllie & Tuttle

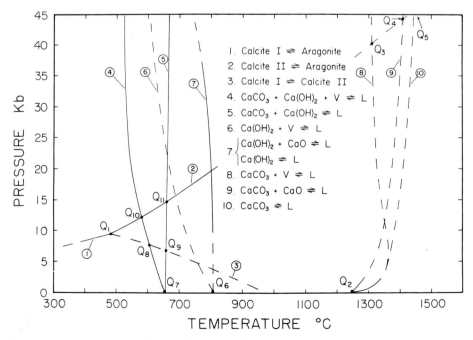

Figure 20.5 Univariant curves showing conditions for melting of carbonates in the system $CaO-CO_2-H_2O$, based on experimental data (solid) and estimates (dashed) of Wyllie & Boettcher (1969). See Figures 20.1A and C for subsequent experimental results in $CaO-CO_2$. Note that the subscripts for invariant points do not correspond.

(1960) reported its composition as CaO plus 19% CO_2 and 16% H_2O at 1 Kb, and Wyllie & Boettcher (1969) illustrated the situation at 25 Kb, with considerable increase in dissolved ($H_2O + CO_2$) and in $H_2O:CO_2$. Figure 20.5 confirms that low-temperature melts capable of precipitating calcite can exist from crustal to mantle pressures, and that these melts would precipitate aragonite rather than calcite at pressures above 13–15 Kb.

20.2.4 Precipitation of dolomite in the system $CaO-MgO-CO_2-H_2O$

Many carbonatites consist largely of dolomite or dolomite plus calcite. Initial attempts to determine the conditions for the precipitation of dolomite from low-temperature synthetic carbonatite magmas were unsuccessful. Addition of MgO to the calcite–portlandite eutectic introduced a liquidus field for periclase or brucite, illustrated in Figure 20.6, with no dolomite in this liquidus area (Walter *et al.* 1962). Wyllie's (1965) analysis suggested that dolomite should appear at higher pressures, but Boettcher *et al.* (1980) followed the solidus to 40 Kb without finding dolomite. The solution to the mystery was discovered by Fanelli *et al.* (1986). These findings are summarized in Figure 20.6, which shows a Cc–Do–Pe–V eutectic.

The experimental approach can be seen in Figures 20.4C and 20.6A. Mixtures were made on the quadrilateral shown in Figure 20.6b and, according to the estimated vapour-saturated liquidus surface drawn in Figure 20.6a, most mixtures on this join when melted should include some free-vapour phase. The phase relationships in the binary system $CaCO_3-MgCO_3$ at 2 Kb correspond to those in Figure 20.4C, with temperatures adjusted to 2 Kb values (Fig. 20.4A) and with a trace of excess CO_2 present as vapour in all phase fields with liquid (Figs 20.1A, 20.2 & 20.6a). Addition of $Ca(OH)_2$ lowered the liquidus temperature below the crest of the calcite–dolomite solvus, and Fanelli & Wyllie (in preparation) confirmed the co-precipitation of calcite and dolomite down to 650 °C. They were able to measure, with electron microprobe, the compositions of magnesian calcite on the solvus limb in reversed runs, with results which have been published by Goldsmith (1983, Fig. 4). Cava & Wyllie (in preparation) followed with experiments using mixtures on the composition join $Ca(OH)_2-MgCO_3-Mg(OH)_2$ (Fig. 20.6b) and located the position of the eutectic for liquid coexisting with calcite, dolomite, periclase, and vapour. Extrapolation of the phase boundaries intersected in Figure 20.6b from the vapour phase join CO_2-H_2O to the estimated vapour-saturated liquidus surface yielded the results shown in Figure 20.6a.

The vapour-saturated field for primary periclase falls steeply from high temperatures with increasing CaO:MgO, and is divided into two regions by a thermal barrier represented by the triangle Cc–Pe–V, which cuts the vapour-saturated surface as the dashed line in Figure 20.6b. A boundary separating the liquidus fields for periclase and calcite passes over this temperature maximum and extends to the two separate regions with low-temperature eutectics. The liquidus surface for calcite rising from one side of the eutectic with dolomite is separated from the liquidus field for periclase by a narrow surface for dolomite extending from 880 °C down to 650 °C. This defines an extensive interval for the possible co-precipitation of

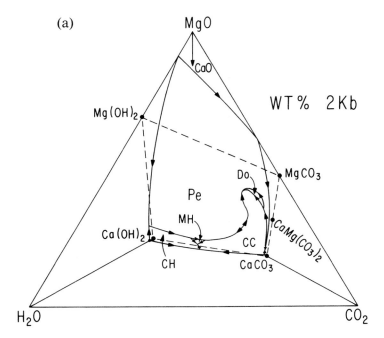
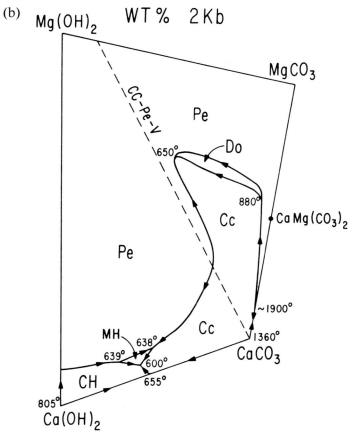

dolomite and calcite, with prospects also for the precipitation of dolomite followed by calcite, or vice versa. Both sequences are known in carbonatites (Zhabin & Cherepivskaya 1965, Hogarth 1966).

In all previous systems investigated the synthetic carbonatite magmas that precipitate calcite terminate at a eutectic with co-precipitation of portlandite or lime, neither of which occurs in carbonatites. The embarrassment of portlandite is removed by discovery of the new eutectic, where periclase (or brucite at somewhat higher pressures) occurs instead. The ubiquitous occurrence in natural carbonatites of magnesioferrite-magnetite with composition very close to Fe_3O_4 suggests that $(Fe + Mg)O$ in the magmas may mimic the role of MgO in the synthetic system. The dearth of experiments in carbonatite systems with Fe is notable. The assemblage calcite–dolomite–periclase has been reported from the Oka carbonatite (Treiman & Essene 1984), and Mariano & Roeder (1983) have described a carbonatite from Kerimasi with euhedral periclase, commonly overgrown on magnesioferrite, in a matrix of calcite and monticellite.

20.2.5 *Precipitation of calcite in the system $CaCO_3$–K_2CO_3–Na_2CO_3*

The eruption at Oldoinyo Lengai of natrocarbonatite whose mineralogy was dominated by Na–K–Ca carbonates (Dawson 1962, 1966; see Dawson, this volume) suggested to many that plutonic carbonatites may have crystallized from a magma containing alkalis that were concomitantly lost to the vapour phase responsible for fenitization of the country rocks. Cooper *et al.* (1975) determined the liquidus phase relationships in the system $CaCO_3$–K_2CO_3–Na_2CO_3 and confirmed that the crystallization history of the lavas could be explained in terms of these ternary phase relationships. Nyerereite, $Na_2Ca(CO_3)_2$, and calcite can be co-precipitated down to 810 °C at 1 Kb without any H_2O present. Cooper *et al.* (1975) concluded that in most carbonatite magmas the alkalis would be removed in fenitizing solutions, and that alkalis would be retained only in dry carbonatite magmas with particularly low silica activity to crystallize as alkali-bearing carbonates.

20.3 ACCESSORY MINERALS IN SYNTHETIC CARBONATITE MAGMAS

One aspect of carbonatites of particular interest is the concentration of trace elements in accessory minerals that can be extracted as ores (Deans 1966, Heinrich 1966). Some of the relevant experimental studies are outlined below.

Figure 20.6 (a) Vapour-saturated liquidus surface in CaO–MgO–CO_2–H_2O at 2 Kb, with field boundaries based on projection from vapour phases (H_2O–CO_2) of phase fields intersected by mixtures on the composition join $CaCO_3$–$MgCO_3$–$Mg(OH)_2$–$Ca(OH)_2$ (dashed line quadrilateral). For temperatures see Figure 20.6b. (b) Vapour-saturated liquidus phase relationships intersected by mixtures on this composition quadrilateral (see dashed lines in Figure 20.6a). For the $CaCO_3$–$MgCO_3$ join at somewhat higher pressure, see Figure 20.4C. Note the eutectic for calcite, portlandite, and vapour on the triangular base CaO–CO_2–H_2O, and the quaternary eutectic involving calcite, dolomite, periclase, and V on the liquidus surface (Fanelli *et al.* 1986). Abbreviations: see Figure 20.4; CH, portlandite; MH, brucite; Pe, periclase (Fanelli *et al.* 1986).

20.3.1 Pyrochlore and the system $CaO-Nb_2O_5-CO_2-H_2O$

Large pyrochlore deposits in some carbonatites have been exploited for Nb. Watkinson (1970) found that about 5% Nb_2O_5 dissolved in the eutectic melt between $CaCO_3$ and $Ca(OH)_2$ at 1 Kb, and that $Ca_2Nb_2O_7$ precipitated along with calcite through a wide temperature interval from higher than 900 °C down to 635 °C. This niobate is related to pyrochlore (see Hogarth, this volume). Other niobate minerals were encountered, analogous to latrappite (perovskite) and fersmite. Watkinson (1970) concluded, from the relatively high solubility of Nb in the carbonate melt, the similarity of experimental and natural reaction relationships of the niobates, and the extensive co-precipitation of calcite and niobates, that Nb minerals in carbonatites are magmatic rather than hydrothermal, although they may be subject to secondary alteration.

20.3.2 Apatite

Biggar (1969) completed a series of studies on carbonate–apatite systems (cf. Wyllie & Biggar 1966) which showed that a carbonatite magma containing a small percentage of P_2O_5 would precipitate calcite and apatite together through a wide temperature interval. Crystal settling in the experimental runs confirmed the ease with which segregation of apatite and flow banding could occur in a carbonatite magma.

20.3.3 Sulphide in the system $CaCO_3-Ca(OH)_2-CaS$

Pyrite is the only common sulphide in carbonatites, occurring usually as disseminated euhedral crystals (Heinrich 1966). At Phalaborwa, however, chalcopyrite is sufficiently abundant to form copper orebodies, and mineralization is probably caused by solutions permeating the carbonatite core (Palabora Mining Company Ltd 1976). Carbonatites tend to contain more S (average 0.62%, Gold 1963; see Woolley & Kempe, this volume, for details) than is typical for other igneous rocks (0.01–0.04%; Parker 1967).

Although CaS (oldhamite) is a rare natural mineral, the system $CaCO_3-Ca(OH)_2-CaS$ was studied by Helz & Wyllie (1979) as a simple model for the solubility and behaviour of sulphides in carbonatite magmas. The field for CaS dominates the ternary liquidus surface at 1 Kb. Only a small percentage of CaS dissolves in the carbonate-rich liquid before a field boundary is reached. Therefore, precipitation of calcite from a carbonatite magma could drive the liquid to the sulphide saturation boundary at a reasonably high temperature, permitting co-precipitation of calcite and a sulphide through a wide temperature interval. Alternatively, for a liquid with lower initial dissolved S, the sulphide could remain stored in the liquid until it reaches a concentration of about 0.9% (CaS = 2.0%) at the final eutectic at 652 °C. The calculated fugacities of O, S, and CO_2 respectively at the ternary eutectic are 1.5×10^{-18}, 6.3×10^{-8}, and $10^{-1.8}$. Helz & Wyllie (1979) reviewed the factors influencing the sulphide solubility in silicate and carbonate melts, and concluded that at comparable fugacities of O, S, and CO_2, natural carbonatite magmas would have a higher sulphide solubility than the eutectic in the synthetic CaS system, because they would contain additional components, such as ferrous iron, which form sulphide complexes in the melt.

The oxidation state of S in carbonatites is variable. There is isotopic evidence that S is present mainly as sulphide in carbonatite magmas with temperatures of 700 °C or higher, with isotopes corresponding to mantle sulphur (see Deines, this volume), but the widespread occurrence of barite in vein carbonatites is indicative of more oxidizing conditions at lower temperatures. The experiments of Kuellmer *et al.* (1966) on the system $BaSO_4$–$CaCO_3$–CaF_2–H_2O confirm that barite can be precipitated from a carbonatite melt, as well as from a hydrothermal solution. This is reviewed below.

20.3.4 REE carbonates and the system $CaCO_3$–$Ca(OH)_2$–$La(OH)_3$

Light REEs are concentrated in carbonatites within three main groups of minerals:

(a) oxides – pyrochlore and perovskite;
(b) phosphates – apatite and monazite; and
(c) carbonates – bastnaesite and related minerals.

Experimental data showing the relatively low solubilities of the niobate oxides (pyrochlore and perovskite) and of apatite were outlined above. The REE carbonate type of carbonatite (Pecora 1956, Heinrich 1966) usually represents the last in a series of carbonatite differentiates, and, under some circumstances, very high concentrations of REE carbonates are produced. In the Mountain Pass carbonatite, for example, bastnaesite constitutes up to 15 vol% of the orebody (Olson *et al.* 1954).

Jones & Wyllie (1986) have studied phase relationships for the simplest model of a REE-carbonatite magma, the composition join $CaCO_3$–$Ca(OH)_2$–$La(OH)_3$ through the system CaO–La_2O_3–CO_2–H_2O. The results at 1 Kb are shown in Figure 20.7. The synthetic carbonatite magma represented by the eutectic E_1 dissolves about 20% $La(OH)_3$ at the ternary eutectic E at 610 °C. With increasing $CO_2:H_2O$ in the liquid, represented by the field boundary along E-a, the solubility of $La(OH)_3$ rises to 40% at the 700 °C piercing point, a.

The solubility of La in the synthetic carbonatite melt is unusually high compared with that of many other accessory components added in experimental studies. The solubility of silicate components (as defined by calcite field boundaries such as E-a in Fig. 20.7) from SiO_2 through Mg_2SiO_4, to feldspars and feldspathoids, is less than 5% (Wyllie 1966*a*). The solubilities of apatite and CaS are similarly low. In these systems, the field boundary limiting the calcite field shows little variation in solubility of the additional components as a function of $CO_2:H_2O$.

Deng & Wyllie (in preparation) continued these experimental studies in the system CaO–La_2O_3–CO_2–H_2O and located a liquidus region for hydroxy-bastnaesite in

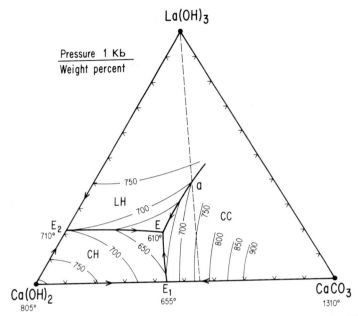

Figure 20.7 Preliminary interpretation of ternary liquidus relationships in part of the join CC–CH–LH at 1 Kb, based on previous results for CC–CH (Figure 20.6) and new runs in the joins CH–LH and the dashed line. No vapour is present. Note the high solubility of La(OH)$_3$ in the carbonate-rich melt (Jones & Wyllie 1986). Abbreviations: see Figure 20.6; LH, La(OH)$_3$.

the volume between the triangle of Figure 20.7 and the edge La$_2$O$_3$–CaO. Hydroxybastnaesite and calcite are co-precipitated from a melt.

Naldrett & Watkinson (1981) pointed out that the later stage, lower-temperature carbonatites, commonly found with hematite, barite, fluorite, and other relatively oxidized minerals, are more favourable for prospecting because they are more likely to be repositories of incompatible elements such as U, Th, and REEs. The most abundant rock in the orebody at the Mountain Pass carbonatite (according to Heinrich 1966) consists of calcite 40–75%, barite 15–50%, and bastnaesite 5–15%. Bastnaesite locally makes up 30% of some dyke rocks. The simple experimental approach to the petrogenesis of REE carbonatites represented in Figure 20.7 is supplemented by the study of synthetic mixtures approximating the composition of the Mountain Pass orebody (Jones & Wyllie 1983, Wyllie & Jones 1985), which requires addition of barite and F to the simple systems. Fluorine is added in the form of CaF$_2$.

20.3.5 Barite and fluorite in the system $CaCO_3–Ca(OH)_2–CaF_2–BaSO_4$

Kuellmer *et al.* (1966) studied the liquidus relationships in the system CaCO$_3$–Ca(OH)$_2$–CaF$_2$–BaSO$_4$ (CC–CH–CF–BS) at 500 bar pressure, and reported a ternary eutectic at 637 °C for CaCO$_3$–Ca(OH)$_2$–BaSO$_4$, with 18% BaSO$_4$ dissolved in the synthetic carbonatite magma. Their results for the system CaCO$_3$–Ca(OH)$_2$–CaF$_2$ with the ternary eutectic liquid containing 18% dissolved CaF$_2$ at

582 °C are consistent with those of Gittins & Tuttle (1964) at 1 Kb. Kuellmer et al. (1966) located a 500 bar ternary eutectic in the anhydrous system $CaCO_3$–CaF_2–$BaSO_4$ at 824 °C containing 22% $BaSO_4$ and 26% CaF_2. They reported a progressive lowering of the solidus temperature with addition of increasing amounts of H_2O, probably due to dissociation of barite or fluorite, increasing the order of the system. In contrast, by adding H_2O in the form of $Ca(OH)_2$, there was no dissociation, and they reported a quaternary eutectic at 576 °C with composition 35% $CaCO_3$, 42% $Ca(OH)_2$, 15% CaF_2 and 8% $BaSO_4$. Kuellmer et al. (1966) discussed the application of their results to magmatic and hydrothermal barite–calcite–fluorite veins that occur in many ore districts.

The shaded area in Figure 20.8 represents the chemical composition of the Mountain Pass orebody without the REEs, and the composition E with $Ca(OH)_2$ added is considered to be close to the quaternary eutectic composition. Calcium hydroxide provides H_2O to bring the liquidus temperature down to appropriate levels, and CaF_2 provides F for the bastnaesite. Jones & Wyllie (1983) and Wyllie & Jones (1985) determined the phase fields intersected by the composition join E–$La(OH)_3$ at 1 Kb. Their results are shown in Figure 20.9. The liquidus has a minimum piercing point at 625 °C, with 18% dissolved $La(OH)_3$. Phase fields including lanthanum-bearing minerals extend from this piercing point. Kutty et al. (1978) showed that a Nd-hydroxybastnaesite was stable to at least 580 °C, well above the solidus shown in Figure 20.9, at 543 °C. Although bastnaesite was not encountered in this composition join, the results suggest that bastnaesite could

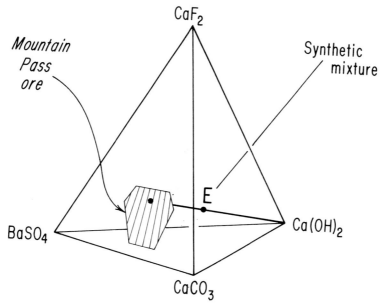

Figure 20.8 Mixture used in experiments of Figure 20.9. The mixture E contains 39% $CaCO_3$, 17% CaF_2, and 13% $BaSO_4$, representing the Mountain Pass ore without REEs, together with 31% $Ca(OH)_2$ to reduce liquidus temperature (Wyllie & Jones 1985).

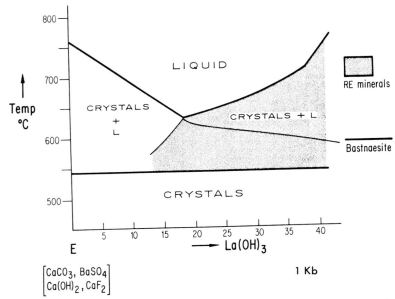

Figure 20.9 Phase fields intersected at 1 Kb by the multicomponent composition join between mixture E in Figure 20.8 and La(OH)$_3$. Note the liquidus piercing point with 18% La(OH)$_3$, the extensive field for the precipitation of hydroxy-REE minerals, and the breakdown temperature for bastnaesite from other subsolidus experiments (Wyllie & Jones 1985).

crystallize along with calcite and barite from a melt of similar composition with suitable proportions of CO_2, H_2O, and F. The results in this system, as well as the model system $CaO-La_2O_3-CO_2-H_2O$, are therefore consistent with a magmatic origin for bastnaesite in the Mountain Pass carbonatite and other carbonatites.

20.4 RELATIONSHIPS BETWEEN HIGH-TEMPERATURE SILICATE MELTS AND LOW-TEMPERATURE SYNTHETIC CARBONATITE MAGMAS

The preceding sections were concerned with the crystallization of carbonates and accessory minerals of carbonatites from melts at the relatively low temperatures appropriate for the emplacement of most carbonatites. We move now to the second major problem, the relationships between the carbonatite magmas and the high-temperature silicate magmas believed by many to be their parents. In this section, crystallization processes are considered at crustal pressures, including fractional crystallization and hypotheses of syntexis and assimilation. In the following section, we consider processes of liquid immiscibility.

20.4.1 $CaO-MgO-SiO_2-CO_2-H_2O$ at crustal pressures

The simplest system combining synthetic carbonatite magmas with silicates is $CaO-SiO_2-CO_2-H_2O$ (Wyllie & Haas 1965, 1966), and a closer approach to realism is provided by the system $CaO-MgO-SiO_2-CO_2-H_2O$ (Franz & Wyllie 1967, Otto & Wyllie 1983, Otto 1984). The synthetic carbonatite magmas dissolve only a few weight per cent of silicate components, and then the liquidus surface rises steeply. From the evidence available in 1966, I concluded in reviews of these systems for shallow crustal pressures (Wyllie 1966a, b) that a persistent thermal barrier on the liquidus is associated with the melting of Mg_2SiO_4. This barrier occurs between the high-temperature alkali-free silicate liquids precipitating minerals such as olivine, enstatite, melilite, and monticellite, and the low-temperature liquids that precipitate, in addition, carbonated and hydrated minerals. According to this evidence, fractionation of the high-temperature liquids at low pressures would not yield the low-temperature synthetic carbonatite magmas. Strongly undersaturated high-temperature liquids with less SiO_2 than in Mg_2SiO_4, however, could follow paths leading to the precipitation of calcite.

The earlier experiments were directed toward the calcite-portlandite eutectic in $CaCO_3-Ca(OH)_2$, but Otto & Wyllie (1983) and Otto (1984) completed a series of runs at 2 Kb connecting the new calcite-dolomite-periclase eutectic (Fig. 20.6) to silicates. The phase fields intersected by one of these composition joins are shown in

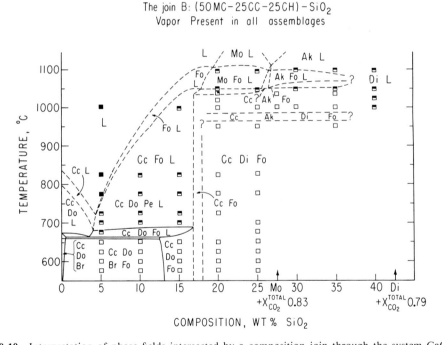

Figure 20.10 Interpretation of phase fields intersected by a composition join through the system $CaO-MgO-SiO_2-CO_2-H_2O$ at 2 Kb (Otto 1984). The mixture B is SiO_2-free with composition near the Do-Cc-Pe-V eutectic in Figure 20.6. Note low-temperature carbonate-rich liquid precipitating calcite, dolomite, forsterite, and periclase, and the high-temperature silicate liquids involving primary akermanite, forsterite, and monticellite. More experiments are needed to determine whether there is a liquidus path passing from the high-temperature liquids, around the temperature maximum corresponding to the calcite-forsterite-vapour join, and down to the low-temperature carbonate liquids. Abbreviations: see Figure 20.6; Mo, monticellite; Ak, akermanite; Fo, forsterite, Di, diopside, Br, brucite.

Figure 20.10. The solidus reaction for the synthetic carbonatite assemblage is:

calcite + dolomite + periclase + forsterite + vapour = liquid.

The high-temperature solidus reaction may be either:

calcite + akermanite + forsterite + vapour = monticellite + liquid,

or:

calcite + monticellite + forsterite + vapour = akermanite + liquid.

The results in Figure 20.10 and the other joins studied by Otto (1984) indicate that there may be peritectic reactions on the high-temperature field boundaries that provide liquidus paths between the high-temperature silicate liquids and the low-temperature carbonate liquids. Whether or not such field boundaries traverse compositions appropriate for natural magmas remains to be established. No evidence was found for liquid immiscibility. Additional experiments at high temperatures are required in this area, of relevance to monticellite- and melilite-bearing magmas as well as to the formation of carbonatites.

20.4.2 Systems including feldspars and feldspathoids with $CaCO_3$

Although alkalis are obviously important in the petrogenesis of carbonatite complexes, they are excluded from the systems discussed so far. Watkinson & Wyllie (1969, 1971) explored the effect of alkalis and alumina by determining at 1 Kb the phase fields intersected by the composition joins $NaAlSi_3O_8$–$CaCO_3$–$Ca(OH)_2$ and $NaAlSiO_4$–$CaCO_3$, with 25% H_2O added. It appears that original liquids precipitating plagioclase feldspar without nepheline cannot yield residual synthetic carbonatite magmas because there is a thermal barrier extending across the liquidus surface related to the barrier for dicalcium silicate in simpler systems (e.g. as illustrated for the join SiO_2–$CaCO_3$; cf. Wyllie 1974, Fig. 3). This barrier is represented in the experimental results by a series of subsolidus assemblages produced by reactions between albite and calcite extending up to temperatures of at least 1100 °C. Results for the albite system also have applications to the limestone syntexis and sial syntexis hypotheses, as described below.

Results for the nepheline system demonstrate that SiO_2-undersaturated alkalic liquids can yield low-temperature melts precipitating calcite. The results in Figure 20.11 show that liquid occurs continuously across the join in equilibrium with vapour and a crystalline phase or phases. Melilite, hauyne, and cancrinite are developed between the end members nepheline and calcite, and the vapour phase becomes enriched in Na as the liquid becomes enriched in $CaCO_3$.

The common ijolite–nephelinite–carbonatite associations are characterized by nepheline–pyroxene assemblages, but many alkalic complexes contain nepheline-, melilite-, and carbonate-bearing rocks. Watkinson & Wyllie (1971) noted that the

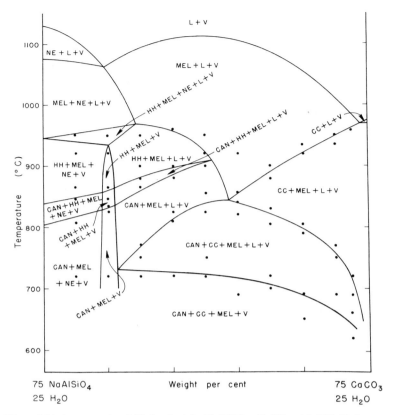

Figure 20.11 Phase fields intersected at 1 Kb by the join $NaAlSiO_4$–$CaCO_3$ with 25% H_2O; vapour coexists with all assemblages. Liquidus paths are continuous from high-temperature nepheline-normative silicate melts to low-temperature carbonate-rich melts (Watkinson & Wyllie 1971). Abbreviations: CC, calcite; NE, nepheline; MEL, melilite; HH, hydroxyhauyne; CAN, cancrinite; L, liquid; V, vapour.

phase fields intersected in Figure 20.11 correspond closely to mineral parageneses in alkalic igneous complexes characterized by melilite. They singled out the Oka complex for comparison because it also contains hauyne. Using a schematic model system nepheline–melilite–calcite–CO_2 to interpret the phase fields in Figure 20.11 in terms of liquidus field boundaries, they showed that with fractional crystallization of an initial liquid precipitating only nepheline, it is possible to derive the whole suite of crystalline assemblages analogous to the okaite series, excluding, of course, the iron–magnesium-rich minerals which are not represented in the synthetic system.

The jacupirangite–melteigite–ijolite–urtite group of pyroxene–nepheline rocks is represented in Figure 20.11 by nepheline alone, nepheline okaite is represented by nepheline + melilite, okaite corresponds to hydroxyhauyne + melilite, and the carbonatites correspond to the assemblages with melilite and calcite. The results in Figure 20.11 demonstrate that this sequence of rocks can be developed by fractional crystallization of a CO_2-bearing nephelinitic parent magma, crystallizing continuously from 1100 °C or more down to less than 700 °C. These results are consistent with field and petrological studies (Davidson 1963, Gold 1966). The sodic, aqueous vapour phase is an excellent candidate for the fenitizing solutions associated with carbonatite complexes. Watkinson (1970) discussed, in addition, the close analogies between textures and mineral reaction relationships in the synthetic systems and natural rocks. The relatively simple experimental system is a remarkably good model for interpretation of the magmatic system.

20.4.3 Limestone syntexis and sial syntexis

Daly (1910) proposed that alkaline magmas were generated by the assimilation of limestone by subalkaline magma, with consequent desilication, and he initiated a debate with advocates and adversaries firmly arguing their cases through many decades. The realization during the 1950s and 1960s that the 'limestones' at so many of the limestone-alkaline rock associations listed by Daly (1910) are in fact carbonatites led most investigators to reject the hypothesis (Heinrich 1966), but Schuiling (1964a) still considered the limestone assimilation hypothesis to be 'the only theory which really has been demonstrated in the field with a degree of certainty, scarcely ever achieved by petrological observation'. He also noted (Schuiling 1964b) that from a physicochemical viewpoint there is no essential difference between the limestone syntexis hypothesis and the hypothesis of sial syntexis by carbonatite magma (von Eckermann 1948, Holmes 1950). The history of the limestone assimilation hypothesis is of particular interest, because it has been so intertwined with the petrogenesis of alkaline rocks and carbonatites, and experimental petrology.

This controversy can be considered in four periods. During the first period, 1910–35, the hypothesis was launched mainly on interpretation of field data. Daly (1910) expressed confidence that future experimental results would favour the idea of limestone control, or at least CO_2 control, in the development of alkaline magmas. Bowen (1922) presented one of the clearest expositions of views opposing the hypothesis, based on paths of crystallization in simple systems, and the deficiency of heat to dissolve enough limestone to make significant changes in composition. Daly (1933) restated the hypothesis, acknowledging arguments presented by Shand (1930), who suggested that alkaline igneous bodies could be maintained at high temperature by associated larger bodies of subalkaline rocks, providing sufficient heat for assimilation.

During the second period, 1936–55, the experimental study of silicate systems revealed a thermal barrier rearing its ugly hump on the liquidus between subalkaline and feldspathoidal liquids (Schairer & Bowen 1935, Schairer 1950). The phase diagram shows that isobaric isothermal desilication of a granitic system causes precipitation of feldspar, with little change in liquid composition (Wyllie 1974). Shand (1943) tried to avoid the thermal barrier by suggesting that perhaps dissolved H_2O and CO_2 released from limestone affected the liquidus, but it took two decades for sufficient experimental data to become available to show that the feldspar barrier persisted in the presence of H_2O under pressure (Fudali 1963, Hamilton & Mackenzie 1965), and that the effect of CO_2 on the liquidus relationships in granitic systems was small (Wyllie & Tuttle 1959). During this second period, up to 1955, increasing attention was being paid to carbonatites. Shand (1945) noted that 'some petrologists preferred to put their faith even in that most strange thing, a carbonate magma'. Von Eckermann (1948) proposed that the Alnö Island carbonatite complex was formed from a primary carbonatite magma, and Holmes (1950) suggested that the strongly alkaline mafic and ultramafic lavas of western Uganda were formed by reaction of primary carbonatite magmas from the mantle with granitic rocks of the crust.

During the third period, 1956–75, the study of carbonatites in the field disrupted the evidence for limestone assimilation. But the prospect that carbonatites might be primary magmas that assimilated sialic material kept debate alive. The search for and study of carbonatites because of their economic potential led most investigators of alkalic–carbonate complexes to conclude that the carbonatites were magmatic (Agard 1956, Pecora 1956, Smith 1956), but the conditions for formation of such magmas at the relatively low temperatures indicated by field studies remained unknown until Wyllie & Tuttle (1960) confirmed Shand's (1945) prediction that H_2O might bring the temperature of a carbonate magma down to a reasonable level. Then followed the series of experimental studies in silicate–carbonate systems outlined above. The sial-syntexis hypothesis was evaluated by Wyllie (1974) on the basis of the phase relationships in $CaO-SiO_2-CO_2-H_2O$ (Wyllie & Haas 1965) and

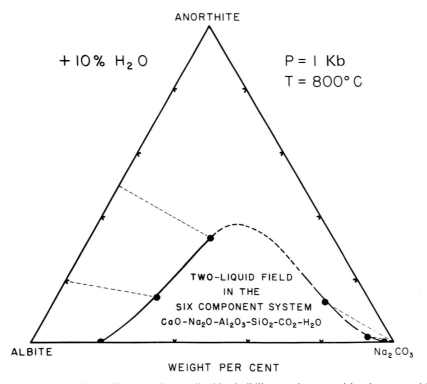

Figure 20.12 The extent of the silicate–carbonate liquid miscibility gap intersected by the composition triangle plagioclase–Na_2CO_3 at 1 Kb, based on experimental results obtained for the two dotted joins (Koster van Groos & Wyllie 1973). See text for analysed melt compositions and comparison with other data.

$NaAlSi_3O_8$–$CaCO_3$–$Ca(OH)_2$ (Watkinson & Wyllie 1969). It appears that the composition of a carbonatitic magma (cafemic variety) assimilating sialic material is limited to SiO_2 contents of a few per cent because of the thermal barrier on the liquidus, and there is no evidence that this kind of syntexis in the crust could produce either alkaline silicate liquids or feldspathoidal rocks.

Roedder's (1965) discovery, of nearly pure CO_2 as inclusions in olivine phenocrysts and peridotite nodules from basalts and kimberlites, directed attention to the possible influence of CO_2 in the mantle, and heralded a fourth period (1975–present) concerned with the prospect that some carbonatites might be derived from mantle sources. This theme was developed from experimental studies by Koster van Groos (1975), Wyllie & Huang (1975), and Eggler (1978), as described in Section 20.6. If alkaline carbonatite magmas can be generated in the mantle or deep crust, the results in Figure 20.12 suggest that immiscible peralkaline magmas would form if a plagioclase-bearing crust were heated sufficiently in contact with the carbonatite melt. The sial-syntexis hypothesis is currently related to liquid immiscibility (§20.5) and mantle-derived magmas (§20.6).

20.5 IMMISCIBILITY BETWEEN SILICATE AND CARBONATE LIQUIDS

The proposal of von Eckermann (1948) that the primary magma at the Alnö alkali complex was a carbonatite composed largely of alkali carbonates was restated by Dawson (1962, 1966) following his account of the remarkable natrocarbonatite lavas from Oldoinyo Lengai. The ideas of von Eckermann (1948) provided the initial stimulus for the experiments of Koster van Groos & Wyllie (1963, 1966) with the composition join $NaAlSi_3O_8$–Na_2CO_3–CO_2, and the relevance of experiments on this exotic join was confirmed by the discovery of natrocarbonatites. At 1 Kb, Koster van Groos & Wyllie (1963, 1966) found a wide field of immiscibility above 870 °C between a silicate liquid with about 20% dissolved Na_2CO_3 and a carbonate-rich liquid with less than 3% dissolved silicate. They determined that the extent of the miscibility gap decreased with decreasing pressure and that it was present at 33 bar, but found no evidence for it at 1 atmosphere. Using these results and those of previous studies in other parts of the system, they sketched the area of immiscible liquids on the CO_2-saturated liquidus surface as a function of pressure. Ellis & Wyllie (in preparation) have determined that comparable miscibility gaps exist at 20 Kb between $NaAlSi_3O_8$–Na_2CO_3 and $KAlSi_3O_8$–K_2CO_3, and similar results were published for the latter join at 5 Kb and 20 Kb by Wendlandt & Harrison (1979).

Further experiments with addition of H_2O to $NaAlSi_3O_8$–Na_2CO_3 at 1 Kb confirmed the persistence of immiscible liquids, with little change in composition until very high H_2O contents are reached (Koster van Groos & Wyllie 1968). The miscibility gap occurs at temperatures above about 725 °C, varying somewhat as a function of composition. There are three coexisting fluid phases in this system, corresponding to a SiO_2-undersaturated silicate magma, sodic carbonatite melts, and aqueous fenitizing solution.

Most carbonatites are composed of calcite with dolomite, and even the natrocarbonatites of Oldoinyo Lengai contain 12.8% CaO (Dawson 1962). Koster van Groos & Wyllie (1973) therefore studied the effect of CaO on the miscibility gap by defining the phase fields intersected by joins between Na_2CO_3 and two plagioclase compositions in the presence of 10% H_2O (Fig. 20.12). The phase relationships are complex, with the formation of nepheline, wollastonite, cancrinite and noselite, but the two-liquid field persists at temperatures above about 750 °C, becoming narrower with increasing anorthite component. Figure 20.12 shows only the extent of the miscibility gap at 800 °C, with an interpretation that it may close off with somewhat greater anorthite component. The compositions of all phases except the plagioclase depart from this triangle.

The products from a mixture on the high-calcium join within the two-liquid field are:

(a) a SiO_2-undersaturated peralkaline silicate liquid with low CaO, and CIPW norm, Ab = 19.8, Ne = 66.5, Cs = 1.1, Ns = 12.6 (H_2O and CO_2 contents of the liquid remain uncertain); and
(b) a carbonate liquid strongly enriched in CaO compared with the silicate liquid, with $CaCO_3$ = 43.3% and Na_2CO_3 = 56.7%, neglecting the silicate content and the estimated 20% dissolved H_2O.

In the less calcic join, the carbonate liquid is correspondingly more sodic, e.g. $CaCO_3$ = 15%, Na_2CO_3 = 85%. These liquids can coexist with (c) a vapour phase strongly enriched in Na_2O and, to a lesser extent, in SiO_2. These three fluid phases correlate well with the rock assemblages in carbonatite complexes, and with the fenites associated with them. Koster van Groos & Wyllie (1973) compared the silicate melt with the nepheline syenites and especially the urtite–melteigite series of carbonatite complexes. The Ca–Na-carbonate liquids would certainly yield calcite cumulates (experiments of Cooper et al. 1975 in §20.2.5), and the residual liquids becoming enriched in Na would evolve vapours capable of fenitization. Note that the range of carbonate liquids produced in these experiments (Fig. 20.12) includes the compositions of the natural natrocarbonatite lavas reported by Dawson (1962).

Koster van Groos (1975) followed up these experiments with more realistic bulk compositions. At pressures up to 10 Kb at temperatures between 650 °C and 725 °C, he reacted CO_2 + H_2O in controlled ratios with materials corresponding to peralkaline rhyolite, nepheline syenite, melteigite, and ijolite. The results confirm that the liquid miscibility gap shown in Figure 20.12 persists with addition of MgO, Fe_2O_3, and K_2O, and through the range of crustal pressures. The products were vapour, silicate liquid, carbonate liquid, silicate minerals, and carbonate minerals in some runs. The carbonate liquid persisted, with little change in composition, below the solidus of the silicate assemblage. The runs involving melteigite and ijolite compositions did not reach high enough temperatures to generate silicate melts. Koster van Groos (1975) concluded that the carbonate melt was produced by reactions:

$$Na_2SiO_3 + CO_2 = Na_2CO_3 + SiO_2$$

and

$$CaSiO_3 + CO_2 = CaCO_3 + SiO_2.$$

Chemical analyses showed that the carbonate liquids (including some carbonate solids) are rich in $CaCO_3$ (80–90%) with Na_2CO_3 next most abundant (up to 15%), and with K_2CO_3, $MgCO_3$, and $FeCO_3$ present in minor amounts (maximum 2.5% $MgCO_3$). The percentage of Na_2CO_3 increases with increasing temperature.

Verwoerd (1978) confirmed the existence of liquid immiscibility at 2 Kb and 10 Kb on a join between a silicate mixture approximating ijolite (acmite, nepheline, and sodium disilicate) and carbonate mixtures of $CaCO_3$ and Na_2CO_3. Silicate and carbonate liquids coexisted at temperatures above about 800 °C. Hamilton et al. (1979) investigated tie lines between immiscible silicate–carbonate liquids in alkali-rich systems, and Kjarsgaard & Hamilton (this volume) review this and more recent work on systems with less alkalies.

Kjarsgaard & Hamilton (1988) defined the miscibility gap at 5 Kb and 1250 °C between mixtures of albite and calcite and anorthite and calcite, corresponding to Figure 20.12 with Na_2CO_3 replaced by $CaCO_3$ and without H_2O. Na_2CO_3 was added to some mixtures. They found, as did Koster van Groos & Wyllie (1973), that CaO is enriched into the carbonate liquid and Na_2O is enriched into the silicate liquid. They followed the miscibility gap all the way to the alkali-free system, and reported the significant discovery of immiscibility between silicate liquid and carbonate liquid with composition almost pure $CaCO_3$.

The projection of the join $NaAlSiO_4$–$CaCO_3$ onto their diagram cuts across the projection of their two-liquid field. Because the results of Watkinson & Wyllie (1971), reproduced here in Figure 20.11, show no liquid immiscibility, Kjarsgaard & Hamilton (1988) concluded that 'the two sets of data are in conflict'. Let us avoid 'conflict' where none is established. The two sets of experiments represent very different conditions: those of Kjarsgaard & Hamilton (1988) were at high temperatures, super-liquidus, and mostly H_2O-free, whereas those of Watkinson & Wyllie (1971) were subliquidus, from 980 °C to 625 °C, and with 25% H_2O. There could be a miscibility gap in Figure 20.11 at high temperatures not reached by Watkinson & Wyllie (1971). The prospect that crystallization paths related to Figure 20.11 could encounter a liquid miscibility gap is clear from the experiments of Koster van Groos (1975) and Kjarsgaard & Hamilton (1988), but not required by any experiments yet published.

Koster van Groos (1975) applied his results to the origin of carbonatites. He concluded that CO_2 in the upper mantle under continental shields will be present mainly in crystalline carbonate, but that under some circumstances the mantle may contain a dispersed carbonate melt, somewhat enriched in alkalis and H_2O. The melt, enriched in alkalis and immiscible with silicate melts, persists down to temperatures below the solidus temperatures for natural hydrous silicate magmas. Migration and concentration of this melt yields primary carbonatite magmas which subsequently react with silicates at shallower levels in the mantle or in the crust, yielding the various rock types present in carbonatite complexes.

20.6 THE FORMATION OF CARBONATE-RICH MELTS IN THE MANTLE

The previous sections have been concerned mainly with the ways by which carbonatite magmas may be developed from silicate magmas, although in the preceding section experimental evidence was outlined for the formation of a somewhat alkaline carbonatite magma by reaction of CO_2 with silicate rocks (Koster van Groos 1975). In this section, evidence is presented from model synthetic systems showing that at high pressures, near-solidus melting of peridotite in the presence of CO_2 produces a carbonatite melt dominated by Mg–Ca. For natural peridotite, this liquid will assuredly be enriched in alkali carbonates to the extent that alkalis are available, but experimental data are few. What happens with reduced oxygen fugacity where volatile components C–H–O do not include CO_2 remains even more uncertain. Evidence from Section 20.5 indicates strongly that any CO_2-bearing alkalic magma generated from peridotite is likely to exsolve a Ca–Na carbonatite melt through a wide range of pressures.

It appears to be generally accepted that most of the CO_2 in carbonatites is derived from silicate magmas rather than from limestones in the crust. If the silicate magmas are derived from the mantle, then the volatile components may be derived from

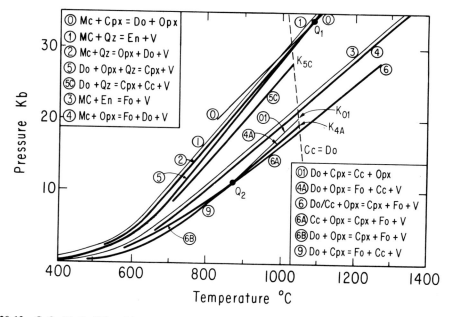

Figure 20.13 $CaO-MgO-SiO_2-CO_2$. Compilation of best estimates of carbonation/decarbonation reactions from Wyllie et al. (1983) and Huang & Wyllie (1984). Heavy lines are quaternary, light lines are degenerate or ternary ($MgO-SiO_2-CO_2$, see Figure 20.14). Compare Figure 20.4A for $CaO-MgO-CO_2$. Note the dashed line for the calcite–dolomite solvus crest, critical end points, K, on this line, and two invariant points, Q_1 and Q_2 (subscripts do not correspond to Q in previous diagrams). Mineral assemblages include representatives of lherzolite, wehrlite, websterite, and harzburgite, coexisting with CO_2 vapour or with carbonate. Abbreviations: see Figures 20.4 and 20.6; Fo, forsterite; En, enstatite; Opx, orthopyroxene; Cpx, clinopyroxene; Qz, polymorphs of SiO_2. Note that MC and En represent end-member minerals and Mc and Opx solid solutions.

primordial deeper mantle, or from subducted material. Huang *et al.* (1980) concluded from phase relationships that some carbonates in subducted limestone or basaltic crust could escape dissociation and melting, be converted to aragonite, and be carried to considerable mantle depths for long-term storage. Calcite should react with adjacent peridotite to form dolomite or magnesite (Fig. 20.13), and, according to many current views on the redox state of the upper mantle, the carbonate should become reduced to graphite or diamond.

20.6.1 The system $CaO-MgO-SiO_2-CO_2$

The system $CaO-MgO-SiO_2$ includes representatives of the mantle minerals olivine, orthopyroxene, and clinopyroxene, and with addition of CO_2 the conditions of carbonation of mantle peridotite and the effect of CO_2 on melting reactions are fairly well modelled. Wyllie & Huang (1975, 1976a) determined the positions of key carbonation reactions to 30 Kb and compiled an array of most of the reactions in the system. Wyllie *et al.* (1983) presented a detailed comparison of most published experimental results and thermodynamic calculations and produced a revised array of reactions. Addition of reaction (9) by Huang & Wyllie (1984) gave the revised and complete sequence of reactions in Figure 20.13. Reactions involving the mineral assemblages representing lherzolite, harzburgite, wehrlite, and websterite are present. Melting reactions have been presented by Wyllie & Huang (1976a), Eggler (1978), and Otto (1984). The relationships are too complex to be covered in this chapter, but the principles can be illustrated in the system with CaO omitted.

20.6.2 The system $MgO-SiO_2-CO_2$ and model harzburgite, $Fo + En + CO_2$

The intersection of the carbonation reactions with a series of solidus reactions for the various mineral assemblages provides the key for understanding the melting relationships in peridotite-CO_2 (Wyllie & Huang 1976a, b, Eggler 1978). The principles can be visualized in the ternary system $MgO-SiO_2-CO_2$, which contains the assemblage Fo + En, representing harzburgite. Figure 20.14 shows an extrapolated diagram using harzburgite reactions (1) and (3) from Figure 20.13, and magnesite dissociation (10) from Figure 20.1. With increasing pressure and excess CO_2, the harzburgite first loses forsterite by reaction (3), and in reaction (1) it is converted completely to magnesite + quartz (the coesite reaction has been omitted for simplicity). Without excess CO_2, each assemblage is partly carbonated. In reaction (3) model harzburgite is converted to MC + Fo + En, and MC + En + Qz is formed in reaction (1). The solidus curves for each silicate mineral assemblage are divided into two parts at the invariant points, Q_3 and Q_1, respectively. In the lower-pressure region below the invariant point, the assemblage melts in the presence of CO_2; the low CO_2 solubility causes only a small depression of the solidus temperature; with increasing pressure, each solidus curve moves rather abruptly to lower temperatures to meet a subsolidus carbonation reaction at an invariant point (Q_3 or Q_1). At higher pressures, the partly carbonated assemblages melt in vapour-absent reactions, producing liquids with much higher percentages of dissol-

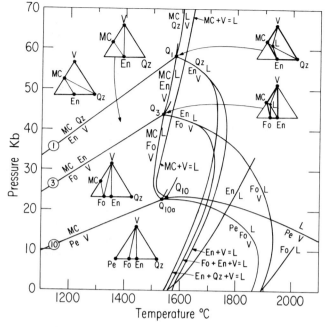

Figure 20.14 Vapour-present reactions in $MgO-SiO_2-CO_2$ (partly schematic) compiled by Wyllie & Huang (1976a, b). Compare Figure 20.13 for reactions (1) and (3), and Figures 20.1B and 20.4 for reaction (10), but note that subscripts change for invariant points, Q. Abbreviations: see Figures 20.4, 20.6, and 20.13.

ved CO_2. These magnesite–silicate solidus curves are represented in Figure 20.14 by the short, unlabelled lines above Q_3 and Q_1.

The composition of the liquid generated from model harzburgite (Fo + En) in the presence of CO_2 with increasing pressure can be traced in the isobaric liquidus diagrams reproduced in Figure 20.15. In Figure 20.15A, at pressures where magnesite dissociates, each of the eutectic liquids in $MgO-SiO_2$ dissolves CO_2 and becomes a ternary, CO_2-saturated eutectic, with low solubility of CO_2. At higher pressures approaching Q_{10} (Fig. 20.14), for liquids with compositions between Fo and MgO, the CO_2 content increases from low levels to about 40%, because the binary liquid in $MgO-CO_2$ approaches $MgCO_3$ in composition (Figs 20.1 & 20.3). With increasing pressure, the general shape of this CO_2-saturated liquidus boundary is preserved, with progressive increase in solubility of CO_2 (Fig. 20.15). The CO_2-saturated liquids coexisting with the silicate assemblages migrate toward the $MgO-CO_2$ side of the triangle with increasing pressure. Consider the point on the CO_2-saturated liquidus boundary giving the composition of the liquid generated from Fo + En in the presence of CO_2. In Figure 20.15A the CO_2 solubility is low. With increased pressure the liquid composition migrates along the field boundary toward the liquidus field for magnesite, as shown in Figure 20.15B, and at pressures above Q_3 (Fig. 20.14), the liquid composition has reached the magnesite field, and migrated to the CO_2-undersaturated liquidus. Here, Fo + En melt together with magnesite (Fig. 20.15C), because at this pressure the harzburgite assemblage cannot

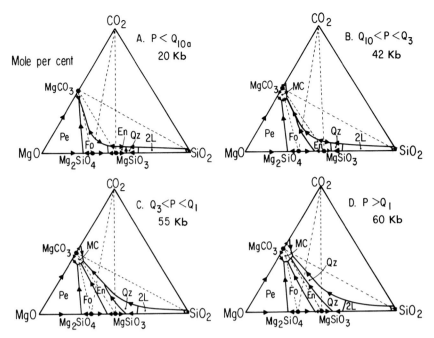

Figure 20.15 Schematic isobaric liquidus diagrams for $MgO-SiO_2-CO_2$ based on Figure 20.14 and other sources cited by Wyllie & Huang (1976b). Of particular interest are (i) the increase in CO_2 solubility with increasing $MgO:SiO_2$, especially in the range Fo–MgO; (ii) the movement of the field boundary between liquidus surfaces for primary Fo and En with increasing pressure; and (iii) the changing composition of the liquid coexisting with Fo + En + V, which reaches the liquidus field for magnesite between Figures (B) and (C) (at Q_3 in Figure 20.14). At higher pressures CO_2-rich liquid is generated from Fo + En + MC. Abbreviations: see Figure 20.14.

exist with free CO_2 (Fig. 20.14). While the composition of the liquid coexisting with model harzburgite exhibits a large increase in CO_2 content with increasing pressure within a pressure interval just below point Q_3, it should be noted that the solubility of CO_2 in the (Fo + En) liquid composition increases only slightly.

20.6.3 Solidus for amphibole–carbonate–lherzolite–C–H–O

Figures 20.14 and 20.15 provide an excellent model for the principles of melting in harzburgite with CO_2, and addition of component CaO brings in clinopyroxene to model lherzolite (Fig. 20.13). However, in order to reach actual reaction temperatures and melt compositions in mantle peridotite with volatile components, we need the additional ingredients associated with amphibole and phlogopite. Phlogopite reactions are not considered further in this treatment, because of their complexity and because K abundances in normal mantle peridotite are so low. In metasomatized mantle, however, the role of phlogopite is significant. Wendlandt & Eggler (1980) have covered in detail the phlogopite–carbonate melting reactions in model peridotite assemblages. The components C–H–O may exist as CO_2–H_2O if conditions are sufficiently oxidizing.

The melting of model lherzolite assemblage Fo + Opx + Cpx in the system

$CaO-MgO-SiO_2-CO_2$ follows the pattern of Fo + En in Figures 20.14 and 20.15, with the CO_2 solubility increasing to high levels in the pressure interval where the carbonation reaction, (6) in Figure 20.13, approaches the solidus curve for Fo + Opx + Cpx + CO_2. The invariant point for model lherzolite corresponding to Q_3 in Figure 20.14 is near 27 Kb and 1200 °C (Wyllie & Huang 1976a, Eggler 1978), as shown by points W and E76, respectively, in Figure 20.16.

The phase relationships for the system lherzolite–CO_2–H_2O, involving the minerals amphibole, phlogopite, dolomite, and magnesite, provide the framework for upper mantle petrology, and for evaluation of the phase relations with reduced oxygen fugacity. Uncertainty remains about the position of the solidus, which also varies as a function of composition. Two experimental studies on amphibole–dolomite–peridotite indicate that the amphibole stability volume overlaps the solidus with dolomite. Figure 20.16 shows a difference of several kilobars between the results of Brey et al. (1983) and those from the detailed investigation of Olafsson & Eggler (1983). I have interpreted these two sets of experimental results in terms of the topology of overlapping phase volumes (Wyllie 1978, 1987a, b). Wyllie & Rutter (1986) have experimental data from lherzolite–CO_2 locating the equivalent of G in Figure 20.16C at 22 Kb and 1090 °C, which fixes H_2 (Fig. 20.16) and Q (Fig. 20.17) at about the same pressure, and slightly lower in temperature (Wyllie 1978, 1987a, b).

Despite differences in experimental detail and interpretation, the existence of a solidus ledge near Q (Fig. 20.17, level 4) where the solidus changes slope and becomes subhorizontal (Eggler 1987) appears to be established. An estimated (but constrained) and extrapolated solidus curve for amphibole–dolomite–peridotite with a high CO_2 : H_2O ratio is compared with solidus curves for peridotite dry and with H_2O in Figure 20.17. This is modified from a previous version (Wyllie 1980) incorporating the new data (Olafsson & Eggler 1983, Wyllie & Rutter 1986, Wyllie 1987b).

There are two sets of experimental data that have not been considered in

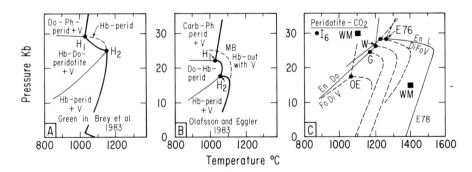

Figure 20.16 (A) and (B) Experimentally determined solidus curves for lherzolite–H_2O–CO_2. Note the different pressures for H_1 and H_2. (C) Locations for the invariant point, I_6 (E76, W, G, and OE) on the solidus for lherzolite–CO_2 based on Figures (A) and (B) and other estimates, deduced and compiled by Wyllie (1987a). Squares WM are two experimental values for the solidus (Wendlandt & Mysen 1980). Abbreviations: see Figure 20.14; Hb, amphibole; Ph, phlogopite; carb, carbonate.

construction of the estimated solidus for lherzolite-volatiles in Figure 20.17. The first is the effect of H_2O on the solidus for carbonates (Figs 20.5 & 20.6), and the second is the effect of CO_2 in extracting alkalis from minerals and generating carbonatite melts at low temperatures (Koster van Groos 1975). These effects may require significant revision of the plotted solidus curve, with the introduction of traces of carbonate-rich liquid persisting to temperatures below the plotted solidus under some circumstances.

It has been assumed in Figure 20.17 that the solidus for lherzolite–CO_2–H_2O remains at temperatures higher than that for lherzolite–H_2O, and this is consistent with the two reported experimental studies on the natural rock compositions (Figs 20.16A, B). At pressures greater than that of Q, however, the lherzolite is converted to a carbonate–lherzolite (see Q_3 in Fig. 20.14 and H_2 in Fig. 20.16), and the effect of H_2O on the melting of carbonates has been neglected. The results plotted in Figure 20.5 show that although calcite melts at temperatures of

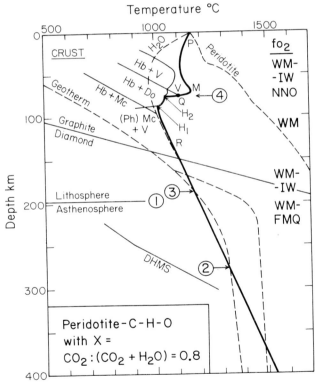

Figure 20.17 Solidus curve (heavy line) for peridotite–CO_2–H_2O (–C–H–O) with defined ratio of CO_2:H_2O (the curve PMQR) extrapolated to high pressure, and compared with the solidus for peridotite–H_2O (dashed line, H_2O) and cratonic geotherms (dashed lines) with and without inflection (Boyd & Gurney 1986). Another dashed line (peridotite) shows the volatile-free solidus. For subsolidus reactions compare Figures 20.13 and 20.16. Oxygen fugacities at various levels taken from Haggerty (1988). With low oxygen fugacity the solidus curve for peridotite–C–H–O may remain close to that for peridotite–H_2O, or it may migrate to higher temperatures (see text). Levels 1, 2, 3, and 4 have significance as described in text (see also Figures 20.18–20.20). Abbreviations: see Figure 20.16; DHMS, dense hydrated magnesium silicates.

1300–1400 °C, it begins to melt in the presence of H_2O at temperatures as low as 600 °C. The amount of liquid produced with small amounts of H_2O is very small (Wyllie & Tuttle 1960). Similarly, the results in Figure 20.6 show that with addition of H_2O at 2 Kb to any calcite–dolomite–magnesite assemblage (see Fig. 20.4 for high melting temperatures) traces of melt are formed at temperatures down to 650 °C, and this situation will persist to high pressures (compare Fig. 20.5). Therefore, carbonate–lherzolite at depths greater than that of Q in Figure 20.17, with a small excess of H_2O over that required to make the maximum amount of phlogopite and amphibole, is likely to coexist with a trace of carbonatite melt down to temperatures several hundred degrees below the plotted solidus.

According to the results of Koster van Groos (1975), CO_2-bearing vapour can react with sodic and calcic minerals in a rock to generate carbonate-rich melts, immiscible with the silicate melts that are generated at higher temperatures. His experiments at pressures corresponding to depths of 40 km produce Ca–Na carbonatite melts at temperatures as low as 650 °C, well below the solidus for peridotite–H_2O in Figure 20.17. The corresponding reactions at pressures greater than that of Q in Figure 20.17, with sufficient CO_2 present in vapour or melt to extract alkalis from the lherzolite minerals, would produce traces of Ca–Mg–Na melt at temperatures well below the peridotite–H_2O solidus, approaching solidus temperatures determined for $CaCO_3$–Na_2CO_3–K_2CO_3 by Cooper et al. (1975). Attempting to represent the low-temperature melting relationships without definitive experimental data is a geometrical problem that would overcomplicate the diagrams that are already too complicated for the comfort of many petrologists.

The traces of carbonatitic melts to be expected at temperatures well below the solidus for lherzolite–H_2O have not been reported in the experiments with natural rocks in the presence of CO_2 and H_2O shown in Figure 20.16A and B, nor in the experiments with dolomite–lherzolite (Wyllie & Rutter 1986). Wendlandt & Mysen (1980) reported traces of melt in lherzolite–CO_2 at 15 Kb and 35 Kb at temperatures below what they considered to be the solidus, and they attributed this liquid to traces of H_2O in the system. Given sufficient CO_2 in an alkali-bearing peridotite, the available experimental data require that a trace of alkali-bearing carbonatite melt be produced at temperatures below the solidus plotted in Figure 20.17, but the amount appears to be below current detection limits in experimental runs, despite a study of SEM photographs of run products (Wyllie & Rutter 1986, in preparation). Despite the small quantities of the alkalic carbonatite melts that may be formed in oxidized mantle, they may have applications to the origin of carbonatites (Koster van Groos 1975) and to trace-element budgets for the mantle and other magmas.

Volatile components could be released within the mantle by activation of a dissociation reaction (Woerman & Rosenhauer 1985), such as that represented in Figure 20.17 by the line, DHMS, giving the boundary for the reaction of olivine to form dense hydrous magnesian silicates. The estimated position of the reaction for the conversion of forsterite to brucite and magnesite in the system MgO–SiO_2–CO_2–H_2O is close to the DHMS curve at 300 km (Ellis & Wyllie 1980).

The redox state of the deeper mantle remains controversial. There are proponents for relatively oxidized gases (Haggerty 1986, Eggler, pers. comm.), but many argue that the components C–H–O exist as C and H_2O or H_2 with CH_4, rather than as CO_2 and H_2O (Deines 1980, Ryabchikov *et al.* 1981, Green *et al.* 1987). Figure 20.17 summarizes Haggerty's (1986, 1988) picture of the possible variations of oxygen fugacity in terms of standard buffers down to 250 km; note the oxidized condition near level 4. According to some investigators, the solidus for lherzolite–C–H–O may remain close to that for peridotite–H_2O, with carbonate ions being generated in the melt when CH_4 or graphite/diamond dissolves (Ryabchikov *et al.* 1981, Eggler & Baker 1982, Woerman & Rosenhauer 1985). In the following discussion, the extrapolated solidus for peridotite–H_2O is adopted as the solidus at depth for lherzolite–C–H–O, with volatile components in the ratio $CO_2 : (CO_2 + H_2O) = 0.8$, if oxidized. With variation in this ratio the solidus temperature changes at pressures less than point Q, but remains buffered and unchanged at higher pressures (Wyllie 1978). According to D. H. Green (pers. comm.), recent experiments with pyrolite–C–H–O at various H_2O activities with reduced oxygen fugacities do move the solidus to higher temperatures. All solidus boundaries in the petrological models can therefore be moved to higher temperature locations as required, once the relationships among oxygen fugacity, water activity, solidus curves, and actual redox conditions in the mantle have been established.

The formation of traces of alkalic carbonatite melt at low temperatures discussed above (Koster van Groos 1975) depends on the presence of H_2O and CO_2. The immiscible melt should not be produced if the redox state of the asthenosphere does not permit the presence of these species in a vapour phase.

20.6.4 *Solubility of CO_2 in mantle magmas, and magma compositions*

There has been debate about the solubility of CO_2 in the liquid along the solidus for model dolomite–lherzolite in the system $CaO–MgO–SiO_2–CO_2$, corresponding to the high-pressure solidus in Figure 20.16 and depths greater than point Q in Figure 20.17. Wyllie & Huang (1976*a*) concluded that the near-solidus liquid contained about 40 wt% CO_2, based on their results on the joins $MgSiO_3–MgCO_3$ (Huang & Wyllie 1974) and $CaMgSi_2O_6–CaMg(CO_3)_2$. Eggler *et al.* (1976) concluded that the corresponding melts at 30 Kb 'contained at most 20 per cent CO_2', and Eggler (1978) concluded that the melt contained 27% CO_2. Wendlandt & Mysen (1980) in their study of peridotite–CO_2 confirmed that 'the melt corresponding to perhaps 100 kilometers in the mantle has a haplocarbonatitic composition', without specifying CO_2 solubility. In an independent approach, Woerman & Rosenhauer (1985) applied Schreinemakers' analysis to the results and phase compositions reported by Eggler (1978) and Wyllie & Huang (1976*a*), and calculated that both sets of data require a minimum concentration of 43 wt% CO_2 in the liquid at the point corresponding to W or E76 in Figure 20.16C.

The reason for the high solubility of CO_2 was illustrated in Figures 20.14 and 20.15 in the system $MgO–SiO_2–CO_2$. As the solidus temperature drops towards the invariant point Q_3 in Figure 20.14, there is a rapid increase in the CO_2 content of the liquid at the solidus. This is caused not by increased solubility as a function of pressure in a liquid of defined composition (with respect to silicate components), but by the change in composition of the liquid coexisting with Fo + En + CO_2 as shown in Figure 20.15. With increasing pressure, the liquid L(Fo, En, V) migrates along a field boundary in the direction of increasing $MgO:SiO_2$, as shown by Figures 20.15A and B. In this direction the proportion of dissolved $MgCO_3$ increases; hence so does the CO_2 dissolved in the liquid. The solubility approaches a maximum of about 45 wt% at Q_3, where the liquid composition reaches the liquidus field for primary magnesite at a pressure between those for Figures 20.15B and C; note the high CO_2 content of liquids around the magnesite field.

The change in CO_2 solubility is associated with invariant points in model systems, such as Q_3 in Figure 20.14, and with corresponding points in whole-rock systems such as H_2 and Q in Figures 20.16 and 20.17. These points occur on near-isobaric ledges on the solidus curves, and there is a marked change in composition of liquids from low-CO_2 moderate-SiO_2, at pressures below that of the ledge, to high-CO_2 low-SiO_2, at pressures above the ledge (Wyllie 1979, Wendlandt & Eggler 1980).

On the basis of experimental data available from model systems, Wyllie (1977) concluded that the near-solidus liquid from peridotite–H_2O–CO_2 at pressures higher than the solidus ledge corresponded to that of a dolomitic carbonatite, with Ca:Mg > 1 (confirmed experimentally by Wyllie et al. 1983), with no more than 10–15% dissolved silicates (5–10% SiO_2), and enriched to some extent in alkali carbonates. He mentioned the probable intersection of a liquid miscibility gap if liquid compositions became sufficiently alkaline. According to the discussion in the preceding section, traces of the immiscible alkali-bearing carbonatite melts should persist down to temperatures much lower than the solidus plotted in Figure 20.17.

20.6.5 Silicate–carbonatite immiscibility in mantle magmas

There have been no reports of immiscible liquids in model mantle systems or in melting experiments with natural peridotite-volatile systems. There is abundant experimental evidence for the occurrence of immiscibility between a variety of silicate and carbonate liquids, and for its persistence to at least 15 Kb (§20.5, and Kjarsgaard & Hamilton, this volume). Nepheline-normative silicate liquids at crustal pressures coexist in equilibrium with carbonate melts with compositions ranging from high to low Ca:Na, and it would be surprising if this miscibility gap did not persist through upper mantle pressures. Under conditions where peridotite yields nepheline-normative silicate magmas, therefore, it is to be expected that the presence of CO_2 would promote the separation of an immiscible carbonatite magma, with Ca:Na and Ca:Mg varying according to conditions.

Three situations for the occurrence of immiscibility in the mantle are as follows:

(a) Partial melting of peridotite could yield two immiscible silicate and carbonatite liquids simultaneously.
(b) A silicate liquid separated from its mantle host could fractionate to a miscibility gap and yield a carbonatite magma.
(c) CO_2 may extract alkalis, especially Na, from mantle peridotite and generate a

Ca–Na carbonatite melt at temperatures below the peridotite–H$_2$O solidus (Koster van Groos 1975); increase of temperature could yield an immiscible silicate magma from the peridotite, or the low-temperature carbonatite melt could migrate to another part of the mantle, encounter a silicate magma, and remain immiscible.

20.7 KIMBERLITES, NEPHELINITES, AND CARBONATITES

Having reviewed the ways in which mantle carbon may become concentrated into CO$_2$-bearing silicate melts or carbonate-rich melts, consideration is now given to the migration of these melts from asthenosphere into lithosphere, and the phase relationships are used as a guide to processes leading to the intrusion of carbonatite magmas.

20.7.1 The asthenosphere and lithosphere

The thickness of the lithosphere is an important variable because of the change in rheology occurring through the asthenosphere–lithosphere boundary layer. This boundary layer is commonly assumed to be near the 1200 °C isotherm, and this temperature is adopted in the following discussion. The layer is depicted as a line, level 1, near 200 km depth in Figures 20.17 and 20.18. Geothermometry and geobarometry of mantle nodules provide a geotherm for cratons consistent with that calculated from heat-loss data (Boyd & Gurney 1986). The geotherm for many kimberlites is inflected to higher temperatures at a depth of about 175 km, somewhat deeper than the graphite-to-diamond transition. Both normal and inflected geotherms are shown in Figure 20.17. Nickel & Green (1985) refined empirical garnet–orthopyroxene geobarometry, and presented a distinctive pattern for South African xenoliths where the high-temperature xenoliths give near-isobaric estimates corresponding to a depth of 150–160 km at 900–1400 °C. If a thermal plume rises below the lithosphere, the asthenosphere–lithosphere boundary rises to shallower levels, causing thinning of the lithosphere, as indicated by the 1200 °C point on the inflected geotherm of Figure 20.17.

Figure 20.17 compares the subcratonic geotherms with and without inflection with the estimated solidus for peridotite–C–H–O. In the absence of volatile components there is no melting. If geotherm and solidus intersect, and if reduced vapours are introduced between the depth levels 2 and 3, then partial melting occurs. The depth interval for melting depends strongly on the geotherm. Furthermore, if the solidus intersects the geotherm, then liquid is the only fluid phase that can exist between levels 2 and 3, because all volatile components would dissolve in the liquid. This limits the depth intervals within which vapour phases can cause metasomatism (Wyllie 1980, 1987b, 1988). Note that if vapour species are sufficiently oxidized, then some of the vapour depicted in Figure 20.18 may be accompanied by traces of low-temperature alkalic carbonatite melt formed at temperatures below the solidus plotted in Figure 20.17 (Koster van Groos 1975; §20.6.3).

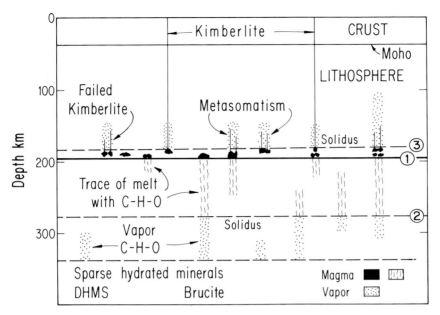

Figure 20.18 Cross-section of a craton, with phase boundaries and normal geotherm transferred from Figure 20.17, showing the depth limits for the migration of deep volatile components, the generation of melts in the asthenosphere, and their entrapment in the lower lithosphere. Release of volatile components as magmas approach the solidus level 3 leads to metasomatism, crack propagation, and intrusion of kimberlites, only some of which reach the surface. Various stages are illustrated, representing events distributed through a long time interval. Levels 1, 2, and 3 are indicated in Figure 20.17.

Petrological structures for the mantle can be constructed if the rock compositions, the phase relationships, and the thermal structure are known. The phase boundaries and the standard geotherm in Figure 20.17 are mapped in terms of pressure and temperature. The phase boundaries can therefore be plotted as a function of depth as in Figure 20.18. Similarly, the phase boundaries can be mapped as lines of variable depth as in Figure 20.19 for the more complex situation with a thermal plume, and lithosphere thinning.

20.7.2 Melts from the asthenosphere

Figure 20.18 shows a stable craton with uninflected geotherm. Phase boundaries are drawn at depths represented by intersections of the uninflected geotherm with the phase boundaries in Figure 20.17. Vapours released from the deep dissociation front by intermittent local disturbances (Woerman & Rosenhauer 1985) would rise through mantle to the solidus at level 2, where they would dissolve in melt. The trace of interstitial melt so developed would migrate through the asthenosphere to the asthenosphere–lithosphere boundary at level 1, where its rate of percolation would be lowered. Small magma chambers could form and remain sealed within the more rigid lithosphere, maintained at temperatures above the solidus for lherzolite–C–H–O. The magmas have little tendency to crystallize or to evolve vapours unless they approach level 3, the solidus, 10–15 km above the asthenosphere–lithosphere boundary.

Those magmas managing to insinuate their way near to level 3 will evolve H_2O-rich vapours. Vapours rich in CO_2 cannot exist in this part of the mantle (Wyllie 1979). The vapours may promote crack propagation, permitting rapid uprise of kimberlite-like magma. Although many intrusions from this level will solidify through thermal death before rising far (Spera 1984), others will enter the crust as kimberlite intrusions (Artyushkov & Sobolev 1984). Alternatively, the solidus with reduced vapours could be too high for the solidus and geotherm to intersect, i.e. points corresponding to levels 2 and 3 in Figures 20.17 and 20.18 would not exist, and no melting would occur unless the vapours became oxidized. If the reduced vapours reached a shallower mantle horizon with higher oxygen fugacity and were oxidized into a condition where the solidus for the system migrated to a lower temperature, this could cause partial melting (Wyllie 1980). Green *et al.* (1987), using data of Taylor & Green (1987), have elaborated the latter process in elegant detail with a major role assigned to 'redox melting' where reduced fluids enter the lithosphere. Haggerty (1986), in contrast, has developed a model with vapours from a relatively oxidized asthenosphere becoming reduced as they enter the lithosphere (Fig. 20.17).

Using the solidus in Figure 20.17 as a guide, aqueous vapours may be expected at level 3. Considering the discussion in Section 20.6.3, however, another possibility is that the oxidized vapour may be accompanied by a trace of carbonatite melt persisting down to lower temperatures.

The depleted, refractory base of the lithosphere, 150–200 km deep, has probably been invaded intermittently by small bodies and dykes of kimberlite through billions of years. Most of these aborted and gave off vapours. The combination of failed kimberlite intrusions and their vapours has contributed to the generation of a heterogeneous layer in this depth interval, characterized by many local volumes metasomatized in diverse ways. The aqueous vapours may percolate upward through many kilometres to shallower levels, perhaps generating the sequence of metasomatized garnet peridotites described in detail by Erlank *et al.* (1987). Schneider & Eggler (1986) suggested on the basis of their experimental data that aqueous vapours and solutions rising through the lithosphere would leach the mantle with minor precipitation until they reached a level of about 70 km where a 'region of precipitation' is associated with the formation of amphibole and consequent change in vapour composition toward CO_2.

If a thermal plume rises below the cratonic lithosphere, the petrological structure of Figure 20.18 becomes modified as the lithosphere thins above the plume (Wyllie 1988). The geotherm rises to a higher, inflected position as depicted in Figure 20.17, and the depth, level 2, at which melting begins is increased considerably. As the plume diverges below the lithosphere, the melt becomes concentrated in layers or chambers in the boundary layer above the plume (Maalöe & Scheie 1982, Ribe & Smooke 1987). Kimberlites may rise either from the early magma accumulating above the plume, or from the lateral magma chambers in the lithosphere base. Other kimberlite-like magmas may be produced by partial melting of previously metasomatized lithosphere, with solidus temperature lower than the neighbouring depleted lithosphere keel (Wyllie et al. 1983). The physical identification of different magma sources and processes should facilitate interpretation of the isotopic characteristics of kimberlites and possibly of carbonatites derived from different silicate parents. Smith et al. (1985) presented evidence that isotopically defined Group I (basaltic) and Group II (micaceous) kimberlites of South Africa have distinctive major- and trace-element signatures. They concluded that Group I kimberlites are derived from asthenospheric sources (similar to ocean island basalts), and that Group II kimberlites originate from sources within ancient lithosphere characterized by time-averaged incompatible element enrichment. Related magmas at depth that are not erupted may yield carbonatites as they are processed through the lithosphere.

20.7.3 Melts in rifted lithosphere

There is evidence that beneath major rifts the lithosphere is thinned. The continued heat flux from a rising plume, and the concentration of hotter magma at the asthenosphere–lithosphere boundary, will promote further thinning of the lithosphere. According to Gliko et al. (1985), it takes only several million years for lithosphere thickness to be halved when additional heat flow of reasonable magnitude is applied to the base of the lithosphere.

Figures 20.19 and 20.20 depict the petrological structure in a craton with rifting crust above a thinning lithosphere. The magma near the boundary layer at levels 1–3 in Figures 20.17 and 20.18 will rise with the layer during lithospheric thinning, either percolating through the newly deformable rock matrix, or rising as diapirs with the amount of liquid increasing as the diapirs extend farther above the solidus for peridotite–C–H–O. The magmas may follow a path a–f in Figure 20.20, or they may follow a path only slightly elevated above the geotherm, represented by a–e (Spera 1984). This magma (with contribution from the lithosphere) approaches the solidus ledge at level 4 in the depth range 90–70 km (Figs 20.17 & 20.20). Magma chambers may be trapped at f or e, and as the magma solidifies, vapours will exsolve causing metasomatism, and intermittent crack propagation which facilitates the escape of magmas through the lithosphere. The release of aqueous vapours at depths greater than Q, and CO_2-rich vapours at shallower depths (Fig. 20.17; Wyllie 1980), is consistent with Haggerty's (1988, this volume) account of two metasomes at these depths. Note that according to the results of Koster van Groos (1975) traces of interstitital low-temperature carbonatite melt could persist at depths shallower than the solidus represented at level 4 in Figures 20.19 and 20.20.

The adiabatic paths for diapirs of higher temperature, rising from greater depths, would miss the solidus at M (Fig. 20.20A) and might reach the solidus for peridotite, with generation of basaltic magmas. Similarly, if the ratio $CO_2 : H_2O$ were lower, the temperature for M would be lower, and again the diapirs could rise past this level without encountering the solidus ledge 4.

The compositions of melts and vapours associated with the solidus in the region of level 4, MQR in Figure 20.17, varies widely for small changes in pressure,

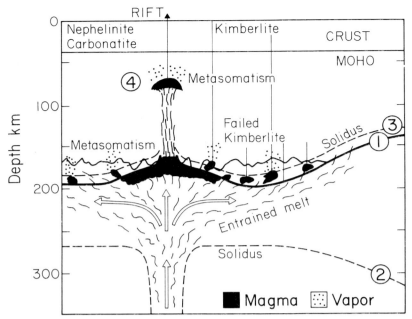

Figure 20.19 Cross-section of a craton corresponding to conditions in Figure 20.18, followed by plume uprise and lithospheric thinning beneath a rift valley. Uprise of the asthenosphere permits upward migration of the magma to level 4 (Figures 20.17 & 20.20) where magma chambers may develop and vapour may be evolved with metasomatic effects. This level is a source of parent magmas for nephelinitic volcanism with associated carbonatites (Wyllie 1987*b*). Levels 1, 2, 3, and 4 are indicated on Figure 20.17.

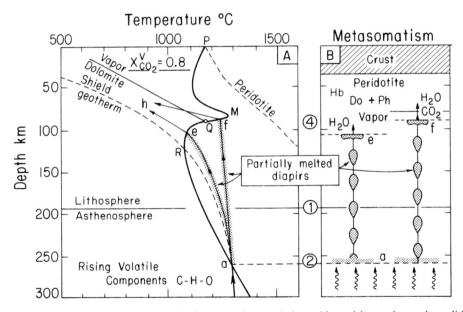

Figure 20.20 The upward migration of diapirs or melt percolation with a rising asthenosphere–lithosphere boundary may follow adiabatic or shallower paths from a region of melt accumulation, a, to the solidus ledge at level 4 (Figures 20.17 & 20.19). Levels 1, 2, and 4 are indicated on Figure 20.17. Crystallization and evolution of vapour if the path approaches the solidus M–Q–R can cause strong metasomatism of the mantle with vapours that may be H_2O-rich or CO_2-rich (B). Note that the paths may be at high enough temperature to miss the solidus Q–M altogether, and rising magmas no longer in equilibrium with peridotite are not affected by this solidus ledge. Modified after Wyllie (1980).

temperature, and $CO_2:H_2O$ (Wyllie 1978, Wendlandt & Eggler 1980, Wendlandt 1984). The igneous sequence associated with the thinning of the lithosphere between the conditions in Figures 20.18 and 20.19 was discussed by Wendlandt & Morgan (1982). Magmas rising from level 4 may include the parents of olivine nephelinites, melilite-bearing magmas, and other igneous associations that include carbonatites.

20.8 THE ORIGIN OF CARBONATITES

The phase relationships incorporated into Figures 20.17, 20.19, and 20.20 indicate the formation of carbonatite magmas and of CO_2-charged silicate magmas at depths ranging from the asthenosphere to level 4 beneath a rifting lithosphere. We consider now the origin of carbonatites emplaced in the crust. This brings us back to the conclusions reached in the earlier part of the chapter, dealing with the carbonate and silicate–carbonate systems.

20.8.1 *Primary carbonatites from the mantle*

Uprise of reduced volatile components in the system C–H–O may induce partial melting of the upper mantle as indicated in Figure 20.18, or melting may be delayed until the vapours reach a more oxidized level in the base of the lithosphere (Green et al. 1987). Kimberlites are erupted from these regions (Figs 20.18 & 20.19). The fact that carbonatite-like magmas are produced by partial melting of peridotite with addition of CO_2 and probably also by reduced gases in the system C–H–O at depths greater than about 75 km (Figs 20.15C, D) supports primary carbonatite magma generation in the mantle. These magmas are enriched in Ca and Mg, unlike the highly alkalic natrocarbonatites of Oldoinyo Lengai (Dawson 1966) and the primary magma proposed by von Eckermann (1948). Such magmas might be expected to develop diatremes similar to those formed by kimberlites, but carbonatites are normally found as small bodies associated with much larger volumes of silicate rocks. The field relationships are more consistent with the formation of carbonatites as residual magmas derived from associated silicate rocks (Barker, Le Bas, this volume). Assuming that primary carbonatite magmas generated directly by partial melting would be similar to kimberlites in terms of eruptive volume, it appears unlikely that intrusive carbonatites could be responsible for the formation of the large volumes of igneous silicate rocks in carbonatite complexes through sial-syntexis. Only if the carbonatite melts were composed essentially of alkali carbonates could they have such a significant effect in modifying sialic rocks. A process for the derivation of such magmas from mantle peridotite has yet to be confirmed, although fractionation at depth of the near-solidus carbonatite melts should yield residual melts resembling natrocarbonatites (Cooper et al. 1975).

At mantle pressures deeper than about 75 km (the solidus ledge in Figs 20.16 & 20.17, near Q), magmas containing dissolved CO_2 that crystallize at depth must yield residual carbonate-rich magmas, with the amount produced depending on the amount of CO_2 present (a model of the crystallization path is offered by Figure

20.15D). Carbonatite magmas could therefore rise independently from the differentiated mantle magma reservoirs postulated at level 4 beneath rifted cratons (Figs 20.17, 20.19, & 20.20). Volume relationships like those outlined above argue against this process.

20.8.2 Immiscible carbonatite magmas from the mantle

Experimental data confirm that a range of silicate magmas containing CO_2 can yield immiscible carbonatite magmas with compositions from alkalic to more calcic, and that these relationships persist to high pressures. Liquid compositions appear to be dominated by Ca + Na at crustal pressures, but Mg probably becomes significant at depths greater than about 75 km (Figs 20.19 & 20.20).

Koster van Groos (1975) developed the theme, based on his experimental data, that reaction of CO_2–H_2O vapours on mantle peridotite would yield a dispersed, immiscible carbonatite melt at temperatures below those where the rock would normally produce a silicate melt. He suggested that this primary melt, enriched in alkalis, would migrate upwards, forming droplets, and would accumulate slowly. Reaction of this carbonatite magma with the mantle or crust at shallower levels would generate the alkalic rocks typical of carbonatite complexes, with the carbonatite melt remaining immiscible with whatever alkalic magmas were generated.

Alternatively, derivative carbonatite magmas could be generated at mantle depths from immiscible fractions formed within deep silicate magma chambers, such as those represented in Figures 20.19 and 20.20 at about 75 km depth. Carbonatite magmas could separate from the silicate magmas for independent intrusion into the crust.

These processes appear to me to suffer from the same problems outlined above for primary carbonatite magmas: most carbonatite complexes have a high ratio of silicate:carbonatite rocks. The prospect of carbonatite magmas derived as primary magmas from mantle depths is real, however, and sufficiently interesting to justify a search for them.

20.8.3 Parent silicate magmas and derivative carbonatites

I now believe that the evidence supports the complex processing of deep, reduced volatile components from the mantle via sparse, kimberlite-like melts (Figs 20.18 & 20.20), followed by the concentration of such melts into a region of rising asthenosphere beneath a rift valley, with the formation of nephelinites, melilitites, and related magmas near the solidus ledge (Figs 20.19 & 20.20). Level 4, near 75 km depth, is a region from which a variety of low-SiO_2, alkalic magmas may rise. These magmas may have remelted mantle previously metasomatized, with some concentration of REEs (Haggerty 1988). Apatite-rich, upper-mantle xenoliths in some locations (Wass *et al.* 1980) suggest that the heterogeneity of metasomatized upper-mantle sources might influence phosphorus depletion and REE-enrichment in parental magmas and derivative carbonatites. The magmas from level 4 include strong candidates for intrusion into the crust with subsequent evolution into carbonatite complexes. Alternatively, separate intrusions of nephelinites and deep-seated immiscible carbonatites from these magma chambers may rise through the same conduit system into shallow complexes.

Experimental results at crustal pressures demonstrate that parent silicate magmas of appropriate compositions can yield carbonatite magmas by fractional crystallization. The magmas must contain dissolved CO_2, obviously, and probably H_2O in addition. Suitable silicate magma compositions are nepheline-normative, undersaturated in SiO_2, and relatively rich in alkalies (Fig. 20.11). Large thermal barriers on the liquidus separate silicate melts with feldspathic components but without normative nepheline from the low-temperature carbonatite melts at crustal pressures (Watkinson 1970, Wyllie 1974). Similar barriers occur in alkali-free systems, but the possibility remains for some initial low-SiO_2 silicate melt compositions to follow liquidus paths, bypassing the barriers (Fig. 20.10).

There is increasing experimental evidence that most alkalic magmas containing CO_2 will evolve to a stage where an immiscible carbonatite fraction must form. Rapid separation of the two melts should occur, yielding carbonatite magma that can be tapped separately for intrusion to shallower levels.

Experimental data confirm that the exsolution of a vapour phase of H_2O–CO_2 with high concentrations of alkalis and other solutes from carbonated silicate magmas, from carbonatite magmas, or from coexisting immiscible magmas is a normal consequence of crystallization within the crust. These vapours are suitable agents for fenitization. Note in Figures 20.18 and 20.20 that vapours and magmas also metasomatize the mantle earlier in the complex magmatic history, with depth ranges for the two levels of metasomatism that are illustrated being controlled by the relative positions of geotherm and solidus in any particular tectonic situation.

20.8.4 *Concluding remarks*

We can expect that CO_2 derived from the mantle may experience a variety of magmatic histories. Gaspar & Wyllie (1984) compared carbonatites with the carbonate-rich rocks associated with some kimberlites, but Mitchell (1979, 1986) has emphasized that the concept of petrogenetic links between kimberlites and carbonatites causes only confusion. Mitchell's arguments are persuasive, and I agree that there is no direct genetic link between these typical occurrences. However, a dominant role for CO_2 is evident in the genesis of both. I think of the petrogenetic links between kimberlites and carbonatites in terms of the liquid paths through the system peridotite–kimberlite–nephelinite–carbonatite–CO_2–H_2O (with varied oxygen fugacities) from high to low pressures, as summarized in a complex petrogenetic scheme proposed by Le Bas (1977). The thermal history experienced by moving masses of rocks, vapours, and magmas controls the compositions of magmas generated, and the proportions of carbonatite-like magma that may reach near-surface levels. A most important aspect of understanding these processes (Figs 20.19 & 20.20) involves thermodynamic and fluid dynamic calculations (e.g.

Spera & Bergman 1980, Spera 1981, 1984, Treiman & Schedl 1983, McKenzie 1984, Marsh 1987, Navon & Stolper 1987, Watson & Brenan 1987).

ACKNOWLEDGEMENTS

I thank the many graduate students and research associates through many years whose collaboration and contributions are evident from the bibliography. This research was supported by the Earth Sciences section of the US National Science Foundation, Grant EAR 84-16583. Caltech Division of Geological and Planetary Sciences Contribution 4545.

REFERENCES

Agard, J. 1956. Les gites minéraux associés aux roches carbonatites. *Sciences de la Terre, Université Nancy* **4**, 105–51.

Ahrens, L. H., J. B. Dawson, A. R. Duncan & A. J. Erlank (eds) 1975. *Physics and chemistry of the Earth*, Vol. 9. Oxford: Pergamon.

Artyushkov, I. V. & S. V. Sobolev 1984. Physics of the kimberlite magmatism. In *Kimberlites I: Kimberlites and related rocks*. Proceedings of the Third International Kimberlite Conference, J. Kornprobst (ed.), 309–22. New York: Elsevier.

Biggar, G. M. 1969. Phase relationships in the join $Ca(OH)_2$–$CaCO_3$–$Ca_3(PO_4)$–H_2O at 1000 bars. *Mineralogical Magazine* **37**, 75–82.

Boettcher, A. L., J. K. Robertson & P. J. Wyllie 1980. Studies in synthetic carbonatite systems: solidus relationships for CaO–MgO–CO_2–H_2O to 40 kilobars and CaO–MgO–SiO_2–CO_2–H_2O to 10 kilobars. *Journal of Geophysical Research* **85**, 6937–43.

Bowen, N. L. 1922. The behavior of inclusions in igneous magmas. *Journal of Geology* **30**, 513–70.

Boyd, F. R. & J. J. Gurney 1986. Diamonds and the African lithosphere. *Science* **232**, 472–7.

Boyd, F. R. & H. O. A. Meyer (eds) 1979a. *Kimberlites, diatremes, and diamonds: their geology, petrology, and geochemistry*. Washington: American Geophysical Union.

Boyd, F. R. & H. O. A. Meyer (eds) 1979b. *The mantle sample: inclusions in kimberlites and other volcanics*. Washington: American Geophysical Union.

Brey, G., W. R. Brice, D. J. Ellis, D. H. Green, K. L. Harris & I. D. Ryabchikov 1983. Pyroxene–carbonate relations in the upper mantle. *Earth and Planetary Science Letters* **62**, 63–74.

Byrnes, A. P. & P. J. Wyllie 1981. Subsolidus and melting relations for the join $CaCO_3$–$MgCO_3$ at 10 kilobars. *Geochimica et Cosmochimica Acta* **45**, 321–8.

Cooper, A. F., J. Gittins & O. F. Tuttle 1975. The system Na_2CO_3–K_2CO_3–$CaCO_3$ at 1 kilobar and its significance in carbonatite petrogenesis. *American Journal of Science* **275**, 534–60.

Daly, R. A. 1910. Origin of alkaline rocks. *Bulletin of the Geological Society of America* **21**, 87–118.

Daly, R. A. 1925. Carbonate dykes of the Premier Diamond Mine, Transvaal. *Journal of Geology* **33**, 659–84.

Daly, R. A. 1933. *Igneous rocks and the depths of the earth*. New York: McGraw-Hill.

Davidson, A. 1963. *A study of okaite and associated rocks near Oka, Quebec*. M.Sc. thesis, University of British Columbia, Vancouver.

Dawson, J. B. 1962. The geology of Oldoinyo Lengai. *Bulletin Volcanologique* **24**, 349–87.

Dawson, J. B. 1966. Oldoinyo Lengai – an active volcano with sodium carbonatite lava flows. In *Carbonatites*, O. F. Tuttle & J. Gittins (eds), 155–68. New York: Wiley.

Dawson, J. B. 1980. *Kimberlites and their xenoliths*. Berlin: Springer.

Dawson, J. B. & J. B. Hawthorne 1973. Magmatic sedimentation and carbonatite differentiation in kimberlite sills at Benfontein, South Africa. *Journal of the Geological Society of London* **129**, 61–85.

Deans, T. 1966. Economic mineralogy of African carbonatites. In *Carbonatites*, O. F. Tuttle & J. Gittins (eds), 385–413. New York: Wiley.

Deines, P. 1980. The carbon isotopic composition of diamonds: relationship to diamond shape, color, occurrence and vapour compositions. *Geochimica et Cosmochimica Acta* **44**, 943–61.

Eckermann, H. von 1948. The alkaline district of Alnö Island. *Sveriges Geoliska Undersökning Ser. Ca.* No. 36.

Eggler, D. H. 1978. The effect of CO_2 upon partial melting of peridotite in the system Na_2O–CaO–Al_2O_3–MgO–SiO_2–CO_2 to 35 kb, with an analysis of melting in a peridotite–H_2O–CO_2 system. *American Journal of Science* **278**, 305–43.

Eggler, D. H. 1987. Discussion of recent papers on carbonated peridotite, bearing on mantle metasomatism and magmatism: an alternative. *Earth and Planetary Science Letters* **82**, 398–400.

Eggler, D. H. & D. R. Baker 1982. Reduced volatiles in the system C–O–H: implications to mantle melting, fluid formation, and diamond genesis. In *High pressure research in geophysics*, S. Akimoto & M. H. Manghnani (eds), 237–50. Tokyo: Centre for Academic Publications.

Eggler, D. H., J. R. Holloway & B. O. Mysen 1976. High CO_2 solubilities in mantle magmas: comment. *Geology* **4**, 198–9.

Ellis, D. & P. J. Wyllie 1980. Phase relations and their petrological implications in the system MgO–SiO_2–CO_2–H_2O at pressures up to 100 kbar. *American Mineralogist* **65**, 540–56.

Erlank, A. J., F. G. Waters, C. J. Hawkesworth, S. E. Haggerty, H. L. Allsop, R. S. Rickard & M. Menzies 1987. Evidence for mantle metasomatism in peridotite nodules from the Kimberley pipes, South Africa. In *Mantle metasomatism*, M. Menzies & C. J. Hawkesworth (eds), 221–311. London: Academic Press.

Fanelli, M. F., N. Cava & P. J. Wyllie 1986. Calcite and dolomite without portlandite at a new eutectic in CaO–MgO–CO_2–H_2O, with applications to carbonatites. In *Morphology and phase equilibria of minerals*, 313–22. Proceedings 13th General Meeting International Mineralogical Association, Bulgarian Academy of Sciences, Sofia.

Franz, G. W. & P. J. Wyllie 1967. Experimental studies in the system CaO–MgO–SiO_2–CO_2–H_2O. In *Ultramafic and related rocks*, P. J. Wyllie (ed.), 323–6. New York: Wiley.

Fudali, R. F. 1963. Experimental studies bearing on the origin of pseudoleucite and associated problems of alkalic rock systems. *Bulletin of the Geological Society of America* **74**, 1101–26.

Gaspar, J. & P. J. Wyllie 1984. The alleged kimberlite–carbonatite relationship: evidence from ilmenite and spinel from Premier and Wesselton Mines and the Benfontein Sill, South Africa. *Contributions to Mineralogy & Petrology* **85**, 133–40.

Gittins, J. & O. F. Tuttle 1964. The system CaF_2–$Ca(OH)_2$–$CaCO_3$. *American Journal of Science* **262**, 66–75.

Gliko, A. L., A. F. Grachev & V. S. Magnitsky 1985. Thermal model for lithospheric thinning and associated uplift in the neotectonic phase of intraplate orogenic activity and continental rifts. *Journal of Geodynamics Research* **3**, 137–8.

Gold, D. P. 1963. Average chemical composition of carbonatites. *Economic Geology* **58**, 988–91.

Gold, D. P. 1966. The minerals of the Oka carbonatite and alkaline complex, Oka, Quebec. In *Papers and Proceedings of the 4th General Meeting, International Mineralogical Association*, IMA Volume, P. R. J. Naidu (ed.), 109–25. Mineralogical Society of India.

Goldsmith, J. R. 1983. Phase relations of rhombohedral carbonates. In *Reviews of mineralogy*. Vol. 11: *Carbonates: mineralogy and chemistry*, R. J. Reeder (ed.), 49–76. Mineralogical Society of America.

Goldsmith, J. R. & H. C. Heard 1961. Sub-solidus phase relationships in the system $CaCO_3$–$MgCO_3$. *Journal of Geology* **69**, 45–74.

Green, D. H., T. J. Falloon & W. R. Taylor 1987. Mantle-derived magmas – roles of variable source peridotite and variable C–H–O fluid compositions. In *Magmatic processes: physicochemical principles*, B. O. Mysen (ed.), 139–54. Geochemical Society Special Publication No. 1.

Haggerty, S. E. 1986. Diamond genesis in a multiply constrained model. *Nature* **320**, 34–8.

Haggerty, S. E. 1988. Source regions for oxides, sulfides and metals in the upper mantle: clues to the stability of diamonds, and the genesis of kimberlites, lamproites and carbonatites. *Proceedings of the Fourth International Kimberlite Conference*, Australian Geological Society, in press.

Hamilton, D. L. & W. S. Mackenzie 1965. Phase equilibrium studies in the system $NaAlSiO_4$ (nepheline)–$KAlSiO_4$ (kalsilite)–SiO_2–H_2O. *Mineralogical Magazine* **34**, 214–31.

Hamilton, D. L., I. C. Freestone, J. B. Dawson & C. H. Donaldson 1979. Origin of carbonatites by liquid immiscibility. *Nature* **279**, 52–4.

Heinrich, E. W. 1966. *The geology of carbonatites*. Chicago: Rand McNally.

Helz, G. R. & P. J. Wyllie 1979. Liquidus relationships in the system $CaCO_3$–$Ca(OH)_2$–CaS and the solubility of sulfur in carbonatite magmas. *Geochimica et Cosmochimica Acta* **43**, 259–65.

Hogarth, D. D. 1966. Intrusive carbonate rock near Ottawa, Canada. In *Papers and Proceedings of the 4th General Meeting, International Mineralogical Association*, IMA Volume, P. R. J. Naidu (ed.), 45–53. Mineralogical Society of India.

Holmes, A. 1950. Petrogenesis of katungite and its associates. *American Mineralogist* **35**, 772–92.

Huang, W. L. & P. J. Wyllie 1974. Eutectic between wollastonite II and calcite contrasted with thermal barrier in MgO–SiO_2–CO_2 at 30 kilobars, with applications to kimberlite–carbonatite petrogenesis. *Earth and Planetary Science Letters* **24**, 305–10.

Huang, W. L. & P. J. Wyllie 1976. Melting relationships in the systems CaO–CO_2 and MgO–CO_2 to 33 kilobars. *Geochimica et Cosmochimica Acta* **40**, 129–32.

Huang, W. L. & P. J. Wyllie 1984. Carbonation reactions for mantle lherzolite and harzburgite. *Proceedings of the 27th International Geological Congress, Moscow* **9**, 455–73. Utrecht: VNU Science Press.

Huang, W. L., P. J. Wyllie & C. E. Nehru 1980. Subsolidus and liquidus phase relationships in the system CaO–SiO_2–CO_2 to 30 kbar with geological applications. *American Mineralogist* **65**, 285–301.

Irving, A. J. & P. J. Wyllie 1973. Melting relationships in CaO–CO_2 and MgO–CO_2 to 36 kilobars, with comments on CO_2 in the mantle. *Earth and Planetary Science Letters* **20**, 220–5.

Irving, A. J. & P. J. Wyllie 1975. Subsolidus and melting relationships for calcite, magnesite, and the join $CaCO_3$–$MgCO_3$ to 36 kilobars. *Geochimica et Cosmochimica Acta* **39**, 35–53.

Jones, A. P. & P. J. Wyllie 1983. Low-temperature glass quenched from a synthetic rare-earth carbonatite: implications for the origin of the Mountain Pass deposit, California. *Economic Geology* **78**, 1721–3.

Jones, A. P. & P. J. Wyllie 1985. Paragenetic trends of oxide minerals in carbonate-rich kimberlites with new analyses from the Benfontein sill, South Africa. *Journal of Petrology* **26**, 210–22.

Jones, A. P. & P. J. Wyllie 1986. Solubility of rare earth elements in carbonatite magmas, indicated by the liquidus surface in $CaCO_3$–$Ca(OH)_2$–$La(OH)_3$ at 1 kbar pressure. *Applied Geochemistry* **1**, 95–102.

King, B. C. & D. L. Sutherland 1960. Alkaline rocks of eastern and southern Africa. *Science Progress* **47**, 298–321, 504–24, 709–20.

Kjarsgaard, B. A. & D. L. Hamilton 1988. Liquid immiscibility and the origin of alkali-poor carbonatites. *Mineralogical Magazine* **52**, 43–55.

Kornprobst, J. (ed.) 1984a. *Kimberlites* I: *Kimberlites and related rocks*. New York: Elsevier.

Kornprobst, J. (ed.) 1984b. *Kimberlites* II: *The mantle and crust–mantle relationships*. New York: Elsevier.

Koster van Groos, A. F. 1975. The effect of high CO_2 pressure on alkalic rocks and its bearing on the formation of alkalic ultrabasic rocks and the associated carbonatites. *American Journal of Science* **275**, 163–85.

Koster van Groos, A. F. & P. J. Wyllie 1963. Experimental data bearing on the role of liquid immiscibility in the genesis of carbonatites. *Nature* **199**, 801–2.

Koster van Groos, A. F. & P. J. Wyllie 1966. Liquid immiscibility in the system Na_2O–Al_2O_3–SiO_2–CO_2 at pressures up to 1 kilobar. *American Journal of Science* **264**, 234–55.

Koster van Groos, A. F. & P. J. Wyllie 1968. Liquid immiscibility in the join $NaAlSi_3O_8$–Na_2CO_3–H_2O and its bearing on the genesis of carbonatites. *American Journal of Science* **266**, 932–67.

Koster van Groos, A. F. & P. J. Wyllie 1973. Liquid immiscibility in the join $NaAlSi_3O_8$–$CaAlSi_2O_8$–Na_2CO_3–H_2O. *American Journal of Science* **273**, 465–87.

Kuellmer, F. J., A. P. Visocky & O. F. Tuttle 1966. Preliminary survey of the system barite–calcite–fluorite at 500 bars. In *Carbonatites*, O. F. Tuttle & J. Gittins (eds.), 353–64. New York: Wiley.

Kutty, T. R. N., M. N. Viswanathiah & J. A. K. Tareen 1978. Hydrothermal equilibria in Nd_2O_3–H_2O–CO_2 system. *Indian Academy of Science Proceedings* **87A**, 69–74.

Le Bas, M. J. 1977. *Carbonatite–nephelinite volcanism*. London: Wiley.

Maalöe, S. & A. Scheie 1982. The permeability controlled accumulation of primary magma. *Contributions to Mineralogy and Petrology* **81**, 350–7.

McKenzie, D. P. 1984. The generation and compaction of partial melts. *Journal of Petrology* **25**, 713–65.

Mariano, A. N. & P. L. Roeder 1983. Kerimasi: a neglected carbonatite volcano. *Journal of Geology* **91**, 449–55.

Marsh, B. 1987. Magmatic processes. *Review of Geophysics* **25**, 1043–53.

Mitchell, R. H. 1979. The alleged kimberlite–carbonatite relationship: additional contrary mineralogical evidence. *American Journal of Science* **279**, 570–89.

Mitchell, R. H. 1986. *Kimberlites: mineralogy, geochemistry, petrology*. New York: Plenum.

Naidu, P. R. J. (ed.) 1966. *Papers and Proceedings of the 4th General Meeting, International Mineralogical Association*, IMA Volume. Mineralogical Society of India.

Naldrett, A. J. & D. H. Wakinson 1981. Ore formation within magmas. In *Mineral resources: genetic understanding for practical applications*, P. B. Barton (ed.), 119. Washington: National Academy.

Navon, O. & E. Stolper 1987. Geochemical consequences of melt percolation: the upper mantle as a chromatographic column. *Journal of Geology* **95**, 285–307.

Nickel, K. G. & D. H. Green 1985. Empirical geothermobarometry for garnet peridotites and implications for the nature of the lithosphere, kimberlites and diamonds. *Earth and Planetary Science Letters* **73**, 158–70.

Nixon, P. H. (ed.) 1973. *Lesotho kimberlites*. Maseru: Lesotho National Development.

Olafsson, M. & D. H. Eggler 1983. Phase relations of amphibole–carbonate, and phlogopite–carbonate peridotite: petrologic constraints on the asthenosphere. *Earth and Planetary Science Letters* **64**, 305–15.

Olson, J. C., D. R. Shawe, L. C. Pray & W. N. Sharp 1954. Rare earth mineral deposits of the Mountain Pass district, San Bernardino County, California. *U.S. Geological Survey Professional Paper* **261**.

Otto, J. 1984. *Melting relations in some carbonate–silicate systems: sources and products of CO_2-rich liquids*. Ph.D. thesis, University of Chicago, Chicago.

Otto, J. W. & P. J. Wyllie 1983. Phase relations on the join calcite–nepheline–albite at 25 kb: subducted basalt limestone as magmatic source. *EOS. Transactions, American Geophysical Union* **64**, 897. (abstract)

Palabora Mining Company Ltd, Mine, Geological and Mineralogical Staff 1976. The geology and the economic deposits of copper, iron, and vermiculite in the Palabora igneous complex, a brief review. *Economic Geology* **71**, 177–92.

Parker, R. L. 1967. Composition of the Earth's crust. In *Data of geochemistry*, 6th edn. U.S. Geological Survey Professional Paper 440-D.

Pecora, W. T. 1956. Carbonatites: a review. *Bulletin of the Geological Society of America* **67**, 1537–56.

Ribe, N. M. & M. D. Smooke 1987. A stagnation point flow model for melt extraction from a mantle plume. *Journal of Geophysical Research* **B7**, 6437–43.

Roedder, E. 1965. Liquid CO_2 inclusions in olivine-bearing nodules and phenocrysts from basalts. *American Mineralogist* **50**, 1746–82.

Ryabchikov, I. D., D. H. Green, W. J. Wall & G. P. Brey 1981. The oxidation state of carbon in the reduced-velocity zone. *Geochemistry International* **1981**, 148–58.

Schairer, J. F. 1950. the alkali-feldspar join in the system $NaAlSiO_4$–$KAlSiO_4$–SiO_2. *Journal of Geology* **58**, 512–17.

Schairer, J. F. & N. L. Bowen 1935. *Preliminary report on equilibrium relations between feldspars, and silica*. Transactions of the American Geophysical Union, 16th annual meeting, 325–8.

Schneider, M. E. & D. H. Eggler 1986. Fluids in equilibrium with peridotite minerals: implications for mantle metasomatism. *Geochimica et Cosmochimica Acta* **50**, 711–24.

Schuiling, R. D. 1964a. The limestone assimilation hypothesis. *Nature* **204**, 1054–5.

Schuiling, R. D. 1964b. Dry synthesis of feldspathoids by feldspar–carbonate reactions. *Nature* **201**, 1115.

Shand, S. J. 1930. Limestone and the origin of feldspathoidal rocks: an aftermath of the Geological Congress. *Geological Magazine* **67**, 415–27.

Shand, S. J. 1943. *Eruptive rocks*, 2nd edn. New York: Wiley.

Shand, S. J. 1945. The present status of Daly's hypothesis of the alkaline rocks. *American Journal of Science* **243-A**, 495–507.

Smith, C. B., J. J. Gurney, E. M. W. Skinner, C. R. Clement & N. Ebrahim 1985. Geochemical character of South African kimberlites: a new approach based on isotopic constraints. *Transactions of the Geological Society of South Africa* **88**, 267–80.

Smith, W. C. 1956. A review of some problems of African carbonatites. *Quarterly Journal of the Geological Society of London* **112**, 189–219.

Spera, F. J. 1981. Carbon dioxide in igneous petrogenesis II: Fluid dynamics of mantle metasomatism. *Contributions to Mineralogy and Petrology* **77**, 56–65.

Spera, F. J. 1984. Carbon dioxide in petrogenesis III: Role of volatiles in the ascent of alkaline magma with special reference to xenolith-bearing mafic lavas. *Contributions to Mineralogy and Petrology* **88**, 217–32.

Spera, F. J. & S. C. Bergman 1980. Carbon dioxide in igneous petrogenesis I: Aspects of dissolution of CO_2 in silicate liquids. *Contributions to Mineralogy and Petrology* **74**, 55–66.

Taylor, W. R. & D. H. Green 1987. The petrogenetic role of methane: effect on liquidus phase relations and the solubility mechanism of reduced C–H volatiles. In *Magmatic processes: physicochemical principles*, B. O. Mysen (ed.), 121–38. Geochemical Society, Special Publication No. 1.

Treiman, A. H. & E. J. Essene 1984. A periclase–dolomite–calcite carbonatite from the Oka complex, Quebec, and its calculated volatile composition. *Contributions to Mineralogy and Petrology* **85**, 149–57.

Treiman, A. H. & A. Schedl 1983. Properties of carbonatite magma and processes in carbonatite magma chambers. *Journal of Geology* **91**, 437–47.

Tuttle, O. F. & J. Gittins 1966. *Carbonatites*. New York: Wiley.

Verwoerd, W. J. 1978. Liquid immiscibility and the carbonatite–ijolite relationship: preliminary data on the join $NaFe^{3+}Si_2O_6$–$CaCO_3$ and related compositions. *Carnegie Institution of Washington Yearbook* **77**, 767–74.

Walter, L. S., P. J. Wyllie & O. F. Tuttle 1962. The system MgO–CO_2–H_2O at high pressures and temperatures. *Journal of Petrology* **3**, 49–64.

Wass, S. Y., P. Henderson & C. J. Elliott 1980. Chemical heterogeneity and metasomatism in the upper mantle: evidence from rare earth and other elements in apatite-rich xenoliths in basaltic rocks from eastern Australia. *Philosophical Transactions of the Royal Society of London* **A297**, 333–46.

Watkinson, D. H. 1970. Experimental studies bearing on the origin of the alkalic rock–carbonatite complex and niobium mineralization at Oka, Quebec. *Canadian Mineralogist* **10**, 350–61.

Watkinson, D. H. & P. J. Wyllie 1969. Phase equilibrium studies bearing on the limestone-assimilation hypothesis. *Bulletin of the Geological Society of America* **80**, 1565–76.

Watkinson, D. H. & P. J. Wyllie 1971. Experimental study of the join $NaAlSiO_4$–$CaCO_3$–H_2O and the genesis of alkalic rock–carbonatite complexes. *Journal of Petrology* **12**, 357–78.

Watson, E. B. & J. B. Brenan 1987. Fluids in the lithosphere, 1. Experimentally determined wetting characteristics of CO_2–H_2O fluids and their implications for fluid transport, host-rock physical properties, and fluid inclusion formation. *Earth and Planetary Science Letters* **85**, 497–515.

Watson, K. D. 1955. Kimberlite at Bachelor Lake, Quebec. *American Mineralogist* **40**, 565–79.

Wendlandt, R. F. 1984. An experimental and theoretical analysis of partial melting in the system $KAlSiO_4$–CaO–MgO–SiO_2–CO_2 and applications to the genesis of potassic magmas, carbonatites and kimberlites. In *Kimberlites I: Kimberlites and related rocks*, Proceedings of the Third International Kimberlite Conference, J. Kornprobst (ed.), 359–69. New York: Elsevier.

Wendlandt, R. F. & D. H. Eggler 1980. The origins of potassic magmas. 1. Melting relations in the systems $KAlSiO_4$–Mg_2SiO_4–SiO_2 and $KAlSiO_4$–MgO–SiO_2–CO_2 to 30 kilobars. 2. Stability of phlogopite in natural spinel lherzolite and in the system $KAlSiO_4$–MgO–SiO_2–H_2O–CO_2 at high pressures and high temperatures. *American Journal of Science* **280**, 385–420, 421–58.

Wendlandt, R. F. & W. J. Harrison 1979. Rare earth partitioning between immiscible carbonate and silicate liquids and CO_2 vapor: results and implications for the formation of light rare earth-enriched rocks. *Contributions to Mineralogy and Petrology* **69**, 409–19.

Wendlandt, R. F. & P. Morgan 1982. Lithospheric thinning associated with rifting in East Africa. *Nature* **298**, 734–6.

Wendlandt, R. F. & B. O. Mysen 1980. Melting phase relations of natural peridotite + CO_2 as a function of degree of partial melting at 15 and 30 kbar. *American Mineralogist* **65**, 37–44.

Woerman, E. & M. Rosenhauer 1985. Fluid phases and the redox stage of the Earth's mantle: extrapolations based on experimental, phase-theoretical and petrological data. *Fortschritte der Mineralogie* **63**, 263–349.

Wyllie, P. J. 1965. Melting relationships in the system CaO–MgO–CO_2–H_2O with petrological applications. *Journal of Petrology* **6**, 101–23.

Wyllie, P. J. 1966a. Experimental studies of carbonatite problems: the origin and differentiation of carbonatite magmas. In *Carbonatites*, O. F. Tuttle & J. Gittins (eds), 311–52. New York: Wiley.

Wyllie, P. J., 1966b. Experimental data bearing on the petrogenetic links between kimberlites and carbonatites. In *Papers and Proceedings of the 4th General Meeting, International Mineralogical Association*, IMA Volume, P. R. J. Naidu (ed.), 67–82. Mineralogical Society of India.

Wyllie, P. J. (ed.) 1967. *Ultramafic and related rocks*. New York: Wiley.

Wyllie, P. J. 1974. Limestone assimilation. In *Alkaline rocks*, H. Sørensen (ed.), 459–74. New York: Wiley.

Wyllie, P. J. 1977. Mantle fluid compositions buffered by carbonates in peridotite–CO_2–H_2O. *Journal of Geology* **85**, 187–207.

Wyllie, P. J. 1978. Mantle fluid compositions buffered in peridotite–CO_2–H_2O by carbonates, amphibole, and phlogopite. *Journal of Geology* **86**, 687–713.

Wyllie, P. J. 1979. Magmas and volatile components. *American Mineralogist* **64**, 469–500.

Wyllie, P. J. 1980. The origin of kimberlites. *Journal of Geophysical Research* **85**, 6902–10.

Wyllie, P. J. 1987a. Discussion of recent papers on carbonated peridotite, bearing on mantle metasomatism and magmatism. *Earth and Planetary Science Letters* **82**, 391–7, 401–2.

Wyllie, P. J. 1987b. Transfer of subcratonic carbon into kimberlites and rare earth carbonatites. In *Magmatic processes: physicochemical principles*, B. O. Mysen (ed.), 107–19. Geochemical Society, Special Publication No. 1.

Wyllie, P. J. 1988. The genesis of kimberlites and some low-SiO_2, high-alkali magmas. *Proceedings of the Fourth International Kimberlite Conference*, Australian Geological Society, in press.

Wyllie, P. J. & G. M. Biggar 1966. Fractional crystallization in the "carbonatite systems" CaO–MgO–CO_2–H_2O and CaO–CaF_2–P_2O_5–CO_2–H_2O. In *Papers and Proceedings of the 4th General Meeting, International Mineralogical Association*, IMA Volume, P. R. J. Naidu (ed.), 92–105. Mineralogical Society of India.

Wyllie, P. J. & A. L. Boettcher 1969. Liquidus phase relationships in the system CaO–CO_2–H_2O to 40 kilobars pressure with petrological applications. *American Journal of Science*, J. Frank Schairer Volume, **267A**, 489–508.

Wyllie, P. J. & J. L. Haas 1965. The system CaO–SiO_2–CO_2–H_2O: I. Melting relationships with excess vapor at 1 kilobar pressure. *Geochimica et Cosmochimica Acta* **29**, 871–92.

Wyllie, P. J. & J. L. Haas 1966. The system CaO–SiO_2–CO_2–H_2O: II. The petrogenetic model. *Geochimica et Cosmochimica Acta* **30**, 525–44.

Wyllie, P. J. & W. L. Huang 1975. Peridotite, kimberlite, and carbonatite explained in the system CaO–MgO–SiO_2–CO_2. *Geology* **3**, 621–4.

Wyllie, P. J. & W. L. Huang 1976a. Carbonation and melting reactions in the system CaO–MgO–SiO_2–CO_2 at mantle pressures with geophysical and petrological applications. *Contributions to Mineralogy and Petrology* **54**, 79–107.

Wyllie, P. J. & W. L. Huang 1976b. High CO_2 solubilities in mantle magma. *Geology* **4**, 21–4.

Wyllie, P. J. & A. P. Jones 1985. Experimental data bearing on the origin of carbonatites, with particular reference to the Mountain Pass rare earth deposit. In *Applied mineralogy*, W. C. Park, D. M. Hausen & R. D. Hagni (eds), 935–49. New York: American Institute of Mining, Metallurgical, and Petroleum Engineers.

Wyllie, P. J. & M. Rutter 1986. Experimental data on the solidus for peridotite–CO_2, with applications to alkaline magmatism and mantle metasomatism. *EOS. Transactions, American Geophysical Union* **67**, 390. (abstract).

Wyllie, P. J. & O. F. Tuttle 1959. Effect of carbon dioxide on the melting of granite and feldspars. *American Journal of Science* **257**, 648–55.

Wyllie, P. J. & O. F. Tuttle 1960. The system CaO–CO_2–H_2O and the origin of carbonatites. *Journal of Petrology* **1**, 1–46.

Wyllie, P. J., W. L. Huang, J. Otto & A. P. Byrnes 1983. Carbonation of peridotites and decarbonation of siliceous dolomites represented in the system CaO–MgO–SiO_2–CO_2 to 30 kbar. *Tectonophysics* **100**, 359–88.

Zhabin, A. G. & G. Ye. Cherepivskaya 1965. Carbonatite dykes as related to ultrabasic-alkalic extrusive igneous activity. *Doklady, Academy of Sciences, Earth Science Section* **160**, 135–8. (English translation)

21
MANTLE METASOMES AND THE KINSHIP BETWEEN CARBONATITES AND KIMBERLITES

S. E. HAGGERTY

ABSTRACT

A model for metasomatic horizons (metasomes) in the upper mantle is developed based on: the mineralogy and chemistry of xenoliths entrained in kimberlites; the prevalent distribution of carbonatite complexes, kimberlites, and closely associated rock types in rifts and cratons, respectively; long-lived mantle heterogeneities; and the fact that enriched upper-mantle source regions are evidently required for the generation of exotic alkali-rich melt compositions. Metasomes are enriched in volatiles (H_2O, CO_2, and S) and high-pressure silicate-incompatible elements (SIEs) such as Ba, K, Na, Nb, Rb, REEs (rare earth elements), Sr, Ti and Zr and are characterized mineralogically by introduced phlogopite, K-richterite, diopside, LIL-titanates, calcite, and zircon into previously depleted harzburgite. Two metasomes are proposed for the mantle lithosphere: one at ~75–100 km in which $K + H_2O$ is dominant; and a second at < 60 km enriched in $Na + CO_2$. A third metasome may be present at the lithosphere–asthenosphere boundary (LAB), represented by glimmerite (mica-rich, amphibole-free) xenoliths. Metasomes form from upwardly migrating fluids (and their dissolved SIEs) released from stagnated melts that abort on intersection with the C–O–H peridotite solidus, or underplate at the LAB. Lithospheric thinning at the edges of cratons and towards adjacent mobile belts and rift zones, with enhanced heat flow and magmatism, leads to a greater production of ponded melts. As a consequence of the combined effects of peridotite solidus intersection, LAB interference and lower crustal underplating, thicker and more enriched and more highly fractionated metasomes are formed. Late-stage emplacement of carbonated kimberlite, and carbonatite into alkali complexes, are accounted for by early melting of deeper upper mantle horizons that tap $K + H_2O$ metasomes, with higher level, later melting and assimilation of $Na + CO_2$ carbonate metasomes.

21.1 INTRODUCTION

Uncertainties in the genesis of carbonatites and kimberlites have led to long-standing debates on the inter-relationship, if any, of these upper-mantle-derived melts (e.g. Dawson 1980, Mitchell 1986). Much emphasis has been placed on opaque mineral compositions, implying a relationship (Gaspar & Wyllie 1984), or demonstrating the absence of a kinship (Mitchell 1979). A major flaw in these discussions is that low-pressure, groundmass crystals of spinel and ilmenite in carbonatites are compared with equivalent minerals of high-pressure origin in upper-mantle xenoliths from kimberlites (Tompkins & Haggerty 1985). On the other hand, globular segregations of calcite are a common feature of kimberlites, and the high Mg contents of beforsites (dolomitic carbonatites) connote ultramafic affinities, consistent with an origin in the upper mantle. Entrained xenoliths in kimberlites exhibit a variety of metasomatic modifications (e.g. Erlank et al. 1987), and the geochemical characteristics of introduced, high-pressure, silicate-incompatible elements (SIEs), specifically Ba, K, LREEs, Na, Sr, Ti, and Zr are remarkably similar to the low-pressure geochemistries of fenite haloes surrounding intrusive carbonatites (e.g. von Eckermann 1966, Currie & Ferguson 1971). Concentrations of SIEs in carbonatites are overwhelmingly greater than in kimberlites or metasomatized upper-mantle-derived xenoliths, so that if kimberlites and carbonatites are linked, a mechanism of selective SIE concentration is required for the formation of carbonatites. Carbonate liquid immiscibility in alkali melts (Hamilton et al. 1979, Kjarsgaard & Hamilton, this volume), and the attendant fractionation of SIEs is a viable process for selective concentration but has not gained widespread acceptance (e.g. Gittins 1986), notwithstanding the fact that all carbonatites are closely associated with alkali intrusives.

Regional tectonic settings for carbonatites in rift zones of high heat flow, and those of kimberlites in stable cratons are accepted norms. The numerous exceptions to these distributions do not, at first sight, make tectonic crustal settings a useful discriminant, particularly because attention has largely focussed on those bodies that are mineralized. An interesting pattern, however, is emerging for long-lived upper-mantle heterogeneities (Bell et al. 1982, Bell & Blenkinsop 1987), in which localized upper mantle has repeatedly been the source of kimberlites for ~1.6 Ga (Bristow et al. 1986), and of carbonatites for ~2.7 Ga (Woolley 1986), implying that localized specific regions of the upper mantle are periodically tapped, and exotic melts are generated in response to heat-flow perturbations and tectonism.

The compositionally exotic melts that comprise carbonatites and kimberlites cannot be derived directly by partial melting of upper-mantle peridotite (Mysen & Kushiro 1977), unless vanishingly small (1% or less) degrees of partial melt are invoked. This is required by the large differences that exist in SIEs and volatiles (H_2O, CO_2) between upper-mantle peridotites and carbonatite and kimberlite magmas. In the model developed here, it is proposed that the upper mantle is preconditioned by SIEs and volatile metasomatism, and that this state is a prerequisite to the genesis of carbonate and alkali-rich magmas in the upper mantle.

21.2 CARBONATE IN KIMBERLITES AND RELATED ROCKS

Primary magmatic calcite (pyrogenic calcite) is widespread in virtually all kimberlites and is included in the definition of kimberlite (Clement et al. 1984). Diapiric calcite segregations and quenched skeletal calcite crystals are present in the Benfontein sill kimberlites (Fig. 21.1a), RSA (Republic of South Africa), and are described in detail by Dawson & Hawthorne (1973), Jones & Wyllie (1985), and McMahon & Haggerty (1984). Similar calcite structures are recognized (4th International Kimberlite Conference Excursion, August 1986) in the Orroroo kimberlites, SE Australia. Kimberlite dykes in the De Beers (Donaldson & Reid 1982) and Wesselton mines (Mitchell 1984) also contain lobe-shaped calcite segregations that are oriented normal to intrusive contacts (Fig. 21.1b). Perhaps the most spectacular association of calcite in kimberlite is at the Premier diamond mine, RSA (Robinson 1975, Jones & Wyllie 1985), where the central core of the pipe, as well as cross-cutting dykes, have high carbonate contents of pyrogenic origin (Fig. 21.1c). Magnesian-titanomagnetite and LREE-enriched perovskite are common minerals at all of these localities, and at Benfontein form layers 1–2 cm thick (Fig. 21.1a).

Although the precise relationship between kimberlites and lamproites is at present obscure, both rock types are alkalic, volatile, and SIE-enriched (e.g. Jaques *et al.* 1984, Fraser *et al.* 1985, Rock 1986), and both may contain diamonds (Atkinson *et al.* 1984). Spherical or oblate segregations of calcite, and more rarely dolomite, are present in diamond-bearing olivine lamproites from Argyle, NW Australia (Fig. 21.1d), and these are commonly accompanied by needles of priderite $(K,Ba)(Ti,Fe)_8O_{16}$, with Sr and LREE-enriched perovskite; rutile is typically present in the groundmass. Kimberlites, carbonatites, and melilitites in the Karasburg Fold Belt, Namibia (Haggerty & Boyd, unpublished data), exhibit a close field relationship between carbonatites and kimberlites, with melilitite in carbonatite dykes,

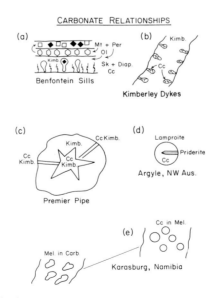

Figure 21.1 Carbonate relationships in kimberlites (a–c), lamproites (d), and melilitites (e). Apparent immiscible carbonate segregations are present in (a), (b), (d), and (e). In (c) the central core complex and cross-cutting dykes are highly carbonated kimberlites intrusive into less carbonated varieties. Mt, magnetite; Per, perovskite; Ol, olivine; Sk, skeletal; Diap, diapiric; Cc, calcite; Kimb, kimberlite; Mel, melilitite; Carb, carbonatite.

and calcite segregations in melilitites (Fig. 21.1e). Magnesian-titanomagnetite and perovskite are abundant in all rock types.

These examples serve to demonstrate:

(a) the presence of pyrogenic calcite in kimberlite;
(b) segregations of calcite in kimberlite, lamproite, and melilitite strikingly similar to experimentally-derived carbonate immiscible liquids in mafic melts (Freestone & Hamilton 1980);
(c) carbonate–LREE affinities in carbonatites, kimberlites, lamproites, and melilitites, consistent with experimental partitioning data (Wendlandt & Harrison 1979, Harrison 1981);
(d) titanium as a ubiquitous component concentrated in Mg-spinel and perovskite; and
(e) relatively high oxidation states in the form of Mg-titanomagnetite (McMahon & Haggerty 1984), consistent with $CO_2 + H_2O$ in the source regions.

21.3 METASOMATISM

Mineralogical and geochemical data on entrained upper-mantle-derived xenoliths from kimberlites (Dawson 1980, Nixon et al. 1981), and alkali basalts (Irving 1980), and models for the generation of ocean-floor basalts and ultramafic peridotites (e.g. Dick & Bullen 1984, Shibata & Thompson 1986), suggest that the upper mantle is layered into a deep fertile asthenosphere and a shallower depleted (i.e. depleted in basaltic components, e.g. Ca, Al, Fe, Si) mantle lithosphere. The lithosphere–asthenosphere boundary (LAB) varies with tectonic setting; it is generally considered (e.g. Haggerty 1986) to be deep (150–200 km) in continental cratonic regions, and shallow in rift zones (10 km or less). Rock types comprising the asthenosphere are lherzolites: garnet-bearing at high $P-T$, spinel-bearing at intermediate $P-T$, and plagioclase-bearing at low $P-T$. Mantle lithosphere is dominated by harzburgite with subsidiary dunite, and is considered to have formed by the extraction of basalt, komatiite, and the crust, from the asthenosphere. In continental subcratonic regions the lower lithosphere is partially desiccated, highly reduced conditions prevail [iron–wüstite to wüstite–magnetite (IW–WM)], and the environment is compatible to the nucleation, growth, and preservation of diamond (e.g. Meyer 1985, Boyd & Gurney 1986, Haggerty 1986). Evidence for metasomatism, however, is documented for garnet inclusions in diamond (Richardson et al. 1984), for fertile garnet lherzolites, depleted harzburgites (Erlank et al. 1987), and for dunites (Boyd et al. 1983). The enriched geochemical signatures produced by metasomatism are manifest in elevated concentrations of SIEs, and in the formation of phlogopite, K-richterite, diopside, and LIL-titanates [lindsleyite, mathiasite (LIMA), armalcolite, rutile, ilmenite, and K and Ba magnetoplumbite structured minerals], described by Haggerty (1983a, 1987, 1988), Haggerty et al. (1983), Erlank et al. (1987), and Grey et al. (1987).

Progressive metasomatism of upper-mantle-derived xenoliths in kimberlites can be traced from garnet peridotites (GP) (i.e. lherzolites and harzburgites), to phlogopite-bearing peridotites (PP), and to phlogopite–K-richterite peridotites (PKP). In the most advanced PKP metasomatic suite (metasomites) exotic arrays of titanate oxides have been recognized in veins and in substrates of olivine and orthopyroxene (i.e. harzburgite) and chromian spinel. Phases contain vestiges of depletion (high MgO + Cr_2O_3), are enriched (Ba, Ca, K, LREEs, Na, Nb, Sr, Ti, Zr) and, from the association with phlogopite and amphibole, it is concluded that previously depleted lithosphere has undergone infiltration fluid metasomatism. Experimental P–T mineral equilibria data for K-richterite, armalcolite, and lindsleyite (summarized in Haggerty 1987), place the zone of metasomatism at 900–1000 °C and 20–25 Kb, equivalent to subcratonic depths of approximately 75–100 km.

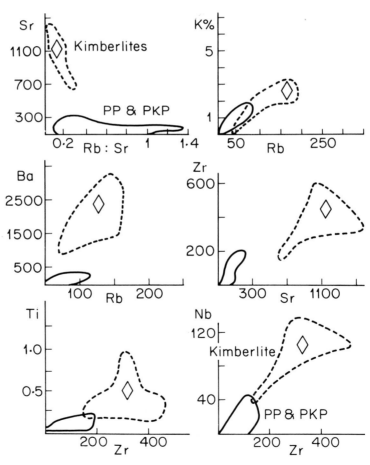

Figure 21.2 Comparisons of silicate incompatible elements (SIEs) between kimberlites (diamond symbol field) and metasomites (PP, phlogopite peridotites; PKP, phlogopite–K-richterite peridotite), Kaapvaal Craton, RSA (Erlank *et al.* 1987). Abundances in p.p.m. unless stated otherwise.

Metasomites containing calcite are restricted to depths of 60 km or less (Woerman & Rosenhauer 1985), and these rocks are characterized by lindsleyite (Ba-specific crichtonite), Nb–Cr-rutile, Mg–Cr-ilmenite, ferro-freudenbergite ($Na_2FeTi_7O_{16}$), and zircon in substrates of olivine + orthopyroxene + clinopyroxene + phlogopite + K-richterite (Haggerty 1983b, Haggerty & Gurney 1984).

Bulk chemical data for carbonate-bearing metasomites are currently unavailable, but a comparison of selected elements in PP and PKP metasomites with kimberlites, both from the Kaapvaal Craton, RSA (Erlank *et al.* 1987), are shown in Figure 21.2. Notwithstanding the enriched nature of these metasomites, it is evident that the abundances of Ba, K, Nb, Sr, Ti, and Zr in kimberlites far exceed the concentrations of PP and PKP metasomatic rocks, implying that kimberlites can only be derived from these sources by selective partial melting. Proto-kimberlite magmas must originate in the asthenosphere because fertile garnet lherzolite is sampled. The SIE and volatile abundances, however, are far too low to satisfy the chemistries of kimberlites. Depleted lithosphere is an even less likely candidate, although it is also sampled along with diamonds. Although small degrees of partial melting of PP and PKP metasomites cannot be unequivocally demonstrated, it is proposed that carbonate-hosted, high-level (i.e. < 60 km), mantle lithospheric metasomatites are also involved. These metasomites are a source of carbonate and elevated contents of SIEs.

21.4 METASOME MODEL

Over 200 kimberlite pipes, dykes, and sills dated between 80 Ma and 1.6 Ga are recognized across the Kaapvaal Craton, extending east to Lesotho, south into the Cape Fold Belt, and west into Namibia (Dawson 1980, Mitchell 1986). Clusters of kimberlites are also present in Zaire, Angola, West Africa, North America, Australia, South America, China, and the USSR. If metasomatism is the key to kimberlite genesis, as assumed here, and favoured by others (e.g. Wyllie 1987), then metasomatism must be regionally extensive within the subcratonic lithosphere. The exotic LIL-titanates described in metasomites from South Africa are also recognized in the Shandong Province of China (Dong *et al.* 1983, Zhou *et al.* 1984); in addition, priderite and jeppeite $(K,Ba)_2(Ti,Fe)_6O_{13}$, which are considered to be the low-pressure geochemical analogues of LIMA and LIL-titanate magnetoplumbites (Haggerty 1987), are present in association with Ti-K-richterite and Ti-ferriphlogopite in olivine and leucite lamproites from NW Australia. The latter rock types may be viewed as the metasomatic equivalents of kimberlites (Haggerty 1987, 1988), with greater proportions of metasome assimilation.

Metasomatism at 60–100 km is thought to result from aborted asthenospheric melts that undergo thermal crisis (Spera 1984) in the lithosphere. This is brought about because the lithosphere is 200–300 °C cooler than the asthenosphere. Melts gel and liberate their fluid and dissolved SIEs on intersection with the C–O–H peridotite solidus (Wyllie 1980, Haggerty 1983, Schneider & Eggler 1984), as

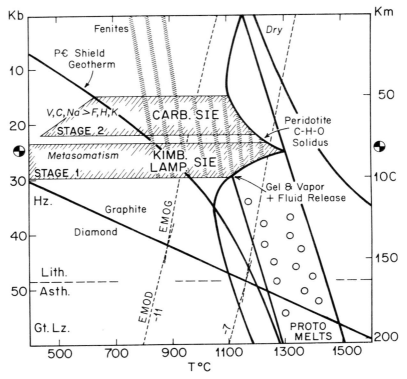

Figure 21.3 *P–T* schematic for the development of metasomes in the interval of ~60–100 km in subcratonic upper mantle lithosphere. Proto-melts originating in the asthenosphere (Asth.) from garnet lherzolite (Gt. Lz.) may gel on intersection with the C–O–H peridotite solidus. Crystallization and fluid and vapour release take place, metasomatizing previously depleted lithosphere (Lith.) which is composed dominantly of harzburgite (Hz). Two metasomes are proposed: a lower horizon from which kimberlite and lamproite SIE signatures are attained; and an upper horizon characterized by calcite. In the latter, vapour (V) + CO_2 (C) + Na_2O (Na) is greater than fluid (F) + H_2O (H) + K_2O (K). Stage 1 and Stage 2 refer to fractional distillation (see text). The 'dry' peridotite solidus is shown to the right of the diagram, and the Precambrian shield geotherm to the left of the diagram. Fenite development from melting of the carbonate metasome is illustrated at low pressures. The near-vertical dashed lines labelled -7, -11, and EMOD-EMOG are oxygen fugacity estimates for the upper mantle (Eggler & Baker 1982), defined by the intersection of two invariant reactions: (1) enstatite + magnesite + carbon = forsterite + vapour; and (2) magnetite + carbon = wüstite + vapour. Other estimates are given in Figure 21.4.

illustrated in Figure 21.3. The position and geometry of the thermal maxima are affected by $CO_2:H_2O$ ratios and chemigraphic interpretation (Eggler 1987a, b, Wyllie 1987), so that the curve illustrated is only one of a family of possibilities. Metasomatism develops by repeated fluxing of fluids from stagnated melts. Given the time-frame over which kimberlites are intruded in restricted regions (~1.6 Ga to 100 Ma), and the antiquity of diamonds (3.2 Ga) with enriched garnet inclusions, melt and fluid release pumping into the lithosphere has seemingly taken place over a long period of time, not continuously but episodically. Adiabatic melts that lead to eruption are only possible if temperatures exceed ~1300 °C at ~75 km (Fig. 21.3), provided that suitable access channels exist or sufficient energy is available for forceful injection; all other melts are impeded by the solidus shoulder. Metasomatism, therefore, is envisaged as being laterally extensive, giving rise to geochemically enriched horizons in the mantle lithosphere defined here as metasomes. Metasome boundaries with depleted lithosphere are unlikely to be sharp because volatile species will migrate along thermal gradients, vertically and horizontally. The less soluble and more refractory components will precipitate closer to the sources of partial melt solidification, while the more soluble and less refractory components will migrate greater distances. This is very likely an oversimplification because elemental solubilities depend also on fluid speciation. Metasomites provide a qualitative insight insofar as those containing carbonate and forming at shallower depths are more enriched in Zr (zircon) and Nb (Nb-Cr-rutile) than those that are carbonate-free (i.e. PKP metasomites). The former also have larger modal contents of amphibole and phlogopite, and are distinguished by the presence of ferro-freudenbergite (Haggerty 1987). Therefore, carbonate metasomes have larger contents of alkalis and volatiles but also $Na + CO_2 > K + H_2O$, relative to PKP metasomes. Injection of melts from the asthenosphere will initiate partial melting of the PKP metasome. These PKP-assimilated or secondary melts will be enriched in SIEs and volatiles but are unlikely to travel great distances because of thermal inertia. The process is perhaps more akin to distillation, and to fractional distillation in particular. A two-stage process for the development of the carbonate metasome is, therefore, proposed. The PKP metasome (Stage 1) results from fluid release of stagnated melts, with moderate fractionation of CO_2 and associated dissolved elements to higher stratigraphic levels, whereas the bulk of the carbonate metasome (Stage 2) results from thermal reworking of the PKP metasome.

21.5 KIMBERLITE AND CARBONATITE GENESIS

Kimberlites are potassium-rich ultramafic rocks with high concentrations of SIEs (e.g. Mitchell 1986). The melt-derived mineralogy is most commonly olivine + phlogopite + spinel + ilmenite + perovskite + calcite + serpentine ± monticellite, clinopyroxene, and apatite. Fragments of upper-mantle-derived xenoliths include lherzolite, harzburgite, dunite, metasomites, and occasional pyroxenites, in association with discrete megacrysts of olivine, clinopyroxene, garnet, spinel, ilmenite, phlogopite, and zircon. Two generations of olivine are standard in the matrix of kimberlites: a coarse-grained xenocrystic population of Fo > 90, of probable origin from depleted lithosphere; and a finer-grained subhedral population that is melt-derived with Fo contents of 80–85. Diamond is an accessory mineral, xenocrystic in origin, and is restricted to cratonic kimberlites. Diamond-bearing olivine lamproites share a common feature with kimberlites in having two generations of olivine, but the melt-derived mineralogy is K–Ti-richterite, Ti-ferriphlogopite, spinel, ilmenite, perovskite, rutile, priderite, and calcite. Bulk compositions are more highly enriched in SIEs, intrusives are into mobile belts, and ages span a range of 1.2 Ga to 22 Ma (Atkinson *et al.* 1984, Jaques *et al.* 1984). Both kimberlite and lamproites are multiple explosive intrusions, and highly carbonated kimberlites are intrusively later than ordinary kimberlite varieties (Fig. 21.1).

Carbonatites are dominated by either calcite (sövite) or dolomite (beforsite) with carbonate minerals > 50 vol%. The most abundant silicates are two generations of olivine in association with clinopyroxene and phlogopite, feldspathoids, perovskite, magnetite, pyrochlore, apatite, ilmenite, and bastnaesite (Kapustin 1980, Hogarth, this volume). With few exceptions, carbonatites are circular alkalic complexes with central cores of sövite or beforsite intruded into marginally concentric ijolite, urtite, jacupirangite, melteigite, and syenite units (e.g. Heinrich 1966). Multiple intrusions are commonly found. Concentrations of SIEs are substantially greater than those of kimberlites and lamproites, with some intrusions attaining ore deposit concentrations of Nb, P, REEs, Ti and Zr. In carbonatites the Na:K ratio is greater than that in kimberlites and lamproites, supporting the proposal for a sodium-enriched carbonate metasome.

Kimberlites are mixtures of products derived from the fertile asthenosphere, the depleted lithosphere, and enriched metasomatized lithosphere. It is argued here and elsewhere (Haggerty 1988) that the SIE signatures of kimberlites and lamproites are derived from PKP and carbonate metasomes (Fig. 21.3) by high-temperature assimilation in melts of asthenospheric origin. Wyllie (1987) favours a zone of metasomatism at the lithosphere–asthenosphere boundary (LAB) for the generation of proto-kimberlite melts. This zone may be represented by glimmerites (Jones 1984), which are composed of sheared phlogopite and accessory clinopyroxene + ilmenite, but lacking in amphibole and the exotic LIL-titanate oxides found in lithospheric PKP and carbonate metasomites. At present little is known about glimmerites but these rocks could be a contributing source for SIE-derived melts underplated at the LAB. That amphibole is absent, places the source of glimmerites at greater depths than PKP metasomites, and shearing is potentially indicative of convective asthenosphere. However, magnesite, which is typically invoked as a deep mantle source for carbonate (e.g. Wyllie 1980), remains elusive.

With reference to Figures 21.1 and 21.2 and discussions in this section, it is noted that carbonate-facies kimberlites post-date carbonate-poor varieties, that carbonatites intrude alkali silicate rock complexes, and that kimberlites (and, therefore, carbonates), are SIE-enriched relative to PKP metasomites. Not shown in Figure 21.2, but quoted in Erlank et al. (1987), are values and distribution patterns for REEs in PKP metasomites and kimberlites. The enriched PKP rocks have lower concentrations of REEs (both are LREE-enriched) than kimberlites, in accord with corresponding distributions for Ba, K, Nb, Rb, Sr, Ti and Zr (Fig. 21.2). It is precisely this array of elements that distinguishes kimberlites (and lamproites) from carbonatites. Following the stratigraphy of metasomes outlined in Figure 21.3, it is proposed that cratonic carbonatites are derived largely from lithospheric metasomes. Lithospheric thinning at the edges of cratons and towards rift zones will result in a thickening of metasomes relative to unaffected depleted lithosphere because of:

(a) enhanced thermal activation and igneous activity;
(b) melt complexing and underplating at the LAB;
(c) melt intersection with the thermal shoulder in the C–O–H peridotite solidus;

and

(d) melt underplating and metasomatism at the lower-crust–upper-mantle boundary.

This regional enhancement is shown schematically in Figure 21.4 with melt migration intrusion paths for kimberlites, lamproites, carbonatites, and alkali rocks through a subcontinental cross-section of craton, mobile belt, and rift zone. The timing of intrusion in single diatremes can be accounted for by early melting of deeper PKP metasomes to form kimberlites and alkalic rocks, and subsequent melting of the shallower carbonate metasome giving rise to later intrusions of carbonated kimberlite and carbonatites.

In this model, the timing and distribution of diamond-bearing kimberlites are understandably a rare event because tapping of the asthenosphere can evidently only

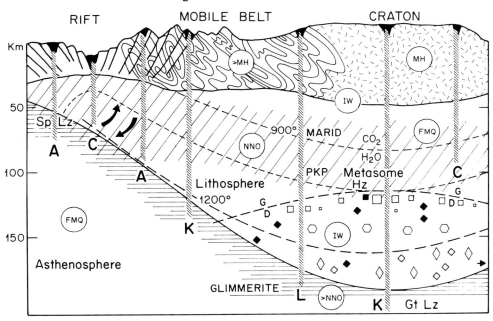

Figure 21.4 Schematic cross-section of subcontinental lithosphere from craton to mobile belt and rift zone. Possible diatreme paths for carbonatites (C), kimberlites (K), lamproites (L), and other alkali-rich (A) melts are shown originating in either the asthenosphere or the lithosphere. Metasomes parallel the lithosphere–asthenosphere boundary and are divided into an upper CO_2-rich, and a deeper H_2O-rich gradational horizon. Curved arrows on lithospheric isotherms indicate possible temperature modifications during active and quiescent phases of activity in the asthenosphere. Note the thickening of metasomes in rift zones inferred from the cumulative effects of solidus intersection and underplating on the lower crust and at the lithosphere–asthenosphere boundary. Symbols below the graphite (G)–diamond (D) stability curve are indicative of diamond morphologies as a function of $T\,°C$ (cubes are at lower T than dodecahedra which are at lower T than octahedra). Abbreviations: Sp Lz, spinel lherzolite; Gt Lz, garnet lherzolite; Hz, harzburgite; MARID, mica–amphibole–rutile–ilmenite–diopside. Circles enclose estimates of oxidation states: MH, magnetite–hematite; IW, iron–wüstite; NNO, nickel–nickel oxide; and FMQ, fayalite–magnetite–quartz.

be accomplished at times of unusually active thermal convection, such as circumstances in the Mesozoic that led to latter-day plate tectonism. Alkalic intrusives and carbonatite complexes are contemporaneous with, or closely follow, such events in rifts and allochogens. This distribution in time and space underscores the relationship between shallow-level, enhanced thermal activity and regional metasomatism. Larger volumes of melts simply produce larger concentrations of SIEs and volatiles by selective fractionation; however, reservoir accumulation is required because single-stage partial melts of fertile asthenosphere are grossly depleted in SIE relative to the abundances present, specifically in carbonatites. Volatile sources remain an issue of debate: some favour deep upper-mantle subduction (e.g. Schultze 1986), whereas others (e.g. Wyllie 1980) call upon primordial degassing. In this chapter the latter is assumed.

Metasome development follows early subcontinental lithospheric depletion. Introduction of fluids and SIEs is a refurbishing and enrichment process, and enrichment is an effective modifier of redox equilibria. Upper-mantle heterogeneities in redox conditions (e.g. Arculus 1985) result, in part, from compositional contrasts between the asthenosphere and the lithosphere (Haggerty & Tompkins 1983) and, in part, from the extent to which these regions have been subsequently reconstituted by metasomatism. Redox estimates for subcontinental mantle lithosphere and adjacent asthenosphere range from equilibria defined by IW (iron–wüstite, i.e. Fe–FeO) to NNO (Ni–NiO), a range of approximately four orders of magnitude that is constrained by the stability of diamond in the lower lithosphere, to carbonates and hydrous minerals in metasomes. Metasomes have evolved with time and upper mantle redox states have, therefore, also evolved.

Finally, in the context of opaque mineral oxides, as outlined in the introduction, and used to determine whether carbonatites and kimberlites are related, it is noteworthy that when comparable crystallization products are considered, a common elemental signature indeed exists. High pyrophanite ($MnTiO_3$) contents are present in carbonatitic, kimberlitic, and lamproitic ilmenites, and high MnO contents are also present in spinel (McMahon & Haggerty 1979, 1984, Gaspar & Wyllie 1984, Jones & Wyllie 1985, Tompkins & Haggerty 1985). The source and behaviour of manganese in the upper mantle are poorly understood, but are undoubtedly related to redox conditions. In common with P, Mn remains elusive in upper-mantle xenoliths and metasomites.

21.6 CONCLUSIONS

A model for the development of laterally extensive metasomatic horizons (metasomes) in the lithospheric upper mantle is supported by:

(a) mineralogy, petrology, mineral chemistry, and bulk chemistry of kimberlites and metasomatized xenoliths (metasomites) in kimberlites;

(b) repeated injections of mantle-derived melts over geologically extended periods of time; and
(c) the fact that an upper mantle source enriched in SIEs and volatiles is required to account adequately for the compositions of carbonatite complexes, kimberlites, lamproites, and related alkali rocks.

Metasomatism develops at restricted zones in the upper mantle and is influenced by a thermal threshold in the C–O–H peridotite solidus at ~75–100 km, and by the discontinuity that is created at the LAB. Melts stagnate and fluid and dissolved solids migrate, enriching previously depleted lithosphere. Metasomes are thickened and more highly fractionated at cratonic boundaries and in rifts because of higher heat flow, greater igneous activity, and through the combined effects of solidus intersection and LAB and lower-crust underplating. Metasomes are stratigraphically zoned with $K + H_2O$ dominant at approximately 75–100 km and $Na + CO_2$ prevalent at depths less than 60 km. If LAB metasomes are present, these are enriched in $K + H_2O$ below cratons, but enriched in $Na + K + H_2O + CO_2$ below rift zones. Mineral repositories for SIEs and volatiles have now been identified in upper mantle metasomites in the form of exotic LIL-titanates, mica, amphibole, and carbonates; previously fictive source regions are, therefore, a reality. High-pressure metasomites share common geochemical and mineral characteristics with fenites surrounding carbonatite complexes, and the similarity extends to olivine lamproites and melilitites. The kinship between carbonatites and kimberlites (and other related alkalic rocks) goes beyond a derivation from the upper mantle – all require the tapping and the assimilation of metasomes. Because the regional development of metasomes differs laterally and vertically, and because melts are initiated at greater depths for kimberlites than carbonatites, differences are expected but vestiges of similarity persist. The preferred spatial distribution of kimberlites in cratons, lamproites in adjacent mobile belts, and carbonatites in rift zones, may be viewed as a progression of increasing mantle metasomatism and metasome fractionation, systematically related to lithospheric thinning, increased heat flow, magma production, stagnation, and reservoir accumulation of SIEs and volatiles. Hence, carbonatite complexes, lamproites, kimberlites, and related alkali rocks are distant relatives rather than brethren, some being more metasomatically mature than others.

ACKNOWLEDGEMENTS

This research was supported by the National Science Foundation under grants EAR83-08297 and EAR86-06496 which are gratefully acknowledged. P. J. Wyllie is thanked for numerous preprints, and P. J. Wyllie, D. Eggler, D. Gold, and K. Bell for discussion. Critical reviews and constructive comments were provided by an anonymous reviewer, R. Vollmer, and K. Bell.

REFERENCES

Arculus, R. J. 1985. Oxidation status of the mantle; Past and present. *Annual Review of Earth and Planetary Sciences* **13**, 75–95.

Atkinson, W. J., F. E. Hughes & C. B. Smith 1984. A review of the kimberlitic rocks of Western Australia. In *Kimberlites* I: *Kimberlites and related rocks*. Proceedings of the Third International Kimberlite Conference, J. Kornprobst (ed.), 195–224. New York: Elsevier.

Bell, K. & J. Blenkinsop 1987. Archean depleted mantle: Evidence from Nd and Sr initial isotopic ratios of carbonatites. *Geochimica et Cosmochimica Acta* **51**, 291–8.

Bell, K., J. Blenkinsop, T. J. S. Cole & D. P. Menagh 1982. Evidence from Sr isotopes for long-lived heterogeneities in the upper mantle. *Nature* **291**, 251–3.

Boyd, F. R. & J. J. Gurney 1986. Diamonds in the African lithosphere. *Science* **232**, 580–4.

Boyd, F. R., R. A. Jones & P. H. Nixon 1983. Mantle metasomatism: The Kimberley dunites. *Carnegie Institution of Washington, Annual Year Book* **82**, 330–6.

Bristow, J. W., C. B. Smith, H. L. Allsopp, S. R. Shee & E. M. N. Skinner 1986. Setting, geochronology and geochemical characteristics of 1600 My kimberlites and related rocks from the Kuruman province, South Africa. 4th International Kimberlite Conference, Extended Abstracts, Geological Society of Australia Abstract Series, No. 16, 112–14.

Clement, C. R., E. M. W. Skinner & B. H. Scott-Smith 1984. Kimberlite redefined. *Journal of Geology* **92**, 223–8.

Currie, K. L. & J. Ferguson 1971. A study of fenitization around the alkaline carbonatite complex at Callander Bay, Ontario, Canada. *Canadian Journal of Earth Science*, **8**, 498–517.

Dawson, J. B. 1980. *Kimberlites and their xenoliths*. Berlin: Springer.

Dawson, J. B. & J. B. Hawthorne 1973. Magmatic sedimentation and carbonatitic differentiation in kimberlite sills at Benfontein, South Africa. *Journal of the Geological Society of London* **129**, 61–85.

Dick, H. J. B. & T. Bullen 1984. Chromian spinel as a petrogenetic indicator in abyssal and alpine-type peridotites and spatially associated lavas. *Contributions to Mineralogy and Petrology* **86**, 54–76.

Donaldson, C. H. & A. M. Reid 1982. Multiple intrusion of a kimberlite dyke. *Bulletin of the Geological Society of South Africa* **85**, 1–12.

Dong, Z., Z. Jianxiong, L. Ol & Z. Peng 1983. Yimengite K (Cr, Ti, Fe, Mg)$_{12}$O$_{19}$ – a new mineral. *Kexue Tongbao (Science Bulletin)* **15**, 932–6.

Eckermann, H. von 1966. Progress of research on the Alnö carbonatite. In *Carbonatites*, O. F. Tuttle & J. Griffins (eds). New York: John Wiley.

Eggler, D. H. 1987a. Discussion of recent papers on carbonated peridotite, bearing on mantle metasomatism and magmatism: an alternative. *Earth and Planetary Science Letters* **82**, 398–400.

Eggler, D. H. 1987b. Discussion of recent papers on carbonated peridotite, bearing on mantle metasomatism and magmatism: final comment. *Earth and Planetary Science Letters* **82**, 403.

Eggler, D. H. & D. R. Baker 1982. Reduced volatiles in the system C–O–H: implications to mantle melting, fluid formation, and diamond genesis. In *High pressure research in geophysics*, S. Akimoto & M. H. Manghnani (eds), 237–50. Tokyo: Centre for Academic Publications.

Erlank, A. J., F. G. Waters, C. J. Hawkesworth, S. E. Haggerty, H. Allsopp, R. S. Rickard & M. Menzies 1987. Evidence for metasomatism in peridotite nodules from the Kimberley pipes, South Africa. In *Mantle metasomatism*, C. J. Hawkesworth & M. Menzies (eds), 221–311. New York: Academic Press.

Fraser, K. J., C. J. Hawkesworth, A. J. Erlank, R. H. Mitchell & B. H. Scott-Smith 1985. Sr, Nd and Pb isotope and minor element geochemistry of lamproites and kimberlites. *Earth and Planetary Science Letters* **76**, 57–70.

Freestone, I. C. & D. L. Hamilton 1980. The role of liquid immiscibility in the genesis of carbonatites – An experimental study. *Contributions to Mineralogy and Petrology* **73**, 105–17.

Gaspar, J. & P. J. Wyllie 1984. The alleged kimberlite–carbonatite relationship: evidence from ilmenite and spinel from Premier and Wesselton Mines and the Benfontein sill, South Africa. *Contributions to Mineralogy and Petrology* **85**, 133–40.

Gittins, J. 1986. Genesis and evolution of carbonatite magmas. *Geological Association of Canada– Mineralogical Association of Canada–Canadian Geophysical Union, Joint Annual Meeting, Ottawa. Program with Abstracts* **11**, 73.

Grey, I. E., I. C. Madsen & S. E. Haggerty 1987. Structure of a new upper mantle magnetoplumbite-type phase, $Ba[Ti_3Cr_4Fe_4Mg]O_{19}$. *American Mineralogist* **72**, 633–6.

Haggerty, S. E. 1983a. The mineral chemistry of new titanates from the Jägersfontein kimberlite, South Africa. Implications for metasomatism in the upper mantle. *Geochimica et Cosmochimica Acta* **47**, 1833–54.

Haggerty, S. E. 1983b. A freudenbergite-related mineral in lower crustal granulites from Liberia. *Neues Jahrbuch für Mineralogie, Abhandlungen* **8**, 375–84.

Haggerty, S. E. 1986. Diamond genesis in a multiply constrained model. *Nature* **320**, 34–8.

Haggerty, S. E. 1987. Metasomatic mineral titanates in upper mantle xenoliths. In *Mantle xenoliths*, P. H. Nixon (ed.), 671–90. New York: Wiley.

Haggerty, S. E. 1988. Upper mantle opaque mineral stratigraphy and the genesis of metasomites and alkali-rich melts. *Proceedings of the Fourth International Kimberlite Conference*, Australian Geological Society, in press.

Haggerty, S. E. & J. J. Gurney 1984. Zircon-bearing nodules from the upper mantle. *EOS. Transactions, American Geophysical Union* **65**, 301. (abstract)

Haggerty, S. E. & L. A. Tompkins 1983. Redox state of Earth's upper mantle from kimberlitic ilmenites. *Nature* **303**, 295–300.

Haggerty, S. E., J. R. Smyth, A. J. Erlank, R. S. Rickard & R. V. Danchin 1983. Lindsleyite (Ba) and mathiasite (K): Two new chromium titanates in the crichtonite series from the upper mantle. *American Mineralogist* **68**, 494–505.

Hamilton, D. L., I. C. Freestone, J. B. Dawson & C. H. Donaldson 1979. Origin of carbonatites by liquid immiscibility. *Nature* **279**, 52–4.

Harrison, W. J. 1981. Partitioning of REE between minerals and coexisting melts during partial melting of a garnet lherzolite. *American Mineralogist* **66**, 242–59.

Heinrich, E. W. 1966. *The geology of carbonatites*. Chicago: Rand McNally.

Irving, A. J. 1980. Petrology and geochemistry of composite ultramafic xenoliths in alkali basalts and implications for magmatic processes within the mantle. *American Journal of Science, Jackson Volume* **280A**, 389–426.

Jaques, A. L., J. D. Lewis, G. P. Gregory, J. Ferguson, B. W. Chappell & M. T. McCulloch 1984. The diamond-bearing ultrapotassic rocks (lamproites) of the West Kimberley region, Western Australia. In *Kimberlites 1: Kimberlites and related rocks*, Proceedings of the Third International Kimberlite Conference, J. Kornprobst (ed.), 225–54. New York: Elsevier.

Jones, A. P. & P. J. Wyllie 1985. Paragenetic trends of oxide minerals in carbonate-rich kimberlites, with new analyses from the Benfontein sill, South Africa. *Journal of Petrology* **26**, 210–22.

Jones, R. A. 1984. *Geochemical and isotopic studies of some kimberlite and included xenoliths from southern Africa*. Ph.D. thesis, University of Leeds, Leeds.

Kapustin, Y. L. 1980. *Mineralogy of carbonatites*. New Delhi: Amerind Publishing.

McMahon, B. M. & S. E. Haggerty 1979. The Oka carbonatite complex: Magnetite compositions and the related role of titanium in pyrochlore. In *Kimberlites, diatremes, and diamonds: their geology, petrology and geochemistry*. Proceedings of the Second International Kimberlite Conference, Vol. 1, F. R. Boyd & H. O. A. Meyer (eds), 382–92. Washington: American Geophysical Union.

McMahon, B. M. & S. E. Haggerty 1984. The Benfontein kimberlite sills: Magmatic reactions and high intrusion temperatures. *American Journal of Science* **284**, 893–941.

Meyer, H. O. A. 1985. Genesis of diamond: a mantle saga. *American Mineralogist* **70**, 344–55.

Mitchell, R. H. 1979. The alleged kimberlite–carbonatite relationship: additional contrary mineralogical evidence. *American Journal of Science* **279**, 570–89.

Mitchell, R. H. 1984. Mineralogy and origin of carbonatite-rich segregations in a composite kimberlite sill. *Neues Jahrbuch für Mineralogie, Abhandlungen* **150**, 185–97.

Mitchell, R. H. 1986. *Kimberlites: mineralogy, geochemistry, and petrology*. New York: Plenum.

Mysen, B. O. & I. Kushiro 1977. Compositional variations of coexisting phases with degree of melting of peridotite in the upper mantle. *American Mineralogist* **62**, 843–56.

Nixon, D. H., N. W. Rogers, I. L. Gibson & A. Grey 1981. Depleted and fertile mantle xenoliths from southern African kimberlites. *Annual Reviews of Earth and Planetary Sciences* **9**, 285–309.

Richardson, S. H., J. J. Gurney, A. J. Erlank & J. W. Harris 1984. Origin of diamonds in old enriched mantle. *Nature* **310**, 180–202.

Robinson, D. N. 1975. Magnetite–serpentine–calcite dykes at Premier Mine and aspects of their relationship to kimberlite and to carbonatite of alkalic carbonatite complexes. *Physics and Chemistry of the Earth* **9**, 61–70.

Rock, N. M. W. 1986. The nature and origin of ultramafic lamprophyres: alnoites and allied rocks. *Journal of Petrology* **27**, 155–96.

Schneider, M. E. & D. H. Eggler 1984. Compositions of fluids in equilibrium with peridotite: Implications for alkaline magmatism–metasomatism. In *Kimberlites 1: Kimberlites and related rocks*, Proceedings of the Third International Kimberlite Conference, J. Kornprobst (ed.), 383–94. New York: Elsevier.

Schultze, D. J. 1986. Calcium anomalies in the mantle and a subducted metaserpentinite origin for diamonds. *Nature* **319**, 483–5.

Shibata, T. & G. Thompson 1986. Peridotites from the mid-Atlantic ridge at 43 °N and their petrogenetic relation to abyssal tholeiites. *Contributions to Mineralogy and Petrology* **93**, 144–59.

Spera, F. J. 1984. Carbon dioxide in petrogenesis. III: Role of volatiles in the ascent of alkaline magma with special reference to xenolith-bearing mafic lava. *Contributions to Mineralogy and Petrology* **88**, 217–32.

Tompkins, L. A. & S. E. Haggerty 1985. Groundmass oxide minerals in the Koidu kimberlite dikes, Sierra Leone, West Africa. *Contributions to Mineralogy and Petrology* **90**, 245–63.

Wendlandt, R. F. & W. J. Harrison 1979. Rare earth partitioning between immiscible carbonate and silicate liquids and CO_2 vapor: results and implications for the formation of light rare earth-enriched rocks. *Contributions to Mineralogy and Petrology* **69**, 409–19.

Woerman, E. & M. Rosenhauer 1985. Fluid phase and the redox state of the Earth's mantle: Extrapolations based on experimental, phase-theoretical and petrological data. *Fortschritte der Mineralogie* **63**, 263–349.

Woolley, A. R. 1986. The distribution of carbonatites in space and time. *Geological Association of Canada–Mineralogical Association of Canada–Canadian Geophysical Union, Joint Annual Meeting, Ottawa. Program with Abstracts* **11**, 147.

Wyllie, P. J. 1980. The origin of kimberlite. *Journal of Geophysical Research* **85**, 6902–10.

Wyllie, P. J. 1987. The genesis of kimberlites and some low-SiO_2 high-alkali magmas. *Proceedings of the Fourth International Kimberlite Conference*, Australian Geological Society, in press.

Zhou, J., G. Yang & J. Zhang 1984. Mathiasite in a kimberlite from China. *Acta Mineralogica Sinica* **9**, 193–200.

22
CARBONATITES, PRIMARY MELTS, AND MANTLE DYNAMICS

D. H. EGGLER

ABSTRACT

Amphibole, amphibole–carbonate, and phlogopite–carbonate peridotite, the volatile-bearing mineral assemblages that exist at the peridotite–H_2O–CO_2 solidus, melt in a regular polybaric trend of primary compositions. Compositions of experimentally determined primary melts can be projected onto the CaO–MgO + FeO–SiO_2 plane with a mantle norm. They reveal that the primary melt trend extends, with increasing pressure, from nephelinites (derived from amphibole peridotite) near the diopside–olivine join through melilitites to dolomite-rich carbonatites (derived from phlogopite–carbonate peridotite). The trend then swings toward olivine-rich compositions, arriving at kimberlite compositions at 55–60 Kb.

Primary carbonate-rich magmas, characterized by high Mg numbers, small but finite silicate components, and modest alkali components, certainly are generated in the mantle. They apparently are rarely erupted, however. Primary melts of 'typical' carbonatite complexes (e.g. Homa Bay, Kenya; Fen, Norway; Kaiserstuhl, F.R.G.) are nephelinitic to melilititic, plotting on or near the primary melt trend at 20–25 Kb pressure (65–80 km). Other mafic to ultramafic alkaline magmas associated with carbonatites (Alnö, Montreal) plot at higher pressures, approaching 50 Kb (160 km). These higher pressures also characterize the off-craton kimberlites of southern Africa that are associated with olivine melilitites and minor carbonatite activity. On-craton kimberlites of southern Africa appear to be derived from highest pressures of about 60 Kb.

Primary melts of 'typical' carbonatite complexes are interpreted to segregate at the boundaries between lithosphere and subcontinental asthenosphere convection cells. These cells represent thinned and partially melted former subcontinental lithospheric material (or lithosphere–asthenosphere mixtures) at temperatures lower than 'normal' asthenosphere. Melt segregation occurs at conditions representing various kinks in the peridotite solidus, depending on temperature and H_2O–CO_2 bulk compositions. If the subcontinental asthenosphere contains 0.3–0.4 wt% H_2O and CO_2, levels characteristic of mildly-depleted OIB mantle, partial melting and fractionation can raise the volatiles to levels expected in carbonatites. Thus no extraordinary fluxing processes are necessary.

Off-craton and on-craton kimberlites plausibly represent primary melts separated from 'normal' asthenosphere near the base of subcontinental lithosphere, the on-craton kimberlites being associated with oldest and thickest lithosphere.

Abbreviations and symbols

Mg number	molar $Mg : (Mg + Fe^{2+})$
CO_2 number	molar $CO_2 : (CO_2 + H_2O)$
l or L	liquid: relatively high-density silicate or silicate–carbonate melt
v or V	vapour (more properly fluid); relatively low-density fluid consisting largely of CO_2 and H_2O with minor amounts of methane and minor to substantial amounts of dissolved silicate components (Eggler 1987c)

22.1 INTRODUCTION

Most carbonatites occur in continental igneous complexes along with alkalic silicate rocks. Such complexes are characteristically associated with and surrounded by fenites. Where both plutonic and volcanic facies of complexes are well exposed (e.g. western Kenya, Le Bas 1977) or where gravity data define hidden portions (e.g. Fen, Ramberg 1973), it appears that mafic alkaline rocks (plutonic ijolites and volcanic nephelinites) are the dominant lithologies of complexes. Carbonatites and differentiated silicate rocks such as phonolites and syenites make up rather small proportions. That is one reason why most petrologists attribute the origin of carbonatites to secondary processes, including crystallization differentiation and liquid immiscibility.

Carbonate-bearing rocks do exist, however, that plausibly have much more direct mantle origin, in particular the kimberlites. Kimberlites have high Mg numbers and contain mantle xenoliths, characteristics commonly attributed to primary or near-primary melts. Experimental petrologists have also shown, over the past 15 years, that dolomite and magnesite can be part of the peridotite mineral assemblage at depths in excess of about 65 km and that peridotite primary melts can be carbonate-rich (e.g. Eggler 1978, Wyllie 1978).

In this paper, relationships will be explored between primary carbonate-rich melts, as defined by experimental petrology, and the melts of typical carbonatite complexes.

22.2 MELTING OF CARBONATED PERIDOTITE

Melting of peridotite–H_2O–CO_2 compositions has been studied in several multi-oxide systems, notably CaO–MgO–SiO_2 (Eggler 1974, 1978, Wyllie & Huang 1976), and $KAlSiO_4$–MgO–SiO_2 (Wendlandt & Eggler 1980), as well as in natural peridotite compositions (Mysen & Boettcher 1975, Green (in Brey *et al.* 1983), Olafsson & Eggler 1983). In addition, Wyllie has published several phase diagrams based on compilations and interpretations (e.g. Wyllie 1978, 1987a, b, this volume, Eggler 1987a, b).

22.2.1 Subsolidus mineralogy

Wyllie (1987a), Green (in Brey *et al.* 1983), and Eggler (1987a) all agree that subsolidus mineralogy of the peridotite–H_2O–CO_2 system consists of facies, with increasing pressure, of amphibole–peridotite, amphibole–dolomite peridotite, and dolomite–phlogopite peridotite (Fig. 22.1). Relatively minor differences concern the pressure ranges of the fields and the nature of the solidus boundaries.

Asthenospheric mantle of ocean-island basalt(OIB)-type contains about 0.05 wt% H_2O and 0.05 wt% CO_2 (§22.4.4). At these levels, the amphibole–dolomite facies will not contain fluid, because volatiles can be contained completely in amphibole and dolomite. The other two facies will contain fluid, but fluid compositions will be buffered (Fig. 22.1), as discussed many times by Eggler and by Wyllie. Fluids are H_2O-rich at high pressures and CO_2-rich at low pressures (Fig. 22.1).

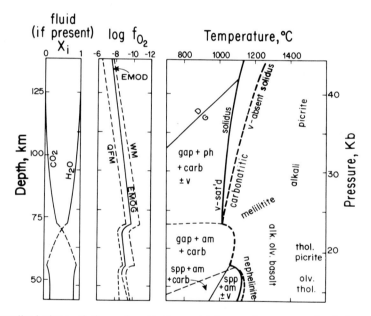

Figure 22.1 Generalized phase relations of peridotite containing small amounts of volatiles, after Olafsson & Eggler (1983) and Eggler (1987c). Diagrams at left show compositions of fluid and f_{O_2}, at the solidus, that are required by subsolidus mineral assemblages in equilibrium with graphite. Fluids were calculated with a modified Redlich–Kwong equation-of-state (Holloway 1981). EMOG/EMOD (enstatite–magnesite–olivine–graphite/ diamond) f_{O_2} is constrained by carbonate-bearing mineral assemblages; f_{O_2}s at pressures below 17 Kb were calculated for the experimentally constrained fluid compositions shown. Supersolidus partial melt compositions are from several sources (see text). Abbreviations: gap and spp, garnet and spinel peridotite (enstatite–diopside–olivine with either garnet or spinel); ph, phlogopite; carb, carbonate; am, amphibole.

22.2.2 Oxidation state and fluids

Oxidation state of the mantle bears upon several petrologic questions, including fluid speciation, partial melting, electrical conductivity, and crustal recycling.

Unfortunately, no consensus exists on oxidation state of various mantle regions. Until about 1980, it was widely held that basalt (and thus mantle) f_{O_2}s are near those of the reference quartz–fayalite–magnetite buffer (e.g. Haggerty 1978) and thus relatively 'oxidized'. Opinion then swung to a 'reduced' view, based on intrinsic f_{O_2} measurements of single-phase xenolithic olivines and spinels (e.g. Arculus & Delano 1981) that placed f_{O_2}s near the iron-wüstite reference buffer. More recently, thermodynamic calculations on multi-phase ilmenite-bearing (Eggler 1983) or spinel-bearing (Mattioli & Wood 1986) peridotite xenoliths have put f_{O_2}s close to or above the magnetite–wüstite (WM) buffer and thus relatively 'oxidized'.

The only evidence that f_{O_2}s may be reduced are the single-phase intrinsic measurements. Inasmuch as such measurements can perturb samples by carbon-induced reduction-exsolution or auto-oxidation (Virgo et al. 1988), it is now unclear how many (if any) such measurements represent either mantle f_{O_2}s or their 1-atm equivalents. In the absence of *positive* evidence for widespread 'reduced' lithospheric f_{O_2}s, it is here assumed that many mantle fluids contain relatively oxidized species (H_2O and CO_2), not reduced species (H_2O and CH_4 with some H_2; Ryabchikov et al. 1981, Eggler & Baker 1982). Relatively 'oxidized' conditions are compatible with, and in fact required by, the peridotite phase diagram (Fig. 22.1); the reference buffer assemblages that include peridotite, carbonate, and graphite (or diamond), such as EMOG (enstatite–magnesite–olivine–graphite) define f_{O_2} near WM at 10–40 Kb pressure. These relatively 'oxidized' conditions are entirely consistent with the existence of graphite or diamond, inasmuch as those minerals are stable over a wide range of f_{O_2}s (e.g. Eggler 1983).

22.2.3 *Compositions of primary melts*

In the effective absence of volatiles, the only melts expected as primary magmas are tholeiitic basalts, alkali olivine basalts, picrites, and komatiites (Basaltic Volcanism Study Project 1981, p. 553). (Even the relatively depleted MORB asthenosphere does contain volatiles. The *effective* absence of volatiles arises from degrees of melting sufficient to dilute volatile effects.) Production of the more exotic (and alkalic) primary magmas results from exotic source regions (mantle metasomatism) or from influence of volatiles at relatively small degrees of melting.

The link between primary basanites–nephelinites and amphibole peridotite was recognized some time ago (Bultitude & Green 1971, Green 1973, Merrill & Wyllie 1975). This link is now attributed to the dominance of the near-solidus melt by an amphibole component (Olafsson & Eggler 1983). At higher degrees of melting, at 15 Kb pressure, melts will move to olivine tholeiitic compositions (Wendlandt & Mysen 1980). Compositions of partial melts from dolomite–amphibole peridotite (between 17.5 and 23 Kb; Fig. 22.1) are not known, but could be carbonatitic. Melts formed at temperatures somewhat above the solidus are likewise unknown, and will depend on bulk CO_2 number. Typically, however, modal amphibole will exceed modal dolomite in such situations, so near-solidus melts will probably grade from nephelinite to melilite nephelinite. Brey & Green (1975, 1976, 1977) demonstrated the plausibility of olivine melilitite as a primary melt at 27 Kb pressure.

Near-solidus melts formed from phlogopite–carbonate peridotite are predicted from simple systems to be carbonatitic. Melts at 30 Kb in phlogopite-absent simple systems (Wyllie & Huang 1976, Eggler 1978) have Ca:Mg ratios somewhat greater than 1; phlogopite–dolomite peridotite would yield near-solidus melts with somewhat higher Mg:Ca ratios (Basaltic Volcanism Study Project 1981) because neither dolomite nor phlogopite is refractory (Holloway & Eggler 1976). With increasing pressure, phlogopite–carbonate peridotite will produce more magnesian melt compositions, in part because carbonate mineralogy changes from dolomite to magnesite at about 32 Kb (Brey *et al.* 1983, Olafsson & Eggler 1983) and in part because melts become picritic (Wendlandt 1984). In fact, high-pressure alkali picrite primary melts occur with or without the influence of volatiles (Jaques & Green 1980). Thus, at 55 Kb, the very magnesian kimberlite composition can be a primary melt (Eggler & Wendlandt 1979).

22.2.4 Primary melts in projection

The display of melting trends in pseudoternary projection has been applied to alkaline mafic magmas by Danchin *et al.* (1975) and McIver & Ferguson (1979). In

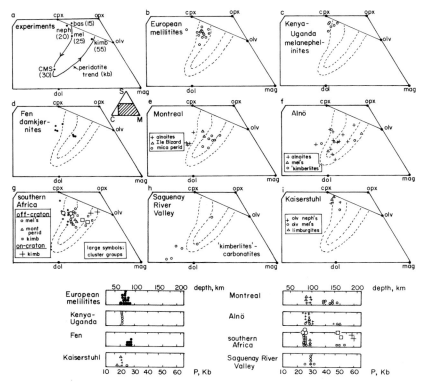

Figure 22.2 Compositions of possible primary melts (high Mg numbers) associated with carbonatites, in CMS projection (Appendix). The calibrating experimental melts include basanite glass (Green 1973), nephelinite (Kakanui kaersutite eclogite of Merrill & Wyllie 1975), olivine melilitite (Brey & Green 1975), the CaO–MgO–SiO$_2$ system (Eggler 1978), and kimberlite (Eggler & Wendlandt 1979). Inferred depths of derivation are summarized in the lower figure.

an approach adopted here, potential projection components, calculated by a mantle norm (Appendix 1), include olivine, pyroxenes, garnet, and phlogopite. Of several projections tested, one that displays a variety of primary mafic alkaline melts is CaO–MgO(+FeO)–SiO$_2$ (CMS). Here the projection is from garnet and phlogopite (as well as ilmenite and apatite), upon the implicit assumption that melts are saturated in those four minerals (the first rule of projections). This scheme is not perfectly suited for basanites and nephelinites, because they plausibly form from amphibole–not phlogopite–peridotite, and saturation in ilmenite or apatite may be questionable. Nevertheless, the scheme seems suited to melilititic and carbonate-rich melts of primary origin. It may be noted that saturation in pyroxenes, olivine, or carbonates is not necessarily assumed, inasmuch as those minerals appear on the CMS plane. Also note that CO_2 or H_2O content of melt is neither used in calculation nor assumed. The CaO and MgO contents of melts that lie on the SiO_2-poor side of the diopside–olivine join (Fig. 22.2a) can be resolved modally into low-pressure melilite or monticellite or high-pressure carbonate.

Experimental melts in projection (Fig. 22.2a) define a polybaric trend representing the peridotite–H_2O–CO_2 solidus. The sweep from the pyroxene–olivine triangle through olivine melilitites toward dolomite composition represents the incoming of subsolidus dolomite and the breakdown of amphibole. The 30–55 Kb sweep back toward kimberlite composition represents the dolomite to magnesite mineralogy change and the increasing picritic nature of high-pressure melts.

22.3 PRIMARY MELTS AND CARBONATITES

22.3.1 *Primary melts: criteria*

The concept of primary melts (in equilibrium with peridotite) was developed in studies of basalt petrogenesis. Candidate silicate rocks should represent near-liquidus melts with appropriate compatible trace-element contents and Mg numbers in excess of about 0.68. These criteria have been applied to mafic silicate rocks associated with carbonatites. Problems with using this approach for carbonatites are several: trace-element data are typically not available; melts are best represented by volcanic rocks, the exception and not the rule in carbonatite complexes; fenitization obscures already complicated geochemical relations; and complexes contain relatively large numbers of cumulate rocks with consequent high Mg numbers. Literature searching is complicated by exotic mineralogy and terminology, a problem somewhat circumvented by the realization that observed low-pressure mineralogies should be expected to be more diverse than hypothesized primary mineralogies.

Application of a Mg number criterion to carbonatites is more difficult than to silicate magmas. Many carbonatites, notably calcitic carbonatites, have low MgO contents and rather high Fe_2O_3 contents due to abundant magnetite; a procedure to massage ferrous/ferric contents to obtain a magmatic Mg number must be rather arbitrary. Moreover, the olivine-melt K_D of 0.30 used for basalt melts (Roeder &

Emslie 1970) may well not apply to carbonatites, even though a nearly identical value, 0.24, applies to the olivine system at 1200 °C. (Olivine melts are more basic than basalt melts, in the sense that they contain more non-bridging oxygens. Olivine melts are therefore somewhat more similar to carbonatites, which have very high ratios of non-bridging to bridging oxygens). Lacking experimental data, a K_D can be calculated for the equilibrium:

$$FeSiO.5O_2 + MgCO_3 = MgSiO.5O_2 + FeCO_3$$

with 298 °K data of Robie et al. (1978). The essentially pressure-independent K_D for Fe–Mg exchange between ideal olivine and carbonate crystalline solutions is 0.27 (1273 °K) to 0.40 (1473 °K). Because the K_D for Fe–Mg exchange between crystalline carbonate and carbonate melt is likely to be near but less than 1, the exchange K_D for olivine–carbonate melt is probably near or somewhat less than 0.30. Thus the same K_D criterion has been applied to both mafic silicate and carbonatite rocks.

22.3.2 *A quest for primary magmas*

A quest for primary magmas in carbonatite literature based on criteria in Section 22.3.1 produced relatively few rocks (Fig. 22.2). Several complexes surveyed (not shown) produced no candidates at all.

European melilitites (Fig. 22.2b) have been included primarily to demonstrate the reasonableness of the projection method. These rocks are near-primary by several geochemical criteria (Alibert et al. 1983) and indeed plot at an expected 20–25 Kb position on the peridotite trend.

The carbonatite–ijolite–nephelinite complexes of western Kenya and eastern Uganda (Le Bas 1977) are well-exposed, partly as volcanics. Melanephelinite and melilitite lithologies constitute 68% of exposed non-fenitized rocks (Le Bas 1977). The majority of exposed melanephelinites and melilitites do not have near-primary Mg numbers. In fact, the weighted average of data from all exposed rocks corresponds to a Mg number = 0.51. Le Bas (1977) took that weighted average, which corresponds to a melanephelinite, to characterize the parent magma. The parent magma is better represented by a small sub-set of melanephelinites and melilitites (Fig. 22.2c) with Mg numbers > 0.64. Carbonatites, which constitute only 3% of the rocks exposed within the complexes, are secondary, formed by liquid immiscibility and crystal fractionation (Le Bas 1977). The non-primary nature of the carbonatites is certainly consistent with relatively low Mg numbers and the position of the carbonatite data in CMS projection.

A similar story emerges from the classic Fen complex (Barth & Ramberg 1966, Griffin & Taylor 1975). Here the only plausible near-primary melts are damkjernites with Mg numbers of 0.61–0.80. These rocks appear to scatter to both sides of a 25-Kb peridotite trend (Fig. 22.2d). Scatter may be due to olivine fractionation or accumulation (which would move compositions to the left or right, respectively) or to the relatively potassic nature of the rocks (§22.3.3). As in Kenya, the average Fen rock is not a plausible parental magma. Indeed, Ramberg (1973) interpreted gravity data to indicate that the major portion of the complex is mafic to ultramafic and buried.

Primary magmas of the Kaiserstuhl, an alkaline complex in the southern Rhine Graben, are olivine nephelinites and olivine melilitites (Keller 1984); limburgites are near-primary but somewhat fractionated. Analyses of such rocks (Wimmenauer 1957, 1963, 1970, Keller 1984), with Mg numbers of 0.66–0.73, plot in the 20–25 Kb range (Fig. 22.2i). Carbonatites of the Kaiserstuhl are small in volume, evolved, and appear in final evolutionary stages (Keller 1984).

The Montreal area contains many alkalic rocks, including the Oka carbonatite complex. Of the analyses listed by Gold *et al.* (1986), none of the Oka rocks appears primary. Groups of alnöites (Mg numbers 0.61–0.70), mica peridotites (Mg numbers 0.77–0.83), and the Ile Bizard aillikite (Mg number = 0.76) are plausibly primary (Fig. 22.2e), with indicated pressures of 25–30 Kb (alnöites) to 38–50 Kb (mica peridotites, in part corrected for olivine fractionation). Note, however, that alnöites and mica peridotites constitute a magmatic series quite separate from the Oka carbonatite complex (Gold *et al.* 1986).

A number of dyke rocks from the Alnö complex (von Eckermann 1948, 1958) have primary Mg numbers. The scattered array in projection (Fig. 22.2f) suggests that the bulk of the rocks were derived at 25–30 Kb, but that a few may be of deeper origin. von Eckermann (1966) contended that the primary mafic melt of the complex was 'kimberlite' or melilitite; a primary alnoite magma was associated in space with the sövites, which represent a magmatic end-product.

The data in Figure 22.2g are from southern African rocks not associated with carbonatite complexes, unlike the rocks discussed above. Rather, they constitute kimberlites from the Kaapvaal Craton and olivine melilititic to 'kimberlitic' rocks, erupted in off-craton mobile belts, that are associated with minor carbonatites. Off-craton occurrences in Namibia and western South Africa include Gibeon (Janse 1975), Gross Brukkaros (Ferguson *et al.* 1975), Saltpetre Kop and Sutherland (McIver & Ferguson 1979), Bitterfontein (McIver 1981), and Garies and Gamoep (Moore & Verwoerd 1985). Also included are cluster-group averages compiled for off-craton and on-craton rocks by Danchin *et al.* (1975) and Ferguson *et al.* (1975). In projection (Fig. 22.2g), olivine melilitites (Mg numbers = 0.67–0.79) and associated cluster groups are plausible mantle melts at 25–28 Kb; on-craton kimberlites (Mg numbers = 0.85–0.86) are plausible melts at 55–60 Kb; off-craton kimberlites (Mg numbers = 0.79–0.86) are plausible melts at 50 Kb; and monticellite peridotites (Mg numbers = 0.72–0.74) are somewhat indeterminate but may represent 30-Kb melts.

Figure 22.2h shows possible (rare) primary carbonatites and 'kimberlites' from the Saguenay River Valley, Quebec (Gittins *et al.* 1975). These rocks, with Mg numbers = 0.67–0.74, may have been derived at about 30 Kb pressure.

22.3.3 *Potassic and sodic rocks*

The experimental melts used to calibrate Figure 22.2 are not potassic ($K:(K + Na) = 0.20–0.25$) except for the kimberlite (0.73). The projection scheme accounts for differences between potassic and sodic rocks only by projecting from phlogopite. It does appear that sodic suites (European melilitites, and samples from Kenya–Uganda and Kaiserstuhl) are less scattered in projection than suites with significant numbers of potassic rocks (K : (K + Na) > 0.5; Montreal, Alnö, Saguenay River Valley, and southern Africa). Yet suites with considerable range in K : (K + Na) (Fen, 0.33–0.62; and Montreal, 0.35–0.75) are reasonably coherent. Coherency does not prove viability of the projection for potassic rocks, but provides no cause for concern.

22.3.4 *Primary carbonatites?*

Based on experimental evidence, it can be concluded that *primary* carbonatites should project near the primary melt trend (Fig. 22.2) and, in particular, should have relatively *high (Mg + Fe) : Ca ratios*, should have *high Mg numbers*, and should contain *finite* amounts of *silicate components* ('S' in CMS projection). Primary carbonatite rocks should also contain at least moderate amounts of *alkalies*, inasmuch as amphibole or phlogopite contribute substantially to the chemistry of near-solidus melts. Primary carbonatite rocks probably also should occur in *isolation* – that is, not associated with extensive suites of allied magma types. Isolation results from the fact that melts, should they rise in the mantle, would encounter ledges in the peridotite solidus (Fig. 22.1). These ledges present a lithospheric thermal structure that encourages interaction with peridotite and freezing, and discourages protracted differentiation processes.

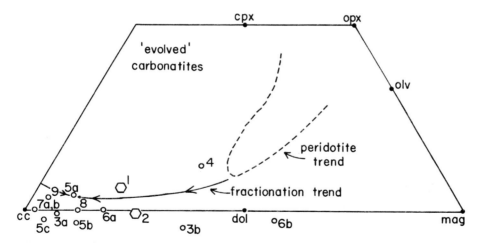

Figure 22.3 Compositions of carbonatites in CMS projection. 1, average carbonatite of Gold (1963); 2, average carbonatite of Heinrich (1966); 3, Fen average sövite (a) and rauhaugite (b) (Saether 1957, Barth & Ramberg 1966); 4, Fort Portal carbonatite lava (von Knorring & DuBois 1961); 5, western Kenya average sövite (a), ferrocarbonatite (b), and alvikite (c) (Le Bas 1977); 6, Oka average sövite (a) and rauhaugite (b) (Gold *et al.* 1986); 7, (a & b), Kerimasi lapilli tuff and porphyritic flow (Mariano & Roeder 1983); 8, Gibeon carbonatite dyke (Janse 1975); 9, Gardiner complex sövite (Nielsen 1980). A fractionation trend was drawn on the assumption that carbonate crystallization differentiation (with diopside and phlogopite but with or without olivine) would proceed to a temperature minimum at the same Ca : Fm ratio as the $CaCO_3$–$MgCO_3$ system at 5 Kb pressure (Irving & Wyllie 1975).

The quest reflected in Figure 22.2 produced virtually no primary carbonatite (carbonate-rich) rocks from 'typical' carbonatite complexes. In fact, the primary magmas of typical carbonatite complexes are olivine melilitites or nephelinites (Fig. 22.2). This conclusion follows that of virtually all carbonatite petrologists (e.g. Le Bas 1977). Typical carbonatites have high $Ca:(Mg + Fe)$ and low silicate components (Fig. 22.3). Among the rare candidates for magmas that contain relatively *large* amounts of carbonate or potential carbonate component may be the Saguenay River Valley carbonatites and possibly the Fort Portal, Uganda carbonatite lavas (Fig. 22.3) with Mg numbers = 0.59–0.60 (Knorring & DuBois 1961).

By contrast, reasonable candidates do exist for primary magmas containing *small* amounts of carbonate or potential carbonate component – the kimberlites and mica peridotites, which fulfil the criteria listed above.

22.3.5 Alkali carbonatites

The only high-alkali extrusive carbonatites occur at Oldoinyo Lengai, Tanzania. Dawson *et al.* (1987) have interpreted older flows at Oldoinyo Lengai to represent alkali carbonatite partially converted to calcite carbonatite by weathering. Conversely, they suggest that many calcitic carbonatites were also alkalic at the magmatic stage. Such reinterpretations do not bear on the question of primary magmas, however. Both the altered and unaltered Oldoinyo Lengai rocks have negligible SiO_2 content and very low $(Mg + Fe):Ca$ and do not appear to be primary.

22.4 MANTLE DYNAMICS

22.4.1 Mantle solidus revisited

The mantle solidus was discussed in Section 22.2.1 with regard to primary melt compositions. The solidus also provides, however, a basis in phase equilibria for discussion of mantle upwelling and magma separation.

The same phase diagram shown in Figure 22.1 appears in Figure 22.4 with two additions. One addition is solidus segment H_2–I_6, an important reaction for peridotite at relatively low temperatures and high CO_2 numbers (Eggler 1987a). The nature of that segment can be seen in the 1150 °C isothermal section in Figure 22.5b.

The second addition is a 'T-max'. In brief, a 'T-max' represents the locus in P-T-X space at which partitioning of CO_2 and H_2O between melt and fluid changes. Thus along segment H_1–N (Fig. 22.4a), the CO_2 number of melt exceeds the CO_2 number of fluid, but along segment H_2–I_6, the CO_2 number of fluid exceeds the CO_2 number of melt. A 'T-max' has not appeared previously in Eggler's diagrams, but Wyllie (e.g. 1987b) is certainly correct that CO_2 numbers of carbonate-rich melts can exceed CO_2 numbers of fluids, at least for H_2O-rich fluids. Here the 'T-max' terminates along the carbonation-melting reaction N–H_1–H_2–I_6. This type of 'T-max', suggested by E. Stolper (pers. comm.), implies that CO_2–H_2O partitioning is a function both of melt composition and of CO_2 number. Along H_1–N fluids are

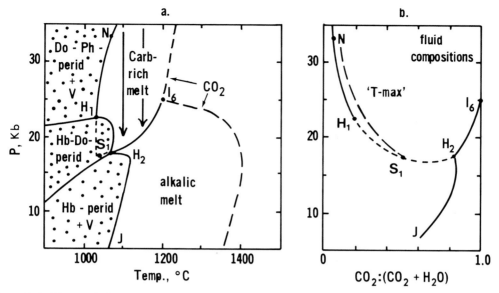

Figure 22.4 Phase relations of peridotite, modified from Eggler (1987a). Additional abbreviations: Hb, amphibole; Do, dolomite. (a) Melting relations. $N-H_1-S_1-H_2-I_6$ is a solidus reaction involving melting of carbonated mineral assemblages. It is metastable between H_1 and H_2, inasmuch as no fluid exists due to amphibole stability. Solidus segment H_2-I_6 applies only to CO_2-rich melts coexisting with dolomite; it represents the 18.5 Kb isobaric line at 1100 °C and the 21 Kb line at 1150 °C in Figure 22.5b. A 'T-max' (see text) terminates at the metastable point S_1. Arrows show the upwelling paths illustrated in Figure 22.5. (b) Compositions of fluids along solidi of (a) and the 'T-max', in part from Olafsson & Eggler (1983) but largely estimated.

H_2O-rich and melts are carbonate-rich, factors that favour a higher CO_2 number in melt. At H_2, fluids are CO_2 rich and melts are not carbonate-rich, factors that favour higher CO_2 number in the fluid. The point of changeover and of termination of the 'T-max', S_1, is actually metastable in Figure 22.4a.

By contrast with Figure 22.4, the 'T-max' of Wyllie (e.g. 1978) terminates on the high-temperature side of I_6, and therefore the carbonation–melting reaction analogous to $N-H_1-H_2-I_6$ enters I_6 at a much different angle. These differences show up rather dramatically in isothermal sections (Fig. 22.5). An 1150 °C section intercepts Wyllie's 'T-max', producing a solidus that sweeps across the diagram (Fig. 22.5a), but an 1150 °C section misses Eggler's 'T-max' (Fig. 22.4a), producing a solidus only at high CO_2 numbers (Fig. 22.5b). [Fig. 22.5a was constructed from Wyllie's projections in his 1978 and 1987a papers for the 'Green' amphibole stability. Rather different isothermal sections would result from the versions for 'MIW' amphibole stability or 'overlapping' amphibole stability. Figure 22.5a appears, however, to be roughly compatible with an unpublished set of projections that was used to construct Figure 20.17 (Wyllie, this volume).]

22.4.2 Lithosphere–asthenosphere boundary

Wendlandt & Morgan (1982) discovered a correlation between chemistry of presumed primary magmas of East Africa (interpreted as depth of origin) and age.

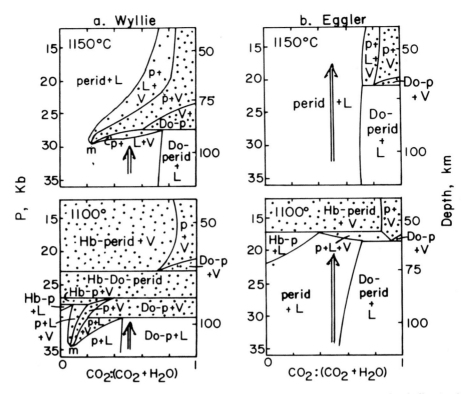

Figure 22.5 Isothermal sections for peridotite compositions containing small amounts of volatiles (< 0.4 wt% H_2O and < 5 wt% CO_2). These sections show the approximate phase relations of upwelling peridotite, and the stippled regions represent petrologic barriers to upwelling; widely spaced stipples show peridotite containing melt and fluid, and closely spaced stipples show subsolidus peridotite. The abscissa is bulk CO_2 number, not fluid composition. Abbreviations as for Figure 22.4, except that p or perid = peridotite. (a) Sections constructed from phase relations of Wyllie (1978, 1987a) with 'Green' amphibole stability. Points m are on the 'T-max'. (b) Sections consistent with phase relations of Figure 22.4.

They proposed that magmas are either generated or segregated at the lithosphere–asthenosphere boundary (LAB), and that age reflects progressive lithospheric thinning.

The Wendlandt–Morgan (1982) hypothesis is attractive in the present context of carbonatitic magmatism. It has been established that parental melts of 'typical' carbonatite complexes plausibly separate at pressures of 20–25 Kb (Fig. 22.2), defining a LAB of 65–80 km depth. Magmatism involving alnöites, mica peridotites, monticellite peridotites, and off-craton 'kimberlites' describes a range of LAB depths (Fig. 22.2) from 80 to 160 km (Montreal, Alnö, and southern Africa). A third type of magmatism, the diamondiferous on-craton kimberlites, defines a LAB of at least 190 km (Fig. 22.2g) and perhaps as much as 550 km, based on included high-pressure garnets (Moore & Gurney 1985).

In the simplest view, the LAB represents a thermal–mechanical boundary. No compositional differences separate lithosphere and asthenosphere, and the LAB can be taken to be essentially equivalent to the peridotite solidus (1050–1200 °C).

Studies of kimberlite-entrained xenoliths show, however, that the bulk of subcontinental lithospheric upper mantle (SCLUM) is more refractory than assumed asthenospheric peridotite, and heavy isotope studies show that SCLUM has existed as a distinct reservoir since Proterozoic or Archaean time. At least part of the LAB must represent, therefore, a *compositional* boundary.

22.4.3 Primary melt separation

Given the variable nature of the LAB, it follows that primary melts can form or segregate at the LAB in different ways (Fig. 22.6).

On-craton and off-craton kimberlites plausibly represent pockets or layers of magma concentrated upon arrest of upwelling plumes at the compositional LAB (Fig. 22.6c, d). Alternatively, magma may separate from peridotite below the LAB and either percolate to the LAB or rise diapirically to the LAB. Kimberlite magmas probably exist at temperatures in excess of 1500 °C, at least in the lower lithosphere (Eggler 1988), and the Group I kimberlites originate in asthenospheric sources (Smith 1983).

Tholeiitic or alkali olivine basaltic volcanism in continental rifts (Fig. 22.6a) probably represents simple lithospheric thinning. Heavy isotope signatures typically are asthenospheric, and eruption temperatures of *c.* 1250 °C are consistent with plume temperatures.

By contrast, nephelinitic–carbonatitic volcanism cannot represent simple thinning. First, isotopic studies demonstrate that magmas parental to carbonatites are generated either entirely within SCLUM, in the case of the Canadian Shield (Bell & Blenkinsop 1987a), or from a mixture of continental lithosphere and asthenosphere, as in East Africa (Bell & Blenkinsop 1987b). Secondly, although 'typical' primary magmas are generated at 65–80 km depth, magmas associated with carbonatites apparently come from a variety of depths. Thirdly, primary nephelinite, melilitite,

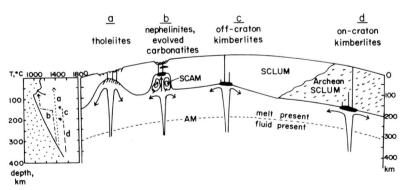

Figure 22.6 Cartoon, after Eggler (1988), illustrating asthenospheric upflow beneath (a) a rift characterized by tholeiitic or alkali olivine basaltic volcanism; (b) a rift characterized by nephelinitic–carbonatitic volcanism; (c) Proterozoic unrifted lithosphere; and (d) thick Archaean lithosphere. The secondary SCAM (subcontinental asthenospheric mantle) convection cell in (b) represents former subcontinental lithospheric upper mantle (SCLUM), or mixtures of SCLUM and asthenospheric mantle (AM), that has been thinned and partially melted. Diagram at left illustrates possible ascent paths of asthenosphere in the four cases, with reference to the solidus (Figure 22.4a).

or carbonatite magmas must separate from peridotite at rather low temperatures (< 1250 °C; Fig. 22.1).

It is possible that the three conditions stated above could be met by models that involve melting of variable portions of SCLUM metasomatic veins and refractory SCLUM keel (e.g. Meen et al., this volume). In such models, convective asthenosphere acts as a heat source, without significant mixing occurring.

However, the relatively consistent depth relations of melt generation (Fig. 22.2) suggest another possibility (Fig. 22.6b). This possibility is a variation on 'convective partial melting' (Mutter et al. 1988), which refers to secondary flow induced by continent–ocean temperature contrasts produced by rifting. Here it is assumed that flow occurs in (former) SCLUM that has been thinned and partially melted (subcontinental asthenospheric mantle or SCAM), or in a mixture of SCAM and 'normal' asthenospheric mantle. In fact, SCAM should be a highly heterogeneous mixture on all size scales of metasomatic SCLUM, refractory SCLUM, and AM. Pieces of metasomatic SCLUM may be preferentially melted (see Meen et al., this volume) or, in the case of Canadian carbonatites (Bell & Blenkinsop 1987a), a more homogeneous SCAM mix may be melted. The SCAM cells are separated from asthenosphere by a thermal boundary layer (Fig. 22.6) and also should be stabilized by their somewhat lower densities due to refractory chemistry. The SCAM model accounts for the SCLUM isotopic characteristics of carbonatites and for the relatively low temperatures required.

Separation of primary melts from SCAM will occur at the LAB, which here (Fig. 22.6b) represents the solidus. The solidus has three segments – relatively vertical segments H_1–N and J–H_2, and a kinked segment H_2–H_1 (Fig. 22.4a). Separation of relatively low-temperature melts at pressures of H_1 to N should produce relatively small amounts of primary carbonatites–mica peridotites (Fig. 22.2a). These low-temperature melts are infrequently erupted because they interact with peridotite upon ascent to shallower depths and thermally die.

Primary nephelinite production probably occurs when convection impinges upon solidus segments H_1–H_2 or H_2–I_6 (Figs 22.4a & 22.6). As developed in Section 22.4.1, the nature of the H_1–H_2–I_6 barrier can be appreciated from isothermal sections (Fig. 22.5). Depending upon temperature and CO_2 number, upwelling mantle or upward-rising melts will tend to arrest at 17.5–25 Kb (at the thickly stippled solidus fields, Fig. 22.5b), partially to arrest by crystallization attendant on fluid evolution (in the thinly stippled peridotite–liquid–vapour fields, Fig. 22.5b), or to avoid the barrier (within the non-stippled peridotite–liquid field, Fig. 22.5b). It is apparent that the 'barrier' is in fact semi-permeable. Melts escaping from arrested diapirs, at 17.5–25 Kb, will be nephelinitic to melilititic and relatively CO_2-rich due to the nature of the arresting process. Such melts are proposed to be parental to 'typical' carbonatite complexes. Transport of melts through lithosphere via cracking will be enhanced by fluid evolution.

Wyllie's (1987a) peridotite solidus presents an analogous barrier or ledge to asthenospheric flow (Wyllie, this volume). In many respects these ledges are nearly identical. Principal differences in the Wyllie barrier that can be appreciated from isothermal sections (Fig. 22.5a), especially the 1150 °C sections, are that the barrier occurs at higher pressures and blankets a wider range of CO_2 numbers. In other words, the Wyllie barrier halts upwelling asthenosphere (or magma that interacts with lithosphere) at deeper levels and over a wider compositional variation. Another petrologic (and possibly undesirable) consequence of Wyllie's model is that, because diapirs are arrested at greater depths (in excess of 25 Kb), primary melts should tend to be more carbonatitic than melilititic or nephelinitic.

22.4.4 Volatiles and lithospheric dynamics

The CO_2 contents of carbonatites are sometimes regarded as extraordinarily high, necessitating an extraordinary process in their origin such as metasomatism or volatile fluxing (i.e. volatiles are decoupled from 'normal' asthenosphere and concentrated).

A useful exercise is to consider SCAM mantle (Fig. 22.6) as analogous to the mildly 'depleted' OIB mantle. In fact, Bell & Blenkinsop (1987a) have pointed out isotopic similarities between OIB and sub-Canadian mantles. The best estimates of volatile contents of OIB mantle come from Hawaiian tholeiitic volcanism. Greenland et al. (1985) have estimated 0.32 wt% H_2O and 0.32 wt% CO_2 in primitive magma from gas emissions and lava volumes. Harris & Anderson (1983) independently estimate 0.46 wt% H_2O and 0.31 wt% CO_2 from melt inclusion data. If tholeiites are produced by 15–25% melting, then the CO_2 content of mantle is 0.05–0.08 wt%. In turn, if that mantle is melted 0.3–2% to produce alkaline magmas, these magmas can contain 2.5–25 wt% CO_2. Reasonable estimates of CO_2 contents of nephelinites or melilitites are 4–8 wt%. Igneous fractionation or immiscibility processes can easily raise those contents by a factor of 5–10 to the contents of carbonatites.

In short, volatiles in carbonatites can easily be accounted for if only mildly 'depleted' peridotite source regions are assumed. No magmatic precursory events that decouple volatiles from more refractory peridotite elements are necessary. In fact, the CO_2 enrichment factors of carbonatites (sample:mantle) are less than those of some trace elements!

22.5 CONCLUSIONS

(a) Primary carbonatites (carbonate-rich melts) should have high Mg numbers, should contain finite amounts of silicate components, should contain alkalies, and should typically occur in magmatic isolation. They should separate from peridotite source regions at pressures of 25–40 Kb. Few, if any, carbonatites fulfil these criteria.

(b) Primary magmas of 'typical' carbonatite complexes are nephelinitic to melilititic, plausibly representing derivation or segregation at 20–25 Kb pressure (65–80 km). Magmas containing small amounts of carbonate or potential carbonate (e.g. mica peridotites and kimberlites) represent derivation at pressures to 60 Kb.

(c) Segregation of primary magmas of 'typical' carbonatite complexes also occurs at relatively low temperatures from subcontinental lithospheric material or from mixed lithosphere–asthenosphere material. These characteristics may represent separation from secondary convection cells developed within rifting lithosphere.

ACKNOWLEDGEMENTS

This paper is a quite different approach to carbonatites than that which I presented at the 1986 Ottawa meeting. Keith Bell has been supportive in these efforts and has been an extremely patient editor. Research is supported by National Science Foundation, Earth Sciences Section, grant EAR-8618919.

REFERENCES

Alibert, C., A. Michard & F. Albarede 1983. The transition from alkali basalts to kimberlites: isotope and trace element evidence from melilitites. *Contributions to Mineralogy and Petrology* **82**, 175–86.

Arculus, R. J. & J. W. Delano 1981. Intrinsic oxygen fugacity measurements: techniques and results for spinels from upper mantle peridotites and megacryst assemblages. *Geochimica et Cosmochimica Acta* **45**, 899–913.

Barth, T. F. W. & I. G. Ramberg 1966. The Fen circular complex. In *Carbonatites*, O. F. Tuttle & J. Gittins (eds), 225–57. New York: Interscience.

Basaltic Volcanism Study Project (BVSP) 1981. *Baslatic volcanism on the terrestrial planets*. Oxford: Pergamon.

Bell, K. & J. Blenkinsop 1987a. Archean depleted mantle: Evidence from Nd and Sr initial isotopic ratios of carbonatites. *Geochimica et Cosmochimica Acta* **51**, 291–8.

Bell, K. & J. Blenkinsop 1987b. Nd and Sr isotopic compositions of East African carbonatites: implications for mantle heterogeneity. *Geology* **15**, 99–102.

Brey, G. & D. H. Green 1975. The role of CO_2 in the genesis of olivine melilitite. *Contributions to Mineralogy and Petrology* **49**, 93–103.

Brey, G. & D. H. Green 1976. Solubility of CO_2 in olivine melilitite at high pressures and role of CO_2 in the earth's upper mantle. *Contributions to Mineralogy and Petrology* **55**, 217–30.

Brey, G. & D. H. Green 1977. Systematic study of liquidus phase relations in olivine melilitite–H_2O + CO_2 at high pressures and petrogenesis of an olivine melilitite magma. *Contributions to Mineralogy and Petrology* **61**, 141–62.

Brey, G., W. R. Brice, D. J. Ellis, D. H. Green, K. L. Harris & I. D. Ryabchikov 1983. Pyroxene-carbonate relations in the upper mantle. *Earth and Planetary Science Letters* **62**, 63–74.

Bultitude, R. J. & D. H. Green 1971. Experimental study of crystal–liquid relationships at high pressures in olivine nephelinite and basanite compositions. *Journal of Petrology* **12**, 121–47.

Danchin, R. V., J. Ferguson, J. R. McIver & P. H. Nixon 1975. The composition of late stage kimberlite liquids as revealed by nucleated autoliths. *Physics and Chemistry of the Earth* **9**, 235–46.

Dawson, J. B., M. S. Garson & B. Roberts 1987. Altered former alkalic carbonatite lava from Oldoinyo Lengai, Tanzania: inferences for calcite carbonatite lavas. *Geology* **15**, 765–8.

Eckermann, H. von 1948. The alkaline district of Alnö Island. *Sveriges Geologiska Undersökning*, Series Ca. No. 36.

Eckermann, H. von 1958. The alkaline and carbonatitic dikes of the Alnö Formation of the mainland north-west of Alnö Island. *Kungliga Svenska Vetenskapakademien Handlingar*, Fjarde serien, v. 7, no. 2.

Eckermann, H. von 1966. Progress of research on the Alnö carbonatite. In *Carbonatites*, O. F. Tuttle & J. Gittins (eds), 3–32. New York: Interscience.

Eggler, D. H. 1974. Effect of CO_2 on the melting of peridotite. *Carnegie Institution of Washington Yearbook* **73**, 215–24.

Eggler, D. H. 1978. The effect of CO_2 upon partial melting of peridotite in the system Na_2O–CaO–Al_2O_3–MgO–SiO_2–CO_2 to 35 kb, with an analysis of melting in a peridotite–H_2O–CO_2 system. *American Journal of Science* **278**, 305–43.

Eggler, D. H. 1983. Upper mantle oxidation state: Evidence from olivine–orthopyroxene–ilmenite assemblages. *Geophysical Research Letters* **10**, 365–8.

Eggler, D. H. 1987a. Discussion of recent papers on carbonated peridotite, bearing on mantle metasomatism and magmatism: an alternative. *Earth and Planetary Science Letters* **82**, 398–400.

Eggler, D. H. 1987b. Discussion of recent papers on carbonated peridotite, bearing on mantle metasomatism and magmatism: final comment. *Earth and Planetary Science Letters* **82**, 403.

Eggler, D. H. 1987c. Solubility of major and trace elements in mantle metasomatic fluids: experimental constraints. In *Mantle metasomatism*, M. A. Menzies & C. J. Hawkesworth (eds), 21–41. London: Academic Press.

Eggler, D. H. 1988. Kimberlites: how do they form? Proceedings of the Fourth International Kimberlite Conference, Australian Geological Society, in press.

Eggler, D. H. & D. R. Baker 1982. Reduced volatiles in the system C–O–H: implications to mantle melting, fluid formation, and diamond genesis. In *Advances in Earth and planetary sciences. High-pressure research in geophysics*, S. Akimoto & M. H. Manghnani (eds), 237–50. Tokyo: Center for Academic Publications.

Eggler, D. H. & R. F. Wendlandt 1979. Experimental studies on the relationship between kimberlite magmas and partial melting of peridotite. In *The mantle sample: inclusions in kimberlites and other volcanics*. Proceedings of the Second International Kimberlite Conference, Vol. 2, F. R. Boyd & H. O. A. Meyer (eds), 213–26. Washington: American Geophysical Union.

Ferguson, J., H. Martin, L. O. Nicolaysen & R. V. Danchin 1975. Gross Brukkaros: a kimberlite–carbonatite volcano. *Physics and Chemistry of the Earth* **9**, 219–34.

Gittins, J., R. H. Hewins & A. F. Laurin 1975. Kimberlitic–carbonatitic dikes of the Saguenay River Valley, Quebec, Canada. *Physics and Chemistry of the Earth* **9**, 137–48.

Gold, D. P. 1963. Average chemical composition of carbonatites. *Economic Geology* **58**, 988–91.

Gold, D. P., G. N. Eby, K. Bell & M. Vallée 1986. Carbonatites, diatremes, and ultra-alkaline rocks in the Oka area, Quebec. *Geological Association of Canada–Mineralogical Association of Canada–Canadian Geophysical Union, Joint Annual Meeting, Ottawa, Field Trip 21, Guidebook*.

Green, D. H. 1973. Conditions of melting of basanite magma from garnet peridotite. *Earth and Planetary Science Letters* **17**, 456–65.

Greenland, L. P., W. I. Rose & J. B. Stokes 1985. An estimate of gas emissions and magmatic gas content from Kilauea volcano. *Geochimica et Cosmochimica Acta* **49**, 125–30.

Griffin, W. L. & P. N. Taylor 1975. The Fen damkjernite: petrology of a "central-complex kimberlite". *Physics and Chemistry of the Earth* **9**, 163–78.

Haggerty, S. E. 1978. The redox state of planetary basalts. *Geophysical Research Letters* **5**, 443–6.

Harris, D. M. & A. T. Anderson, Jr 1983. Concentrations, sources, and losses of H_2O, CO_2, and S in Kilauean basalt. *Geochimica et Cosmochimica Acta* **47**, 1139–50.

Heinrich, E. W. 1966. *The geology of carbonatites*. Chicago: Rand McNally.

Holloway, J. R. 1981. Compositions and volumes of supercritical fluids in the Earth's crust. In *Fluid inclusions: applications to petrology*, L. S. Hollister & M. L. Crawford (eds), 13–38. Calgary: Mineralogical Association of Canada.

Holloway, J. R. & D. H. Eggler 1976. Fluid-absent melting of peridotite containing phlogopite and dolomite. *Carnegie Institution of Washington Yearbook* **75**, 636–9.

Irving, A. J. & P. J. Wyllie 1975. Subsolidus and melting relationships for calcite, magnesite, and the join $CaCO_3$–$MgCO_3$ to 36 kilobars. *Geochimica et Cosmochimica Acta* **39**, 35–53.

Janse, A. J. A. 1975. Kimberlite and related rocks from the Nama Plateau of South-West Africa. *Physics and Chemistry of the Earth* **9**, 81–94.

Jaques, A. L. & D. H. Green 1980. Anhydrous melting of peridotite at 0–15 kb pressure and the genesis of tholeiitic basalts. *Contributions to Mineralogy and Petrology* **73**, 287–310.

Keller, J. 1984. Geochemie und magmenentwicklung im Kaiserstuhl. *Fortschritte der Mineralogie* **62**, 116–18.

Knorring, O. von & C. G. B. DuBois 1961. Carbonatitic lava from Fort Portal area in western Uganda. *Nature* **192**, 1064–5.

Le Bas, M. J. 1977. *Carbonatite–nephelinite volcanism*. London: Wiley.

McIver, J. R. 1981. Aspects of ultrabasic and basic alkaline intrusive rocks from Bitterfontein, South Africa. *Contributions to Mineralogy and Petrology* **78**, 1–11.

McIver, J. R. & J. Ferguson 1979. Kimberlitic, melilititic, trachytic and carbonatite eruptives at Saltpetre Kop, Sutherland, South Africa. In *Kimberlites, diatremes, and diamonds: their geology, petrology, and geochemistry*. Proceedings of the Second International Kimberlite Conference, Vol. 1, F. R. Boyd & H. O. A. Meyer (eds), 111–28. Washington: American Geophysical Union.

Mariano, A. N. & P. L. Roeder 1983. Kerimasi: a neglected carbonatite volcano. *Journal of Geology* **91**, 449–55.

Mattioli, G. S. & B. J. Wood 1986. Upper mantle oxygen fugacity recorded by spinel lherzolites. *Nature* **322**, 626–9.

Merrill, R. B. & P. J. Wyllie 1975. Kaersutite and kaersutite eclogite from Kakanui, New Zealand – water-excess and water-deficient melting relations to 30 kilobars. *Geological Society of America Bulletin* **86**, 555–70.

Moore, A. E. & W. J. Verwoerd 1985. The olivine melilitite–'kimberlite'–carbonatite suite of Namaqualand and Bushmanland, South Africa. *Transactions of Geological Society of South Africa* **88**, 281–94.

Moore, R. O. & J. J. Gurney 1985. Pyroxene solid solution in garnets included in diamond. *Nature* **318**, 553–5.

Mutter, J. C., W. R. Buck & C. M. Zehnder 1988. Convective partial melting I. A model for the formation of thick basaltic sequences during the initiation of spreading. *Journal of Geophysical Research* **93**, 1031–48.

Mysen, B. O. & A. L. Boettcher 1975. Melting of a hydrous mantle: I. Phase relations of natural peridotite at high pressures and temperatures with controlled activities of water, carbon dioxide, and hydrogen. *Journal of Petrology* **16**, 520–48.

Nielsen, T. F. D. 1980. The petrology of a melilitolite, melteigite, carbonatite and syenite ring dike system, in the Gardiner complex, East Greenland. *Lithos* **13**, 181–97.

Olafsson, M. & D. H. Eggler 1983. Phase relations of amphibole, amphibole-carbonate, and phlogopite-carbonate peridotite: petrologic constraints on the asthenosphere. *Earth and Planetary Science Letters* **64**, 305–15.

Ramberg, I. B. 1973. Gravity studies on the Fen Complex, Norway, and their petrological significance. *Contributions to Mineralogy and Petrology* **38**, 115–34.

Robie, R. A., B. S. Hemingway & J. R. Fisher 1978. Thermodynamic properties of minerals and related substances at 298.15 °K and 1 bar pressure and at higher temperatures. *United States Geological Survey Bulletin*, No. 1452.

Roeder, P. L. & R. F. Emslie 1970. Olivine–liquid equilibrium. *Contributions to Mineralogy and Petrology* **29**, 275–89.

Ryabchikov, I. D., D. H. Green, V. J. Wall & G. Brey 1981. The oxidation state of carbon in the environment of the low velocity zone. *Geokhimiya* **2**, 221–32.

Saether, E. 1957. The alkaline rock province of the Fen area in southern Norway. *Det Kongelige Norske Videnskabers Selskab Skrifter* No. 1.

Smith, C. B. 1983. Pb, Sr, and Nd isotopic evidence for sources of southern African Cretaceous kimberlites. *Nature* **304**, 51–4.

Virgo, D., R. W. Luth, M. A. Moats & G. C. Ulmer 1988. Constraints on the oxidation state of the mantle: an electrochemical and ^{57}Fe Mossbauer study of mantle-derived ilmenites. *Geochimica et Cosmochimica Acta*, in press.

Wendlandt, R. F. 1984. An experimental and theoretical analysis of partial melting in the system $KAlSiO_4-CaO-MgO-SiO_2-CO_2$ and applications to the genesis of potassic magmas, carbonatites and kimberlites. In *Kimberlites I: Kimberlites and related rocks*. Proceedings of the Third International Kimberlite Conference, J. Kornprobst (ed.), 359–70. New York: Elsevier.

Wendlandt, R. F. & D. H. Eggler 1980. The origins of potassic magmas: 2. Stability of phlogopite in natural spinel lherzolite and in the system $KAlSiO_4-MgO-SiO_2-H_2O-CO_2$ at high pressures and high temperatures. *American Journal of Science* **280**, 421–58.

Wendlandt, R. F. & P. Morgan 1982. Lithospheric thinning associated with rifting in East Africa. *Nature* **298**, 734–6.

Wendlandt, R. F. & B. O. Mysen 1980. Melting phase relations of natural peridotite + CO_2 as a function of degree of partial melting at 15 and 30 Kbar. *American Mineralogist* **65**, 37–44.

Wimmenauer, E. 1957. Beitrage zur petrographie des Kaiserstuhls. *Neues Jahrbuch für Mineralogie. Abhandlungen* **91**, 131–50.

Wimmenauer, E. 1963. Contribution to the petrography of the Kaiserstuhl. *Neues Jahrbuch für Mineralogie, Abhandlungen* **99**, 231–76.

Wimmenauer, E. 1970. Zur petrologie der magmatite des Oberrheingrabens. *Fortschritte der Mineralogie* **47**, 242–62.

Wyllie, P. J. 1978. Mantle fluid compositions buffered in peridotite–CO_2–H_2O by carbonates, amphibole, and phlogopite. *Journal of Geology* **86**, 687–713.

Wyllie, P. J. 1987a. Discussion of recent papers on carbonated peridotite, bearing on mantle metasomatism and magmatism. *Earth and Planetary Science Letters* **82**, 391–7.

Wyllie, P. J. 1987b. Discussion of recent papers on carbonated peridotite, bearing on mantle metasomatism and magmatism. *Earth and Planetary Science Letters* **82**, 401–2.

Wyllie, P. J. & H. L. Huang 1976. Carbonation and melting reactions in the system CaO–MgO–SiO_2–CO_2 at mantle pressures with geophysical and petrological applications. *Contributions to Mineralogy and Petrology* **54**, 79–107.

APPENDIX: MANTLE NORM AND CMS PROJECTION

1. Set total Fe = FeO + (0.9 × Fe_2O_3); FeO = 0.9 × (total Fe); and Fe_2O_3 = 0.11 × (total Fe).
2. Convert oxide weight amounts to single cation amounts with appropriate molecular weights. Al = Al + Cr + Fe^{3+}; Fe^{2+} = Fe^{2+} + Mn.
3a. If Fe^{2+} > Ti, then IL = Ti; Fe^{2+} = Fe^{2+} − Ti; proceed to 4.
3b. If Fe^{2+} < Ti, then IL = Fe^{2+}; PF = Ti − Fe^{2+}; Fe^{2+} = 0; Ca = Ca − PF.
4. AP = P/2 and Ca = Ca − (3.33 × AP).
5. Fm = Mg + Fe^{2+}; PH = K/2; Fm = Fm − (6 × PH); Al = Al − (2 × PH); Si = Si − (6 × PH).
6. If Al > Na, then JD = Na; Al = Al − JD; Si = Si − (2 × JD); proceed to 7.
6a. If Na > Al, then JD = Al; Al = 0; Si = Si − (2 × JD); Na = (Na − JD)/2.
7. SP = Al/2; Al = 0; Fm = Fm − SP; proceed to 9 (step 7 is omitted for the CMS projection from garnet).
8a. GAR = Al/2; Al = 0; Si = Si − (3 × GAR).
8b. If Ca > (5.45 × Fm) then Ca = Ca − (0.465 × GAR); Fm = Fm − (2.535 × GAR); proceed to 9 (Ca content of this garnet is typical of mantle garnets).
8c. If Ca > (5.45 × Fm) then Ca = Ca − {3 × GAR × [Ca/(Ca + Fm)]}; Fm = Fm − ⟨3 × GAR × {1 − [Ca/(Ca + Fm)]}⟩.
9. Ca = Ca + Na + JD; Fm = Fm + JD; Si = Si + (2 × JD) (this step adds jadeite to diopside and, in the infrequent case of excess Na, adds Na to Ca as potential Na_2CO_3).
10a. For projection into CMS, the sum of Ca, Fm, and Si is normalized to 100.
10b. For a norm, Ca, Fm, and Si are apportioned among diopside, orthopyroxene, olivine, monticellite, akermanite, and carbonates by an additional algorithm.

23
THE ORIGIN AND EVOLUTION OF CARBONATITE MAGMAS

J. GITTINS

ABSTRACT

Published schemes of carbonatite magma genesis involve either crystal fractionation at crustal pressures of a mantle-derived parental magma (usually a 'carbonated nephelinite') or separation of carbonate and silicate liquids by immiscibility.

Fractionation of 'carbonated nephelinite' is rejected because:

(a) it is unlikely that such magma contains enough CO_2 to allow abundant crystallization of carbonate; Ca and Mg will continue to form olivine and clinopyroxene rather than carbonate;
(b) rocks intermediate between nephelinite and carbonatite, required by such a scheme, are unknown;
(c) fractionation of nephelinite magma could not generate the Nb and REE (rare earth element) concentrations characteristic of carbonatites.

Although certain silicate and carbonate liquids are immiscible at crustal pressures, no liquid has yet been synthesized that exsolves into nephelinitic and carbonatitic liquids. Immiscibility fails to account for the high Nb, P, REE, and Sr content of carbonatites; it is a late-stage process with only local effects.

Genesis of carbonatite magma by direct melting of partially carbonated mantle is proposed, in which Na_2CO_3 is formed from the jadeitic component of clinopyroxene as well as calcite–dolomite solid solutions from the diopsidic component and olivine. This carbonated mantle is metasomatized by fluorine-rich fluids that supply Nb, P, REEs, and Sr, and melts to form carbonate magma that dissolves enough olivine and pyroxene to provide Al, Fe, and Si necessary for crystallization of silicate minerals. Primitive carbonatite parent magma is dominantly calcitic with varied Ca:Mg ratio, and contains alkalies as well as Al, Fe, P, and Si. Its composition approximates olivine–calcite carbonatite or olivine–calcite–dolomite carbonatite.

Extensive differentiation in the crust can follow a normal or alkalic trend. The normal trend involves:

(a) alkali enrichment followed by alkali depletion; and

(b) Mg depletion coupled with either Fe enrichment followed by Fe depletion, or continuous Fe enrichment to form ferrocarbonatite.

Alkalies are lost as chlorides in an aqueous fluid whose separation from carbonatite magma is controlled by magmatic differentiation and rate of magma ascent. The alkalic trend is rare and occurs only in dry magmas with high initial Cl and F contents. Aqueous fluids do not form and alkalies accumulate in the magma. The concept of alkalic carbonatite magma as the parent of all carbonatites is rejected.

Carbonatite lavas and pyroclastic rocks are rarely as alkalic as is often claimed. Fluorine and modest alkali contents allow calcite to crystallize at atmospheric pressure from magma with as little as 8% Na_2CO_3.

Carbonatite fluids are silicon-deficient brines containing Cl and F and a solute composed of Ca, Fe^{3+}, Na, and Mg. Fenitizing fluids, derived principally from carbonatite rather than ijolite (nephelinite) magma, may be evolved episodically depending on rate of magma ascent and degree of differentiation.

23.1 INTRODUCTION

Knowledge of carbonatite genesis has lagged far behind that of the commoner silicate rocks. It is only now beginning to emerge from the formative period during which essential features were established, such as the range of petrographic types, field relations (both within the individual complex and the large-scale tectonic setting), major- and minor-element and isotopic compositions, and detailed mineralogy. Eventually the igneous nature of carbonatite complexes, and hence also the existence of carbonatite magmas, became accepted, but only after a lengthy period varying from scepticism to complete disbelief.

The nature of carbonatite magmas and their possible parents has remained rather vague and no unified theory of petrogenesis has yet become generally accepted. The tendency has been to equate particular rock types to particular magmas and there is still no clear picture of whether carbonatite magmas are of primary or secondary derivation, or whether mantle or crustal processes are dominant. The problem has been exacerbated by lack of agreement over whether the carbonate rocks are cumulates and, hence, unrepresentative of the parent magma. The tendency to study the mineralogical exotica at the expense of the major rock-forming minerals has not helped.

Against this confused background it is not surprising that no clear picture has emerged of what carbonatite magmas are, what chemical compositions they have, where they develop, and how they evolve. Carbonatites have now begun to enjoy a resurgence of interest, and a vast amount of information is beginning to appear, as is very evident from the preceding chapters of this book. For this reason it is worth trying to create some general framework into which the new work might fit, and to suggest some of the bases from which future work needs to proceed. Caution is also expressed at some of the approaches taken and at some of the conclusions that are being accepted uncritically.

23.2 WHERE DO CARBONATITES COME FROM?

Prior to about 1960 there was no clearer understanding of carbonatite genesis than there was of the origin of alkalic rocks in general. Little was known of the mantle, and few people looked deeper than the crust for any igneous rocks. Moreover, there was a very influential period in the history of alkalic rock petrology when the Daly–Shand limestone syntexis hypothesis held sway. It is not particularly surprising that in such a context Shand (1927) saw no reason to consider carbonatite complexes in any different light. The associated alkalic (feldspathoidal) rocks he accepted as igneous but he tended to dismiss the carbonatite itself as sedimentary limestone. Thus, the Phalaborwa complex became a massive raft of Transvaal dolomite that has floated upward in a feldspathoidal magma.

During the 1960s, as the mantle gradually was recognized as the source of most magmas, carbonatites came to be seen as having their origins there as well. However, a curious dichotomy developed between those who saw carbonatites as developing essentially within the crust, or only slightly below it, by modification of a mantle-derived magma, and those who emphasized the vexed question of an alleged kimberlite–carbonatite association, and took a quantum leap into the bowels of the Earth to seek the birthplace of carbonatites.

At about the same time the first systematic $^{87}Sr : ^{86}Sr$ ratios became available and, although crude by modern standards, suggested that carbonatites were indeed of mantle origin. This much, then, was settled. What was not settled, and still remains unclear, is whether carbonatite magmas originate by direct mantle melting or whether they evolve from a silicate, mantle-derived magma during low-pressure fractionation in the crust.

Writings on carbonatite magma genesis consider essentially two possibilities:

(a) fractionation at low, crustal pressure, of a mantle-derived parental magma (usually a 'carbonated nephelinite'); and
(b) immiscible separation of a carbonate liquid and a silicate liquid (nephelinitic or phonolitic) from a mantle-derived parent magma that formed within the mantle. The immiscible separation might be at high or low pressure and, hence, occur within mantle or crust.

A third possibility proposed here considers direct melting of partly carbonated, metasomatized mantle to produce separate, primary carbonate and silicate magmas.

23.3 ASSESSMENT OF MAGMA DERIVATION SCHEMES

23.3.1 Fractionation of a 'carbonated nephelinite' magma at crustal pressure

What is a carbonated nephelinite magma? Proponents of the scheme have not really explained, but what is probably meant is a high content of dissolved CO_2 that will stabilize carbonates and allow them to crystallize. The scheme, whereby a carbonatite magma is generated from a CO_2-rich nephelinite magma fails, however, on several scores.

There is nothing to indicate high CO_2 solubility in nephelinite magma and there is no reason why, at crustal pressures, Ca, Fe^{2+}, and Mg should favour CO_2 over SiO_2 or Al_2O_3 to form carbonates instead of olivine and clinopyroxene.

A further problem with fractional crystallization of a nephelinite magma is that clinopyroxene, and sometimes olivine, is the principal mafic mineral and crystallizes in abundance. This would reduce rather than concentrate CaO and MgO. Furthermore, the production of carbonatite magma from nephelinite magma by fractional crystallization would require the generation of rock types intermediate between nephelinite and carbonatite, but such rocks are not found. It might be claimed that silicocarbonatites fill this role, but their silicate mineralogy is generally not that of nephelinites; many silicocarbonatites are probably mechanical hybrids generated by the intrusion of carbonatite magma into already solid ijolite or alkalic pyroxenite.

It is also unlikely that fractional crystallization could generate the Nb and REE concentrations that characterize carbonatites, even though nephelinites are significantly enriched in LREEs.

There are additional problems. How did sufficient CO_2 become incorporated into the magma in the first place, and where did it come from? The source of CO_2 must surely be in the mantle because any known crustal reservoirs of CO_2 would also be rich in ^{87}Sr so that the Sr isotope ratio of the nephelinite magma would be altered beyond what is observed in carbonatites. Carbonates can exist in the mantle as a result of carbonation of clinopyroxene at pressures > 27 Kb (Wyllie, this volume) and might be considered as a source of CO_2 for a 'carbonated nephelinite magma'. It seems more logical, however, that this mantle carbonate might have melted to produce a carbonatite magma directly.

There is no reason to suppose that a nephelinite magma can dissolve much CO_2, but even if it could there is little reason to suppose that such a magma would crystallize calcite at crustal pressures rather than exsolving gaseous CO_2. Watkinson & Wyllie (1971) counter this by referring to a pseudobinary section through the multi-component system $NaAlSiO_4$–$CaCO_3$ at 25% H_2O which shows that a sequence of mineral assemblages involving nepheline, melilite, hydroxyhauyne, cancrinite, and calcite can crystallize continuously from 1100 °C to less than 700 °C at 1 Kb. It is difficult to equate this system to the crystallization of a CO_2-bearing nephelinite magma. These assemblages are not found in substantial amounts in carbonatite complexes but do occur in late-stage, rather rare, small-volume rocks such as the okaites (nepheline–melilite–perovskite rocks) of Oka. Such a system is more likely to develop and operate in the late stages of crystallization of ijolitic or carbonatite magmas that have become hybridized.

The existence of nephelinite magma rich in dissolved CO_2 remains speculative. There is no evidence that it exists in nature, and the degree to which it has dominated discussions of carbonatite magma genesis is unwarranted.

23.3.2 *Liquid immiscibility*

The popularity of liquid immiscibility as a petrological process has waxed and waned throughout much of this century, but its application to carbonatite problems dates from experimental studies in the 1960s when some of Wyllie's students investigated the possibility of its existence in silicate–carbonate systems. Koster van Groos & Wyllie (1966, 1973) discovered it in the system $NaAlSi_3O_8$–Na_2CO_3–CO_2 and in $NaAlSi_3O_8$–$CaAl_2Si_2O_8$–Na_2CO_3–H_2O. Wyllie showed the persistence of this immiscibility at up to 20 Kb pressure, and in the presence of water at 1 Kb (Koster van Groos & Wyllie 1968, 1973). Ferguson & Currie (1971) argued for immiscible separation of carbonate from silicate melts based on ocelli in ultrabasic dykes at the Callander Bay complex in Ontario, and on the basis of subsequent melting experiments. Immiscibility was demonstrated by Verwoerd (1978), in $NaFeSi_2O_6$–Na_2CO_3 and in ijolite–carbonatite lava; by Wendlandt & Harrison (1978), in $KAlSi_3O_8$–K_2CO_3 at pressures up to 20 Kb; and by Freestone & Hamilton (1980), in experiments using Oldoinyo Lengai alkalic carbonatite and silicate lavas.

The immiscible origin of carbonatite magma has been popularized by Le Bas (1977, 1981, 1987, this volume) and the general idea has developed that 'There is now a consensus that their parental magmas originate by the separation of an immiscible liquid phase from a CO_2-saturated nephelinite or phonolite magma' (Fitton & Upton 1987). A similar statement in a petrology textbook (Hall 1987) reads that 'Immiscible separation of carbonatitic and nephelinitic liquids from a common parent is now a widely favoured hypothesis for the origin of carbonatite magmas.' I venture to suggest that the case for immiscibility is greatly exaggerated and bears re-examination.

Immiscibility between certain carbonate and silicate liquids does exist; the experiments prove that. What has to be decided is the significance of immiscibility in carbonatite petrogenesis and, far more importantly, the relationship between experimental results and the actual separation of two conjugate liquids in nature from some credible parent magma. Field data are notoriously difficult to interpret, particularly 'ocelli', and immiscibility that goes to completion destroys the evidence of its having occurred.

In assessing the role of immiscibility, the principal concerns are:

(a) whether natural magmas evolve to compositions at which immiscibility occurs;
(b) the stage of magmatic evolution at which immiscibility might occur and, hence, whether it is an early process of fundamental importance, or a late, relatively low-pressure event of minor importance; and
(c) the range of compositions over which immiscibility might occur.

Freestone & Hamilton (1980) established that molten nephelinite and phonolite are each immiscible with molten alkalic carbonatite. Powdered nephelinite mixed with synthetic alkalic carbonatite lava was heated to a temperature above the liquidus of both members, and then quenched to establish texturally that two immiscible liquids had coexisted. Experiments with phonolite produced similar results; both sets were conducted over a pressure range from 0.7 to 7.6 Kb.

It was concluded that the silicate and carbonate liquids separate from a common parent. The experiments established that the two liquids are immiscible but not that they came into being together. The experiments have not created a single liquid from two immiscible ones, and without doing so it can not be said that the immiscible liquids have a common nephelinitic or phonolitic parent. The liquids simply have the property of being immiscible. Nothing more has been proved. Kjarsgaard & Hamilton (1988) have extended the range of immiscibility in plagioclase systems to albite–calcite and anorthite–calcite at 5 Kb but the same problem remains.

Another major weakness with liquid immiscibility as a means of generating carbonatite magma is that its proponents fail to explain how the characteristic carbonatite element concentrations are produced from either nephelinitic or phonolitic magma. Immiscibility probably occurs locally as a late-stage process, but is unlikely to have played a major role in developing carbonatite magmas.

23.3.3 *Direct melting within the mantle*

Alternatives to differentiation of silicate magmas within the crust began to emerge as long ago as 1975 with the establishment of carbonation reactions at mantle pressures. Carbon dioxide can not exist as a gas at pressures greater than about 27 Kb because of a series of reactions, involving olivine and pyroxene, that were discovered largely by Wyllie, Eggler, and their associates. These reactions open up the possibility of direct development of carbonatite magma by partial melting of carbonated peridotite. Reactions of particular interest for carbonatite magma genesis are the ones that generate solid solutions of calcite, dolomite, and magnesite. These are:

$$Fo + Cpx + CO_2 = 3Opx + CC \qquad (23.1)$$

$$2Fo + Cpx + 2CO_2 = 4Opx + Do. \qquad (23.2)$$

At high pressure Do + CC form a solid solution series, and the two reactions are combined as:

$$3Fo + 2Cpx + 3CO_2 = 7Opx + Do/CC_{ss}. \qquad (23.3)$$

Other carbonation reactions are:

$$Fo + Do + CO_2 = Opx + MC_{ss} \text{ (2MC.CC)}, \qquad (23.4)$$

and

$$Opx + Do + CO_2 = Coes/Qtz + MC_{ss} \text{ (2MC.CC)}. \qquad (23.5)$$

Abbreviations are: Cpx, clinopyroxene; Opx, orthopyroxene; Fo, forsterite; CC calcite; Do, dolomite; Do/CC$_{ss}$, dolomite/calcite solid solution; MC, magnesite; Coes/Qtz, silica minerals.

Reaction 23.3 requires about 5% CO_2 for completion, but 23.4 and 23.5 require about 23% because of the high CO_2-buffering capacity of olivine. It is far more likely that 5% of CO_2 will be available in the mantle than 23%, and so reaction 23.3

is a reasonable possibility, whereas reactions 23.4 and 23.5 seem very unlikely to occur.

An important question is whether reaction 23.3 can be modified to permit development of a realistic carbonatite magma. In its unmodified form the reaction offers no obvious means by which a liquid generated from direct melting of the carbonate in a carbonated peridotite would contain the elements necessary to permit the crystallization of pyroxenes, amphiboles, and micas as well as carbonates, and give carbonatites their characteristic geochemical signature. The problem can be divided into two parts:

(a) the major elements (Al, Fe, K, Na, Si);
(b) the minor elements, such as Nb, P, REEs, and Sr.

The first problem can be examined in the light of reaction 23.3. Experimental determinations have considered only the ideal compositions of Fo as Mg_2SiO_4 and Cpx as $CaMgSi_2O_6$, yet most mantle Cpx contains a significant jadeite component which, if carbonated, will produce Na_2CO_3. Thus, a further (as yet hypothetical) carbonation reaction may be written:

$$Fo + Jd + CO_2 = Opx + NC + AS, \qquad (23.6)$$

where Jd = jadeite, NC = Na_2CO_3, and AS = aluminosilicate. Combination with reaction 23.3 yields:

$$Fo + Jd/Di_{ss} + CO_2 = Opx + Do/CC_{ss} + NC + AS, \qquad (23.7)$$

where Jd/Di_{ss} is a jadeite–diopside solid solution.

Although the reaction is written as producing aluminosilicate, it is more likely that Al will enter Opx and that Si will convert more Fo into Opx. A somewhat similar approach was proposed by Treiman & Essene (1983) who suggested the melting of carbonate-bearing eclogite. Their proposal involved the carbonation of diopside to dolomite, but instead of considering the possible carbonation of jadeite they looked at melting in the system jadeite–calcite–CO_2. Since this involves CO_2 gas it is a process that would operate at depths less than those at which mantle carbonation reactions occur, and so is rather different from what is proposed here.

The principal effect of producing Na_2CO_3 as well as calcite and dolomite is to reduce drastically the solidus temperature of the carbonate component of the mantle peridotite, and hence the temperature at which melting can begin (Cooper *et al.* 1975). Even at 1 Kb the dry system NC–Do has a eutectic at 575–600 °C (Beckett 1987), but at mantle temperatures of 1200–1500 °C the carbonates must melt completely, even allowing for substantial increase of solidus temperature at 27 Kb in the dry system. Such a liquid should be capable of dissolving enough olivine and orthopyroxene to give the silica and alumina levels required by a carbonatite magma. The amount of Na that might be introduced in this way would depend on the jadeite content of the mantle clinopyroxene.

It is not known whether jadeite can be carbonated as suggested but it seems possible that a calcite/dolomite melt containing alkalies, Al, and Si might be produced in this way. Any phlogopite present in the mantle peridotite is likely to be an early contributor to the melt and would add K. Clinopyroxene would also contribute Fe^{2+}.

The problem remains of producing the characteristic minor-element suite of carbonatites. Bedson (1983) showed that CO_2 can transport the REEs, and the very fact that carbonation has taken place points to CO_2 migration. However, if CO_2 is the agent which transports the characteristic minor element suite to the mantle melting site and carbonates the mantle peridotite, it is difficult to understand why all of the mantle is not similarly enriched in the carbonatite minor-element suite. Furthermore, CO_2 can not move freely at depths greater than those at which carbonation reactions occur, so its efficacy is severely restricted.

It is, of course, possible that the mantle is not uniformly carbonated. Carbonatites are found almost exclusively under stable cratons, and this raises two possibilities. One is that carbonation of peridotite occurs only in such localities, and the other is that carbonation is widespread but that some other factor is responsible for inducing melting. Because CO_2 does not seem to be a likely agent for transporting the required minor-element suite to the melting site, another agent must be found, and it is conceivable that this agent will also be responsible for inducing melting of an already carbonated peridotite.

It is tempting to turn to F and Cl because we know that these elements reach high concentrations in the alkali carbonatite magma of Oldoinyo Lengai, and there is abundant evidence that they are present in most carbonatite magmas. Rare earth elements, and probably also Nb and Sr, readily form fluoride complexes. Chlorine may play a similar role to F. There is no certainty that the mantle is extensively carbonated; indeed, it seems likely that if the amount of CO_2 available is fairly small a correspondingly small proportion of peridotite will be carbonated and subsequently melt. Under these conditions very little F and Cl are needed at the melting site to allow the carbonatite magma to acquire a few per cent of each element, assuming that both elements dissolve preferentially in carbonatite melts. If there is only a small volume of carbonatite melt, and F and Cl dissolve in it, then very small amounts of Nb, REEs and Sr carried as fluoride could be concentrated into the magma.

Experimental studies in progress at the University of Toronto (B. Jago, unpublished data) have shown that F drastically lowers the solidus and liquidus of most carbonate systems. It is, therefore, reasonable to expect that its presence in the mantle would also lower the solidus of the carbonate portion of the mantle and further assist in the formation of carbonatite magma. There is the further possibility that the halogens are the joint agent both of mantle metasomatism and of inducing melting of a partly carbonated mantle peridotite; this would explain why carbonate magma formation is not a generalized mantle phenomenon.

Some doubt might be cast on a direct mantle origin for carbonatites by the Ni content of carbonatite olivines that is usually below the detection limits of the energy-dispersive electron microprobe, and stands in sharp contrast to the olivines of associated silicate rocks where values of 0.1–0.6% NiO are common. But this is not necessarily a problem. Nothing is known about the behaviour of Ni during olivine carbonation, but given its compatibility with the olivine structure it seems probable that it would diffuse into remaining olivine in lherzolite, rather than enter carbonate. In this way the carbonate melt would have a very low Ni content from the outset. Olivines in carbonatites are generally more magnesian than those of associated silicate rocks and sometimes more magnesian even than mantle olivines (e.g. Sokli, Kovdor, and Jacupiranga; Twyman 1983). The high Mg content of carbonatite olivines is attributed to the rather low Fe content of the initial carbonatite melt which, in turn, is controlled by the highly magnesian composition of the mantle olivine and clinopyroxene.

If carbonatite magma is generated directly by mantle melting, the associated mafic alkalic silicate magma, largely nephelinitic, must be developed separately. There seems to be little reason why nephelinitic and carbonatitic magma can not form at similar pressures (Brey & Geen 1976). If mantle carbonation is of limited extent, and the triflingly small volume of carbonatite in the crust–mantle system suggests that it is, then silicate magma that developed after carbonatite magma would form a mantle that was distinctly depleted in the carbonatite element suite relative to the F-metasomatized, carbonated mantle. It would crystallize olivine that was more magnesian and richer in Ni.

The scheme envisaged is of nephelinite and carbonatite magmas forming sequentially, or possibly simultaneously, from carbonated mantle. Migrating halogens (principally Cl and F) carrying low concentrations of Nb, Sr, and REEs dissolve mainly in the carbonatite magma. Variation in the Nb and REE content of different carbonatites is probably attributable to variation in the amounts of these elements carried as migrating halides. The general character of the magma is established by the carbonation reactions, the alkali content by the amount of jadeite component that was present or became carbonated, and the Si and Al content by the extent to which mantle silicates dissolved into the carbonatite magma.

Although part of the variation of CC:Do of carbonatites is, no doubt, due to fractionation and magmatic differentiation, some at least is probably due to distinctly different magmas that result from variations in the degree of mantle carbonation. Sharp contacts between dominantly calcitic and dominantly ankeritic to dolomitic carbonatites probably represent the intrusion of separate magmas.

In summary, then, it is worth considering that carbonatite magmas develop by direct melting following partial carbonation of mantle peridotite in which the jadeitic as well as the diopsidic component has been affected allowing the formation of NC in addition to CC–Do. Subsequent metasomatism by F, and possibly Cl, generates the characteristic trace-element suite of carbonatites and also further induces melting of the carbonate fraction. The halogens play an essential role in developing carbonatite magma and may explain why carbonatites are restricted to certain regions of the Earth's crust.

23.3.4 *The primitive, parental carbonatite magma*

There is no reason to suppose that only one type of parental carbonatite magma exists. If carbonatite magma is produced directly in the mantle by partial melting of carbonated peridotite, then variations in alkali, Al, Fe, and Si content and f_{O_2} are likely because of local variations in mantle composition, extent of carbonation, and concentration of the halogens, REEs and P. It is possible, therefore, for the composition of carbonatite parent magmas to vary.

It seems likely, however, that the carbonatite parental magma which reaches the crust is dominantly calcitic (although with wide variation in Ca : Mg ratio), modestly alkalic (at least several per cent of $Na_2O + K_2O$), and contains enough Al, Fe, P, and Si to allow the crystallization of amphibole, apatite, clinopyroxene, magnetite, mica, and olivine. It also contains sufficient Nb, REEs, and Sr to generate rocks with Nb minerals, abundant total REEs relative to chondrites, and with very strong enrichment in the LREEs as well as Sr contents usually of the order of several thousand p.p.m. The F and Cl abundances are low and probably only a few tenths of a per cent.

23.4 CARBONATITE MAGMA EVOLUTION IN THE CRUST

Although a case has been made for the production of carbonatite parent magma by direct melting of a carbonated mantle and accompanying metasomatism, there is overwhelming evidence for subsequent differentiation within the crust. Rocks of progressively varied composition can commonly be mapped, individual minerals are usually zoned, and there is sympathetic variation in chemical composition between mineral assemblages. Fenitization is itself an indication of relatively long crustal residence time.

Differentiation in the crust follows several possible directions, which may be grouped as the 'normal' and the 'alkali enrichment' trends.

23.4.1 The 'normal' carbonatite trend

Petrographic and experimental evidence suggest that calcite and dolomite are among the first minerals to crystallize and continue to do so throughout most of the crystallization history.

Olivine is relatively rare and is probably restricted to both a very early stage of crystallization and to fairly deep levels. It does not seem to have a long crystallization history; its composition is initially Fo_{85-90}, although some carbonatites show extensive differentiation to about Fo_{65}. Carbonatites that appear to have crystallized at deep crustal levels generally contain olivine and sometimes titanoclinohumite. The combined effect of dolomite and olivine crystallization is Mg depletion and a trend to more calcitic carbonatite.

The solubility of P_2O_5 in most carbonate liquids is low, and apatite will start to crystallize very early. In spite of this, the continued crystallization of carbonates in abundance will concentrate P_2O_5 and tend to maintain saturation so that apatite is probably able to crystallize throughout most of the magmatic history. In doing so it tends to remove F because most carbonatite apatites are fluorapatites. The efficiency of F removal, therefore, depends on the P content of the magma.

Mineralogical evolution of carbonatites is discussed by Gittins (1978) in a review of the crystallization history of the Cargill intrusion of northern Ontario, Canada, where the course of differentiation began deep in the crust. The complex was intruded into a granulite terrain and consists of carbonatites accompanied by layered ultramafic rocks but it lacks feldspathoidal rocks and fenites. It displays evidence of quiescent crystallization. Ultramafic rocks (olivine clinopyroxenites, hornblende clinopyroxenites, hornblendites, and late veins of aegirine augite–calcite) appear to represent crystallization of an alkalic magma under progressively increasing pressure of H_2O that eventually stabilized amphiboles spanning the range from pargasite to hastingsite and containing 20% normative nepheline and leucite. They are probably the deep-seated equivalents of olivine leucite nephelinite or olivine phlogopite nephelinite.

Calcite, dolomite, and calcite–dolomite carbonatites occur at Cargill and contain combinations of olivine, titanoclinohumite, amphibole, and biotite, with and without apatite and magnetite. The assemblage amphibole + biotite displays a sympathetic Fe enrichment with essentially constant Al : (Fe + Mg) ratio. However, in the assemblage amphibole + biotite + titanoclinohumite the latter mineral buffers the amphibole causing it to maintain constant composition; titanoclinohumite progressively increases its Fe : Mg ratio, and biotite displays extensive Fe enrichment by substitution of Fe^{3+} for Al^{3+}. The result is zonation from normal biotite cores to reversely pleochroic rims of tetraferriphlogopite. Such differentiation required substantial crustal residence time. Progressive enrichment of Fe and alkalis is documented by Twyman & Gittins (1987) for several other carbonatite complexes.

These mineralogical trends indicate the directions of carbonatite magma evolution but need to be interpreted cautiously. For example, zonation from phlogopite cores to tetraferriphlogopite rims, with almost total replacement of Al^{3+} by Fe^{3+}, does not necessarily imply a sharp rise in f_{O_2} of the magma, although there is probably a small progressive increase from the primitive parental magma; the accompanying oxide mineral remains magnetite. The original magma probably contained enough Al to initiate phlogopite crystallization but not enough to sustain it until the available Si and K were exhausted. Because Fe^{3+}–Al^{3+} is a readily available substitution, mica crystallization continues as long as K and Si are available, using Fe^{3+} instead of Al^{3+}. This may limit the rate at which f_{O_2} might otherwise increase.

Many carbonatites contain micas that display the effect of the magma becoming depleted in K and Al. In thin section these micas have the normal brown colour of titanian phlogopite but change toward the rim through pale green to an intense deep green, almost opaque, hydro-mica. There is little or no K at the rim and total Fe, as FeO, ranges from 12.9% to 28.5%. One interesting aspect of these micas is the indication that K can be depleted to zero before water saturation has been reached; it merely re-emphasizes that Na is generally vastly more abundant than K in carbonate magmas but that considerable variation in K content is still possible.

Iron enrichment or iron depletion? It is commonly stated that the normal end-product of carbonatite magma differentiation is ferrocarbonatite, but this is an oversimplified generalization.

As Twyman & Gittins (1987) showed, it is common for carbonatite silicate minerals to show both Fe and alkali enrichment, and for the ankerite component of dolomite to increase from core to rim. Whether carbonatite magma is a primary mantle melt or of secondary derivation, it is thought to have an initial f_{O_2} possibly as low as Ni–NiO (see Wyllie, this volume, Fig. 20.17) which will prevent magnetite from crystallizing. Iron enrichment will then occur until the f_{O_2} has increased to a value at which magnetite can crystallize, and at this stage there will be rapid Fe depletion to an almost iron-free composition, a feature that is seen in some calcite carbonatites.

Other factors that complicate this simple trend, however, are the effects of declining temperature and rising CO_2 content. Yang (1987) has shown that titanomagnetite crystallization is terminated in the presence of CO_2 at about 450 °C and 3 Kb, 400 °C and 2 Kb, or 370 °C at 1 Kb. At lower temperatures, still in the presence of a CO_2-rich fluid, the equilibrium assemblage is siderite–hematite–rutile. Siderite will most likely enter solid solution as ankerite or ferroan dolomite, generating an ankerite–hematite ± calcite rock, or what is commonly known as ferrocarbonatite. Rutile, not generally reported in ferrocarbonatites, may be present in solid solution in the hematite.

Iron enrichment is not, then, a simple matter; its course is dictated by changing f_{O_2} as magma fractionation proceeds. The low initial f_{O_2} causes Fe enrichment but this ceases when f_{O_2} becomes high enough to form Fe^{3+} and magnetite begins to crystallize. Extensive crystallization of magnetite at high temperature will then lead to rapid depletion of Fe. Cooling of the magma, on the other hand, may terminate depletion by diverting Fe into ankerite and hematite at a stage where the magma is close to its solidus. Rapid intrusion of carbonatite magma into the cooler, upper levels of a volcanic structure will produce correspondingly rapid cooling of the magma to a temperature at which magnetite crystallization is impossible. This may explain why ferrocarbonatites are so common as dykes at the volcanic stage of carbonatite activity.

At the same time, some caution is necessary in interpreting ferrocarbonatites. Some rocks that are called ferrocarbonatites may not warrant the name; although they have a messy, rather altered appearance with abundant visible hematite, this appearance may lead to an exaggerated estimate of their Fe content. Many such carbonatites may contain no more Fe, and possibly less, than some of the seemingly pure carbonatites that contain coarse magnetite.

Iron enrichment generally occurs in carbonatite magma evolution but it is not a simple, uninterrupted trend. It may cease at an early stage or continue until ferrocarbonatite is produced. It is controlled by evolving f_{O_2}, which in turn is controlled by the cooling history and rate of ascent of the magma.

Alkali enrichment or alkali depletion? Evolution of the alkali content is also complex. Mineralogical evidence indicates initial alkali enrichment, but such enrichment must eventually be terminated because only one carbonatite (Oldoinyo Lengai) is known to have produced an extreme alkalic composition. Initial enrichment is a logical consequence of differentiation because, although crystallization of amphiboles, micas, and pyroxenes removes alkalies, the low Si and Al content of the magma severely limits the amount that can be removed in this way.

Alkali enrichment or depletion is also affected by the behaviour of Cl, F, H_2O, and P. Crystallization of apatite may remove some of the F but not necessarily all, while Cl and H_2O are concentrated. The dominance of anhydrous carbonates in the crystallization of carbonatite magma eventually concentrates H_2O to the point where saturation occurs and a fluid forms that removes alkalies from the magma, probably as chlorides in aqueous solution. The efficiency of this process depends largely on the Cl content of the magma. If Cl is insufficient to remove all of the alkalies, then, clearly, some alkalies will remain in the magma.

The normal magmatic evolutionary trend may, therefore, be taken as alkali enrichment followed by alkali depletion. The fate of magma, following exsolution of the aqueous phase, depends upon a number of factors. It will still be water-saturated, although the solubility of H_2O in the now largely calcitic or dolomitic liquid is quite low, and how much longer it can remain liquid will depend mainly upon F and alkali contents because both can greatly lower the liquidus temperature. In general, alkali enrichment is probably terminated by separation of the aqueous fluid phase.

Water saturation may occur, however, at various stages in the evolution of the magma and, hence, at different alkali contents. The solubility of water in carbonatite magma increases with alkali content, which is largely controlled by the degree of fractionation that has occurred. This, in turn, is related to the rate at which the magma has risen, which is a major factor governing the length of time available for crystal fractionation to occur. In a rapidly rising magma there is relatively little time available for fractionation to develop a very alkalic composition; consequently, water saturation occurs at a low alkali level and the exsolved aqueous fluid has a correspondingly low alkali content. In a slowly rising magma, fractionation is extensive and higher alkali enrichment occurs, thus allowing greater solubility of water, so that when water saturation eventually occurs the aqueous fluid is more alkalic.

In this way it is possible for various degrees of alkali enrichment to occur according to how rapidly or slowly the primitive parent magma has risen. The extent of alkali enrichment will always be limited, however, by the solubility of water in the melt, and its Cl content. Precise limits can not be set, for the data on water solubility are known only in general outline, but it does seem that water saturation and alkali loss will always occur before an Oldoinyo Lengai type magma is developed. A very pressing need is to determine the solubility of Cl and H_2O in carbonate melts of varied $(Na,K):(Ca,Mg)$ ratios, and to determine the partition coefficients of Cl and F between carbonate melts and aqueous fluid at crustal temperature and pressure.

In summary, the 'normal' carbonatite trend involves:

(a) initial alkali enrichment followed by alkali depletion;
(b) Mg depletion; and either
(c) Fe enrichment followed by Fe depletion; or
(d) continued Fe enrichment to ferrocarbonatite.

Plutonic crystallization generally ends with Fe depletion, whereas volcanic crystallization commonly leads to the production of ferrocarbonatite.

23.4.2 *The alkalic trend (or the Oldoinyo Lengai trend)*

The extreme form of alkali-rich carbonatite magma is represented by the lavas of Oldoinyo Lengai. The discussion so far has focussed on the role of H_2O, alkalies and the halogens (F and Cl) in maintaining carbonatite liquids at reasonable temperature and pressure. But H_2O is essentially denied in the development of extreme alkali enrichment because it must ultimately separate as an aqueous phase and terminate the process, and because the lavas show no evidence of abundant H_2O having been present. They are composed entirely of anhydrous minerals, and specimens that have been carefully protected from the air since collection have very low H_2O contents. If fractionation has occurred under water-free conditions, some agent other than H_2O must be responsible for lowering the liquidus of the primitive parental magma in order to permit fractionation to continue. A reasonable alternative to H_2O is F and Cl, because the Oldoinyo Lengai magma contained at least 4% Cl and 8% F. If Cl partitions into an aqueous phase in preference to carbonatite magma, as it does in silicate magmas, then the presence of so much Cl is further evidence that the magma was dry, because an aqueous fluid would have removed it.

The effect of 5% of F is to reduce the liquidus temperatures in the system Na_2CO_3 (NC)–$CaCO_3$(CC) at 1 Kb by 150–200 °C, depending on the NC:CC ratio, over those of the volatile-free system (B. Jago, Ph.D. thesis in preparation), an effect distinctly greater than that produced by the same percentage of H_2O. The ability of F to play a role similar to that of H_2O in the normal carbonatite magma evolutionary trend is thus established.

It is proposed that in the absence of H_2O and the presence of F, a calcitic magma with no more than 8% combined Na_2O and K_2O is completely liquid and undergoes fractional crystallization. Because the magma is dry, no aqueous fluid develops; alkalis, therefore, are not removed but are progressively concentrated in the magma by continued fractionation of essentially alkali-free minerals. A compositional range from primitive, parental calcitic carbonatite to the Oldoinyo Lengai type is possible. A further constraint, however, is a low P content in the initial magma; otherwise, crystallization of apatite would remove the F which is principally responsible for maintaining the magma in a liquid state.

Why should some parental magma be water-free and phosphorus-poor? The answer is unknown but the anhydrous nature of the Oldoinyo Lengai lavas, coupled with high Cl and F contents, is proof that such magmas exist. Development of an alkali-rich magma at Oldoinyo Lengai must be a special event because ejected fenite blocks suggest the presence of a 'normal' magmatic trend earlier in the history of the volcano.

23.5 ALKALIC CARBONATITE MAGMA AS A PARENTAL MAGMA

In 1981 Le Bas promoted the Oldoinyo Lengai magma-type to a parent from which the full diversity of carbonatite rock types could be derived. He accepted the primacy of liquid immiscibility and proposed a scheme in which a 'carbonated nephelinite magma' separated at 1000–1100 °C into an ijolitic (nephelinitic) silicate liquid and an alkalic carbonate liquid, the latter being at least 500 °C above its liquidus. The pressure at which immiscibility occurs was not stated but lower crustal rather than mantle conditions were implied. Le Bas envisaged development of a calcite–dolomite magma by loss of alkalis during cooling through the 500 °C of superheat; this was then followed by crystallization of calcitic carbonatite and later by dolomitic and ankeritic carbonatites. It was further suggested that the proportions of Ca:Na:K in the carbonatite magma depend upon the temperature and pressure at which immiscible separation occurred, a conclusion drawn from the experimental work of Freestone & Hamilton (1980).

A detailed rebuttal of this parental scheme is given by Twyman & Gittins (1987) who concluded that such magma, rather than being parental, is a late-stage fractionation product developed under special conditions as described above. They emphasized that liquidus temperatures are far higher than the 500–600 °C proposed by Le Bas and can be as high as 1100 °C.

A further point, not developed in Twyman & Gittins (1987), concerns the major-element composition of the Oldoinyo Lengai magma, which is notable for its extremely low content of Al, Fe, Mg, and Si. Alkali loss from such a magma could not concentrate these elements sufficiently to permit crystallization of the silicate and oxide minerals that are common in most carbonatites.

The parental status scheme is not supportable. The litany of its inadequacies is legion and it would be most profitable to discard it.

23.6 CARBONATITE LAVAS AND PYROCLASTICS: THEIR BEARING ON CARBONATITE MAGMA EVOLUTION

All carbonatite lavas and pyroclastics are calcitic, with the single exception of the lava at Oldoinyo Lengai. Many contain tabular calcite, some in relatively large single crystals, and some apparently as former single crystals now pseudomorphed by small calcite crystals. There is a general consensus that the single crystals are primary magmatic calcite and the aggregates are nyerereite 'calcitized' by groundwater (Hay & O'Neil 1983, Deans & Roberts 1984, Clarke & Roberts 1986, Dawson *et al.* 1987, Turner 1988). Both are presumed to have crystallized from a magma on the calcite side of the nyerereite–fairchildite join in the system Na_2CO_3–K_2CO_3–$CaCO_3$ (Cooper *et al.* 1975) and hence to represent a more calcitic magma than that of Oldoinyo Lengai, but still alkali-rich. Calcite crystals are presumed to be phenocrysts that crystallized in the magma chamber underlying the volcano, and were carried in suspension during lava eruption.

Many geologists have had difficulty reconciling field evidence for the extrusive character of these lavas and pyroclastics with the presumed impossibility of calcite crystallizing at atmospheric pressure; the presumption of former alkali-rich composition avoids this problem. Such an explanation, however, implies that alkali-rich carbonatite magma is developed far more commonly than was once thought. It is possible, however, that these rocks were not nearly so alkalic as the explanation assumes.

Crystallization of calcite at atmospheric pressure is limited by its dissociation temperature of 900 °C. In the system NC–CC this temperature corresponds to a composition on the liquidus at NC35 CC65 and 1 Kb, and at 1 bar will probably be NC30 CC70. Thus, any carbonatite magma with at least 30% of NC will be capable of crystallizing calcite at 1 bar. Liquids more calcitic than CC70 could crystallize calcite in the higher pressure of a magma chamber and continue to crystallize calcite on surface if erupted after the magma composition had changed to at least CC70.

Another factor, however, is F. As was stated previously, with 5% F the liquidus in the system NC–CC is depressed by 150–200 °C at 1 Kb, depending on the NC : CC ratio. The Oldoinyo Lengai magma contained at least 8% F, and most carbonatite magmas contain significant amounts of F. Thus, a liquidus depression of 200 °C at 1 Kb is very likely, and a temperature of 900 °C then corresponds to a composition on the liquidus of approximately NC15 CC85 (Fig. 23.1). Given the high solubility of F in carbonate melts and the fact that it remains in solution at low pressure (as shown by its presence in the Oldoinyo Lengai lava) it may be reasonable to assume that the 1 bar liquidus temperature occurs at about the same composition. The effect of F is to compensate the increase in liquidus temperature caused by the increase in pressure to 1 Kb. A magma with only 15% NC thus can crystallize calcite as a liquidus phase at atmospheric pressure if it contains > 5% F. When crystallization is complete the rock will consist of 71% calcite and 29% nyerereite. Although such lavas have yet to be found, it seems likely that they exist. Of significance is the fact that calcite can crystallize as the liquidus phase, to be joined at lower temperatures by nyerereite.

The example cited was for a magma extruded at its liquidus temperature of 900 °C. But 1100 °C is not an unreasonable temperature for some carbonatite magmas (Gittins 1979, Twyman & Gittins 1987), and with at least 5% F this temperature corresponds to a liquidus composition at 1 Kb of NC8 CC92. If such magma crystallizes calcite and differentiates until it has cooled to 900 °C, it will, when erupted, continue to crystallize calcite as before. Such a magma, corresponding to 84% calcite and 16% nyerereite, is capable of crystallizing calcite after its eruption as a lava so long as it has undergone a pre-eruption period of fractionation. These lavas still crystallize nyerereite, but the amount is so small that it could easily be leached by groundwater and calcite could be deposited in its place. The important point is that calcitic lavas need not have been alkali-rich.

A further consideration is the role of phosphorus in alkalic carbonatite lavas. Study of the system Na_2CO_3–$CaCO_3$–apatite, both volatile-free and under fluorine pressure, shows that sodium-carbonate-rich liquids crystallize complex sodium calcium carbonate phosphates and only minute amounts of apatite (B. Jago,

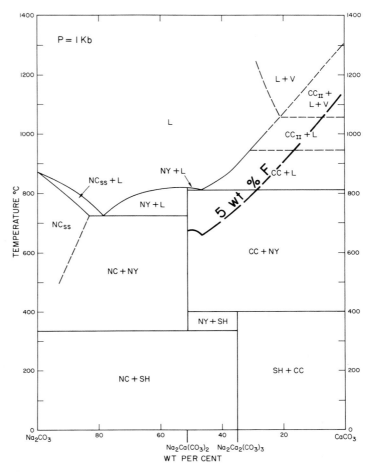

Figure 23.1 Isobaric equilibrium diagram for the system Na_2CO_3–$CaCO_3$ at 1 Kb, showing the liquidus with 5 wt% F as a heavy dashed line. NC, Na_2CO_3; NC_{ss}, sodium carbonate solid solutions; NY, $Na_2Ca(CO_3)_2$ (nyerereite); SH, $Na_2Ca_2(CO_3)_3$ (shortite); CC, $CaCO_3$; L, liquid; V, vapour. Adapted from Cooper *et al.* (1975) and incorporating data from B. Jago (Ph.D. research in progress at the University of Toronto).

unpublished data). Apatite is not present in the lavas of Oldoinyo Lengai, except for very rare grains that are probably xenocrystic. Thus, the abundance of apatite in some calcite-bearing carbonatite lavas is strong evidence that the rock was not formerly rich in nyerereite.

In conclusion, it seems likely that very calcitic magmas are capable of giving rise to lava flows and pyroclastic eruptions as long as they contain 8–10% NC and about 5% F, and eruption occurs below 900 °C. They will crystallize some nyerereite but the amount can be so little as to be readily lost by groundwater-leaching. It is highly likely that lavas of varied alkali content have been erupted, but there is no need to assume that all presently calcitic carbonatite lavas were once alkali-rich. The current tendency to do so is serving to inflate the number of alleged examples of alkali-rich carbonatite magma. Considerable caution must be exercised in interpreting calcitic carbonatite lavas or pyroclastic rocks as former alkali-rich (nyerereite-bearing) lavas. The extreme alkali-rich lava of Oldoinyo Lengai remains the only one whose existence we can be sure of, and its uniqueness tends to strengthen the argument that its development requires distinct conditions that may operate only rarely.

23.7 CARBONATITE FLUIDS AND FENITIZATION

Fluids are commonly associated with carbonatite magmas and are responsible for fenitization, late-stage development of iron-rich, pipe-like bodies enriched in REEs and U, and such minor features as barite veins and localized dolomitization of earlier calcitic carbonatite. The composition of these fluids can be estimated from fluid inclusions, the mineralogy of fenites, and from mineral associations in some carbonatites.

Wyllie and his co-workers chose CO_2 and H_2O as the principal volatiles in their carbonate melting experiments of the 1960s (e.g. Wyllie & Tuttle 1960), and this led to the concept of a 'carbothermal fluid' of varied $CO_2:H_2O$ ratio. Because H_2O allows calcite to melt at petrologically acceptable temperatures and pressures it has dominated discussion of carbonatite fluids ever since and is now widely assumed to be the dominant component. This view, however, needs to be modified because there is evidence to suggest that H_2O is neither the sole, nor necessarily the dominant component of carbonatite fluids. Treiman & Essene (1984) have, for example, deduced from theoretical considerations that the fluid in equilibrium with a carbonatite at Oka '... was essentially a H_2O-CO_2 fluid with $X(H_2O) + X(CO_2) > 0.99$ and $X(CO_2)$ near 0.1.' Unfortunately, the deductions assumed incorrectly that the rock is a eutectic assemblage. Studies at the University of Toronto (B. Jago, unpublished data) show that the particular rock does not melt in the presence of excess H_2O even at 950 °C and 1 Kb. It has been successfully melted in the presence of both F and Na_2CO_3 but over a considerable temperature interval. The rock is not, therefore, a eutectic mineral assemblage, and any fluid phase present with the magma was not simply H_2O.

The Oldoinyo Lengai carbonatite lava shows that Cl, F, and S (or sulphate) are probable constituents of carbonatite fluids, and the mineralogy of fenites requires that the fluid must contain Ca, Fe^{3+}, Mg, K, and Na. Because such a fluid dissolves SiO_2 during fenitization it must also be strongly deficient in, and a powerful solvent for, SiO_2. The abundance of acmitic pyroxene, riebeckitic amphibole, and hematite in fenites further suggests that at the depth where fenitization occurs the fluid has a relatively high f_{O_2} (probably close to the hematite–magnetite buffer).

In carbonatite magmas, Cl will probably partition strongly into an accompanying aqueous fluid, with F remaining in the carbonate liquid. Thus, in the 'normal' carbonatite evolutionary trend, the aqueous fluid that separates from a water-saturated magma will be a supercritical brine. Such a brine is perfectly capable of removing alkalies, Ca, Fe, and Mg from carbonatite magma as chlorides, and it will contain virtually no Al or Si since neither is present in the carbonatite magma in other than triflingly small amounts. Currie & Ferguson (1971) also concluded that the fenitizing fluid at Callander Bay, Ontario, Canada, was a NaCl brine.

Uncertainty has sometimes been expressed about whether fenitizing fluids are derived from carbonatite magma or from associated silicate magma (ijolite or nephelinite), but silicate magmas can generally be eliminated as a source in light of the preceding discussion. Neither ijolite nor nephelinite magma contains much H_2O, but an aqueous fluid produced by the small amount that is present would contain substantial amounts of Al, Na, K, and Si because of the relatively high solubility of nepheline in supercritical H_2O. The fluid would be unable to leach SiO_2 efficiently from the country rocks but would also have too low a Na:K ratio to change drastically the compositions of their feldspars. Currie & Ferguson (1971) have made the same point.

Fenitizing fluids generally come from carbonatite magma, and an obvious relationship between the course of magmatic evolution and the nature of fenitization is suggested. Earlier it was postulated that a magma which rises slowly fractionates to a greater extent, and so produces a more alkali-rich aqueous fluid, than one which rises rapidly and so reaches water-saturation before fractionation has time to increase the alkali content substantially. This, in turn, must affect the type of fenitization that occurs. It also suggests that there is a critical depth within the crust below which fenitization is unlikely to occur. An aqueous fluid can not separate before carbonatite magma reaches water-saturation, and it will not accomplish much fenitization unless, at the time the fluid separates, the magma has become sufficiently enriched in Cl to allow the fluid to dissolve the necessary agents of fenitization. Because the extent of fractionation is a function of the rate at which the magma rises, the composition suitable for such a brine to form is unlikely to develop in the lower crust or upper mantle. Support for this view is seen in the absence of fenites around some carbonatites thought to have crystallized at considerable depth in a granulite terrain of Northern Ontario, Canada (Gittins 1978). There are some minor metasomatic effects but nothing like the classic fenite zone of high-level carbonatites.

There is also much uncertainty about the identity of the gaseous species present at deeper crustal–mantle levels. Wyllie (this volume) thinks that fluids at mantle depth consist of C, H_2O, and CH_4, and this may be another factor that prevents fenitization from occurring in deep crust.

It is, of course, possible for a magma that has risen rapidly into the lower crust, and produced an alkali-poor fluid, to pause in its ascent at this depth; fractionation will then lead to the development of a second fluid that will, by this time, have become a fenitizing agent. Thus, fenitizing fluids may evolve at several stages in the ascent of a carbonatite magma, and give rise to the episodic fenitization seen in a number of carbonatite complexes.

Progressive evolution of alkali chloride fluids during magma fractionation is supported by fluid inclusion studies of Fen apatite crystals (Andersen 1986). Two types of fluid were found. One (type A), in equilibrium with a very early stage of magmatic evolution, has the composition: H_2O 88–90%, CO_2 5–27%, NaCl 5%; and another later fluid (type B) has a range of NaCl from 1 to 24% in a CO_2-free, aqueous solution. A third group of inclusions, containing aqueous fluids with up to 35% NaCl and solid halite crystals, may represent very late fluids derived in part from groundwater. Andersen puts it very clearly when he writes: 'Type B inclusions define a continuous trend from low towards higher salinities and densities and formed as a result of cooling and partitioning of alkali chloride components in the carbonatite system into the fluid phase.'

Fenitizing fluid is generally derived from carbonatite rather than silicate magmas. It is dominantly a brine carrying Ca, Fe^{3+}, Mg, Na, and K as chlorides; it has relatively high f_{O_2} (close to the hematite–magnetite buffer), and its composition depends on the rate of ascent and degree of fractionation of the carbonatite magma.

23.8 EPILOGUE

This volume indicates the breadth of carbonatite studies that are in progress, and the extent of interest that carbonatites have engendered. They are a treasure trove of the rarer metals and of commoner industrial and agriculturally important elements. They are, in addition, a unique window on the Earth's mantle. This chapter has sought to draw together some of the disparate threads in the complex history of carbonatite studies, to create a petrogenetic scheme that might be tested experimentally, and to guide future studies while cautioning against some interpretations that are likely to mislead.

REFERENCES

Andersen, T. 1986. Magmatic fluids in the Fen carbonatite complex, S.E. Norway. Evidence of mid-crustal fractionation from solid and fluid inclusions in apatite. *Contributions to Mineralogy and Petrology* **93**, 491–503.

Beckett, M. F. 1987. *Phase relations in synthetic alkali-bearing dolomite carbonatites and the effect of alkalinity and fluorine content on the solubility of pyrochlore and the formation of niobium deposits in carbonatites.* M.Sc. thesis, University of Toronto, Toronto.

Bedson, P. 1983. *The origin of the carbonatites and their relations to other rocks by liquid immiscibility.* Ph.D. thesis, University of Manchester, Manchester.

Brey, G. & D. H. Green 1976. The solubility of CO_2 in olivine melilitite at high pressures and the role of CO_2 in the Earth's upper mantle. *Contributions to Mineralogy and Petrology* **55**, 217–30.

Clarke, M. G. C. & B. Roberts 1986. Carbonated melilitites and calcitized alkali carbonatites from Homa Mountain, western Kenya: A reinterpretation. *Geological Magazine* **123**, 683–92.

Cooper, A. F., J. Gittins & O. F. Tuttle 1975. The system Na_2CO_3–K_2CO_3–$CaCO_3$ at 1 kilobar and its significance in carbonatite petrogenesis. *American Journal of Science* **275**, 534–60.

Currie, K. L. & J. Ferguson 1971. A study of fenitization around the alkaline complex at Callander Bay, Ontario, Canada. *Canadian Journal of Earth Sciences* **8**, 498–517.

Dawson, J. B., M. S. Garson & B. Roberts 1987. Altered former alkalic carbonatite lava from Oldoinyo Lengai, Tanzania: Inferences for calcite carbonatite lavas. *Geology* **15**, 765–8.

Deans, T. & B. Roberts 1984. Carbonatite tuffs and lava clasts of the Tinderet foothills, western Kenya: a study of calcified natrocarbonatites. *Journal of the Geological Society of London* **141**, 563–80.

Ferguson, J. & K. L. Currie 1971. Evidence of liquid immiscibility in alkaline ultrabasic dykes at Callander Bay, Ontario. *Journal of Petrology* **12**, 561–86.

Fitton, J. G. & B. G. J. Upton 1987. Introduction. In *Alkaline igneous rocks*, J. A. Fitton & B. G. J. Upton (eds). Geological Society Special Publication No. 30.

Freestone, I. C. & D. L. Hamilton 1980. The role of liquid immiscibility in the genesis of carbonatites – An experimental study. *Contributions to Mineralogy and Petrology* **73**, 105–17.

Gittins, J. 1978. Some thoughts on the current status of carbonatite studies. In *Proceedings of the First International Carbonatite Symposium, Poços de Caldas, Brazil*, June 1976, 105–15. Brasilia: Brasil Departamento Nacional da Produção Mineral.

Gittins, J. 1979. Problems inherent in the application of calcite–dolomite geothermometry to carbonatites. *Contributions to Mineralogy and Petrology* **69**, 1–4.

Hall, A. 1987. *Igneous petrology*. London: Longman.

Hay, R. L. & J. R. O'Neil 1983. Carbonatite tuffs in the Laetolil Beds of Tanzania and the Kaiserstuhl in Germany. *Contributions to Mineralogy and Petrology* **82**, 403–6.

Kjarsgaard, B. A. & D. L. Hamilton, 1988. Liquid immiscibility and the origin of alkali-poor carbonatites. *Mineralogical Magazine* **52**, 43–55.

Koster van Groos, A. F. & P. J. Wyllie 1966. Liquid immiscibility in the systems Na_2O–Al_2O_3–SiO_2–CO_2 at pressures up to 1 kilobar. *American Journal of Science* **264**, 234–55.

Koster van Groos, A. F. & P. J. Wyllie 1968. Liquid immiscibility in the join $NaAlSi_3O_8$–Na_2CO_3–H_2O and its bearing on the genesis of carbonatites. *American Journal of Science* **266**, 932–67.

Koster van Groos, A. F. & P. J. Wyllie 1973. Liquid immiscibility in the join $NaAlSi_3O_8$–$CaAl_2Si_2O_8$–Na_2CO_3–H_2O. *American Journal of Science* **273**, 465–87.

Le Bas, M. J. 1977. *Carbonatite–nephelinite volcanism*. Chichester: Wiley.

Le Bas, M. J. 1981. Carbonatite magmas. *Mineralogical Magazine* **44**, 133–40.

Le Bas, M. J. 1987. Nephelinites and carbonatites. In *Alkaline igneous rocks*, J. G. Fitton & B. G. J. Upton (eds), 53–83. Geological Society Special Publication No. 30.

Shand, S. J. 1927. *Eruptive rocks*. London: Murby.

Treiman, A. H. & E. J. Essene 1983. Mantle eclogite and carbonate: a possible source of sodic carbonatites and alkaline magmas. *Nature* **302**, 700–4.

Treiman, A. H. & E. J. Essene 1984. A periclase–dolomite–calcite carbonatite from the Oka complex, Quebec, and its calculated volatile composition. *Contributions to Mineralogy and Petrology* **85**, 149–57.

Turner, D. C. 1988. Volcanic carbonatites of the Kaluwe complex, Zambia. *Journal of the Geological Society* **145**, 95–106.

Twyman, J. D. 1983. *The generation, crystallization, and differentiation of carbonatite magmas: evidence from the Argor and Cargill complexes, Ontario*. Ph.D. thesis, University of Toronto, Toronto.

Twyman, J. D. & J. Gittins 1987. Alkalic carbonatite magma: parental or derivative? In *Alkaline igneous rocks*, J. G. Fitton & B. G. J. Upton (eds), 85–94. Geological Society Special Publication No. 30.

Verwoerd, W. J. 1978. Liquid immiscibility and the carbonatite–ijolite relationship: preliminary data on the join $NaFe^{3+}Si_2O_6$–$CaCO_3$ and related compositions. *Carnegie Institution of Washington Yearbook* **77**, 767–74.

Watkinson, D. H. & P. J. Wyllie 1971. Experimental study of the join $NaAlSiO_4$–$CaCO_3$–H_2O and the genesis of alkalic rock-carbonatite complexes. *Journal of Petrology* **12**, 357–78.

Wendlandt, R. F. & W. J. Harrison 1978. Phase equilibria and rare earth element partitioning between coexisting immiscible carbonate and silicate liquids and CO_2 vapor in the system K_2O–Al_2O_3–SiO_2–CO_2. *Annual Report of the Director of the Geophysical Laboratory, Carnegie Institution of Washington Yearbook* **77**, 695–703.

Wyllie, P. J. & O. F. Tuttle 1960. The system CaO–CO_2–H_2O and the origin of carbonatites. *Journal of Petrology* **1**, 1–46.

Yang, H. Y. 1987. Stability of ilmenite and titanomagnetite in the presence of carbon dioxide – a thermodynamic evaluation. *Contributions to Mineralogy and Petrology* **95**, 202–6.

SUBJECT INDEX*

aa lava 76, 268
accumulation 248
ACM plot 213, *9.8*
actinolite 135, 136, 205
adcumulate texture 440
aegirine 2, 20, 168, 182, 204, 205, 206, 209, 229, 241, 261, 262, 266, 438, 445, 590, Table 6.2
aegirine-augite 168, 267, 440, Table 6.2
African Plate 255, 288, 289
agglomerate 45, 73, 77, 258, 264, 274, *9.5*
agpaite 138
aillikite 216, 568
akermanite 516, *20.10*
alabandite 42, 268
albite 56, 182, 209, 211, 393, 516, 522, Table 6.2, *9.8*
albitite 60, 182, 209, 211
Aldan Shield, USSR 28
alkaline complexes 19–20, 22, 23, 27–8, 31, 79, 177, 181, 192, 193, 362, 369, 370–4, 376, 378, 379, 382, 383, 554, 568
alkaline rocks 23, 53, 55, 154, 159, 164, 179, 202, 207, 215, 216, 234, 249, 361, 364, 373, 466, 489, 518, 555, 562, 568, 582
alkalinity index 132
allanite 128, 205
allochogen 556
allochthon 206
almandine 313
alnoite 54, 568, 572, *13.6*, *22.2*
Alpine Fault, New Zealand 31
alstonite Table 6.2
alteration
 contact 201
 deuteric 58, 60, 62, 157
 hydrothermal 60, 62, 235
 metasomatic 182, 201–2, 211, 224
 post-magmatic 301
 seawater 382
 secondary 323, 510
aluminium (Al) 10, 90, 97, 155, 234, 236, 241, 248, 263, 267, 268, 272, 398, 435–6, 453, 473, 474, 477–8, 549, 586, 588, 589, 590, 592, 594, 598, *6.9*
aluminosilicate anions 95, 97, 101
alvikite 2, 3, 4, 6, 72, 76, 79, 80, 398–9, *1.2*, *22.3*
amphibole 57, 58, 60, 109, 140, 184, 202, 204, 205, 209, 211, 238, 438, 445, 450, 451, 453, 459, 468, 470, 473, 478, 479, 494, 526–30, 534, 550, 553, 554, 557, 563, 564, 566, 569, 571, 586, 589, 590, 592, 597, Table 6.2, Table 6.8, Table 6.9, **6.4**, *6.8–6.11*, *9.8*, *22.1*
amphibolite 160, 162, 182, 204, 208, 382
analcime 61, 204
anatase 161, 162, 163, 164, 172
ancylite 153, 171
andesite 438, 489, 491

andradite 265, 266, 267, 272–3, 313, Table 6.2, *13.13*
ankaramite 46
ankerite 3, 4, 5, 10, 11, 41, 42, 46, 49, 50, 57, 58, 59, 71, 108, 153, 184, 442, 443, 588, 591, Table 6.2, *1.2*, *17.7*
anorthite 393, 522
anti-correlation plot (ACP) 245, 280–1, 286–90, 291–2, 295, *10.11*, *11.8*, *12.3–7*, *19.5*
apatite 2, 6, 20, 40, 42, 43, 44, 47, 49, 50, 55, 57, 60, 61, 109, 6.3, 123–4, 131–2, 153, 154, 155, 157, 161, 162, 163, 164, 166, 167, 169, 171, 172, 184, 187, 189, 192, 204, 205, 209, 211, 224, 226, 227, 228, 238, 240, 241, 247, 248, 262, 265, 271, 272, 293, 368, 371, 402, 429, 431, 436, 440, 441, 444, 445, 456, 478, 491, 510, 511, 538, 553, 554, 566, 589, 590, 592, 596, 598, Table 6.2, Table 6.5, Table 6.6, Table 6.7, **7.4**, *6.5–6.7*
Appalachian Orogen 25, *2.5*
aragonite 506–7, 524, *20.4*, *20.5*
arfvedsonite 132, 138, 206, 438, 440, 445
argon (Ar) 327
armalcolite 549, 550
arsenic (As) 129
ash 44, 77, 78, 211, 257, 259, 260, 261, 274, *4.4*, *4.8*, *4.9*
assimilation 241, 292, 514, 518, 554
assimilation-fractional-crystallization model (AFC) 292
asthenosphere 245, 273, 294, 297, 383, 465, 488, 502, 532–3, 533–5, 538, 549, 551–3, 554, 555–6, 556, 563, 564, 575
augite 229, 258, 266, 267, 367, 590, *13.14*
auto-oxidation 564
autometasomatism 228, 318

baddeleyite 114, 161, 171, 205, 226, 228, 230, 231, 241, 402, 425, 479, Table 19.5, *19.3*
banakite 489
banding (layering) 98, 167, 170, 205, 209, 224, 227, 228, 237, 238, 373
bariopyrochlore 157, 158, 161
barite (baryte) 42, 50, 58, 59, 60, 61, 109, 124, 153, 161, 169, 171, 204, 209, 342, 348, 443, 444, 511, 512–14, 597, **7.9**, *13.50*
barium 9, 11, 125, 139, 186, 444, 445
barium (Ba) 9, 11, 52, 57, 60, 78, 94, 121, 129, 150, 153, 157, 166, 189, 192, 211, 240, 241, 263, 267, 270, 271, 407, 412, 413, 414–15, 419, 422–4, 431, 453, 456, 457, 459, 461, 465, 468, 491, 547, 549, 551, 554, Table 3.1, Table 6.2, *8.7*, *16.1*, *16.5*, *16.6*, *16.10*, *16.11*, *21.2*
barytocalcite Table 6.2
basalt 49, 61, 89, 98, 245, 256, 286, 310, 312, 314, 328–31, 330, 332, 343, 375, 383, 398, 438, 459, 466, 468, 477, 480, 482, 487, 489, 520, 549, 564,

*Page numbers in italics refer to line diagrams. Page numbers in bold refer to Sections.

566, 573, *12.6*, *13.12*, *13.30–13.32*, *13.45*, *13.36*, *14.2*, *14.8*, *19.9*, *22.1*, *22.2*, *22.6*
basanite 564, 566, *13.12*, *13.30–13.32*
base metals 189, 192
base surge deposits 72
bastnaesite 59, 94, 150, 153, 154, 155, 156, 157, 167, 168, 171, 204, 425, 443, 511, 511–12, 513–14, 554, Table 6.2, *20.9*
beforsite 2, 3, 6, 7, 155, 547, 554, *1.3*
belovite 128
belovite scheme 131
bergalite 79
berthierine 204
beryllium (Be) 129
binary mixing 244–5, 292, 294, 297, 368, 487, 494
biotite 19, 20, 45, 51, 57, 58, 157, 166, 167, 184, 204, 205, 206, 235, 236, 267, 322, 349, 431, 440, 444, 445, 590, *13.8*, *13.14*
blister mounds 45
blocks 83, 264–7
bombs 46, 77, 83, 85, *4.10–4.12*
bornite 170, 226, 227, 228, 229
boron (B) 550
breccia 49, 61, 75, 123, 124, 129, 130, 167, 179, 185, 229, 501, *3.5*, *3.7*, *8.5*
    diatreme 52, 75, 77, 200, 215
    explosion 80, 167
brecciation 47, 58, 157, 187, 228, 229
breunnerite 3, 5, *1.2*
britholite 128, 153
britholite scheme 131
bromine (Br) 484
brookite 160, 162, 163, 166
brucite 507, 509, 529, *20.6*, *20.10*
buffers
    hematite-magnetite 597, 599
    iron-wüstite 549, 556, 564
    magnetite-wüstite 549, 564
    quartz-fayalite-magnetite 564
bulk Earth 280, 281, 285, 286, 288, 292, 293, 297, 362, 364, 367, 375, 378, 381, 449, 453, 467, 479, 480, 489, 492, 493, Table 14.4, **14.3**, *14.7*, *14.8*
Bushveld Igneous Complex, South Africa 231

calcification 78
calciocarbonatite 6, 10, 11, 12, 13, 169, 435, 436, 438, *1.2*, *1.5*, *17.7*
calcite 2, 3, 5, 6, 11, 40, 41, 42, 43, 44, 45, 46, 47, 49, 50, 55, 56, 57, 58, 59, 60, 61, 71, 79, 80, 83, 85, 100, 101, 108–9, 127, 153, 155, 162, 163, 164, 166, 168, 171, 184, 204, 209, 211, 226, 227, 258, 272, 273–4, 293, 307, 313, 315–16, 317, 320, 322, 325, 331, 332, 339, 340, 367, 368, 369, 388, 390, 402, 425, 431, 433, 434, 442, 443, 504, 505, 506–7, 509, 511–12, 513, 516, 521, 522, 529, 547, 548, 549, 553, 554, 583, 585, 590, 594, 595, Table 6.2, *13.5–13.8*, *13.13–13.17*, *13.23–13.25*, *13.33*, *13.34*, *20.1*, *20.4–20.6*, *20.11*
    polycrystalline 73, 78
    primary 75, 78, 271, 433, 547, 594
    sparry 80
    tabular 42, 46, 71, 80, 85, 594

calcium (Ca) 9, 82, 91, 93, 101, 154, 155, 157, 164, 234, 263, 270, 398, 444, 445, 453, 456, 477, 484, 507, 521, 522, 538, 549, 550, 565, 566, 570, 583, 589, 594, 597, 599, *1.1*, *6.2*, *6.9*, *6.10*, *8.7*, *9.8*, *18.6*
calcrete 257, 274
calderas 72
Caledonian Orogen 25, 27, *2.5*, *2.6*
Canadian Shield 23, 284, 288, 361, 369, 372, 375, 378, 379, 382, 465, 466, Table 12.1, Table 14.1, *2.5*, *14.1*, *19.10*
cancrinite 53, 262, 516, 521, 583, *20.11*
Cape Fold Belt, South Africa 195, 551
carbon (C) 39, 52, 58, 60, 62, 80, 83, 130, 204, 326–7, 431, 523, 526, 530, 598
    juvenile 38, 39, 60
    see also isotopic tracers
carbon dioxide 9, 11, 54, 263, 270, 317, 323, 331, 398, 399, 460, 467, 468, 470, 473, 475, 478, 479, 483, 484, 493, 504, 507, 510, 511, 518, 520, 521, 522, 523, 524, 525, 526, 527, 528, 529, 530–1, 534, 535, 537, 539, 549, 553, 557, 566, 582, 587, 597, *1.1*, *22.1*
carbonated alkaline melts (hyperalkaline, hydrated carbonate alkaline) 294, 467, 468–9, 479, 493–4
carbonated nephelinite 582–5, 594
carbonated peridotite 431, 501, 585, 586, 587, 588, 589
carbonation 585, 586, 587, 588, 594, *20.3*
carbonatite activity
    increase in with time **2.5**
    repetition of **2.4**
carbonatites
    classification of 2–3
    extrusive 9, 23, 71, 85, 202, 206, 207, 209–11, 213, 216
    intrusive 206, 208–9
    metamorphosed **3.8**
    metasomatic 41
    minerals in Table 6.1
    naming of 5–6
    origin of **20.8**
    pyroclastic 44, 44–6, 56, 71, 76, 77, 179, 209, 257, 434, 594, 595, 596, 597, *4.4*, *4.5*
    source of **14.7**
    source region **12.5**
    stages of 57–8, 72, 78, 109, 125, 132, 186, 347–8, Table 6.2, Table 13.3
    volcanological classification of **4.3**
cathodoluminescence (CL) 85, 128, 131, 154, 156, 160, 166, 167, 224, 229, *4.3*, *4.12*
cation diffusivity Table 5.3
celestite 169, 342, 521
cement 171
    ankerite 46
    calcite 45, 46, 80
cerianite 156, 163, 166
cerium (Ce) 13, 78, 154, 156, 407, 426, 459, 461, Table 6.2, *8.7*, *16.1*
cerium (Ce) anomalies 118, 120, 127–8, 154, 158
CFM plot *1.2–1.4*, *8.2*, *8.8*, *9.6*, *17.7*
chalcocite 170, 226
chalcopyrite 170, 209, 226, 228, 342, 510
chemical compositions Table 1.1, Table 4.1, Table

8.3–8.6, Table 9.2, Table 11.1–11.3, Table 16.1, Table 16.3, Table 19.1, Table 19.4, *1.1*
chemical elements and oxides *see under* individual names
chlorine (Cl) 11, 130, 263, 431, 587, 588, 589, 592, 593, 597
chlorite 43, 58, 59, Table 6.2
chondritic meteorites 310, 374, 589
Chondritic Uniform Reservoir (CHUR) 280, 281, 285, 466
chondrodite 226, 228
chromium (Cr) 11, 139, 163, 234, 235, 248, 263, 407, 412, 414–15, 416, 419, 422–4, 450–1, 451, 453, 460, 551, 553, *16.1–16.5, 16.6, 16.10, 16.11*
churchite 156, 166
clasts 49, 80, 83, 211
clinohumite 57, 226, 227, 228, 589, Table 6.2
CMS projection *22.2, 22.5*
CO_2 number 564, 570, 571, 574, 575, *22.4, 22.5*
cobalt 11
coesite 585
columbite 159, 204, 205
comb structures 100, 265, 399, 440–1, *5.1*
comb-quenching 80
combeite 261, 263
cone sheets 47, 48, 50, 57, 71
cones
  ash 260, *11.5*
  parasitic 259
  pyroclastic 45, 258
  tephra 42, 45
  tuff 71, 258, 273, 274, *11.1, 11.2*
conjugate melt 128, 399, 434, 435, 436, 437, 443, 584, *17.4, 17.5*
contamination 243, 245, 292, 329, 343, 349, 362, 373, 383, 402, 430
  post-crystallization 377
  secondary 330–1
  wall-rock 439–40
continental margins 201, 202, 215, 216, *2.4*
convection 98, 102
copper (Cu) 12, 94, 129, 150, 161, 171, 192, 222, 237, 247, 263, 267, 405, 407, 412, 414–15, 416, 419, 422–4, 425, 431, **7.13**, *16.1, 16.5, 16.6, 16.10, 16.11*
coupled substitution 130–1, 139–40
crandallite-group minerals 108, 155, 156, 163, 166, 171
craters 258, 259, 260, 261, 268, *11.3–11.6*
cratons 17, 23, 30, 54, 196, 201, 215, 537, 547, 549, 554, 555, 557, 572, 587, *8.1, 8.12, 20.18, 20.19, 21.4, 22.6*
crichtonite 451, 453, 479, 551, *18.2*
crustal shortening 195
cubanite 170, 228
cumulate minerals 400
cumulates 54, 98, 102, 162–3, 224, 231, 238, 249, 265, 272, 402, 433, 440–2, 441, 453, 501, 503, 521, 566, 581
cuprite 170

dacite 438
damkjernite 567, *22.2*
davidite 451, 452, *18.3*
dawsonite 108
decalcification 163–4, 170, 172
decarbonation reactions 505, *20.13*
delta notation 301
dendritic crystallization 47, 90, 98, 100, 102, 184, 187, 192, *5.2, 5.3*
density 46, 53, 96, Table 5.3
depleted mantle source 281, 293, 294, 295, 296, 297, 373–4, 375–9, 380, 382, 442, 487, 491, 492, 535, 549, 551, 553–6, 557, 575
depolymerization 478
desilication 518
diabase 215, 230, 395, *15.7*
diamond 46–7, 189, 192, 327, 328, 332, 348, 465, 466, 524, 530, 548, 549, 551, 552, 553, 555, 556, 564, 572, *13.29, 13.33*
diapirs 61, 395, 535, 547–8, 573, 575, *15.7, 20.20*
diatremes 47, 49, 52, 61, 215, 216, 537
differentiation index 263
diffusion coefficient (D) 486
dilatant bodies 58
diopside 205, 233, 235, 267, 272, 450, 451, 453, 456, 457, 459, 468, 480, 566, 586, 588, *11.8, 13.13, 13.14, 18.5, 20.10, 20.16*
dissociation curves
  calcite *20.1*
  magnesite 503, 524, *20.1*
dissociation front 533
dissociation temperature 595
distribution (partion) coefficients (KD) 240, 382
distribution (partition) coefficients (KD) 235, 237, 240, 293, 382, 405, 411, 434, 480, 567
  effect of composition **16.3**
  effect of pressure **16.5**
  temperature **16.4**
dolerite 229–30
dolomite 2, 3, 4, 5, 6, 11, 40, 41, 47, 50, 56, 57, 58, 59, 71, 93, 108, 153, 155, 157, 168, 171, 204, 209, 227, 273, 293, 307, 315, 315–16, 318, 325, 331, 332, 339, 340, 395, 402, 425, 433, 439, 442, 443, 505–6, 507, 509, 516, 521, 524, 527, 529, 530–1, 548, 554, 563, 564, 565, 585, 586, 588, 590, 594, Table 6.2, *1.2, 13.5–13.8, 13.14–13.16, 13.23, 13.25, 13.33, 13.34, 13.41–13.43, 15.7, 17.7, 20.4, 20.10, 20.13, 20.16*
dolomitization 435, 597
domes/doming (up-doming) 16, 17, 23, 27, 31, 34, 162, 178, 194, 206, 256
drainage basins 256
dunite 52, 240, 456, 484, 549, 553
Dupal anomaly 288
dykes (dikes) 7, 17, 27, 40, 48, 49, 50, 52, 55, 56, 57, 58, 59, 60–1, 62, 71, 98, 179, 182, 184, 195, 204, 216, 223, 229–30, 230, 322–3, 325, 466, 484, 485, 548, 551, 568, 584, 591, Table 10.3, *3.3, 13.8, 13.26, 15.7, 21.1, 22.3*
  alvikite 72, 79, 80, 441
  ankerite 58, 204
  diabase 215, 395
  lamprophyre 30–1, 59, 293
  radial 42, 47
  sövite 50, 204
dysprosium (Dy) 166

East African Rift 17, 18, 19, 31, 32, 179, 194, 195, 255, *2.1*, *8.1–8.3*, *8.12*, *12.5*, *12.6*, *13.36*
eckermanite 132
eclogite 453, 586, *13.11*, *18.4*
edenite 132, 135, 136
edenitic hornblende 132
edingtonite 204
element distribution 10–13
eluvial enrichment 121
EM 1295 *12.5*
enriched source 281, 296, 379–81, 383, 449, 465, 466, 484, 554, 556
enstatite 515, *13.13*, *20.13–20.16*
epeirogenesis 178
epsilon notation (Nd and Sr) 280, Table 12.2
eruptions, phreatomagmatic 72, 75
eucolite 265, 266
euectic, sodium carbonate–magnesium carbonate 586
Eurasian Plate 29
europium (Eu) 166, 407, *16.1*
europium (Eu) anomalies 120, 127–8
eutectic
   $CaCO_3$-$Ca(OH)_2$-$BaSO_4$ 513
   $CaCO_3$-$Ca(OH)_2$-CaF_2 513
   $CaCO_3$-$Ca(OH)_2$-CaF_2-$BaSO_4$ 513
   calcite-dolomite-periclase 515
   calcite-portlandite 507, 515
extension 215
extensional basins 216

failed arm 21
fairchildite 594, *17.6*
faults/faulting 16, 17, 31, 32, 34, 49, 182, 184, 192, 216, 256. *2.1*, *2.6*, *2.7*, *8.2*, *8.3*, *8.12*, *9.3*
fayalite 564
feldspar 162, 164, 167, 182, 204, 209, 213, 241, 262, 367, 368, 371, 436, 438, 501, 511, 516, 518, 598, Table 6.2, *9.8*
feldspathization 187, *17.6*
feldspathoids 211, 501, 511, 554, 582
femit 597
fenites/fenitization 44, 47, 50, 54, 56, 58, 60, 78, 128, 138, 154, 162, 164, 166, 167, 168, 182, 184, 185, 192, 204, 205, 205–6, 208, 209, 211, 213, 214, 216, 217, 230, 233, 235, 267, 271, 273, 309, 319, 320, 399, 403, 405, 429, 430, 431, 432, 434, 435–6, 438, 438–9, 442, 449, 509, 518, 521, 562, 566, 589, 593, Table 3.1, **23.7**, *3.4*, *8.4*, *8.6*, *8.7*, *10.1–10.3*, *13.19*, *17.6*, *21.3*
fenitizing fluids *see* fenites/fenitization
ferri-kataphorite 132
ferri-winchite *6.8*
ferriphlogopite 553
ferro-hornblende 132
ferroan pargasite 132
ferrocarbonatite 2, 3, 6, 10, 11, 12, 13, 169, 184, 185, 186, 211, 402, 443–4, 445, 590, 591, 592, 593, *1.4*, *1.5*, *8.4*, *8.7*, *8.8*, *9.6*, *17.7*, *22.3*
ferrocolumbite 156, 157, 159, 171
ferrohornblende 132
fersmite 157, 159, 171, 204, 510
fertile mantle 554, 556
fervanite 168
fiamme 77

fission tracks 161, 453
fissure eruptions 259
florencite 156
flow banding (structures) 47, 50, 510, *3.3*, *4.1*
flow deformation 440
flow differentiation 224, 248
flow (movement) 56, 60, 61, 77, 98, 102, 486
flow structure 47, 50
fluid inclusions 40, 41, 44, 51, 78, 121, 167, 226, 229, 235, 248, 271, 429, 493, 597, 598
fluids 43, 62, 153, 154, 155, 157, 166, 168, 170, 187, 243, 286, 317, 318, 320, 322, 342, 348, 349, 399, 403, 438, 443, 444, 449, 450, 457, 459, 460, 461, 467, 468, 470, 478, 483, 556, 562–5, 570, 574, 592, 597, 598, **23.7**
fluorapatite 11, 121, 129–30, 182, 192, 589
fluoride 587
fluorine anomalies 211, 213
fluorine (F) 11, 60, 78, 90, 109, 130, 137–8, 164, 192, 217, 425, 431, 444, 478, 512, 587, 588, 589, 593, 595, 597, Table 6.2, *6.7*, *6.11*, *23.1*
fluorite 11, 50, 58, 59, 150, 153, 155, 157, 162, 171, 189, 192, 201, 204, 443, 444, 445, 512–14, **7.8**
fold belts 25, 27, 29, 32, 34, 182, 192, 195, 548, 555, 557, 568, *8.12*, *21.4*
forsterite 441, 444, 516, 524, 529, 585, Table 6.2, *13.13*, *20.10*, *20.13–20.16*
fractional crystallization (crystal settling) 40, 43, 56, 62, 70, 80, 97–8, 128, 248, 272, 290, 291, 292, 301, 348, 388, 401, 402, 402–3, 403, 405, 429, 435, 438, 439, 441, 442–3, 444, 445, 501, 510, 514, 517, 539, 556, 567, 575, 582, 583, 588, 592, 593, *22.2*
fractional distillation 553, *21.3*
fractionation (stable isotopes) 70, 305, *13.13–13.18*, *13.33*, *13.42*
fracture (crack) propagation 484, 534, 535, 574
francolite 123, 124
free energy of mixing 92
freudenbergite 551, 553
fugacity 43, 59, 62, 233, 238, 247–8, 249, 265, 343, 402, 441, 442, 445, 477, 510, 511, 523, 527, 530, 534, 539, 564, 590–1, 597, *20.17–21.3*, *21.4*, *22.1*
fumaroles 331
fused salts *see* ionic liquids
fusion curves (calcite, magnesite) 503–4

gadolinium (Gd) 407, 418, *16.1*
galena 207, 342, 348, 367, 369, 377, 378, *13.50*
garnet 46, 49, 261, 262, 313, 382, 453, 454, 460, 466, 486, 493, 549, 552, 553, 566, *18.5*, *22.1*
geobarometry 532
geotherm 532, 533, 534, *20.17–20.20*, *21.3*, *21.4*
  inflected 532, 534
geothermometry 40, 41, 130, 532
glass 83, 263, 265, 390, 393, 396, 475, *15.5*, *15.7*
glaucophane 132
glimmerite 128, 134, 137, 161, 162, 166, 167, 224, 231, 554, *10.1*, *10.2*, *21.4*
globules 56, 61, 396, 429
gneiss 182, 184, 202, 208
gold (Au) 171, 222, 228
Gondwanaland 177, 195, *8.14*

SUBJECT INDEX

gorceixite 156
goyazite 156
graben 27, 31
grain-edge infiltration 486
granite 19, 89, 159, 222, 228–9, 229, 241, 438, Table 10.3
granulite 373, 590, 598
graphite 524, 530, 564, *13.33*
graphite-diamond transition 532, *20.17, 21.3, 21.4, 22.1*
gravity anomalies 50–1, 163, 369, 567–8
gregoryite 42, 71, 77, 85, 94, 108, 268, 274, 431, 442, *17.6*
Grenville Province, Canada 23, 284, 284–5
grossular 313, *13.13*
groundwater leaching 58, 60, 167, 168, 596

hafnium (Hf) 382, 407, 424, 436, 454, 457, *17.4*
halite 598
halogens 468, 587, 589, 593
harzburgite 456, 473, 524, 524–6, 549, 553, Table 19.1
hastingsite 132, 590
hauyne 79, 342, 516, 517, 583, *20.11*
heat capacity 96, Table 5.3
heat of fusion Table 5.3
heat of melting 91–2, 93, 96
heat of mixing 93–5
hedenbergite 266
helium (He) 327
hematite 41, 43, 58, 59, 153, 171, 443, 512, 591, 597
hematite ore 171
hewettite 168
HIMU 295–6, *12.5*
hornblende 132, 136, 137, 202, 590, *20.16*
hornblendite 590
hot springs 167
hydration 121, 468
hydrobiotite 235
hydrogen (H) 448, 523, 526, 530, 532, 537, 551
hydrothermal activity 38, 41, 60, 62, 153, 154, 155, 157, 161, 162, 166, 167, 168, 169, 170, 171, 218, 243, 308, 316, 317, 319, 348, 444, 510, 513, **3.9**
hydrous silicate melt 467
hydroxybastnaesite *see* bastnaesite
hypabyssal intrusions **3.5**

ignimbrite 77
ijolite 2, 6, 42, 43, 44, 128, 159, 161, 194, 200, 204, 235, 258, 265, 266, 267, 291, 309, 320, 321, 322, 324, 400, 429, 434, 436, 470, 516, 517, 521, 554, 562, 567, 594, 598, Table 3.1, Table 11.2, Table 19.1, **19.50**, *13.9, 13.16, 13.19–13.21, 15.8, 17.2, 17.4*
ilmenite 55, 56, 59, 61, 164, 204, 205, 209, 211, 226, 227–8, 247, 450, 456, 476, 479, 494, 547, 549, 551, 553, 554, 556, 564, 566, Table 19.5, *19.3*
immiscibility 43, 46, 54, 55–6, 60–1, 62, 70, 92, 94, 95, 101, 128, 138, 271, 290, 291, 292, 348, 388–403, 405, 412, 424, 425, 426, 428, 429, 431, 432–4, 436, 437, 438, 439, 443, 459, 501, 502, 514, 516, 531–2, 538, 539, 547, 549, 562, 567, 583–5, 594, Table 16.2, Table 15.1, **20.5**, *15.1–15.4, 15.7–15.9, 17.2, 17.5, 17.6*

incompatible elements 405, 451, 456, 459, 460, 461
Indian plate 29, 30
ionic liquids 40, 90–1, 92, 95, 97
ionic species
  CO_3 155
  H_3O^+ 130
  OH^- 431
  SO_4 431, 444
iron (Fe) 4, 9, 10, 50, 62, 94, 114, 128, 137, 154, 163, 167, 214, 236, 241, 263, 267, 272, 431, 435, 443, 445, 449, 450–1, 451, 466, 468, 476, 478, 484, 509, 511, 517, 521, 522, 549, 556, 566, 569, 570, 583, 586, 587, 589, 590, 591, 594, 599, Table 6.2, *1.1, 6.10, 8.7, 9.8, 17.7*
Irumides, East Africa *8.1, 8.12, 8.13*
isotopes, stable 244–5
isotopic ages 18, 30, 33, 53, 179–81, 193, 202, 205, 207, 215, 229, 367, 368, 370, 371, 372, 374, 375, 465, 466, Table 8.1, *2.1–2.9, 10.4, 10.12, 14.3–14.7*
isotopic tracers
  carbon 39, 45, 55, 58, 60, 80, 83, 242, 243, 244, 245, 270, Table 3.1, Table 8.1, **13.3, 13.4**, *10.10, 13.19, 13.20, 13.22–13.37, 13.39–13.43, 13.49*
  lead 246, 270, 271, 296, 364, 370, 373–4, 379, 380–1, 465, 466, 482, 483, 487, 489, 493, 494, Table 14.1–14.4, Table 19.6, **14.4**, *10.12, 14.2–14.10, 19.6*
  neodymium ($^{143}Nd:^{144}Nd$ ratios) 245, 246, 270, 279–81, 285–97, 364, 465, 466, 480, 488, 489, 492, 493, 494, Table 12.2, Table 19.6, *10.11, 11.8, 12.2–12.7, 19.5, 19.8–19.10*
  oxygen 45, 55, 58, 60, 80, 83, 243, 244, 245, 270, 301, Table 3.1, Table 9.2, **3.4, 13.2**, *10.10, 13.1–13.21, 13.39–13.43*
  strontium ($^{87}Sr:^{88}Sr$ ratios) 55, 60, 224, 241, 242, 243, 246, 270, 271, 279–97, 364, 370, 373–4, 380, 381, 383, 465, 466, 480, 488, 489, 492, 494, 582, Table 3.1, Table 10.1, Table 12.2, Table 14.2, Table 19.6, **14.4**, *10.8–10.11, 11.8, 12.1, 12.3–12.7, 14.8, 19.5, 19.8–19.10*
  sulphur 247, 248, 301, Table 13.3, **13.5**, *13.44–13.50*

jacupirangite 204, 265, 266, 267, 517, 554, Table 11.2, *13.16*
jadeite 586, 588, *13.14*
jeppeiite 551
joins
  albite–calcite 390, 394, *15.1*
  $CaCO_3$-$Ca(OH)_2$-$La(OH)_3$ 511, *20.7*
  $CaCO_3$-$MgCO_3$ 40–1, 505
  $CaCO_3$-$MgCO_3$-$Mg(OH)_2$-$Ca(OH)_2$ *20.6*
  calcite-forsterite-vapour *20.10*
  $CaMgSi_2O_6$-$CaMg(CO_3)_2$ 530
  $Ca(OH)_2$-$La(OH)_3$ *20.7*
  $Ca(OH)_2$-$MgCO_3$-$Mg(OH)_2$ 507
  CO_2-H_2O 507
  diopside-olivine 566
  ijolite-harzburgite 470, 471–4, Table 19.2, Table 19.3, *19.2*
  $KAlSi_3O_8$-K_2CO_3 520
  $MgCO_3$-$CaCO_3$-$Ca(OH)_2$-SiO_2 *20.10*
  $MgSiO_3$-$CaMg(CO_3)_2$ 530

Na_2CO_3-$CaCO_3$ *17.1*
$NaAlSi_3O_4$-$CaCO_3$ 516
$NaAlSi_3O_8$-$CaCO_3$-$Ca(OH)_2$ 516, 520
$NaAlSi_3O_8$-Na_2CO_3 520
$NaAlSi_3O_8$-Na_2CO_3-CO_2 520
$NaAlSiO_4$-$CaCO_3$ 390, 394, 516, 522, *15.1, 20.11*
nyerereite-fairchildite 594
SiO_2-$CaCO_3$ 516

Kaapvaal craton, South Africa 17–18, 551, 568
kaersutite 132, 138, 258, 468
kalsilite 473
kamphorite 121
Kapuskasing Structural Zone 23, 31–2, 296, 297, 369, *2.5, 14.1*
Karasburg Fold Belt, Namibia 548, *21.2*
katophorite 132
Kibaran Fold Belt, Africa *8.1, 8.12, 8.13*
kimberlite 39, 52, 60, 192, 194, 200, 215–16, 233, 235, 236, 240, 242, 245–6, 270, 273, 289, 295, 297, 306–7, 323, 328, 396, 431, 448, 449, 450, 453, 456, 457, 458, 459, 460, 465, 466, 468, 520, 537, 538, 539, 546, 549–50, 552, 553, 556, 557, 562, 566, 568, 570, 572, 573, 575, 582, Table 3.1, **3.7, 18.4, 20.7, 21.2,** *3.8, 11.8, 12.7, 13.4, 15.7, 17.2, 18.4–18.8, 20.18, 20.19, 21.1–21.4, 22.2, 22.6*
kimberlitic signature **18.4**
komatiite 40, 100, 361, 372, 378, 487, 549, 564
KREEP 453
Kursk Magnetic Anomaly, USSR 137
kyanite 57, 205

lahar deposits 258
laminar flow 61, 98
lamproite 54, 245, 288, 449, 450, 468, 548, 549, 551, 553, 554, 555, 556, 557, *10.11, 21.1, 21.3*
lamprophyre 30–1, 52, 58, 61, 162, 215, 293, 324, 466, 491, *13.9, 13.19, 19.9*
lanthanum (La) 13, 78, 94, 118, 120, 128, 154, 241, 407, 451, 459, 461, 511, 513, *6.4, 6.7, 16.1*
lapilli 45, 46, 71, 72, 73, 75, 77, 79–80, 82, 85, 399, 433, 434, *4.1, 4.6–4.9, 22.3*
latent heat of fusion 97
laterite 20, 155, 156, 157, 158, 161, 164, 168, 170, 171, Table 7.3, **7.6**
latrappite 157, 159, 160, 510
lava flows (lavas) 9, 19, 41–4, 42, 45, 50, 52, 71, 76, 109, 179, 195, 241, 258, 259, 260, 261–2, 262–3, 267, 268, 268–71, 271–2, 272–4, 273–4, 306, 322–3, 325, 326, 333, 334, 396, 399, 400, 406, 433, 436, 439, 441, 519, 520, 521, 593, 595, 596, Table 11.1, Table 11.3, Table 11.4, Table 13.2, *11.3, 11.4, 11.6, 13.3, 13.27, 22.3*
lavas, potassium-rich 241, 244, 248
layering (rythmic) 51, 98, 168, 204, 209, *3.6*
lead (Pb) 30, 58, 129, 192, 205, 207, 231, 263, 270, 290, 362, 381–2, 453, 466, 467, 480, 483, 485, 487, 489, 493, 494
   see also isotopic tracers
lead (Pb) ratios 362
leucite 42, 46, 551, 590
leucitite 46, *10.11*

lherzolite 46, 273, 294, 382, 470, 478, 479, 480, 484, 492, 493, 494, 524, 526–30, 530–1, 549, 553, 588
LIL elements 245, 279, 281, 286, 288, 361, 367, 369, 370, 371, 382, 451, 466, 467, 468, 479, 480, 485, 486, 494, 549, 551
LIMA 451, 549, 551
limburgite 568
limestone 52, 60, 152, 284, 331, 518–20, 524
   isotopic composition 284
limestone assimilation (syntexis) 502, 582
lindsleyite 451, 479, 549, 550, 551
lineaments 17, 19, 20, 27, 30, 31–2
lithium (Li) 263
lithosphere 17, 32, 33, 273, 297, 448, 449, 450, 451, 460, 461, 466, 484, 492, 502, 532–3, 534, 535, 537, 549, 551–4, 556, 572, **12.5**
   reduced 564
lithosphere-asthenosphere boundary (LAB) 532, 533, 534, 535, 549, 554, 557, 571–3, 574, *20.18–20.20, 21.3, 21.4, 22.6*
lithospheric keel 293, 535, 574
lithospheric thinning 533, 535, 554, 557, 572, 573, *20.19, 22.6*
LoNd line 295
loparite 157, 162, 263
loveringite 451, 452, 453, *18.2, 18.3*
lutetium (Lu) 13, 382, 407, *16.1*

madupite 235
magma mixing 292
magmas, primary carbonatite 522, 537, 564, 568, 570, 575
magmas see rock types
magnesio-arfvedsonite 132, 134, 135, 136, 137, 138, 140, 141, *6.8, 6.9*
magnesio-hastingsite 132, 134, 136, 160
magnesio-katophorite 132, 134, 135, 136, 137, *6.9*
magnesio-riebeckite 132, 135, 136, 137, 138, 141, *6.8, 6.9*
magnesiocarbonatite 3, 6, 10, 11, 12, 13, 169, 184, 186, 211, 431, 435, 438, 441, 443, 445, *1.3, 1.5, 8.4, 8.7, 8.8, 9.6, 17.5, 17.7*
magnesioferrite 509
magnesite 3, 5, 505, 524–6, 527, 529, 554, 562, 565, 585, Table 6.2, Table 10.2, *20.1, 20.4, 20.13–20.15*
magnesium index 132
magnesium (Mg) 9, 41, 50, 62, 91, 93, 101, 155, 214, 241, 263, 267, 272, 431, 445, 453, 477, 478, 501, 503, 505–6, 507, 509, 511, 514–16, 517, 521, 522, 524, 525, 527, 529, 530, 531, 538, 547, 550, 551, 565, 566, 569, 570, 583, 588, 589, 590, 592, 594, 599, Table 6.2, *1.1, 6.10, 8.7, 9.8, 17.7*
magnesium number see Mg numbers
magnetic (aeromagnetic) anomalies 137, 163, 369
magnetite 2, 10, 40, 49, 50, 55, 57, 58, 59, 60, 85, 121, 157, 161, 162, 162–3, 164, 166, 167, 168, 171, 204, 205, 209, 211, 224, 226, 227, 228, 235, 236, 237, 238, 247, 248, 272, 402, 431, 440, 441, 443, 444, 445, 549, 554, 564, 589, 590, 591, 597, *13.8, 13.14*
magnetoplumbite 549, 551
majorite 460

malachite 170
malignite 159
manganese (Mn) 9, 11, 42, 50, 57, 78, 108, 124, 139, 159, 166, 211, 263, 412, 413, 414–15, 416, 419, 422–4, 435, 451, 556, Table 6.2, *6.5, 16.1, 16.10, 16.11*
mantle, redox state of 529–30, 556
mantle array 281, 286, 489, 491, 493, 494, *19.5*
mantle evolution model 361–2
mantle metasomatism 286, 294, 295, 448–9, 465–95, 557, 587
mantle norm 566, 579
mantle plumes 22, 32, 294, 297, 449, 533, 573, *20.19*
marble 52, 60, 152, 206, 208, 284
MARID suite 450–1, 454, 456–8, 460–1, *18.1, 18.5–18.8, 21.4*
mathiasite 451, 549
melanephelinite 43, 259, 271, 400, 567
melanite 262
melilite 42, 44, 45, 51, 52, 57, 58, 60, 62, 73, 78, 229, 258, 261, 266, 322, 515, 516, 517, 564, 566, 583, Table 3.1, *13.8, 13.14, 15.9, 20.11*
melilitite 45, 50, 53, 54, 55, 79, 258, 263, 399, 400, 501, 502, 538, 548, 549, 557, 564, 566, 567, 568, 569, 570, 573–4, 575, Table 3.1, *17.2, 21.1, 22.1, 22.2*
melt inclusions 101
melteigite 42, 159, 205, 235, 241, 265, 266, 517, Table 3.1, 517, 521, 554, Table 11.2
melts, primary 53, 62, 77, 562, 564–5, 565–6, 566–7, 567–8, 591
metamorphism 56, 202, 204, 206
metasomatism (metasomites, metasomatic activity) 41, 43, 54, 62, 166, 167, 204, 223, 233, 236, 245, 249, 258, 265, 267, 273, 286, 292, 293–4, 294, 309, 320, 349, 383, 399, 428, 431, 448, 449, 450, 451, 453, 455, 456, 457, 459, 460, 461, 466–7, 467, 480, 487, 501, 526, 532, 534, 535, 547, **21.3**, 550, 551, 551–3, 553, 554, 555, 556, 557, 575, 598, *13.9, 13.16, 18.1, 18.2, 18.6, 19.4–19.9, 20.18–20.20, 21.2, 21.4*
meteoric waters 44, 45, 55, 58, 78, 80, 164, 317, 319–20, 431
meteorites 310, 312, 314, 327, 328, 332, 343, 348, 375, *13.10, 13.28, 13.45*
methane 530, 564, 598
Mg numbers 235, 240, 562, 566, 567, 568, 569, 570, 575, Table 10.3, *22.2*
mica 224, 235–7, 258, 450, 451, 492, 557, 568, 570, 572, 574, 575, 586, 589, 590, 592, Table 10.2, *10.7*
microcline 205, 229
mineralization 123, 189, 189–92, 195, 222, 228, 229, 431, 443–4, 510, **7.2**
minerals
  in carbonatites Table 6.1
  see under individual minerals
minette 215, 466
miogeocline 205, 216, 217
miscibility gap 389, 390, 391, 395, 403, 468, 520, 521, 522, 531
mobile belts *see* fold belts
molar properties Table 5.1
molybdenite 209

monazite 23, 128, 153, 154, 155, 156, 163, 166, 167, 170, 171, 205, 209, 227, 511
monchiquite 492
monticellite 42, 44, 49, 51, 57, 58, 60, 515, 516, 553, 566, 568, 572, Table 6.2, *13.8, 20.10*
montrosite 168
monzonite 202, 205
MORB 245, 286, 294, 295, 331, 363, 364, 369, 374, 382, 383, 467, 489, 564, *12.3, 13.30, 13.31, 13.45, 14,2, 19.9*
Mozambique Belt, East Africa *8.1*
muscovite Table 6.2

natrocarbonatite 2, 6, 41, 42, 43, 44, 54, 71, 80, 85, 258, 259, 260, 262, 267–72, 273, 274, 291–2, 388, 399, 402, 403, 429, 431, 432, 433, 434, 436, 438, 439, 440, 442, 444, 509, 520, 521, 537, *11.1–11.8, 17.2, 17.3, 17.5, 17.6, 23.1*
natrocarbonatite *vs.* calcite carbonatite **4.4**
natrolite 204
navajoite 168
neodymium (Nd) 58, 272–3, 278, 279, 280, 362, 367, 375, 381, 382, 465, 466, 480, 487, 488, 491, 494, Table 12.1, Table 12.2, *6.4, 8.7*
see also isotopic tracers
nepheline 45, 52, 57, 58, 60, 205, 258, 261, 262, 265, 266, 267, 322, 388, 390, 402, 431, 435, 459, 473, 516, 521, 531, 573, 583, 590, Table 3.1, *13.14, 15.6, 15.9, 20.11*
nepheline syenite 6, 19, 25, 50, 52, 56, 62, 120, 157, 161, 162, 164, 265, 267, 400, 429, 521, *15.8, 17.2*
nephelinite 39, 41, 42, 49, 50, 53, 54, 79, 194, 257, 258, 262, 263, 267, 271, 272, 288, 292, 294, 296, 400, 406, 411, 414–15, 419, 424, 426, 429, 440, 453, 487, 491, 502, 516, 539, 562, 564, 566, 567, 568, 570, 573–4, 574, 575, 582, 584, 585, 588, 598, Table 3.1, Table 11.1, Table 16.1, **17.4, 20.7**, *15.6–15.8, 16.1, 16.4, 16.6, 16.9, 16.11, 17.2–17.5, 22.1, 22.2, 22.6*
nickel (Ni) 11, 228, 237, 263, 308, 407, 453, 587–8, 591
niobium (Nb) 9, 12, 50, 55, 57, 58, 60, 61, 94, 101, 129, 139–40, 150, 155, 160, 161, 162, 163, 166, 171, 183, 189, 204, 211, 213, 217, 240, 267, 405, 451, 456, 457, 458, 468, 510, 550, 551, 553, 554, 583, 586, 587, 588, 589, Table 3.1, Table 7.3, **7.3**, *6.1, 18.7*
niobium (Nb) anomalies 166
niocalcite 157, 159, 171
nodules 451, 466–7, 468, 520, 532
non-bridging oxygens (NBO) 411, 412, 567
nordmarkite 19
nosean 342
nuée ardente 77
nyerereite 42, 44, 45, 46, 71, 73, 77, 78, 80, 94, 108, 268, 272, 274, 399, 431, 432, 433, 442, 509, 594, 595

ocelli 56, 395, 396, 584, *15.7*
Ohre rift 27
OIB 245, 286, 288, 294, 295, 297, 361, 363, 364, 369, 382, 383, 489, 535, 563, 575, *12.3, 12.6, 13.30–13.32, 14.2, 19.9*

okaite 79, 309, 322, 324, 517, 583, Table 3.1, *13.9, 13.16, 13.19*
oldhamite 510, 511
olivine 40, 42, 43, 44, 46, 50, 55, 57, 58, 60, 79, 100, 205, 224, 226, 227, 235, 236, 237, 248, 258, 263, 266, 267, 271, 313, 330, 431, 440, 441, 466, 470, 473, 477, 478, 480, 487, 501, 515, 520, 524, 529, 551, 553, 554, 564, 566, 567, 568, 570, 573, 583, 585, 586, 587–8, 589, 590, Table 10.2, *15.9, 18.5*
olivinite 441
orogenesis (orogenic activity) 15, 23, 25, 27, 29, 32, 33, 34, 177, 178, 194, 195, 201, 202, 204, 216
orogenic belts *see* fold belts
orthopyroxene 470, 473, 478, 487, 494, 524
orthosilicate anions 97
oscillatory zoning 229
overlapping phase volume 527
oxidation 141
oxygen (O) 58, 60, 80, 83, 431, 510, 511, 523, 526, 530
   see also fugacity, isotopic tracers

pahoehoe lava 42, 76, 262, 268
palaeomagnetism 229, 231
palagonitization 263
Pan-African event 193, 195
Pangaea 18, 33
pantellerite *15.1*
parent magma 271–2, 398, 399
pargasite 132, 137, 468, 590
parisite 94, 153, 154, 156, 168, 171, Table 6.2
partition coefficients 480
pegmatite 159, 170, 224, 308, 460, *13.8*
Pele's tears 40, 71, 75, 80, 82, 85, 399
pentlandite 226
periclase 40, 42, 49, 60, 507, 509, 516, *20.1, 20.6, 20.10, 20.14, 20.15*
peridotite 2, 43, 54, 163, 273, 313, 348, 431, 450, 451, 453, 455, 456, 457, 459, 461, 466, 467–9, 470, 474, 493, 494, 502, 520, 523, 524, 526, 527, 530, 534, 550, 562, 564, 565, 566, 567, 568, 569, 570, 572, 574, 575, 586, *19.50, 11.8, 13.13, 13.18, 18.6, 18.7, 19.1, 22.1, 22.3*
peridotite solidus 54, 294, 448, 551, 554, 557, 566, 572, 574
perovskite 42, 44, 58, 60, 61, 156, 160, 161, 162, 162–4, 169, 170, 171, 204, 265, 272, 436, 441, 456, 458, 476, 510, 511, 548, 553, 554, 583, *15.6*
phenocrysts 42, 44, 45, 46, 47, 50, 71, 75, 80, 83, 85, 98, 132, 138, 227, 262, 268, 271, 431, 433, 436, 441, 520, 594, *4.6, 4.10–4.12, 15.16*
phlogopite 27, 40, 42, 45, 50, 55, 56, 57, 58, 60, 137, 157, 162, 166, 167, 182, 187, 204, 205, 209, 211, 224, 226, 227, 231, 235, 236, 237, 247, 248, 249, 258, 266, 267, 382, 440, 444, 445, 451, 453, 455, 456, 457, 459, 526, 527, 529, 550, 551, 553, 554, 563, 565, 566, 569, 587, 590, Table 6.2, *18.1, 18.5, 20.16*
phonolite 39, 41, 48, 49, 53, 54, 58, 79, 256, 257, 262, 263, 267, 271, 272, 292, 388, 399, 401, 402, 403, 406, 411, 412, 413, 414, 415, 417–18, 424, 426, 429, 433, 434, 435, 437, 438, 440, 584, 585, Table 3.1, Table 16.1, *15.8, 16.1–16.3, 16.5, 16.7, 16.8, 16.10, 17.2, 17.3, 17.5, 17.6*

phoscorite 9, 121, 128, 130, 134, 170, 222, 224, 226, 227, 228, 231, 235, 237, 240, 241, 248, *10.1–10.3*
phosphate 150, 153, 154, 160, 162, 189, 489, 491, 502, 511
phosphorite 160–1, 172, *6.5*
phosphorus (P) 9, 11, 60, 61, 90, 130, 154, 161, 162, 163, 192, 222, 238, 263, 270, 272, 405, 425, 440, 474–8, 476, 478, 491, 501, 510, 538–9, 554, 586, 589, 592, 593, 595–6, Table 19.5, *1.1, 8.7, 19.3*
phreatomagmatic activity 72, 75, 76, 77
picrite 564, 565, 566
pipes 47, 77, 170, 222, 228–9, 248, 551
pirrsonite 268, 274
plagioclase 50, 60, 209, 211
platinum (Pt) group elements 171, 222, 228
pleochroism, reverse 209, 226, 227, 235, 236, 590
plugs 47, 50, 57, 98, 162, 179, 184, 223, 273
plumes *see* mantle plumes
plutons 53, 58, 371, 373–4, 379, **3.6**
podolite 123
polymerization 90, 95, 97, 101, 411, 413, 425, 478
portlandite 78, 506, 507, 509, *20.16*
potassium (K) 11, 62, 91, 121, 137–8, 213, 240, 241, 248, 263, 267, 270, 271, 272, 431, 434, 438, 439, 442, 444, 445, 453, 456, 458, 468, 474, 478, 521, 522, 526, 547, 549, 550, 551, 553, 554, 557, 568, 586, 587, 589, 590, 592, 593, 594, 598, 599
prehnite Table 6.2
priderite 548, 551, 553
primitive liquids 263, 272, 478
pseudocarbonates **3.10**
pseudomorphs 55, 73, 78
pumice 77
pyrite 58, 60, 204, 209, 211, 342, 368, 511, *13.50*
pyrochlore 2, 6, 12, 28, 49, 54, 55, 57, 59, 60, 101, 156, 157–9, 160, 164, 166, 169, 171, 182, 189, 204, 205, 209, 402, 425, 445, 458, 510, 511, 554, Table 3.1, Table 6.2–6.5, Table 7.2, **6.2**, *6.1–6.4*
pyroclastic flow deposits 45, 56, 76, 77, 594, 595, 596
pyrope 313, *13.13*
pyrophanite 556
pyroxene 42, 44, 45, 49, 50, 51, 52, 54, 55, 57, 58, 60, 62, 100, 184, 202, 205, 211, 223, 226, 229, 231, 233, 233–5, 240, 242, 244–5, 262, 265, 266, 272–3, 313, 322, 390, 445, 453, 460, 466, 467, 473, 478, 480, 484, 487, 494, 501, 524, 551, 553, 554, 566, 583, 585, 586, 587, 589, 592, 597, Table 3.1, Table 10.2, *9.8, 10.5, 10.6, 10.10, 13.8, 15.6, 15.9, 20.13*
pyroxenite 52, 128, 159, 161, 162, 163, 166, 167, 170, 171, 182, 187, 222, 223, 224, 226, 227, 228–9, 231, 235, 236, 238, 240, 241, 248, 249, 265, 266, 267, 272, 291, 468, 470, 474, 480, 492, 493, 553, 590, Table 10.3, *10.1–10.3*
pyrrhotite 57, 209, 261, 265, 342, 442

quartz 50, 58, 59, 153, 155, 157, 162, 166, 167, 168, 171, 524, 564, 585, Table 6.2, *20.13–20.15*
quartz syenite 228, 229, 438
quartzite 154, 206
quench crystals 90, 265, 433, 434, 440, 548
quenched droplets 40, 80
quenched liquids 39, 80, 85, 90, 472, 475, Table 15.2
quenching, volcanic 85

radiometric anomalies 169, 170
radium (Ra) 54
rauhaugite 2, 3, 6, 157, 168, 204, 205, 308, 309, 336, 431, *1.3*, *22.3*
Rayleigh fractionation 331, 335, 336, 337, 340, 341, 348, *13.38*, *13.40*
Rayleigh number 98
recrystallization 78, 80, 228, 279, 291, 294
redox melting 534
redox state, mantle 529, 556
reduction-exsolution 564
REEs 13, 52, 54–5, 57, 59, 60, 78, 109, 117–21, 118, 121, 123, 125, 128, 131, 140, 150, 156–7, 157, 158, 161, 163, 164, 166, 167, 169, 170, 171, 186, 187, 188, 189, 204, 213, 217, 240, 241, 248, 263, 270, 271, 291, 294, 405, 413, 415, 416, 417–19, 424, 425, 426, 431, 436, 437, 441, 443, 444, 445, 449, 450, 451, 452, 453, 457, 459, 460, 461, 465, 466, 492, 502, 511–12, 513, 538–9, 547, 548, 549, 550, 554, 583, 586, 587, 589, 597, Table 1.2, Table 3.1, Table 6.5, Table 6.7, **7.2**, *1.5*, *6.3*, *6.6*, *8.9*, *8.10*, *8.11*, *9.7*, *11.4*, *11.7*, *16.1*, *16.2–16.4*, *16.7–16.9*, *20.6*, *20.9*
regular solution parameters 92, 94, 95, Table 5.2
replacement bodies 58, 59
rhabdophane 156
rhyodacite 477
rhyolite 96, 521
rhythmic layering 51, 204, *52*
richterite 57, 132, 134, 135, 137, 140, 205, 382, 440, 451, 456, 457, 468, 550, 551, 553, *6.8*, *18.1*
riebeckite 132, 134, 138, 597
rifts/rifting (rift zones, palaeorifts) 16, 18, 19, 27, 28–9, 29, 31, 34, 177, 178, 194, 195, 196, 215, 216, 217, 255, 256, 258, 263, 334, 535, 537, 538, 547, 548, 555, 557, 574, *2.1*, *2.6*, *8.1*, *8.3*, *8.12–8.14*, *11.1*, *11.2*, *12.5*, *12.6*, *14.1*, *20.19*, *21.4*, *22.6*
  see also East African Rift
ring dykes 20, 42, 47, 48, 50
rödbergite 58, 171, 443
rubidium (Rb) 279
rubidium (Rb) 20, 23, 30, 181, 241, 242, 263, 270, 281, 381, 453, 456, 459, 461, 465, 468, 479, 480, 494, 554, *21.2*
rutile 58, 59, 156, 160, 162, 166, 204, 205, 450, 456, 460, 476, 548, 549, 551, 553, 591

sag structures 46
salite 162, 163, 233, 262
samarium (Sm) 166, 279, 381, 407, 465, 466, 480, 487, 488, 494, Table 1.2, *16.1*
sanidine 168, 262, 265, 267, 473
saprolitic cover 168
schlieren 184, 223, 440, 441
schorlomite 261, 262
scoria 46, 268
sedimentation basins 256
segregations, calcite-rich 60–1
selenium (Se) 228
sericite Table 6.2
serpentine 55, 224, 553, Table 6.2
serpentinization 226, 227
settling velocities 98

shear zones 57, 181, 184, 192, 193, 195, 196, 228, *8.2*
shonkinite 23
shortite *23.1*
shoshonite 489
siderite 3, 5, 11, 109, 204, 591, *1.2–1.4*, *13.14*
silicate incompatible elements (SIEs) 547, 548, 549, 551, 553, 554, 556, 557, *21.2*, *21.3*
silicification 162, 187, 443, 444
silicocarbonatite 2, 139, 159, 161, 440, 445, 583, *3.5*
silicon (Si) 2, 7, 10, 41, 62, 90, 97, 129, 241, 263, 267, 268, 320, 398, 435, 468, 478, 501, 511, 523–4, 525, 531, 538–9, 549, 566, 583, 586, 588, 589, 590, 592, 594, 597, 598, *1.1*, *6.9*
sillimanite 57, 208
sills (sill-like bodies) 49, 56, 60–1, 98, 159, 202, 205, 215, 343, 396, 551, *13.45*, *15.7*, *21.2*
silver (Ag) 222, 228
skeletal crystals 90, 98, 100, 102, 184, 187, 192
slags 323, 324, 325, *13.3*, *13.27*
sodalite 200, 201, 205, 267
sodium (Na) 11, 62, 71, 82, 85, 91, 128, 131, 157, 164, 213, 234, 241, 248, 263, 267, 270, 271, 272, 434, 438, 439, 442, 444, 445, 453, 468, 473, 474, 478, 521, 522, 538, 547, 550, 553, 557, 568, 586, 589, 592, 593, 594, 597, 598, 599, Table 6.2, *6.7*, *6.9*, *6.10*, *9.8*, *17.3*, *17.6*
soils, residual 189, 192
solidus ledge (shoulder) 527, 531, 535, 537, 538, 552, 569, 574
solubility, CO_2 504
solution cavities 167
solvus 138, 247, 400, 401, 402, 432, 505, 507, *15.7–15.9*, *17.2*, *17.3*
sövite 2, 3, 4, 6, 41, 42, 43, 44, 46, 49, 50, 53, 79, 80, 109, 111, 113, 114, 115, 139, 155, 157, 159, 160, 161, 167, 168, 184, 185, 186, 187, 192, 205, 211, 258, 265, 271, 272, 308, 309, 320, 322, 324, 432, 440, 441, 443, 444, 445, 554, *1.2*, *8.4*, *8.7*, *9.6*, *13.16*, *13.21*, *22.3*
spatter 71, 73, 77, 82, *4.2*, *4.3*, *11.3*
spherules 40, 56, 390, 396
spinel 46, 55, 56, 61, 95, 453, 547, 549, 550, 553, 556, 564
spurrite 42, 95
staffelite 123
stockworks 58, 492
stoping 52
strontianite 58, 59, 124, 153, 155, 168, 169, 171, 209, Table 6.2, *13.14*
strontium (Sr) 11, 20, 23, 30, 52, 57, 58, 60, 61, 62, 109, 111, 124, 139, 150, 153, 155, 166, 181, 186, 211, 240, 241, 242–4, 263, 267, 270, 271, 272–3, 278, 279, 280, 281, 362, 367, 375, 381, 382, 431, 442, 456, 459, 461, 465, 468, 479, 480, 484, 487, 494, 547, 550, 551, 554, 586, 587, 589, Table 3.1, Table 6.2, Table 12.1, Table 12.2, Table 19.6, **7.11**, *6.5*, *6.7*, *8.7*, *10.9*, *18.6*, *21.2*
  see also isotopic tracers
sub-continental upper mantle (SCUM) 244, 278, 296, 382, 383, 450, 451, 465, 466, 467, 484, 485, 488, 491, 492, 493, 494, 549, 556, 573, 574, 576
subduction 216, 294, 296, 311–12, 331, 382, 383, 524, 556
subsolidus assemblages 516

SUBJECT INDEX

sulphides 153, 226–8, 237, 241, 247, 248–9, 502, 510–11, *13.44*
sulphur (S) 9, 11, 101, 129, 153, 263, 270, 431, 444, 468, 510, 511, 597, **13.5**
supercritical brine 597
supergene 155–6, 166, 169, 171
superheat 399, 402
Superior Province, Canada 247, 284–5, 293, 296, 297
surge deposits 45, 72, 73, 76, 77
swells *see* domes/doming
syenite 19, 29, 53, 59, 62, 114, 120, 161, 181, 200, 201, 202, 204, 205, 206, 211, 213, 215, 222, 223, 228–9, 229, 231, 240, 241, 248, 265, 285, 291, 368, 371, 429, 554, Table 3.1, Table 10.3, Table 11.2, *10.1*
sylvite 268
synchysite 94, 153, 156
syntexis 502, 514, 518–20, 537, 582
systems
  albite-calcite 585
  albite-calcite-H_2O 390
  albite-sodium carbonate-H_2O 389
  anorthite-calcite 403, 585
  $BaSO_4$-$CaCO_3$-CaF_2-$Ca(OH)_2$ *20.8*
  $BaSO_4$-$CaCO_3$-CaF_2-$Ca(OH)_2$-$La(OH)_2$ *20.9*
  $BaSO_4$-$CaCO_3$-CaF_2-H_2O 511
  C-H-O 537
  $CaCO_3$-CaF_2 93
  $CaCO_3$-CaF_2-$BaSO_4$ 512
  $CaCO_3$-CaF_2-$Ca(OH)_2$-$BaSO_4$ 39
  $CaCO_3$-CaF_2-$Ca(OH)_2$-$BaSO_4$-$La(OH)_2$ 39
  $CaCO_3$-$Ca(OH)_2$ 93
  $CaCO_3$-$Ca(OH)_2$-CaF_2 512
  $CaCO_3$-$Ca(OH)_2$-CaF_2-$BaSO_4$ 512–14
  $CaCO_3$-$Ca(OH)_2$-CaS 510–11
  $CaCO_3$-$Ca(OH)_2$-$La(OH)_3$ 39, 511–12
  $CaCO_3$-K_2CO_3 93
  $CaCO_3$-$MgCO_3$ 507, *20.4*
  $CaCO_3$-Na_2CO_3 93, 431, 595, *23.1*
  $CaCO_3$-Na_2CO_3-apatite 595–6
  $CaCO_3$-Na_2CO_3-K_2CO_3 39, 41, 76, 268, 529, 594
  CaO-CO_2 503–6, *20.2*
  CaO-CO_2-H_2O 41, 71, 76, 78, 501, 506–7, *20.5*
  CaO-La_2O_3-CO_2-H_2O 511, 512, 514
  CaO-MgO-CO_2 505–6
  CaO-MgO-CO_2-H_2O 40, 41, 507, 509, *20.6*
  CaO-MgO-SiO_2 524–6, 562
  CaO-MgO-SiO_2-CO_2 524, 526–7, 530, *20.13*
  CaO-MgO-SiO_2-CO_2-H_2O 41, 514–16, *20.10*
  CaO-Na_2O-Al_2O_3-SiO_2-CO_2-H_2O *20.12*
  CaO-Nb_2O_5-CO_2-H_2O 510
  CaO-SiO_2-CO_2-H_2O 514–15, 519
  CO_2-CaO-P_2O_5 154
  jadeite-calcite-CO_2 586
  K_2CO_3-$MgCO_3$ 39, 93
  $KAlSi_3O_8$-K_2CO_3 584
  $KAlSiO_4$-MgO-SiO_2 562
  lherzolite-C-H-O 526–30, 533
  lherzolite-H_2O-CO_2 529, *20.1*, *20.16*
  MgO-CO_2 503–6, *20.3*
  MgO-SiO_2 524–6
  MgO-SiO_2-CO_2 530–1, *20.14*, *20.15*
  MgO-SiO_2-CO_2-H_2O 529
  $NaAlSi_3O_8$-$CaAl_2Si_2O_8$-Na_2CO_3-H_2 584
  $NaAlSi_3O_8$-Na_2CO_3-CO_2 584
  $NaAlSiO_4$-$CaCO_3$ 583
  $NaFeSi_2O_6$-Na_2CO_3 584
  nepheline-calcite-H_2O 390
  peridotite-C-H-O 532, 535
  peridotite-CO_2 530
  peridotite-H_2O-CO_2, 478, 484, 493–4, 562, 563, **19.3**, *19.1*, *19.4*, *20.17*, *22.1*, *22.4*, *22.5*
  peridotite-kimberlite-nephelinite-carbonatite-CO_2-H_2O 539
  $(SiO_2 + Al_2O_3)$-Na_2O-CaO-CO_2 *15.4*
  $(SiO_2 + TiO_2 + Al_2O_3) - (Na_2O + K_2O)$-$CaO + MgO + FeO) - (CO_2)$ *15.9*

tantalum (Ta) 94, 113, 115, 129, 139, 150, 158, 159, 170, 171, 213, 407, 412, 414–15, 416, 419, 422–4, 436, *6.1*, *6.2*, *16.1*, *16.5*, *16.6*, *16.10*, *16.11*, *17.4*
Tanzanian Shield *8.1*, *8.12*
taphrogenesis 177
tellurium (Te) 228
temperature of crystallization 40, 247–8, 400, 437
temperature, liquidus 76, 268
tephra 41, 42, 43, 50, 52, 209, 211, 271, 274, 434, **3.4**
tephrite 42, 53, 54, 79
ternary eutectic 506, 512–14
tetraferriphlogopite Table 6.2, 440, 590
tetranatrolite 204
thermal boundary layer 574
thermal conductivity Table 5.3
thermal death 484, 534, 551, 574
thermal diffusivity 96, Table 5.3
thermal expansion 96, Table 5.3
tholeiite 245, 564, 575
thorite 169
thorium (Th) 13, 54, 60, 113–14, 128, 129, 150, 153, 163, 228, 270, 361, 362, 368, 443, 444, 453, 465, 467, 480, 483, 487, 489, 491, 493, 494, 512, Table 3.1, Table 14.1, **7.12**, **14.6**
tie lines
  albite–calcite *15.3*
  nephelinite–calcic natrocarbonatite *17.2*
  phonolite–natrocarbonatite 399, *17.2*
  silicate–carbonate 390, 396, *15.7*
tilleyite 95
tin (Sn) 114
titanate minerals 449, 450, 451, 453, 457, 458, 459, 460, 461, 550, 551, 554, 557, Table 18.1, **18.3**, *18.2–18.7*
titanaugite *13.14*
titanite 49, 128, 156, 160, 162, 202, 209, 262, 265, 371, 452, 476
titanium (Ti) 10, 55, 94, 135, 138, 150, 158, 159, 162, 163, 171, 228, 234, 267, 272, 431, 441, 449, 450, 451, 456, 458, 460, 466, 468, 474–8, 494, 547, 549, 550, 551, 553, 554, Table 19.5, **7.5**, *6.1*, *18.1*, *18.7*, *19.3*, *21.2*
titanoclinohumite 589, 590
titanomagnetite 42, 46, 163, 262, 265, 267, 548, 549, 591
trachyte 45, 58, 228–9, 292
transform faults 31, 32
tremolite 132, 135, 136, 205, *6.8*, *6.9*
triple junctions 21, 22

trona 45, 268
tuff 9, 44, 45, 46, 71, 72, 73, 75, 77, 79–80, 82, 83, 85, 109, 114, 138, 202, 211, 258, 259, 273–4, 306, 322–3, 325, 326, 334, 433, *4.7*, *4.8*, *9.4*, *9.5*, *13.3*, *13.27*
tuff rings 77
turjaite 79, Table 3.1

Ubendian belt, East Africa 182, *8.1–8.3*, *8.12*, *8.13*
ultramafic rocks (bodies, lavas, liquids) 46, 50–4, 161, 167, 168, 201, 202, 237, 243, 312–13, 314, 320, 396, 547, 549, 553, 568, 590, *13.11*, *15.7*
ulvospinel 226
uncompahgrite Table 3.1
uranium (U) 13, 30, 113, 114, 115, 150, 158, 169, 205, 207, 222, 228, 231, 270, 290, 361, 362, 371, 443, 444, 451, 453, 454, 459, 461, 465, 466, 467, 480, 483, 487, 489, 493, 494, 512, Table 14.1, **7.12**, *6.2*
uranoan thorianite 228, 230, 231, 246
urtite 128, 159, 202, 204, 205, 470, 517, 521, 554, Table 3.1, *13.16*

valleriite 170, 228
vanadium (V) 11, 163, 263, **7.10**
vapour phase 399, 521, 532, 539
vapour phase transfer 291
vein carbonate 184, 187, 320, 321–2, 322–3, 398, 443, 511, *8.7*, *13.2*, *13.8*, *13.20*, *13.21*, *13.26*
veins
  dilatant 50, 60, 62, **3.9**
  metasomatic 479, 480, 483–6, 487, 491–4, 574, *19.5*, *19.7*
  mineralized 153, 154, 157, 168, 170, 171, 192, *13.2*
  replacement 50, 153, 171
vent 77, 80
vermiculite 49, 150, 161, 235, **7.7**
vesicles 60–1, 82, 83, 263, 268, *15.5*
viscosity of melts 42, 46, 53, 61, 80, 96, 97, 98, 102, 268, 402, 441, Table 3.1
vishnevite 262, 266
volatile components 53, 217, 399, 428, 448, 457, 468, 477, 502, 523, 529, 547, 551, 553, 556, 557, 564, 575, *20.18*
volcanism 573
  alkaline 179
  basaltic 194, 449
volcanoes, carbonatite **4.2**

wall-rock reaction 189, 263, 428, 439–40, 444–5, 450–1, 468, 469, 478, 486, 487, 494

Walvis Ridge 295, 365, 375–6, 377, 493, *12.6*, *14.2*, *14.8–14.10*, *19.9*
water 9, 54, 95–6, 97, 101, 131, 154, 166, 263, 399, 402, 425, 429, 449, 453, 460, 467, 468, 469–70, 473, 484, 502, 503, 507, 511, 513, 518, 520, 522, 526–9, 530, 535, 537, 538, 539, 547, 549, 551, 553, 557, 562, 563, 566, 570–1, 575, 583, 592, 593, 597, 598, *22.1*
weathering 121, 153, 167
  chemical 155, 168
  lateritic 163, 164, 166, 171
weathering profile 121, 165–6
websterite 470, 524
wehrlite 493, 524
weloganite 108
whitlockite 296, 476, 479, 482, 484, 491, 494, Table 19.5, *19.3*
winchite 132
witherite *13.14*
wöhlerite 156, 157, 159–60, 171
wollastonite 261, 262, 265, 521, Table 6.2
wüstite 549, 556

xenocrysts 45, 136, 140, 226, 237, 242, 262, 553, 596
xenoliths 46, 50, 58, 137, 167, 184, 229, 233, 235, 258, 295, 310, 311, 312, 314, 383, 440, 449, 450, 451, 459, 460, 466, 467, 468, 480, 484, 491, 492, 493, 532, 538, 547, 549–50, 553, 556, 562, 564, 573, *3.8*, *11.8*, *13.11*, *19.9*
xenon (Xe) 327
xenotime 156, 166

ytterbium (Yb) 186, 407, *16.1*
yttrium (Y) 78, 156, 166, 188, 451

Zambezi Rift 17
zeolites 45, 58, 204
zinc (Zn) 192, 207, 263, 270, 451
zircon 114, 166, 181, 202, 204, 205, 207, 209, 231, 367, 371, 372, 442, 454, 476, 479, 551, 553, Table 19.5, *19.3*
zirconium (Zr) 55, 60, 78, 94, 114, 129, 150, 159, 222, 228, 240–1, 241, 248, 263, 405, 407, 412, 414–15, 416, 419, 422–4, 436, 451, 454, 456, 457, 474–8, 494, 547, 550, 551, 553, 554, Table 3.1, Table 19.5, *16.1*, *16.5*, *16.10*, *16.11*, *19.3*, *21.2*
zoned crystals 42, 85, 116–17, 120, 131–2, 140, 229, 266, 466, 590

LOCATION INDEX

Adiounedj (Mali) 157
Afghanistan 29–30, *2.8*, 44, 49, 106, 274
Africa *2.1*, 16–20, *2.2*, 21, *2.4*, 23, 27, 31, 32, 33, 82, 106, 121, 129, 150, 164, 177–96, 221–49, 255–74, 278, Table 12.2, *12.4*, 288, *12.6*, 294, 295, 296, 297, 323, 325, *13.29*, 328, 365, 561, *22.2*, 568, 572
Afrikanda (USSR) Table 6.9, Table 13.1
Ahaggar (Algeria) *12.6*
Ailagyrsk (Turkmen, USSR) *2.7*
Akjoujt (Mauritania) Table 13.1
Albany Forks (Ontario, Canada) Table 13.1
Alberta (Canada) *9.1*
Aldan (USSR) 28
Aley (British Columbia, Canada) 113, *9.1*, 201, Table 9.1, 204, *9.6*, 213, *9.7*, *9.8*
Alice Springs (Northern Territory, Australia) 56
Alnö (Sweden) 7, 10, *2.6*, 27, 48, 125, 273, Table 13.1, 315, *13.35*, 434, *17.5*, 440, 441, 519, 520, 561, *22.2*, 568, 569, 572
Alto Paranaiba (Brazil) 20, *2.3*, *2.4*
Amazon (Brazil) 20, *2.3*, 164, 170, 171
Amba Dongar (India) 29, *2.8*, 46, 49, Table 7.1, 167, 284, 438, 444
Angola 16, 17, 18, 19, 22, 23, 31, 284, 551
Anitapolis (Brazil) 20, *2.3*, Table 7.1
Anstey Arm (British Columbia, Canada) *9.3*
Araxá (Brazil) 20, *2.3*, 111, Table 6.4, 149, Table 7.1, 150, 154, 155, 156, 157, Table 7.2, 158, 160, 161, Table 7.3, 168, 169, 171, 172, 444
Arbarastakh (USSR) *2.7*, 28, 114
Area Zero (Araxá, Brazil) 155
Argentina *2.4*, 22
Argor (Ontario, Canada) Table 6.8, *6.8*, *6.9*, *6.10*
Argyle (Western Australia) *21.1*, 548
Arkansas (USA) 138, 140, 163, *10.11*, 364, Table 14.2, 375
Armykon Hill (Tanzania) *11.2*, 258
Arusha (Tanzania) *8.1*, 178, *11.1*, 257, 273
Arvida (Quebec, Canada) 24, 25
Arvika Bay (Sweden) 27
Ataq (South Yemen) 466, *19.9*, 491
Atlantic Ocean 17, 19, 31, 32, 286, 430
Atlas Mts (Morocco) *see* High Atlas Mts (Morocco)
Austral Islands *19.9*, 493
Australia 15, 30, 150, 245, Table 12.2, *12.4*, 492, 551
Azov (Ukraine, USSR) Table 13.1, *13.25*

Babati (Tanzania) *11.1*
Badloch (Kaiserstuhl, FRG) *6.3*
Baikal, Lake (USSR) *2.6*, 69
Bailundo (Angola) 113, 161, Table 13.1, *13.35*
Bancroft (Ontario, Canada) *6.3*
Barberton (South Africa) Table 14.3, 377
Barchinsk (Kazakh, USSR) *2.7*
Basotu (Tanzania) *11.1*, 257
Bayan Obo (China) 149, Table 7.1, 155, 167, 438, 439

Bear Lodge (Wyoming, USA) 169
Bear Paw (Montana, USA) *13.47*, Table 13.1, 489, *19.9*, 492
Bearpaw Ridge (British Columbia, Canada) *9.1*, 201
Beemerville (New Jersey, USA) *2.5*, 25
Belaya Zima (East Sayan, Russia, USSR) Table 13.1, 337
Beloziminsky (East Sayan, Russia, USSR) Table 13.1
Benfontein (South Africa) 61, 270, 395, 396, *15.7*, *21.1*, 548
Betafo (Madagascar) *6.3*, 118
Bews Creek (British Columbia, Canada) *9.3*
Big Beaver House (Ontario, Canada) Table 12.2, Table 13.1
Big Spruce Lake (Northwest Territories, Canada) *2.5*, 23, 370–1
Bingo (Zaire) 164
Bitterfontein (South Africa) 568
Blackburn (Ontario, Canada) Table 6.5, 132
Blue River (British Columbia, Canada) Table 6.5, 116, *6.2*, 117, *6.3*, Table 6.5, 120, 130, Table 7.2, 159, 160, 169, *9.1*, 205, *9.6*, 213, *9.7*, *9.8*
Bol'shaya Tagna (East Sayan, USSR) Table 13.1
Bol'shetaginskiy (East Sayan, USSR) Table 13.1
Bol'shoy (USSR) Table 13.1, *13.35*, *13.47*, 347, *13.49*, Table 13.3
Bone Creek (British Columbia, Canada) Table 7.2, 159, 160, *9.1*, 202
Bonga (Angola) 155, 157, Table 7.3, 169
Borden (Ontario, Canada) 129, Table 12.2, *14.1*, 364, 369–70, *14.5*, Table 14.2
Brava (Cape Verde Islands) 20, 429, 434, 438
Brazil 17, 20, *2.3*, 21, *2.4*, 121, 160, 161, 164, 171
Breivikbotn (Norway) 56
British Columbia (Canada) 23, 32, 200–17, Table 9.1, *9.3*, Table 9.2, *9.7*, *9.8*
Bufumbira (Uganda) *10.11*
Bukusu (Uganda) Table 12.2, Table 13.1
Bultfontein (South Africa) 450, *18.2*, *18.6*, 459
Bunyaruguru Volcanic Field (Uganda) 45–6
Burritt Island (Ontario, Canada) Table 12.2
Burundi 181, 192
Busumbu (Uganda) Table 6.6, Table 13.1

Calgary (Alberta, Canada) *9.1*
Callandar Bay (Ontario, Canada) 584, 597
Callander Bay (Ontario, Canada) 24
Canada 27, 111, 121, 194, 200–17, 245, *10.11*, 247, 278, *12.1*, 284–5, 285, *12.2*, 286, 288, 291, 293, 297, 465–6, 491–2, 573, 574, 575
Canary Islands *2.1*, 17, *12.6*, 294, 430
Cane Valley (Utah, USA) 61
Cantley (Quebec, Canada) 138
Cape Verde Islands *2.1*, 17, 20, 46, 52, 75, 273, 294, 429, *17.1*, 430, 430–1, 434, *17.4*, 436, *17.5*
Cappelen Quarry (Fen, Norway) *3.10*
Carb Lake (Ontario, Canada) *2.5*, 23
Cargill (Ontario, Canada) Table 6.8, *6.8*, *6.9*, *6.10*,

LOCATION INDEX

Table 7.1, 161, 167, Table 12.2, Table 13.1, *13.35*, *14.1*, 369–70, *14.5*, Table 14.2, 590
Caspian Sea (USSR) *2.7*
Castignon Lake (Quebec, Canada) *2.5*, 23, 25
Catalão (Brazil) 20, *2.3*, 149, Table 7.1, 154, 156, 157, 160, 161, 162, 163, 167, 168, 169, 170, 171
Catanda (Angola) 109, 139
Cerro Impacto (Venezuela) 149, 155, 164, 166, 168
Cerro Manomo (Bolivia) 46
Cerro Sarambí (Paraguay) *2.3*, 20
Chadobetsk (USSR) *2.7*
Changit (Maimecha-Kotui, USSR) Table 13.1
Chasweta (Rufunsu, Zambia) 72
Chernigov (Ukraine, USSR) 125, Table 6.7, *6.6*, *13.25*, Table 13.1
Chigwakwalu-Hill, Malawi Table 13.1
Chilwa (Malawi) *2.1*, 17, *3.7*, *17.7*
China 31, 551
Chiriguelo (Paraguay) *2.3*, 20, 155, 157, 168, 169, 170, 171
Chishanya (Zimbabwe) 193
Clay-Howells (Ontario, Canada) Table 12.2, *12.4*
Coldwell (Ontario, Canada) see Port Coldwell (Ontario, Canada)
Colombia 20
Colorado (USA) 215, 323
Columbia R. (Canada/USA) *9.3*, 383
Comores 295
Cook Islands *19.9*, 493
Coola (Angola) Table 13.1
Cranbrook (British Columbia, Canada) *9.1*
Crater Highlands (Tanzania) *11.1*, 256
Crazy Mts (Montana, USA) 489, *19.9*, 491
Crevier (Quebec, Canada) 113, 114, 115, 149, 159, 169
Czechoslovakia 27

Dalbykha (Maimecha-Kotui, USSR) 113, Table 13.1, *13.47*
Damaraland (Namibia) *2.4*, 22
Dorowa (Zimbabwe) 435

East Africa 15, 16, 17, 18, 19, 31, 32, 177–96, *8.1*, 245, *10.11*, 246, 255–74, *11.8*, *12.5*, 288, 289, *12.7*, 291, 292, 293, 296, *13.6*, 333, *14.2*, 430, 433, *17.4*, 437, 440, 441, *19.9*, 491, *19.10*, 492, 571–2, 573
East Griqualand (South Africa) 46
Eastern Desert (Egypt) 53
Eastern Ghats (India) 15, 29, *2.8*
Egypt 17, 19
Eledoi (Tanzania) *11.2*, 258, *11.8*, 273
Elk Creek (Nebraska, USA) *2.5*, 24
Elkford (British Columbia, Canada) *9.1*
Embagai (Tanzania) *11.2*
Enesei Ridge (USSR) 28–9
Engaruka (Tanzania) *11.2*
Eppawala (Sri Lanka) 29, *2.8*
Essei Maimecha-Kotui (USSR) 130
 see also Yessey (Maimecha-Kotui, Russia, USSR)
Europe *2.2*, *2.6*, 26–7, 33, 150, 161, 171, Table 12.2, *12.4*, 294, *22.2*, 567, 569
Eyasi, Lake (Turkey) *11.1*

Fazenda Boa Vista (Tapira, Brazil) 163

Fen (Norway) *2.6*, 27, 41, 48, 50–1, 54, 58, *3.10*, 149, 157, 169, Table 7.2, 159, 161, 169, 171, 241, 247, Table 12.2, Table 13.1, *13.15*, 315, 316, *13.35*, 431, 444, *19.10*, 492, 561, 562, *22.2*, 567, *22.3*, 598
Finland 15, 26–7, 27, 29, 123
Finnmark (Norway) 49
Finsch (South Africa) *13.29*, 480
Firesand River (Ontario, Canada) 137, Table 6.9, 139, Table 12.2, Table 13.1, *14.1*, 364, *14.3*, Table 14.1, 367, 369, Table 14.2
Fogo (Cape Verde Islands) Table 13.1, *13.35*
Fort Portal (Uganda) 42, *3.2*, 44, 45, 71, 76, 109, 114, 138, Table 13.1, 399, *17.1*, 434, *22.3*, 570
Foskar (Phalaborwa, South Africa) 224, *10.2*, 230
Frederikshaabs Isblink (Greenland) 25
Frenchman Cap (British Columbia, Canada) 23, 201, 202, Table 9.1, 206, *9.3*, 207, *9.4*, Table 9.2, *9.6*, 213, *9.8*, 214, 215
Fuerteventura (Canary Islands) 20, Table 13.1, *13.35*

Gakara (Burundi) Table 7.1
Gallapo (Tanzania) 179
Gamoep (South Africa) 568
Gardar (Greenland) *2.5*, 25, 42, 426
Garies (South Africa) 568
Garub (Namibia) *2.1*, 19
Gatineau (Quebec, Canada) 109, Table 6.6, 123, 124, 125, 127, 128, 129, *6.7*, 131, Table 6.8, Table 6.9, *6.8*, *6.10*, 137, *6.11*, 138, 139, 140
Gebel Tarbti (Egypt) *2.1*, 19
Gek (Siberia, USSR) Table 13.1
Gelai (Tanzania) *11.1*, 256, *11.2*
Gem Park (Colorado, USA) 159, 169, 215
Germany 106, 343, *13.46*
Geronimo Volcanic Field (Arizona, USA) Table 19.9
Gibeon (South-West Africa) 568, *22.3*
Glenover (South Africa) 121
Goias (Brazil) 20, 21, 149, Table 7.1, 162, 163
Golden (British Columbia, Canada) *9.1*
Goldray (Ontario, Canada) 41, Table 6.8, *6.9*, *6.10*, *14.1*, 369–70, *14.5*
Gornoye Ozero (Sette Deban, USSR) Table 13.1
Got Chiewo (Homa Mountain, Kenya) *4.2*, 73, *4.4*, 76, Table 4.1
Goudini (South Africa) *12.2*, Table 13.1
Greece *13.46*
Greenland 15, *2.2*, *2.5*, 25–7, 27, 32, 77, 114, 138, 273, 297, Table 13.1
Griqualand see East Griqualand (South Africa)
Gross Brukkaros (Namibia) 568
Gula (Maimecha-Kotui, USSR) Table 13.1
Gula-North (Maimecha-Kotui, USSR) Table 13.1
Gula-South (Maimecha-Kotui, USSR) Table 13.1
Guli (Siberia, USSR) *3.3*
Gulinsk (USSR) *2.7*, 28
Gunnison County (Colorado, USA) Table 7.1, 163
Guyana 20

Hanang (Tanzania) 179, *11.1*, 257, 274
Hawaii (USA) 327, *13.30*, *13.31*, 330, 331, 375–6, 377, 575
Hegau (FRG) 46, 75, *4.6*
Henkenberg (Kaiserstuhl, FRG) 71, *4.6*, Table 4.1, 79–80, 82

Hick's Dome (USA) 167
High Atlas Mountains (Morocco) 17, 20
Himalayas *2.8*
Homa Bay (Kenya) 50, 72, Table 6.8, *6.8*, *6.9*, *6.10*, 274, Table 12.2, 288, 296, 561
Homa Mountain (Kenya) 46, 48, 72, *4.2*, 73, *4.4*, Table 13.1, *13.35*, 338, 438
Homa, Mount (Kenya) *see* Homa Mountain (Kenya)
Howard Creek (British Colombia, Canada) 160, 162, *9.1*, 202, *9.2*
Hst River (New Zealand) Table 13.1
Hualalai (Hawaii, USA) 327

Ice River (British Columbia, Canada) *9.1*, 201, 202, Table 9.1, 204, *9.6*
Iceland *13.30*, 330
Igaliko (Greenland) 25
Ile Bizard (Quebec, Canada) 568
Ile de Fogo (Cape Verde Islands) *see* Fogo (Cape Verde Islands)
Ile de Fuerte Ventura (Canary Islands) *see* Fuerteventura (Canary Islands)
Ilimaussaq (Greenland) 235
Illinois (USA) 167
Il'men (USSR) Table 13.1
Ilmenogorsk (USSR) *2.7*
Ilomba Hill (Malawi) Table 8.1, 181
Iluilarssuk (Greenland) *2.5*, 25
In Imanal (Mali) 160, 161
Independence Volcano (Montana, USA) 465, 489, *19.9*, 491
India 29–30, *2.8*, 49
Ingilil (USSR) *2.7*, 28
Invermere (British Columbia, Canada) *9.1*
Ipanema (Brazil) Table 7.1
Iran 31
Iron Hill (Colorado, USA) 109, 111, 125, 129, Table 6.8, *6.8*, *6.9*, *6.10*, *6.11*, Table 6.9, 139, 215, Table 13.1, *13.35*
Iron Island (Ontario, Canada) Table 12.2
Ishkulusk (USSR) Table 13.1
Isua (Greenland) Table 14.3, 377
Italy 31
Itapirapuá (Brazil) 154, 157

Jacupiranga (Brazil) 20, *2.3*, 50, Table 7.1, 167, 171, Table 12.1, 292, *14.2*, 364, 365, *14.9*, *14.10*, 588
Jägersfontein (South Africa) *13.29*, *18.2*
James Bay (Canada) 23
Japan *13.46*, 343
Jebel Dumbeir (Sudan) *2.1*, 19
Jodipatti (India) 29
Jordan River (British Columbia, Canada) *9.3*, 207
Jwaneng (Botswana) *13.29*

Kaiserstuhl (FRG) 27, 31, 40, 46, 47, 71, *4.1*, 75, 78, Table 4.1, 79–85, *4.7*, 85, 111, *6.3*, Table 6.5, Table 6.6, Table 6.7, *6.6*, 128, 130, 131, 139, 270, 273, 274, Table 12.2, Table 13.1, 315, Table 13.2, 326, *13.35*, 338, 399, *17.1*, 431, 434, *17.5*, 438, 441, 561, *22.2*, 568, 569
Kakanui (New Zealand) *22.2*
Kalix (Sweden) *2.6*, 27
Kalkfeld (Namibia) 171

Kalyango (Uganda) 42, Table 12.2, Table 13.1, Table 13.2
Kamchatka (USSR) 431
Kaminak Lake (Northwest Territories, Canada) *2.5*, 23, 26
Kamploops (British Columbia, Canada) *9.1*
Kangankunde (Malawi) 153, 169, 170, Table 12.2
Karonge (Burundi) 149, Table 7.1, 154–5, 171
Kasakere (Uganda) 71
Katwe-Kikorongo (Uganda) 45–6
Kavirondo Gulf (Kenya) 19
Kayanza (Burundi) *2.1*, 19
Kentucky (USA) 167
Kenya 16, 17, 19, 78, 128, 195, 256, 263, *12.6*, 429, 431, *17.4*, 437, 561, 562, 567, *22.3*, 569, *22.2*
Kerimasi (Tanzania) 42, 45, 49, 73, 85, 123, 127, 130, 170, *11.1*, 257, 271, 274, Table 13.1, *17.1*, 433, *17.5*, 509, *22.3*
Kgopoeloe (South Africa) 221, 229, 241
Khanneshin (Afghanistan) *2.8*, 29, 42, 46, 76, 109, 139
Kharla, Sanilen Highlands (Tuva, USSR) Table 13.1
Khibina Massif (Kola Peninsula, USSR) 161
Kiama (New South Wales, Australia) *19.9*, 493
Kibuye (Rwanda) Table 13.1
Kiisk (USSR) *2.7*, 29
Kilauea (Hawaii, USA) 331, 332
Kilimanjaro (Tanzania) *11.1*, 256
Killala (Ontario, Canada) *14.1*, *14.3*, Table 14.1, 367, 368, 369
Kimberley (South Africa) 451, 455, *18.6*, *18.7*, 467, *21.2*
Kimsey Quarry, Magnet Cove (Arkansas, USA) 160
Kirbyville Creek (British Columbia, Canada) *9.3*, *9.5*
Kirchberg (Kaiserstuhl, FRG) Table 4.1, 80–3, *4.7*, *4.8*, *4.9*, *4.10*, *4.11*
Kirumba (Zaire) *2.1*, 19, 181, Table 13.1
Kisete (Tanzania) *11.2*, 258
Kisingiri (Kenya) 288, 296
Kitchener's Kop (Phalaborwa, South Africa) 229
Kitumbeine (Tanzania) *11.1*, 256
Kiya-Shaltysk (USSR) Table 13.1
Koffiefontein (South Africa) *13.29*
Koga (Pakistan) 29
Koksharovsk (USSR) *2.7*
Kola (USSR) *2.6*, 27, *2.7*, 29, 31, 45, 46, 76–7, 113, 125, 127, 129, Table 6.9, 161, 162, 273, 431, *17.5*, 440, 441, 444
Kolyano (Uganda) Table 13.1
Kontozero District, Kola Peninsula (USSR) 46
Koratti (India) 29, 113
Kortejarvi (Finland) *2.6*, 26
Kotamu (Malawi) Table 13.1
Kovdor (Kola-Karelia, USSR) 4, 27, 116, 121, Table 6.6, 124, 125, 129, 130, 135, Table 13.1, 161, *13.35*, *13.47*, *13.49*, Table 13.3, 347, 441, 588
Kovdor Massif (USSR) 111, 113
Krasnomaisk (USSR) *2.7*
Krasnyye Obryvy (Maimecha-Kotui, USSR) Table 13.1
Kruidfontein (South Africa) 431, 433, 438, *17.1*, 444
Kugda (Maimecha-Kotui, USSR) Table 13.1
Kunja Area, Panda Hill (Tanzania) *8.5*, Table 8.4
Kursk (USSR) 137

Kwaraha (Tanzania) Table 13.1

Laacher See (FRG) Table 13.1, *13.35*, Table 13.1, *13.35*, 342, 343
Lackner Lake (Ontario, Canada) Table 13.1
Laetolil (Tanzania) Table 13.1, 431
Laivajoki (Finland) *2.6*, 26
Lajes (Brazil) 20, *2.3*
Lalarasi (Tanzania) *11.2*, 258, 263, *11.7*
Lashaine (Tanzania) 46, *11.8*, 274
Lelyaki, Dneiper-Donets (Ukraine, USSR) Table 13.1
Lemitar Mts (New Mexico, USA) 215
Lesnaya Varaka (USSR) Table 13.1
Lesotho 551
Leucite Hills (Wyoming, USA) 234, 241, *10.11*, 489, *19.9*
Libby (Montana, USA) Table 7.1, 167, 168
Lillebukt (Norway) 56
Lobo Hills (New Mexico, USA) 215
Loch Roag, Isle of Lewis (Scotland) *19.9*, 492
Loe Shilman (Afghanistan/Pakistan) 29, 49, 111, 438, *17.7*, 444
Loluni (Tanzania) *11.2*, 258
Longido (Tanzania) *11.1*
Lonnie (British Columbia, Canada) 56, *3.9*, Table 9.1
Lovozero Massif (Kola Peninsula, USSR) 162
Lower Sayansk (USSR) Table 6.9, 139
Luangwa (Zambia) 17
Lueshe (Zaire) *1.3*, *2.1*, 19, 53, 111, Table 6.4, 164, Table 7.3, 181, Table 13.1, *13.35*

McCloskey's Field (Quebec, Canada) 47, *3.4*
Mackenzie (British Columbia, Canada) *9.1*
McLure Mountain (Colorado, USA) 215
Madagascar *8.14*
Magadi, Lake (Kenya) 261
Magan (USSR) *2.7*, 28
Magnet Cove (Arkansas, USA) *2.5*, 41, 44, 100, 101, 155, 160, 162, 168, 169, Table 12.2, Table 13.1, *13.35*, *13.47*, *13.49*, 347, *14.2*, 364, 376, 380
Maicuru (Pará, Brazil) Table 7.1, 164
Maimecha-Kotui (USSR) *2.7*, 28, 29, 47
Maio (Cape Verde Islands) 20, 437
Maji ya Weta (Tanzania) 179
Makome (Uganda) Table 13.1
Makonde (Tanzania) 178, *8.2*, Table 8.1, 180, 182, Table 8.2, 189, *8.11*
Malakand (Afghanistan/Pakistan) 29
Malawi 18, 31, 180, 181, 192, 193, *8.14*
Mali 30
Maly Sayanski (East Sayan, USSR) Table 13.1, *13.35*, *13.47*, *13.49*, Table 13.3, 347
Manitou Islands (Ontario, Canada) 24
Manomo (Bolivia) *2.3*, 21
Manson Creek (British Columbia, Canada) *9.1*, 202, Table 9.1, 205–6, *9.6*, *9.7*, *9.8*
Mansouri (Egypt) 53
Manyara, Lake (Tanzania) *11.1*, 256
Marongwe-Hill (Malawi) Table 13.1
Martison (Ontario, Canada) Table 7.1
Mato Preto (Brazil) Table 7.1, 167
Matongo (Burundi) 181
Matongo-Bandaga (Burundi) 154, 161, Table 7.3, 168
Mauna Loa (Hawaii, USA) 332

Mauritania 20
Mawenzi (Tanzania) 256
Mbalizi (Tanzania) Table 6.5, 120, *8.2*, Table 8.1, 180, *8.3*, 182, Table 8.2, 186–8, *8.10*, Table 8.5, 192
Mbeya (Tanzania) 41, 116, *8.1*, 178, *8.2*, 180, *8.3*, Table 13.1, *13.15*, 316, 332–3, *13.35*, *13.42*
Mbozi (Tanzania) *8.2*, 181
Mbuga (Uganda) Table 13.1
Meech Lake (Quebec, Canada) Table 6.4, 111, 116, Table 6.5, 120, 137
Melkfontein (South Africa) 46
Meru (Tanzania) *11.1*, 257
Mezenskyi (USSR) *2.6*
Mica Creek (British Columbia, Canada) Table 9.1
Minarot Valley (South Ruri, Kenya) 73
Minas Gerais (Brazil) 20, 21, 149, Table 7.1, 162, 163, Table 7.3
Monashee Mts (British Columbia, Canada) 206, *9.3*
Mongolia 27–9, *2.7*, 32, 273
Montana 489
Montana (USA) 464, 489
Monte Verde (Angola) Table 13.1, *13.35*
Monteregian Hills (Quebec, Canada) 24, 31
Montreal (Quebec, Canada) Table 6.1, 108, 561, *22.2*, 568, 569, 572
Morogoro (Tanzania) *8.1*, 178, 179
Morro dos Seis Lagos (Brazil) 155, 166, 168
Mosonik (Tanzania) *11.1*
Mount Copeland (British Columbia, Canada) *9.1*, 202, Table 9.1, *9.3*, 207, 215
Mount Elgon (Kenya/Uganda) 274
Mount Grace (British Columbia, Canada) 23, 46, 49, 56–7, *9.1*, 202, Table 9.1, *9.3*, 207, *9.4*, 209–11, *9.5*, *9.6*, 213, *9.7*, *9.8*, 215
Mount Royal (Quebec, Canada) 480
Mount Shaheru (Zaire) 263
Mountain Pass (California, USA) 23, *2.5*, 109, 149, Table 7.1, 150, 153, 157, 168, 169, 170, 171, Table 13.1, 344, Table 13.3, *13.50*, 348, 349, 425, 511, 512, *20.8*, 513, 514
Mozambique 16, 18, 31, 180
Mozambique Belt 32, *8.1*
Mrima Hill (Kenya) 111
Mrima (Kenya) 149, 155, 164, 168, 169, 170, Table 13.1
Mud Lake (British Columbia, Canada) *9.1*, 202
Mud Tank (Northern Territory, Australia) 30, 56, 167, Table 13.1, *13.35*
Muri Mts (Guyana) 160, 164
Murun (USSR) *2.7*, Table 13.1
Musensi (Tanzania) *8.2*, Table 8.1, 180, *8.3*, 182, Table 8.2, 185
Mush Kuduk (Outer Mongolia) 123, 125
Mutum (Brazil/Guyana) 20, *2.3*

Nachenedzwaya (Tanzania) *8.2*, Table 8.1, 180, 181, 182, Table 8.2, 189, Table 8.6, Table 12.2
Namibia 17, 18, 19, *2.4*, 31, 548, 551, 568
Napak (Uganda) 17, 49, 111, 274, Table 12.2, Table 13.1
Natron, Lake (Tanzania) 255, *11.1*, 256, *11.2*, 257
Ndeke (Uganda) Table 13.1
Nemegos (Ontario, Canada) 121

LOCATION INDEX

Nemegosenda Lake (Ontario, Canada) Table 12.2, *12.4*, Table 13.1, *13.35*, *14.1*, *14.3*, 365, Table 14.1, 368–9
New England Seamounts 295
New South Wales (Australia) *10.11*
New Zealand 31, Table 12.2, *12.4*, 293, 364, 365, *14.9*, 380, *14.10*, 381, 466, 467, *19.9*, 491, 493
Newania (India) 29, *2.8*, *17.5*, 438, 439
Newlands (South Africa) *13.29*
Ngorongoro (Tanzania) *11.1*, 256
Ngualla (Tanzania) *2.1*, *8.9*, 18, 47, 149, Table 7.1, 159, Table 7.3, 169, 177, *8.2*, Table 8.1, 180, 182, *8.4*, Table 8.2, 184, 185, Table 8.3, 186, *8.7*, *8.8*, 189, *8.11*, 192, 193
Nipissing, Lake (Ontario, Canada) 24, 31, 32
Nipissing (Ontario, Canada) 32
Nissikkatch Lake (Saskatchewan, Canada) 23, *2.5*
Nizhnesayan Massif (USSR) 164
Nkombwa Hill (Zambia) *2.1*, 18, 111, Table 6.6, 181
Nor'ilsk (USSR) 343
North America *2.2*, 23–5, *2.5*, 27, 33, 159, 200, 202, 215, 245, 246, 272, Table 12.2, *12.4*, 294, 360, 362–74, *14.7*, *14.8*, 467, 551
North Ruri (Nyanza, Kenya) Table 13.1
Novaya Poltava (Ukraine, USSR) Table 13.1
Nunivak (Alaska, USA) 466
Nyasa, Lake (Africa) *8.1*, 178, 180, 182, *8.5*, *13.36*

Octobre (USSR) *2.7*
Odikhincha (Maimecha-Kotui, USSR) Table 13.1
Ohre River (Czechoslovakia) *2.6*, 27
Oka (Quebec, Canada) *2.5*, 24, 31, 32, 40, 41, 44, 51, *3.6*, *5.1*, *5.2*, 100, 108, 111, Table 6.4, 113, 114, 115, 116, 117, *6.3*, Table 6.5, *6.4*, 120, 121, Table 6.6, 124, 125, Table 6.7, *6.6*, 127, 129, 130, 131, Table 7.1, 155, 157, 159, 160, 161, 169, Table 12.2, 285, 291, 301, Table 13.1, 306, 308, *13.7*, *13.8*, *13.9*, *13.16*, 316, *13.19*, 319, 320–2, *13.20*, *13.21*, *13.24*, 324, *13.35*, 335–7, *13.37*, *13.38*, *13.39*, 338, *13.42*, 340, 342, *13.47*, *13.49*, 347, 348, 349, *14.1*, *14.2*, 364, Table 14.2, 376, 380, 441, 466, 471, *19.9*, 491–2, 509, 517, 568, *22.3*, 583, 597
Okorusu (Namibia) Table 7.1, 167
Oldoinyo Lengai (Tanzania) 2, 9, *2.1*, 19, 41, 42, 44, 54, 70, 71, 76, 77, 85, 108, 179, 255–74, *11.1*–*11.8*, Tables 11.1–11.4, Table 12.2, 291–2, Table 13.1, Table 13.2, 333, *13.35*, 338, 400, 402, 403, Table 16.1, 406, *16.2*, 426, 429, 431, 432, 433, 436, 439, 441, 509, 521, 537, 570–1, 584, 587, 591, 592–4, 595, 596, 597
Oldoinyo Loolmurwak (Tanzania) *11.2*, 258, 263, Table 11.1, *11.7*, 272
Olduvai Gorge (Tanzania) *11.1*, 256, 258, 274
Olenek (USSR) *2.7*, 29
Ondurakorume (Namibia) 169
Ontario (Canada) 25, *14.1*, *14.3*, *14.6*, 598
Oppelo (Arkansas, USA) Table 6.9
Orapa (Botswana) *13.29*
Orroroo (South Australia) 548
Otago (New Zealand) Table 12.2
Ottawa (Ontario, Canada) 47, 52, *14.1*
Ozernaya Varaka (Kola-Karela, USSR) 130, Table 13.1

Ozernyi (Siberia, USSR) *2.7*, 28, Table 13.1
Pajarito Mountain (New Mexico, USA) 161
Pakistan 29–30, *2.8*, 32, 49, 442–3
Panda Hill (Tanzania) 111, 157, 169, *8.2*, Table 8.1, 180, *8.3*, 182, Table 8.2, 184–5, *8.5*, *8.6*, 186, Table 8.4, *8.8*, 189, 192, Table 12.2
Pará (Brazil) Table 7.1, 164
Paradise (British Columbia, Canada) 160, *9.1*, 202, Table 9.1
Paraguay *2.4*, 164
Paraná (Brazil) 20, *2.4*, Table 7.1, 160, 162, 167
Patrocinio (Brazil) Table 7.1
Pello Hill (Tanzania) *11.2*, 258, *11.8*, 274
Perkins (Quebec, Canada) Table 6.7, 130, 228
Perry River (British Columbia, Canada) *9.1*, 207, *9.4*, 211, *9.6*, *9.7*, *9.8*, Table 9.1, *9.3*
Peshawar Plain (Pakistan/Afghanistan) 29, *2.8*
Pesochnaya Varaka (USSR) Table 13.1
Phalaborwa (South Africa) *2.1*, 12, 17, 100, 121, 123, 130, Table 7.1, 150, 160, 161, 167, 169, 170–1, 172, 221–49, 288, *12.6*, 289, 294, Table 13.1, *13.35*, 338, *13.47*, *13.49*, 347, 425, 431, 510, 582
Pilbara (Western Australia) Table 14.3, 377
Pokrovo-Kirieevskie (Ukraine, USSR) *2.7*
Pollen (Norway) 49
Ponta Grossa Arch (Brazil) *2.3*, 20, *2.4*, 22
Poohbah Lake (Ontario, Canada) *12.1*, 285, *12.2*, *12.4*, *14.1*, 371, *14.6*, Table 14.2
Port Coldwell (Ontario, Canada) *14.1*, *14.3*, 365, Table 14.1, 366, 369
Potash Sulphur Springs (Arkansas, USA) 168
Povorotnyy (Siberia, USSR) Table 13.1
Powderhorn (Ontario, USA) Table 7.1, 163, 170
Prairie Lake (Ontario, Canada) 53, 111, 160, 169, 170, Table 12.2, Table 13.1, *14.1*, *14.3*, Table 14.1, 367, *14.4*, 369
Premier Mine (South Africa) 55

Qagssiarssuk (Greenland) 42, 46
Qaqarssuk (Greenland) *2.5*, 25
Quebec (Canada) 46, 65, *14.1*
Quigussaq (Greenland) *2.5*, 25
Quilengues (Angola) Table 7.1

Rainy Creek (Montana, USA) Table 7.1, 167, 236
Rangwa (Nyanza, Kenya) 72, Table 6.6, 274, Table 13.1
Ratchford Creek (British Columbia, Canada) *9.1*, *9.2*, 207–8, *9.4*, *9.6*, *9.7*
Red Sea 17, 19
Ren (British Columbia, Canada) 208, 213
Revelstoke (British Columbia, Canada) *9.1*, *9.3*
Rhine Graben (FRG) *2.6*, 27, 46, 568
Richat Dome (Mauritania) *2.1*, 17
Rio Grande do Sul Arch (Brazil) *2.4*, 22
Rison (Arkansas, USA) 163
Rock Canyon Creek (British Columbia, Canada) *9.1*, 201, 202, Table 9.1, 213, *9.7*
Rocky Mts (British Columbia, Canada) 200, 201, 202, Table 9.1, 211, 215
Rufiji River (Tanzania) 179
Rufunsa (Zambia) 46, 71, 72, 121
Rukwa (Tanzania) *8.2*

616

Rukwa Trough (Tanzania) *8.3*, 182
Ruri (Kenya) 72, 73
Rusekere (Uganda) Table 12.2
Russia *see* USSR

Saddle Mts (USA) 383
Sadiman (Tanzania) *11.1*
Saguenay River Valley (Quebec, Canada) *22.2*, 568, 569, 570
St André (Quebec, Canada) 111, Table 13.1
St Helena *12.6*, 295, *19.9*, 493
St Honoré (Quebec, Canada) 24–5, 25, Table 7.1, 169, Table 12.2, Table 13.1, 306, 307, 315, 316, 318, *13.23*, 324, *13.35*, *13.39*, 338, *13.41*, *13.42*, 349
St Michel (Quebec, Canada) 108
Saka (Fort Portal, Uganda) 71, Table 13.1
Salitre (Brazil) 20, *2.3*, Table 7.1, 160, 161, 162, 163, 170
Sallanlatva (USSR) 4, Table 6.6, Table 13.1
Salmogorsk (USSR) 27
Saltpetre Kop (South Africa) 55, 111, 568
Samoa Islands *19.9*, 493
San Felix 295
San Vicente (Cape Verde Islands) 437
Sangu-Ikola (Tanzania) 178, *8.2*, Table 8.1, 180, 182, Table 8.2, 184, 189, *8.11*, Table 8.6, 192
Santa Catarina (Brazil) 20, Table 7.1
Santiago (Cape Verde Islands) 20, 433, 437
São Francisco Craton (Brazil) 20
São Paolo (Brazil) 20, Table 7.1, 154, 166
Sarfartok (Greenland) *2.5*, 25, 40, 117, 140, Table 7.1
Saudi Arabia *13.46*
Sayan (USSR) Table 6.4
Sayan-Tuva Province (USSR) 28, 29
Schryburt Lake (Ontario, Canada) Table 12.2
Scotland 492
Seabrook Lake (Ontario, Canada) *3.5*, Table 12.2, Table 13.1
Sebl'yarvi (USSR) 111, 113, Table 13.1
Seiland (Norway) Table 13.1
Selkirk Mts (British Cloumbia, Canada) 206, *9.3*
Sengeri Hill (Tanzania) *8.2*, Table 8.1, 180, *8.3*, 182, Table 8.2, 185, *8.8*, 192, Table 12.2, Table 8.4
Serengeti Plains (Tanzania) *11.1*, 257, 261, 274
Serra Negra (Brazil) Table 7.1, 160, 162, 163, 170
Sevathur (India) 29, Table 13.1
Seymour River (British Columbia, Canada) *9.3*
Shandong Province (China) 551
Shawa (Zimbabwe) 193, *8.14*, 435, *17.5*, 441
Shira (Tanzania) 256
Shombole (Kenya) *11.1*, 396, *15.6*, *15.7*, 429
Shonkin Sag (Montana, USA) 235, 240
Siberia (USSR) Table 6.4, 111, 113, 115, Table 6.5, 120, 125, 127, 129, Table 6.9, 139, 140, Table 13.1
Siilinjarvi (Finland) *2.6*, 26, 56, 137, *14.6*, 372, *14.8*
Silai Patti (Pakistan) 49
Silver Crater (Ontario, Canada) 118
Sinai (Egypt) 395, *15.7*
Smedsgården (Alnö, Sweden) 440
Smoky Butte (Montana, USA) *19.9*, 489
Snake River (USA) 383

Sokli (Finland) 27, 113, Table 6.8, *6.9*, *6.10*, Table 13.1, *13.35*, *13.47*, *13.49*, 347, 440, 588
Sol'bel'der (Tuva, USSR) 115
Songhor (Tinderet, Kenya) 75
Songwe (Malawi) Table 8.1, 181
Songwe (Tanzania) 49, *8.2*, Table 8.1, 180, *8.3*, 182, Table 8.2, *8.5*, 187–8, 189, *8.10*, Table 8.5, 438
Söröy (Norway) 56
South Africa 16, 17, 23, 31, 171, 221–49, 448, 450, 460, 466, 467, 468, 532, 535, 551, 568, *18.5*
South America 17, *2.2*, 20–3, *2.3*, *2.4*, 150, 170, 278, Table 12.2, *12.4*, 551
South Island (New Zealand) 58
South Ruri (Kenya) 73, *4.5*, Table 4.1, Table 13.1
South Westland (South Island, New Zealand) 30–1
Spain *10.11*, 245
Spanish River (Ontario, Canada) Table 12.2, Table 13.1, 370
Spitzbergen (Norway) 396, *15.7*
Spitzkop (South Africa) 221, 229, Table 13.1
Srednezimiskiy (East Sayan, USSR) Table 13.1
Stjernøy (Norway) 56, Table 13.1
Strangways Range (Northern Territory, Australia) 30, *12.2*
Sturgeon Lake (Ontario, Canada) *14.1*, 371, *14.6*, Table 14.2
Sudbury (Ontario, Canada) 343
Sukulu (Uganda) 121, Table 7.1, 171, Table 12.2, Table 13.1, 309, *13.35*
Sumbadzi River (Tanzania) 179
Superior, Lake (North America) 23
Sutherland (South Africa) 568
Sweden 27
Syanskiy (USSR) Table 13.1

Tadhak (Mali) *2.1*, 17, 19
Tagna (USSR) *2.7*
Tajno (Poland) *2.6*, 27
Tamazert (Morocco) 19, 20, Table 13.1, *2.1*
Tamil Nadu (India) 15, 29, *2.8*
Tanga (East Sayan, USSR) Table 13.1
Tanganyika, Lake (Africa) *8.1*, 178, *8.2*, 181, 182, *13.36*
Tanginski (Sayan, USSR) Table 13.1
Tanzania 17, 18, 19, 45, 49, 72, 77, 177–96, 255–74, 431
Tapira (Brazil) 20, *2.3*, Table 7.1, 160, 162, 163, 167, 170
Tatarsky (Yenisei Mts, USSR) Table 13.1, *13.35*
Tchivira (Angola) Table 7.1, 160, 292, Table 13.1
Three Valley Gap (British Columbia, Canada) *9.1*
Thumb (Colorado Plateau, USA) 466
Tien-Shan (USSR) 113
Timan (USSR) 337
Timmins (Ontario, Canada) 124
Tinderet (Kenya) 46, 73, 75, 139, 274, 429, *17.1*, 431, 434
Tororo (Uganda) 41, 75, Table 12.2, Table 13.1, *13.35*
Transvaal (South Africa) 582
Trident Mountain (British Columbia, Canada) *9.1*, 202, Table 9.1, 205
Tristan da Cunha *12.6*
Tubuaii 295, *19.9*, 493

Tundulu (Malawi) 17, Table 13.1, *17.7*
Tupertalik (Greenland) *2.5*, 25, 26, 56
Turii Peninsula (USSR) 27, 113, 115, 129, Table 6.8, *6.8*, *6.9*, *6.10*, *6.11*, Table 13.1
Turkey 31
Twareitau Mountain (Guyana) 155, 164, Table 7.3, 165, 166, 169

Ufiome (Tanzania) *11.1*, 257
Uganda 19, 195, 235, 248, 263, 288, 467, 519, *22.2*, 567, 569
Ukraine (USSR) 28
Uluguru Mountains (Tanzania) 179
Umanak (Greenland) 25
Union of Soviet Socialist Republics *see* USSR
United States of America *see* USA
Upper Ruvubu (Burundi) 181
Upper Sayansk Massif (USSR) Table 6.4
Ural Mts (USSR) 115, 117
Uruguay 17, *2.4*, 22
USA 215, 245, 465, 468
USSR 27–9, *2.7*, 32, 52, 106, 108, 114, Table 6.5, 121, 123, 157, 163, *13.29*, 328, 551
Uweinat (Libya) *2.1*, 19

Valentine Township (Ontario, Canada) 138, Table 6.9
Vancouver (British Columbia, Canada) *9.1*
Venezuela 20
Verity (British Columbia, Canada) 56, *6.2*, *6.3*, 160, *9.1*, 202, Table 9.1
Victoria, Lake (Africa) *8.1*, *13.36*
Vipeto (Fen, Norway) 159
Virunga (Zaire) *10.11*
Vishnev Hills (USSR) *13.1*
Vishnevogorsk (USSR) *2.7*, Table 13.1

Voin (USSR) *2.7*
Vuorijarvi (Kola-Karelia, USSR) Table 13.1, 337, *13.47*, Table 13.3, 441
Vuoyarvi (Kola-Karelia, USSR) 114

Walloway (Australia) Table 12.2
Walvis Ridge 295, 365, 375–6, 377, 493, *12.6*, *14.2*, *14.8–14.10*, *19.9*
Wasaki (Kenya) 438
Weld, Mount (Western Australia) 30, 149, 155, 161, Table 7.3, 166, 167, 169, Table 12.2
Wesselton (South Africa) 548
West Africa 551
West African Craton 17, 30
West Eifel (FRG) 467
Western Australia 127, 129–30, *10.11*
Westland (New Zealand) Table 12.2, 466
Wet Mountains (Colorado, USA) *2.5*, 24, 59, Table 13.1, *13.35*
Wigu Hill (Tanzania) 149, 154, 157, 170, 171, 179
Wyoming (USA) 215, 248, 465, 489

Yellowknife (Northwest Territories, Canada) 370–1
Yemen 31, 491
Yessey (Maimecha-Kotui, USSR) Table 13.1, *13.47* *see also* Essei (Maimecha-Kotui, USSR)
Yrass (Maimecha-Kotui, USSR) Table 13.1, *13.47*, 344
Yukon Territories (Canada) 215

Zaire 181, 192, 551
Zambia 44, 49, 180, 181
Zhidoiskii (USSR) *2.7*, 28
Zimbabwe 18, 180, *8.12*, 193
Ziminskyi (USSR) *2.7*